Safety Symbols

Safety symbols in the following table are used in the lab activities to indicate possible hazards. Learn the meaning of each symbol. **It is recommended that you wear safety goggles and apron at all times in the lab. This might be required in your school district.**

Safety Symbols		Hazard	Examples	Precaution	Remedy
Disposal		Special disposal procedures need to be followed.	certain chemicals, living organisms	Do not dispose of these materials in the sink or trash can.	Dispose of wastes as directed by your teacher.
Biological		Organisms or other biological materials that might be harmful to humans	bacteria, fungi, blood, unpreserved tissues, plant materials	Avoid skin contact with these materials. Wear mask or gloves.	Notify your teacher if you suspect contact with material. Wash hands thoroughly.
Extreme Temperature		Objects that can burn skin by being too cold or too hot	boiling liquids, hot plates, dry ice, liquid nitrogen	Use proper protection when handling.	Go to your teacher for first aid.
Sharp Object		Use of tools or glassware that can easily puncture or slice skin	razor blades, pins, scalpels, pointed tools, dissecting probes, broken glass	Practice common-sense behavior and follow guidelines for use of the tool.	Go to your teacher for first aid.
Fume		Possible danger to respiratory tract from fumes	ammonia, acetone, nail polish remover, heated sulfur, moth balls	Be sure there is good ventilation. Never smell fumes directly. Wear a mask.	Leave foul area and notify your teacher immediately.
Electrical		Possible danger from electrical shock or burn	improper grounding, liquid spills, short circuits, exposed wires	Double-check setup with teacher. Check condition of wires and apparatus. Use GFI-protected outlets.	Do not attempt to fix electrical problems. Notify your teacher immediately.
Irritant		Substances that can irritate the skin or mucous membranes of the respiratory tract	pollen, moth balls, steel wool, fiberglass, potassium permanganate	Wear dust mask and gloves. Practice extra care when handling these materials.	Go to your teacher for first aid.
Chemical		Chemicals that can react with and destroy tissue and other materials	bleaches such as hydrogen peroxide; acids such as sulfuric acid, hydrochloric acid; bases such as ammonia, sodium hydroxide	Wear goggles, gloves, and an apron.	Immediately flush the affected area with water and notify your teacher.
Toxic		Substance may be poisonous if touched, inhaled, or swallowed.	mercury, many metal compounds, iodine, poinsettia plant parts	Follow your teacher's instructions.	Always wash hands thoroughly after use. Go to your teacher for first aid.
Flammable		Flammable chemicals may be ignited by open flame, spark, or exposed heat.	alcohol, kerosene, potassium permanganate	Avoid open flames and heat when using flammable chemicals.	Notify your teacher immediately. Use fire safety equipment if applicable.
Open Flame		Open flame in use, may cause fire.	hair, clothing, paper, synthetic materials	Tie back hair and loose clothing. Follow teacher's instruction on lighting and extinguishing flames.	Notify your teacher immediately. Use fire safety equipment if applicable.

 Eye Safety Proper eye protection should be worn at all times by anyone performing or observing science activities.

 Clothing Protection This symbol appears when substances could stain or burn clothing.

 Animal Safety This symbol appears when safety of animals and students must be ensured.

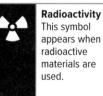 **Radioactivity** This symbol appears when radioactive materials are used.

 Handwashing After the lab, wash hands with soap and water before removing goggles.

Inspire
Earth Science

Mc Graw Hill

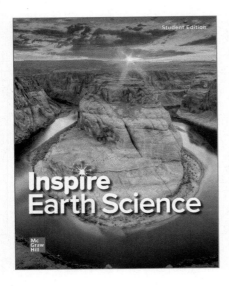

Student Edition

Inspire Earth Science

McGraw Hill

Phenomenon: Canyon formation

This spectacular shot is only possible because of a change in Earth's surface leading to a process called rejuvenation.

Fun Fact

To cause rejuvenation, Earth's surface must rise above its base level. The river will then cut through the rock to go back to the original base level.

FRONT COVER: ronnybas/Shutterstock. **BACK COVER:** ronnybas/Shutterstock.

mheducation.com/prek-12

Send all inquiries to:
McGraw-Hill Education
STEM Learning Solutions Center
8787 Orion Place
Columbus, OH 43240

ISBN: 978-0-02-145261-3
MHID: 0-02-145261-X

Printed in the United States of America.

2 3 4 5 6 7 8 9 QVS 23 22 21 20 19

McGraw-Hill is committed to providing instructional materials in Science, Technology, Engineering, and Mathematics (STEM) that give all students a solid foundation, one that prepares them for college and careers in the 21st century.

Welcome to

Inspire
Earth Science

Explore Our Phenomenal World

The Inspire High School Series brings phenomena to the forefront of learning to engage and inspire students to investigate key science concepts through their three-dimensional learning experience.

Start exploring now!

Inspire Curiosity   •   **Inspire Investigation**   •   **Inspire Innovation**

Owning Your Learning

1 **Encounter the Phenomenon**

Every day, you are surrounded by natural phenomena that makes you wonder.

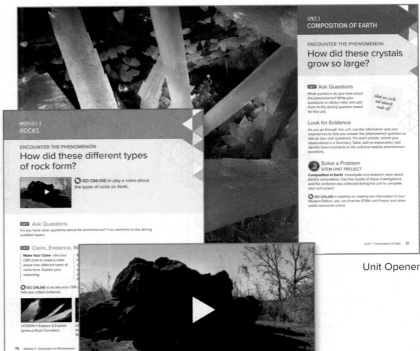

Unit Opener

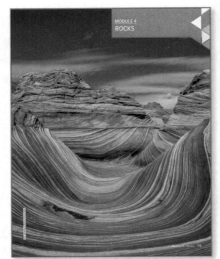

Module Opener

Phenomenon Video

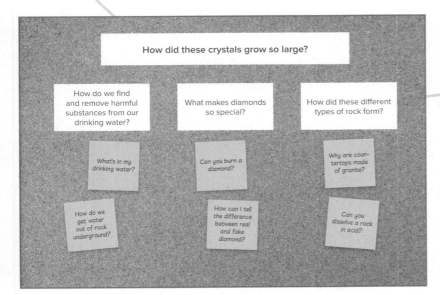

How did these crystals grow so large?

How do we find and remove harmful substances from our drinking water?

What makes diamonds so special?

How did these different types of rock form?

What's in my drinking water?

Can you burn a diamond?

Why are countertops made of granite?

How do we get water out of rock underground?

How can I tell the difference between real and fake diamond?

Can you dissolve a rock in acid?

2 **Ask Questions**

At the beginning of each unit and module, make a list of the questions you have about the phenomenon. Share your questions with your classmates.

3 Claim, Evidence, Reasoning

As you investigate each phenomenon, you will write your claim, gather evidence by performing labs and completing reading assignments and Applying Practices, and explain your reasoning to answer the unit and module phenomena.

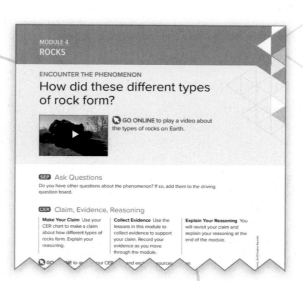

MODULE 4
ROCKS

ENCOUNTER THE PHENOMENON
How did these different types of rock form?

GO ONLINE to play a video about the types of rocks on Earth.

SEP Ask Questions
Do you have other questions about the phenomenon? If so, add them to the driving question board.

CER Claim, Evidence, Reasoning

Make Your Claim Use your CER chart to make a claim about how different types of rocks form. Explain your reasoning.

Collect Evidence Use the lessons in this module to collect evidence to support your claim. Record your evidence as you move through the module.

Explain Your Reasoning You will revisit your claim and explain your reasoning at the end of the module.

GO ONLINE to your CER and explore resources

SUMMARY TABLE

Activity Model	Observation Evidence	Explanation Reasoning	Connection to Phenom	Questions Answered	New Questions
Virtual Investigation: Mineral Properties	Minerals have distinct properties that can be used to identify them.	A mineral's chemical composition and crystal structure give it unique properties, by which it can be identified.	Unit: There is no limit the size a crystal can grow, because the chemical patterns can repeat indefinitely. Module: Diamond is special because of its chemical make-up and crystal structure.	How can I tell the difference between real and fake diamond?	How is diamond made in nature?

4 Summarize Your Work

When you collect evidence, you can record your data in a summary table and use the data to collaborate with others to answer the questions you had.

5 Apply Your Evidence and Reasoning

At the end of the unit, modules, and lessons, you can use all of the data you collected to help complete your STEM Unit Project.

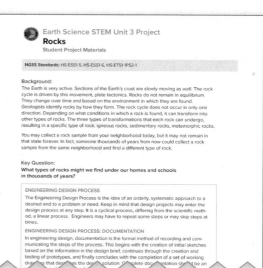

Earth Science STEM Unit 3 Project
Rocks
Student Project Materials

NGSS Standards: HS-ESS1-5, HS-ESS1-6, HS-ETS1-1PS2-1

Background:
The Earth is very active. Sections of the Earth's crust are slowly moving as well. The rock cycle is driven by this movement, plate tectonics. Rocks do not remain in equilibrium. They change over time and based on the environment in which they are found. Geologists identify rocks by how they form. The rock cycle does not occur in only one direction. Depending on what conditions in which a rock is found, it can transform into other types of rocks. The three types of transformations that each rock can undergo, resulting in a specific type of rock: igneous rocks, sedimentary rocks, metamorphic rocks.

You may collect a rock sample from your neighborhood today, but it may not remain in that state forever. In fact, someone thousands of years from now could collect a rock sample from the same neighborhood and find a different type of rock.

Key Question:
What types of rocks might we find under our homes and schools in thousands of years?

ENGINEERING DESIGN PROCESS
The Engineering Design Process is the idea of an orderly, systematic approach to a desired end to a problem or need. Keep in mind that design projects may enter the design process at any step. It is a cyclical process, differing from the scientific method, a linear process. Engineers may have to repeat some steps or may skip steps at times.

ENGINEERING DESIGN PROCESS: DOCUMENTATION
In engineering design, documentation is the formal method of recording and communicating the steps of the process. This begins with the creation of initial sketches based on the information in the design brief, continues through the creation and testing of prototypes, and finally concludes with the completion of a set of working drawings that describe the design solution. Complete documentation should be an

ADVISORS AND CONSULTANTS

Teacher Advisory Board

The Teacher Advisory Board gave the editorial staff and design team feedback on the content and design of both the Student Edition and Teacher Edition. We thank these teachers for their hard work and creative suggestions.

Bill Brown
Grandview Heights High School
Columbus, OH

Carmen S. Dixon
East Knox High School
Howard, OH

Joel Heuberger
Waite High School
Toledo, OH

Jane Karabaic
Steubenville City Schools
Steubenville, OH

Terry Stephens
Edgewood High School
Trenton, OH

Content Consultants

Content consultants each reviewed selected chapters of *Inspire Earth Science* for content accuracy and clarity.

Anastasia Chopelas, PhD
Research Professor of Earth and Space Sciences
University of Washington
Seattle, WA

Diane Clayton, PhD
University of California at Santa Barbara
Santa Barbara, CA

Sarah Gille, PhD
Associate Professor
Scripps Institution of Oceanography
   and Department of Mechanical and Aerospace
   Engineering
University of California San Diego
San Diego, CA

Alan Gishlick, PhD
National Center for Science Education
Oakland, CA

Janet Herman, PhD
Professor and Director of Program of Interdisciplinary
Research in Contaminant Hydrogeology
University of Virginia
Charlottesville, VA

David Ho, PhD
Storke-Doherty Lecturer & Doherty
Associate Research Scientist
Lamont-Doherty Earth Observatory
Columbia University
New York, NY

Jose Miguel Hurtado, PhD
Associate Professor of Geology
University of Texas at El Paso
El Paso, TX

Monika Kress, PhD
Assistant Professor of Physics and Astronomy
San Jose State University
San Jose, CA

Amy Leventer, PhD
Associate Professor of Geology
Colgate University
Hamilton, NY

Amala Mahadevan, PhD
Associate Research Professor
Department of Earth Sciences
Boston University
Boston, MA

Nathan Niemi, PhD
Assistant Professor of Geological Sciences
University of Michigan
Ann Arbor, MI

Anne Raymond, PhD
Professor of Geology and Geophysics
Texas A&M University
College Station, TX

Safety Consultant

The safety consultant reviewed labs and lab materials for safety and implementation.

Kenneth Russell Roy, PhD
Director of Environmental Health and Safety
Glastonbury Public Schools
Glastonbury, CT

Smithsonian

Smithsonian

Following the mission of its founder James Smithson for "an establishment for the increase and diffusion of knowledge," the Smithsonian Institution today is the world's largest museum, education, and research complex. To further their vision of shaping the future, a wealth of Smithsonian online resources are integrated within this program.

SpongeLab Interactives

SpongeLab Interactives is a learning technology company that inspires learning and engagement by creating gamified environments that encourage students to interact with digital learning experiences.

Students participate in inquiry activities and problem-solving to explore a variety of topics using games, interactives, and video while teachers take advantage of formative, summative, or performance-based assessment information that is gathered through the learning management systems.

PhET Interactive Simulations

The PhET Interactive Simulations project at the University of Colorado Boulder provides teacher and students with interactive science and math simulations. Based on extensive education research, PhET sims engage students through an intuitive, game-like environment where students learn through exploration and discovery.

ABOUT THE AUTHORS

Dr. Francisco Borrero, PhD

Research Associate
Cincinnati Museum Center
Cincinnati, OH
Dr. Borrero's research examines the relationship between physical habitat characteristics and the diversity and distribution of natural populations of mollusks.

Dr. Frances Scelsi Hess, Ed.D.

Educational Consultant
Formerly: Teacher
Cooperstown High School
Cooperstown, NY
Dr. Hess taught Earth science and advanced placement environmental science

Dr. Chia Hui (Juno) Hsu, PhD

Associate Project Scientist
University of California
Irvine, CA
Dr. Hsu's research interests include the dynamics of monsoons, climate regime shifts, and modeling global-cale atmospheric chemistry.

Dr. Gerhard Kunze, PhD

Professor Emeritus of Geology
University of Akron
Akron, OH
Dr. Kunze's current research interests include engineering geophysical surveys and digital modeling of geophysical anomalies.

Dr. Stephen A. Leslie, PhD

Professor and Department Head of Geology and Environmental Science
James Madison University
Harrisburg, VA
Dr. Leslie's areas of research include paleontology, stratigraphy, and the evolution of early life on Earth.

Stephen Letro

Meteorologist-in-Charge
National Weather Service Local Forecast Office
Jacksonville, FL
Mr. Letro received his B.S. in meteorology from Florida State University with an emphasis on tropical meteorology. He is a member of National Hurricane Center's Hurricane Liaison Team.

Dr. Michael Manga, PhD

Professor of Earth and Planetary Science
University of California Berkeley
Berkley, CA
Dr. Magna's areas of research include volcanology, the internal evolution and dynamics of planets, and hydrogeology

Len Sharp

Formerly: Teacher of Earth Science
Liverpool High School
Liverpool, NY
Mr. Sharp was president of the Science Teachers Association of New York (1991–1992) and president of the National Earth Science Teachers Association (1992–1994).

Dr. Theodore Snow, PhD

Professor of Astronomy
University of Colorado
Boulder, CO
Dr. Snow's research examines the gas and dust between the stars, called diffuse interstellar bands (DIBs).

Dinah Zike

President and Founder
Dinah-Might Adventures, L.P.
San Antonio, TX
Dina Zike is an inventor who has developed educational products and three-dimensional, interactive graphic organizers for over 30 years. Dinah Zike's Foldables are an exclusive feature of McGraw-Hill textbooks.

UNIT 1
COMPOSITION OF EARTH

ENCOUNTER THE PHENOMENON

How did these crystals grow so large?

 STEM UNIT 1 PROJECT 31

UNIT 2
SURFACE PROCESSES ON EARTH

ENCOUNTER THE PHENOMENON

How did wind, water, and ice shape this landscape?

Coppee Audrey/Shutterstock

UNIT 3
THE ATMOSPHERE AND THE OCEANS

ENCOUNTER THE PHENOMENON

How are animals in the ocean affected by things that happen in the sky?

 STEM UNIT 3 PROJECT 195

UNIT 4
THE DYNAMIC EARTH

ENCOUNTER THE PHENOMENON

Why are the rock layers sideways?

Ian_Redding/iStock/Getty Images

UNIT 5
GEOLOGIC TIME

ENCOUNTER THE PHENOMENON

What can these fossils tell us about Earth millions of years ago?

 STEM UNIT 5 PROJECT 441

Zens photo/Moment/Getty Images

UNIT 6
RESOURCES AND THE ENVIRONMENT

ENCOUNTER THE PHENOMENON

How does coal mining affect both Earth and human communities?

 STEM UNIT 6 PROJECT

David T. Stephenson/Shutterstock

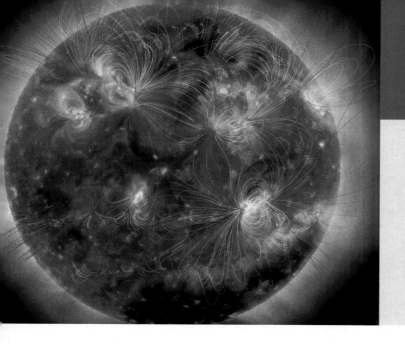

UNIT 7
BEYOND EARTH

ENCOUNTER THE PHENOMENON

How is space different from the planet we call home?

ENCOUNTER THE PHENOMENON

What can maps tell us about our world?

GO ONLINE to play a video about the importance of cartography.

SEP Ask Questions

Do you have other questions about the phenomenon? If so, add them to the driving question board.

CER Claim, Evidence, Reasoning

Make Your Claim Use your CER chart to make a claim about what maps tell us about our world. Explain your reasoning.

Collect Evidence Use the lessons in this module to collect evidence to support your claim. Record your evidence as you move through the module.

Explain Your Reasoning You will revisit your claim and explain your reasoning at the end of the module.

GO ONLINE to access your CER chart and explore resources that can help you collect evidence.

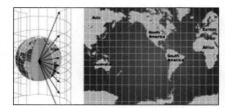

LESSON 2: Explore & Explain: Projections

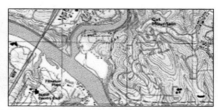

LESSON 3: Explore & Explain: The Geographic Information System

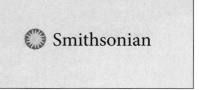

Additional Resources

Video Supplied by BBC Worldwide Learning

WHAT IS EARTH SCIENCE?

FOCUS QUESTION

Why is understanding Earth important?

The Scope of Earth Science

From the maps you use when traveling to the weather report you use when deciding whether or not to carry an umbrella, Earth science is a part of your everyday life. The scope of Earth science is vast. It is partly for this reason that the broad field of Earth science is often broken into five major areas of specialization: astronomy, meteorology, geology, oceanography, and environmental science.

Astronomy

The study of objects beyond Earth's atmosphere is called **astronomy.** Prior to the invention of sophisticated instruments, such as the telescope shown in **Figure 1,** many astronomers merely described the locations of objects in space in relation to each other. Today, Earth scientists study the universe and everything in it, including galaxies, stars, planets, moons, and other bodies they have identified. Astronomers focus on the movement, composition, and structure of bodies in space.

Meteorology

The study of forces and processes that cause the atmosphere to change and produce weather is **meteorology.** Meteorologists also try to forecast the weather and learn how changes in weather over time might affect Earth's climate. They use technology ranging from satellites to thermometers to make forecasts and to analyze trends.

Figure 1 This telescope is one of 13 located on Mauna Kea in Hawaii.

 3D THINKING   **DCI** Disciplinary Core Ideas   **CCC** Crosscutting Concepts   **SEP** Science & Engineering Practices

COLLECT EVIDENCE

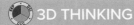 Use your Science Journal to record the evidence you collect as you complete the readings and activities in this lesson.

INVESTIGATE

🧭 **GO ONLINE** to find these activities and more resources.

🥽 **Virtual Investigation: The Nature of Science**
Plan and carry out an investigation using a scientific method to optimize a design solution.

🥽 **Investigation Lab: Observing and Analyzing Stream Flow**
Analyze and interpret data to determine the effect water has on earth materials.

Geology

The study of materials that make up Earth, the processes that form and change these materials, and the history of the planet and its life-forms since its origin is the branch of Earth science known as **geology.** Geologists identify rocks and fossils, study glacial movements, interpret clues to Earth's 4.6-billion-year history, and determine how forces change our planet.

Oceanography

The study of Earth's oceans, which cover nearly three-fourths of the planet, is called **oceanography.** Oceanographers study the creatures that inhabit salt water, measure different physical and chemical properties of the oceans, and observe various processes in these bodies of water. When oceanographers are conducting field research, they often have to dive into the ocean to gather data, as shown in **Figure 2.**

Figure 2 Oceanographers study the life and properties of the ocean.

Investigate *What kind of training would this Earth scientist need?*

Environmental science

The study of interactions among organisms and their surroundings is called **environmental science.** Environmental scientists study how organisms impact the environment both positively and negatively. The topics an environmental scientist might study include the use of natural resources, the effects of pollution, alternative energy sources, and the impact of humans on the atmosphere.

Subspecialties

The study of our planet is a broad endeavor, and, as such, each of the five major areas of Earth science consists of a variety of subspecialties. These include climatology, paleontology, and environmental chemistry. The descriptions of several subspecialties of Earth science are listed in **Table 1.**

Table 1 Subspecialties of Earth Science

Major Area of Study	Subspecialty	Subjects Studied
Astronomy	astrophysics	physics of the universe, including the physical properties of objects in space
	planetary science	planets of the solar system and the processes that form them
Meteorology	climatology	patterns of weather over a long period of time
	atmospheric chemistry	chemistry of Earth's atmosphere and the atmospheres of other planets
Geology	paleontology	remains of organisms that once lived on Earth; ancient environments
	geochemistry	Earth's composition and the processes that change it
Oceanography	physical oceanography	physical characteristics of oceans, such as salinity, waves, and currents
	marine geology	geologic features of the ocean floor, plate tectonics of the ocean
Environmental science	environmental soil science	interactions between humans and the soil, such as the impact of farming practices; effects of pollution on soil, plants, and groundwater
	environmental chemistry	chemical alterations to the environment through pollution and natural means

Figure 3 These Earth scientists are monitoring water quality to ensure that the water is safe for human consumption.

Importance of Earth Science

The study of science, including Earth science, has led to many discoveries that have been applied to address society's needs and problems. The application of scientific discoveries is called technology. Technology is transferable, which means that it can be applied to new situations. Freeze-dried foods, ski goggles, laptops, and the ultralight materials used to make many pieces of sports equipment were created from technologies used in our space program. Smoke detectors were also invented as part of the space program and were adapted for use in everyday life.

Impacts on society

In addition to making life easier, technology can make life safer. For example, Earth scientists monitor and analyze natural disasters ranging from earthquakes to hurricanes. They help predict when these disasters will occur, allowing people to seek shelter or evacuate the area. Earth Scientists also aid in the aftermath of a disaster. Mapping areas affected by natural disasters with satellite and aerial images helps relief workers gain safe access. Relief workers are better able to prepare for the changes in local geography, destruction of buildings, and other physical challenges in the disaster zone.

Earth scientists also work to improve farming methods, helping to produce more food for a growing human population. Precision farming provides farmers with detailed data about soil composition, topography, and pest infestations. The data, gathered by satellites and computer simulations, allow farmers to increase crop yields, reduce waste, and protect natural resources.

In addition to an adequate food supply, people need clean water and clean air. Earth scientists such as those in **Figure 3** monitor water quality and air quality and develop strategies to reduce environmental degradation. Through computer simulations and other studies, important discoveries are being made about how Earth's systems are modified in response to human activities.

Impacts on worldview

Ancient people had a limited view of their surroundings. They were not sure of the shape of Earth or its landmasses. Through exploration and advances in mapping technology, we now have a clear image of Earth and can navigate around the world. **Figure 4** shows some advances in Earth science over time.

Figure 4

Major Events in Earth Science

Many discoveries during the twentieth and early twenty-first centuries revolutionized our understanding of Earth and its systems.

1 **1907** Scientists begin using radioactive decay to determine that Earth is billions of years old. This method will be used to develop the first accurate geological time scale.

2 **1913** French physicists discover the ozone layer in Earth's upper atmosphere and propose that it protects Earth from the Sun's ultraviolet radiation.

3 **1925** Cecilia Payne's analysis of the spectra of stars reveals that hydrogen and helium are the most abundant elements in the universe.

4 **1936** Inge Lehmann proposes that Earth's center consists of a solid inner core and a liquid outer core based on her studies of seismic waves.

5 **1962** Harry Hess's seafloor spreading hypothesis, along with discoveries made about the ocean floor, lays the foundation for plate tectonic theory.

6 **1990** The Hubble Space Telescope goes into orbit, exploring Earth's solar system, measuring the expansion of the universe, and providing evidence of black holes.

7 **2004** A sediment core retrieved from the ocean floor discloses 55-million-year Earth's atmospheric and climatic history. The sample reveals that the North Pole once had a warm climate.

8 **2015** Scientists observe gravitational waves for the first time. The existence of gravitational waves was predicted by Albert Einstein in 1916.

Earth's Systems

Scientists who study Earth have identified four main Earth systems: the geosphere, atmosphere, hydrosphere, and biosphere. Each system is unique, yet each interacts with the others. When investigating Earth's systems, scientists define the boundaries of each system. The inputs and outputs of the systems can be analyzed using models.

Geosphere

The area from the surface of Earth down to its center is called the **geosphere.** The geosphere is divided into three main parts: the crust, mantle, and core. These three parts are illustrated in **Figure 5.**

The rigid outer shell of Earth is called the crust. There are two kinds of crust—continental crust and oceanic crust. Continental crust can be billions of years old. It is generally older than oceanic crust, which is less than 200 million years old. Just below the crust is Earth's mantle. The mantle differs from the crust both in composition and behavior. The mantle ranges in temperature from 100°C to 4000°C—much warmer than the temperatures found in Earth's crust. Below the mantle is Earth's core. Temperatures in the core may be as high as 7000°C.

Atmosphere

The blanket of gases that surrounds our planet is called the **atmosphere.** Earth's atmosphere contains about 78 percent nitrogen and 21 percent oxygen. The remaining 1 percent of gases in the atmosphere include water vapor, argon, carbon dioxide, and other trace gases. Earth's atmosphere provides oxygen for living things, protects Earth's inhabitants from harmful radiation from the Sun, and helps to keep the planet at a temperature suitable for life.

Hydrosphere

All the water on Earth, including the water in the atmosphere, makes up the **hydrosphere.** About 97 percent of Earth's water exists as salt water, while the remaining 3 percent is freshwater contained in lakes and rivers, beneath Earth's surface as groundwater, and in glaciers. The region of permanently frozen water on Earth is called the **cryosphere.** Only a fraction of Earth's total amount of freshwater is in lakes, ponds, streams, and rivers.

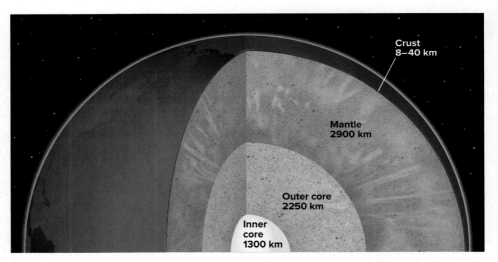

Figure 5 Earth's geosphere is composed of everything from the crust to the center of Earth. Notice how thin the crust is in relation to the rest of the geosphere's components.

CCC CROSSCUTTING CONCEPTS

Systems and Models Make observations of the natural world. Cite evidence of interactions among Earth's systems. Describe how you could model these interactions.

SCIENCE USAGE v. COMMON USAGE

crust

Science usage: the thin, rocky outer layer of Earth

Common usage: the hardened exterior or surface part of bread

Biosphere

The **biosphere** includes all organisms on Earth and the environments in which they live. Most organisms live within a few meters of Earth's surface, but some exist deep within the vast ocean or high atop rugged mountain peaks.

As illustrated in **Figure 6,** Earth's systems are dynamic and interactive. Feedbacks between the biosphere and other Earth systems cause a continual co-evolution of Earth's surface and living things. For example, the atmosphere of early Earth did not contain oxygen. Roughly 2.5 billion years ago, photosynthesizing organisms in Earth's biosphere helped form Earth's present atmosphere by releasing oxygen as a by-product of photosynthesis. The addition of oxygen in the atmosphere greatly influenced the evolution of complex life-forms.

Such interactions among Earth's systems cause feedback effects that can increase or decrease the original changes. For example, when trees in the biosphere are cut down, rates of photosynthesis decrease, and the atmosphere is affected.

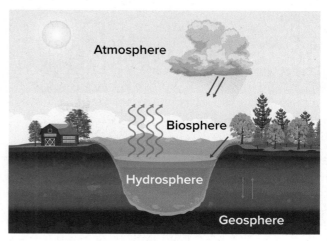

Figure 6 All of Earth's systems are interdependent. Notice how water from the hydrosphere enters the atmosphere, falls on the biosphere, and soaks into the geosphere.

Check Your Progress

Summary

- Earth is divided into four main systems: the geosphere, hydrosphere, atmosphere, and biosphere.
- Earth systems are all dynamic and interactive.
- Identifying the interrelationships among Earth systems leads to specialties and subspecialties of Earth science.
- Earth science has contributed to society and to the development of many items used in everyday life.

Demonstrate Understanding

1. **Explain** why it is helpful to identify specialties and subspecialties of Earth science.
2. **Apply** What are three items you use on a daily basis that have come from research in Earth science?
3. **Hypothesize** how organisms in the biosphere, including humans, can cause the co-evolution of one other Earth system.
4. **Compare and contrast** geology and the geosphere.
5. **Differentiate** between the hydrosphere and the biosphere.

Explain Your Thinking

6. **Predict** what would happen if the composition of the atmosphere changed. How might this affect the biosphere?
7. **WRITING** **Connection** Research an interaction among Earth's systems. Make a flowchart that shows how the interaction causes feedback effects that can increase or decrease the original changes.

LEARNSMART Go online to follow your personalized learning path to review, practice, and reinforce your understanding.

UNDERSTANDING MAPS

FOCUS QUESTION

How do we use maps to describe Earth?

Latitude

Maps are flat models of three-dimensional objects. For thousands of years, people have used maps to define borders and to find places. The science of mapmaking is called **cartography.**

Cartographers use an imaginary grid of parallel lines to locate exact points on Earth. In this grid, the **equator** horizontally circles Earth halfway between the North and South Poles. The equator separates Earth into two equal halves called the northern hemisphere and the southern hemisphere.

Lines on a map running parallel to the equator are called lines of **latitude.** Latitude is the distance in degrees north or south of the equator, as shown in **Figure 7.** The equator, which serves as the reference point for latitude, is numbered 0° latitude. The poles are each numbered 90° latitude. Latitude is thus measured from 0° at the equator to 90° at the poles.

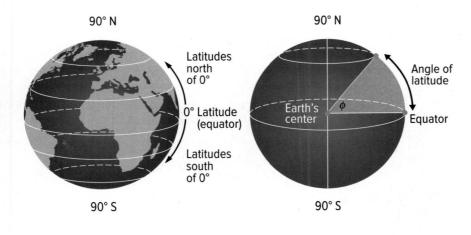

Figure 7 Lines of latitude are parallel to the equator. The value in degrees of each line of latitude is determined by measuring the imaginary angle created between the equator, the center of Earth, and the line of latitude, as seen in the globe on the right.

 **3D THINKING**   **DCI** Disciplinary Core Ideas   **CCC** Crosscutting Concepts   **SEP** Science & Engineering Practices

COLLECT EVIDENCE

 Use your Science Journal to record the evidence you collect as you complete the readings and activities in this lesson.

INVESTIGATE

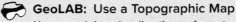 **GO ONLINE** to find these activities and more resources.

GeoLAB: Use a Topographic Map
Use a model to visualize the surface structure of Earth.

Investigation Lab: Interpreting Political and Landform Maps
Develop and use models at different scales to optimize the design of a map.

Locations north of the equator are referred to by degrees north latitude (N). Locations south of the equator are referred to by degrees south latitude (S). For example, Syracuse, New York, is located at 43°N, and Christchurch, New Zealand, is located at 43°S.

Degrees of latitude

Each degree of latitude is equivalent to about 111 km on Earth's surface. How did cartographers determine this distance? Earth is a sphere and can be divided into 360°. The circumference of Earth is about 40,000 km. To find the distance of each degree of latitude, cartographers divided 40,000 km by 360°.

To locate positions on Earth more precisely, cartographers break down degrees of latitude into 60 smaller units, called minutes. The symbol for a minute is '. The actual distance on Earth's surface of each minute of latitude is 1.85 km, which is obtained by dividing 111 km by 60'.

A minute of latitude can be further divided into seconds, which are represented by the symbol ". Longitude is also divided into degrees, minutes, and seconds.

Longitude

To locate positions in east and west directions, cartographers use lines of longitude, also known as meridians. As shown in **Figure 8, longitude** is the distance in degrees east or west of the prime meridian, which is the reference point for longitude.

The **prime meridian** represents 0° longitude. In 1884, astronomers decided that the prime meridian should go through Greenwich, England, home of the Royal Naval Observatory. Points west of the prime meridian are numbered from 0° to 180° west longitude (W); points east of the prime meridian are numbered from 0° to 180° east longitude (E).

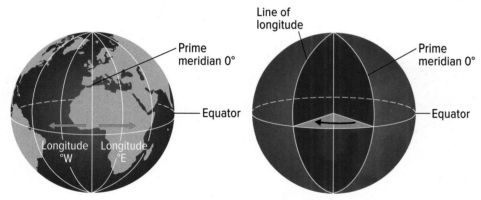

Figure 8 The reference line for longitude is the prime meridian. The degree value of each line of longitude is determined by measuring the imaginary angle created between the prime meridian, the center of Earth, and the line of longitude, as seen on the globe on the right.

SCIENCE USAGE v. COMMON USAGE

minute

Science usage: a unit used to indicate a portion of a degree of latitude

Common usage: a unit of time comprised of 60 seconds

Semicircles

Unlike lines of latitude, lines of longitude are not parallel. Instead, they are large semicircles that extend vertically from pole to pole. For instance, the prime meridian runs from the North Pole, through Greenwich, England, to the South Pole.

The line of longitude on the opposite side of Earth from the prime meridian is the 180° meridian. There, east lines of longitude meet west lines of longitude. This meridian, also known as the International Date Line, will be discussed later in this lesson.

Degrees of longitude

Degrees of latitude cover relatively consistent distances. So, the distance between any two latitude lines is approximately the same everywhere on Earth. The distances covered by degrees of longitude, however, vary with location. Lines of longitude converge at the poles into a point. Thus, one degree of longitude varies from about 111 km at the equator to 0 km at the poles.

 Get It?

Compare degrees of latitude and degrees of longitude.

Using coordinates

Both latitude and longitude are needed to locate positions on Earth precisely. For example, it is not sufficient to say that Charlotte, North Carolina, is located at 35°14′N because that measurement includes any place on Earth located along the 35°14′ line of north latitude.

The same is true of the longitude of Charlotte; 80°50′W could be any point along that longitude from pole to pole. To locate Charlotte, you must use its complete coordinates—latitude and longitude—as shown in **Figure 9.**

Figure 9 The precise location of Charlotte, North Carolina, is 35°14′N, 80°50′W. Note that latitude comes first in reference to the coordinates of a particular location.

Time zones

Earth is divided into 24 time zones. Why 24? Earth takes about 24 hours to rotate once (360°) on its axis. Therefore, there are 24 times zones, each representing a different hour. Every hour Earth spins approximately 15°, so each time zone is 15° wide, corresponding roughly to lines of longitude. To avoid confusion, however, time zone boundaries have been adjusted in local areas so that cities and towns are not split into different time zones.

For example, all of Morton County, North Dakota, operates within the central time zone, even though the western part of the county is within the mountain-time-zone boundary. As shown in **Figure 10,** there are six time zones in the United States. The majority of states within these time zones recognize Daylight Saving Time (DST), wherein clocks are set forward one hour in the spring and back one hour in the fall. This system was put in place to extend daylight hours during warmer months, and so save energy.

International Date Line Each time you travel through a time zone, you gain or lose time until, at some point, you gain or lose an entire day. The **International Date Line,** which is the 180° meridian, serves as the transition line for calendar days. This imaginary line runs through the Pacific Ocean, as shown in **Figure 10.** If you were traveling west across the International Date Line, you would advance your calendar one day. If you were traveling east, you would move your calendar back one day. Note that the International Date Line runs in a general north-south direction, but it tends to follow political boundaries so that countries are not split into different calendar days.

 Get It?

Estimate the time difference between your home and places that are 60° east and west longitude of your home.

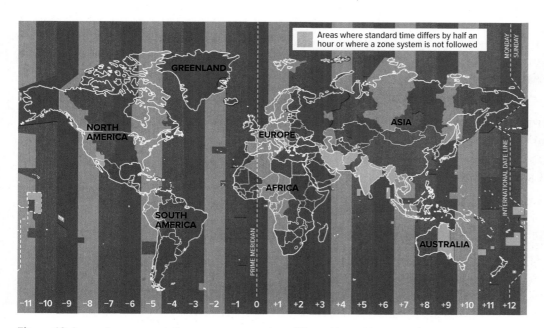

Figure 10 In most cases, each time zone represents a different hour. However, there are some exceptions. **Identify** *two areas where the time zone is not standard.*

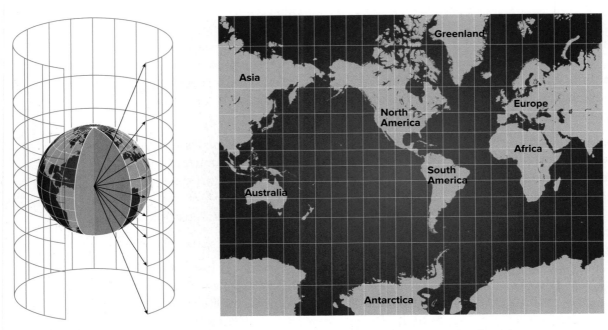

Figure 11 In a Mercator projection, points and lines on a globe are transferred onto cylinder-shaped paper. Mercator projections show true direction but distort the size of the regions near the poles.

Projections

Because Earth is spherical, it is difficult to represent on a piece of paper. For this reason, all flat maps distort to some degree either the shapes or the areas of landmasses. Cartographers use projections to make maps. A map projection is made by transferring points and lines on a globe's surface onto a sheet of paper.

Mercator projections

A **Mercator projection** is a map that has parallel lines of latitude and longitude. Recall that lines of longitude meet at the poles. When lines of longitude are projected as being parallel on a map, landmasses near the poles are exaggerated. So, in a Mercator projection, the shapes of the landmasses are correct, but their areas are distorted.

Figure 11 shows a Mercator projection. As you can see, Greenland appears much larger than Australia. In reality, Greenland is much smaller than Australia. Because Mercator projections show the correct shapes of landmasses and also clearly indicate direction in straight lines, they are most commonly used for navigating ships.

 Get It?

Determine On a Mercator projection, where does most of the distortion occur? Why?

STEM CAREER Connection

GIS Technician

Would you like to create a map that shows how much wood is in the world's forests? Then you may want to explore a career as a Geographic Information System (GIS) technician. GIS technicians have backgrounds in cartography and computer science. They analyze data and create detailed, layered maps that can show data about Earth from the top of its atmosphere to deep underground.

Conic projections

A **conic projection** is made by projecting points and lines from a globe onto a paper cone, as shown in **Figure 12.** The cone touches the globe at a particular line of latitude. There is little distortion in the areas or shapes of landmasses that fall along this line of latitude. Distortion is evident, however, near the top and bottom of the projection. In **Figure 12,** the landmass at the top of the map is distorted.

Because conic projections have a high degree of accuracy for limited areas, they are excellent for mapping small regions. Hence, they are used to make road maps and weather maps.

Gnomonic projections

A **gnomonic (noh MAHN ihk) projection** is made by projecting points and lines from a globe onto a piece of paper that touches the globe at a single point. At the single point where the map is projected, there is no distortion. But outside of this single point, great amounts of distortion are visible both in direction and landmass, as shown in **Figure 13.**

Because Earth is a sphere, it is difficult to plan long travel routes on a flat projection with great distortion, such as a conic projection. To plan such a trip, a gnomonic projection is often used. Although the direction and landmasses on the projection are distorted, it is useful for navigation. A straight line on a gnomonic projection is the straightest route from one point to another when traveled on Earth. This straight line is called a "great circle route" because it represents a segment of the largest circle that can be drawn on a globe.

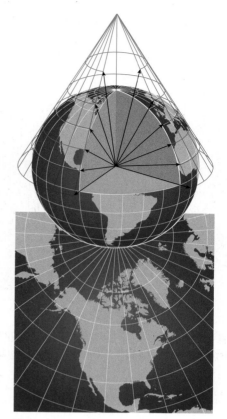

Figure 12 In a conic projection, points and lines on a globe are projected onto cone-shaped paper. There is little distortion along the line of latitude touched by the paper.

 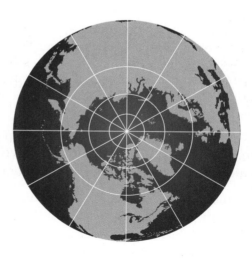

Figure 13 In a gnomonic projection, points and lines from a globe are projected onto paper that touches the globe at a single point.

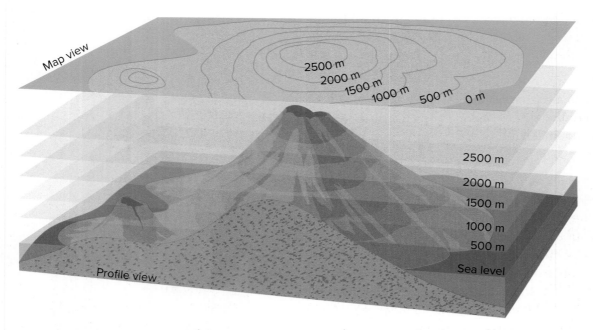

Figure 14 Points of elevation on Earth's surface are projected onto paper to make a topographic map.

Interpret *How many meters high is the highest point on the map?*

Topographic Maps

Detailed maps showing the hills and valleys of an area are called topographic maps. **Topographic maps** show changes in elevation of Earth's surface, as shown in **Figure 14.** They also show mountains, rivers, forests, and bridges, among other features. Topographic maps use lines, symbols, and colors to represent changes in elevation and features on Earth's surface. These types of maps are used for a wide variety of purposes, from hiking to surveying. Because topographic maps include symbols for roads, buildings, and other structures, they are useful for city and county planning.

Contour lines

Elevation on a topographic map is represented by a contour line. Elevation refers to a location's distance above or below sea level. A **contour line** connects points of equal elevation. Because contour lines connect points of equal elevation, they never cross. If they did, it would mean that the point where they crossed had two different elevations, which is impossible.

Contour intervals As **Figure 14** shows, topographic maps use contour lines to show changes in elevation. The difference in elevation between two side-by-side contour lines is called the **contour interval.** The contour interval is dependent on the terrain.

For mountains, the contour lines might be very close together, and the contour interval might be as great as 100 m. This would indicate that the land is steep because there is a large change in elevation between lines. For flat areas, such as plains, the contour lines would be far apart, and the contour interval would be small because there is not much change in elevation over a large area.

 Get It?

Describe how you could use a contour map to determine the best route to take when hiking up a mountain.

Index contours

To aid in the interpretation of topographic maps, some contour lines are marked by numbers representing their elevations. These contour lines are called index contours, and they are used hand-in-hand with contour intervals to help determine elevation.

If you look at a map with a contour interval of 5 m, you can determine the elevations represented by other lines around the index contour by adding or subtracting 5 m from the elevation indicated on the index contour.

Get It?

Analyze If you were looking at a topographic map with a contour interval of 50 m and the contour lines were far apart, would this indicate a rapid increase or slow increase in elevation? Explain your answer.

Depression contour lines

The elevations of some features, such as volcanic craters and mines, are lower than that of the surrounding landscape. Depression contour lines are used to represent such features.

On a map, depression contour lines look like regular contour lines, but they have hachures—short lines at right angles to the contour line—to indicate depressions. As shown in **Figure 15,** the hachures point toward lower elevations. Any point inside a depression contour line is at a lower elevation than the contour line. Any point outside a depression contour line is at a higher elevation than the contour line. This is the opposite of regular contour lines, where points inside a regular contour line are at higher elevations than the contour line, and points outside a regular contour line are at lower elevations than the contour line.

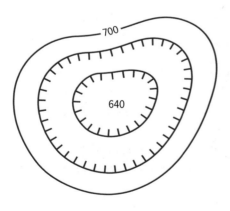

Figure 15 The depression contour lines shown here indicate that the center of the area has a lower elevation than the outer portion of the area. The short lines pointing inward are called hachures and indicate the direction of the elevation change.

Geologic Maps

A useful tool for a geologist is a geologic map. A **geologic map** is used to show the distribution, arrangement, and type of rocks located below the soil. A geologic map can also show features such as fault lines, bedrock, and geologic formations.

Using the information contained on a geologic map, combined with data from visible rock formations, geologists can infer how rocks might look below Earth's surface. They can also gather information about geologic trends, based on the type and distribution of rock shown on the map.

Geologic maps are most often superimposed over topographic maps and color coded by the type of rock formation, as shown in the map of the Grand Canyon in **Figure 16.** Each color corresponds to the type of rock present in a given area. Symbols are used to represent mineral deposits and other structural features. The key under the map shows what the colors and symbols represent.

Study the map shown in **Figure 16** closely. Notice the abundance of Older Precambrian rock formations in the Grand Canyon. These rock formations formed relatively early in Earth's geologic past. Extensive uplift and erosion by water and wind have resulted in the exposure of these ancient rocks.

Three-dimensional maps

Topographic and geologic maps are two-dimensional models of Earth's surface. They are flat, so symbols, numbers, and lines must be used to interpret Earth's features. Sometimes, scientists need to visualize Earth three-dimensionally. For example, they may want to study the path of a river down a mountain. To do this, scientists often rely on computers to digitize features such as rivers, mountains, valleys, and hills, creating three-dimensional maps. Refer to **Table 2** on the following page to compare three-dimensional maps to the other maps you have learned about in this module.

Geologic Map of Grand Canyon

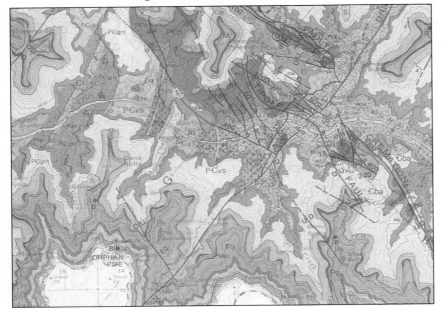

Figure 16 Geologic maps show the distribution of surface geologic features. This map of the Grand Canyon uses colors and symbols to distinguish different types of rock.

QUATERNARY
- S Landslides and rockfalls
- r River sediment

PERMIAN
- Pk Kaibab Limestone
- Pt Toroweap Formation
- Pc Coconino Sandstone
- Ph Hermit Shale
- Pe Esplanade Sandstone

PENNSYLVANIAN
- Ps Supai Formation

MISSISSIPPIAN
- Mr Redwall Limestone

DEVONIAN
- Dtb Temple Butte
       Limestone

CAMBRIAN
- Cm Muav Limestone
- Cba Bright Angel Shale
- Ct Tapeats Sandstone

YOUNGER PRECAMBRIAN
- PCi Diabase sills and dikes
- PCs Shinumo Quartzite
- PCh Hakatai Shale
- PCb Bass Formation

OLDER PRECAMBRIAN
- PCgr Zoroaster Granite
- PCgnt Trinity Gneiss
- PCvs Vishnu Schist

Table 2 Types of Maps and Projections

Map or Projection	Common Uses	Distortions
Mercator projection	ship navigation	The land near the poles is distorted.
Conic projection	road and weather maps	The areas at the top and the bottom of the map are distorted.
Gnomonic projection	great circle routes	The direction and distance between land-masses are distorted.
Topographic map	show elevation changes on a flat projection	It depends on the type of projection used.
Geologic map	show the types of rocks present in a given area	It depends on the type of projection used.
Three-dimensional map	aid in conceptualizing geologic structures and processes	It depends on the type of projection used.

Map Legends

Most maps include both human-made and natural features located on Earth's surface. These features are represented by symbols, such as black dotted lines for trails, solid red lines for highways, and small black squares and rectangles for buildings. A **map legend,** such as the one shown in **Figure 17,** explains what the symbols represent.

 Get It?

Identify You see a blue line on a map. Which feature on Earth's surface does this represent?

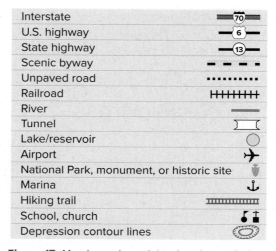

Interstate	
U.S. highway	
State highway	
Scenic byway	
Unpaved road	
Railroad	
River	
Tunnel	
Lake/reservoir	
Airport	
National Park, monument, or historic site	
Marina	
Hiking trail	
School, church	
Depression contour lines	

Figure 17 Map legends explain what the symbols on maps represent.

Map Scales

When using a map, you need to know how to measure distances. This is accomplished by using a map scale. A **map scale** is the ratio between distances on a map and actual distances on the surface of Earth. Normally, map scales are measured in SI units, such as kilometers and centimeters, but sometimes they are measured in different units, such as miles and inches. There are three types of map scales: verbal scales, graphic scales, and fractional scales.

Verbal scales

To express distance as a statement, such as "one centimeter is equal to one kilometer," Earth scientists use verbal scales. The verbal scale, in this example, means that one centimeter on the map represents one kilometer on Earth's surface.

Graphic scales

Instead of stating the map scale in words, graphic scales consist of a line that represents a certain distance, such as 5 km or 5 miles. The line is labeled and divided into sections with hash marks, with each section representing a distance on Earth's surface. For instance, a graphic scale of 5 km might be divided into five sections, with each section representing 1 km. Graphic scales are the most common type of map scale.

 Get It?

Infer why an Earth scientist might use different types of scales on different types of maps.

Fractional scales

Fractional scales express distance as a ratio, such as 1:63,500. This means that one unit of distance on the map represents 63,500 units of distance on Earth's surface. One centimeter on a map, for instance, would be equivalent to 63,500 centimeters on Earth's surface. Any unit of distance can be used in fractional scales, but the units on each side of the ratio must always be the same.

A large ratio indicates that the map represents a large area, while a small ratio indicates that the map represents a small area. Therefore, a map with a large fractional scale, such as 1:100,000 km, would show less detail than a map with a small fractional scale, such as 1:1000 km.

Check Your Progress

Summary

- Latitude lines run parallel to the equator. Longitude lines run vertically, wrapping around Earth and meeting at the North and South Poles.
- Both latitude and longitude lines are necessary to locate exact places on Earth.
- Earth is divided into 24 time zones, each 15° wide.
- Different types of map projections are used for different purposes.
- Geologic maps help Earth scientists study large-scale patterns in geologic formations.
- Maps often include a map legend for interpreting map symbols and a map scale for determining distances.

Demonstrate Understanding

1. **Explain** why it is important to include both latitude and longitude when giving coordinates.
2. **Describe** how the distance of a degree of longitude varies from the equator to the poles.
3. **Compare and contrast** Mercator and gnomonic projections. What are these projections commonly used for?
4. **Describe** what it would be like to fly from where you live to Paris, France. How many time zones would you cross? What would it be like to adjust to the time difference?

Explain Your Thinking

5. **Evaluate** If you were flying directly south from the North Pole and reached 70°N, how many degrees of latitude would be between you and the South Pole?
6. **Model** how a conic projection is made. Why is this type of projection best suited for small areas?
7. **WRITING** ▸**Connection** Suppose you are a city planner. Explain how a geologic map could help you decide where to build a city park.

LEARNSMART Go online to follow your personalized learning path to review, practice, and reinforce your understanding.

REMOTE SENSING

FOCUS QUESTION

What tools do we use to map Earth?

Landsat Satellite

Advanced technology has changed the way maps are made. The process of gathering data about Earth using instruments mounted on satellites, airplanes, or ships is called **remote sensing.**

One form of remote sensing uses satellites. Features on Earth's surface, such as rivers and forests, radiate warmth at slightly different frequencies. **Landsat satellites** record reflected wavelengths of energy from Earth's surface. To obtain images such as the one in **Figure 18,** each Landsat satellite is equipped with a moving mirror that scans Earth's surface. This mirror has rows of detectors that measure the intensity of energy received from Earth. Computers then convert this information into digital images that show landforms in great detail.

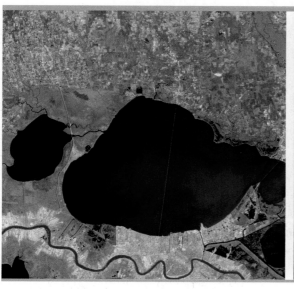

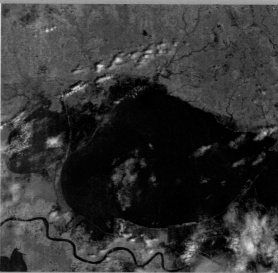

Figure 18 Landsat satellites are used to study pollution, the movements of Earth's plates, and the melting of glaciers and ice caps. They are also used to aid in natural disaster relief planning. Notice the differences between the two Landsat photos of New Orleans.

Interpret *Which image was taken after Hurricane Katrina in 2005? Explain.*

3D THINKING **DCI** Disciplinary Core Ideas **CCC** Crosscutting Concepts **SEP** Science & Engineering Practices

COLLECT EVIDENCE

 Use your Science Journal to record the evidence you collect as you complete the readings and activities in this lesson.

INVESTIGATE

GO ONLINE to find these activities and more resources.

 Review the News
Obtain information from a current news story about current remote sensing research. Evaluate your source and communicate your findings to your class.

 Revisit the Encounter the Phenomenon Question
What information from this lesson can help you answer the Unit and Module questions?

OSTM/Jason Satellites

One satellite that uses radar to measure and map sea surface height is the *OSTM/Jason-3* satellite. **OSTM** stands for **O**cean **S**urface **T**opography **M**ission and is a follow-on to the *TOPEX/Poseidon, Jason-1,* and *Jason-2* satellites. Radar uses high-frequency signals that are transmitted from the satellite to the surface of the ocean. A receiving device then picks up the returning echo as it is reflected off the water.

The distance to the water's surface is calculated using the known speed of light and the time it takes for the signal to be reflected. These data are used to make images, such as the one in **Figure 19,** that are important for measuring variations in sea level and monitoring changes in global ocean currents and heat transfer. A rise in regional or global sea level is often caused by the melting of ice sheets and glaciers, which is associated with climate change. A rise in ocean surface temperatures is also indicative of climate change.

Using *OSTM/Jason* satellite data, scientists are able to accurately estimate global sea level to within a few millimeters. Scientists can use these data combined with other existing data to create maps of ocean-floor features. For instance, ocean water bulges over sea-floor mountains and forms depressions over seafloor valleys.

SeaBeam

SeaBeam technology is also used to map the ocean. **Figure 20** shows an example of a map created with information gathered with SeaBeam technology. To map ocean-floor features, SeaBeam relies on **sonar,** which is the use of sound waves to detect and measure objects underwater.

First, to gather the information needed to map the seafloor, a sound wave is sent from a ship toward the ocean floor. A receiving device then picks up the returning echo when it bounces off the seafloor.

Computers on the ship calculate the distance from the ship to the ocean floor using the speed of sound in water and the time it takes for the sound to be reflected. SeaBeam technology is used by fishing fleets; deep-sea drilling operations; and scientists such as oceanographers, volcanologists, and archaeologists. See **Figure 21** for a history of mapping technology.

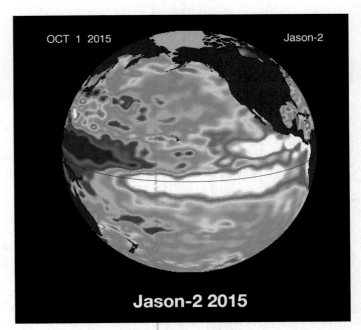

Figure 19 This image, which focuses on the Pacific Ocean, was created with data from *OSTM/Jason-2*. The red color along the equator shows the rise in ocean depth and temperature (relative to normal) during an El Niño event.

Figure 20 This offshore image of an area near San Diego, CA, was created with data from SeaBeam. The change in color indicates a change in elevation. The red-orange colors are the highest elevations, and the blue colors are the lowest.

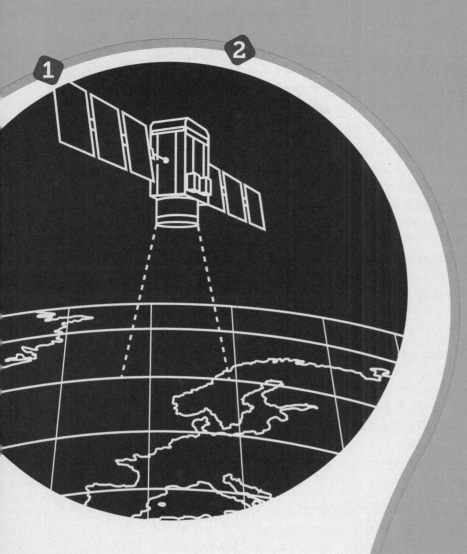

Figure 21

Mapping Technology

Advances in mapping have relied on technological developments.

1 **1300 B.C.** An ancient Egyptian scribe draws the oldest surviving topographical map.

2 **150 B.C.** The ancient Greek scientist Ptolemy creates the first map using a coordinate grid. It depicts Earth as a sphere and includes Africa, Asia, and Europe.

3 **A.D. 1154** Arab scholar Al-Idrisi creates a world map used by European explorers for several centuries. Earlier medieval maps showed Jerusalem as the center of a flat world.

4 **1569** Flemish geographer Gerhardus Mercator devises a way to project the globe onto a flat map using lines of longitude and latitude.

5 **1752** A French cartographer first uses contour lines to represent elevation and marine depth for sailors exploring the New World.

6 **1875** American governess Ellen Eliza Fitz invents a method to mount a globe that shows the position of the Sun and the length of nights and days.

7 **1966** Harvard University researchers develop the first computerized grid-based mapping system, the forerunner of GIS.

8 **2000** Space shuttle *Endeavour* collects the most complete topographical data of Earth, mapping almost 80 percent of Earth's land surface.

9 **2016** *OSTM/Jason-3* launches. The satellite measures the height of the ocean's surface to help scientists study climate change.

The Global Positioning System

The **Global Positioning System (GPS)** is a satellite navigation system that allows users to locate their approximate position on Earth. There are at least 24 satellites orbiting Earth for use with GPS units. The satellites are positioned around Earth and are constantly orbiting so that signals from at least three or four satellites can be picked up at any given moment by a GPS receiver.

You need a GPS receiver to find your location on Earth. The receiver calculates your approximate latitude and longitude—usually within 10 m—by processing the signals emitted by the satellites. If enough information is present, these satellites can also relay information about elevation, direction of movement, and speed. With signals from three satellites, a GPS receiver can calculate locations on Earth without elevation, while four satellite signals will allow a GPS receiver to also calculate elevation. For more information on how the satellites are used to determine location, see **Figure 22** on the next page.

Uses for GPS technology GPS technology is used extensively for navigation by airplanes and ships. However, it is also used to help detect earthquakes, create maps, and track wildlife.

GPS technology also has many applications for everyday life. GPS receivers are often placed in cars to help navigate to preprogrammed destinations such as restaurants, hotels, and homes. Portable, handheld GPS systems are also used in hiking, biking, and other travels. They allow for finding destinations more quickly and can help in determining specific locations on a map. Cell phones may also contain GPS systems that aid in finding locations.

The Geographic Information System

The **Geographic Information System (GIS)** combines many of the traditional types and styles of mapping described in this module. GIS mapping uses a database of information gathered by scientists, professionals, and students like you from around the world to create layers, or "themes," of information that can be placed one on top of the other to create a comprehensive map. These "themes" are often maps that were created with information gathered by remote sensing.

Scientists from many disciplines use GIS technologies. A geologist might use GIS mapping when studying a volcano to help track historical eruptions. An ecologist might use GIS mapping to track pollution or to follow animal or plant population trends of a given area.

 Get It?

Compare applications for GPS and GIS technology. Give an example of how you would use each type of technology in a scientific investigation.

ACADEMIC VOCABULARY

comprehensive
covering completely or broadly
The teacher gave the students a comprehensive study guide for the final exam.

Figure 22 Visualizing GPS Satellites

GPS receivers detect signals from a core group of at least 24 GPS satellites orbiting Earth. Using signals from at least three satellites, the receiver can calculate location to within 10 m.

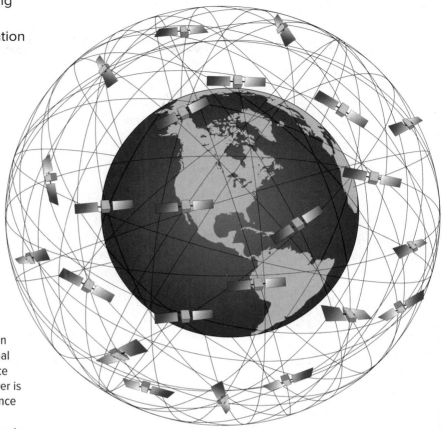

First, a GPS receiver, located in New York City, receives a signal from one satellite. The distance from the satellite to the receiver is calculated. Suppose the distance is 20,000 km. This limits the possible location of the receiver to anywhere on a sphere 20,000 km from the satellite.

Next, the receiver measures the distance to a second satellite. Suppose this distance is calculated to be 21,000 km away. The location of the receiver has to be somewhere within the area where the two spheres intersect, shown here in orange.

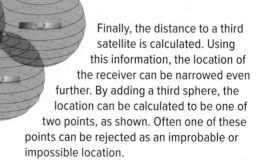

Finally, the distance to a third satellite is calculated. Using this information, the location of the receiver can be narrowed even further. By adding a third sphere, the location can be calculated to be one of two points, as shown. Often one of these points can be rejected as an improbable or impossible location.

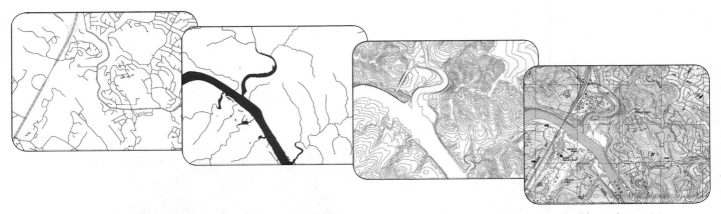

Figure 23 GIS mapping involves layering one map on top of another. In this image, you can see how one layer builds on the next.

GIS maps might contain many layers of information compiled from different types of maps, such as a geologic map and a topographic map. As shown in **Figure 23,** maps of rivers, topography, roads, cities, and other notable landforms from the same geographic area can be layered on top of each other to create one comprehensive map.

In addition, GIS can interpret aerial photographs and data from spreadsheets, such as population trends. Digital images from satellites can also be incorporated into the system to produce a comprehensive map. GIS and other technologies increase the ability of scientists to model, predict, and manage current and future impacts of human activities.

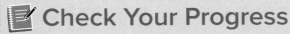

Check Your Progress

Summary

- Remote sensing is an important part of modern cartography.
- Satellites are used to gather data about features on Earth's surface.
- Sonar is also used to gather data about features on Earth's surface.
- GPS is a navigational tool that is now used for many everyday applications.
- GIS mapping uses different databases to create comprehensive maps.

Demonstrate Understanding

1. **Describe** how remote sensing works and why it is important to cartography.

2. **Apply** Why is GPS navigation important to Earth scientists?

3. **Compare and contrast** SeaBeam images with *OSTM/Jason-2* images and how each might be used.

4. **Predict** why it might be important to be able to add and subtract map layers, as with GIS mapping.

Explain Your Thinking

5. **Infer** How could GIS mapping be helpful in determining where to build a housing development?

6. **Explain** why it is important to have maps of the ocean floor, such as those gathered with SeaBeam technology.

7. **WRITING Connection** The impact of human activities on Earth's systems is greater than ever. Explain how remote sensing can help scientists reduce the impact of human activities.

U.S. Geological Survey

LEARNSMART Go online to follow your personalized learning path to review, practice, and reinforce your understanding.

A Bird's-Eye View of Disasters

Uncrewed aerial vehicles (UAVs), more commonly known as drones, are changing how governments respond to disasters such as hurricanes and earthquakes. Drones can be deployed quickly, can fly into places that are not safe or accessible to humans, and are less expensive than traditional aircraft. The data that drones gather are used to coordinate disaster relief and plan rebuilding.

After Hurricane Maria hit Puerto Rico in 2017, a drone captured this image of a destroyed road.

Assistance from Above

One might wonder how such small aircraft can make a difference in a large-scale disaster. The key is their versatility and flexibility. Drones do not need an airstrip to take off. They can be in the air almost immediately after a disaster, and they can fly long distances on battery power. Drones are equipped with cameras that first responders use to locate obstacles—such as crumpled roads and bridges—that could impede them as they move into areas to help survivors.

Some drones are operated by pilots on the ground, while others use autonomous flight and coordination capability. Drones can gather information with neural network image recognition. To identify damage, they compare pre-disaster images of the affected area to what they are "seeing" now. Imaging drones can be deployed in a grid pattern so that no part of an area is missed. In addition, drones can concentrate on important sites, such as bridges, railways, and hospitals.

Drones Around the World

China uses drones to help first responders find survivors in collapsed buildings after earthquakes. In Africa, Malawi is testing the use of drones to deliver medical supplies to remote areas, and Rwanda transports blood supplies to hospitals via a drone network.

The United States used drones to survey damage in Texas, Florida, and Puerto Rico during the intense hurricane season of 2017. In that same year, Mexico used drones to assist its recovery from a 7.1-magnitude earthquake.

In the future, areas that experience frequent disasters could have sets of drones with secure power sources ready to use. After a disaster, these drones could be employed to survey the damage and to locate survivors.

ASK QUESTIONS TO CLARIFY

Brainstorm and write two questions you have about how drones can aid first responders in disaster areas. Use print or online sources to find answers. Present your research to your class.

Ricardo Arduengo/AFP/Getty Images

GO ONLINE to study with your Science Notebook.

Lesson 1 WHAT IS EARTH SCIENCE?

- Earth is divided into four main systems: the geosphere, hydrosphere, atmosphere, and biosphere.
- Earth systems are all dynamic and interactive.
- Identifying the interrelationships among Earth systems leads to specialties and subspecialties.
- Earth science has contributed to society and to the development of many items used in everyday life.

- astronomy
- meteorology
- geology
- oceanography
- environmental science
- geosphere
- atmosphere
- hydrosphere
- cryosphere
- biosphere

Lesson 2 UNDERSTANDING MAPS

- Latitude lines run parallel to the equator. Longitude lines run east and west of the prime meridian.
- Both latitude and longitude lines are necessary to locate exact places on Earth.
- Earth is divided into 24 time zones, each 15° wide.
- Different types of map projections are used for different purposes.
- Geologic maps help Earth scientists study large-scale patterns in geologic formations.
- Maps often include a map legend for interpreting map symbols and a map scale for determining distances.

- cartography
- equator
- latitude
- longitude
- prime meridian
- International Date Line
- Mercator projection
- conic projection
- gnomonic projection
- topographic map
- contour line
- contour interval
- geologic map
- map legend
- map scale

Lesson 3 REMOTE SENSING

- Remote sensing is an important part of modern cartography.
- Satellites and sonar are used to gather data about features on Earth's surface.
- GPS is a navigational tool that is now used for many everyday applications.
- GIS mapping uses different databases to create comprehensive maps.

- remote sensing
- Landsat satellite
- sonar
- Global Positioning System (GPS)
- Geographic Information System (GIS)

Module Wrap-Up

REVISIT THE PHENOMENON

What can maps tell us about our world?

CER Claim, Evidence, Reasoning

Explain Your Reasoning Revisit the claim you made when you encountered the phenomenon. Summarize the evidence you gathered from your investigations and research and finalize your Summary Table. Does your evidence support your claim? If not, revise your claim. Explain why your evidence supports your claim.

GO FURTHER

SEP Data Analysis Lab
How can you analyze changes in elevation?

Gradient refers to the steepness of a slope. To measure gradient, divide the change in elevation between two points on a map by the distance between the two points.

Data and Observations Use the map scale to determine the distance from Point A to Point B on the map. Convert your answers to SI units. Record the change in elevation.

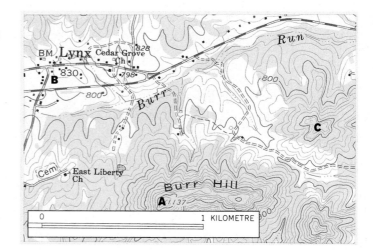

CER Analyze and Interpret Data

1. **Claim** If you were to hike the distance from Point A to Point B, what would be the gradient of your climb?

2. **Reasoning** Would it be more difficult to hike from Point A to Point B or from Point B to Point C? Explain.

3. **Claim, Evidence** Between Point A and Point C, where is the steepest part of the hike? How do you know?

ENCOUNTER THE PHENOMENON

How did these crystals grow so large?

SEP Ask Questions

What questions do you have about the phenomenon? Write your questions on sticky notes and add them to the driving question board for this unit.

What are rocks and minerals made of?

Look for Evidence

As you go through this unit, use the information and your experiences to help you answer the phenomenon question as well as your own questions. For each activity, record your observations in a Summary Table, add an explanation, and identify how it connects to the unit and module phenomenon questions.

Solve a Problem
STEM UNIT PROJECT

Composition of Earth Investigate and research more about Earth's composition. Use the results of these investigations and the evidence you collected during the unit to complete your unit project.

GO ONLINE In addition to reading the information in your Student Edition, you can find the STEM Unit Project and other useful resources online.

MATTER AND CHANGE

ENCOUNTER THE PHENOMENON

How do we find and remove harmful substances from our drinking water?

GO ONLINE to play a video about changing the pH of mine water.

SEP Ask Questions

Do you have other questions about the phenomenon? If so, add them to the driving question board.

CER Claim, Evidence, Reasoning

Make Your Claim Use your CER chart to make a claim about how we find and remove harmful substances from our drinking water. Explain your reasoning.

Collect Evidence Use the lessons in this module to collect evidence to support your claim. Record your evidence as you move through the module.

Explain Your Reasoning You will revisit your claim and explain your reasoning at the end of the module.

GO ONLINE to access your CER chart and explore resources that can help you collect evidence.

LESSON 2: Explore & Explain: Chemical Reactions

LESSON 2: Explore & Explain: Mixtures and Solutions

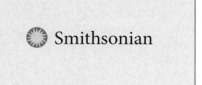

Smithsonian

Additional Resources

FOCUS QUESTION

What is water made of?

Atoms

Matter is anything that has volume and mass. Everything in the physical world that surrounds you is composed of matter. On Earth, matter usually occurs as a solid, a liquid, or a gas. All matter is made of substances called elements. An **element** is a substance that cannot be broken down into simpler substances by physical or chemical means. For example, gold, which is often used in jewelry, is so soft that it can be molded, hammered, sculpted, or drawn into wire. Whatever its size or shape, the gold is still gold. Gold is a type of element.

Each element has unique physical and chemical properties. Although aluminum has different properties from gold, both aluminum and gold are elements that are made up of atoms. All atoms consist of even smaller particles—protons, neutrons, and electrons. **Figure 1** shows one method of representing an atom. The center of an atom is called the nucleus (NEW klee us) (plural, *nuclei*). The **nucleus** of an atom is made up of protons and neutrons. A **proton** (p) is a tiny particle that has mass and a positive electric charge. A **neutron** (n) is a particle with approximately the same mass as a proton, but it is electrically neutral; that is, it has no electric charge. All atomic nuclei have positive charges because they are composed of protons with positive electric charges and neutrons with no electric charge.

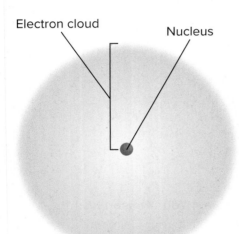

Electron cloud

Nucleus

Atom

Figure 1 In this representation of an atom, the fuzzy area surrounding the nucleus is referred to as an electron cloud.

🌀 **3D THINKING**   **DCI** Disciplinary Core Ideas   **CCC** Crosscutting Concepts   **SEP** Science & Engineering Practices

COLLECT EVIDENCE

📝 Use your Science Journal to record the evidence you collect as you complete the readings and activities in this lesson.

INVESTIGATE

🌐 **GO ONLINE** to find these activities and more resources.

🥽 **Quick Investigation: Identify Elements**
Obtain and communicate information about the function and use of natural resources in your environment.

🥽 **Virtual Investigation: Properties of Elements**
Analyze and interpret data to identify elements by their properties.

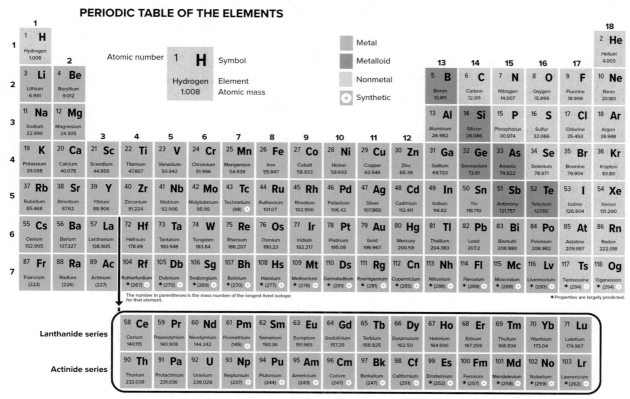

Figure 2 The periodic table of the elements is arranged so that a great deal of information about all of the known elements is provided in a small space.

Surrounding the nucleus of an atom are smaller particles called electrons. An **electron** (e⁻) has little mass, but it has a negative electric charge that is exactly the same magnitude as the positive charge of a proton. When an atom has an equal number of protons and electrons, the electric charge of an electron cancels the positive charge of a proton, resulting in an atom that has no overall charge. Notice that the electrons in **Figure 1** are shown as a cloudlike region surrounding the nucleus. This is because electrons are in constant motion around an atom's nucleus, and their exact positions at any given moment cannot be determined.

 Get It?

Describe the charges of the three atomic particles in a neutral atom.

Symbols for elements

There are 92 elements that occur naturally on Earth and in the stars. Other elements have been produced in laboratory experiments. Generally, each element is identified by a one-, two-, or three-letter abbreviation known as a chemical symbol. For example, the symbol H represents the element hydrogen, C represents carbon, and O represents oxygen. Elements identified in ancient times, such as gold and mercury, have symbols of Latin origin. For example, gold is identified by the symbol Au for its Latin name, *aurum,* and mercury is identified by the symbol Hg for its Latin name, *hydrargyrum.* All elements are classified and arranged according to their chemical properties in the periodic table of the elements, shown in **Figure 2.**

Mass number

The number of protons and neutrons in atoms of different elements varies widely. The lightest of all atoms is hydrogen, which has only one proton in its nucleus. The heaviest naturally occurring atom is uranium. Uranium-238 has 92 protons and 146 neutrons in its nucleus. The number of protons in an atom's nucleus is its **atomic number.** The sum of the protons and neutrons is its **mass number.** Because electrons have little mass, they are not included in determining mass number. For example, the atomic number of uranium is 92, and its mass number is 238 (92 protons + 146 neutrons). **Figure 3** illustrates how atomic numbers and mass numbers are listed in the periodic table of the elements.

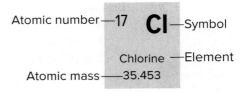

Atomic number—17 **Cl** —Symbol

Chlorine —Element

Atomic mass—35.453

Figure 3 The element chlorine is atomic number 17.

Isotopes

Recall that all atoms of an element have the same number of protons. However, the number of neutrons of an element's atoms can vary. For example, all chlorine atoms have 17 protons in their nuclei, but they can have either 18 or 20 neutrons. This means that a chlorine atom could have a mass number of 35 (17 protons + 18 neutrons) or 37 (17 protons + 20 neutrons). Atoms of the same element that have the same number of protons but a different number of neutrons, and thus different mass numbers, are called **isotopes.** The element chlorine has two isotopes: Cl-35 and Cl-37. Because the number of electrons in a neutral atom equals the number of protons, isotopes of an element have the same chemical properties.

Look again at the periodic table in **Figure 2.** Scientists have measured the mass of atoms of elements. The atomic mass of an element is the average of the mass numbers of the isotopes of an element. Most elements are mixtures of isotopes. For example, notice in **Figure 2** that the atomic mass of chlorine is 35.453. This number is the average of the mass numbers of the naturally occurring isotopes of chlorine-35 and chlorine-37.

Radioactive isotopes

The nuclei of some isotopes are unstable and tend to break down. When this happens, the isotope also emits energy in the form of radiation. Radioactive decay is the spontaneous process through which unstable nuclei emit radiation. In the process of radioactive decay, a nucleus will either lose protons and neutrons, change a proton to a neutron, or change a neutron to a proton. Because the number of protons in a nucleus identifies an element, decay also changes the identity of an element. For example, the isotope polonium-218 decays at a steady rate over time into bismuth-214. The polonium originally present in a rock is gradually replaced by bismuth. The process of radioactive decay is often used to calculate the ages of rocks.

HISTORY Connection Some isotopes have played an important role in deciphering the history of ancient people. For example, carbon-14 (C-14), an isotope of carbon, can be used to date organic remains up to 60,000 years old. Testing of clothing, dried food, and even mummies have provided date ranges for the lifetimes of ancient people, plants, and animals. Carbon-14 is used for remains that are much more recent as well. In recent years, carbon-14 has been used to solve modern missing persons cases by dating found remains and comparing the ages to missing persons cases in the area. The technique is becoming so refined that some researchers claim that date of death and even year of birth, can be determined using carbon-14.

Electrons in Energy Levels

Although the exact position of an electron cannot be determined, scientists have discovered that electrons occupy areas called energy levels. Look again at **Figure 1** and the relative size of the nucleus compared to the electron cloud. The volume of an atom is mostly empty space. However, the size of an atom depends on the number and arrangement of its electrons.

Filling energy levels

Figure 4 presents a model to help you visualize atomic particles. Note that electrons are distributed over one or more energy levels in a predictable pattern. Keep in mind that the electrons are not sitting still in one place. Each energy level can hold only a limited number of electrons. For example, the smallest, innermost energy level can hold only two electrons, as illustrated by the oxygen atom in **Figure 4.** The second energy level is larger, and it can hold up to eight electrons. The third energy level can hold up to 18 electrons, and the fourth energy level can hold up to 32 electrons. Depending on the element, an atom might have electrons in as many as seven energy levels surrounding its nucleus.

 Get It?

Analyze Look at **Figure 4.** How can you use the number of electrons in an aluminum atom to determine how many protons and neutrons are in its nucleus?

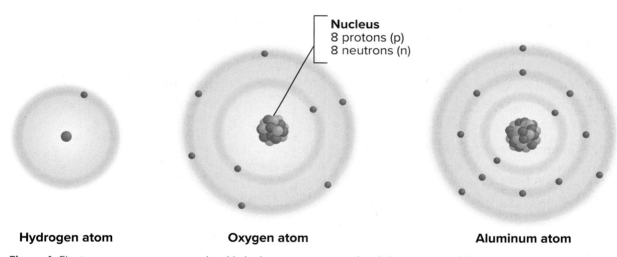

Nucleus
8 protons (p)
8 neutrons (n)

Hydrogen atom   **Oxygen atom**   **Aluminum atom**

Figure 4 Electrons occupy one energy level in hydrogen, two energy levels in oxygen, and three energy levels in aluminum.

Valence electrons

The electrons in the outermost energy level determine the chemical behavior of an element. These outermost electrons are called valence electrons. Elements with the same number of valence electrons have similar chemical properties. For example, on the left side of the periodic table are sodium and potassium; they have different atomic numbers, but each has one valence electron. Thus, sodium and potassium exhibit similar chemical behavior. These elements are highly reactive metals, which means that they combine easily with many other elements. For example, sodium and potassium both combine easily with chlorine, making sodium chloride (NaCl), also known as table salt, and potassium chloride (KCl), a salt sometimes used to treat low levels of potassium in the blood.

On the right side of the periodic table are the noble gases. These elements, including helium and argon, have full outermost energy levels. For example, an argon atom, shown in **Figure 5,** has 18 electrons, with two electrons in the first energy level and eight electrons in the second and third (outermost) energy levels. Elements that have full outermost energy levels are highly unreactive. Compare this to the elements oxygen and hydrogen. These are also gases at room temperature, but they are highly flammable. Their outermost electron shells are not full so they readily react with other elements.

Ions

Sometimes atoms gain or lose electrons from their outermost energy levels. Recall that atoms are electrically neutral when the number of electrons, which have negative charges, balances the number of protons, which have positive charges. An atom that gains or loses an electron has a net electric charge and is called an **ion.** In general, an atom in which the outermost energy level is less than half-full—that is, it has fewer than four valence electrons—tends to lose its valence electrons. When an atom loses valence electrons, it becomes positively charged. In chemistry, a positive ion is indicated by a superscript plus sign. For example, a sodium ion is represented by Na^+. If more than one electron is lost, that number is placed before the plus sign. For example, a magnesium ion, which forms when a magnesium atom has lost two electrons, is represented by Mg^{2+}.

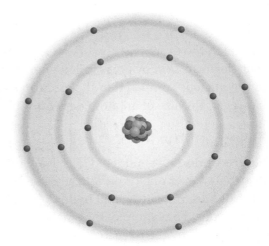

Argon atom

Figure 5 The inert nature of argon makes it an ideal gas to use inside an incandescent lightbulb because it does not react with the extremely hot filament.

 Get It?

Explain what makes an ion positive.

An atom in which the outermost energy level is more than half-full—that is, it has more than four valence electrons—tends to fill its outermost energy level. Such an atom forms a negatively charged ion. Negative ions are indicated by a superscript minus sign. For example, a nitrogen atom that has gained three electrons is represented by N^{3-}. Some substances contain ions that are made up of groups of atoms—for example, silicate ions $(SiO_4)^{4-}$. These complex ions are important constituents of most rocks and minerals.

CCC CROSSCUTTING CONCEPTS

Patterns The periodic table of the elements is based on the very orderly nature of atoms and the particles that form them. Patterns can be observed across all of the periodic table. Using **Figure 5** as your guide, predict what the next atom will be if one more electron energy level is filled. How many protons, neutrons, and electrons will there be?

ACADEMIC VOCABULARY

exhibit

to show or display outwardly
The dog exhibited aggression by baring its teeth and growling.

Abundance of Elements

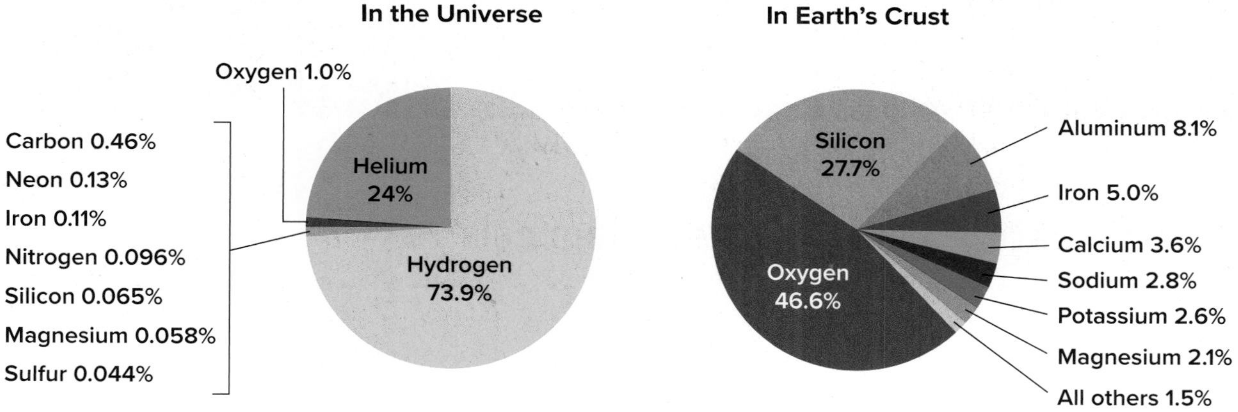

In the Universe

Oxygen 1.0%

Carbon 0.46%
Neon 0.13%
Iron 0.11%
Nitrogen 0.096%
Silicon 0.065%
Magnesium 0.058%
Sulfur 0.044%

Helium 24%

Hydrogen 73.9%

In Earth's Crust

Silicon 27.7%

Oxygen 46.6%

Aluminum 8.1%
Iron 5.0%
Calcium 3.6%
Sodium 2.8%
Potassium 2.6%
Magnesium 2.1%
All others 1.5%

Figure 6 The most abundant elements in the universe are greatly different from the most abundant elements on Earth.

What elements are most abundant?

Astronomers have identified the two most abundant elements in the universe as hydrogen and helium. All other elements account for less than 1 percent of all atoms in the universe, as shown in **Figure 6.** Analyses of the composition of rocks and minerals on Earth indicate that the percentages of elements in Earth's crust differ from the percentages in the universe. As shown in **Figure 6,** 98.5 percent of Earth's crust is made up of only eight elements. Two of these elements, oxygen and silicon, account for almost 75 percent of the crust's composition. This means that most of the rocks and minerals on Earth's crust contain oxygen and silicon. Minerals that contain oxygen and silicon are called silicates. The next two most abundant elements in Earth's crust are aluminum and iron.

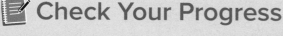 Check Your Progress

Summary

- Atoms consist of protons, neutrons, and electrons.

- An element consists of atoms that have a specific number of protons in their nuclei.

- Isotopes of an element differ by the number of neutrons in their nuclei.

- Elements with full outermost energy levels are highly unreactive.

- Ions are electrically charged atoms or groups of atoms.

Demonstrate Understanding

1. **Differentiate** among the three particles of an atom in terms of their location, charge, and mass.

2. **Infer** why the elements magnesium and calcium have similar properties.

3. **Illustrate** how a neutral atom becomes an ion.

4. **Compare and contrast** these isotopes: uranium-239, uranium-238, and uranium-235.

Explain Your Thinking

5. **Illustrate** a model of a calcium atom, including the number and position of protons, neutrons, and electrons in the atom.

6. **Interpret** the representation of magnesium in the periodic table. Explain why the atomic mass of magnesium is not a whole number.

7. **MATH > Connection** As the radioactive isotope radium-226 decays, it emits two protons and two neutrons. How many protons and neutrons are now left in the nucleus? What is the atom's new atomic number? What is the name of this element?

LEARNSMART Go online to follow your personalized learning path to review, practice, and reinforce your understanding.

FOCUS QUESTION

Why does water readily mix with other chemicals?

Compounds

Look at **Figure 7.** How can two dangerous elements combine to form a material that you sprinkle on your popcorn? Table salt is a compound, not an element. A **compound** is a substance that is composed of atoms of two or more different elements that are chemically combined. Water is another example of a compound because it is composed of two elements—hydrogen and oxygen. Most compounds have different properties from the elements of which they are composed. For example, both oxygen and hydrogen are highly flammable gases at room temperature, but in combination they form water—a liquid.

Chemical formulas

Compounds are represented by chemical formulas. These formulas include the symbol for each element followed by a subscript number that stands for the number of atoms of that element in the compound. If there is only one atom of an element, no subscript number follows the symbol. Thus, the chemical formula for table salt is NaCl. The chemical formula for water is H_2O.

Figure 7 Chlorine is a green, poisonous gas. Sodium is a silvery metal that is soft enough to cut with a knife. When they react, they produce sodium chloride, a white solid.

(l)Joe Franek/McGraw-Hill Education, (c)sciencephotos/Alamy Stock Photo, (r)Fuse/Getty Image

---

3D THINKING   **DCI** Disciplinary Core Ideas   **CCC** Crosscutting Concepts   **SEP** Science & Engineering Practices

COLLECT EVIDENCE

 Use your Science Journal to record the evidence you collect as you complete the readings and activities in this lesson.

INVESTIGATE

 GO ONLINE to find these activities and more resources.

GeoLAB: Precipitate Salt
Carry out an investigation to determine which factors affect salt solubility and precipitation.

Design Your Own: Rates of Chemical Reactions
Carry out an investigation to determine which factors affect a chemical reaction.

Covalent Bonds

Recall that an atom is chemically stable when its outermost energy level is full. A state of stability is achieved by some elements by forming chemical bonds. A **chemical bond** is the force that holds together the elements in a compound. One way in which atoms fill their outermost energy levels is by sharing electrons. For example, individual atoms of hydrogen each have just one electron. Each atom becomes more stable when it shares its electron with another hydrogen atom so that each atom has two electrons in its outermost energy level. **Figure 8** shows an example of this bond. How do these two atoms stay together? The nucleus of each atom has one proton with a positive charge, and the two positively charged protons attract the two negatively charged electrons. This attraction of two atoms for a shared pair of electrons that holds the atoms together is called a **covalent bond.**

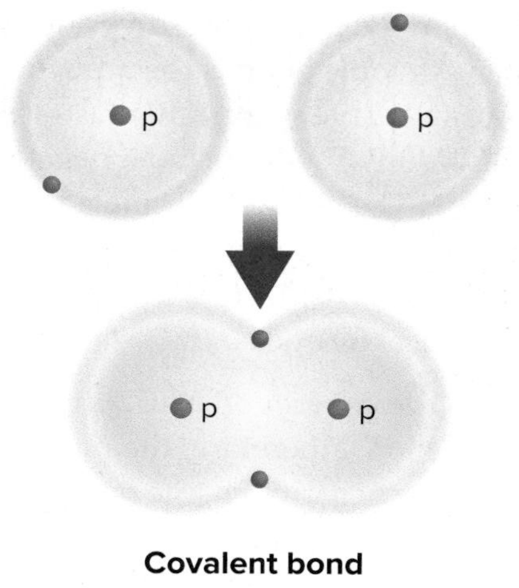

Molecules

A **molecule** is composed of two or more atoms held together by covalent bonds. Molecules have no overall electric charge because the total number of electrons equals the total number of protons. Water is an example of a compound whose atoms are held together by covalent bonds, as illustrated in **Figure 9.** The chemical formula for a water molecule is H_2O because, in this molecule, two atoms of hydrogen, each of which need to gain an electron to become stable, are combined with one atom of oxygen, which needs to gain two electrons to become stable. A compound comprised of molecules is called a molecular compound.

Polar molecules

Although water molecules are held together by covalent bonds, the atoms do not share the electrons equally. As shown in **Figure 9,** the shared electrons in a water molecule are attracted more strongly by the oxygen atom than by the hydrogen atoms. As a result, the electrons spend more time near the oxygen atom than they do near the hydrogen atoms. This unequal sharing of electrons results in polar molecules. A polar molecule has a slightly positive end and a slightly negative end.

Covalent bond

Figure 8 In this covalent bond example, notice the positions of the electrons in the outermost energy levels. They can now be considered as part of each atom.

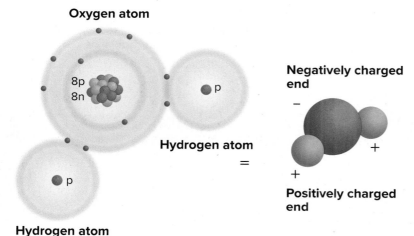

Oxygen atom

8p
8n

Hydrogen atom

Hydrogen atom

Negatively charged end

Positively charged end

Figure 9 Polar molecules are similar to bar magnets. At one end of a water molecule, the hydrogen atoms have a positive charge, while at the opposite end, the oxygen atom has a negative charge.

Ionic Bonds

As you might expect, positive and negative ions attract each other. An **ionic bond** is the attractive force between two ions of opposite charge. **Figure 10** illustrates an ionic bond between a positive ion of sodium and a negative ion of chlorine, called chloride. The chemical formula for common table salt is NaCl, which consists of equal numbers of sodium ions (Na^+) and chloride ions (Cl^-). Note that positive ions are always written first in chemical formulas.

Within the compound NaCl, there are as many positive ions as negative ions; therefore, the positive charge on the sodium ion equals the negative charge on the chloride ion, and the net electric charge of the compound NaCl is zero. Magnesium and oxygen ions combine in a similar manner to form the compound magnesium oxide (MgO)—one of the most common compounds on Earth. Compounds formed by ionic bonding are called ionic compounds. Other ionic compounds have different proportions of ions. For example, oxygen and sodium ions combine in the ratio shown by the chemical formula for sodium oxide (Na_2O), in which there are two sodium ions to each oxygen ion.

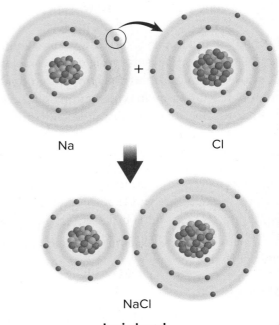

Na   +   Cl

NaCl

Ionic bond

Figure 10 The single valence electron in a sodium atom is used to form an ionic bond with a chlorine atom. Once an ionic bond is formed, the negatively charged ion is slightly larger than the positively charged ion.

Metallic Bonding

Most compounds on Earth are held together by ionic or covalent bonds, or by a combination of these bonds. Another type of bond is shown in **Figure 11**. In metals, the valence electrons are shared by all the atoms, not just by adjacent atoms, as they are in covalent compounds. You could think of a metal as a group of positive ions surrounded by a sea of freely moving negative electrons. The positive ions of the metal are held together by the attraction to the negative electrons between them. This type of bond, known as a **metallic bond,** allows metals to conduct electricity because the electrons can move freely throughout the entire solid metal.

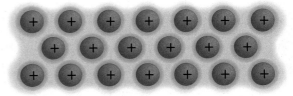

Metallic bond

Figure 11 Metallic bonds are formed when valence electrons are shared equally among all the positively charged atoms. Because the electrons flow freely among the positively charged ions, you can visualize electricity flowing through electrical wires.

 Get It?

Explain the difference between covalent, ionic, and metallic bonds.

SCIENCE USAGE v. COMMON USAGE

polar

Science usage: the unequal sharing of electrons

Common usage: locations of or near the North or South Pole, or the ends of a magnet

Metallic bonding also explains why metals are so easily deformed. When a force, such as the blow of a hammer, is applied to a metal the electrons are pushed aside. This allows the metal ions to move past each other, thus deforming or changing the shape of the metal. **Figure 13** summarizes how valence electrons are used to form the three different types of bonds.

Chemical Reactions

You have learned that atoms gain, lose, or share electrons to become more stable and that these atoms form compounds. Sometimes, compounds break down into simpler substances. The change of one or more substances into other substances, such as those in **Figure 12,** is called a **chemical reaction.** Chemical reactions are described by chemical equations. For example, water (H_2O) is formed by the chemical reaction between hydrogen gas (H_2) and oxygen gas (O_2). The formation of water can be described by the following chemical equation.

$$2H_2 + O_2 \rightarrow 2H_2O$$

You can read this chemical equation as "two molecules of hydrogen and one molecule of oxygen react to yield two molecules of water." In this reaction, hydrogen and oxygen are the reactants, and water is the product. When you write a chemical equation, you must balance the equation by showing an equal number of atoms for each element on each side of the equation. Therefore, the same amount of matter is present both before and after the reaction. Note that there are four hydrogen atoms on each side of the above equation ($2 \times 2 = 4$). There are also two oxygen atoms on each side of the equation.

Another example of a chemical reaction, one that takes place between iron (Fe) and oxygen (O), is represented by the following chemical equation.

$$4Fe + 3O_2 \rightarrow 2Fe_2O_3$$

Chemical reactions also occur when materials dissolve. Water is often called the universal solvent because so many of Earth's materials dissolve readily in it. For example, NaCl is a common mineral in Earth's crust. It dissolves quickly in water, which often leads to changes in the rock and mineral composition at Earth's surface.

Figure 12 When a copper wire is placed in a solution of silver nitrate in the beaker, a chemical reaction occurs in which silver replaces copper in the wire, and an aqua-colored copper nitrate solution forms.

Stephen Frisch/McGraw-Hill Education

STEM CAREER Connection

Electroplate Technician

No one can turn straw into gold. But electroplate technicians can make substances look like gold. When an electrical current is passed through a solution of carefully chosen chemicals, a chemical reaction occurs, and a thin metal layer is attached to the surface of the object being coated. In the case of jewelry, the piece could appear to be made out of gold or silver. Electroplating is also used to protect metals from corrosion, to prevent rust from forming, and, in the case of space suit helmets, to protect astronauts' eyes from solar radiation.

Figure 13 Visualizing Bonds

Atoms gain stability by sharing, gaining, or losing electrons to form ions and molecules. The properties of metals can be explained by metallic bonds.

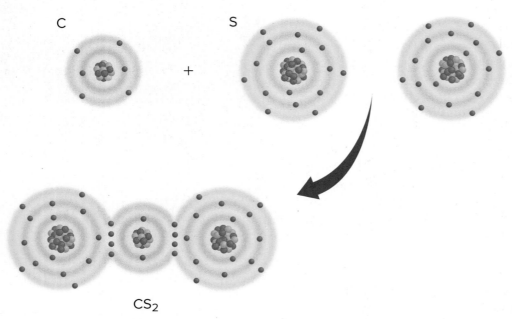

C   S

+

CS_2

Covalent bond Shared electrons fill outermost energy levels and make stable molecular compounds.

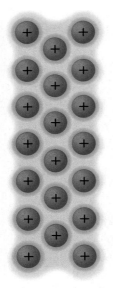

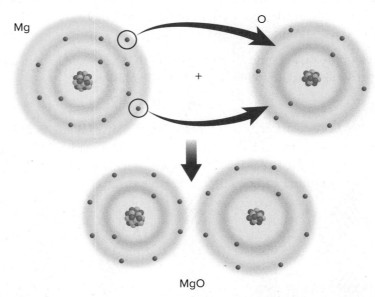

Mg   O

+

MgO

Metallic bond Within metals, valence electrons move freely around positively charged ions.

Ionic bond Once valence electrons are gained or lost to fill outermost energy levels and form stable ions, the oppositely charged ions are attracted to each other.

Mixtures and Solutions

Unlike a compound, in which the atoms combine and lose their identities, a mixture is a combination of two or more components that retain their identities. When a mixture's components are easily recognizable, it is called a heterogeneous mixture. For example, beach sand, shown in **Figure 14,** is a heterogeneous mixture because its components—shells, small pieces of broken shells, grains of minerals, and so on—are still recognizable. In a homogeneous mixture, which is also called a **solution,** the component particles cannot be distinguished from each other, even though they still retain their original properties.

A solution can be liquid, gaseous, or solid. Seawater is a solution consisting of water molecules and ions of many elements that exist on Earth. Molten rock is also a liquid solution; it is composed of ions representing all atoms that were present in the solid rock before it melted. Air is a solution of gases, mostly nitrogen and oxygen molecules mixed with other atoms and molecules. Metal alloys, such as bronze and brass, are also solutions. Bronze is a homogeneous mixture of copper and tin atoms; brass is a similar mixture of copper and zinc atoms. Such solid homogeneous mixtures are called solid solutions.

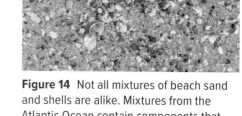

Figure 14 Not all mixtures of beach sand and shells are alike. Mixtures from the Atlantic Ocean contain components that are different from mixtures that form in the Pacific Ocean.

Get It?

Describe three examples of solutions.

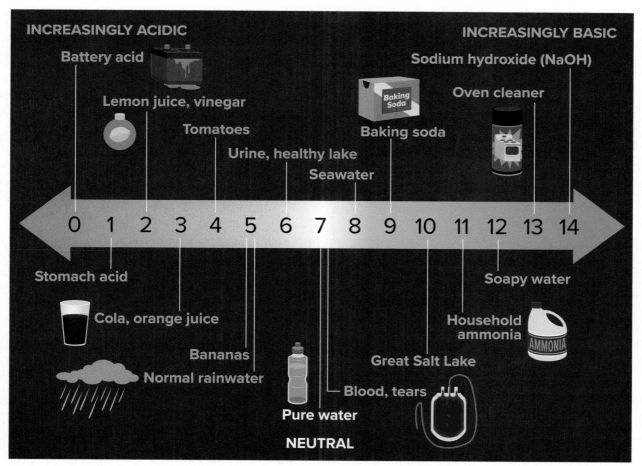

Figure 15 The pH scale is not only relevant to science class. All substances have a pH value, as you can see by the common household substances shown here.

Acids

Many chemical reactions on Earth involve acids and bases. An **acid** is a solution containing a substance that produces hydrogen ions (H^+) in water. A hydrogen atom has one proton and one electron. When a hydrogen atom loses its electron, it becomes a hydrogen ion (H^+). The pH scale, shown in **Figure 15,** is based on the amount of hydrogen ions in a solution, referred to as the concentration. A value of 7 is considered neutral; distilled water usually has a pH of 7. A solution with a pH value below 7 is considered to be acidic. The lower the number, the more acidic the solution.

The most common acid on Earth is carbonic acid (H_2CO_3). It is produced when carbon dioxide (CO_2) is dissolved in water (H_2O) by the following reaction.

$$H_2O + CO_2 \rightarrow H_2CO_3$$

Some of the carbonic acid (H_2CO_3) in the water ionizes, or breaks apart, into hydrogen ions (H^+) and bicarbonate ions (HCO_3^-), as shown by the equation

$$H_2CO_3 \rightarrow H^+ + HCO_3^-$$

These two equations play a major role in the dissolution of limestone and the formation of caves. In addition, rainwater is slightly acidic, with a pH of 5.0 to 5.6. Because many of the reaction rates involved in geological processes are very slow, it takes thousands of years for enough carbonic acid mixed with groundwater to dissolve limestone and form a cave.

Bases

A **base** is a substance that produces hydroxide ions (OH^-) in water. A base can neutralize an acid because hydrogen ions (H^+) from the acid react with the hydroxide ions (OH^-) from the base to form water through the following reaction.

$$H^+ + OH^- \rightarrow H_2O$$

Figure 15 shows the pH values of some basic substances. A solution with a reading above 7 is considered to be basic. The higher the number, the more basic the solution. Many household cleaning products are basic, with pH values from 11 to 13.

Check Your Progress

Summary

- Atoms of different elements combine to form compounds.
- Covalent bonds form from shared electrons between atoms.
- Ionic compounds form from the attraction of positive and negative ions.
- There are two types of mixtures—heterogeneous and homogeneous.
- Acids are solutions containing hydrogen ions. Bases are solutions containing hydroxide ions.

Demonstrate Understanding

1. **Explain** why molecules do not have electric charges.
2. **Differentiate** between molecules and compounds.
3. **Calculate** the number of atoms needed to balance the following equation: $CaCO_3 + HCl \rightarrow CO_2 + H_2O + CaCl$
4. **Diagram** how an acid can be neutralized.
5. **Compare and contrast** mixtures and solutions by using specific examples of each.

Explain Your Thinking

6. **Design** a procedure to demonstrate whether whole milk, which consists of microscopic fat globules suspended in a solution of nutrients, is a homogeneous or heterogeneous mixture.
7. **Predict** what kind of chemical bond forms between nitrogen and hydrogen atoms in ammonia (NH_3). Sketch this molecule.
8. **WRITING ▸Connection** Antacids are used to relieve indigestion and upset stomachs. Write an advertisement for a new antacid product. Explain how the product works in terms that people who are not taking a science class will understand.

LEARNSMART Go online to follow your personalized learning path to review, practice, and reinforce your understanding.

STATES OF MATTER

FOCUS QUESTION

What forms does water take?

Solids

Solids are substances with densely packed particles, which can be ions, atoms, or molecules. Most solids are **crystalline structures,** with particles arranged in regular geometric patterns. Examples of crystals are shown in **Figure 16.** Because of their crystalline structures, solids have both a definite shape and volume.

Perfectly formed crystals are rare. When many crystals form in the same space at the same time, crowding prevents the formation of smooth, well-defined crystal faces. The result is a mass of intergrown, randomly arranged crystals. **Figure 16** shows the crystalline nature of the rock granite.

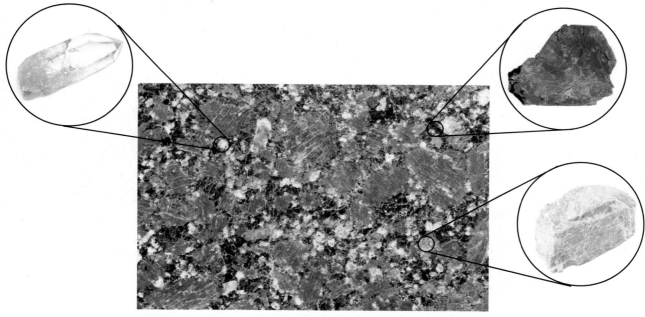

Figure 16 This granite is composed of mineral crystals that fit together like interlocking puzzle pieces. The minerals that make up the rock are composed of individual atoms and molecules that are aligned in a repeating pattern.

3D THINKING   **DCI** Disciplinary Core Ideas   **CCC** Crosscutting Concepts   **SEP** Science & Engineering Practices

COLLECT EVIDENCE

Use your Science Journal to record the evidence you collect as you complete the readings and activities in this lesson.

INVESTIGATE

GO ONLINE to find these activities and more resources.

Design Your Own: Changes in State
Plan and carry out an investigation to demonstrate and compare the properties of matter.

? Revisit the Encounter the Phenomenon Question
What information from this lesson can help you answer the Unit and Module questions?

Some solid materials have no regular internal patterns. **Glass** is a solid that consists of densely packed atoms arranged randomly. Glasses form when molten material is chilled so rapidly that atoms do not have enough time to arrange themselves in a regular pattern. These solids do not form crystals. Window glass consists mostly of disordered silicon and oxygen (SiO_2).

Figure 17 Each of these containers has the same volume of liquid in it.

Liquids

At any temperature above absolute zero ($-273°C$), the atoms in a solid vibrate. Because these vibrations increase with increasing temperature, they are called thermal vibrations. At the melting point of the material, these vibrations become vigorous enough to break the forces holding the solid together. The particles can then slide past each other, and the substance becomes liquid. Liquids take the shape of the container they are placed in, as you can see in **Figure 17.** However, liquids do have definite volume.

Gases

The particles in liquids vibrate vigorously. As a result, some particles can gain sufficient energy to escape the liquid. This process of change from a liquid to a gas at temperatures below the boiling point is called **evaporation.** When any liquid reaches its boiling point, it vaporizes quickly as a gas. In gases, the particles are separated by relatively large distances, and they travel at high speeds in one direction until they bump into another gas particle or the walls of a container. Gases, like liquids, have no definite shape. Gases also have no definite volume unless they are restrained by a container or a force such as gravity. For example, Earth's gravity keeps gases in the atmosphere from escaping into space.

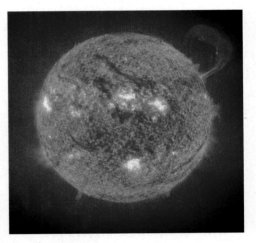

Figure 18 The Sun's temperature is often expressed in kelvins; 273 K is equal to 0°C. The Sun's corona, which is a plasma, has a temperature of about 15,000,000 K.

(t)John Evans/McGraw-Hill Education, (b)Digital Vision/Getty Images

Plasma

When matter is heated to a temperature greater than 5000°C, the collisions between particles are so violent that electrons are knocked away from atoms. Such extremely high temperatures exist in stars and, as a result, the gases of stars consist entirely of positive ions and free electrons. These hot, highly ionized, electrically conducting gases are called **plasmas. Figure 18** shows the plasma that forms the Sun's corona. You have seen matter in the plasma state if you have ever seen lightning or a neon sign.

CCC CROSSCUTTING CONCEPTS
Energy and Matter Scientists have determined that during any chemical reaction, no energy or matter is created or destroyed. Lighting a candle and letting it burn involves chemical reactions. Describe the chemical reactions you observe. What energy changes do you notice? Where do the candle wax and the candle wick go during burning? Share your observations with your class.

Changes of State

Solids melt when they absorb enough thermal energy to cause their orderly internal crystalline structure to break down. This happens at the melting point. When liquids are cooled, they solidify at that same temperature and release thermal energy. This temperature is called the freezing point. Freezing is the reverse of melting. When a liquid is heated to the boiling point and absorbs enough thermal energy, vaporization occurs, and the liquid becomes a gas. When a gas is cooled to the boiling point, it becomes a liquid in a process called **condensation,** shown in **Figure 19.** Condensation is the reverse of vaporization. Energy that was absorbed during vaporization is released upon condensation. A special type of vaporization, called sublimation, can occur below the boiling point. You might have noticed that even on winter days with temperatures below freezing, snow gradually disappears. This slow change of state from a solid (ice crystals) to a gas (water vapor) without an intermediate liquid state is called **sublimation.**

Figure 19 As hot, moist air from the shower encounters the cool glass of the mirror, water vapor in the air condenses on the glass.

Conservation of Energy

Matter can be changed through chemical reactions and nuclear processes, and its state can be changed under different thermal conditions. A chemical equation must be balanced because matter cannot be created or destroyed. This fundamental fact is the law of conservation of matter. Like matter, energy cannot be created or destroyed, but it can be changed from one form to another. For example, electric energy can be converted into light energy. This law, called the conservation of energy, is also known as the first law of thermodynamics.

✏️ Check Your Progress

Summary

- Changes of state involve thermal energy.
- The law of conservation of matter states that matter cannot be created or destroyed.
- The law of conservation of energy states that energy is neither created nor destroyed.

Demonstrate Understanding

1. **Explain** how thermal energy is involved in changes of state.
2. **Evaluate** the nature of the thermal vibrations in each of the four states of matter.
3. **Apply** what you know about thermal energy to compare evaporation and condensation.

Explain Your Thinking

4. **Infer** how the boiling point of water (100°C) would change if water molecules were not polar molecules.
5. **Consider** glass and diamond—two clear, colorless solids. Why does glass shatter more easily than diamond?
6. **MATH > Connection** Refer to **Figure 18.** Calculate the corona's temperature in degrees Celsius. Remember that 273 K is equal to 0°C.

LEARNSMART Go online to follow your personalized learning path to review, practice, and reinforce your understanding.

Seeing the Light

Device screens have come a long way since the days of cathode-ray tube (CRT) displays, used in twentieth-century televisions, and early liquid crystal display (LCD) technology, used in computer screens of the 1980s. Today's LCD screens offer picture quality that was unimaginable in the earliest days of consumer technology.

The picture quality of IPS LCD screens is favorably compared with the picture quality of any other type of screens.

How LCD Technology Works

LCD technology is made using two sheets of glass, with a row of light-emitting diode (LED) lights, called backlights, behind the bottom piece of glass. Layers of materials distribute this light equally over all parts of the glass. Each sheet of glass is lined with polarizing film. Sandwiched between the layers are a type of liquid crystal called twisted nematics (TN). Liquid crystals are a state of matter that have properties of both a liquid and a solid. They have some of the order of a solid and also some of the fluidity of a liquid. Their molecules can move around.

The amount and arrangement of light we see depends on the liquid crystals' orientation. Applying electric current causes the crystals to untwist, which controls the passage of light. The light that passes through the crystals looks white when the crystals are in their twisted state. But when electric current is applied, the crystals untwist and line up in such a way that light cannot pass through them.

Over the front sheet of glass, there are millions of tiny, colored blocks called pixels. Each pixel is controlled by a thin-film transistor (TFT) on the back sheet of glass. The pixels are switched on and off by the TFTs regulating the electricity flowing through the liquid crystals. This updates the screen from top to bottom at a speed so rapid that people cannot see the change happening.

LCD versus IPS LCD

In-plane switching (IPS) LCD improves on older LCD technology by adding a stronger backlight and by controlling each pixel with two TFTs instead of just one.

IPS LCD screens use more power than older LCD screens, but they also have better color, can be viewed from a wider angle, and are useful in touch-screen technology because they do not show when a screen has been touched. IPS LCD screens are found in a range of smartphones, televisions, and tablets.

USE A MODEL TO ILLUSTRATE

Use print or online sources to find diagrams and more information about how these technologies work. Then work with a partner to create a model of an LCD or IPS LCD screen. Label the parts of your model.

Justin Sullivan/Getty Images News/Getty Images

GO ONLINE to study with your Science Notebook.

Lesson 1 MATTER

- Atoms consist of protons, neutrons, and electrons.
- An element consists of atoms that have a specific number of protons in their nuclei.
- Isotopes of an element differ by the number of neutrons in their nuclei.
- Elements with full outermost energy levels are highly unreactive.
- Ions are electrically charged atoms or groups of atoms.

- matter
- element
- nucleus
- proton
- neutron
- electron
- atomic number
- mass number
- isotope
- ion

Lesson 2 COMBINING MATTER

- Atoms of different elements combine to form compounds.
- Covalent bonds form from shared electrons between atoms.
- Ionic compounds form from the attraction of positive and negative ions.
- There are two types of mixtures—heterogeneous and homogeneous.
- Acids are solutions containing hydrogen ions. Bases are solutions containing hydroxide ions.

- compound
- chemical bond
- covalent bond
- molecule
- ionic bond
- metallic bond
- chemical reaction
- solution
- acid
- base

Lesson 3 STATES OF MATTER

- Changes of state involve thermal energy.
- The law of conservation of matter states that matter cannot be created or destroyed.
- The law of conservation of energy states that energy is neither created nor destroyed.

- crystalline structure
- glass
- evaporation
- plasma
- condensation
- sublimation

REVISIT THE PHENOMENON

How do we find and remove harmful substances from our drinking water?

CER Claim, Evidence, Reasoning

Explain your Reasoning Revisit the claim you made when you encountered the phenomenon. Summarize the evidence you gathered from your investigations and research and finalize your Summary Table. Does your evidence support your claim? If not, revise your claim. Explain why your evidence supports your claim.

STEM UNIT PROJECT
Now that you've completed the module, revisit your STEM unit project. You will summarize your evidence and apply it to the project.

GO FURTHER

SEP Data Analysis Lab
How do compounds form?

Data and Observations Many atoms gain or lose electrons in order to have eight electrons in the outermost energy level. In the diagram, energy levels are indicated by the circles around the nucleus of each element. The colored spheres in the energy levels represent electrons, and the spheres in the nucleus represent protons and neutrons.

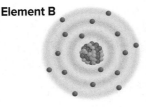

CER Analyze and Interpret Data

1. **Claim** Different atoms have different numbers of electrons. How many electrons are present in atoms of Element A? Element B?
2. **Claim** How many protons are present in the nuclei of these atoms?
3. **Evidence, Reasoning** Use a periodic table to determine the name of and symbol for Element A and Element B.
4. **Reasoning** Decide if these elements can form ions. If so, what would be the electric charges (magnitude and sign) of and chemical symbols for these ions?
5. **Reasoning** Formulate a compound from these two elements. What is the chemical formula for the compound?

rgbdigital/iStock/Getty Images

MINERALS

ENCOUNTER THE PHENOMENON

What makes diamonds so special?

GO ONLINE to play a video to learn more about diamonds.

SEP Ask Questions

Do you have other questions about the phenomenon? If so, add them to the driving question board.

CER Claim, Evidence, Reasoning

Make Your Claim Use your CER chart to make a claim about what makes diamonds so special. Explain your reasoning.

Collect Evidence Use the lessons in this module to collect evidence to support your claim. Record your evidence as you move through the module.

Explain Your Reasoning You will revisit your claim and explain your reasoning at the end of the module.

GO ONLINE to access your CER chart and explore resources that can help you collect evidence.

LESSON 1: Explore & Explain: Mineral Characteristics

LESSON 2: Explore & Explain: Mineral Groups

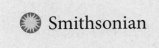

Additional Resources

WHAT IS A MINERAL?

FOCUS QUESTION

What properties does diamond have?

Mineral Characteristics

Earth's crust is composed of about 3000 types of minerals. Minerals play important roles in forming rocks and in shaping Earth's surface. A select few have helped shape civilization. For example, great progress in prehistory was made when early humans began making tools from iron.

A **mineral** is a naturally occurring, inorganic solid with a specific chemical composition and a definite crystalline structure. This crystalline structure is often exhibited by the crystal shape itself. Examples of two mineral crystal shapes, pyrite and calcite, are shown in **Figure 1**.

 Get It?

Identify the requirements a substance must meet in order to be classified as a mineral.

Naturally occurring and inorganic

Minerals are naturally occurring, meaning that they are formed by natural processes. Thus, synthetic diamonds and other substances developed in labs are not minerals. All minerals are inorganic. They are not alive and never were alive. Based on these criteria, salt is a mineral, but sugar, which is harvested from plants, is not. What about coal? According to the scientific definition of minerals, coal is not a mineral because millions of years ago, it formed from organic materials.

Pyrite

Calcite

Figure 1 The shapes of these mineral crystals reflect the internal arrangements of their atoms.

(t)Doug Sherman/Geofile, (b)Jiri Vaclavek/Shutterstock

3D THINKING **DCI** Disciplinary Core Ideas   **CCC** Crosscutting Concepts   **SEP** Science & Engineering Practices

COLLECT EVIDENCE

 Use your Science Journal to record the evidence you collect as you complete the readings and activities in this lesson.

INVESTIGATE

GO ONLINE to find these activities and more resources.

 Quick Investigation: Recognize Cleavage and Fracture
Carry out an investigation to visualize the structure found in mineral crystal shapes.

 GeoLAB: Make a Field Guide for Minerals
Obtain, evaluate, and communicate information about how to identify minerals by their properties.

Definite crystalline structure

The atoms in minerals are arranged in regular geometric patterns that are repeated. This regular pattern results in the formation of a crystal. A **crystal** is a solid in which the atoms are arranged in repeating patterns. Sometimes, a mineral will form in an open space and grow into one large crystal. The well-defined crystal shapes shown in **Figure 1** are rare. More commonly, the internal atomic arrangement of a mineral is not apparent because the mineral formed in a restricted space. **Figure 2** shows a sample of quartz that formed in a restricted space.

Figure 2 This piece of quartz most likely formed in a restricted space, such as within a crack in a rock.

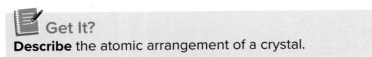

Get It?

Describe the atomic arrangement of a crystal.

Solids with specific compositions

The fourth characteristic of minerals is that they are solids, like the quartz and aquamarine in **Figure 3**. Recall that solids have definite shapes and volumes, while liquids and gases do not. Because of this, no gas or liquid can be considered a mineral.

Each type of mineral has a chemical composition unique to that mineral. This composition might be specific, or it might vary within a set range of compositions. A few minerals, such as copper, silver, and sulfur, are composed of single elements. The vast majority, however, are made from compounds. The mineral quartz (SiO_2), for example, is a combination of two atoms of oxygen and one atom of silicon. Although other minerals might contain silicon and oxygen, the arrangement and proportion of these elements in quartz are unique to quartz.

Quartz   Aquamarine

Figure 3 Minerals must be solids, each with its own unique chemical composition.

ACADEMIC VOCABULARY

restricted

small space; to have limits

The room was so small that it felt very restricted.

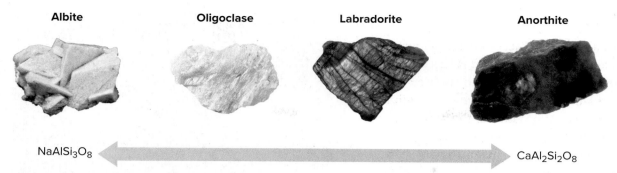

| Albite | Oligoclase | Labradorite | Anorthite |

$NaAlSi_3O_8$ ←——————————————————→ $CaAl_2Si_2O_8$

Figure 4 The mineral albite is a sodium-rich feldspar, while anorthite is calcium-rich. Oligoclase and labradorite contain both sodium and calcium in varying compositions.

Variations in composition

In some minerals, chemical composition can vary slightly depending on the temperature at which the mineral crystallizes. The plagioclase feldspar, shown in **Figure 4,** ranges from sodium-rich albite (AHL bite) at low temperatures to calcium-rich anorthite (uh NOR thite) at high temperatures. The difference in the minerals' appearance is due to a slight change in chemical composition and a difference in growth pattern as the temperature changes. At intermediate temperatures, both calcium and sodium are incorporated into the crystal structure. This builds up alternating layers that allow light to interfere with itself, producing a range of colors, as shown in the labradorite in **Figure 4.**

Rock-Forming Minerals

Although about 3000 minerals occur in Earth's crust, only about 30 of these are common. Eight to ten of these minerals are referred to as rock-forming minerals because they make up most of the rocks in Earth's crust. They are primarily composed of the eight most common elements in Earth's crust, which are listed in **Table 1**.

Table 1 Most Common Rock-Forming Minerals

Quartz	Feldspar	Mica*	Pyroxene*
SiO_2	$NaAlSi_3O_8 -$ $CaAl_2Si_2O_8$ $KAlSi_3O_8$	$K(Mg,Fe)_3(AlSi_3O_{10})$ $(OH)_2$ $KAl_2(AlSi_3O_{10})(OH)_2$	$MgSiO_3$ $Ca(Mg,Fe)Si_2O_6$ $NaAlSi_2O_6$
Amphibole*	**Olivine**	**Garnet***	**Calcite**
$Ca_2(Mg,Fe)_5Si_8O_{22}(OH)_2$ $Fe_7Si_8O_{22}(OH)_2$	$(Mg,Fe)_2SiO_4$	$Mg_3Al_2Si_3O_{12}$ $Fe_3Al_2Si_3O_{12}$ $Ca_3Al_2Si_3O_{12}$	$CaCO_3$

*representative mineral compositions

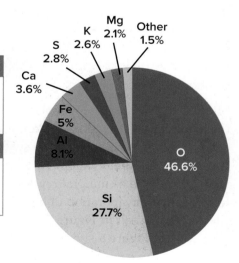

Minerals from magma

Molten material that forms and accumulates below Earth's surface is called **magma.** Magma is less dense than the surrounding solid rock, so it can rise upward into cooler layers of Earth's interior. Here, the magma cools and crystallizes. The type and number of elements present in the magma determine which minerals will form. The rate at which the magma cools determines the size of the mineral crystals. If the magma cools slowly within Earth's heated interior, the atoms have time to arrange themselves into large crystals. If the magma reaches Earth's surface, comes in contact with air or water, and cools quickly, the atoms do not have time to arrange themselves into large crystals. Thus, small crystals form from rapidly cooling magma, and large crystals form from slowly cooling magma. The mineral crystals in the granite shown in **Figure 5** are the result of slowly cooling magma.

 Get It?

Explain how the rate of cooling of magma affects the size of the crystals that form.

Minerals from solutions

Minerals are often dissolved in water. For example, the salts that are dissolved in ocean water make it salty. When a liquid becomes full of a dissolved substance and it can dissolve no more of that substance, the liquid is saturated. If more solute is added, the solution is called supersaturated, and conditions are right for minerals to form. At this point, individual atoms bond together and mineral crystals precipitate, which means that they form into solids from the solution.

Minerals also crystallize when the solution in which they are dissolved evaporates. You might have experienced this if you have ever gone swimming in the ocean. As the water evaporated off your skin, the salts were left behind as mineral crystals. Minerals that form from the evaporation of liquid are called evaporites. The rock salt in **Figure 5** was formed from evaporation. **Figure 6** shows Mammoth Hot Springs, a large evaporite complex in Yellowstone National Park.

 Get It?

Identify two ways minerals can form from solutions.

Granite

Rock salt

Figure 5 The crystals in these two samples formed in different ways.

Describe *the differences you see in these rock samples.*

Figure 6 This large complex of evaporite minerals is in Yellowstone National Park. The variation in color is a result of the variety of elements that are dissolved in the water.

(t)Doug Sherman/Geofile, (c)mariusFM77/iStock/Getty Images, (b)ablokhin/iStock/Getty Images

Identifying Minerals

Geologists rely on several simple tests to identify minerals. These tests are based on a mineral's physical and chemical properties: crystal form, luster, hardness, cleavage, fracture, streak, color, texture, density, specific gravity, and special properties. Scientists usually use a combination of tests instead of just one to identify minerals.

Crystal form

Some minerals form such distinct crystal shapes that they are immediately recognizable. Halite always forms perfect cubes. Quartz crystals, with their double-pointed ends and six-sided structure, are also readily recognized. However, as you learned earlier in this lesson, perfect crystals are not always formed, so identification based only on crystal form is rare.

Luster

The way that a mineral reflects light from its surface is called **luster.** There are two types of luster—metallic luster and nonmetallic luster. Silver, gold, copper, and galena have shiny surfaces that reflect light, like the chrome trim on cars. Thus, they are said to have a metallic luster. Not all metallic minerals are metals. If their surfaces have shiny appearances like metals, they are considered to have a metallic luster. Pyrite, for example, is a mineral with a metallic luster, but it is not a metal.

Minerals with nonmetallic lusters, such as calcite, gypsum, sulfur, and quartz, do not shine like metals. Nonmetallic lusters might be described as dull, pearly, waxy, silky, or earthy. Differences in luster, shown in **Figure 7,** are caused by differences in the chemical compositions of minerals. Describing the luster of nonmetallic minerals is a subjective process. A mineral that appears waxy to one person might not appear waxy to another. When identifying a mineral, luster should be used in combination with other physical characteristics.

| Talc | Kaolinite |

Figure 7 The flaky and shiny nature of talc gives it a pearly luster. Another white mineral, kaolinite, contrasts sharply with its dull, earthy luster.

STEM CAREER Connection
Mineralogist

A mineralogist studies minerals. Some mineralogists work for mining companies, analyzing economically important minerals and the best way to extract them. Others teach at universities, work in laboratories, or conduct surveys in the field.

Hardness

One of the most useful and reliable tests for identifying minerals is hardness. **Hardness** is a measure of how easily a mineral can be scratched. German geologist Friedrich Mohs developed a scale by which an unknown mineral's hardness can be compared to the known hardness of ten minerals. The minerals in the Mohs scale of hardness were selected because they are easily recognized and, with the exception of diamond, readily found in nature.

Talc is one of the softest minerals and can be scratched by a fingernail; therefore, talc represents 1 on the Mohs scale of hardness. In contrast, diamond is so hard that it can be used as a sharpener and cutting tool, so diamond represents 10 on the Mohs scale of hardness. The scale, shown in **Table 2,** is used in the following way: a mineral that can be scratched by your fingernail has a hardness less than 2.5. A mineral that cannot be scratched by your fingernail and cannot scratch glass has a hardness value between 5.5 and 2.5. Finally, a mineral that scratches glass has a hardness greater than 5.5.

Table 2 Mohs Scale of Hardness

Mineral	Hardness	Hardness of Common Objects
Diamond	10	
Corundum	9	
Topaz	8	
Quartz	7	streak plate = 7
Feldspar	6	steel file = 6.5
Apatite	5	glass = 5.5
Fluorite	4	iron nail = 4.5
Calcite	3	piece of copper = 3.5
Gypsum	2	fingernail = 2.5
Talc	1	

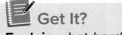 **Get It?**

Explain what *hardness* means.

Using other common objects, such as those listed in the table, can help you determine a more precise hardness and provide you with more information with which to identify an unknown mineral. Sometimes more than one mineral is present in a sample. If this is the case, it is a good idea to test more than one area of the sample. This way, you can be sure that you are testing the hardness of the mineral you are studying. **Figure 8,** on the next page, shows two minerals that have different hardness values.

 Get It?

Analyze What is the hardness of a mineral that can be scratched by a streak plate but not by glass?

Figure 8 The mineral on the left can be scratched with a fingernail. The mineral on the right easily scratches glass.

Determine *Which mineral has greater hardness?*

Cleavage and fracture

Atomic arrangement also determines how a mineral will break. Minerals break along planes where atomic bonding is weak. A mineral that splits relatively easily and evenly along one or more flat planes is said to have **cleavage.** To identify a mineral according to its cleavage, geologists count the number of cleaved planes and study the angle or angles between them. For example, mica has perfect cleavage in one direction. It breaks in sheets because of weak atomic bonds. Halite, shown in **Figure 9,** has cubic cleavage, which means that it breaks in three directions along planes of weak atomic attraction. Quartz and flint do not have natural planes of separation. They fracture instead of cleave.

| Halite | Quartz | Flint |

Figure 9 Halite, common table salt, breaks apart into pieces that have 90° angles. The strong bonds in quartz prevent cleavage from forming. Conchoidal fractures are characteristic of certain minerals that do not cleave, such as the microcrystalline mineral flint.

Quartz, shown in **Figure 9,** breaks unevenly along jagged edges because of its tightly bonded atoms. Minerals that break with rough or jagged edges are said to have **fracture.** Flint, jasper, and chalcedony (kal SEH duh nee) (microcrystalline forms of quartz) exhibit a unique fracture with arclike patterns resembling clamshells, also shown in **Figure 9.** This fracture is called conchoidal (kahn KOY duhl) fracture and is diagnostic in identifying the rocks and minerals that exhibit it.

 Get It?

Differentiate between cleavage and fracture.

Streak

A mineral rubbed across an unglazed porcelain plate will sometimes leave a colored powdered streak on the surface of the plate. **Streak** is the color of a mineral when it is broken up and powdered. The streak of a nonmetallic mineral is usually white. Streak is most useful in identifying metallic minerals.

Sometimes, a metallic mineral's streak does not match its external color, as shown in **Figure 10.** The mineral hematite occurs in two different forms, resulting in two distinctly different appearances. Hematite that forms from weathering and exposure to air and water is a rusty red color and has an earthy luster.

Figure 10 Despite the fact that these pieces of hematite appear remarkably different, their chemical compositions are the same. Thus, the streak that each makes is the same color.

Hematite that forms from crystallization of magma can be silver and metallic in appearance. However, both forms make a reddish-brown streak when tested. The streak test can be used only on minerals that are softer than a porcelain plate. This is another reason why streak cannot be used to identify all minerals.

 Get It?

Explain which type of mineral can be identified using streak.

Color

One of the most noticeable characteristics of a mineral is its color. Color is sometimes caused by the presence of trace elements or compounds within a mineral. For example, quartz occurs in a variety of colors, as shown in **Figure 11.** These different colors are the result of different trace elements in the quartz samples. Red jasper, purple amethyst, and orange citrine contain different amounts and forms of iron. Rose quartz contains manganese or titanium. However, the appearance of milky quartz is caused by the numerous bubbles of gas and liquid trapped within the crystal. In general, color is one of the least reliable clues of a mineral's identity.

| Red Jasper | Amethyst | Citrine | Rose Quartz |

Figure 11 These varieties of quartz all contain silicon and oxygen. Trace elements determine their colors.

Table 3 Special Properties of Minerals

Property	Double refraction occurs when a ray of light passes through the mineral and is split into two rays.	Effervescence occurs when a reaction with hydrochloric acid causes the mineral calcite in limestone to fizz.	Magnetism occurs between minerals that contain iron; only magnetite and pyrrhotite are strongly magnetic.	Iridescence is a play of colors, caused by light rays interfering with each other.	Fluorescence occurs when some minerals are exposed to ultraviolet light, which causes them to glow in the dark.
Mineral	Calcite—Variety Iceland Spar	Calcite	Magnetite Pyrrhotite	Labradorite	Calcite
Example					

Special properties

Several special properties of minerals can also be used for identification purposes. Some of these properties, shown in **Table 3,** are double refraction, effervescence with hydrochloric acid, magnetism, iridescence, and fluorescence. For example, Iceland spar is a form of calcite that exhibits double refraction. The arrangement of atoms in this type of calcite causes light to be bent in two directions when it passes through the mineral. The refraction of the single ray of light into two rays creates the appearance of two images.

Texture

Texture describes how a mineral feels to the touch. This property, like luster, is subjective. Therefore, texture is often used in combination with other tests to identify a mineral. The texture of a mineral might be described as smooth, rough, ragged, greasy, or soapy. For example, fluorite, shown in **Figure 12,** has a smooth texture, while the texture of talc, shown in **Figure 7,** is greasy.

Figure 12 Textures are interpreted differently by different people. The texture of this fluorite is usually described as smooth.

Density and specific gravity

Sometimes, two minerals of the same size have different weights. Differences in weight are the result of differences in density, which is defined as mass per unit of volume. Density is expressed as follows.

$$D = \frac{M}{V}$$

In this equation, D = density, M = mass, and V = volume. For example, pyrite has a density of 5.2 g/cm³, and gold has a density of 19.3 g/cm³. If you had a sample of gold and a sample of pyrite of the same size, the gold would have greater weight because it is denser.

Density reflects the atomic mass and structure of a mineral. Because density is not dependent on the size or shape of a mineral, it is a useful identification tool. Often, however, differences in density are too small to be distinguished by lifting different minerals. Thus, for accurate mineral identification, density must be measured. The most common measure of density used by geologists is **specific gravity,** which is the ratio of the mass of a substance to the mass of an equal volume of water at 4°C. For example, the specific gravity of pyrite is 5.2. The specific gravity of pure gold is 19.3.

 Get It?

Explain the relationship among density, mass, and volume.

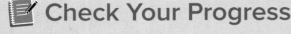

Check Your Progress

Summary

- A mineral is a naturally occurring, inorganic solid with a specific chemical composition and a definite crystalline structure.
- A crystal is a solid in which the atoms are arranged in repeating patterns.
- Minerals form from magma, supersaturated solutions, or evaporation of solutions in which they are dissolved.
- Minerals can be identified based on physical and chemical properties.
- The most reliable way to identify a mineral is by using a combination of several tests.

Demonstrate Understanding

1. **List** two reasons why petroleum is not a mineral.
2. **Define** *naturally occurring* in terms of mineral formation.
3. **Contrast** the formation of minerals from magma and their formation from solution.
4. **Differentiate** between subjective and objective mineral properties.

Explain Your Thinking

5. **Develop** a plan to test the hardness of a sample of feldspar using the following items: glass plate, copper penny, and streak plate.
6. **Predict** the success of a lab test in which students plan to compare the streak colors of fluorite, quartz, and feldspar.
7. **MATH ▸Connection** Calculate the volume of a 5-g sample of pure gold.

TYPES OF MINERALS

FOCUS QUESTION

Why are diamond and graphite so different when they are made of the same element?

Mineral Groups

You have learned that elements combine in many different ways and proportions. One result is the thousands of different minerals present on Earth. In order to study these minerals and understand their properties, geologists have classified them into groups. Each group has a distinct chemical nature and specific characteristics.

Silicates

Oxygen is the most abundant element in Earth's crust, followed by silicon. Minerals that contain silicon and oxygen, and usually one or more additional elements, are known as **silicates.** Silicates make up approximately 96 percent of the minerals present in Earth's crust. The two most common minerals, feldspar and quartz, are silicates. The basic building block of the silicates is the silicon-oxygen tetrahedron, shown in **Figure 13.** A **tetrahedron** (plural, *tetrahedra*) is a geometric solid having four sides that are equilateral triangles, resembling a pyramid. Recall that the electrons in the outermost energy level of an atom are called valence electrons. The number of valence electrons determines the type and number of chemical bonds an atom will form. Because silicon atoms have four valence electrons, silicon has the ability to bond with four oxygen atoms. As shown in **Figure 14,** silicon-oxygen tetrahedra can share oxygen atoms. This structure allows tetrahedra to combine in a number of ways, which accounts for the large diversity of structures and properties of silicate minerals.

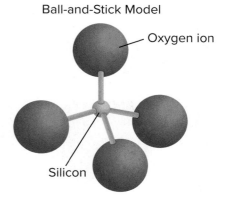

Ball-and-Stick Model

— Oxygen ion

Silicon

Space-Filling View

Oxygen (O^{2-})

Silicon (Si)

Figure 13 The silicate polyatomic ion SiO_4^{4-} forms a tetrahedron in which a central silicon atom is covalently bonded to oxygen atoms.

Specify *How many atoms are in one tetrahedron?*

 3D THINKING   **DCI** Disciplinary Core Ideas   **CCC** Crosscutting Concepts   **SEP** Science & Engineering Practices

COLLECT EVIDENCE

 Use your Science Journal to record the evidence you collect as you complete the readings and activities in this lesson.

INVESTIGATE

 GO ONLINE to find these activities and more resources.

Investigation Lab: Growing Crystals
Carry out an investigation to classify the results of evaporating solutions.

((📡)) **Review the News**
Obtain information from a current news story about current use of crystals in technology. Evaluate your source and **communicate** your findings to your class.

Figure 14 Visualizing the Silicon-Oxygen Tetrahedron

A silicon-oxygen tetrahedron contains four oxygen atoms bonded to a central silicon atom. Chains, sheets, and complex structures form when tetrahedra share oxygen atoms. These structures and the types of metal ions bonded to them determine the numerous silicate minerals that are present on Earth.

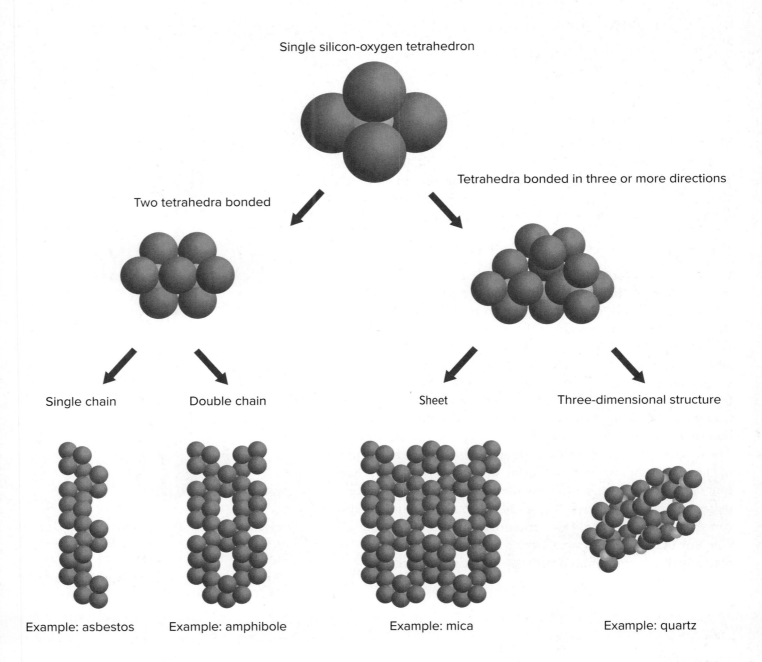

Single silicon-oxygen tetrahedron

Tetrahedra bonded in three or more directions

Two tetrahedra bonded

Single chain

Double chain

Sheet

Three-dimensional structure

Example: asbestos

Example: amphibole

Example: mica

Example: quartz

| Asbestos | Mica |

Figure 15 The differences in silicate minerals are due to differences in the arrangement of their silicon-oxygen tetrahedra. Certain types of asbestos consist of weakly bonded double chains of tetrahedra, while mica consists of weakly bonded sheets of tetrahedra.

Individual tetrahedron ions are strongly bonded. They can bond together to form sheets, chains, and complex three-dimensional structures. The bonds between the atoms help determine several mineral properties, including a mineral's cleavage or fracture. Mica, shown in **Figure 15,** is a sheet silicate, also called a phyllosilicate, in which positive potassium or aluminum ions bond the negatively charged sheets of tetrahedra together. Mica separates easily into sheets because the attraction between the tetrahedra and the aluminum or potassium ions is weak. Asbestos, also shown in **Figure 15,** consists of double chains of tetrahedra that are weakly bonded together. This results in a fibrous nature.

Carbonates

Oxygen combines easily with many elements and forms other mineral groups, such as carbonates. Carbonates are minerals composed of one or more metallic elements and the carbonate ion CO_3^{2-}. Examples of carbonates are calcite, dolomite, and rhodochrosite. Carbonates are the primary minerals found in rocks such as limestone and marble. Some carbonates have distinctive colorations, as exhibited by the calcite and rhodochrosite shown in **Figure 16.**

| Calcite | Rhodochrosite |

Figure 16 Carbonates such as calcite and rhodochrosite occur in distinct colors due to trace elements found in them.

SCIENCE USAGE v. COMMON USAGE

phyllo

Science usage: sheets of silica tetrahedra

Common usage: sheets of dough used to make pastries and pies

Oxides

Oxides are compounds of oxygen and a metal. Hematite (Fe_2O_3) and magnetite (Fe_3O_4) are common iron oxides and good sources of iron. The mineral uraninite (UO_2) is valuable because it is the major source of uranium, which is used to generate nuclear power.

Other Groups

Other major mineral groups are sulfides, sulfates, halides, and native elements. Sulfides, such as pyrite (FeS_2), are compounds of sulfur and one or more elements. Sulfates, such as anhydrite ($CaSO_4$), are composed of elements and the sulfate ion SO_4^{2-}. Halides, such as halite ($NaCl$), are made up of chloride or fluoride along with calcium, sodium, or potassium. A native element such as silver (Ag) or copper (Cu), is made up of one element only.

 Get It?

Compare carbonates and oxides.

Economic Minerals

Minerals are important to society. They are used to make computers, cars, televisions, desks, roads, buildings, jewelry, beds, paints, sports equipment, and medicines, in addition to many other things. You can learn about how the availability and use of minerals has guided the development of human society by examining **Figure 17,** on the next page.

Ores

Many of the items just mentioned are made from ores. A mineral is an **ore** if it contains a valuable substance that can be mined at a profit. Hematite, for instance, is an ore that contains the element iron. Consider your classroom. If any items are made of iron, their original source might have been the mineral hematite. If there are items in the room made of aluminum, their original source was the ore bauxite. A common use of the metal titanium, obtained from the mineral ilmenite, is shown in **Figure 18.**

Figure 18 Parts of this athlete's wheelchair are made of titanium. Its light weight and extreme strength make it an ideal metal for this purpose.

Mines Ores that are located deep within Earth's crust are removed by underground mining. Ores that are near Earth's surface are obtained from large, open-pit mines. When a mine is excavated, unwanted rock and minerals, known as gangue, are dug up along with the valuable ore. The overburden must also be removed before the ore can be used. Removing the overburden can be expensive and, in some cases, harmful to the environment. If the cost of removing the overburden or separating the gangue becomes higher than the value of the ore itself, the mineral will no longer be classified as an ore. It would no longer be economical to mine.

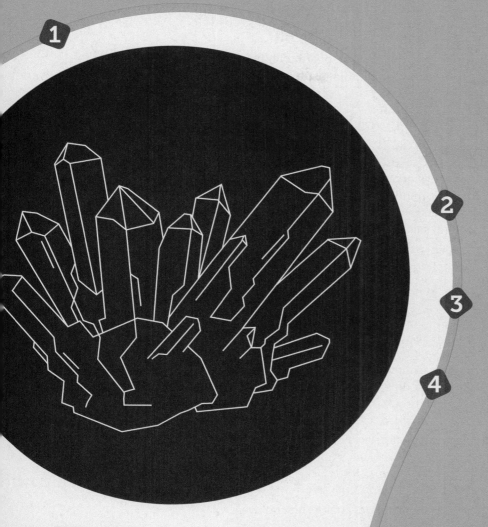

Figure 17

Mineral Use Through Time

While the values and uses of minerals have changed over time, some are consistently important. For example, quartz has always been important—from the first trade routes for flint over 10,000 years ago to the labs that turn quartz into computer chips today.

1 **3300–3000 B.C.** Bronze weapons and tools become common in the Near East as large cities and powerful empires arise.

2 **1200–1000 B.C.** In the Near East, bronze becomes scarce and is replaced by iron in tools and weapons.

3 **800 B.C.** Diamonds spread from India to other parts of the world to be used for cutting, for engraving, and in ceremonies.

4 **506 B.C.** Rome takes over the salt industry at Ostia. The word *salary* comes from *salarium argentums*, the salt rations paid to Roman soldiers.

5 **A.D. 200–400** Iron farming tools and weapons allow people to migrate across Africa, clearing and cultivating land for agricultural settlement and driving out hunter-gatherer societies.

6 **800–900** Chinese alchemists combine saltpeter with sulfur and carbon to make gunpowder, which is first used for fireworks and later used for weapons.

7 **1546** South American silver mines help establish Spain as a global trading power, supplying silver needed for coinage.

8 **2010** Over 60 individual minerals, including quartz, bauxite, and halite, contribute to the modern computer.

Classification of Minerals The classification of a mineral as an ore can also change if the supply of or demand for that mineral changes. Consider a mineral that is used to make computers. Engineers might develop a more efficient design or decide to use a less costly alternative material. In either of these cases, the mineral would no longer be used in computers. Demand for the mineral would drop, so it would not be profitable to mine. Therefore, the mineral would no longer be considered an ore. **Table 4** summarizes the mineral groups and their major uses.

Table 4 **Major Mineral Groups**

Group	Examples	Economic Use
Silicates	mica (biotite) olivine (Mg_2SiO_4) quartz (SiO_2) vermiculite	furnace windows gem (as peridot) timepieces potting soil additive (swells when wet)
Sulfides	pyrite (FeS_2) marcasite (FeS_2) galena (PbS) sphalerite (ZnS)	used to make sulfuric acid; often mistaken for gold (fool's gold) jewelry lead ore zinc ore
Oxides	hematite (Fe_2O_3) corundum (Al_2O_3) uraninite (UO_2) ilmenite ($FeTiO_3$) chromite ($FeCr_2O_4$)	iron ore; red pigment abrasive, gem (as in ruby or sapphire) uranium source titanium source; pigment (replaced lead in paint) chromium source, plumbing fixtures, auto accessories
Sulfates	gypsum ($CaSO_4 \cdot _2H_2O$) anhydrite ($CaSO_4$)	plaster, drywall (slows drying in cement) plaster (name indicates absence of water)
Halides	halite (NaCl) fluorite (CaF_2) sylvite (KCl)	table salt, stock feed, weed killer, food preparation and preservative steel manufacturing, enameling cookware fertilizer
Carbonates	calcite ($CaCO_3$) dolomite ($CaMg(CO_3)_2$)	Portland cement, lime, chalk Portland cement, lime; source of calcium and magnesium in vitamin supplements
Native elements	gold (Au) copper (Cu) silver (Ag) sulfur (S) graphite (C)	monetary standard, jewelry coinage, electrical wiring, jewelry coinage, jewelry, photography sulfa drugs and chemicals; match heads; fireworks pencil lead, dry lubricant

Gems

What makes a ruby more valuable than mica? Rubies are rarer and more visually pleasing than mica. Rubies are thus considered gems. **Gems** are valuable minerals that are prized for their rarity and beauty. They are very hard and scratch resistant. Gems such as rubies, emeralds, and diamonds are cut, polished, and used for jewelry. Because of their rareness, rubies and emeralds are more valuable than diamonds. **Figure 19,** on the next page, shows a rough diamond (on the left) and a polished diamond (on the right).

Figure 19 The real beauty of gemstones is revealed once they are cut and polished.

In some cases, the presence of trace elements can make one variety of a mineral more colorful and more prized than other varieties of the same mineral. Amethyst, for instance, is the gem form of quartz. Amethyst contains traces of iron, which gives the gem a purple color. The mineral corundum, which is often used as an abrasive, also occurs as rubies and sapphires. Red rubies contain trace amounts of chromium, while blue sapphires contain trace amounts of cobalt or titanium. Green emeralds are a variety of the mineral beryl and are colored by trace amounts of chromium or vanadium.

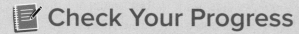

📝 Check Your Progress

Summary

- In silicates, one silicon atom bonds with four oxygen atoms to form a tetrahedron.

- Major mineral groups include silicates, carbonates, oxides, sulfides, sulfates, halides, and native elements.

- An ore contains a valuable substance that can be mined at a profit.

- Gems are valuable minerals that are prized for their rarity and beauty.

Demonstrate Understanding

1. **Formulate** a statement that explains the relationship between chemical elements and mineral properties.

2. **List** the two most abundant elements in Earth's crust. What mineral group do these elements form?

3. **Hypothesize** some environmental consequences of mining ores.

Explain Your Thinking

4. **Hypothesize** why the mineral opal is often referred to as a mineraloid.

5. **Evaluate** which metal is better to use in sporting equipment and medical implants: titanium—specific gravity = 4.5, contains only Ti; or steel—specific gravity = 7.7, contains Fe, O, Cr.

6. **WRITING ▸ Connection** Design an advertisement for a mineral of your choice. You might choose a gem or industrially important mineral. Include any information that you think will help your mineral sell.

(l)Zbynek Burival/iStock/Getty Images, (r)Mark Evans/E+/Getty Images

LEARNSMART® Go online to follow your personalized learning path to review, practice, and reinforce your understanding.

Better Than the "Real" Thing?

Synthetic and natural diamonds come in many different shapes, sizes, and colors. They are used in everything from jewelry to industrial drills.

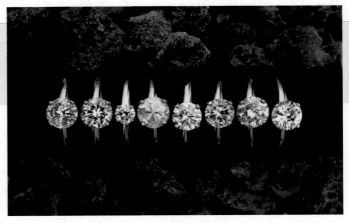

Can you tell which diamonds are synthetic? Synthetic diamonds are visually and chemically identical to natural diamonds.

Diamonds

When you think about diamonds, expensive jewelry and unethical mining practices might come to mind. But the uses of diamonds extend far beyond fashion, such as drill bits and optics for lasers. With growing accessibility and customization options for synthetic diamonds, materials scientists are exploring new uses for them.

Synthetic diamonds are made in a lab instead of being mined from a chunk of billion-year-old magma. They have the same characteristics as natural diamonds, and usually cost much less.

While scientists have been making diamonds in labs since 1954, modern influences—including a bigger market for sustainable, conflict-free diamonds—have helped to grow the synthetic diamond industry.

How to Make a Diamond

Scientists use two processes to make diamonds in the lab: chemical vapor deposition (CVD) and high pressure, high temperature (HPHT).

For wearable, jewelry-grade gems, CVD is the preferred process. In this process, a chemical vapor, or plasma, of carbon and hydrogen atoms are created by superheating methane and hydrogen in a closed chamber. Diamond seeds, each made of a repeating lattice of carbon atoms, are placed on a disk and added to the chamber.

The temperature is adjusted depending on the qualities the scientists want in that batch of diamonds. Over the course of a few weeks, the diamond seed grows, atom by atom, until the process is complete. The diamonds are then cut and polished, just as a natural diamond would be.

The HPHT process is a little closer to how natural diamonds form. Natural diamonds form when carbon is trapped deep in Earth's interior and subjected to the intense heat and pressure there. Scientists think that most natural diamonds formed billions of years ago.

Using the HPHT process, graphite, a soft form of carbon used to make pencil lead, is placed inside a machine that applies intense heat and pressure, similar to the conditions found in Earth's interior. After just a few days of this treatment, the graphite is changed into a high-quality diamond.

DEVELOP AND USE MODELS TO ILLUSTRATE

Research more about the differences between the CVD and HPHT processes, including costs, characteristics of diamonds produced, and the history of each process. Develop an illustration that describes the differences.

MODULE 3
STUDY GUIDE

 **GO ONLINE** to study with your Science Notebook.

Lesson 1 WHAT IS A MINERAL?

- A mineral is a naturally occurring, inorganic solid with a specific chemical composition and a definite crystalline structure.
- A crystal is a solid in which the atoms are arranged in repeating patterns.
- Minerals form from magma, supersaturated solutions, or the evaporation of solutions in which they are dissolved.
- Minerals can be identified based on physical and chemical properties.
- The most reliable way to identify a mineral is based on a combination of properties.

- mineral
- crystal
- magma
- luster
- hardness
- cleavage
- fracture
- streak
- specific gravity

Lesson 2 TYPES OF MINERALS

- In silicates, one silicon atom bonds with four oxygen atoms to form a tetrahedron.
- Major mineral groups include silicates, carbonates, oxides, sulfides, sulfates, halides, and native elements.
- An ore contains a valuable substance that can be mined at a profit.
- Gems are valuable minerals that are prized for their rarity and beauty.

- silicate
- tetrahedron
- ore
- gem

(t)Jiri Vaclavek/Shutterstock.com; (b)Zbynek Burival/iStock/Getty Images

REVISIT THE PHENOMENON

What makes diamonds so special?

CER Claim, Evidence, Reasoning

Explain your Reasoning Revisit the claim you made when you encountered the phenomenon. Summarize the evidence you gathered from your investigations and research and finalize your Summary Table. Does your evidence support your claim? If not, revise your claim. Explain why your evidence supports your claim.

STEM UNIT PROJECT

Now that you've completed the module, revisit your STEM unit project. You will summarize your evidence and apply it to the project.

GO FURTHER

SEP Data Analysis Lab

What information should you include in a mineral identification chart?

The table shows some of the properties of different minerals.

Data and Observations Copy the data table and use the *Reference Handbook* to complete the table. Expand the table to include the names of the minerals, other properties, and uses.

CER Analyze and Interpret Data

1. **Claim, Evidence** Which of these minerals will scratch glass? Explain.
2. **Claim, Evidence** Which of these minerals might be present in both a painting and your desk? Explain.
3. **Reasoning** What other information could you include in the table? Why?

Mineral Identification Chart

Mineral Color	Streak	Hardness	Breakage Pattern
copper red		3	hackly, fracture
	red and reddish brown	6	irregular fracture
pale to golden yellow	yellow		
	colorless	7.5	conchodial fracture
gray, green, or white			two cleavage planes

*Data obtained from: Klein, C. 2002. *The Manual of Mineral Science*.

rgbdigital/iStock/Getty Images

Putt Sakdhnagool/EyeEm/Getty Images

ENCOUNTER THE PHENOMENON

How did these different types of rock form?

GO ONLINE to play a video about the types of rocks on Earth.

SEP Ask Questions

Do you have other questions about the phenomenon? If so, add them to the driving question board.

CER Claim, Evidence, Reasoning

Make Your Claim Use your CER chart to make a claim about how different types of rocks form. Explain your reasoning.

Collect Evidence Use the lessons in this module to collect evidence to support your claim. Record your evidence as you move through the module.

Explain Your Reasoning You will revisit your claim and explain your reasoning at the end of the module.

GO ONLINE to access your CER chart and explore resources that can help you collect evidence.

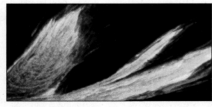

LESSON 1: Explore & Explain: Igneous Rock Formation

LESSON 3: Explore & Explain: Recognizing Metamorphic Rocks

 Smithsonian

Additional Resources

(t)Video Supplied by BBC Worldwide Learning, (b)Frizi/Shutterstock, (br)©Stephen Reynolds

IGNEOUS ROCKS

FOCUS QUESTION

How do rocks form from magma?

Igneous Rock Formation

Igneous rocks form when lava or magma cools and crystallizes. **Lava** is magma that flows out onto Earth's surface. In the laboratory, most rocks must be heated to temperatures of 800°C to 1200°C before they melt. In nature, these temperatures are present in the upper mantle and lower crust. Scientists theorize that the remaining energy from Earth's molten formation and the heat generated from the decay of radioactive elements are the sources of Earth's thermal energy.

Composition of magma

The type of igneous rock that forms depends on the composition of the magma. Magma is often a slushy mix of molten rock, dissolved gases, and mineral crystals. The common elements present in magma are the same major elements that are in Earth's crust: oxygen (O), silicon (Si), aluminum (Al), iron (Fe), magnesium (Mg), calcium (Ca), potassium (K), and sodium (Na). Of all the compounds present in magma, silica is the most abundant and has the greatest effect on magma characteristics. **Table 1** shows that magma is classified as basaltic, andesitic, or rhyolitic, based on the amount of silica it contains. Silica content affects melting temperature and impacts a magma's viscosity, or resistance to flow. Rhyolitic magma has a higher viscosity than basaltic magma. Once magma is free of the overlying pressure of the rock layers around it, dissolved gases are able to escape into the atmosphere. Thus, the chemical composition of lava is slightly different from the chemical composition of the magma from which it developed.

Table 1 Types of Magma

Group	Silica Content	Example Location
Basaltic	45–52%	Hawaiian Islands
Andesitic	52–66%	Cascade Mountains, Andes Mountains
Rhyolitic	more than 66%	Yellowstone National Park

 **3D THINKING**   **DCI** Disciplinary Core Ideas   **CCC** Crosscutting Concepts   **SEP** Science & Engineering Practices

COLLECT EVIDENCE

Use your Science Journal to record the evidence you collect as you complete the readings and activities in this lesson.

INVESTIGATE

GO ONLINE to find these activities and more resources.

Mapping Lab: Locating Igneous Rocks on Earth
Develop and use a model to visualize the patterns of global igneous rock locations.

Quick Investigation: Compare Igneous Rocks
Carry out an investigation to compare the properties of igneous rocks from different locations.

Magma formation

Magma can be formed either by melting of Earth's crust or by melting within the mantle. The four main factors involved in the formation of magma are temperature, pressure, water content, and the mineral content of the crust or mantle.

Temperature generally increases with depth in Earth's crust. This temperature increase, known as the geothermal gradient, is plotted in **Figure 1.** Drill bits, such as the one shown in **Figure 2,** can encounter temperatures in excess of 200°C when drilling deep oil wells.

Pressure also increases with depth. This is a result of the weight of overlying rock. Laboratory experiments show that as pressure on a rock increases, its melting point also increases. Thus, a rock that melts at 1100°C at Earth's surface will melt at 1400°C at a depth of 100 km.

The third factor that affects the formation of magma is water content. Rocks and minerals often contain small percentages of water, which changes the melting point of the rocks. As water content increases, the melting point decreases.

Get It?

Explain how water content changes the melting point of rocks.

Mineral content

In order to better understand how the types of elements and compounds present give magma its overall character, it is helpful to discuss this fourth factor in more detail. Different minerals have different melting points. For example, rocks such as basalt, which are formed of olivine, calcium feldspar, and pyroxene (pi RAHK seen), melt at higher temperatures than rocks such as granite, which contain quartz and potassium feldspar. Granite has a melting point that is lower than basalt's melting point because granite contains more water and minerals that melt at lower temperatures. In general, rocks that are rich in iron and magnesium melt at higher temperatures than rocks that contain higher levels of silicon.

Get It?

Explain how levels of iron and magnesium in a rock affect its melting temperature.

Earth's Geothermal Gradient

Figure 1 The average geothermal gradient in the crust is about 25°C/km, but scientists think that it drops sharply in the mantle to as low as 1°C/km.

Figure 2 The temperature of Earth's upper crust increases with depth by about 25°C for each 1 km. At a depth of 3 km, this drill bit will encounter rock that is close to the temperature of boiling water.

Lowell Georgia/Corbis Documentary/Getty Images

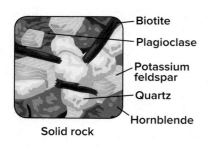

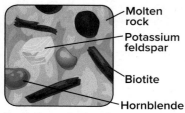

Solid rock

Partially melted rock

Figure 3 As the temperature increases in an area, minerals begin to melt.

Determine *What can you infer about the melting temperature of quartz based on this diagram?*

Partial melting

Suppose you freeze melted candle wax and water in an ice cube tray. You take the tray out of the freezer and leave it at room temperature. The ice melts, but the candle wax does not. The two substances have different melting points. Rocks melt in a similar way because the minerals they contain have different melting points. Not all parts of a rock melt at the same temperature. This explains why magma is often a slushy mix of crystals and molten rock. The process whereby some minerals melt at relatively low temperatures while other minerals remain solid is called **partial melting.** Partial melting is illustrated in **Figure 3.** As each group of minerals melts, different elements are added to the magma mixture, changing its composition. If temperatures are not high enough to melt the entire rock, the resulting magma will have a different composition from that of the original rock. This is one way in which different types of igneous rocks form.

 Get It?

Summarize the formation of magma that has a different chemical composition from the original rock.

Bowen's Reaction Series

In the early 1900s, Canadian geologist N. L. Bowen demonstrated that as magma cools and crystallizes, minerals form in predictable patterns in a process now known as **Bowen's reaction series. Figure 4** illustrates the relationship between cooling magma and the formation of minerals that make up igneous rock. Bowen discovered two main patterns, or branches, of crystallization. The left-hand branch is characterized by an abrupt change of mineral type in the iron-magnesium group. A continuous, gradual change of mineral composition characterizes the right-hand branch.

Figure 4 On the left side of Bowen's reaction series, minerals rich in iron and magnesium change abruptly as the temperature of the magma decreases.

Compare *How does this compare to the feldspars on the right side of the diagram?*

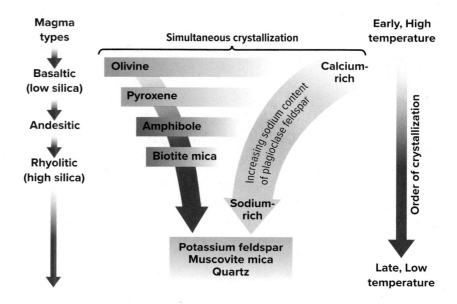

Iron-rich minerals

The left branch of Bowen's reaction series represents the iron-rich minerals. These minerals undergo abrupt changes as magma cools and crystallizes. For example, olivine is the first mineral to crystallize when magma that is rich in iron and magnesium begins to cool. When the temperature decreases enough for a completely new mineral, pyroxene, to form, the olivine that previously formed reacts with the magma and is converted to pyroxene. As the temperature decreases further, similar reactions produce the minerals amphibole and biotite mica.

Figure 5 Plagioclase feldspars undergo a continuous change of composition in Bowen's reaction series.

Feldspars

In Bowen's reaction series, the right branch represents the plagioclase feldspars, shown in **Figure 5.** These undergo a continuous change of composition. As magma cools, the first feldspars to form are rich in calcium. As cooling continues, these feldspars react with magma, and their calcium-rich compositions change to sodium-rich compositions. In some instances, such as when magma cools rapidly, the calcium-rich crystals are unable to react completely with the magma. The result is a zoned crystal with a calcium-rich core and sodium-rich outer layers.

Fractional Crystallization

When magma cools, it crystallizes in the reverse order of partial melting. That is, the first minerals that crystallize from magma are the last minerals to melt. This process, called **fractional crystallization,** is similar to partial melting in that the composition of magma can change. In this case, however, early formed crystals are removed from the magma and cannot react with it. As minerals form and their elements are removed from the remaining magma, it becomes concentrated in silica, as shown in **Figure 4.**

As is often the case with scientific inquiry, the discovery of Bowen's reaction series led to more questions. For example, if olivine converts to pyroxene during cooling, why is olivine present in some rocks? Geologists hypothesize that, under certain conditions, newly formed crystals are separated from the cooling magma, and the chemical reactions between the magma and the minerals stop. This can occur when crystals settle to the bottom of the magma body and when liquid magma is squeezed from the crystal mush. This results in the formation of two distinct igneous bodies with different compositions. **Figure 6,** on the next page, illustrates this process and the concept of fractional crystallization with an example from the Hudson River Valley in New York and New Jersey. This is one way in which the magmas listed in **Table 1** are formed.

Figure 7 Quartz veins are deposited from the quartz that was dissolved by hot water from a magma body that cooled and crystallized.

As fractional crystallization continues, and more and more crystals are separated from magma, the magma becomes more concentrated in silica, aluminum, and potassium. For this reason, the last two minerals to crystallize out of a cooling body of magma are potassium feldspar and quartz. Potassium feldspar is one of the most common feldspars in Earth's crust. Quartz often occurs in veins, as shown in **Figure 7,** because hot water dissolved the quartz that crystallizes from the magma. The water then flowed through the cracks, cooled down, and deposited the quartz.

(t)Doug Sherman/Geofile, (b)John A. Karachewski

Figure 6 Visualizing Fractional Crystallization and Crystal Settling

The Palisades Sill in the Hudson River Valley of New York and New Jersey is a classic example of fractional crystallization and crystal settling. When magma cools and solidifies before reaching the surface, the rock that forms is called an intrusion. In this basaltic intrusion, small crystals formed in the chill zone as the outer areas of the intrusion cooled more quickly than the interior.

Sandstone
Chill zone—small crystals

Mostly plagioclase:
no olivine

Plagioclase
and pyroxene:
no olivine

Olivine layer
Chill zone—small crystals
Sandstone

Basaltic intrusion

As magma in an intrusion begins to cool, crystals form and settle to the bottom. This layering of crystals is fractional crystallization.

June Marie Sobrito/Shutterstock.com

Gabbro

Granite

Diorite

Figure 8 Differences in magma composition can be observed in the rocks that form when the magma cools and crystallizes.

Observe *Describe the differences you see in these rocks.*

Mineral Composition of Igneous Rocks

Igneous rocks are broadly classified as intrusive or extrusive. When magma cools and crystallizes below Earth's surface, **intrusive rocks** form. If the magma is injected into the surrounding rock, it is called an igneous intrusion. Crystals of intrusive rocks are generally large enough to see without magnification. Magma that cools and crystallizes on Earth's surface forms **extrusive rocks.** These are sometimes referred to as lava flows or flood basalts. The crystals that form in these rocks are small and difficult to see without magnification. Geologists classify igneous rocks by their mineral compositions. In addition, physical properties such as grain size and texture serve as clues for the identification of various igneous rocks.

Igneous rocks are classified according to their mineral compositions. **Basaltic rocks,** also called mafic rocks, are dark-colored, have lower silica contents, and contain mostly plagioclase and pyroxene. **Granitic rocks,** or felsic rocks, are light-colored, have high silica contents, and contain mostly quartz and feldspar. Rocks that have a composition of minerals that is somewhere in between basaltic and granitic are called intermediate rocks. They consist mostly of plagioclase feldspar and hornblende. **Figure 8** shows examples from these three main compositional groups of igneous rocks: gabbro is basaltic, granite is granitic, and diorite is intermediate. A fourth category called ultrabasic, or ultramafic, contains rocks with only iron-rich minerals such as olivine and pyroxene and are always dark. **Figure 9** summarizes igneous rock identification.

Igneous Rock Identification

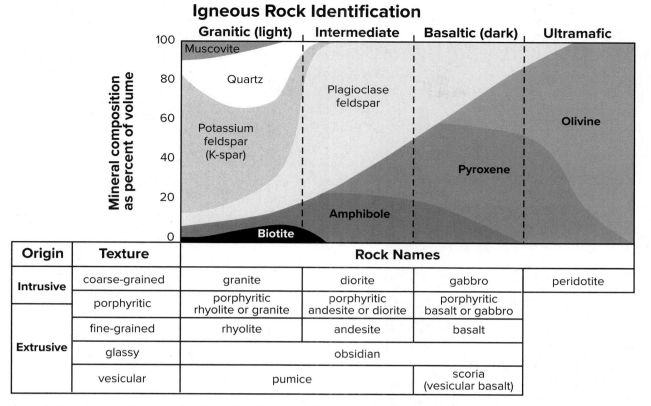

Origin	Texture	Rock Names			
Intrusive	coarse-grained	granite	diorite	gabbro	peridotite
Extrusive	porphyritic	porphyritic rhyolite or granite	porphyritic andesite or diorite	porphyritic basalt or gabbro	
	fine-grained	rhyolite	andesite	basalt	
	glassy	obsidian			
	vesicular	pumice		scoria (vesicular basalt)	

Figure 9 Rock type can be determined by estimating the relative percentages of minerals in the rocks.

Texture

In addition to differences in their mineral compositions, igneous rocks differ in the sizes of their grains or crystals. **Texture** refers to the size, shape, and distribution of the crystals or grains that make up a rock. For example, as shown in **Figure 10,** the texture of rhyolite can be described as fine-grained, while granite can be described as coarse-grained. The difference in crystal size can be explained by the fact that one rock is extrusive and the other is intrusive.

Rhyolite   Granite   Obsidian

Figure 10 Rhyolite, granite, and obsidian have different textures because they formed in different ways. Obsidian's glassy texture is a result of rapid cooling.

Porphyry

Vesicular basalt

Pumice

Figure 11 Rock textures provide information about a rock's formation. Evidence of the rate of cooling and the presence or absence of dissolved gases is preserved in the rocks shown here.

Crystal size and cooling rates When lava flows on Earth's surface, it cools quickly and there is not enough time for large crystals to form. The resulting extrusive igneous rocks, such as rhyolite, which is shown in **Figure 10,** have crystals so small that they are difficult to see without magnification. Sometimes, cooling occurs so quickly that crystals do not form at all. The result is volcanic glass, called obsidian, also shown in **Figure 10.** In contrast, when magma cools slowly beneath Earth's surface, there is sufficient time for large crystals to form. Thus, intrusive igneous rocks, such as granite, diorite, and gabbro, can have crystals larger than 1 cm.

Porphyritic rocks Look at the textures of the rocks shown in **Figure 11.** The top photo shows a rock with different crystal sizes. This rock has a **porphyritic** (por fuh RIH tihk) **texture,** which is characterized by large, well-formed crystals surrounded by finer-grained crystals of the same mineral or different minerals.

What causes minerals to form both large and small crystals in the same rock? Porphyritic textures indicate a complex cooling history during which a slowly cooling magma suddenly began cooling rapidly. Imagine a magma body cooling slowly, deep in Earth's crust. As it cools, the first crystals to form grow large. If this magma were to be suddenly moved higher in the crust, or if it erupted onto Earth's surface, the remaining magma would cool quickly and form smaller crystals.

Vesicular rocks Magma contains dissolved gases that escape when the pressure on the magma lessens. If the lava is thick enough to prevent the gas bubbles from escaping, holes called vesicles are left behind. The rock that forms looks spongy. This spongy appearance is called **vesicular texture.** Pumice and vesicular basalt are examples shown in **Figure 11.**

 Get It?

Explain what causes holes to form in igneous rocks.

Thin Sections

It is usually easier to observe the sizes of mineral grains than it is to identify the mineral. To identify minerals, geologists examine samples that are called thin sections. A thin section is a slice of rock, generally 2 cm × 4 cm and only 0.03 mm thick. Because it is so thin, light is able to pass through it.

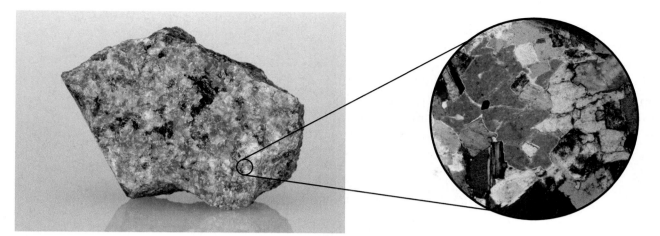

Figure 12 The minerals that make up this piece of granite can be identified in a thin section.

When viewed through a special microscope, called a petrographic microscope, mineral grains exhibit distinct properties. These properties allow geologists to identify the minerals present in the rock. For example, feldspar grains often show a distinct banding called twinning. Quartz grains might appear wavy as the microscope stage is rotated. Calcite crystals become dark, or extinguish, as the stage is rotated. **Figure 12** shows the appearance of a thin section of granite under a petrographic microscope.

Igneous Rocks as Resources

The cooling and crystallization history of igneous rocks sometimes results in the formation of unusual but useful minerals. These minerals can be used in many fields, including construction, energy production, and jewelry making.

Veins

As you have learned, ores are minerals that contain a useful material that can be mined for a profit. Valuable ore deposits often occur within igneous intrusions. In other cases, ore minerals are found in the rocks surrounding intrusions. These types of deposits sometimes occur as veins. Recall from Bowen's reaction series that the fluid left during magma crystallization contains high levels of silica and water. This fluid also contains any leftover elements that were not incorporated into the common igneous minerals. Some important metallic elements that are not included in common minerals are gold, silver, lead, and copper. These elements are released at the end of magma crystallization in a hot, mineral-rich fluid that fills cracks and voids in the surrounding rock. This fluid deposits metal-rich quartz veins, such as the gold-bearing veins in the Sierra Nevada. An example of gold formed in a quartz vein is shown in **Figure 13.**

Figure 13 Gold and quartz are extracted from mines together. The two are later separated.

Infer *What can you determine from this photo about the melting temperature of gold?*

 Get It?

Explain why veins have high amounts of quartz.

Pegmatites

Vein deposits can contain other valuable resources in addition to metals. Igneous rocks that are made of extremely large-grained minerals are called **pegmatites** and are usually found as igneous intrusions or veins. Ores of rare elements, such as lithium (Li) and beryllium (Be), often form in pegmatites. In addition to ores, pegmatites can produce beautiful crystals. Because veins fill cavities and fractures in rock, minerals grow into voids and retain their shapes. Some of the world's most beautiful minerals have been found in pegmatite veins. A famous pegmatite is the source rock for the Mount Rushmore National Memorial, located near Keystone, South Dakota. A close-up view of President Thomas Jefferson, shown in **Figure 14,** reveals the huge mineral veins that run through the rock.

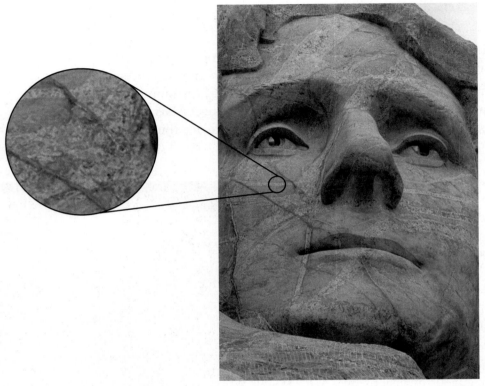

Figure 14 Pegmatite veins cut through much of the rock from which Mount Rushmore National Memorial is carved. You can see the veins running across Thomas Jefferson's face.

Kimberlites

Diamond is a valuable mineral found in rare, ultrabasic rocks known as **kimberlites,** named after Kimberly, South Africa, where the intrusions were first identified. These unusual rocks are a variety of peridotite. They most likely form in the mantle at depths of 150 to 300 km. This is because diamonds and other minerals present in kimberlites can form only under very high pressure.

Geologists hypothesize that kimberlite magma is intruded rapidly upward toward Earth's surface, forming long, narrow, pipelike structures. These structures extend many kilometers into the crust, but they are only 100 to 300 m in diameter. For this reason, they are often called kimberlite pipes.

Most of the world's diamonds come from South African mines, such as the one shown in **Figure 15.** Many kimberlites have been discovered in the United States, but diamonds have been found only in Arkansas and Colorado. The diamond mine in Colorado is the only diamond mine currently in operation in the United States.

 Get It?

Explain how kimberlites form and why diamonds are found in kimberlites.

Igneous rocks in construction

Igneous rocks have several characteristics that make them especially useful as building materials. The interlocking grain textures of igneous rocks make them strong. In addition, many of the minerals present in igneous rocks are resistant to weathering.

Granite is among the most durable of igneous rocks. You have probably seen many items, such as countertops, floors, and statues, made from the wide variety of granite that has formed on Earth.

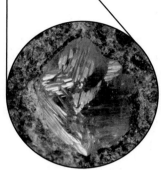

Figure 15 Diamonds are mined from kimberlite in mines like this one in Richtersveld, Northern Cape, South Africa.

Check Your Progress

Summary

- Magma consists of molten rock, dissolved gases, and mineral crystals.
- Magma is classified as basaltic, andesitic, or rhyolitic, based on the amount of silica it contains.
- Different minerals melt and crystallize at different temperatures.
- Bowen's reaction series defines the order in which minerals crystallize from magma.
- Igneous rocks are either ultramafic, basaltic, intermediate, or granitic.
- The rate of cooling determines crystal size.
- Some igneous rocks are used as building materials.

Demonstrate Understanding

1. **Predict** the appearance of an igneous rock that formed as magma cooled quickly and then more slowly.
2. **List** the eight major elements present in most magmas. Include the chemical symbol for each element.
3. **Summarize** the factors that affect the formation of magma.
4. **Identify** mineral resources and how they are used.

Explain Your Thinking

5. **Predict** If the temperature increases toward the center of Earth, why is the inner core solid?
6. **Infer** the silica content of magma derived from partial melting of an igneous rock. Would it be higher, lower, or about the same as the rock itself? Explain.
7. **WRITING Connection** A local rock collector claims that she has found the first example of pyroxene and sodium-rich feldspar in the same rock. Write a commentary about her claim for publication in a rock collector society newsletter.

LEARNSMART Go online to follow your personalized learning path to review, practice, and reinforce your understanding.

FOCUS QUESTION

How do rocks become sediment?

Weathering and Erosion

Wherever rock is exposed at Earth's surface, it is continuously being broken down by weathering—a set of physical and chemical processes that breaks rock into smaller pieces. **Sediments** are small pieces of rock that are moved and deposited by water, wind, glaciers, and gravity. When sediments become glued together, they form sedimentary rocks. The formation of sedimentary rocks begins when weathering and erosion produce sediments.

Weathering

Weathering produces rock and mineral fragments, known as sediments, that range in size from huge boulders to microscopic particles. Chemical weathering occurs when the minerals in a rock are dissolved or otherwise chemically changed. What happens to more resistant minerals during weathering? While the less stable minerals are chemically broken down, the more resistant grains are broken off of the rock as smaller grains. During physical weathering, however, minerals remain chemically unchanged. Rock fragments break off of the solid rock along fractures or grain boundaries. The rock in **Figure 16** has been chemically and physically weathered.

Figure 16 When exposed to both chemical and physical weathering, granite eventually breaks apart and might look like the decomposed granite shown here.

Infer *which of the three common minerals—quartz, feldspar, and mica—is most resistant to chemical weathering.*

 **3D THINKING**   **DCI** Disciplinary Core Ideas   **CCC** Crosscutting Concepts   **SEP** Science & Engineering Practices

COLLECT EVIDENCE

 Use your Science Journal to record the evidence you collect as you complete the readings and activities in this lesson.

INVESTIGATE

 GO ONLINE to find these activities and more resources.

👓 **Investigation Lab: Comparing Chemical Sedimentary Rocks and Modeling Their Formation**
Carry out an investigation into the changes that occur during the formation of sedimentary rocks.

👓 **Quick Investigation: Model Sediment Layering**
Develop and use a model to visualize the result of sediment settling.

Mark Dierker/Bear Dancer Studios

Erosion

The removal and transport of sediment is called erosion. **Figure 17** shows the four main agents of erosion: wind, moving water, gravity, and glaciers. Visible signs of erosion are all around you. You can observe erosion in action when a gust of wind blows soil across the infield at a baseball park. The force of the wind removes the soil and carries it away. In another example, water in streams becomes muddy after a storm because eroded silt and clay-sized particles have been mixed in it. Glaciers are large masses of ice that move across land. As they move, they drag sediments with them.

After rock fragments and sediments have been weathered out of the rock, they often are transported to new locations through the process of erosion. Eroded material is almost always carried downhill. Although wind can sometimes carry fine sand and dust to higher elevations, particles transported by water are almost always moved downhill. Eventually, even windblown dust and fine sand are pulled downhill by gravity.

 Get It?

Summarize what occurs during erosion.

Wind

Moving water

Gravity

Glaciers

Figure 17 Rocks and sediment are weathered and transported by the main agents of erosion—wind, moving water, gravity, and glaciers.

(t)MarketPlace/Media Bakery, (tr)Kai Honkanen/PhotoAlto, (bl)Matthew | Thomas/Shutterstock.com, (br)Photograph by Bruce F. Molnia, U.S. Geological Survey

Deposition

When transported sediments are deposited on the ground or sink to the bottom of a body of water, deposition occurs. Sediments in nature are deposited when transport stops. Perhaps the wind stops blowing or a river enters a quiet lake or an ocean. In each case, the particles being carried will settle out, forming layers of sediment, with the largest grains at the bottom and the smallest grains at the top.

Energy of transporting agents

Fast-moving water can transport larger particles better than slow-moving water. As water slows down, the largest particles settle out first, then the next largest, and so on, so that different-sized particles are sorted into layers. Such deposits are characteristic of sediment transported by water and wind. Wind, however, can move only small grains. For this reason, sand dunes are commonly made of fine, well-sorted sand, as shown in **Figure 18.** Not all sediment deposits are sorted. Glaciers, for example, move all materials with equal ease. Large boulders, sand, and mud are all carried along by the ice and dumped in an unsorted pile as the glacier melts. Landslides create similar deposits when sediment moves downhill in a jumbled mass.

Figure 18 These sand dunes at White Sands National Monument in New Mexico were formed by windblown sand that has been transported and redeposited. Notice the uniform size of the sand grains.

Lithification

Most sediments are ultimately deposited on Earth in low areas such as valleys and ocean basins. As more sediment is deposited in an area, the bottom layers are subjected to increasing pressure and temperature. These conditions cause **lithification,** the physical and chemical processes that transform sediments into sedimentary rocks. *Lithify* comes from the Greek word *lithos,* which means "stone."

()rucasrucas/Shutterstock, (r)Frank Krahmer/Radius Images/Corbis

STEM CAREER Connection

Sedimentologist

Do you wonder if sediments contain pollutants or other chemicals? Do you like learning about human history by examining layers of sediment? If so, you may enjoy working as a sedimentologist. Sedimentologists work both in the field, where they collect samples, and in the lab, where they analyze samples.

CCC CROSSCUTTING CONCEPTS

Structure and Function Beach dunes are formed when wind blows grains of sand into an area where the sand accumulates. You may have noticed that on the side of a dune facing the wind, the sand grains are blown up the side. The side away from the wind is usually smooth compared to the side facing the wind.

Compaction

Lithification begins with compaction. The weight of overlying sediments forces the sediment grains closer together, causing the physical changes shown in **Figure 19.** Layers of mud can contain up to 60 percent water, and these shrink as excess water is squeezed out. Sand does not compact as much as mud during burial, partly because individual sand grains, usually composed of quartz, do not deform under normal burial conditions. Grain-to-grain contacts in sand form a supporting framework that helps maintain open spaces between the grains. Groundwater, oil, and natural gas are commonly present in these spaces in sedimentary rocks.

Cementation

Compaction is not the only force that binds the grains together. **Cementation** occurs when mineral growth glues sediment grains together into solid rock. This occurs when a new mineral, such as quartz (SiO_2), calcite ($CaCO_3$), or iron oxide (Fe_2O_3), grows between sediment grains as dissolved minerals precipitate out of groundwater. This process is illustrated in **Figure 20.**

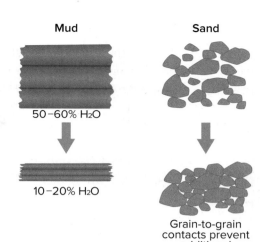

Mud   Sand

50–60% H₂O

10–20% H₂O

Grain-to-grain contacts prevent additional compaction.

Figure 19 The high water content and flat shape of particles in mud cause it to compact greatly when subjected to the weight of overlying sediments.

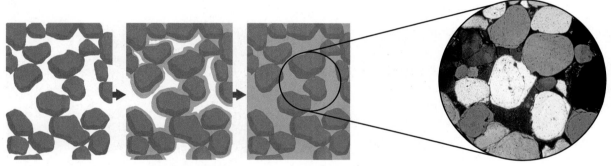

Figure 20 Minerals precipitate out of water as it flows through pore spaces in the sediment. These minerals form the cement that glues the sediments together.

Sedimentary Features

Just as igneous rocks contain information about the history of their formation, sedimentary rocks also have features and characteristics that help geologists interpret how they formed and the history of the area in which they formed.

Bedding

The primary feature of sedimentary rocks is horizontal layering called **bedding.** This feature results from the way sediment settles out of water or wind. Individual beds can range in thickness from a few millimeters to several meters. There are two different types of bedding, each dependent upon the method of transport. However, the size of the grains and the material within the bedding depend upon many other factors.

Graded bedding

Bedding in which the particle sizes become progressively finer and lighter toward the top layers is called **graded bedding.** Graded bedding is often observed in marine sedimentary rocks that were deposited by underwater landslides. As the sliding material slowly came to rest under water, the largest and heaviest material settled out first and was followed by progressively finer material. An example of graded bedding is shown in **Figure 21.**

Cross-bedding

Another characteristic feature of sedimentary rocks is cross-bedding. **Cross-bedding,** shown in **Figure 22,** is formed as layers of sediment are deposited at an incline across a horizontal surface. When these deposits become lithified, the cross-beds are preserved in the rock. Small-scale cross-bedding forms on sandy beaches and along sandbars in streams and rivers. Most large-scale cross-bedding is formed by migrating sand dunes.

 Get It?

Contrast graded bedding and cross-bedding.

Ripple marks When sediment is moved into small ridges by wind or wave action or by a river current, ripple marks form. The back-and-forth movement of waves on a shore pushes the sand on the bottom into symmetrical ripple marks, where grain size is evenly distributed. When a current flows in one direction, such as in a river, it pushes the sediment on the bottom into asymmetrical ripple marks. The ripple marks are steeper upstream and contain coarser sediment on the upstream side. If a rippled surface is buried gently by more sediment without being disturbed, it might be preserved in solid rock. The formation of ripple marks is illustrated in **Figure 23.**

 Get It?

Explain how the two types of ripple marks form.

Figure 21 The graded bedding (shown between the arrows in this photo) of the Furnace Creek Formation in Death Valley, California, records an episode of deposition during which the water that carried these sediments slowed and lost energy.

Figure 22 The large-scale cross-beds in these ancient dunes at Zion National Park were deposited by wind.

SCIENCE USAGE v. COMMON USAGE

grade

Science usage: a position in a scale of ranks or qualities

Common usage: a mark indicating a degree of knowledge or completion in school

(t.b)©Doug Sherman/Geofile

Figure 23 Visualizing Cross-Bedding and Ripple Marks

Moving water and loose sediment result in the formation of sedimentary structures such as cross-bedding and ripple marks.

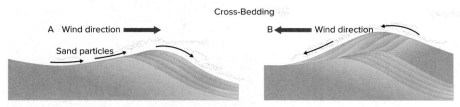

Sand carried by wind gets deposited on the downwind side of a dune. As the wind changes direction, cross-bedding is formed that records this change in direction.

Sediment on the river bottom gets pushed into small hills and ripples by the current. Additional sediment gets deposited at an angle on the downcurrent side of these hills forming cross-beds. Eventually, it levels out or new hills form and the process begins again.

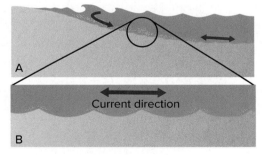

The back-and-forth wave action on a shore pushes the sand on the bottom into symmetrical ripple marks. Grain size is evenly distributed.

Current that flows in one direction, such as that of a river, pushes sediment on the bottom into asymmetrical ripple marks. They are steeper upstream and contain coarser sediment on the upstream side.

Angular vs. rounded

Some individual sediment grains have jagged, angular edges, and some are rounded. When a rock breaks apart, the pieces are initially angular in shape. As the sediment is transported away from its source, individual pieces knock into each other. The edges are broken off and, over time, the pieces become rounded. The amount of rounding is influenced by how long the sediment has been in transport and how far the sediment has traveled. Also, harder minerals with little to no cleavage have a better chance of becoming rounded before they break apart. As shown in **Figure 24,** quartz sand on beaches is nearly round, while carbonate sand, which is made up of seashells and calcite, is usually more angular because it is deposited closer to the source of the sediment.

Evidence of past life

Probably the best-known features of sedimentary rocks are fossils. Fossils are the preserved remains or impressions or any other evidence, of once-living organisms. When an organism dies, if its remains are buried without being disturbed, it might be preserved as a fossil. During lithification, parts of the organism can be replaced by minerals and turned into rock, such as shells that have been mineralized.

Clastic Sedimentary Rocks

The most common sedimentary rocks, **clastic sedimentary rocks,** are formed from the abundant deposits of loose sediments that accumulate on Earth's surface. The word **clastic** comes from the Greek word *klastos,* meaning "broken." These rocks are further classified according to the sizes of their particles. As you read about each rock type, refer to **Table 2** on the next page, which summarizes the classification of sedimentary rocks based on grain size, mode of formation, and mineral content.

Coarse-grained rocks Sedimentary rocks consisting of gravel-sized rock and mineral fragments are classified as coarse-grained rocks, samples of which are shown in **Figure 25.** Conglomerates have rounded, gravel-sized particles. Because of its relatively large mass, gravel is transported by high-energy flows of water, such as those generated by mountain streams, flooding rivers, some ocean waves, and glacial meltwater. During transport, gravel becomes abraded and rounded as the particles scrape against one another. This is why beach and river gravels are often well rounded. Lithification turns these sediments into conglomerates.

In contrast, breccias are composed of angular, gravel-sized particles. The angularity indicates that the sediments from which they formed did not have time to become rounded. This suggests that the particles were transported only a short distance and deposited close to their source. Refer to **Table 2** to see how these rocks are named.

Quartz Sand

Carbonate Sand

Figure 24 The carbonate sand has sharp, jagged pieces and is not as rounded and smooth as the quartz sand.

Conglomerate

Breccia

Figure 25 Conglomerates and breccias are made of coarse sediments that have been transported by high-energy water.

Infer *the circumstances that might cause the types of transport necessary for each to form.*

Table 2 Classification of Sedimentary Rocks

Classification	Texture/Grain Size	Composition	Rock Name
Clastic	coarse (> 2 mm)	Fragments of any rock type—quartz, chert and quartzite common } rounded angular	conglomerate breccia
	medium (1/16 mm to 2 mm)	quartz and rock fragments quartz, potassium feldspar and rock fragments	sandstone arkose
	fine (1/256 mm–1/16 mm)	quartz and clay	siltstone
	very fine (< 1/256 mm)	quartz and clay	shale
Biochemical	microcrystalline with conchoidal fracture	calcite ($CaCO_3$) quartz (SiO_2)	micrite chert
	abundant fossils in micrite matrix	calcite ($CaCO_3$)	fossiliferous limestone
	shells and shell fragments loosely cemented	calcite ($CaCO_3$)	coquina
	microscopic shells and clay	calcite ($CaCO_3$)	chalk
	variously sized fragments	highly altered plant remains, some plant fossils	coal
Chemical	ooids (small spheres of calcium carbonate)	calcite ($CaCO_3$)	oolitic limestone
	fine to coarsely crystalline	calcite ($CaCO_3$)	crystalline limestone
	fine to coarsely crystalline	dolomite ($(Ca,Mg)CO_3$ (will effervesce if powered)	dolostone
	very finely crystalline	quartz (SiO_2)—light colored; dark colored calcite ($CaCO_3$)	chert; flint micrite
	fine to coarsely crystalline	gypsum ($CaSO_4 \cdot 2H_2O$)	rock gypsum
	fine to coarsely crystalline	halite (NaCl)	rock salt

Medium-grained rocks Stream and river channels, beaches, and deserts often contain abundant sand-sized sediments. Sedimentary rocks that contain sand-sized rock and mineral fragments are classified as medium-grained clastic rocks. Refer to **Table 2** for a listing of rocks with sand-sized particles. Sandstone usually contains several features of interest to scientists. For example, because ripple marks and cross-bedding indicate the direction of current flow, geologists use sandstone layers to map ancient stream and river channels.

Another important feature of sandstone is its relatively high porosity. **Porosity** is the percentage of open spaces between grains in a rock. Loose sand can have a porosity of up to 40 percent. Some of these open spaces are maintained during the formation of sandstone, often resulting in porosities as high as 30 percent. When pore spaces are connected to one another, fluids can move through sandstone. This feature makes sandstone layers valuable as underground reservoirs of oil, natural gas, and groundwater.

ACADEMIC VOCABULARY

reservoir
a subsurface area of rock that has enough porosity to allow for the accumulation of oil, natural gas, or water
The newly discovered reservoir contained large amounts of natural gas and oil.

Fine-grained rocks Sedimentary rocks consisting of silt- and clay-sized particles, such as siltstone and shale, are called fine-grained rocks. These rocks represent environments like swamps, ponds, and deep oceans which have still or slow-moving waters. In the absence of strong currents and wave action, these sediments settle to the bottom where they accumulate in thin horizontal layers. Shale often breaks along thin layers, as shown in **Figure 26.** Unlike sandstone, fine-grained sedimentary rock has low porosity and often forms barriers that hinder the movement of groundwater and oil. **Table 2** shows how these rocks are named.

Figure 26 The very fine-grained sediment that formed this shale was deposited in thin layers in still waters.

Get It?

Identify the types of environments in which fine-grained rocks form.

Chemical and Biochemical Sedimentary Rocks

The formation of chemical and biochemical rocks involves the processes of evaporation and precipitation of minerals. During weathering, minerals can be dissolved and carried into lakes and oceans. As water evaporates from the lakes and oceans, the dissolved minerals are left behind. In arid regions, high evaporation rates can increase the concentration of dissolved minerals in bodies of water. The Great Salt Lake, shown in **Figure 27**, is an example of a lake that has high concentrations of dissolved minerals.

Chemical sedimentary rocks When the concentration of dissolved minerals in a body of water reaches saturation, crystals can precipitate out of solution and settle to the bottom. As a result, layers of chemical sedimentary rocks form, most of which are called **evaporites.** Evaporites primarily form in arid regions, drainage basins on continents that have low water flow, and in coastal settings. Because these areas usually have minimal freshwater input and high rates of evaporation, the concentration of dissolved minerals remains high. Over time, thick layers of evaporite minerals can accumulate on basin floors, as illustrated in **Figure 27**.

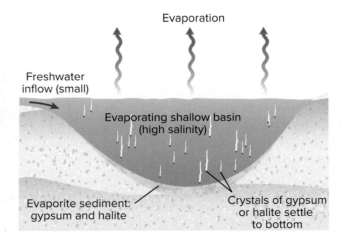

Figure 27 The constant evaporation from a body of salt water results in precipitation of large amounts of salt. This process has been occurring in the Great Salt Lake in Utah for approximately 18,000 years.

Biochemical sedimentary rocks Biochemical sedimentary rocks are formed from the remains of once-living organisms. The most abundant of these rocks is limestone, which is composed primarily of calcite. Some organisms that live in the ocean use the calcium carbonate that is dissolved in seawater to make their shells. When these organisms die, their shells settle to the bottom of the ocean and can form thick layers of carbonate sediment. During burial and lithification, calcium carbonate precipitates out of the water, crystallizes between the grains of carbonate sediment, and forms limestone.

Limestone is common in shallow water environments, such as those in the Bahamas, where coral reefs thrive in 15 to 20 m of water just offshore. The skeletal and shell materials that are currently accumulating there will someday become limestone as well. Many types of limestone contain evidence of their biological origin in the form of abundant fossils. As shown in **Figure 28,** these fossils can range from large-shelled organisms to microscopic, unicellular organisms. However, not all limestone contains fossils or is biochemical in origin. Some limestone has a crystalline texture or consists of tiny spheres of carbonate sand called ooids. These are listed in **Table 2.**

Other organisms make their shells out of silica, or microcrystalline quartz. After these organisms die and settle to the bottom of the ocean, their shells can form sediment that is often referred to as siliceous ooze because it is rich in silica. Siliceous ooze becomes lithified into the sedimentary rock chert, which is also listed in **Table 2.**

Figure 28 Limestone can contain many different fossil organisms. Geologists can interpret where and when the limestone formed by studying the fossils within the rock.

Check Your Progress

Summary

- The processes of weathering, erosion, deposition, and lithification form sedimentary rocks.
- Sediments are lithified into rock by the processes of compaction and cementation.
- Clastic rocks form from sediments and are classified by particle size and shape.
- Chemical rocks form primarily from minerals precipitated from water.
- Biochemical rocks form from the remains of once-living organisms.

Demonstrate Understanding

1. **Describe** how sediments are produced by weathering and erosion.
2. **Sequence** Use a flowchart to show why sediment deposits tend to form layers.
3. **Compare** temperature and pressure conditions at Earth's surface and below Earth's surface, and relate them to the process of lithification.

Explain Your Thinking

4. **Determine** whether you are walking upstream or downstream along a dry mountain streambed if you notice that the shape of the sediment is getting more angular as you continue walking. Explain.
5. **WRITING ›Connection** Imagine you are designing a museum display based on a sedimentary rock that contains fossils of corals and other marine animals. Draw a picture of what this environment might have looked like, and write the accompanying description that will be posted next to the display.

LEARNSMART Go online to follow your personalized learning path to review, practice, and reinforce your understanding.

METAMORPHIC ROCKS

FOCUS QUESTION

How can we tell when a rock has been transformed?

Recognizing Metamorphic Rock

The rock layers shown in **Figure 29** have been metamorphosed (meh tuh MOR fohzd)—this means that they have been changed. How do geologists know that this has happened? Pressure and temperature increase with depth. When temperature and pressure become high enough, rocks melt and form magma. But what happens if the rocks do not reach the melting point? When temperature and pressure combine and change the texture, mineral composition, or chemical composition of a rock without melting it, a metamorphic rock forms. The word *metamorphism* is derived from the Greek words *meta*, meaning "change," and *morphé*, meaning "form." During metamorphism, a rock changes form while remaining solid.

The high temperatures required for metamorphism are ultimately derived from Earth's internal heat, either through deep burial or from nearby igneous intrusions. The high pressures required for metamorphism come from deep burial or from compression during mountain building.

Figure 29 Strong forces were required to bend these rock layers into the shape they are today.

Hypothesize *the changes that occurred to the sediments after they were deposited.*

©Stephen Reynolds

 **3D THINKING**　**DCI** Disciplinary Core Ideas　**CCC** Crosscutting Concepts　**SEP** Science & Engineering Practices

COLLECT EVIDENCE

 Use your Science Journal to record the evidence you collect as you complete the readings and activities in this lesson.

INVESTIGATE

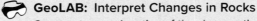 **GO ONLINE** to find these activities and more resources.

　GeoLAB: Interpret Changes in Rocks
　Construct an explanation of the changes that occur to a rock during metamorphosis.

　?　**Revisit the Encounter the Phenomenon Question**
　What information from this lesson can help you answer the Unit and Module questions?

Metamorphic minerals

How do minerals change without melting? Think back to the concept of fractional crystallization. Bowen's reaction series shows that all minerals are stable at certain temperatures, and they crystallize from magma along a range of different temperatures. Scientists have discovered that these stability ranges also apply to minerals in solid rock. During metamorphism, the minerals in a rock change into new minerals that are stable under the new temperature and pressure conditions. Minerals that change in this way are said to undergo solid-state alterations. Scientists have conducted experiments to identify the metamorphic conditions that create specific minerals. When the same minerals are identified in rocks, scientists are able to interpret the conditions inside the crust during the rocks' metamorphism. **Figure 30** shows some common metamorphic minerals.

Figure 30 Metamorphic minerals, such as mica, staurolite, garnet, and talc (shown above, clockwise from top left), occur in many colors, shapes, and crystal sizes. Colors can be dark or light, and crystal form can be unique.

 Get It?

Explain what metamorphic minerals are.

Metamorphic textures

Metamorphic rocks are classified into two textural groups: foliated and nonfoliated. Geologists use metamorphic textures and mineral composition to identify metamorphic rocks. **Figure 31** shows how these two characteristics are used in the classification of metamorphic rocks.

Foliated rocks Layers and bands of minerals characterize **foliated** metamorphic rocks. High pressure during metamorphism causes minerals with flat or needlelike crystals to form with their long axes perpendicular to the pressure, as shown in **Figure 32.** This parallel alignment of minerals creates the layers observed in foliated metamorphic rocks.

Metamorphic Rock Identification Chart

Texture		Composition	Rock Name
Foliated	Layered	Fine-grained		Slate
		Fine-to-medium-grained	Chlorite / Mica / Quartz / Feldspar / Amphibole / Pyroxene	Phyllite
		Coarse-grained		Schist
	Banded	Coarse-grained		Gneiss
Nonfoliated		Fine- to coarse-grained	Quartz	Quartzite
			Calcite or dolomite	Marble

Figure 31 Increasing grain size parallels changes in composition and development of foliation. Grain size is not a factor in nonfoliated rocks.

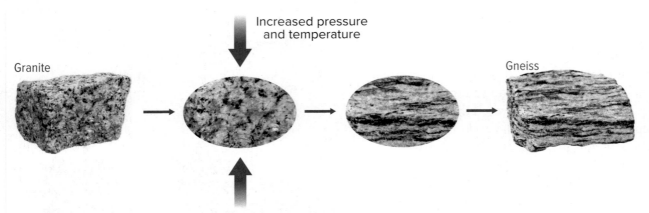

Increased pressure and temperature

Granite

Gneiss

Figure 32 Foliation develops when pressure is applied from opposite directions. The foliation develops perpendicular to the pressure direction.

Nonfoliated rocks

Unlike foliated rocks, **nonfoliated** metamorphic rocks are composed mainly of minerals that form with blocky crystal shapes. Two common examples of nonfoliated rocks, shown in **Figure 33,** are quartzite and marble. Quartzite is a hard, often light-colored rock formed by the metamorphism of quartz-rich sandstone. Marble is formed by the metamorphism of limestone or dolomite. Some marbles have smooth textures that are formed by interlocking grains of calcite. These marbles are often used in sculptures. Fossils are rarely preserved in metamorphic rocks.

Under certain conditions, new metamorphic minerals can grow large while the surrounding minerals remain small. The large crystals, which can range in size from a few millimeters to a few centimeters, are called porphyroblasts. Although these crystals resemble the very large crystals that form in pegmatite granite, they are not the same. Instead of forming from magma, they form in solid rock through the reorganization of atoms during metamorphism. Garnet, shown in **Figure 33,** is a mineral that commonly forms porphyroblasts.

Marble

Quartzite

Garnet porphyroblasts in schist

Figure 33 As a result of the extreme heat and pressure during metamorphism, marble rarely contains fossils. Metamorphism does not, however, always destroy cross-bedding and ripple marks, which can be seen in some quartzites. Garnet porphyroblasts can grow to be quite large in some rocks, as shown by the garnets in this sample of schist.

Grades of Metamorphism

Different combinations of temperature and pressure result in different grades of metamorphism. Low-grade metamorphism is associated with low temperatures and pressures and a particular suite of minerals and textures. High-grade metamorphism is associated with high temperatures and pressures and a different suite of minerals and textures. Intermediate-grade metamorphism is in between low- and high-grade metamorphism.

Figure 34 shows the minerals present in metamorphosed shale. Note the change in composition as conditions change from low-grade to high-grade metamorphism. Geologists can create metamorphic maps by plotting the location of metamorphic minerals. Knowing the temperatures that certain areas experienced when rocks were forming helps geologists locate valuable metamorphic minerals such as garnet and talc. Studying the distribution of metamorphic minerals helps geologists to interpret the metamorphic history of an area.

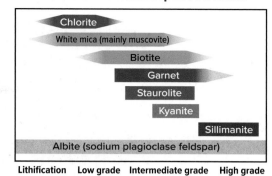

Minerals in Metamorphosed Shale

Chlorite
White mica (mainly muscovite)
Biotite
Garnet
Staurolite
Kyanite
Sillimanite
Albite (sodium plagioclase feldspar)

Lithification   Low grade   Intermediate grade   High grade

Figure 34 Metamorphism of shale results in the formation of minerals that provide the wide variety of color observed in slate.

Types of Metamorphism

The effects of metamorphism can be the result of contact metamorphism, regional metamorphism, or hydrothermal metamorphism. The minerals that form and the degree of change in the rocks provide information as to the type and grade of metamorphism that occurred.

Regional metamorphism

When temperature and pressure affect large regions of Earth's crust, they produce large belts of **regional metamorphism.** The metamorphism can range in grade from low to high. Results of regional metamorphism include changes in minerals and rock types, foliation, and folding and deforming of the rock layers that make up the area. The folded rock layers shown in **Figure 29** experienced regional metamorphism.

Contact metamorphism

When molten material, such as that in an igneous intrusion, comes in contact with solid rock, a local effect called **contact metamorphism** occurs. High temperature and moderate-to-low pressure form mineral assemblages that are characteristic of contact metamorphism.

SCIENCE USAGE v. COMMON USAGE
intrusion
Science usage: the placement of a body of magma into preexisting rock
Common usage: the act of joining or coming into without being invited

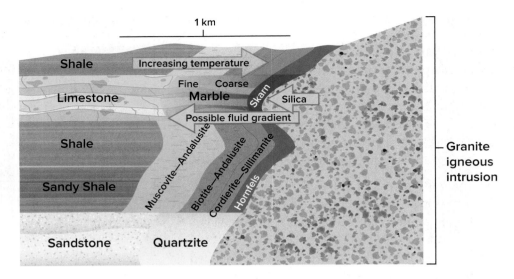

Figure 35 Contact metamorphism from the intrusion of this granite batholith has caused zones of metamorphic minerals to form.

Figure 35 shows zones of different minerals surrounding an intrusion. Because temperature decreases with distance from an intrusion, metamorphic effects also decrease with distance. Recall that minerals crystallize at specific temperatures. Metamorphic minerals that form at high temperatures occur closest to the intrusion, where it is hottest. Because lava cools too quickly for the heat to penetrate far into surface rocks, contact metamorphism from extrusive igneous rocks is limited to thin zones.

Hydrothermal metamorphism

When very hot water reacts with rock and alters its chemical and mineral composition, **hydrothermal metamorphism** occurs. The word *hydrothermal* is derived from the Greek words *hydro*, meaning "water," and *thermal*, meaning "heat." As hot fluids migrate in and out of the rock during metamorphism, the original mineral composition and texture of the rock can change. Chemical changes are common during contact metamorphism near igneous intrusions and active volcanoes. Valuable ore deposits of gold, copper, zinc, tungsten, and lead are formed in this manner.

Economic Importance of Metamorphic Rocks and Minerals

The modern way of life is made possible by a great number of naturally occurring Earth materials. We use salt for cooking; gold for trade; other metals for construction and industrial purposes; fossil fuels for energy; and rocks and various minerals for construction, cosmetics, and more. **Figure 36,** on the next page, shows two examples of how metamorphic rocks are used in construction. Many economic mineral resources are produced by metamorphic processes. Among these are the metals gold, silver, copper, and lead, as well as many significant nonmetallic resources.

Figure 36 Marble and slate are metamorphic rocks that have been used in construction for centuries.

Metallic mineral resources

Metallic resources occur mostly in the form of metal ores, although deposits of pure metals are occasionally discovered. Many metallic deposits are precipitated from hydrothermal solutions and are either concentrated in veins or spread throughout a rock mass. Native gold, silver, and copper deposits tend to occur in hydrothermal quartz veins near igneous intrusions or in contact metamorphic zones. However, most hydrothermal metal deposits are in the form of metal sulfides such as galena (PbS) or pyrite (FeS_2). The iron ores magnetite and hematite are oxide minerals often formed by precipitation from iron-bearing hydrothermal solutions.

Get It?

State the resources that hydrothermal metamorphism produces.

Nonmetallic mineral resources

Metamorphism of ultrabasic igneous rocks produces the minerals talc and asbestos. Talc, with a hardness of 1, is used as a dusting powder, as a lubricant, and to provide texture in paints. Because it is not combustible and has low thermal and electric conductivity, asbestos has been used in fireproof and insulating materials. Prior to the recognition of its cancer-causing properties, it was also widely utilized in the construction industry. Many older buildings still have asbestos-containing materials. Graphite, the main ingredient of the lead in pencils, is formed by the metamorphism of organic material.

Get It?

Explain why asbestos is no longer used in the construction industry.

CCC CROSSCUTTING CONCEPTS

Energy and Matter The rock cycle provides another example of the concept that neither matter nor energy can be created or destroyed. In the rock cycle, older rocks are recycled to form younger rocks. Energy in the rock cycle comes from the Sun and Earth's interior.

The Rock Cycle

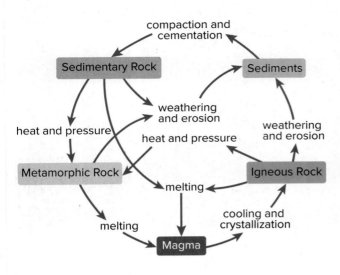

Figure 37 Rocks are continually being changed above and beneath Earth's surface. The rock cycle shows some of the series of changes rocks undergo.

Metamorphic rocks form when other rocks change. The three types of rock—igneous, sedimentary, and metamorphic—are grouped according to how they form. Igneous rocks crystallize from magma underneath Earth's surface or from lava on Earth's surface; sedimentary rocks form from cemented or precipitated minerals and sediments; and metamorphic rocks form from changes in temperature and pressure. Once a rock forms, does it remain the same type of rock always? Possibly, but it most likely will not.

Heat and pressure can change an igneous rock into a metamorphic rock. A metamorphic rock can be changed into another metamorphic rock or melted to form an igneous rock. Alternately, metamorphic rock can be weathered and eroded into sediments that might become cemented into a sedimentary rock. In fact, any rock can be changed into any other type of rock. The continuous changing and remaking of rocks is called the **rock cycle.** The rock cycle is summarized in **Figure 37.** The arrows represent the different processes that change rocks into different types.

Check Your Progress

Summary

- The three main types of metamorphism are regional, contact, and hydrothermal.

- The texture of metamorphic rocks can be foliated or nonfoliated.

- During metamorphism, new minerals form that are stable under the increased temperature and pressure conditions.

- The rock cycle is the set of processes through which rocks continuously change into other types of rocks.

Understand Main Ideas

1. **Summarize** how temperature increases can cause metamorphism.

2. **Summarize** what causes foliated metamorphic textures to form.

3. **Apply** the concept of the rock cycle to explain how the three main types of rocks are classified.

4. **Compare and contrast** the factors that cause the three main types of metamorphism.

Think Critically

5. **Infer** which steps in the rock cycle are skipped when granite metamorphoses to gneiss.

6. **Predict** the location of an igneous intrusion based on the following mineral data. Muscovite and chlorite were collected in the northern portion of the area of study; garnet and staurolite were collected in the southern portion of the area.

7. **MATH** ▶**Connection** Gemstones often form as porphyroblasts. Gemstones are described in terms of carat weight. A carat is equal to 0.2 g, or 200 mg. A large garnet discovered in New York in 1885 weighs 4.4 kg and is 15 cm in diameter. What is the carat weight of this gemstone?

LEARNSMART Go online to follow your personalized learning path to review, practice, and reinforce your understanding.

A Rock by Any Other Name

Geologists study Earth, the materials that compose it, and its processes. Many geologists study the different types of rocks on Earth. But did you know that some geologists also study rocks from outer space?

Geologists study Earth rocks as well as rocks of extraterrestrial origin.

Extraterrestrial rocks

The extraterrestrial rocks that geologists study come from the Moon, Mars, and asteroids. These rocks were collected in different ways. Rocks and other materials from the Moon were collected during the Apollo missions. About 124 Martian rocks have landed on Earth as meteorites. Most rocks from asteroids that geologists study—more than 60,000 of them—also fell to Earth in the form of meteorites. However, we do have one sample, taken by Japanese Aerospace Exploration Agency (JAXA), from an asteroid in orbit around the Sun.

Similarities to Earth rocks

Geologists have found that extraterrestrial rocks are similar to Earth rocks in several ways. For example, some Martian rocks show signs of having formed in the presence of water, as many sedimentary rocks on Earth are formed. Mars and Earth both have mudstone, sandstone, shale, and conglomerate rocks. Mars and the Moon also have basalt, a volcanic rock that shows evidence of Earth-like volcanic activity.

Differences from Earth rocks

Geologists have also found a few differences between Earth rocks and extraterrestrial rocks. Moon rocks have tiny pockmarks called zap pits because the Moon's atmosphere is so thin that even micrometeorites impact its surface. Earth's rocks do not have zap pits because its atmosphere burns up most micrometeorites. Most Martian rocks have few zap pits; the planet's atmosphere is thin, but it still protects the surface from micrometeorites.

Another difference relates to oxidation. On Earth, rocks and minerals undergo chemical weathering when oxygen in the atmosphere reacts with them. However, because the Moon's atmosphere does not contain free oxygen, its rocks do not show the effects of oxidation. In years to come, geologists will likely find more similarities and differences between Earth, Martian, and lunar rocks.

ENGAGE IN ARGUMENT FROM EVIDENCE

Work with a small group to find more information on this topic from print or online sources. Then use evidence to construct an argument about which rocks are most similar. Debate the issue with another group.

Print Collector/Hulton Archive/Getty Images

MODULE 4
STUDY GUIDE

 GO ONLINE to study with your Science Notebook.

Lesson 1 IGNEOUS ROCKS

- Magma consists of molten rock, dissolved gases, and mineral crystals.
- Magma is classified as basaltic, andesitic, or rhyolitic based on the amount of silica it contains.
- Different minerals melt and crystallize at different temperatures.
- Bowen's reaction series defines the order in which minerals crystallize from magma.
- Igneous rocks are either ultramafic, basaltic, intermediate, or granitic.
- The rate of cooling determines crystal size.
- Some igneous rocks are used as building materials.

- igneous rock
- lava
- partial melting
- Bowen's reaction series
- fractional crystallization
- intrusive rock
- extrusive rock
- basaltic rock
- granitic rock
- texture
- porphyritic texture
- vesicular texture
- pegmatite
- kimberlite

Lesson 2 SEDIMENTARY ROCKS

- The processes of weathering, erosion, deposition, and lithification form sedimentary rocks.
- Sediments are lithified into rock by the processes of compaction and cementation.
- Sedimentary rocks might contain features such as horizontal bedding, cross-bedding, and ripple marks.
- Clastic rocks form from sediments and are classified by particle size and shape.
- Chemical rocks form primarily from minerals precipitated from water.
- Biochemical rocks form from the remains of once-living organisms.

- sediment
- lithification
- cementation
- bedding
- graded bedding
- cross-bedding
- clastic sedimentary rocks
- clastic
- porosity
- evaporite

Lesson 3 METAMORPHIC ROCKS

- The three main types of metamorphism are regional, contact, and hydrothermal.
- The texture of metamorphic rocks can be foliated or nonfoliated.
- During metamorphism, new minerals form that are stable under the increased temperature and pressure conditions.
- The rock cycle is the set of processes through which rocks continuously change into other types of rocks.

- foliated
- nonfoliated
- regional metamorphism
- contact metamorphism
- hydrothermal metamorphism
- rock cycle

REVISIT THE PHENOMENON

How did these different types of rock form?

CER Claim, Evidence, Reasoning

Explain Your Reasoning Revisit the claim you made when you encountered the phenomenon. Summarize the evidence you gathered from your investigations and research and finalize your Summary Table. Does your evidence support your claim? If not, revise your claim. Explain why your evidence supports your claim.

STEM UNIT PROJECT
Now that you've completed the module, revisit your STEM unit project. You will apply your evidence from this module and complete your project.

GO FURTHER

SEP Data Analysis Lab

Which metamorphic minerals will form?

The minerals that form in metamorphic rocks depend on the metamorphic grade and composition of the original rock. The diagram on the right and **Figure 23** show the mineral groups that form under different metamorphic conditions. Use both diagrams to answer the questions below.

CER Analyze and Interpret Data

1. **Claim, Evidence** What minerals are formed when shale and basalt are exposed to low-grade metamorphism? Use evidence to explain your answer.
2. **Claim, Evidence** Identify the mineral groups that you would expect to form from intermediate-grade metamorphism of shale and basalt. Support your answer with evidence.
3. **Reasoning** Explain the major compositional differences between shale and basalt. How are these differences reflected in the minerals formed during metamorphism?

Minerals in Metamorphosed Basalt

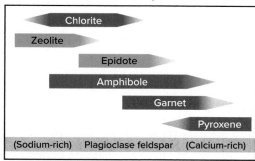

Chlorite
Zeolite
Epidote
Amphibole
Garnet
Pyroxene
(Sodium-rich)   Plagioclase feldspar   (Calcium-rich)

Lithification   Low grade   Intermediate grade   High grade

ENCOUNTER THE PHENOMENON

How did wind, water, and ice shape this landscape?

SEP Ask Questions

What questions do you have about the phenomenon? Write your questions on sticky notes and add them to the driving question board for this unit.

What is the difference between weathering and erosion?

Look for Evidence

As you go through this unit, use the information and your experiences to help you answer the phenomenon question as well as your own questions. For each activity, record your observations in a Summary Table, add an explanation, and identify how it connects to the unit and module phenomenon questions.

Solve a Problem
STEM UNIT PROJECT

Surface Processes on Earth Investigate and research more about Earth's surface processes. Use the results of these investigations and the evidence you collected during the unit to complete your unit project.

GO ONLINE In addition to reading the information in your Student Edition, you can find the STEM Unit Project and other useful resources online.

WEATHERING, EROSION, AND SOIL

ENCOUNTER THE PHENOMENON

How does ice carve large valleys out of mountain rock?

GO ONLINE to play a video about glacial ice.

SEP Ask Questions

Do you have other questions about the phenomenon? If so, add them to the driving question board.

CER Claim, Evidence, Reasoning

Make Your Claim Use your CER chart to make a claim about how ice can carve large valleys out of mountain rock. Explain your reasoning.

Collect Evidence Use the lessons in this module to collect evidence to support your claim. Record your evidence as you move through the module.

Explain Your Reasoning You will revisit your claim and explain your reasoning at the end of the module.

GO ONLINE to access your CER chart and explore resources that can help you collect evidence.

LESSON 1: Explore & Explain: Mechanical Weathering

LESSON 2: Explore & Explain: Glacial Erosion

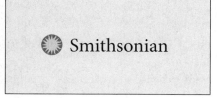

Additional Resources

(t)Video Supplied by BBC Worldwide Learning, (b)Richard Myers/Alamy Stock Photo

FOCUS QUESTION

How does ice help to break down rock?

Mechanical Weathering

Weathering is the process in which materials on or near Earth's surface break down and change. **Mechanical weathering** is a type of weathering in which rocks and minerals break down into smaller pieces. This process is also called physical weathering. Mechanical weathering does not involve any change in a rock's composition, only changes in the size and shape of the rock. A variety of factors are involved in mechanical weathering, including changes in temperature and pressure.

Effect of temperature

Temperature plays a role in mechanical weathering. When water freezes, it expands and increases in volume by 9 percent. You have observed this increase in volume if you have ever frozen water in an ice cube tray. In many places on Earth's surface, water collects in the cracks of rocks and rock layers. If the temperature drops to the freezing point, water freezes, expands, exerts pressure on the rocks, and can cause the cracks to widen, as shown in **Figure 1**. When the temperature increases, the ice melts in the cracks of rocks and rock layers. The freeze-thaw cycles of water in the cracks of rocks is called **frost wedging.** Frost wedging is responsible for the formation of potholes in many roads in the northern United States, where winter temperatures vary frequently between freezing and thawing.

Figure 1 Frost wedging begins in hairline fractures of a rock. Repeated cycles of freeze and thaw cause the crack to expand over time.

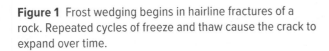

3D THINKING   **DCI** Disciplinary Core Ideas   **CCC** Crosscutting Concepts   **SEP** Science & Engineering Practices

COLLECT EVIDENCE

 Use your Science Journal to record the evidence you collect as you complete the readings and activities in this lesson.

INVESTIGATE

 GO ONLINE to find these activities and more resources.

Investigation Lab: Chemical Weathering and Temperature
Conduct an investigation into the effects of temperature on the chemical weathering of limestone.

GeoLAB: Model Mineral Weathering
Conduct an investigation to determine how time and composition affect the weathering of minerals.

H.R.Photos/Shutterstock

Effect of pressure

Another factor involved in mechanical weathering is pressure. Roots of trees and other plants can exert pressure on rocks when they wedge themselves into the cracks in rocks. As the roots grow and expand, they exert increasing amounts of pressure, which often causes the rocks to split, as shown in **Figure 2**.

On a much larger scale, pressure also functions within Earth. Bedrock at great depths is under tremendous pressure from the overlying rock layers. A large mass of rock, such as a batholith, may originally form under great pressure from the weight of several kilometers of rock above it. When the overlying rock layers are removed by processes such as erosion or mining, the pressure on the bedrock is reduced. The bedrock surface that was buried expands, and long, curved cracks can form. These cracks, also known as joints, occur parallel to the surface of the rocks. Reduction of pressure also allows existing cracks in the bedrock to widen. For example, when several layers of overlying rocks are removed from a deep mine, the sudden decrease of pressure can cause large pieces of rocks to explode off the walls of the mine tunnels.

After a rock body is uplifted as a result of geological processes such as mountain building, fine cracks may develop in the rock due to a decrease in overlying pressure. Over time, the outer layers of rock can be stripped away in succession, similar to the way an onion's layers can be peeled. The process by which outer rock layers are stripped away is called **exfoliation.** Exfoliation often results in dome-shaped formations, such as Moxham Mountain in New York and Half Dome in Yosemite National Park in California, shown in **Figure 3**.

Figure 2 Tree roots can grow within the cracks and joints in rocks and eventually cause the rocks to split.

Figure 3 The rock that makes up Half Dome in Yosemite National Park fractures along its outer surface in a process called exfoliation. Over time this has resulted in the dome shape of the outcrop.

Chemical Weathering

Chemical weathering is the process by which rocks and minerals undergo changes in their composition. Agents of chemical weathering include water, oxygen, carbon dioxide, and acid precipitation. The interaction of these agents with rock can cause some substances to dissolve and some new minerals to form. The new minerals have properties different from those that were in the original rock. For example, iron often combines with oxygen to form iron oxide, such as in the mineral hematite.

Get It?

Express in your own words the effect that chemical weathering has on rocks.

The composition of a rock determines the effects that chemical weathering will have on it. Some minerals, such as calcite, which is composed of calcium carbonate, can decompose completely in acidic water. Limestone and marble are made almost entirely from calcite and are, therefore, greatly affected by chemical weathering. Buildings and monuments made of these rocks usually show signs of wear as a result of chemical weathering. The statue in **Figure 4** shows an example of chemical weathering from acid precipitation.

Figure 4 This statue has been chemically weathered by acidic water and atmospheric pollutants.

Effect of water

Water is an important agent in chemical weathering because it can dissolve many kinds of minerals and rocks. Water also plays an active role in many reactions by serving as a medium in which the reactions can occur. Water can also react directly with minerals in a chemical reaction. In one common reaction with water, large molecules of the mineral break down into smaller molecules. This reaction decomposes and transforms many silicate minerals. For example, potassium feldspar decomposes into kaolinite, a fine-grained clay mineral common in soils.

Effect of oxygen

An important element in chemical weathering is oxygen. The chemical reaction of oxygen with other substances is called **oxidation.** Approximately 21 percent of Earth's atmosphere is oxygen gas. Iron in rocks and minerals combines with the oxygen in water and air to form minerals with the oxidized form of iron. A common mineral that contains the oxidized form of iron is hematite.

ACADEMIC VOCABULARY

process
a natural phenomenon marked by gradual changes that lead toward a particular result
The process of growth changes a seedling into a tree.

Effect of carbon dioxide

Another atmospheric gas that contributes to the chemical weathering process is carbon dioxide. Carbon dioxide is a gas that occurs naturally in the atmosphere as a product of living organisms. When carbon dioxide combines with water in the atmosphere, it forms a very weak acid called carbonic acid, which falls to Earth's surface as precipitation.

Precipitation includes rain, snow, sleet, and fog. Natural precipitation has a pH of 5.6. The slight acidity of precipitation causes it to dissolve certain rocks, such as limestone.

Decaying organic matter and respiration produce high levels of carbon dioxide. When slightly acidic water from precipitation seeps into the ground and combines with carbon dioxide in the soil, carbonic acid becomes a stronger agent in the chemical weathering process. Carbonic acid slowly reacts with minerals such as calcite in limestone and marble to dissolve rocks. After many years, limestone caverns can form where the carbonic acid flowed through cracks in limestone rocks and reacted with calcite.

Effect of acid precipitation

Another agent of chemical weathering is acid precipitation, which is caused by sulfur dioxide, carbon dioxide, and nitrogen oxides. These compounds are released into the atmosphere, often by human activities. Sulfur dioxide and carbon dioxide are primarily the product of burning fossil fuels. Motor vehicle exhaust contributes to the emissions of nitrogen oxides. These three gases combine with oxygen and water in the atmosphere and form strong sulfuric, nitric, and carbonic acids.

Recall that the acidity of a solution is described using the pH scale. Acid precipitation is precipitation that has a pH value below 5.6—the pH of normal rainfall. Because strong acids can be harmful to many organisms and destructive to human-made structures, acid precipitation often creates problems. Many plant and animal populations, such as the forest shown in **Figure 5,** cannot survive even slight changes in acidity. Acid precipitation is a serious issue in New York, West Virginia, and much of Pennsylvania.

Figure 5 Forests around the world have been damaged by the effects of acid precipitation. Acid precipitation can make forests more vulnerable to disease and damage by insects.

Piotr Zawisza/E+/Getty Images

Rate of Weathering

The natural weathering of earth materials occurs slowly. For example, it can take 2000 years to weather 1 cm of limestone, and most rocks weather at even slower rates. Certain conditions and interactions can accelerate or slow the weathering process.

Effects of climate on weathering

Climate is the major influence on the rate of weathering of earth materials. Precipitation, temperature, and evaporation are factors of climate. The interaction between temperature and precipitation in a given climate determines the rate of weathering in a region.

 Get It?

Explain why different climates have different rates of weathering.

Rates of chemical weathering Chemical weathering is rapid in climates with warm temperatures, abundant rainfall, and lush vegetation. These climatic conditions can produce soils that are rich in organic matter. Water from heavy rainfalls combines with the carbon dioxide in soil's organic matter and produces high levels of carbonic acid. The resulting carbonic acid accelerates the weathering process. The effects of chemical weathering are greatest along the equator, where rainfall is plentiful and the temperature tends to be high, as shown in **Figure 6**.

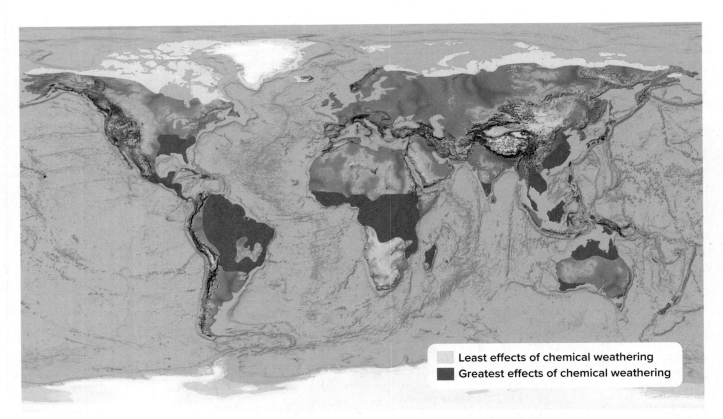

Least effects of chemical weathering
Greatest effects of chemical weathering

Figure 6 The impact of chemical weathering is related to a region's climate. Warm, lush areas such as the tropics experience the fastest chemical weathering.

Infer *what parts of the world experience less chemical weathering.*

Rates of physical weathering Conversely, physical weathering can break down rocks more rapidly in cool climates. Physical weathering rates are highest in areas where water in cracks within the rocks undergoes repeated freezing and thawing. Conditions in such climates do not favor chemical weathering because cool temperatures slow or inhibit chemical reactions. Little or no chemical weathering occurs in areas that are frigid year-round.

The different rates of weathering caused by different climatic conditions can be emphasized by a comparison of Asheville, North Carolina, and Phoenix, Arizona. Phoenix has warm, dry conditions; temperatures do not drop below the freezing point of water, and humidity is low. In Asheville, temperatures can drop below freezing during the winter months, and Asheville has more monthly rainfall and higher levels of humidity than Phoenix. Because of these differences in climates, rocks and human-made structures in Asheville experience higher rates of mechanical and chemical weathering than those in Phoenix.

Figure 7 shows how rates of weathering are dependent on climate. Both Egyptian obelisks were carved from granite more than 3500 years ago. They stood in Egypt's dry climate, showing few effects of weathering. In 1881, Cleopatra's Needle was transported from Egypt to New York City. After the move, it had been thought that acid precipitation and the repeated cycles of freezing and thawing in New York City accelerated the processes of chemical and physical weathering. In comparison, the obelisk that remains in Egypt appears unchanged. Recent studies, however, hypothesize that Cleopatra's needle and other monuments that have been moved from Egypt's dry climate may have begun to weather before leaving their original home. This would mean that New York City's climate has simply continued a process that was already in progress and is not totally to blame.

Rock type and composition

Not all rocks in the same climate weather at the same rate. The effects of climate on the weathering of rock also depends on rock type and composition. For example, rocks containing mostly calcite, such as limestone and marble, are more easily weathered than rocks containing mostly quartz, such as granite and quartzite.

Figure 7 The climate of New York City may have caused the obelisk on the left to weather rapidly. The obelisk on the right has been preserved by Egypt's dry, warm climate.

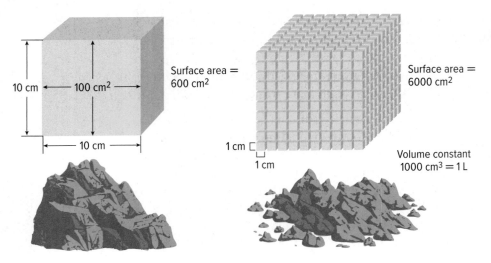

Figure 8 When an object is broken into pieces, the surface area increases. The large cube has a volume of 1000 cm³. When it is broken into 1000 pieces, the volume is unchanged, but the surface area is increased 1000 times.

Surface area

The rate of weathering also depends on the surface area that is exposed. Mechanical weathering breaks rocks into smaller pieces. As the pieces get smaller, their surface area increases, as illustrated in **Figure 8.** When this happens, there is more total surface area available for chemical weathering. The result, as you learned in the Launch Lab, is that weathering has a greater effect on multiple, smaller particles than it does on a single, large rock.

Topography

The slope of a landscape also determines the rate of weathering. Rocks on level areas are likely to remain in place over time, whereas the same rocks on slopes tend to move downslope as a result of gravity. Steep slopes, therefore, promote erosion and continually expose more rocks to mechanical and chemical weathering.

Check Your Progress

Summary

- Mechanical weathering changes a rock's size and shape.
- Frost wedging and exfoliation are forms of mechanical weathering.
- Chemical weathering changes the composition of a rock.
- The rate of chemical weathering depends on the climate, rock type, surface area, and topography.

Demonstrate Understanding

1. **Distinguish** between the characteristics of an unweathered rock and of a highly weathered rock.
2. **Describe** the factors that control the rates of chemical weathering and those of physical weathering.
3. **Compare** chemical weathering to mechanical weathering.
4. **Analyze** the relationship between surface area and weathering.

Explain Your Thinking

5. **Infer** which headstone engravings would last longer—those in one made of marble or those in one made of granite.
6. **MATH Connection** Infer the relationship between rate of weathering and the surface area of a material. Create a graph that illustrates the relationship.

LEARNSMART Go online to follow your personalized learning path to review, practice, and reinforce your understanding.

FOCUS QUESTION

How does erosion done by ice differ from erosion done by water?

Gravity's Role

Recall that the process of weathering breaks rock and soil into smaller pieces, but never moves it. The removal of weathered rock and soil from its original location is a process called **erosion.** Erosion can remove material through a number of different agents, including running water, glaciers, wind, ocean currents, and waves. These agents of erosion can carry rock and soil thousands of kilometers away from their source. After the materials are transported, they are dropped in another location in a process known as **deposition.**

Gravity is associated with many erosional agents because the force of gravity tends to pull all materials downslope. Without gravity, neither streams nor glaciers would flow. In the process of erosion, gravity pulls loose rocks and other material downslope. Additionally, the effects of gravity on erosion by running water can often produce dramatic landscapes with steep valleys, such as that shown in **Figure 9.**

Get It?

How is gravity helping to widen the valley of the Colorado River through the Grand Canyon?

Figure 9 Boch Hollow Nature Preserve in Ohio is home to waterfalls such as this one, aptly named Corkscrew Falls.

Kenneth Keifer/Shutterstock

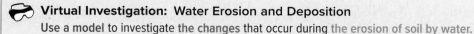

3D THINKING **DCI** Disciplinary Core Ideas   **CCC** Crosscutting Concepts   **SEP** Science & Engineering Practices

COLLECT EVIDENCE

 Use your Science Journal to record the evidence you collect as you complete the readings and activities in this lesson.

INVESTIGATE

🌐 **GO ONLINE** to find these activities and more resources.

⚙ **Applying Practices: Investigate Stream Erosion**
HS-ESS2-5. Plan and conduct an investigation of the properties of water and its effects on Earth materials and surface processes.

🥽 **Virtual Investigation: Water Erosion and Deposition**
Use a model to investigate the changes that occur during the erosion of soil by water.

Figure 10 Rill erosion (left) can occur in an agricultural field. Gully erosion (right) often develops from rills.

Suggest *land management practices that can slow or prevent the development of gully erosion.*

Erosion by Water

Moving water is perhaps the most powerful agent of erosion. Stream erosion can reshape entire landscapes. Stream erosion is greatest when a large volume of water is moving rapidly, such as during spring thaws and torrential downpours. Water flowing down steep slopes has additional erosive potential resulting from gravity, causing it to cut downward into the slopes, carving steep valleys and carrying away rock and soil. Water that flows swiftly or in large volumes can independently carry more material. The Mississippi River carries more than 400,000 metric tons of sediment each day from thousands of kilometers away due to the volume of water in the river.

 Get It?

Predict what time of year water has the most potential for erosion.

Erosion by water can have destructive results. For example, water flowing downslope can carry away fertile soil, thus affecting agricultural areas. **Rill erosion** develops when running water cuts small channels into the side of a slope, as shown in **Figure 10.** When a channel becomes deep and wide as a result of further erosion, rill erosion evolves into **gully erosion,** also shown in **Figure 10.** The channels formed in gully erosion can transport much more water and, consequently, more soil than rills. Gullies can be more than 3 m deep and can cause major problems in farming and grazing areas.

Rivers and streams

Each year, streams carry billions of metric tons of sediments and weathered material to coastal areas. Once a river enters the ocean, the current slows down, which reduces the potential of the stream to carry sediment. As a result, rivers deposit large amounts of sediments in the region where they enter the ocean. The buildup of sediments over time forms deltas, such as the Nile River Delta, shown in **Figure 11.** The volume of river flow and the action of tides determine the shape of deltas, most of which contain fertile soil. The Nile River Delta has the classic fan or triangle shape associated with many deltas.

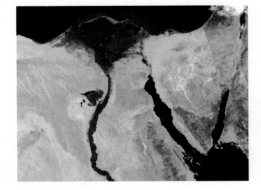

Figure 11 Rivers slow down when they meet the ocean. In these regions, sediments are deposited by the river, resulting in the development of a delta.

Wave action

Erosion of materials also occurs along the ocean floor and at continental and island shorelines. The work of ocean currents, waves, and tides carves out cliffs, arches, and other features along the continents' edges. In addition, sand particles accumulate on shorelines and form dunes and beaches. The constant movement of water and the availability of accumulated weathered material result in a continuous erosional process, especially along ocean shorelines. Sand along a shoreline is repeatedly picked up, moved, and deposited by ocean currents. As a result, sandbars form from offshore sand deposits. If sandbars continue to be built up with sediments, they can develop into barrier islands. Many barrier islands, such as Hatteras Island off the coast of North Carolina, shown in **Figure 12,** have formed along both the Gulf and Atlantic Coasts of the United States.

Just as shorelines are built by the process of deposition in some areas, they are reduced by the process of coastal erosion in other areas. Changing tides and conditions associated with coastal storms can also have a great impact on coastal erosion. Human development and population growth along shorelines have led to attempts to control the erosion of sand. However, efforts to keep the sand on one beachfront disrupt the natural migration of sand along the shore, depleting sand from another area.

Figure 12 Hatteras Island is a well-known barrier island off the coast of North Carolina. It has been built over time by deposition of sand and sediments.

National Geographic/SuperStock

CCC CROSSCUTTING CONCEPTS

Stability and Change Weathering, erosion, and deposition occur in a nonstop series of processes on Earth's surface. A place that seems stable one day may suddenly change overnight. Use what you have learned about the agents of erosion to explain why many public beaches need to add more sand to the shoreline in the spring.

Glacial Erosion

Although glaciers currently cover about 10 percent of Earth's surface, they have covered more than 30 percent of Earth's surface in the past. The erosional effects of glaciers are large-scale and dramatic. Glaciers scrape, scratch, and gouge out large sections of Earth's landscape. Because they can move as dense, enormous rivers of slowly flowing ice, glaciers have the capacity to carry huge rocks and piles of debris over great distances, all the while grinding the rocks beneath them into flour-sized particles. The features left in the wake of glacial movements include steep-sided, U-shaped valleys and deep, steep-sided lakes such as the one shown in **Figure 13.**

The effects of glaciers on the landscape also include deposition. For example, soils in the northern parts of the United States are formed from material that was transported and deposited by glaciers. Although the last ice age ended about 12,000 years ago, glaciers continue to affect erosional processes on Earth.

Figure 13 This lake in Glacier National Park, Montana, was formed by glaciers.

Wind Erosion

Wind can be a major erosional agent, especially in arid and coastal regions. Such regions tend to have little vegetation to hold soil in place. Wind can easily pick up and move fine, dry particles. The effects of wind erosion can be both dramatic and devastating. The abrasive action of wind-blown particles can damage both natural features and human-made structures. Winds can even blow against the force of gravity and easily move fine-grained sediments and sand uphill. Some types of sand dunes—such as those in large, sand-filled deserts—form in this way.

Wind barriers One farming method that can reduce the effects of wind erosion is the planting of wind barriers, also called windbreaks, shown in **Figure 14.** Windbreaks are trees or other vegetation planted perpendicular to the direction of the wind. A wind barrier might also be a row of trees along the edge of a field. In addition to reducing erosion, wind barriers can trap blowing snow, conserve moisture, and protect crops from the effects of wind.

Figure 14 A windbreak can reduce the speed of the wind for distances up to 30 times the height of the trees. These windbreaks are in North Dakota.

Calculate *If these trees are 10 m tall, what is the distance over which they can serve as a windbreak?*

Figure 15 In this construction project, the landscape was considerably altered.

Analyze *the effects of this alteration on both the landscape and the organisms that live here.*

Erosion by Living Things

Plants and animals also play a role in erosion. As plants and animals carry out their life processes, they frequently move Earth's surface materials from one place to another. For example, rocks and sediments are moved by plants' roots, or by animals burrowing into soil. Humans also play a role in erosion when excavating large areas and moving soil from one location to another, as shown in **Figure 15.** Planting a garden, developing a new athletic field, and building a highway are all examples of human activities that result in moving earth materials from one place to another.

📝 Check Your Progress

Summary

- The processes of erosion and deposition have shaped Earth's landscape in many ways.
- Gravity is the driving force behind major agents of erosion.
- Agents of erosion include running water, waves, glaciers, wind, and living things.

Demonstrate Understanding

1. **Discuss** how weathering and erosion are related.
2. **Describe** how gravity is associated with many erosional agents.
3. **Classify** the type of erosion that could move sand along a shoreline.
4. **Compare and contrast** rill erosion and gully erosion.

Explain Your Thinking

5. **Analyze** how geologic processes and features are expressed in your area.
6. **Diagram** a design for a wind barrier to prevent wind erosion.
7. **WRITING** **Connection** Research how a development in your area has alleviated or contributed to erosion. Present your findings to the class, including which type of erosion occurred and where the eroded materials will eventually be deposited.

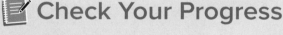 **LEARNSMART®** Go online to follow your personalized learning path to review, practice, and reinforce your understanding.

FOCUS QUESTION

How does eroded rock become soil?

Soil Formation

What is soil? It is found almost everywhere on Earth's surface. Weathered rock alone is not soil. **Soil** is the loose covering of weathered rock particles and decaying organic matter, called humus, overlying the bedrock of Earth's surface, and it serves as a medium for the growth of plants. Soil is the product of thousands of years of chemical and mechanical weathering and biological activity.

Soil development

The soil-development process often begins when weathering breaks solid bedrock into smaller pieces. These pieces of rock continue to weather and break down into smaller pieces. Worms and other organisms help break down organic matter, add nutrients to the soil, and create passages for air and water, as shown in **Figure 16.** As nutrients are added to the soil, its texture changes, and the soil's capacity to hold water increases. While all soil contains some organic matter in various states of decay, the amount varies widely among different types of soil. For example, as much as 5 percent of the volume of prairie soils is organic matter, while most desert soils have almost no organic matter.

Figure 16 Organisms in the soil change the soil's structure over time by adding nutrients and passages for air.

 3D THINKING   **DCI** Disciplinary Core Ideas   **CCC** Crosscutting Concepts   **SEP** Science & Engineering Practices

COLLECT EVIDENCE

 Use your Science Journal to record the evidence you collect as you complete the readings and activities in this lesson.

INVESTIGATE

GO ONLINE to find these activities and more resources.

Mapping Lab: Global Soils and Climate
Use models to visualize the patterns of climate and soils in different regions.

Review the News
Obtain information from a current news story about soil science. Evaluate your source and communicate your findings to your class.

Soil Layers

During the process of soil formation, layers develop in the soil. Most of the volume of soil is formed from the weathered products of a source rock, called the parent material. The parent material of a soil is often the bedrock. As the parent material weathers, the weathering products rest on top of the parent material. Over time, a layer of the smallest pieces of weathered rock develops above the parent material. Eventually, living organisms such as plants and animals become established and use nutrients and shelter available in the material. Rainwater seeps through this top layer of materials and dissolves soluble minerals, carrying them into the lower layers of the soil.

A soil whose parent material is the local bedrock is called **residual soil.** Kentucky's bluegrass soil is an example of residual soil, as are the red soils in Georgia. Not all soil develops from local bedrock. **Transported soil,** shown in the valley in **Figure 17,** is soil that develops from parent material that has been moved far from its original location. Agents of erosion transport parent material from its place of origin to new locations. For example, glaciers have transported sediments from Canada to many parts of the United States. Streams and rivers, especially during times of flooding, also transport sediments downstream to floodplains. Wind also carries sediments to new locations. Over time, processes of soil formation transform these deposits into mature, well-developed soil layers.

 Get It?

Explain how residual soils are different from transported soils.

Figure 17 In a stream valley, transported soils are often found in the floodplain. Residual soils are often found in the higher, mountainous regions.

Pixtal/age fotostock

Figure 18 An undeveloped soil (left) has few, if any, distinct layers, while mature soil (right) is characterized by several soil horizons that have developed over time.

Soil profiles

Digging a deep hole in the ground can reveal a soil profile. A **soil profile** is a vertical sequence of soil layers. Some soil profiles have more distinct layers than others. Relatively new soils that have not yet developed distinct layers are called undeveloped soils, as shown in **Figure 18.** It can take tens of thousands of years for distinct layers to form creating a mature soil. An example of a mature soil is also shown in **Figure 18.**

Soil horizons A distinct layer within a soil profile is called a **soil horizon.** There are typically four major soil horizons in mature soils—O, A, B, and C. The O-horizon is the top layer of organic material, which is made of humus and leaf litter. Below that, the A-horizon is a layer of weathered rock combined with a rich concentration of dark brown to black organic material. The B-horizon, also called the zone of accumulation, is a red or brown layer that has been enriched over time by clay and minerals that are deposited by water flowing in the layers above or that percolate upward from layers below. Usually the clay gives a blocky structure to the B-horizon. Accumulations of certain minerals can result in a hard layer called hardpan. Hardpan can be so dense that it allows little or no water to pass through it. The C-horizon contains little or no organic matter and is often made of broken-down bedrock. The development of each horizon depends on the factors of soil formation.

HISTORY ▶ Connection **Sod busters and hardpan** Pioneers that settled in the North American midwest and Great Plains earned a nickname—sod busters. This name comes from the fact that the original prairie soil was covered in grass whose roots grew so long and deep that the pioneers literally had to break or bust up the sod in order to prepare the soil for planting. Added to this difficulty was the layer of hardpan beneath the A-horizon, or topsoil. Nearly impenetrable to plant roots, it too had to be broken up to allow water to flow freely through the soil and for plants to extend their roots deep below the surface.

Factors of Soil Formation

Five factors influence soil formation: climate, topography, parent material, biological activity, and time. These factors combine to produce different types of soil, called soil orders, that differ by region. Soil taxonomy (tak SAH nuh mee) is the system that scientists use to classify soils into orders and other categories. The five factors of soil formation result in 12 different soil orders.

Climate

Climate is the most significant factor controlling the development of soils. Temperature, wind, and the amount of rainfall determine the type of soil that can develop.

Recall from Lesson 1 that rocks tend to weather rapidly under humid, temperate conditions, such as those present in climates along the eastern United States. Weathering results in soils that are rich in aluminum and iron oxides. Water from abundant rainfall moves downward, carrying dissolved minerals into the B-horizon. In contrast, the soils of arid regions are so dry that water from below ground moves up through evaporation and leaves an accumulation of white calcium carbonate in the B-horizon. Tropical areas experience high temperatures and heavy rainfall. These conditions lead to intensely weathered soils from which all but the most insoluble minerals have been flushed out.

Topography

Topography, which includes the slope and orientation of the land, affects the type of soil that forms. On steep slopes, weathered rock is carried downhill by agents of erosion. As a result, hillsides tend to have shallow soils, while valleys and flat areas develop deeper soils with more organic material. The orientation of slopes also affects soil formation. In the northern hemisphere, slopes that face south receive more sunlight than other slopes. The extra sunlight allows more vegetation to grow. Slopes without vegetation tend to lose more soil to erosion. **Figure 19** shows how the orientation and slope of a landscape can affect the formation of soil.

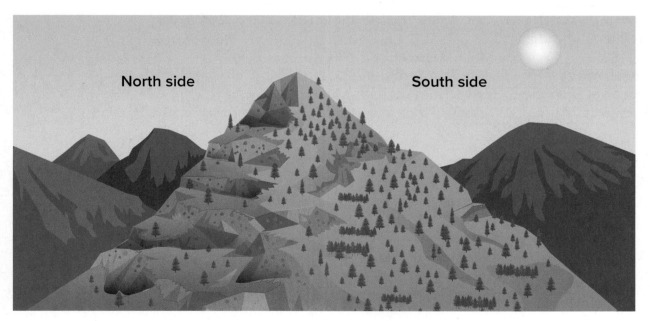

Figure 19 The slope on the right side faces south, and the slope on the left side faces north.

Interpret *why one slope has more vegetation than the other.*

Parent material

Recall that a soil can be either residual or transported. If the soil is residual, it will have the same chemical composition as the local bedrock. For example, in regions near volcanoes, the soils form from weathered products of lava and ash. Volcanic soils tend to be rich in the minerals that were present in the lava. If the soil is transported, the minerals in the soil are likely to be different from those in the local bedrock.

Biological activity

Organisms including fungi and bacteria, as well as plants and animals, interact with soil. Microorganisms decompose dead plants and animals. Plant roots can open channels, and when they decompose, they add organic material to the soil. Different types of living organisms in a soil can result in different soil orders. Mollisols (MAH lih sawlz), which are called prairie soils, and alfisols (AL fuh sawlz), also called woodland soils, both develop from the same climate, topography, and parent material. Different sets of organisms result in two soils with entirely different characteristics. For example, the activity of prairie organisms in mollisols produces a thick A-horizon, rich in organic matter. Some of the most fertile agricultural lands in the Great Plains region are mollisols.

 Get It?

Describe how microorganisms affect soil formation.

Time

The effects of time alone can determine the characteristics of a soil. New soils, such as entisols (EN tih sawlz), are often found along rivers, where sediment is deposited by periodic flooding. This type of soil is shown as a light blue color in **Figure 20.** These soils have had little time to weather and develop soil horizons. The effects of time on soil can be easy to recognize. After tens of thousands of years of weathering, most of the original minerals in a soil are changed or washed away. Minerals containing aluminum and iron remain, which can give older soils, such as ultisols (UL tih sawlz), a red color. **Figure 21** shows the locations of the 12 soil orders in the United States.

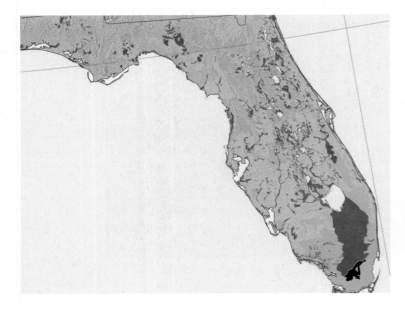

Figure 20 Soil types vary widely from one area to the next depending on the local climate, topography, parent material, organisms, and age of the soil. Entisols are shown in light blue, and ultisols are shown in orange on this map.

Infer *how differences in topography have affected the types of soils in Florida.*

State Soil Geographic Database (STATSGO)/NRCS/USDA

Figure 21 Visualizing Soil Orders

The five factors of soil formation determine how the soil orders are distributed across the United States. Soil profiles of three soil orders from different parts of the country are shown. Each soil profile has soil horizons expressed differently.

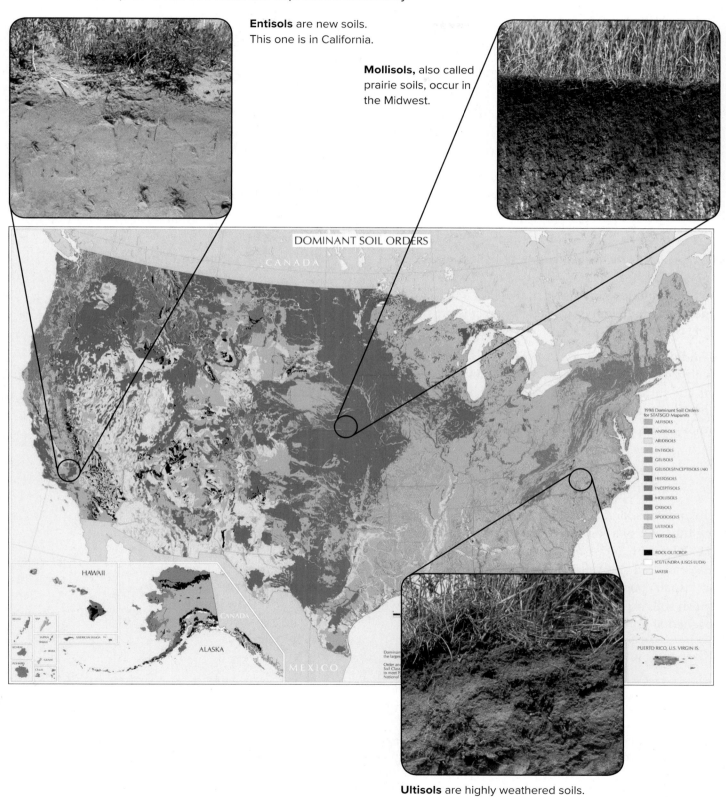

Entisols are new soils. This one is in California.

Mollisols, also called prairie soils, occur in the Midwest.

DOMINANT SOIL ORDERS

1998 Dominant Soil Orders for STATSGO Mapunits
ALFISOLS
ANDISOLS
ARIDISOLS
ENTISOLS
GELISOLS
GELISOLS/INCEPTISOLS (AK)
HISTOSOLS
INCEPTISOLS
MOLLISOLS
OXISOLS
SPODOSOLS
ULTISOLS
VERTISOLS

ROCK OUTCROP
ICE/TUNDRA (USGS LUDA)
WATER

HAWAII

ALASKA

PUERTO RICO, U.S. VIRGIN IS.

Ultisols are highly weathered soils. This one is in North Carolina.

Soil Texture

Particles of soil are classified according to size as clay, silt, or sand, with clay being the smallest and sand being the largest. The relative proportions of particle sizes determine a soil's texture, as shown in **Figure 22.** Soil texture affects a soil's capacity to accept and retain moisture and, therefore, its ability to support plant growth. Soil texture also varies with depth.

 Get It?

What soil classification would you assign a sample that contains 40% sand, 40% silt, and 20% clay?

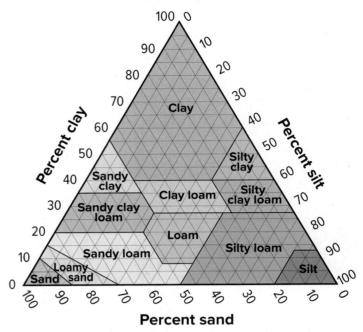

USDA Soil Classification

Figure 22 A soil classification triangle is used to determine a soil's texture.

Soil Fertility

Soil fertility is the measure of how well a soil can support the growth of plants. Factors that affect soil fertility include topography, availability of minerals and nutrients, number of microorganisms present, amount of precipitation available, and level of acidity.

Conditions necessary for growth vary with plant species. Farmers use natural and commercially produced fertilizers to replace minerals and maintain soil fertility. Commercial fertilizers add nitrates, potassium, and phosphorus to soil. The planting of legumes, such as beans and clover, allows bacteria to grow on plant roots and replace nitrates in the soil. Pulverized limestone is often added to soil to reduce acidity and enhance crop growth.

Soil Color

The minerals, organic matter, and moisture in each soil horizon determine its color. Examining the color of a soil can reveal many of its properties. For example, the layers that compose the O-horizon and A-horizon are usually dark-colored because they are rich in organic material. Red and yellow soils might be the result of oxidation of iron minerals. Yellow soils are usually poorly drained and are often associated with environmental problems. Grayish or bluish soils are common in poorly drained regions, where soils are constantly wet, might be leached of minerals, and lack oxygen.

STEM CAREER Connection

Soil Scientist
Do you love working outdoors? Are you interested in the land and how it is used? You could be destined to become a soil scientist. Professionals in this field determine where to build structures, evaluate groundwater quality, predict the fertility of the soil, and develop ways to prevent topsoil erosion such as that which happened during the Dust Bowl of the 1930s.

Figure 23 Hue, value, and chroma can be determined using the Munsell System of Color Notation.

Scientists use the Munsell System of Color Notation, shown in **Figure 23,** to describe soil color. This system consists of three parts: hue (color), value (lightness or darkness), and chroma (intensity). Each color is shown on a chip from a soil book. Using the components of hue, value, and chroma, a soil's color can be precisely described.

 Get It?

Evaluate which is better to use—a dry sample or a wet sample—when evaluating hue, value, and chroma.

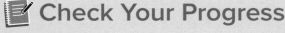 **Check Your Progress**

Summary

- Soil consists of weathered rock and humus.
- Soil is either residual or transported.
- A typical soil profile has an O-horizon, A-horizon, B-horizon, and C-horizon.
- Five factors influence soil formation: climate, topography, parent material, biological activity, and time.
- Characteristics of soil include texture, fertility, and color.

Demonstrate Understanding

1. **Describe** how soil forms.
2. **Summarize** the features of each horizon of soil.
3. **Classify** a soil profile based on whether it is mature or immature.
4. **Generalize** the effect that topography has on soil formation.

Explain Your Thinking

5. **Infer** Soil scientists discover that a soil in a valley has a deep C-horizon of glacial till above a bedrock of granite. Is this a transported soil or a residual soil? Explain.
6. **Hypothesize** what type of soil exists in your area, and describe how you would determine whether your hypothesis is correct.
7. **WRITING** **Connection** Soil in a portion of a garden is found to be claylike and acidic. Design a plan for improving the fertility of this soil.

LEARNSMART Go online to follow your personalized learning path to review, practice, and reinforce your understanding.

The Latest Dirt on Healthy Soil

Many farms routinely apply chemical fertilizers, pesticides, and herbicides on their crops, but these substances may come with a price. Evidence suggests that the chemicals commonly used to increase crop yield may, over time, actually do the opposite. An increasing number of professionals in a range of STEM fields are discovering that the solution to growing more food might just be right under our feet.

Changing the way farmers treat soil may be the answer to growing more and healthier food.

Go Organic

To build proteins, plants must draw elements like potassium, nitrogen, and phosphorus from the soil. These elements are not always plentiful since they usually come from the slow decay of organic matter. Often, farmers add chemical nitrogen and phosphorous to increase the rate of plant growth. In the short term, it works. But in the long term, it's destructive to the environment. Among the many problems caused by over-application of chemicals are an increase in plant diseases, disruption of surrounding ecosystems by eutrophication, and a decrease in the crops' ability to withstand drought.

How can farmers provide plants with necessary nutrients without causing new problems? Agricultural scientists have found that the answer may lie in a healthy soil ecosystem. Like all ecosystems, a healthy soil ecosystem requires a balance between its abiotic and biotic components.

Chemical fertilizers upset this balance, changing abiotic factors like porosity and pH and decreasing the number of bacteria and other microbes. Research suggests that adding organic materials like compost and manure to soil, along with pure minerals, can restore this balance. The result? Increased crop yield.

Agricultural engineers design ways to add organic materials efficiently, at low cost, and at minimal hazard to the environment. They develop new technology to determine how to maximize crop yield without harming the soil ecosystem. Computer programmers play a role, too, by designing models to predict the soil's ability to retain water and resist erosion.

We'll need plentiful food in the future to feed a growing human population. Scientists, farmers, engineers, and other professionals are working together to address this challenge and those yet to come.

COMMUNICATE SCIENTIFIC AND TECHNICAL INFORMATION

Research a scientist or other STEM professional who is working to improve soil quality. Create a brochure about this professional and his or her work.

Stephen R Ausmus USDA-ARS

MODULE 5
STUDY GUIDE

 GO ONLINE to study with your Science Notebook.

Lesson 1 WEATHERING

- Mechanical weathering changes a rock's size and shape.
- Frost wedging and exfoliation are forms of mechanical weathering.
- Chemical weathering changes the composition of a rock.
- The rate of chemical weathering depends on the climate, rock type, surface area, and topography.

- weathering
- mechanical weathering
- frost wedging
- exfoliation
- chemical weathering
- oxidation

Lesson 2 EROSION AND DEPOSITION

- The processes of erosion and deposition have shaped Earth's landscape in many ways.
- Gravity is the driving force behind major agents of erosion.
- Agents of erosion include running water, waves, glaciers, wind, and living things.

- erosion
- deposition
- rill erosion
- gully erosion

Lesson 3 SOIL

- Soil consists of weathered rock and humus.
- Soil is either residual or transported.
- A typical soil profile has an O-horizon, A-horizon, B-horizon, and C-horizon.
- Five factors influence soil formation: climate, topography, parent material, biological activity, and time.
- Characteristics of soil include texture, fertility, and color.

- soil
- residual soil
- transported soil
- soil profile
- soil horizon

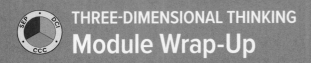

REVISIT THE PHENOMENON

How does ice carve large valleys out of mountain rock?

CER Claim, Evidence, Reasoning

Explain your Reasoning Revisit the claim you made when you encountered the phenomenon. Summarize the evidence you gathered from your investigations and research and finalize your Summary Table. Does your evidence support your claim? If not, revise your claim. Explain why your evidence supports your claim.

STEM UNIT PROJECT

Now that you've completed the module, revisit your STEM unit project. You will summarize your evidence and apply it to the project.

GO FURTHER

SEP Data Analysis Lab

How can examining the components of a soil be used to determine a soil's texture?

Soils can be classified with the use of a soil textural triangle. Soil texture is determined by the percentages of sand, silt, and clay that make up the soil. These percentages vary with depth, from one soil horizon to another. Below are data from three horizons of a soil in North Carolina.

Data and Observations

Soil Sample	Percent Clay	Percent Silt	Percent Sand	Texture
A	11	48		Loam
B	67		5	
C		53	38	

CER Analyze and Interpret Data

1. **Claim, Evidence** Use the soil texture triangle to complete the data table.
2. **Claim, Evidence** Infer from the data table which soil sample has the greatest percentage of the smallest-sized particles.

3. **Claim, Evidence** Identify the maximum percentage of clay in clay loam.
4. **Reasoning** If water passes quickly through sand particles, what horizon will have the most capacity to hold soil moisture?

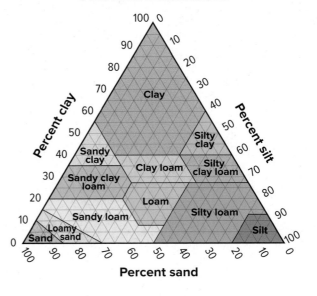

USDA Soil Classification

PK289/Shutterstock

MASS MOVEMENTS, WIND, AND GLACIERS

ENCOUNTER THE PHENOMENON

What dangers do mass movements and erosion pose to people?

GO ONLINE to play a video about landslides and landslide prediction.

SEP Ask Questions

Do you have other questions about the phenomenon? If so, add them to the driving question board.

CER Claim, Evidence, Reasoning

Make Your Claim Use your CER chart to make a claim about how mass movements and erosion pose dangers to people. Explain your reasoning.	**Collect Evidence** Use the lessons in this module to collect evidence to support your claim. Record your evidence as you move through the module.	**Explain Your Reasoning** You will revisit your claim and explain your reasoning at the end of the module.

GO ONLINE to access your CER chart and explore resources that can help you collect evidence.

LESSON 1: Explore & Explain: Mass Movements

LESSON 3: Explore & Explain: Glacial Erosion

Additional Resources

MASS MOVEMENTS

FOCUS QUESTION

What can we do to reduce the risks that mass movements pose to people?

Mass Movements

How do landforms, such as mountains, hills, and plateaus, wear down and change? Landforms can change through processes involving wind, ice, and water, and sometimes through the force of gravity alone. The downslope movement of soil and weathered rock resulting from the force of gravity is called **mass movement.** Recall that weathering processes weaken and break rock into smaller pieces. Mass movements often carry the weathered debris downslope. Because climate has a major effect on the weathering activities that occur in a particular area, climatic conditions determine the extent of mass movement.

All mass movements, such as the one shown in **Figure 1,** occur on slopes. Because few places on Earth are completely flat, most of Earth's surface undergoes mass movement. Mass movements range from motions that are barely detectable to sudden slides, falls, and flows. The earth materials that are moved range in size from fine-grained mud to large boulders.

Get It?

Describe how gravity causes a mass movement.

Figure 1 In this example of mass movement, tree trunks curved in order to continue growing opposite the pull of gravity, which is toward the center of Earth.

saraporn/Shutterstock

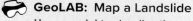 **3D THINKING**   **DCI** Disciplinary Core Ideas   **CCC** Crosscutting Concepts   **SEP** Science & Engineering Practices

COLLECT EVIDENCE

Use your Science Journal to record the evidence you collect as you complete the readings and activities in this lesson.

INVESTIGATE

GO ONLINE to find these activities and more resources.

GeoLAB: Map a Landslide
Use a model to visualize the result of a landslide.

((( ))) **Review the News**
Obtain information from a current news story about a recent mass movement. Evaluate your source and communicate your findings to your class.

Factors That Influence Mass Movements

Several factors influence the mass movement of Earth's materials. One factor is the material's weight, which pulls the material downslope. A second factor is the material's resistance to sliding or flowing, which depends on the amount of friction, how cohesive the material is, and whether the material is anchored to bedrock. A third factor is a trigger, such as an earthquake, that shakes material loose. Mass movement occurs when the forces pulling material downslope are stronger than the material's resistance to sliding, flowing, or falling.

Water is a fourth variable that influences mass movements. The landslide shown in **Figure 2** occurred after days of heavy rain. Saturation by water greatly increases the weight of soils and sediments. In addition, as water fills the tiny, open spaces between grains, it acts as a lubricant, reducing the friction between the grains.

Figure 2 Mass movements like the one shown here can significantly alter landscapes.

Summarize *the factors that might have been involved in this mass movement.*

Types of Mass Movements

Mass movements are classified as creep, flows, slides, and rockfalls. Different types of materials are moved in various ways.

Creep

The slow, steady, downhill flow of loose, weathered earth materials, especially soils, is called **creep.** Because movement might be as little as a few centimeters per year, the effects of creep are usually noticeable only after long periods of time. One way to tell whether creep has occurred is to observe the positions of structures and objects. As illustrated in **Figure 3,** creep can cause once-vertical utility poles and fences to tilt and cause trees and walls to break. Loose materials on almost all slopes undergo creep.

Creep that usually occurs in regions of permafrost, or permanently frozen soil, is called solifluction (SOH luh fluk shun). The material moved in solifluction is a mudlike liquid that is produced when water is released from melting permafrost during the warm season. The water saturates the surface layer of soil and is unable to move downward. As a result, the entire surface layer can slide slowly downslope.

Tilted fence posts, trees, and poles

Figure 3 All slopes undergo creep to some extent. Tilting of vertical objects is often the result.

Figure 4 The city of Armero, in Colombia, was covered in mud and debris by a lahar that contained snowmelt and volcanic material.

Describe *the effect of the lahar on the city shown above.*

Figure 5 Mudflows, like this one in Asakura, Japan, can be extremely destructive and can result in severe property damage, road closures, and power outages.

Flows

In some mass movements, earth materials flow as if they were a thick liquid. The materials might move as slowly as a few centimeters per year or as rapidly as 100 kilometers per hour. Earth flows are moderately slow movements of soils, whereas **mudflows** are swiftly moving mixtures of mud and water. Mudflows can be triggered by earthquakes or similar vibrations and are common in volcanic regions, where the heat from a volcano melts snow on nearby slopes that have fine sediment and little vegetation. Meltwater fills the spaces between the small particles of sediment, allowing them to slide over one another and move downslope.

A lahar (LAH har) is a type of mudflow that occurs after a volcanic eruption. Often a lahar results when a snow-topped volcanic mountain erupts and melts the snow on top of the mountain. The melted snow mixes with ash and flows downslope. **Figure 4** shows how a lahar that originated from Nevado del Ruiz, one of the volcanic mountains in the Andes, devastated a town. The Nevado del Ruiz is 5389 m high and covered with 25 km² of snow and ice, which melted when it erupted. Four hours after Nevado del Ruiz erupted, lahars had traveled more than 100 km downslope. As a result of these lahars, which occurred in 1985, approximately 23,000 people were killed, 5000 people were injured, and 5000 homes were destroyed.

 Get It?

Determine what triggers a lahar.

Mudflows are also common in sloped, semiarid regions that experience intense, short-lived rainstorms. In such areas, periods of drought and forest fires leave the slopes with little protective vegetation. When heavy rain falls in these areas, it can cause massive, destructive mudflows because there is little vegetation to anchor the soil. Mudflows are especially destructive in areas where urban development has spread to the bases of mountainous areas. These mudflows can bury homes, as shown in **Figure 5.**

 Get It?

Describe ways in which the risk of mudflows like the one in **Figure 5** could be reduced.

(t)Jacques Langevin/Sygma/Getty Images, (b)The Asahi Shimbun/Getty Images

Figure 6 Typical of landslides, this soil moved in a large block.

Slides

A rapid, downslope movement of earth materials that occurs when a relatively thin block of soil, rock, and debris separates from the underlying bedrock is called a **landslide,** shown in **Figure 6.** The material rapidly slides downslope as one block, with little internal mixing. A landslide mass eventually stops and becomes a pile of debris at the bottom of a slope, sometimes damming rivers and causing flooding. Landslides are common on steep slopes, especially when soils and weathered bedrock are fully saturated by water. This destructive form of mass movement causes almost 2 billion dollars in damage and 25 to 50 associated deaths per year in the United States alone.

A rockslide is a type of landslide that occurs when a mass of broken rock moves downhill on a sloped surface. During a rockslide, some blocks of rock are broken into smaller blocks as they move downslope, as shown in **Figure 7.** Often triggered by earthquakes, rockslides can move large amounts of material.

 Get It?

Describe the relationship between a landslide and a rockslide.

Figure 7 During this rockslide, blocks of rock were broken into smaller blocks as they moved downslope.

(t)E.L. Harp/U.S. Geological Survey, (b)Lloyd Cluff/Corbis Documentary/Getty Images

Slumps

When the mass of material in a landslide moves along a curved surface, a **slump** results. Material at the top of the slump moves downhill and slightly inward, while the material at the bottom of the slump moves outward. Slumps can occur in areas that have thick soils on moderate to steep slopes. Sometimes, slumps occur along highways where the slopes of soils are extremely steep. Slumps are common after rain, when water reduces the frictional contact between grains of soil and acts as a lubricant between surface materials and underlying layers. The weight of the additional water pulls material downhill. As with other types of mass movement, slumps can be triggered by earthquakes. Slumps leave crescent-shaped scars on slopes, as shown in **Figure 8.**

Figure 8 Slumps leave distinct crescent-shaped scars on hillsides as the soil rotates downward.

Avalanches

Landslides that occur in mountainous areas with thick accumulations of snow are called **avalanches.** About 10,000 avalanches occur each year in the mountains of the western United States. Radiation from the Sun can melt surface snow, which then refreezes at night into an icy crust. Snow that falls on top of this crust can eventually build up, become heavy, slip off, and slide downslope as an avalanche. Avalanches can happen in early winter when snow accumulates on the warm ground. The snow in contact with the warm ground melts and then refreezes into a layer of jagged, slippery snow crystals.

Avalanches of dangerous size, like the one shown in **Figure 9,** occur on slope angles between 30° and 45°. When the angle of a slope is greater than 45°, snow cannot accumulate enough to create a large avalanche. At angles less than 30°, the slope is not steep enough for snow to begin sliding. A vibrating trigger, even from a single skier, can send this unstable layer sliding down a mountainside. Avalanches pose significant risks in places such as Switzerland, where more than 50 percent of the population lives in avalanche terrain.

Figure 9 Vibrations from a single skier can trigger an avalanche.
Identify *the conditions that make a landscape more vulnerable to avalanches.*

Figure 10 This rockfall in Topanga Canyon, California, was unusual in that it involved mainly one large rock.

Rockfalls

On high cliffs, rocks are loosened by physical weathering processes, such as freezing and thawing, and by plant growth. As rocks break up and fall directly downward, they can bounce and roll, ultimately producing a cone-shaped pile of coarse debris, called talus, at the base of the slope. Rockfalls, such as the one shown in **Figure 10,** commonly occur at high elevations, in steep road cuts, and on rocky shorelines. Rockfalls are less likely to occur in humid regions, where the rock is typically covered by a thick layer of soil, vegetation, and loose materials. On human-made rock walls, such as road cuts, rockfalls are particularly common.

Mass Movements Affect People

While mass movements are natural processes, human activities often contribute to the factors that cause them. For example, the construction of buildings, roads, and other structures can make slopes unstable. In addition, poor maintenance of septic systems, which often leak, can trigger slides. In 2006, mudslides in the Philippines, shown in **Figure 11,** were triggered after ten days of torrential rain delivered 5 m of precipitation. Four years later, a massive storm once again caused mudslides, affecting thousands of people.

Figure 11 The mudslide on the island of Luzon, in the Philippines, occurred after many days of rain.

Reducing the risks

Catastrophic mass movements are most common on slopes greater than 25° that experience annual rainfall over 90 cm. Risk increases if that rainfall tends to occur over a short period of time. Humans can minimize the destruction caused by mass movement by not building structures on or near the base of steep or unstable slopes.

Although preventing mass-movement disasters is not easy, some actions can help reduce the risks. For example, a series of trenches can be dug to divert running water around a slope and control its drainage. Landslides and rockfalls can be controlled by covering steep slopes with materials such as steel nets, shown in **Figure 12,** and constructing fences along highways in areas where mass movements are common. Another approach is to install retaining walls that support the bases of weakened slopes. Most of these efforts at slope stabilization and mass-movement prevention are only temporarily successful.

The best way to reduce the number of disasters related to mass movements is to continue to monitor the movements and to educate people about the problems of building on steep slopes.

Figure 12 Covering hillsides with steel nets can reduce risks of mass movements and harm to humans.

Identify *the type of mass movement that these steel nets help prevent.*

Check Your Progress

Summary

- Mass movements are classified in part by how rapidly they occur.
- Factors involved in the mass movement of earth materials include the material's weight, its resistance to sliding, the trigger, and the presence of water.
- Mass movements are natural processes that can affect human lives and activities.
- Human activities can increase the potential for the occurrence of mass movements.

Demonstrate Understanding

1. **Compare** the speed of creep to the speeds of the other types of mass movements.
2. **Identify** the underlying force behind all forms of mass movement.
3. **Analyze** how water affects mass movements. Use two examples of mass movement in your analysis.
4. **Appraise** the effects of one type of mass movement on humans.

Explain Your Thinking

5. **Generalize** in which regions of the world mudflows are most common.
6. **Evaluate** how one particular human activity can increase the risk of mass movement, and suggest a solution to the problem.
7. **WRITING Connection** Make a poster that compares and contrasts solifluction and a slump. Consider the way soil moves and the role of water.

LEARNSMART Go online to follow your personalized learning path to review, practice, and reinforce your understanding.

FOCUS QUESTION

How do human activities affect wind erosion?

Wind Erosion and Transport

A current of rapidly moving air can pick up and carry sediment in the same way that water does. However, except in extreme cases, wind generally cannot carry particles as large as those transported by moving water. Regardless, wind is a powerful agent of erosion, which is a problem in many parts of the United States, as shown in **Figure 13.**

Wind transports materials by causing their particles to move in different ways. For example, wind can move sand on the ground in a rolling motion. A method of wind transport, called saltation, causes a bouncing motion of larger particles. This bouncing kicks up other particles that then join in on the saltation motion. Saltation accounts for most sand transport by wind. Another method of transport by which strong winds cause small particles to stay airborne for long distances is called suspension.

Wind Erosion in the United States

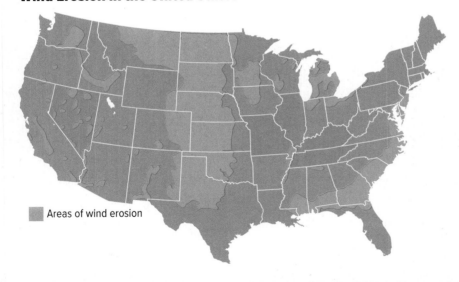

Areas of wind erosion

Figure 13 Wind erosion does not affect all areas of the United States equally.

Determine *if wind erosion is problematic in your area, and describe an example.*

3D THINKING   **DCI** Disciplinary Core Ideas   **CCC** Crosscutting Concepts   **SEP** Science & Engineering Practices

COLLECT EVIDENCE
Use your Science Journal to record the evidence you collect as you complete the readings and activities in this lesson.

INVESTIGATE
GO ONLINE to find these activities and more resources.

Investigation Lab: How does wind erosion take place?
Develop and use a model to visualize the result of wind transporting sand.

Review the News
Obtain information from a current news story about wind erosion and farming. Evaluate your source and communicate your findings to your class.

Figure 14 Through deflation, wind can create a bowl-shaped blowout.

Limited precipitation leads to increased wind erosion because precipitation holds down sediments and allows plants to grow. Thus, wind transport and erosion primarily occur in areas with little vegetative cover, such as deserts, semiarid areas, seashores, and some lakeshores.

Deflation

The lowering of the land surface that results from the wind's removal of surface particles is called **deflation.** During the 1930s, portions of the Great Plains region, which stretches from Montana to Texas, experienced severe drought. The area was already suffering from the effects of poor agricultural practices, in which large areas of natural vegetation were removed to clear the land for farming. Strong winds readily picked up the dry surface particles, which lacked any protective vegetation. Severe dust storms resulted in daytime skies that were often darkened, and the region became known as the Dust Bowl.

Today, the Great Plains are characterized by thousands of shallow depressions known as deflation blowouts. Many result from the removal of surface sediment by wind erosion during the 1930s. The depressions range in size from a few meters to hundreds of meters in diameter. Deflation blowouts are also found in areas that have sandy soil, as shown in **Figure 14.** Wind erosion continues today throughout the world, as shown by the duststorm in **Figure 15.**

 Get It?

Explain the causes and results of deflation.

Figure 15 A dust storm in a desert region fills the air with dust.

Deflation is a major problem in many agricultural areas of the world as well as in deserts, where wind has been consistently strong for thousands of years. In areas of intense wind erosion, coarse gravel and pebbles are usually left behind as the finer surface material is removed by winds. The coarse surface left behind is called desert pavement.

 Get It?

Explain why desert pavement results in low crop yields and why these areas might be prone to famine.

Abrasion

Another process of erosion, called **abrasion,** occurs when particles rub against the surface of rocks or other materials. Abrasion occurs as part of the erosional activities of winds, streams, and glaciers. In wind abrasion, wind picks up materials such as sand-sized particles and blows them against anything in their path. Because sand is often made of quartz, a hard mineral, wind abrasion can be an effective agent of erosion. Windblown sand particles eventually wear away rocks. Over long periods of time, wind erosion can produce unique structures such as those shown in **Figure 16.** Structures such as telephone poles can also be worn away or undermined by wind abrasion, and paint and glass on homes and vehicles can be damaged by windblown sand.

Materials that are exposed to wind abrasion show unique characteristics. For example, windblown sand causes rocks to become pitted and grooved. With continued abrasion, rocks become polished on the windward side and develop smooth surfaces with sharp edges. In areas of shifting winds, abrasion patterns correspond to wind shifts, and different sides of rocks become polished and smooth. Rocks that have been shaped by windblown sediments are called **ventifacts.** They range in size from pebbles to boulders.

 Get It?

Identify the unique characteristics of materials shaped by abrasion.

Arch

Caprock and Pillars

Figure 16 Arches and pillars with caprock form in different types of environments, but they most commonly occur in arid climates where wind is the dominant erosional force.

Figure 17 When there is a lot of sand, Barchan dunes merge to form a series of ridges call transverse dunes. The wind blows perpendicular to the ridge.

Identify *the dominant direction of wind in the figure.*

Wind Deposition

Wind deposition occurs in areas where wind velocity decreases. As the wind velocity slows down, some of the windblown sand and other materials cannot stay airborne. They drop out of the airstream, forming a deposit on the ground.

Dunes

In windblown environments, sand particles tend to accumulate where an object, such as a rock, landform, or piece of vegetation, blocks the forward movement of the particles. Sand continues to be deposited as long as winds blow in one general direction. Over time, the pile of windblown sand develops into a **dune,** as shown in **Figure 17.** All dunes have a characteristic profile. The gentler slope of a dune, located on the side from which the wind blows, is called the windward side. The steeper slope, on the side protected from the wind, is called the leeward side. The conditions under which a dune forms determine its shape. These conditions include the availability of sand, wind velocity, wind direction, and the amount of vegetation present. The different types of dunes are shown in **Table 1** on the next page.

Dune migration As long as winds continue to blow, dunes will migrate. As shown in **Figure 18,** dune migration is caused when prevailing winds continue to move sand from the windward side of a dune to its leeward side, causing the dune to move slowly over time.

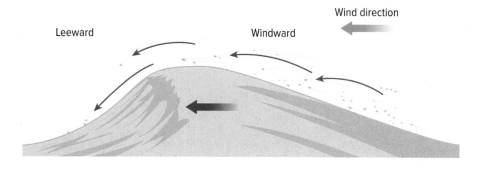

Leeward   Windward   Wind direction

Figure 18 Dune migration is caused by wind.

ACADEMIC VOCABULARY

migrate
to move from one location to another
Dunes migrate as wind blows over sand.

CCC CROSSCUTTING CONCEPTS

Stability and Change Review Figures 16 and 17 and construct explanations of how these land features were changed over time.

Table 1 Types of Dunes

Example of Dune	Description
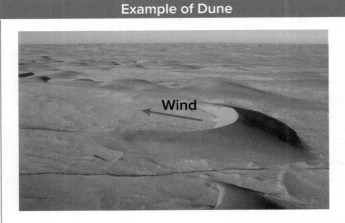	**Barchan Dunes** • form solitary crescent shapes • form from a small amount of sand • when there is a lot of sand, they merge to form long ridges called transverse dunes, as shown in **Figure 17** • covered by minimal or no vegetation • form in flat areas of constant wind direction • have crests that point downwind
	**Star Dunes** • form three or more irregularly shaped "arms" • form from a large amount of sand • covered by minimal or no vegetation • form in areas where the wind patterns are complex and changable
	**Parabolic Dunes** • form U shapes • form from a large amount of sand • covered by minimal vegetation • form in humid areas with moderate winds • crests point upwind
	**Longitudinal Dunes** • form series of ridge shapes • form from small or large amounts of sand • covered by minimal or no vegetation • form parallel to variable wind direction

Loess

Wind can carry fine, lightweight particles such as silt and clay in great quantities and for long distances. Many parts of Earth's surface are covered by thick layers of yellow-brown, wind-blown silt, which are thought to have accumulated as a result of thousands of years of dust storms. The source of these silt deposits might have been the fine sediments that were exposed when glaciers melted after the last ice age, more than 10,000 years ago. These thick, windblown silt deposits are known as **loess** (LUSS). Loess soils are very fertile because they contain abundant minerals and nutrients. **Figure 19** shows the states in which agriculture has benefited from loess deposits.

Distribution of Loess Deposits in the United States

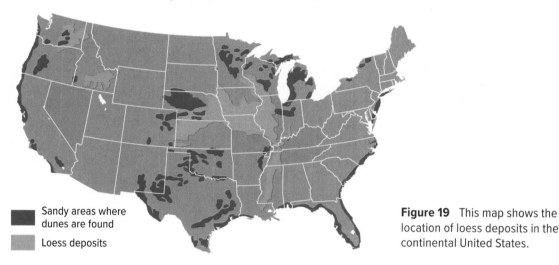

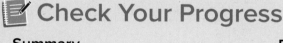

■ Sandy areas where dunes are found

Loess deposits

Figure 19 This map shows the location of loess deposits in the continental United States.

Check Your Progress

Summary

- Wind is a powerful agent of erosion.
- Wind can transport sediment in two ways, suspension and saltation.
- Dunes form when wind velocity slows down and windblown sand is deposited.
- Dunes migrate as long as winds continue to blow.

Demonstrate Understanding

1. **Distinguish** the various types of landforms formed by wind and how these landforms are created.
2. **Identify** conditions that can contribute to an increase in wind erosion.
3. **Examine** why the particles that form loess can travel much greater distances than sand.
4. **Analyze** how wind erosion and other factors affect dune development.

Explain Your Thinking

5. **Infer** how the movement of sand grains by saltation affects the overall movement of dunes.
6. **Evaluate** why wind erosion is an effective agent of erosion.
7. **WRITING** **Connection** Write a paragraph that explains how human activities directly affect wind erosion on coastlines.

LEARNSMART Go online to follow your personalized learning path to review, practice, and reinforce your understanding.

FOCUS QUESTION

What happens to the rock that is carved out of landscapes by glaciers?

Moving Masses of Ice

A large mass of moving ice is called a **glacier.** Glaciers form near Earth's poles and in mountainous areas at high elevations. They currently cover about 10 percent of Earth's surface, as shown in **Figure 20.** In the past, glaciers were more widespread than they are today. During the last ice age, which began about 1.6 mya and ended about 11,700 years ago, ice covered over 30 percent of Earth's land surface.

Areas at extreme northern and southern latitudes, such as Greenland and Antarctica, and areas of high elevations, such as the Alps, have temperatures near 0°C year-round. Cold temperatures keep fallen snow from completely melting, and each year the snow that has not melted accumulates in an area called a snowfield.

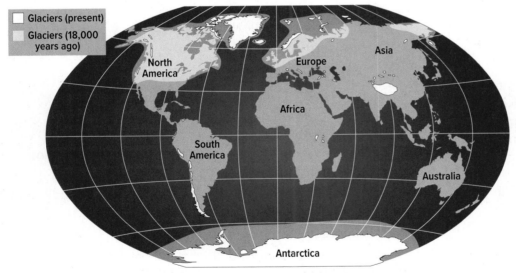

Figure 20 Glaciers around the world have changed in distribution throughout geologic time.

Infer *what changes have occurred in the distribution of glaciers around the world.*

🌐 **3D THINKING**   **DCI** Disciplinary Core Ideas   **CCC** Crosscutting Concepts   **SEP** Science & Engineering Practices

COLLECT EVIDENCE

 Use your Science Journal to record the evidence you collect as you complete the readings and activities in this lesson.

INVESTIGATE

 GO ONLINE to find these activities and more resources.

 Design Your Own: Analysis of Glacial Till
Plan and carry out an investigation to determine the properties that differentiate glacial till and outwash.

Quick Investigation: Model Glacial Deposition
Use a model to visualize the result of a glacier melting.

Thus, the total thickness of the snow layer increases as years pass. The weight of the top layers of snow eventually exerts enough downward pressure to force the accumulated snow below to recrystallize into ice. In this way, the accumulated snow develops into the ice of a glacier. A glacier can develop in any location that provides the necessary conditions. Glaciers can be classified as one of two types—valley glaciers or continental glaciers.

Valley glaciers

Glaciers that form in valleys in high, mountainous areas are called **valley glaciers.** Movement of a valley glacier occurs when the growing ice mass becomes so heavy that the ice maintains its rigid shape and begins to flow downhill. For most valley glaciers, flow begins when the accumulation of snow and ice exceeds 40 m in thickness. As a valley glacier moves, deep cracks in the surface of the ice, called crevasses, can form.

The slope of the valley floor, the temperature and thickness of the ice, and the shape of the valley walls affect the speed of a valley glacier's movement. The sides and bottom of a valley glacier move more slowly than the middle because friction slows down the sides and bottom, where the glacier comes in contact with the ground. Movement downslope is usually slow— less than a few millimeters per day. Over time, as valley glaciers flow downslope, their powerful carving action transitions V-shaped stream valleys into U-shaped glacial valleys.

Continental glaciers

Glaciers that cover broad, continent-sized areas are called **continental glaciers.** These glaciers form in cold climates where snow accumulates over many years. A continental glacier is thickest at its center. The weight of the center forces the rest of the glacier to flatten in all directions. In the past, when Earth experienced colder average temperatures than it does today, continental glaciers covered huge portions of Earth's surface. Today, they are confined to Greenland and Antarctica.

Glacial movement

Both valley glaciers and continental glaciers move outward when snow gathers at the zone of accumulation, a location in which more snow falls than melts, evaporates, or sublimates. For valley glaciers, the zone of accumulation is at the top of mountains, while for continental glaciers, the zone of accumulation is the center of the ice sheet. Both types of glaciers recede when the ends melt faster than the zone of accumulation builds up snow and ice.

Glacial Erosion

Of all the erosional agents, glaciers are the most powerful because of their great size, weight, and density. When a valley glacier moves, it breaks off pieces of rock through a process called plucking. When glaciers with embedded rocks move over bedrock, they act like the grains on a piece of sandpaper, grinding parallel scratches into the bedrock. Small scratches are called striations, and larger ones are called grooves. Striations and grooves provide evidence of a glacier's history and indicate its direction of movement.

STEM CAREER Connection

Glaciologist
What do you think it would be like to live on a glacier for several months? Recording temperatures and other data is part of a glaciologist's job. Glaciologists must love adventure and traveling to interesting places all over the world.

| Cirque | Horn | Hanging valley |

Figure 21 Glacial erosion by valley glaciers creates features such as cirques, horns, and hanging valleys.

Glacial erosion by valley glaciers can create features like those shown in **Figure 21.** At high elevations where snow accumulates, valley glaciers also scoop out deep, bowl-shaped depressions called **cirques.** When there are glaciers on three or more sides of a mountaintop, the carving action creates a steep, pyramid-shaped peak called a horn. The most famous example of this feature is Switzerland's Matterhorn.

Valley glaciers can also leave hanging valleys in the glaciated landscape. Hanging valleys are formed when higher tributary glaciers converge with the lower primary glaciers and later retreat. The primary glacier is so thick that it meets the height of the smaller tributary glacier. When the glaciers melt, the valley is left hanging high above what is now a river in the primary valley floor. Hanging valleys are often characterized by waterfalls.

Glacial Deposition

Glacial till is the unsorted rock, gravel, sand, and clay that glaciers deposit. The sediment is embedded in a glacier's ice and on their tops, sides, and front edges. The sediment is formed from the grinding action of the glacier on underlying rock. Glaciers deposit unsorted ridges of till called **moraines** when the glacier melts. Terminal moraines are found along the edge where the retreating glacier melts. Ground moraines form hilly areas instead of ridges and usually form underneath the glacier as a blanket of till.

Outwash

When the farthest ends of a glacier melt and the glacier begins to recede, meltwater flows into the valley below. Meltwater contains gravel, sand, and fine silt. When this sediment is sorted and deposited by meltwater and carried away from the glacier, it is called outwash. The area where the meltwater flows and deposits outwash is the **outwash plain.**

Drumlins, eskers, and kames

Continental glaciers that move over older ground moraines form the material into elongated landforms called **drumlins.** A drumlin's steeper slope faces the direction from which the glacier came. Streams flowing under melting glaciers leave long, winding ridges of layered sediments called **eskers.** A **kame** is a mound of layered sediment that forms when sediment gets washed into depressions or openings in the melting ice. When the ice finally melts, a cone-shaped hill or mound is left. Drumlins, eskers, and kames are all shown in **Figure 22.**

Figure 22 Visualizing Continental Glacial Features

Continental glaciers alter vast regions of landscape, leaving behind distinctive features such as kames, eskers, drumlins, and moraines.

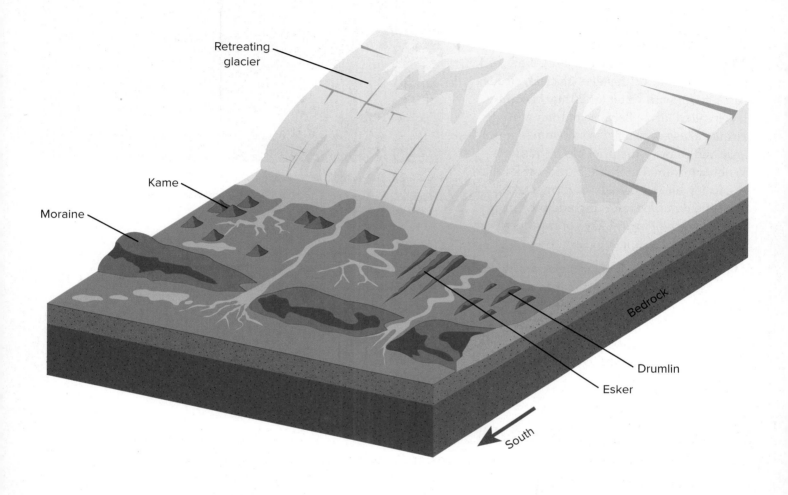

Kames are cone-shaped mounds of sand and gravel. They form when material is washed into a depression or hole in the ice. The material settles into the distinct shape as the ice beneath it melts.

Eskers are long ridges of sorted deposits. They are shaped from outwash deposited by water flowing through tunnels in a glacier.

Drumlins are shaped as a glacier moves over old moraines.

Glacial lakes

Sometimes a large block of ice breaks off a continental glacier, and the surrounding area is covered by sediment. When the ice block melts, it leaves behind a depression called a kettle hole. After the ice block melts, the kettle hole fills with water from precipitation and runoff, forming a kettle lake. **Kettles,** or kettle lakes, such as those shown in **Figure 23,** are common in New England, New York, and Wisconsin. With valley glaciers, cirques can also fill with water, and they become lakes called tarns. Because of their altitude, cirque lakes are usually frozen much of the year and thaw in the brief alpine summer. When a terminal moraine blocks off a valley, the valley fills with water to form a lake. Moraine-dammed lakes include the Great Lakes and the Finger Lakes of northern New York, which are long and narrow.

Figure 23 These kettle lakes in South Dakota are a result of glacial retreat.
Describe *how you might be able to locate kettles on a topographic map.*

Mass movements, wind, and glaciers all contribute to the changing of Earth's surface. These processes erode landforms constantly and, in many ways, impact human populations and activities.

Check Your Progress

Summary

- Glaciers are large, moving masses of ice that form near Earth's poles and in mountainous areas.

- Glaciers can be classified as valley glaciers or continental glaciers.

- Glaciers modify the landscape by erosion and deposition.

- Features formed by glaciers include U-shaped valleys, hanging valleys, moraines, drumlins, and kettles.

Demonstrate Understanding

1. **Describe** two examples of how glaciers modify landscapes.
2. **Explain** how glaciers form.
3. **Compare and contrast** the characteristics of valley glaciers and continental glaciers.
4. **Differentiate** among different glacial depositional features.

Explain Your Thinking

5. **Evaluate** the evidence of past glaciers that can be found on Earth today.
6. **Infer** whether valley glaciers or continental glaciers have shaped more of the landscape of the United States.
7. **WRITING** **Connection** Draw a picture of each glacial feature, and write a description to accompany each picture.

LEARNSMART Go online to follow your personalized learning path to review, practice, and reinforce your understanding.

Going, Going, Gone!

Glaciers shape Earth's surface, carving out landforms and moving rock and sediment from place to place. The amount of Earth's surface covered by glaciers varies over time. However, climate change has caused an extraordinary amount of glacial melt over the past 100 years. Not only does this rapid change affect Earth's surface, it affects society as well.

Changes in Water Availability

Glaciers are an important water source. Melting glaciers feed rivers and maintain watersheds. When the rate of glacial melt changes, water availability can change. For example, millions of people in Peru use water from glaciers. Runoff from high-altitude glaciers is used for household needs, agriculture, and generating electricity. Some of these glaciers have lost about 90% of their mass due to climate change, decreasing the quantity of available water. Other areas of Peru are facing flooding due to an increased rate of glacial melting.

Changes in Sea Level

Much of the water from melting glaciers flows to the ocean. Along with several other factors, melting ice has caused Earth's sea levels to rise at increasing rates. Scientists project that the water level in the world's oceans might rise between 28 and 98 centimeters by 2100.

Coastal erosion and storm surges become greater threats as sea levels rise. And as sea levels rise, the shoreline moves inland. Entire areas become uninhabitable because of flooding.

Grinnell Glacier, 1900
F. E. Matthes photo, courtesy of GNP Archives

Grinnell Glacier, 2008
Lisa McKeon photo, USGS

These photos of the same location show how Grinnel Glacier in Glacier National Park in Montana changed between 1900 and 2008.

Changes in Tourism

Glaciers are a tourist attraction. However, the number of glaciers has drastically decreased in the past two centuries. It is predicted that the last glaciers have a lifespan of only 10 to 30 years.

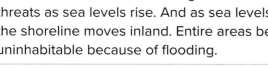 EVALUATE EVIDENCE SUPPORTING CLAIMS

Find a reliable source of information about the impact of climate change on glaciers. Write a statement explaining whether the source supports the claim that the rate at which glaciers are melting has an impact on society.

MODULE 6
STUDY GUIDE

 GO ONLINE to study with your Science Notebook.

Lesson 1 MASS MOVEMENTS

- Mass movements are classified in part by how rapidly they occur.
- Factors involved in the mass movement of earth materials include the material's weight, its resistance to sliding, the trigger, and the presence of water.
- Mass movements are natural processes that can affect human lives and activities.
- Human activities can increase the potential for the occurrence of mass movements.

- mass movement
- creep
- mudflow
- landslide
- slump
- avalanche

Lesson 2 WIND

- Wind is a powerful agent of erosion.
- Wind can transport sediment in two ways, suspension and saltation.
- Dunes form when wind velocity slows down and windblown sand is deposited.
- Dunes migrate as long as winds continue to blow.

- deflation
- abrasion
- ventifact
- dune
- loess

Lesson 3 GLACIERS

- Glaciers are large, moving masses of ice that form near Earth's poles and in mountainous areas.
- Glaciers can be classified as valley glaciers or continental glaciers.
- Glaciers modify the landscape by erosion and deposition.
- Features formed by glaciers include U-shaped valleys, hanging valleys, moraines, drumlins, and kettles.

- glacier
- valley glacier
- continental glacier
- cirque
- moraine
- outwash plain
- drumlin
- esker
- kame
- kettle

REVISIT THE PHENOMENON

What dangers do mass movements and erosion pose to people?

CER Claim, Evidence, Reasoning

Explain Your Reasoning Revisit the claim you made when you encountered the phenomenon. Summarize the evidence you gathered from your investigations and research and finalize your Summary Table. Does your evidence support your claim? If not, revise your claim. Explain why your evidence supports your claim.

STEM UNIT PROJECT

Now that you've completed the module, revisit your STEM unit project. You will summarize your evidence and apply it to the project.

GO FURTHER

SEP Data Analysis Lab

How much radioactivity is in ice cores?

Glaciologists have found that ice cores taken from the arctic region contain preserved radioactive fallout. Data collected from the study of these ice cores have been plotted on the graph.

Data and Observations The graph shows a subset of the data obtained.

CER Analyze and Interpret Data

1. **Determine** the depth in the ice cores where the highest and lowest amounts of radioactivity were found.
2. **Claim, Evidence** Describe what happened to the amount of radioactivity in the ice cores between the pre-test ban and Chernobyl.
3. **Reasoning** Infer what happened to the amount of radioactivity in the ice cores after Chernobyl.
4. **Reasoning** Explain what information or material, other than radioactive fallout, ice cores might preserve within them.

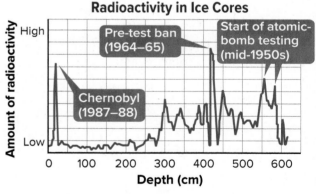

Radioactivity in Ice Cores

Pre-test ban (1964–65)

Start of atomic-bomb testing (mid-1950s)

Chernobyl (1987–88)

*Data obtained from: Mayewski, et al. 1990. Beta radiation from snow. *Nature* 345:25.

ENCOUNTER THE PHENOMENON

How does rain water get into this underground cave?

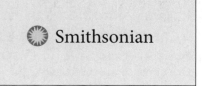 **GO ONLINE** to play a video about freshwater stored in an aquafer.

SEP Ask Questions

Do you have other questions about the phenomenon? If so, add them to the driving question board.

CER Claim, Evidence, Reasoning

Make Your Claim Use your CER chart to make a claim about how rain water gets into underground caves. Explain your reasoning.

Collect Evidence Use the lessons in this module to collect evidence to support your claim. Record your evidence as you move through the module.

Explain Your Reasoning You will revisit your claim and explain your reasoning at the end of the module.

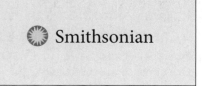 **GO ONLINE** to access your CER chart and explore resources that can help you collect evidence.

LESSON 2: Explore and Explain: Origins of Lakes

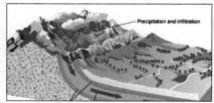

LESSON 3: Explore and Explain: Groundwater Movement

Additional Resources

FOCUS QUESTION

What happens to surface water that does not soak into the ground?

The Hydrosphere

The water on and in Earth's crust makes up the hydrosphere, named after *hydros*, the Greek word for *water*. You have learned about the hydrosphere in the context of Earth's systems, including the geosphere, atmosphere, and biosphere. About 97 percent of the hydrosphere is contained in the oceans. The water contained by landmasses—nearly all of it freshwater—makes up only about 3 percent of the hydrosphere.

Freshwater is one of Earth's most abundant and important renewable resources. However, of all the freshwater, about 70 percent is held in polar ice caps and glaciers. All the rivers, streams, and lakes on Earth represent only a small fraction of Earth's liquid freshwater. The distribution of the world's water is shown in **Table 1.** Recall that water in the hydrosphere moves through the water cycle.

 Get It?

Identify Where is the majority of Earth's water located?

Table 1 Estimated World's Water Supply

Location	Percentage of Total Water	Water Volume (km³)	Estimated Average Residence Time of Water
Oceans	97	1,316,700,000	thousands of years
Ice caps and glaciers	2.1	28,700,000	tens of thousands of years
Groundwater	0.31	4,200,000	hundreds to many thousands of years
Lakes	0.15	2,100,000	tens of years
Atmosphere	0.1	1,140,000	nine days
Rivers and streams	0.003	40,000	two weeks

 **3D THINKING**   **DCI** Disciplinary Core Ideas   **CCC** Crosscutting Concepts   **SEP** Science & Engineering Practices

COLLECT EVIDENCE

Use your Science Journal to record the evidence you collect as you complete the readings and activities in this lesson.

INVESTIGATE

GO ONLINE to find these activities and more resources.

 Investigation Lab: Analyzing Watersheds
Evaluate and communicate information about the effects of human activity on the health of watersheds.

 Review the News
Obtain information from a current news story about the effect of flooding on human populations. Evaluate your source and communicate your findings to your class.

The Water Cycle

Earth's water supply is recycled in a continuous process called the water cycle, shown in **Figure 1.** Water molecules move continuously through the water cycle, following many pathways. For example, they evaporate from a body of water or the surface of Earth, or they evaporate through the pores in plant leaves, a process called transpiration. As water vapor rises in the atmosphere, it cools and condenses into cloud droplets. Eventually, water falls back to Earth's surface as precipitation, including rain, snow, and sleet. On Earth's surface, water may run off into rivers and oceans or infiltrate the ground.

As part of a continuous cycle, water molecules eventually evaporate back into the atmosphere, form clouds, fall as precipitation, and repeat the cycle. Understanding the mechanics of the water cycle will help you understand the reasons for variations in the amount of freshwater that is available throughout the world.

Often, a water molecule's pathway includes time spent within a living organism or as part of a snowfield, glacier, lake, or ocean. Although water molecules might follow a number of different pathways, the overall process is one of repeated evaporation and condensation powered by the Sun's energy.

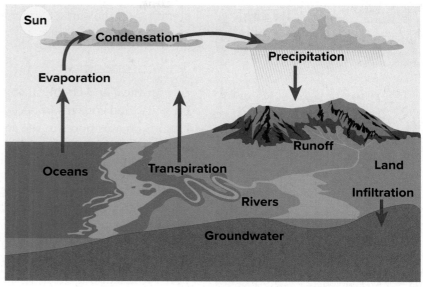

Figure 1 The water cycle, also referred to as the hydrologic cycle, is a never-ending, natural circulation of water through Earth's systems.

Identify *the driving force for the water cycle.*

 Get It?

Explain What happens once water reaches Earth's surface?

Runoff

Water flowing downslope along Earth's surface is called **runoff.** Runoff might reach a stream, river, or lake; it might evaporate; or it might accumulate as puddles in small depressions and infiltrate the ground. During and after heavy rains, you can observe these processes in your yard or local park. Water that infiltrates Earth's surface becomes groundwater.

SCIENCE USAGE v. COMMON USAGE

infiltrate

Science usage: to pass into or through a substance by filtering or permeating through pores or gaps

Common usage: to gradually become established within a group, usually for the purpose of gathering information or undermining authority

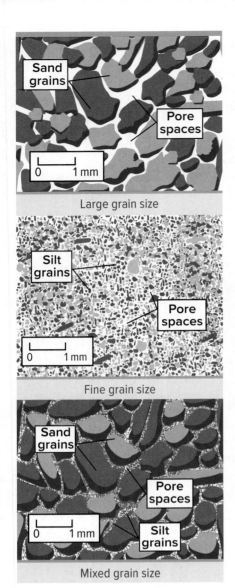

Large grain size

Fine grain size

Mixed grain size

Figure 2 Soil that has open surface pores allows water to infiltrate. The particle size that makes up a soil helps determine the pore space of the soil.

A number of conditions determine whether water on Earth's surface will infiltrate the ground or become runoff. For water to enter the ground, there must be large enough pores or spaces in the soil and rock to accommodate the water's volume, as in the loose soil illustrated in **Figure 2.** If the pores already contain water, the newly fallen precipitation will either remain in puddles on top of the ground or, if the area has a slope, run downhill. Water standing on the surface of Earth eventually evaporates, flows away, or slowly enters the groundwater.

Soil composition

The physical and chemical composition of soil affects its water-holding capacity. Soil consists of decayed organic matter, called humus, and minerals. Humus creates pores in the soil, thereby increasing a soil's ability to retain water. The minerals in soil have different particle sizes, which are classified as sand, silt, or clay.

Recall that the percentages of particles of each size vary from soil to soil. Soil with a high percentage of coarse particles, such as sand, has relatively large pores between its particles, which allows water to enter and pass through the soil quickly. In contrast, soil with a high percentage of fine particles, such as clay, clumps together and has few or no spaces between the particles. Small pores restrict both the amount of water that can enter the ground and the ease of movement of water through the soil.

Get It?

Predict Which type of soil would you expect to hold more water after a heavy rain—sand or clay? Explain your reasoning.

Rate of precipitation

Light, gentle precipitation can infiltrate dry ground. However, the rate of precipitation might temporarily exceed the rate of infiltration. For example, during heavy precipitation, water falls too quickly to infiltrate the ground and becomes runoff. Thus, a gentle, long-lasting rainfall is more beneficial to plants and causes less erosion by runoff than a torrential downpour. If you have a garden, remember that more water will enter the ground if you water your plants slowly and gently.

Get It?

Argue Make an argument that light precipitation is better for plants and soil than is heavy precipitation.

ACADEMIC VOCABULARY

accommodate

to hold without crowding or inconvenience

The teacher said she could accommodate three more students in her classroom.

Vegetation

Soils that contain grasses or other vegetation allow more water to enter the ground than do soils with no vegetation. Precipitation falling on vegetation slowly flows down leaves and branches and eventually drops gently to the ground, where the plants' root systems help maintain the pore space needed to hold water, as shown in **Figure 3**. In contrast, precipitation strikes with far more force on barren land. In such areas, soil particles clump together and form dense aggregates with little space between them. The force of falling rain can then push the soil clumps together, thereby closing pores and allowing less water to enter.

Grasses slow the movement of runoff water.

Figure 3 Vegetation can slow the rate of surface water runoff. Raindrops are slowed when they strike tree leaves or blades of grass, and they trickle down slowly.

Slope

The slope of a land area plays a significant role in determining the ability of water to enter the ground. Water from precipitation falling on slopes flows to areas of lower elevation. The steeper the slope, the faster the water flows. There is also greater potential for erosion on steep slopes. In areas with steep slopes, much of the precipitation is carried away as runoff.

 Get It?

Describe the relationship between the steepness of a slope and the ability of water to enter the ground.

Stream Systems

Precipitation that does not enter the ground usually runs off the surface quickly. Some surface water flows in thin sheets and eventually collects in small channels, which are the physical areas where streams flow. As the amount of runoff increases, the channels widen, deepen, and become longer. Although these small channels often dry up after precipitation stops, the channels fill with water each time it rains, and they become larger and longer.

Tributaries

All streams flow downslope to lower elevations. However, the path of a stream can vary considerably, depending on the slope and the type of material through which the stream flows. Some streams flow into lakes, while others flow directly into the ocean. Rivers that flow into other streams are called tributaries. For example, as shown in **Figure 4** on the next page, the Missouri River is a tributary of the Mississippi River.

Watersheds and divides

All of the land area whose water drains into a stream system is called the system's **watershed.** Watersheds can be relatively small or extremely large in area. A **divide** is an elevated land area that separates one watershed from another. In a watershed, the water flows away from the divide, as this is the high point of the watershed.

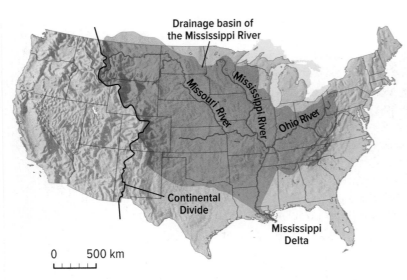

Drainage basin of
the Mississippi River

Missouri River

Mississippi River

Ohio River

Continental
Divide

Mississippi
Delta

0   500 km

Figure 4 The watershed of the Mississippi River includes many stream systems, including the Mississippi, Missouri, and Ohio Rivers. The Continental Divide marks the western boundary of the watershed.

Identify *what portion of the continental United States eventually drains into the Mississippi River.*

Each tributary in a stream system has its own watershed and divides, but they are all part of the larger stream system to which the tributary belongs. The watershed of the Mississippi River, shown in **Figure 4,** is the largest in North America.

Stream Load

The material that a stream carries is known as stream load. Stream load is carried in three ways.

Materials in suspension

Suspension is the method of transport for all particles small enough to be held up by the turbulence of a stream's moving water. Particles such as silt, clay, and sand are part of a stream's suspended load. The stream's suspended load varies based on the volume and velocity of the stream. Rapidly moving water carries larger particles in suspension than slowly moving water.

Bed load

Sediment that is too large or heavy to be held up by turbulent water is transported by streams as the bed load. A stream's **bed load** consists of sand, pebbles, and cobbles that the stream's water can roll or push along the bed of the stream. The faster the water moves, the larger the particles it can carry. As the particles move, they rub against one another or the solid rock of the streambed, which can erode the surface of the streambed, as shown in **Figure 5.**

Materials in solution

Minerals that are dissolved in a stream's water are called materials in solution. When water runs through or over rocks with soluble minerals, it dissolves small amounts of the minerals and carries them in solution. Groundwater adds the majority of the dissolved load to streams. The amount of dissolved material that water carries is often expressed in parts per million (ppm). For example, a measurement of 10 ppm means that there are 10 parts of dissolved material for every 1 million parts of water. The total concentration of materials in solution in streams averages 115–120 ppm, although some streams carry as little dissolved material as 10 ppm.

Figure 5 Particles rub, scrape, and grind against one another in a streambed, which can create potholes and other erosional features.

Stream Carrying Capacity

The ability of a stream to transport material, referred to as its carrying capacity, depends on both the velocity and the amount of water moving in the stream. The channel's slope, depth, and width all affect the speed and direction the water moves within it. A stream's water moves more quickly where there is less friction; consequently, deep and smooth-sided channels with steep slopes allow water to move the most rapidly. The total volume of moving water also affects a stream's carrying capacity. **Discharge,** shown in **Figure 6,** is the measure of the volume of stream water that flows past a particular location within a given period of time. Discharge is commonly expressed in cubic meters per second (m³/s). The following formula is used to calculate the discharge of a stream.

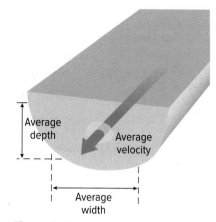

Figure 6 Stream discharge is the product of a stream's average width, average depth, and average velocity.

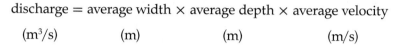

$$\text{discharge} = \text{average width} \times \text{average depth} \times \text{average velocity}$$

$$\text{(m}^3\text{/s)} \qquad \text{(m)} \qquad \text{(m)} \qquad \text{(m/s)}$$

 Get It?

Calculate the discharge of a stream with an average width of 8 m, an average depth of 2 m, and an average velocity of 0.9 m/s.

The largest river in North America, the Mississippi River, has a huge average discharge of about 17,000 m³/s. The Amazon River, the largest river in the world, has a discharge of about ten times that amount. The discharge from the Amazon River over a two-hour period would supply New York City's water needs for an entire year!

As a stream's discharge increases, its carrying capacity also increases. Both water velocity and volume increase during times of heavy precipitation, rapid melting of snow, and flooding. In addition to increasing a stream's carrying capacity, these conditions heighten a stream's ability to erode the land over which it passes. As a result of an increase in erosional power, a streambed can widen and deepen, adding to the stream's carrying capacity. Streams shape the landscape during both periods of normal flow and during floods.

 Get It?

Explain how heavy precipitation, rapid melting of snow, and flooding can affect a stream's erosional power.

Floods

The amount of water being transported in a particular stream at any given time varies with weather conditions. For example, a severe thunderstorm can lead to increased amounts of water in a stream. Heavy rains from hurricanes can also dump huge amounts of water into streams. Sometimes, more water pours into a stream than the banks of the stream channel can hold. A **flood** occurs when water spills over the sides of a stream's banks and onto the adjacent land. The broad, flat area that extends out from a stream's bank and is covered by excess water during times of flooding is known as the stream's **floodplain.**

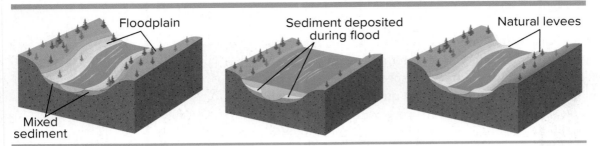

Figure 7 When rivers overflow their banks, the floodwater deposits sediment. Over time, sediment accumulates along the edges of a river, resulting in natural levees.

Floodwater carries along with it a great amount of sediment eroded from Earth's surface and the sides of the stream channel. As floodwater recedes and its volume and speed decrease, the water drops its sediment load onto the stream's floodplain. After repeated floods over time, sediments deposited by floods tend to accumulate along the banks of the stream. These develop into continuous ridges, called natural levees, along the sides of a river, as shown in **Figure 7**. Floodplains develop highly fertile soils as more sediment is deposited with each subsequent flood. The fertile soils of floodplains make some of the best croplands in the world. **Figure 8** details how floods have shaped the landscape and affected human lives.

 Get It?

Describe what happens when floodwaters recede.

Stages of floods

Floods are a common hazard in flat and low-lying areas of land. This makes it difficult for water to drain off the land.

All floods have common stages. First, water enters a stream. The water level rises and might reach its highest point, called its crest, days after precipitation ends. When the water level rises above the stream's banks, the river is at flood stage. A small, local, upstream flood can occur. But the most intensive flooding often occurs downstream. The tremendous volume of water in a downstream flood can cause extensive damage, as shown in **Figure 9.**

In some coastal areas, floods also result from storm surges associated with hurricanes. During a storm surge, a large volume of ocean water sweeps over the land.

Flood Monitoring and Warning Systems

As one of the most destructive natural disasters, floods affect almost 100 million people worldwide annually. The resulting damage is costly as well, at almost 14 billion dollars per year. Floods displace people from their homes, often leave them without clean water to drink, and can lead to water-borne infections such as cholera.

Figure 9 This flood was caused by heavy rainfall upstream. Notice the farm fields that have been covered in floodwater.

Analyze *What long-term effects might this flood have on the crops grown in this area?*

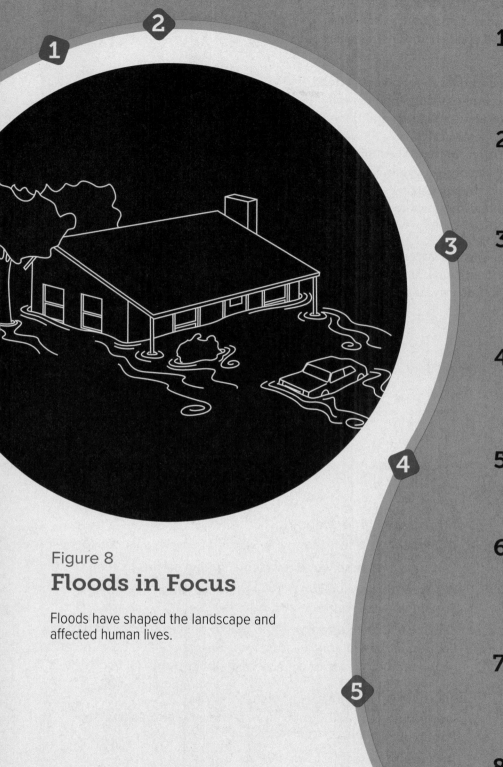

Figure 8
Floods in Focus

Floods have shaped the landscape and affected human lives.

1 **1927** Heavy rains flood the Mississippi River from Illinois to Louisiana, leaving more than 600,000 people homeless.

2 **1931** China's Yellow River floods when heavy rain causes the river's large silt deposits to shift and block the channel.

3 **1958** Following a flood that claimed almost 2000 lives, Holland begins creating a vast network of dams, dikes, and barriers, shortening its coastline by 700 km.

4 **1974** The United Kingdom begins building the Thames Barrier to protect London from rising tide levels as the city sinks and sea levels rise.

5 **1988** Monsoon rains in Bangladesh flood two-thirds of the country, affecting 45 million people.

6 **1996** Volcanic eruptions in Iceland release meltwater from under the Vatnajökull glacier, washing away power lines, major roads, and bridges.

7 **2010** China's Three Gorges Dam, which plays a key role in the flood control of the Yangtze River, reaches full capacity.

8 **2017** Hurricane Harvey hits eastern Texas and drops at least 50 inches of rain over 4 days, causing catastrophic flooding. This is the most expensive natural disaster in the U.S. to date.

Since floods are so destructive, government agencies, such as the National Weather Service, work to track rainfall and large storms that can cause flooding. By monitoring potential flood conditions, agencies can provide warnings for people at risk.

Several types of technology are used to collect data about potential flood conditions. For example, Earth-orbiting weather satellites photograph Earth and collect and transmit information about weather conditions, storms, and streams. In addition, the U.S. Geological Survey (USGS) has established approximately 9200 gaging stations in the United States, like the one shown in **Figure 10.** These gaging stations provide a continuous record of the water level in each monitored stream. Gaging systems often transmit data to satellites and telephone lines, where the information is then sent to the local monitoring office.

In areas that are prone to severe flooding, warning systems are the first step in implementing emergency management plans. Flood warnings and emergency plans often allow people to safely evacuate an area in advance of a flood.

Figure 10 Gaging stations, like this one, can send data to meteorologic stations. There, scientists can process the information and alert the public to potential floods.

 Get It?

Describe technology used to monitor potential flood conditions.

Check Your Progress

Summary

- Infiltration of water into the ground depends on the number of open pores.
- All of the land area that drains into a stream system is the system's watershed.
- Elevated land areas called divides separate one watershed from another.
- A stream's load is the material the stream carries.
- Flooding occurs as upstream floods in small, localized areas or as large downstream floods.

Demonstrate Understanding

1. **Analyze** ways in which moving water can carve a landscape.
2. **Describe** the three ways in which a stream carries its load.
3. **Analyze** the relationship between the carrying capacity of a stream and its discharge and velocity.
4. **Describe** Use **Figure 8** and the text to describe how floods have shaped human history. Include measures that have been taken to reduce flooding and to warn people about potential floods.

Explain Your Thinking

5. **Determine** how a floodplain forms and why people live on floodplains.
6. **Analyze** how natural levees form.
7. **MATH Connection** Design a data table that compares how silt, clay, sand, and large pebbles settle to the bottom of a stream as the water's velocity decreases.

LEARNSMART Go online to follow your personalized learning path to review, practice, and reinforce your understanding.

U.S. Geological Survey

STREAMS, LAKES, AND WETLANDS

FOCUS QUESTION

How does surface water interact with the land?

Stream Development

The region where water first accumulates to supply a stream is the headwaters. Often a stream's headwaters are high in the mountains. Falling precipitation accumulates in small gullies at these higher elevations and forms streams. As surface water begins its flow, its path might not be well defined. In time, the moving water carves a narrow pathway called the **stream channel** into the sediment or rock. The channel widens and deepens as more water accumulates and cuts into Earth's surface. **Stream banks** hold the moving water within them.

When small streams erode away the rock or soil at the head of a stream, it is known as headward erosion. These streams move swiftly over rough terrain and often form waterfalls and rapids as they flow over steep inclines. Sometimes a stream erodes the high area separating two drainage basins, joining another stream. It then draws water away from the other stream in a process called stream capture, as shown in **Figure 11.**

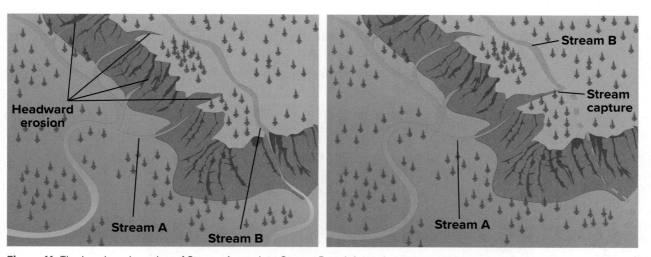

Figure 11 The headward erosion of Stream A cuts into Stream B and draws its water away into one stream.

3D THINKING **DCI** Disciplinary Core Ideas   **CCC** Crosscutting Concepts   **SEP** Science & Engineering Practices

COLLECT EVIDENCE

Use your Science Journal to record the evidence you collect as you complete the readings and activities in this lesson.

INVESTIGATE

GO ONLINE to find these activities and more resources.

Applying Practices: Investigate Stream Erosion
HS-ESS2-5. Plan and conduct an investigation of the properties of water and its effects on Earth materials and surface processes.

Mapping Lab: Interpreting a River's Habits
Interpret data to determine the effects of surface water on the surface of Earth.

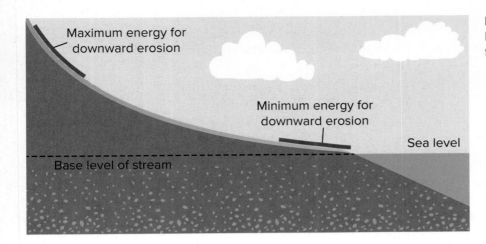

Figure 12 The height of a stream above its base level determines how much downcutting energy the stream will have.

Maximum energy for downward erosion

Minimum energy for downward erosion

Sea level

Base level of stream

Formation of Stream Valleys

The driving force of a stream is the force of gravity on water. This means that the energy of a stream comes from the movement of water down a slope called a stream gradient. When the gradient of a stream is steep, water in the stream moves downhill rapidly, cutting steep valleys. The gradient of the stream depends on its **base level,** which is the elevation at which it enters another stream or body of water. The lowest base level possible for any stream is sea level, the point at which the stream enters the ocean, as shown in **Figure 12.**

Far from its base level, a stream actively erodes a path through the sediment or rock, and a V-shaped channel develops. V-shaped channels have steep sides and sometimes form canyons or gorges. The Yellowstone River in Wyoming flows through an impressive example of this type of narrow, deep gorge carved by a stream. **Figure 13** shows the classic V-shaped valley. As a stream approaches its base level, it has less energy for downward erosion. Instead, streams that are near their base level tend to erode at the sides of the stream channel and, over time, result in broader valleys with gentle slopes, as shown in **Figure 13.**

Figure 13 A V-shaped valley is formed by the downcutting of a stream. A broad valley is a result of stream erosion over a long period of time.

Identify *which river is closer to its base level.*

Meanders

As stream channels develop into broader valleys, the volume of water and sediment that they are able to carry increases. In addition, a stream's gradient decreases as it nears its base level, and the channel gets wider as a result. The decrease in gradient causes an increase in the volume of water the stream channel can carry. Sometimes, the water begins to erode the sides of the channel in such a way that the overall path of the stream starts to bend or wind. A bend or curve in a stream channel caused by moving water is called a **meander,** shown in **Figure 14.**

Water in the straight parts of a stream flows at different velocities, depending on its location in the channel. In a straight length of a stream, water in the center of the channel flows at the maximum velocity. Water along the bottom and sides of the channel flows more slowly because it experiences friction as it moves against the land.

In contrast, the water moving along the outside of a meander curve experiences the greatest velocity within the meander. The water that flows along this outside part of the curve continues to erode the sides of the streambed, thus making the meander larger. Along the inside of the meander, the water moves more slowly, and deposition is dominant, forming point bars, as shown in **Figure 15** on the next page. The differences in the velocity within a meander causes it to become more accentuated over time.

Oxbow lakes Meanders continue to develop and become larger and wider over time. After enough winding, however, it is common for a stream to cut off a meander and once again flow along a straighter path. The stream then deposits material along the adjoining meander and eventually blocks off its water supply, as shown in **Figure 14.** The blocked-off meander becomes an oxbow lake, which eventually dries up.

As a stream approaches a larger body of water or its endpoint—the ocean—the streambed's gradient flattens out, and its channel becomes very wide. The area of the stream that leads into the ocean or another large body of water is called the mouth.

Figure 14 As the path of the stream bends and winds, it creates meanders and, eventually, oxbow lakes.

SCIENCE USAGE v. COMMON USAGE

meander

Science usage: a bend or curve in a stream channel caused by moving water

Common usage: to follow a winding path or course

CCC CROSSCUTTING CONCEPTS

Structure and Function Using evidence from the text, explain how the properties of a stream valley can be inferred from the overall structure of the stream and its components, such as meanders and oxbow lakes.

Figure 15 Visualizing Erosion and Deposition in a Meander

As water travels down a meander, the area of maximum velocity changes. As shown in cross section A, when the meander is straight, the maximum velocity is located near the center. When the meander curves, the maximum velocity shifts to the outside of the curve, as shown in cross section B. As the meander travels around to cross section C, the maximum velocity shifts again to the outside of the curve. Erosion occurs around curves in the meander in areas of high velocity. The water's high velocity carries sediment downstream and deposits it where the velocity decreases, on the inside of a curve. The area where erosion occurs is called a cutbank, and the area where deposition occurs is called a point bar.

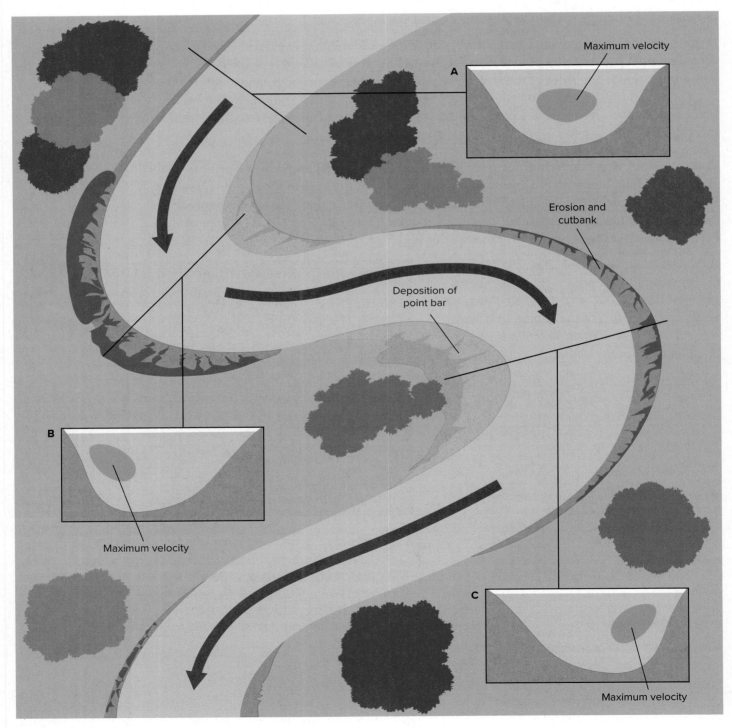

A — Maximum velocity

Erosion and cutbank

Deposition of point bar

B — Maximum velocity

C — Maximum velocity

Deposition of Sediment

The velocity of a stream determines how much sediment it can transport. When streams lose velocity, they lose some of the energy needed to transport sediment, and deposition of sediment occurs.

Alluvial fans

A stream's velocity lessens and its sediment load is deposited when its gradient abruptly decreases. In dry regions such as the North American Southwest, mountain streams flow intermittently down steep, rocky slopes and then flatten out onto expansive dry lake beds. In areas such as these, a stream's gradient suddenly decreases, causing the stream to drop its sediment at the base of the mountain in a fan-shaped deposit called an alluvial fan. Alluvial fans are formed at the bases of slopes and are composed mostly of sand and gravel. An example of an alluvial fan is shown in **Figure 16.**

Figure 16 An alluvial fan is a fan-shaped depositional feature.

 Get It?

Describe how an alluvial fan is formed.

Deltas Streams also lose velocity and some of their capacity to carry sediment when they join larger bodies of quiet water. The often triangular deposit that forms where a stream enters a large body of water is called a **delta.** Delta deposits usually consist of layers of silt and clay particles. As a delta develops, sediments build up and slow the stream water, sometimes even blocking its movement. Smaller distributary streams then form to carry the stream water through the developing delta. Deltas, such as the Mississippi River Delta, are normally areas where the stream flow changes direction frequently.

Over the course of thousands of years, the Mississippi River Delta has frequently changed. Today, any small change in the drainage channels can result in catastrophic flooding. To prevent floods, an extensive system of dams and levees is in place to help protect people and economic activities. Flood control, however, can result in a decrease in the regular deposition of sediment throughout the delta. In the absence of regular deposition throughout the delta, normal processes of coastal erosion have caused the delta to shrink over time, as shown in **Figure 17.**

1973

2003

Figure 17 The Mississippi River Delta formed from the deposition of river sediments. The area in the top left of both images is a marshland used for both recreation and business. Since 1973, waters upstream of the Mississippi River have been dammed, reducing the sediment flow. Over the course of 30 years, the area of the marshland decreased as the sediment input from upstream decreased.

 Get It?

Describe how a delta is formed.

(t)totajla/Shutterstock, (c, b)U.S. Geological Survey

Rejuvenation

During the process of stream formation, downcutting can occur. Downcutting is the wearing away of the streambed and is a major erosional process that influences the stream until it reaches its base level. If the base level drops as a result of geologic processes, the stream undergoes rejuvenation.

Rejuvenation means "to make young again." During **rejuvenation,** a stream actively resumes the process of downcutting toward its base level. This causes an increase in the stream's velocity, and the stream's channel once again cuts downward into the existing meanders. Rejuvenation can cause deep-sided canyons to form.

A well-known example of rejuvenation is the Grand Canyon. Millions of years ago, the Colorado River was near its base level, like much of the Mississippi River today. Then the land was uplifted compared to the level of the ocean, which caused the base level of the Colorado River to drop. This caused the process of rejuvenation, in which renewed erosion occurred as the river began cutting downward into the existing meanders. The result is the 1.6-km-deep canyons, which attract millions of visitors each year from all over the world. Rejuvenation is still occurring today in the Grand Canyon as the Colorado River continues downcutting toward its base level.

Origins of Lakes

Natural **lakes,** bodies of water surrounded by land, form in different ways in surface depressions and in low areas. As you have previously read, oxbow lakes form when streams cut off meanders and leave isolated channels of water. Lakes also form when stream flow becomes blocked by sediment from landslides or other sources. Still other lakes have glacial origins. The basins of these lakes formed when glaciers gouged out the land during the ice ages. Many of the lakes in the northernmost parts of Europe and North America are in recently glaciated areas. Glacial moraines originally dammed some of these depressions and restricted the outward flow of water. The lakes that formed as a result are known as moraine-dammed lakes. In another process, cirques carved high in the mountains by valley glaciers filled with water to form cirque lakes. Other lakes formed as blocks of ice left on the outwash plain ahead of melting glaciers eventually melted, leaving depressions called kettles. When these depressions filled with water, they formed kettle lakes such as those shown in **Figure 18.**

Figure 18 Lakes such as these in Minnesota were formed from blocks of ice that melted after glaciers retreated.

Lakes Undergo Change

Water from precipitation, runoff, and underground sources can maintain a lake's water supply. Some lakes contain water only during times of heavy rain or excessive runoff from spring thaws. A depression that receives more water than it loses to evaporation or use by humans will exist as a lake for a long period of time. However, most lakes are temporary water-holding areas; over hundreds of thousands of years, lakes usually fill in with sediment and become part of a new landscape.

Eutrophication

The process by which the surrounding watershed enriches bodies of water with nutrients that stimulate excessive plant growth is called **eutrophication.** Although eutrophication is a natural process, it can be sped up with the addition of nutrients, such as fertilizers, that contain nitrogen and phosphorus. Other major sources of nutrients that concentrate in lakes are animal wastes and phosphate detergents.

When eutrophication occurs, the animal and plant communities in a lake can change rapidly. Algae growing at the surface of the water can suddenly multiply very quickly. The excessive algae growth in a lake or pond appears as a green blanket, as shown in **Figure 19.** Other organisms that eat the algae can multiply in numbers as well. In addition, the population of algae on the surface can block sunlight from penetrating to the bottom of the lake, causing sunlight-dependent plants and other organisms below the surface to die. The resulting overpopulation and, later, the decay of a large number of plants and animals, depletes the water's dissolved oxygen supply. Fish and other sensitive organisms might die due to the lack of dissolved oxygen in the water. In some cases, the algae can also release into the water toxins that are harmful to other organisms. Scientists use the amount of dissolved oxygen present in a body of water to assess the water's overall quality. A body of water must have dissolved oxygen in order to support life.

 Get It?

Identify the effects of eutrophication on the aquatic animals in an affected lake system.

Figure 19 Eutrophication is a natural process that can be accelerated with the addition of nitrogen and phosphorus to a body of water. Once the process begins, it can cause rapid changes in the plant and animal communities in the affected body of water.

Wetlands

A **wetland** is any land area that is covered with water for a part of the year. Wetlands include environments commonly known as bogs, marshes, and swamps. Each wetland has a certain soil type and supports specific plant species. Soil type depends on the degree of water saturation.

Bogs Bogs are not stream-fed; instead, they receive their water from precipitation. The waterlogged, acidic soil supports unusual plant species, including insect-eating pitcher plants such as sundew and Venus flytrap.

Marshes Freshwater marshes often form along the mouths of streams and near deltas. The constant supply of water and nutrients allows for the lush growth of marsh grasses. Grasses, reeds, and rushes, along with abundant wildlife, are common in marshes.

Swamps Swamps are low-lying areas often located near streams. Swamps can develop from marshes that have filled in sufficiently to support the growth of shrubs and trees. As these larger plants grow and begin to shade the marsh plants, the marsh plants die.

Wetlands and water quality Wetlands play a valuable role in improving water quality. They serve as a filtering system that traps pollutants, sediments, and pathogenic bacteria contained in water sources. Wetlands also provide vital habitats for migratory waterbirds and homes for an abundance of other wildlife. In the past, it was common for wetland areas to be filled in to create more land on which to build. Government data reveal that from the late 1700s to the mid-1980s, the continental United States lost 50 percent of its wetlands. By 1985, it was estimated that 50 percent of the wetlands in Europe were drained. Now, however, the restoration and preservation of existing wetland areas, as well as the creation of new wetlands to replace those that were drained or filled, has become a global concern.

Check Your Progress

Summary

- Stream water flows in channels confined by the stream's banks.
- Alluvial fans and deltas form when stream velocity decreases and sediment is deposited.
- Lakes form in a variety of ways when depressions on land fill with water.
- Wetlands are low-lying areas that are periodically saturated with water.

Demonstrate Understanding

1. **Describe** how a V-shaped valley is formed.
2. **Identify** four changes that a stream undergoes before it reaches the ocean.
3. **Compare** the velocity on the inside of a meander curve with that on the outside of the curve.
4. **Explain** the transformation process that a lake might undergo as it changes to dry land.

Explain Your Thinking

5. **Infer** how you can tell that rejuvenation has modified the landscape.
6. **Organize** a data table to compare various types of lakes and their origins.
7. **WRITING ▸ Connection** Write an essay explaining why proper management of wetlands is important.

FOCUS QUESTION

How does water flow through rock?

Groundwater and Precipitation

The ultimate source of all water on land is the oceans. Evaporation of seawater cycles water into the atmosphere in the form of invisible water vapor and visible clouds. Winds and weather systems move this atmospheric moisture all over Earth, with much of it concentrated over the continents. Precipitation brings atmospheric moisture back to Earth's surface. Some of this precipitation falls directly into the oceans, and some falls on land.

Infiltration is the process by which precipitation that has fallen on land trickles into the ground and becomes groundwater. Only a small portion of precipitation becomes runoff and is returned directly to the oceans through streams and rivers. Groundwater slowly moves through the ground, eventually returns to the surface through springs and seepage into wetlands and streams, and then flows back to the oceans.

 Get It?

Identify the ultimate source of all water on land.

Groundwater Storage

Puddles of water that are left after it rains quickly disappear, partly by infiltrating the ground. On sandy soils, rain soaks into the ground almost immediately. Where does that water go? The water seeps into small openings within the ground. Remember from lesson 1, although Earth's crust appears solid, it is composed of soil, sediment, and rock that contain countless small openings, called pore spaces.

Pore spaces make up large portions of some of these materials. The amount of pore space in a material is its porosity. The greater the porosity, the more water can be stored in the material. Subsurface materials have porosities ranging from 2 percent to more than 50 percent.

 **3D THINKING**   **DCI** Disciplinary Core Ideas   **CCC** Crosscutting Concepts   **SEP** Science & Engineering Practices

COLLECT EVIDENCE

Use your Science Journal to record the evidence you collect as you complete the readings and activities in this lesson.

INVESTIGATE

GO ONLINE to find these activities and more resources.

Design Your Own: Analysis of Drinking Water
Evaluate and communicate information about the effect of human activity on water resources.

Investigation Lab: Measuring Permeability Rate
Analyze and interpret data to visualize how the structure of rocks influences the movement of water through them.

Well-sorted, large sand grains Unsorted sand grains Well-sorted, small sand grains

Figure 20 Porosity depends on the size and variety of particles in a material.

For example, the porosity of well-sorted sand is 30 percent; however, in poorly sorted sediment, smaller particles occupy some of the pore spaces and reduce the overall porosity of the sediment, as shown in **Figure 20.** Because of the enormous volume of sediment and rock beneath Earth's surface, enormous quantities of groundwater are stored in the pore spaces.

The Zone of Saturation

The region below Earth's surface in which groundwater completely fills all the pores of a material is called the **zone of saturation.** The upper boundary of the zone of saturation is the **water table,** shown in **Figure 21.** In the **zone of aeration,** above the water table, materials are moist, but because they are not saturated with water, air occupies much of the pores.

Water movement

Water in the zone of saturation and zone of aeration can be classified as either gravitational water or capillary water. Gravitational water is water that trickles downward as a result of gravity. Capillary water is water that is drawn upward through capillary action above the water table and is held in the pore spaces of rocks and sediment because of surface tension.

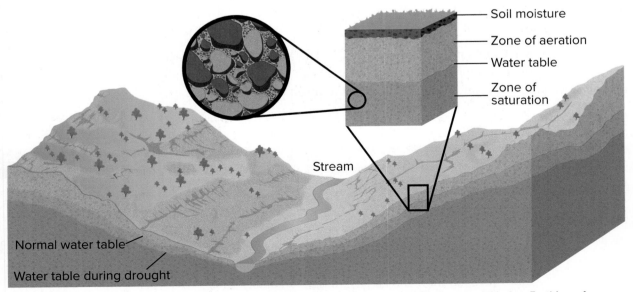

Soil moisture
Zone of aeration
Water table
Zone of saturation

Stream

Normal water table

Water table during drought

Figure 21 The zone of saturation is where groundwater completely fills all the pores of a material below Earth's surface.

The water table

The depth of the water table often varies depending on local conditions. For example, in stream valleys, groundwater is relatively close to Earth's surface, and thus the water table can be only a few meters deep. In swampy areas, the water table is at Earth's surface, whereas on hilltops or in arid regions, the water table can be tens to hundreds of meters or more beneath the surface. As shown in **Figure 21,** the topography of the water table generally follows the topography of the land above it. For example, the slope of the water table corresponds to the shape of valleys and hills on the surface above.

Because of its dependence on precipitation, the water table fluctuates with seasonal and other weather conditions. It rises during wet seasons, usually in spring, and drops during dry seasons, often in late summer.

Groundwater Movement

Groundwater flows downhill in the direction of the slope of the water table. Usually, this downhill movement is slow because the water has to flow through numerous tiny pores in the subsurface material. The ability of a material to let water pass through it is its **permeability.** Materials with large, connected pores, such as sand and gravel, have high permeability and permit relatively high flow velocities up to hundreds of meters per hour. Other permeable subsurface materials include highly fractured bedrock, sandstone, and limestone.

Permeability

Groundwater flows through permeable sediment and rock, called **aquifers,** such as the one shown in **Figure 22.** In aquifers, the pore spaces are large and connected. Fine-grained materials have low permeabilities because their pores are small. These materials are said to be impermeable. Groundwater flows so slowly through impermeable materials that the flow is often measured in millimeters per day. Some examples of impermeable materials include silt, clay, and shale. Clay is so impermeable that a clay-lined depression will hold water. For this reason, clay is often used to line artificial ponds and landfills. Impermeable layers, called **aquicludes,** are barriers to groundwater flow.

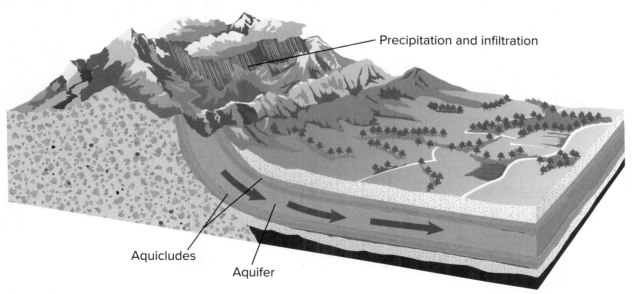

Figure 22 An aquifer is a layer of permeable subsurface material that is saturated with water. This aquifer is located between two impermeable layers called aquicludes.

Flow velocity

The flow velocity of groundwater depends on the slope of the water table and the permeability of the material through which the groundwater is moving. The force of gravity pulling the water downward is greater when the slope of the water table surface is steeper. Water also flows faster through a large opening than through a small opening. The flow velocity of groundwater is proportional to both the slope of the water table and the permeability of the material through which the water flows.

Springs

Groundwater moves slowly but continuously through aquifers and eventually returns to Earth's surface. In most cases, groundwater emerges wherever the water table intersects Earth's surface. Such intersections commonly occur in areas that have sloping surface topography. The exact places where groundwater emerges depend on the arrangement of aquifers and aquicludes in an area.

 Get It?

Explain the relationship between the slope of the land and where groundwater emerges.

Recall that aquifers are permeable underground layers through which groundwater flows easily, and aquicludes are impermeable layers. Aquifers are commonly composed of layers of sand and gravel, sandstone, and limestone. In contrast, aquicludes, such as layers of clay or shale, block groundwater movement. As a result, groundwater tends to discharge at Earth's surface where an aquifer and an aquiclude are in contact. These natural discharges of groundwater, shown in **Figure 23,** are called **springs.** Florida has the most springs of any state in the United States. This is mainly due to its plentiful rainfall and its limestone formations, which form underground pathways for water.

Emergence of springs

The volume of water that is discharged by a spring might be a trickle or a rushing stream. In places called karst regions, an entire river might emerge from the ground. Such a superspring is called a karst spring. Karst springs occur in northwest Florida and other limestone regions where springs discharge water from underground pathways. In regions of nearly horizontal sedimentary rocks, springs often emerge on the sides of valleys near the base of an aquifer, as shown in **Figure 24.** Springs might also emerge at the edges of perched water tables. In a perched water table, a zone of saturation that overlies an aquiclude separates it from the main water table below. Other areas where springs tend to emerge are along faults, when huge blocks of rock shift and block aquifers.

Figure 23 Springs are natural discharges of groundwater.

 Get It?

Identify areas where springs may emerge.

John A. Karachewski

Figure 24 Visualizing Springs

A spring is the result of groundwater that emerges at Earth's surface. Springs can be caused by a variety of conditions.

Compare and contrast *the origins of the four types of springs.*

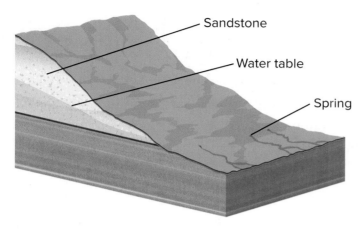

A spring forms where a permeable layer and an impermeable layer come together.

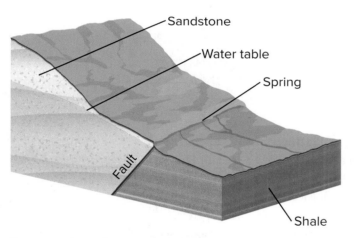

Some springs form where a fault has brought together two different types of bedrock, such as a porous rock and a nonporous rock.

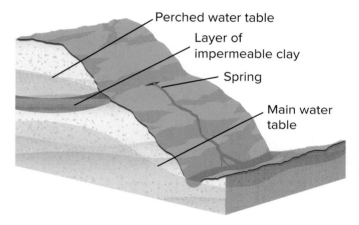

A layer of impermeable rock or clay can create a perched water table. Springs can result where groundwater emerges from a perched water table.

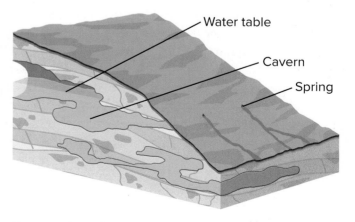

Karst springs form where groundwater weathers through limestone bedrock, and water in the underground caverns emerges at Earth's surface.

Temperature of springs

The temperature of groundwater that is discharged through a spring is generally the average annual temperature of the region in which it is located. Thus, springs in New England have temperatures of about 10°C, while springs in the Gulf states have temperatures of about 20°C. Compared to air temperatures, groundwater is generally colder in the summer and warmer in the winter.

However, in some regions around the world, springs discharge water that is much warmer than the average annual temperature. These springs are called warm springs or **hot springs,** depending on their temperatures. Hot springs are springs that have a temperature above 36.6°C. Most hot springs are located in areas where the subsurface is still hot from nearby igneous activity.

Among the most spectacular features produced by Earth's underground thermal energy in volcanic regions are geysers, shown in **Figure 25. Geysers** are explosive hot springs. In a geyser, water is heated past its boiling point, causing it to vaporize. The resulting water vapor builds up tremendous pressure, which fuels eruptions.

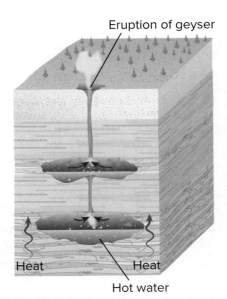

Figure 25 A geyser is a type of hot spring from which very hot water and vapor erupt at the surface.

Wells

Wells are holes dug or drilled into the ground to reach an aquifer. There are two main types of wells: ordinary wells and artesian wells.

Ordinary wells

The simplest wells are those that are dug or drilled below the water table, into what is called a water-table aquifer, as shown in **Figure 26.** In a water-table aquifer, the level of the water in the well is the same as the level of the surrounding water-table. As water is drawn out of a well, it is replaced by surrounding water in the aquifer.

Overpumping occurs when water is drawn out of the well at a rate that is faster than that at which it is replaced. Overpumping of the well lowers the local water level and results in a cone of depression around the well, as shown in **Figure 26.** The difference between the original water-table level and the water level in the pumped well is called the **drawdown.**

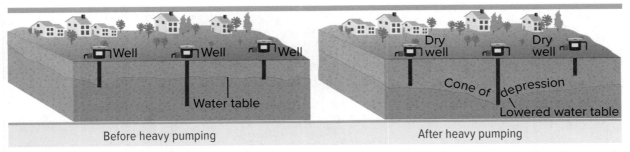

Figure 26 Overpumping from one well or multiple wells can result in a cone of depression and a general lowering of the water table.

If many wells withdraw water from a water-table aquifer, the cones of depression can overlap and cause an overall lowering of the water table, causing shallow wells to become dry. Water from precipitation replenishes the water content of an aquifer in the process of **recharge.** Groundwater recharge from precipitation and runoff sometimes replaces the water withdrawn from wells. However, if withdrawal of groundwater exceeds the aquifer's recharge rate, the drawdown increases until all wells in the area become dry.

Artesian wells

An aquifer's area of recharge is often at a higher elevation than the rest of the aquifer. An aquifer located between aquicludes, called a confined aquifer, can contain water that is under pressure. This is because the water at the top of the slope exerts gravitational force on the water downslope. An aquifer that contains water under pressure is called an artesian aquifer. When the rate of recharge is high enough, the pressurized water in a well drilled into an artesian aquifer can spurt above the land surface in the form of a fountain known as an **artesian well.** The level to which water in an open well can rise is called its pressure surface, as shown in **Figure 27.** Similarly, a spring that discharges pressurized water is called an artesian spring. The name *artesian* is derived from the French province of Artois, where such wells were first drilled almost 900 years ago.

Figure 27 An artesian aquifer contains water under pressure.

Identify *the features that cause the primary difference between an ordinary well and an artesian well.*

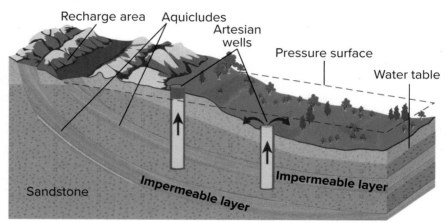

Threats to Our Water Supply

Freshwater is Earth's most precious natural resource. It is essential for life. Water is also used extensively in agriculture and industry. Groundwater supplies much of this water.

Estimates of water supplies are the result of a dynamic equilibrium between various factors. These factors include amounts of precipitation and infiltration, surface drainage, porosity and permeability of subsurface rock or sediment, and volume of groundwater naturally discharged back to the surface. These factors can be affected by natural processes or human activities, which can lead to issues such as those shown in **Table 2** on the next page.

Table 2 Processes and Activities That Affect Water Supply

Process or Activity	Effect on Water Supply
Overuse	If groundwater use exceeds the recharge rate, the groundwater supply will decrease and the water table will drop. Groundwater supplies can be depleted.
Subsidence	The volume of water underground helps support the weight of the soil, sediment, and rock above. When the height of the water table drops, the weight of the overlying material is increasingly transferred to the aquifer's mineral grains, which then squeeze together more tightly. Subsidence, the sinking of land, can result.
Pollution	In general, the most easily polluted groundwater reservoirs are water-table aquifers, which lack a confining layer above them. Confined aquifers are affected less frequently by local pollution because they are protected by impermeable barriers. When the recharge areas of confined aquifers are polluted, however, those aquifers can also become contaminated. Sources of groundwater pollution include sewage from faulty septic tanks and farms, landfills, and other waste disposal sites. Pollutants usually enter the ground above the water table, but they infiltrate to the water table. In highly permeable aquifers, pollutants can spread quickly in a specific direction, such as toward a well.
Chemicals	Because chemicals dissolved and transported with groundwater are submicroscopic, they can travel through the smallest pores of fine-grained sediment. For this reason, chemicals such as arsenic can contaminate any type of aquifer. When chemical and biological contaminants enter groundwater, they flow through the aquifer at the same rate as the rest of the groundwater. Over time, an entire aquifer can become toxic. Once chemical contaminants have entered groundwater, they cannot be easily removed. Aquifers are particularly vulnerable to pollution in humid areas where the water table is shallow and can more easily come in contact with waste.
Salt	Ordinary table salt is used to season food, but water is undrinkable when its salt content is too high. In a similar way, groundwater is unusable after the intrusion of salt water. Salt pollution is one of the major threats to groundwater supplies, especially in coastal areas. In coastal areas, salty seawater, which is denser than freshwater, underlies the groundwater near Earth's surface. Overpumping of wells can cause underlying salt water to rise into the wells and contaminate the freshwater aquifer.
Radon	Another source of natural groundwater pollution is radon gas. Radon is one of the products of the radioactive decay of uranium in rocks and sediment, and it usually occurs in very low concentrations in all groundwater. However, some rocks, especially granite and shale, contain more uranium than others. Therefore, the groundwater in areas where these rocks are present contains higher levels of radon.

Protecting Our Water Supply

There are a number of ways by which groundwater resources can be protected and restored. First, major pollution sources, many of which are listed in **Table 3,** need to be located, identified, and eliminated. Pollution can enter groundwater resources via runoff and infiltration or directly from underground. Pollution plumes that already exist can be monitored with observation wells and other techniques. Most pollution plumes spread slowly, providing adequate time for alternate water supplies to be found. In some cases, pollution plumes can be stopped by building impermeable underground barriers around the polluted area. Sometimes, polluted groundwater can be pumped out for chemical treatment on the surface.

While these measures can have limited success, they alone cannot save Earth's water supply. Humans must be aware of how their activities impact the groundwater system so that they can protect the water supply.

Table 3 Groundwater Pollution Sources

Infiltration from fertilizers
Leaks from storage tanks
Drainage of acid from mines
Seepage from faulty septic tanks
Saltwater intrusion into aquifers near shorelines
Leaks from waste disposal sites
Radon

Get It?

Identify the different ways pollution can enter groundwater.

Check Your Progress

Summary

- Groundwater is stored below the water table in pore spaces of rocks and sediment.

- Groundwater moves through permeable layers called aquifers and is trapped by impermeable layers called aquicludes.

- Groundwater emerges from the ground where the water table intersects Earth's surface.

- Wells are drilled into the zone of saturation to obtain water.

- Sources of groundwater pollution include sewage, landfills, and other waste disposal sites.

Demonstrate Understanding

1. **Explain** how the movement of groundwater is related to the water cycle.
2. **Illustrate** how the relative positions of an aquifer and aquiclude can result in the presence of a spring.
3. **Analyze** the factors that determine flow velocity.
4. **Evaluate** the problems associated with overpumping wells.
5. **Illustrate** the difference between an artesian well and an ordinary well.

Explain Your Thinking

6. **Infer** why it is beneficial for a community to have an aquiclude located beneath the aquifer from which it draws its water supply.
7. **Formulate** an experiment that would test for impermeable barriers around a polluted area.
8. **WRITING** **Connection** Choose one process or activity that can negatively affect the water supply, and describe one way a community can reduce or stop the effect entirely. Explain why is it important to manage water resources responsibly.

LEARNSMART® Go online to follow your personalized learning path to review, practice, and reinforce your understanding.

GROUNDWATER WEATHERING AND DEPOSITION

FOCUS QUESTION

How does groundwater interact with the land?

Carbonic Acid

Acids are aqueous solutions that contain hydrogen ions. Most groundwater is slightly acidic due to carbonic acid. Carbonic acid forms when carbon dioxide gas dissolves in water and combines with water molecules. This happens when precipitation falls through the atmosphere and interacts with carbon dioxide gas or when groundwater infiltrates the products of decaying organic matter in soil. As a result of these processes, groundwater is usually slightly acidic and attacks carbonate rocks, especially limestone. Limestone mostly consists of calcite, also called calcium carbonate, which reacts with any kind of acid.

Dissolution by Groundwater

The process by which carbonic acid forms and dissolves calcite can be described by three simple chemical reactions.

In the first reaction, carbon dioxide (CO_2) and water (H_2O) combine to form carbonic acid (H_2CO_3), as represented by the following equation.

$$CO_2 + H_2O \rightarrow H_2CO_3$$

In the second reaction, carbonic acid splits into hydrogen ions (H^+) and bicarbonate ions (HCO_3^-). This process is represented by the following equation.

$$H_2CO_3 \rightarrow H^+ + HCO_3^-$$

In the third reaction, the hydrogen ions (H^+) react with calcite ($CaCO_3$) and form calcium ions (Ca_2^+) and bicarbonate ions (HCO_3^-).

$$CaCO_3 + H^+ \rightarrow Ca_2^+ + HCO_3^-$$

3D THINKING **DCI** Disciplinary Core Ideas   **CCC** Crosscutting Concepts   **SEP** Science & Engineering Practices

COLLECT EVIDENCE

 Use your Science Journal to record the evidence you collect as you complete the readings and activities in this lesson.

INVESTIGATE

GO ONLINE to find these activities and more resources.

? **Revisit the Encounter the Phenomenon Question**
What information from this lesson can help you answer the Unit and Module questions?

CCC **Identify Crosscutting Concepts**
Create a table of the **crosscutting concepts** and fill in examples you find as you read.

The resulting calcium ions (Ca_2^+) and bicarbonate ions (HCO_3^-) are then carried away in the groundwater. Eventually, they precipitate, which means they crystallize out of the solution, somewhere else. Precipitation occurs when the groundwater evaporates or when the carbon dioxide gas leaves the water. The processes of dissolving, called dissolution, and precipitation of calcite both play a major role in the formation of limestone caves, such as those shown in **Figure 28.**

Caves

A natural underground opening with a connection to Earth's surface is called a **cave** or a cavern. Some caves form three-dimensional mazes of passages, shafts, and chambers that stretch for many kilometers. Some caves are dry, while some contain underground streams or lakes. Others are totally flooded and can be explored only by cave divers. Mammoth Cave in Kentucky, shown in **Figure 28,** is composed of a series of connected underground passages.

Most caves are formed when groundwater dissolves limestone. The development of most caves begins in the zone of saturation, just below the water table. As groundwater infiltrates the cracks and joints of limestone formations, it gradually dissolves the adjacent rock and enlarges these passages to form an interconnected network of openings. As the water table is lowered, the cave system becomes filled with air. New caves then form beneath the lowered water table. If the water table continues to drop, the thick limestone formations eventually become honeycombed with caves. This is a common occurrence in limestone regions that have been uplifted by tectonic forces.

 Get It?

Explain how most caves form.

Carlsbad Caverns, New Mexico | Mammoth Cave, Kentucky

Figure 28 The dissolution and precipitation of ions into and out of groundwater results in a variety of features in caves.

Identify *which chemical reactions might be at work.*

WORD ORIGINS

cave

comes from the Latin word *cavus* meaning "hollow"

Karst topography

Figure 29 shows some of the surface features produced by the dissolution of limestone bedrock. One of the main features is a **sinkhole**—a depression in the ground caused by the collapse of a cave or by the direct dissolution of limestone by acidic water. Another type of feature, called a disappearing stream, forms when a surface stream drains into a cave system and continues flowing underground, leaving a dry valley above. Disappearing streams sometimes reemerge on Earth's surface as karst springs.

Limestone regions that have sinkholes and disappearing streams are said to have **karst topography.** The word *karst* comes from the name of a region in Croatia where these features are especially well developed. Prominent karst regions in the United States are located in Florida, Kentucky, Indiana, and Missouri.

Figure 29 Karst topography is characterized by sinkholes formed by dissolution of limestone. This karst topography is on South Island, New Zealand.

Identify *what controls the rate of dissolution of bedrock in karst topography.*

Florida's karst topography is related to its geologic past—the state was underwater millions of years ago. The limestone that formed during this time contains the fossils of marine organisms. The rocks were exposed and later dissolved by chemical weathering. Sinkholes proliferated and grew.

In karst areas, rates of weathering depend on factors such as humidity and soil composition. In humid Florida, more water infiltrates the porous soil, and rates of weathering are high.

 Get It?

Define What is a sinkhole?

Groundwater Deposits

Calcium ions eventually precipitate from groundwater and form new calcite minerals. These minerals create spectacular natural features.

Dripstones

The most remarkable features produced by groundwater are the rock formations called dripstone that decorate many caves above the water table, as shown in **Figure 30.** These formations are built over time as water drips through caves. Each drop of water hanging on the ceiling of a cave loses some carbon dioxide and precipitates some calcite. A form of dripstone called a **stalactite** hangs from the cave's ceiling like an icicle and forms gradually. As the water drips to the floor of the cave, it may also slowly build mound-shaped dripstones called **stalagmites.**

Figure 30 These stalactites formed from buildup of minerals precipitated from groundwater.

Over time, stalactites and stalagmites can meet and grow into one another to form dripstone columns. Increasingly, researchers are finding abundant and varied microorganisms associated with dripstone formations in caves. It is possible that these organisms play an important role in the deposition of at least some of the materials found in caves.

 Get It?

Explain how dripstones and dripstone columns form.

Hard water

You are probably aware that tap water contains various dissolved solids. While some of these materials are added by water treatment facilities, others come from the dissolution of minerals in soils and subsurface rock and sediment. Water that contains high concentrations of calcium, magnesium, or iron is called hard water. Hard water is common in areas where the subsurface rock is limestone. Because limestone is made of mostly calcite, the groundwater in these areas contain significant amounts of dissolved calcite. Hard water used in households can sometimes cause problems. Just as calcite precipitates in caves, it can also precipitate in water pipes, as shown in **Figure 31,** and on the heating elements of appliances. Over time, deposits of calcite can clog water pipes and render some electrical appliances useless.

Figure 31 Hard water contains high concentrations of minerals, which leave precipitates such as these in household water pipes.

Identify *one of the likely precipitates in these pipes.*

Check Your Progress

Summary

- Groundwater dissolves limestone and forms underground caves.
- Sinkholes form at Earth's surface when bedrock is dissolved or when caves collapse.
- Irregular topography caused by groundwater dissolution is called karst topography.
- The precipitation of dissolved calcite forms stalactites and stalagmites in caves.

Demonstrate Understanding

1. **Analyze** how limestone is weathered, and identify the features that are formed as a result of this dissolution.
2. **Identify** the acid that is most common in groundwater.
3. **Illustrate** in a series of pictures how caves are formed.
4. **Examine** Why is hard water more common in some areas than others?

Explain Your Thinking

5. **Compare and contrast** the formation of stalactites and stalagmites.
6. **Describe** how you might identify an area of karst topography on a topographic map.
7. **WRITING Connection** Explain how, in a karst area, an increase in humidity could affect rates of weathering.

LEARNSMART Go online to follow your personalized learning path to review, practice, and reinforce your understanding.

Sipping Seawater, Chugging Clouds

Earth is in the midst of a global water crisis. With a population of more than 7 billion—and growing—scientists and engineers are working on innovative solutions to provide clean drinking water for every person on the planet.

Innovations in technology, such as these filters that can produce clean water, will help to solve the global water crisis.

Getting the Salt Out

According to the World Health Organization (WHO), at least 2 billion people on Earth drink contaminated water. Experts predict that by 2025, over half of the world's population will live in a region where clean drinking water is scarce. How can science and engineering help solve this problem?

One potential solution is Earth's largest source of liquid water—the ocean. Through the process of desalination, salt and other minerals can be removed from seawater. Most desalination plants use high-tech membranes to filter out impurities from seawater. The process is effective, but it's costly and it comes with an added environmental price. Desalination consumes a lot of energy and releases highly concentrated salt back into marine ecosystems. Engineers continue to work on ways to improve the membranes used in the process, making them more efficient at removing salt. Other solutions include using clean energy to power desalination plants.

Looking High and Low for Solutions

Some engineers have looked to another Earth system as a source of water—the atmosphere. Among the technologies being developed for this purpose are billboards and towers that condense humidity and specialized nets that catch and condense dense mountain fog. The nets can be as large as 600 square meters and can generate nearly 64 L of clean water for each square meter of net.

In areas where water is plentiful but not clean, filters are one answer to the problem. Filters that look like large drinking straws are placed directly into a river or other water source. As a person sucks the water up, it moves through the filters and is cleaned. Although small and light, these filters can remove a host of harmful particles like viruses and bacteria.

While new kinds of technology are key to solving the global water crisis, innovation is only part of the answer. Education, water and waste management, and good policies must also play a role. Scientists and engineers are hopeful that by working together, they can help ensure that everyone will have access to safe, clean drinking water.

COMMUNICATE TECHNICAL INFORMATION

Choose a type of clean-water technology and research it further. Create an infographic to show how it works to provide clean drinking water.

The Disappearance of Lake Chad

Straddling four international borders in west-central Africa, Lake Chad is important to the countries of Niger, Nigeria, Cameroon, and Chad. It is a large, shallow, freshwater lake that has no outflow, meaning the waters do not leave the area to reach the ocean. Despite not having an outflow, the lake has been disappearing.

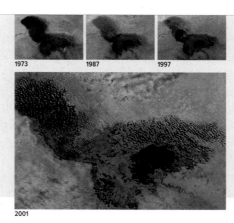

1973   1987   1997

2001

Vanishing act

The lake has always fluctuated in size due to the seasons, but from about 1963 to 1998, Lake Chad shrank overall by almost 95 percent. Much of the northern part of the lake has become wetlands. The big question is why. Increased demand due to population growth is thought to be half the reason. Studies of the area surrounding the lake say the lake's reduction is caused by overgrazing, which causes desertification, and inefficient damming and irrigation methods. Changing climate patterns are another cause. The thought is that air pollution from Europe has shifted rainfall patterns farther south, causing the region to become drier.

Hope for the future?

Now that the European Union has implemented regulations concerning air pollutants, some of the rainfall has started to return. This has led to a slight increase in the size of the lake from 2001 to 2007.

The upper photos have vegetation colored red, the lower photo has vegetation colored green. The lake shrank by almost 95 percent in 35 years.

However, this slight increase is not enough to help the Lake Chad region, so international organizations have been discussing ways to help. One initiative is educating the local farmers and herders on efficient ways of using the water. There is also talk of building canals to pump water from the Congo River to the Chari River, the main source of water for the lake. This project was first suggested in 1982 but met with resistance due to its size and the political tension in the area. Drought and extreme poverty have led to almost continuous civil unrest and war. Many people have become displaced refugees due to the disappearance of the lake. The political tension has only worsened, but the need is so great, governments and aid organizations hope that something can be done soon.

Goddard Space Flight Center Scientific Visualization Studio/NASA

OBTAIN AND COMMUNICATE INFORMATION

Investigate the current initiatives created to mitigate the Lake Chad crisis. Write a short description of one of these initiatives.

GO ONLINE to study with your Science Notebook.

Lesson 1 SURFACE WATER MOVEMENT

- Infiltration of water into the ground depends on the number of open pores.
- All the land area that drains into a stream system is the system's watershed.
- Elevated land areas called divides separate one watershed from another.
- A stream's load is the material the stream carries.
- Flooding occurs as upstream floods in small, localized areas or as large downstream floods.

- runoff
- watershed
- divide
- suspension
- bed load
- discharge
- flood
- floodplain

Lesson 2 STREAMS, LAKES, AND WETLANDS

- Stream water flows in channels confined by the stream's banks.
- Alluvial fans and deltas form when stream velocity decreases and sediment is deposited.
- Lakes form in a variety of ways when depressions on land fill with water.
- Wetlands are low-lying areas that are periodically saturated with water.

- stream channel
- stream bank
- base level
- meander
- delta
- rejuvenation
- lake
- eutrophication
- wetland

Lesson 3 GROUNDWATER

- Groundwater is stored below the water table in pore spaces of rocks and sediment.
- Groundwater moves through permeable layers called aquifers and is trapped by impermeable layers called aquicludes.
- Groundwater emerges from the ground where the water table intersects Earth's surface.
- Wells are drilled into the zone of saturation to obtain water.
- Sources of groundwater pollution include sewage, landfills, and other waste disposal sites.

- infiltration
- zone of saturation
- water table
- zone of aeration
- permeability
- aquifer
- aquiclude
- spring
- hot spring
- geyser
- well
- drawdown
- recharge
- artesian well

Lesson 4 GROUNDWATER WEATHERING AND DEPOSITION

- Groundwater dissolves limestone and forms underground caves.
- Sinkholes form at Earth's surface when bedrock is dissolved or when caves collapse.
- Irregular topography caused by groundwater dissolution is called karst topography.
- The precipitation of dissolved calcite forms stalactites and stalagmites in caves.

- cave
- sinkhole
- karst topography
- stalactite
- stalagmite

REVISIT THE PHENOMENON

How does rain water get into this underground cave?

CER Claim, Evidence, Reasoning

Explain Your Reasoning Revisit the claim you made when you encountered the phenomenon. Summarize the evidence you gathered from your investigations and research and finalize your Summary Table. Does your evidence support your claim? If not, revise your claim. Explain why your evidence supports your claim.

STEM UNIT PROJECT
Now that you've completed the module, revisit your STEM unit project. You will apply your evidence from this module and complete your project.

GO FURTHER

SEP Data Analysis Lab
How do sediments move in a stream?

The critical velocity of water determines the size of particles that can be moved. The higher the stream velocity, the larger the particles that can be transported.

CER Analyze and Interpret Data

1. **Claim, Evidence** At what velocity would flowing water pick up a pebble? Use evidence from the graph to support your claim.

2. **Claim, Evidence** At what range of velocities would flowing water carry a pebble? Use evidence from the graph to support your claim.

3. **Reasoning** Which of the following objects would not fall into the same size range as a pebble: an egg, a baseball, a golf ball, a table tennis ball, a volleyball, and a pea. How would you test your conclusions?

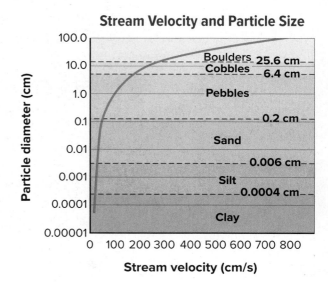

Stream Velocity and Particle Size

Boulders 25.6 cm
Cobbles 6.4 cm
Pebbles
0.2 cm
Sand
0.006 cm
Silt
0.0004 cm
Clay

Particle diameter (cm): 100.0, 10.0, 1.0, 0.1, 0.01, 0.001, 0.0001, 0.00001

Stream velocity (cm/s): 0, 100, 200, 300, 400, 500, 600, 700, 800

ENCOUNTER THE PHENOMENON

How are animals in the ocean affected by things that happen in the sky?

SEP Ask Questions

What questions do you have about the phenomenon? Write your questions on sticky notes and add them to the driving question board for this unit.

What influences climate?

Look for Evidence

As you go through this unit, use the information and your experiences to help you answer the phenomenon question as well as your own questions. For each activity, record your observations in a Summary Table, add an explanation, and identify how it connects to the unit and module phenomenon questions.

Solve a Problem
STEM UNIT PROJECT

The Atmosphere and the Oceans Investigate and research more about Earth's atmosphere and oceans. Use the results of these investigations and the evidence you collected during the unit to complete your unit project.

GO ONLINE In addition to reading the information in your Student Edition, you can find the STEM Unit Project and other useful resources online.

ENCOUNTER THE PHENOMENON

How do clouds and mist form?

GO ONLINE to play a video about the formation of clouds.

SEP Ask Questions

Do you have other questions about the phenomenon? If so, add them to the driving question board.

CER Claim, Evidence, Reasoning

Make Your Claim Use your CER chart to make a claim about how clouds and mist form. Explain your reasoning.

Collect Evidence Use the lessons in this module to collect evidence to support your claim. Record your evidence as you move through the module.

Explain Your Reasoning You will revisit your claim and explain your reasoning at the end of the module.

GO ONLINE to access your CER chart and explore resources that can help you collect evidence.

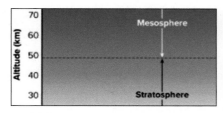

LESSON 2: Explore and Explain: Air Pressure

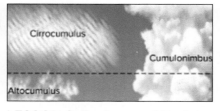

LESSON 3: Explore and Explain: Types of Clouds

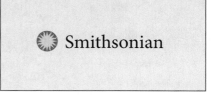

Additional Resources

FOCUS QUESTION

Where in the atmosphere do clouds form?

Atmospheric Composition

Air is a combination of gases, such as nitrogen and oxygen, and particles, such as dust, water droplets, and ice crystals. These gases and particles form Earth's atmosphere, which surrounds Earth and extends from Earth's surface to outer space.

Permanent atmospheric gases

About 99 percent of the atmosphere is composed of nitrogen (N_2) and oxygen (O_2). The remaining 1 percent consists of argon (Ar), carbon dioxide (CO_2), water vapor (H_2O), and other trace gases, as shown in **Figure 1.** Over Earth's history, the composition of the atmosphere has changed greatly. For example, Earth's early atmosphere probably contained mostly helium (He), hydrogen (H_2), methane (CH_4), and ammonia (NH_3). Today, oxygen and nitrogen are continually recycled among Earth's systems.

Variable atmospheric gases

The concentrations of some atmospheric gases are not constant over time. Gases such as water vapor and ozone (O_3) can vary significantly from place to place. Water vapor and carbon dioxide in particular play important roles in regulating the amount of energy the atmosphere absorbs and emits back to Earth's surface.

Water vapor Water vapor is the invisible, gaseous form of water. The amount of water vapor in the atmosphere can vary greatly over time and from one place to another. For example, the amount of water vapor in the air over a desert is much less than the amount of water vapor in the air over a tropical rain forest. At a given place and time, the concentration of water vapor can be as much as 4 percent or as little as nearly zero.

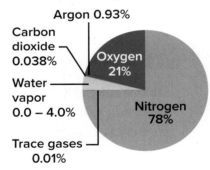

Composition of Earth's Atmosphere

Argon 0.93%
Carbon dioxide 0.038%
Water vapor 0.0 – 4.0%
Trace gases 0.01%
Oxygen 21%
Nitrogen 78%

Figure 1 Earth's atmosphere consists mainly of nitrogen (78 percent) and oxygen (21 percent).

 3D THINKING   **DCI** Disciplinary Core Ideas   **CCC** Crosscutting Concepts   **SEP** Science & Engineering Practices

COLLECT EVIDENCE

 Use your Science Journal to record the evidence you collect as you complete the readings and activities in this lesson.

INVESTIGATE

 GO ONLINE to find these activities and more resources.

Design Your Own: What is in the air?
Plan and carry out an investigation to discover the effect of weather and atmospheric changes on the concentrations of particulates in the air.

Virtual Investigation: Earth's Atmosphere
Carry out an investigation to discover the structure of the atmosphere and how the layers interact.

In addition to varying with location, the concentration of water vapor in the atmosphere varies with the seasons, with the altitude of a body of air, and with the properties of the surface beneath the air.

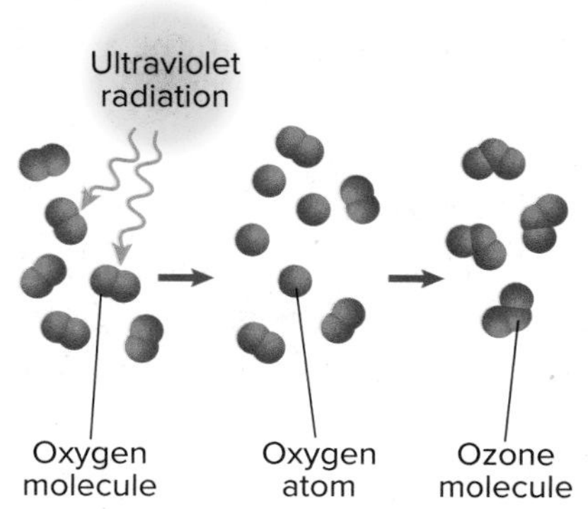

Carbon dioxide

Carbon dioxide, another variable gas, currently makes up about 0.038 percent of the atmosphere. During the past 150 years, measurements have shown that the concentration of atmospheric carbon dioxide has increased from about 0.028 percent to its present value. Carbon dioxide is also cycled among Earth's systems.

The recent increase in atmospheric carbon dioxide is due primarily to the burning of fossil fuels. These fuels are burned to heat buildings, produce electricity, and power vehicles. Burning fossil fuels can also produce other gases, such as sulfur dioxide and nitrogen oxides, that can cause respiratory illnesses as well as environmental problems.

Ozone

Molecules of ozone are formed by the addition of an oxygen atom to an oxygen molecule, as shown in **Figure 2.** Most atmospheric ozone is present in the ozone layer, 20 km to 50 km above Earth's surface, as shown in **Figure 3.** The maximum concentration of ozone in this layer—9.8×10^{12} molecules/cm^3—is only about 0.0012 percent of the atmosphere.

The ozone concentration in the ozone layer varies seasonally at higher latitudes, reaching a minimum in the spring. The greatest seasonal changes occur over Antarctica. During the past several decades, measured ozone levels over Antarctica in the spring have dropped significantly. This decrease is due to the presence of chemicals called chlorofluorocarbons (CFCs) that react with ozone and break it down in the atmosphere.

Atmospheric particles Earth's atmosphere also contains variable amounts of solids in the form of tiny particles, such as dust, salt, and ice. Dust and soil are carried into the atmosphere by wind. Winds also pick up salt particles from ocean spray.

Ultraviolet radiation

Oxygen molecule · Oxygen atom · Ozone molecule

Figure 2 Molecules of ozone are formed by the addition of an oxygen atom to an oxygen molecule.

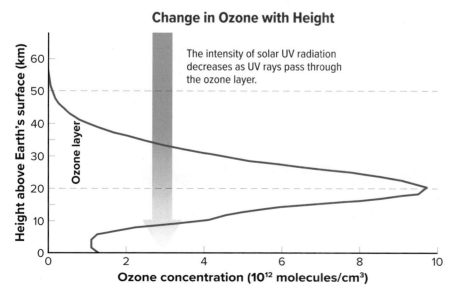

Change in Ozone with Height

The intensity of solar UV radiation decreases as UV rays pass through the ozone layer.

Ozone layer

Height above Earth's surface (km)

Ozone concentration (10^{12} molecules/cm^3)

Figure 3 The ozone layer prevents harmful ultraviolet rays from reaching Earth's surface. Ozone concentration is highest at about 20 km above Earth's surface, in the ozone layer.

Airborne microorganisms, such as fungi and bacteria, can also be found attached to microscopic dust particles in the atmosphere.

Atmospheric Layers

The atmosphere is divided into five different layers, as shown in **Table 1** and **Figure 4.** These layers are the troposphere, stratosphere, mesosphere, thermosphere, and exosphere. Each layer differs in composition and temperature profile.

Troposphere

The layer closest to Earth's surface, the **troposphere,** contains most of the mass of the atmosphere. Weather occurs in the troposphere. In the troposphere, air temperature decreases as altitude increases. The altitude at which the temperature stops decreasing is called the tropopause. The height of the tropopause varies from about 16 km above Earth's surface in the tropics to about 9 km above it at the poles. Temperatures at the tropopause can be as low as –60°C.

Stratosphere

Above the tropopause is the **stratosphere,** a layer that contains the ozone layer and in which the air temperature mainly increases with altitude. In the lower stratosphere below the ozone layer, the temperature stays constant with altitude. However, starting at the bottom of the ozone layer, the temperature in the stratosphere increases as altitude increases. This heating is caused by ozone molecules, which absorb ultraviolet radiation from the Sun.

At the stratopause, air temperature stops increasing with altitude. The stratopause is about 50 km above Earth's surface. About 99.9 percent of the mass of Earth's atmosphere is below the stratopause.

Mesosphere

Above the stratopause is the **mesosphere,** which is about 50 km to 85 km above Earth's surface. In the mesosphere, air temperature decreases with altitude, as shown in **Figure 4.** This temperature decrease occurs because very little solar radiation is absorbed in this layer. The top of the mesosphere, where temperatures stop decreasing with altitude, is called the mesopause.

Thermosphere

The **thermosphere** is the layer between about 85 km and 600 km above Earth's surface. This layer is hotter than the others because it absorbs most of the high-energy radiation, like x-rays, from the Sun. Temperatures in this layer can be up to 2000°C. The ionosphere, which is made of electrically charged particles, is part of the thermosphere.

Table 1 Components of the Atmosphere

Atmospheric Layer	Components
Exosphere	outermost layer of Earth's atmosphere, transitional space between Earth's atmosphere and outer space
Thermosphere	layer above the mesosphere, absorbs solar radiation
Mesosphere	layer above the stratosphere, ends at the mesopause
Stratosphere	layer above the troposphere, contains the ozone layer and ends at the stratopause
Troposphere	layer closest to Earth's surface, ends at the tropopause

Figure 4 Visualizing the Layers of the Atmosphere

Earth's atmosphere is made up of five layers. Each layer is unique in composition and temperature. As shown, air temperature changes with altitude. When you fly in a plane, you might be flying at the top of the troposphere, or you might enter into the stratosphere.

In the exosphere, gas molecules can be exchanged between the atmosphere and space.

Noctilucent clouds are shiny clouds that can be seen in the twilight in the summer around 50°–60° latitude in the northern and southern hemispheres. These are the only clouds that form in the mesosphere.

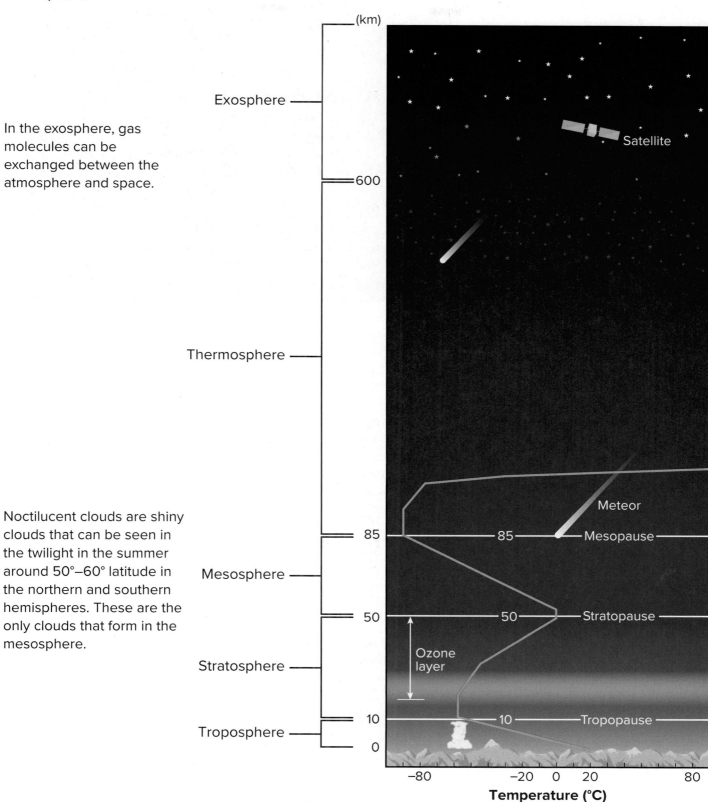

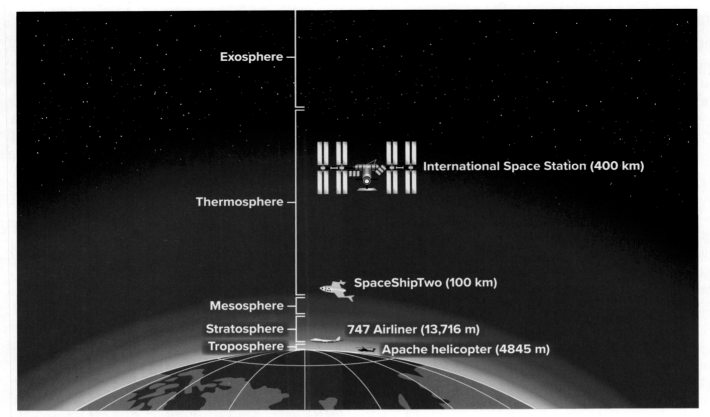

Figure 5 Different spacecraft can traverse the various layers of the atmosphere.

Compare *the number of atmospheric layers each spacecraft can reach in its flight path.*

Exosphere

The **exosphere** is the outermost layer of Earth's atmosphere, as shown in **Figure 5.** The exosphere extends from about 600 km to more than 10,000 km above Earth's surface. There is no clear boundary at the top of the exosphere. Instead, the exosphere can be thought of as the transitional region between Earth's atmosphere and outer space. The number of atoms and molecules in the exosphere becomes very small as altitude increases.

In the exosphere, atoms and molecules are so far apart that they rarely collide with each other. In this layer, some atoms and molecules move fast enough that they are able to escape into outer space.

Energy Transfer in the Atmosphere

All materials are made of particles, such as atoms and molecules. These particles are always moving, even if the object is not moving. The particles move in all directions with various speeds—a type of motion called random motion. A moving object has a form of energy called kinetic energy. As a result, the particles moving in random motion have kinetic energy. The total energy of the particles in an object due to their random motion is called thermal energy.

Heat is the transfer of thermal energy from a region of higher temperature to a region of lower temperature. In the atmosphere, thermal energy can be transferred by radiation, conduction, and convection.

Radiation

Just as the Sun warms Earth, the heat lamp shown in **Figure 6** uses the process of radiation to warm the iguana. **Radiation** is the transfer of thermal energy by electromagnetic waves. The heat lamp emits visible light and infrared waves that travel from the lamp and are absorbed by the iguana. The thermal energy carried by these waves increases the body temperature of the animal. In the same way, thermal energy is transferred from the Sun to Earth by electromagnetic radiation. The solar radiation that reaches Earth is absorbed and reflected by the atmosphere and Earth's surface.

Figure 6 A heat lamp transfers thermal energy by radiation. Here, the thermal energy helps to keep the iguana warm.

Absorption and reflection Most of the solar energy that reaches Earth is in the form of visible light waves and infrared waves. Almost all of the visible light waves pass through the atmosphere and strike Earth's surface, where they are absorbed. As the surface absorbs these visible light waves, it also emits infrared waves. The atmosphere absorbs some infrared waves from the Sun and emits infrared waves with different wavelengths, as shown in **Figure 7.** About 30 percent of solar radiation is reflected into space by Earth's surface, the atmosphere, or clouds. Another 20 percent is absorbed by the atmosphere and clouds. About 50 percent of solar radiation is absorbed directly or indirectly by Earth's surface and keeps Earth's surface warm.

Rates of absorption The rate of absorption for any particular area depends on the physical characteristics of the area and the amount of solar radiation it receives. Different areas absorb energy and heat at different rates. For example, water heats and cools more slowly than land.

Figure 7 Incoming solar radiation is either reflected back into space or absorbed by Earth's atmosphere or its surface.
Trace *the pathways by which solar radiation is absorbed and reflected.*

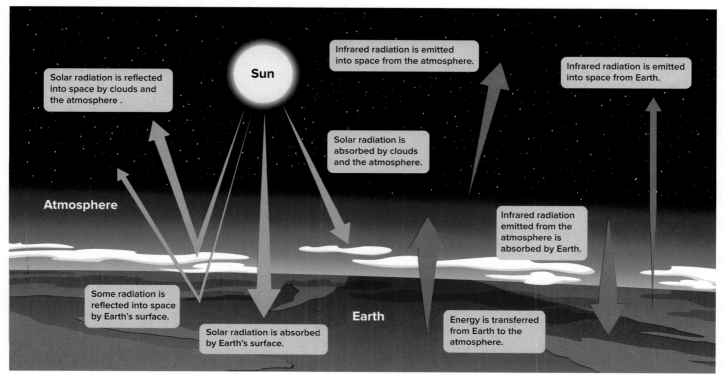

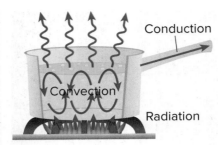

Figure 8 Thermal energy is transferred to the burner from the heat source by radiation. The burner transfers the energy to the atoms in the bottom of the pot. The atoms collide, transferring energy by conduction to other parts of the pot, including the handle.

Also, as a general rule, darker objects absorb energy faster than light-colored objects. For instance, a black asphalt driveway heats faster on a sunny day than a light-colored concrete driveway.

Conduction

Energy transfer can also occur when two objects at different temperatures are in contact. **Conduction** is the transfer of thermal energy between objects when their atoms or molecules collide, as shown in **Figure 8.** Conduction can occur more easily in solids and liquids, where particles are close together, than in gases, where particles are farther apart.

Because air is a mixture of gases, it is a poor conductor of thermal energy. In the atmosphere, conduction occurs between Earth's surface and the lowest part of the atmosphere. Conduction requires molecular contact. Energy is transferred through direct contact between Earth's surface and the atmosphere.

 Get It?

Infer How is conduction related to cold air temperatures at the poles, which are covered with ice and snow?

Convection

Throughout much of the atmosphere, thermal energy is transferred by a process called convection. **Convection** is the transfer of thermal energy by the movement of heated material from one place to another. This process occurs mainly in liquids and gases. **Figure 8** illustrates the process of convection in a pot of water.

A similar process occurs in the atmosphere. Parcels of air near Earth's surface are heated, become less dense than the surrounding air, and rise. As the warm air rises, it cools and its density increases. When it cools below the temperature of the surrounding air, the air parcel becomes denser than the air around it and sinks. As it sinks, it warms again, and the process repeats. These movements of air, called convection currents, are the main mechanism for energy transfer in the atmosphere. Convection is also involved in the redistribution of thermal energy among the atmosphere, oceans, and land.

The transfer of energy by convection distributes energy through the movement of a fluid until equilibrium is reached within that fluid. Earth's atmosphere is such a large volume of fluid that equilibrium is never attained.

 Get It?

Model How could you make a model to show the process of convection? What materials would you use? What would each component of your model represent?

How the Atmosphere Retains Heat

Earth's atmosphere significantly influences its temperature, weather, and climate. Solar radiation that is not reflected by clouds passes freely through the atmosphere. It is then absorbed by Earth's surface and released as infrared radiation. This radiation is absorbed by atmospheric gases. Some of this absorbed energy is reradiated back to Earth's surface.

This process—the absorption and radiation of energy in the atmosphere—results in the greenhouse effect, shown in **Figure 9**. The **greenhouse effect** is the natural heating of Earth's surface caused by certain atmospheric gases called greenhouse gases, including water vapor, methane, and carbon dioxide. The greenhouse effect keeps some of the radiation and heat from the Sun from radiating back into space. This process warms Earth's surface by more than 30°C. Without the greenhouse effect, life as it currently exists on Earth would not be possible.

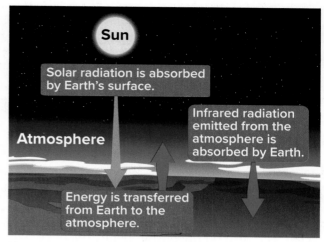

Figure 9 The greenhouse effect helps keep temperatures on Earth relatively stable and at optimal levels for life.

 Get It?

Synthesize What would conditions be like on Earth if the greenhouse effect did not exist? Mention different Earth systems in your answer.

Check Your Progress

Summary

- Earth's atmosphere is composed of several gases, primarily nitrogen and oxygen, and also contains small particles.
- Earth's atmosphere is divided of five layers that differ in their compositions and temperatures.
- Solar energy reaches Earth's surface in the form of visible light and infrared waves.
- Solar energy absorbed by Earth's surface is transferred as thermal energy throughout the atmosphere.
- Earth's surface is naturally heated by the greenhouse effect.

Demonstrate Understanding

1. **Rank** the gases in the atmosphere in order from most abundant to least abundant.
2. **Name** the four types of particles found in the atmosphere.
3. **Compare and contrast** the five layers that make up the atmosphere.
4. **Explain** why temperature increases with height in the stratosphere.
5. **Describe** how solar radiation reaches Earth and is absorbed and emitted by Earth's atmosphere and surface.

Explain Your Thinking

6. **Predict** whether a pot of water heated from the top would boil more quickly than a pot of water heated from the bottom. Explain your answer.
7. **MATH Connection** In the troposphere, temperature decreases with height at an average rate of 6.5°C/km. If the temperature at 2.5 km altitude is 7.0°C, what is the temperature at 5.5 km altitude?

LEARNSMART Go online to follow your personalized learning path to review, practice, and reinforce your understanding.

PROPERTIES OF THE ATMOSPHERE

FOCUS QUESTION

How do water vapor and the atmosphere interact?

Temperature

When you turn on the burner beneath a pot of water, thermal energy is transferred to the water, and the temperature increases. Recall that particles in any material are in random motion. Temperature is a measure of the average kinetic energy of the particles in a material. Particles have more kinetic energy when they are moving faster, so the higher the temperature of a material, the faster the particles are moving.

Measuring temperature

Temperature is usually measured using one of two common temperature scales: the Fahrenheit (°F) scale, used mainly in the United States, and the Celsius (°C) scale. The SI temperature scale used in science is the Kelvin (K) scale. **Figure 10** shows the differences among these temperature scales. The Fahrenheit and Celsius scales are based on the freezing point and boiling point of water. The zero point of the Kelvin scale is absolute zero—the lowest temperature that any substance can have.

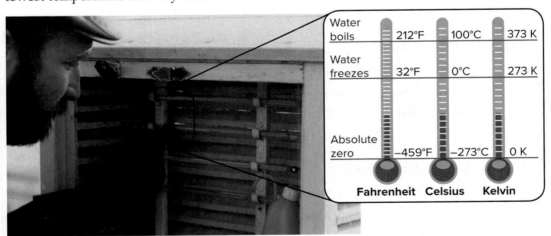

	Fahrenheit	Celsius	Kelvin
Water boils	212°F	100°C	373 K
Water freezes	32°F	0°C	273 K
Absolute zero	−459°F	−273°C	0 K

Figure 10 Temperature can be measured in degrees Fahrenheit, degrees Celsius, or in Kelvin. The Kelvin scale starts at 0 K, which corresponds to −273°C and −459°F.

3D THINKING   **DCI** Disciplinary Core Ideas   **CCC** Crosscutting Concepts   **SEP** Science & Engineering Practices

COLLECT EVIDENCE

 Use your Science Journal to record the evidence you collect as you complete the readings and activities in this lesson.

INVESTIGATE

 GO ONLINE to find these activities and more resources.

GeoLAB: Interpret Pressure-Temperature Relationships
Develop and use a model to visualize the effect of the dynamic interaction between temperature and pressure on the atmosphere.

Investigation Lab: Temperature Inversion
Analyze and interpret data to visualize the effect of temperature inversions on air quality and humans.

Air Pressure

If you hold your hand out in front of you, Earth's atmosphere exerts a downward force on your hand due to the weight of the atmosphere above it. The force exerted on your hand is air pressure. Air pressure is the pressure exerted on a surface by the weight of the atmosphere above the surface. Pressure is equal to force divided by area, so the units for pressure are Newtons per square meter (N/m^2). Air pressure is often measured in units of millibars (mb), where 1 mb equals 100 N/m^2. At sea level, the atmosphere exerts a pressure of approximately 100,000 N/m^2, or 1000 mb. As you go higher in the atmosphere, air pressure decreases as the mass of the air above you decreases. **Figure 11** shows how pressure in the atmosphere changes with altitude.

Density of air

The density of a material is the mass of material in a unit volume. Atoms and molecules become farther apart in the atmosphere as altitude increases. This means that the density of air decreases with increasing altitude, as shown in **Figure 11.** Near sea level, the density of air is about 1.2 kg/m^3. At the average altitude of the tropopause, or about 12 km above Earth's surface, the density of air is about 25 percent of its sea-level value. At the stratopause, about 48 km above Earth's surface, air density decreases to about 0.2 percent of sea-level value.

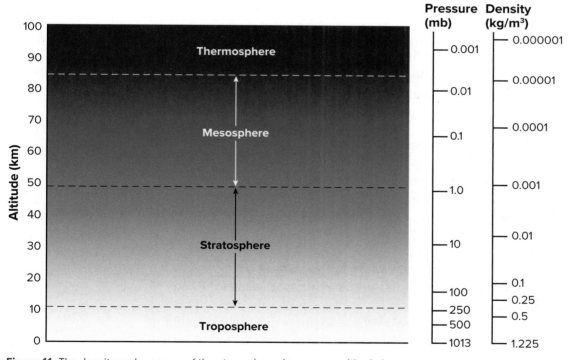

Figure 11 The density and pressure of the atmosphere decrease as altitude increases.

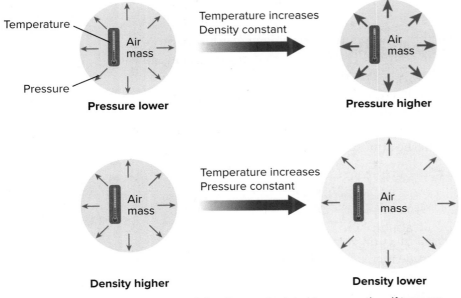

Temperature

Air mass

Pressure

Pressure lower

Temperature increases
Density constant

Air mass

Pressure higher

Air mass

Density higher

Temperature increases
Pressure constant

Air mass

Density lower

Figure 12 Temperature, pressure, and density are all related to one another. If temperature increases and density is constant, pressure increases. If temperature increases and pressure is constant, density decreases.

Pressure-temperature-density relationship

In the atmosphere, the temperature, pressure, and density of air are related to each other, as shown in **Figure 12.** Imagine a sealed container that holds only air. The pressure exerted by the air inside the container is related to the air temperature inside the container and the air density. Now imagine that the air temperature or density change. How will this affect air pressure inside the container?

Air pressure and temperature The pressure exerted by air in the container is due to the collisions of gas particles in the air with the sides of the container. When these particles move faster because of an increase in temperature, they exert a greater force when they collide with the sides of the container. Therefore, air pressure inside the container increases. This means that for air with the same density, warmer air is at a higher pressure than cooler air.

Air pressure and density Imagine that the temperature of the air does not change, but that more air is pumped into the container. Now there are more gas particles in the container; therefore, the mass of the air in the container has increased. Because the volume has not changed, the density of the air has increased. Now there are more gas particles colliding with the walls of the container, so more force is being exerted by the particles on the walls. This means that at the same temperature, air with a higher density exerts more pressure than air with a lower density.

 Get It?

Deduce why air pressure does not crush a human.

Temperature and density Heating a balloon makes the air inside the balloon move faster, causing the balloon to expand and increase in volume. As a result, the air density inside the balloon decreases. The same is true for air masses in the atmosphere. At the same pressure, warmer air is less dense than cooler air.

Temperature inversion

In the troposphere, air temperature decreases as height increases. However, over a localized region in the troposphere, a temperature inversion can sometimes occur. A **temperature inversion** is an increase in temperature with height in an atmospheric layer. In other words, when a temperature inversion occurs, warmer air is on top of cooler air. The temperature-altitude relationship is inverted, or turned upside down, as shown in **Figure 13.**

Causes of temperature inversion A temperature inversion in the troposphere can occur when land rapidly cools on a cold, clear, winter night with calm air. Under these conditions, the land does not radiate thermal energy to the lower layers of the atmosphere. As a result, the lower layers of air become cooler than the air above them. Temperature increases with height, forming a temperature inversion.

Effects of temperature inversion If the sky is very hazy, there is probably an inversion somewhere in the lower atmosphere. A temperature inversion can lead to fog or low-level clouds. Fog is a significant factor in lowering visibility in many coastal cities, such as San Francisco. In some cities, such as the one shown in **Figure 14,** a temperature inversion can worsen air-pollution problems. Because air rises as long as it is warmer than the air above it, the cool pollution-filled air becomes trapped under the warm layer. Pollutants are consequently unable to be lifted from Earth's surface.

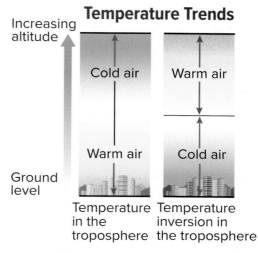

Temperature Trends

Increasing altitude

Cold air / Warm air

Warm air / Cold air

Ground level

Temperature in the troposphere

Temperature inversion in the troposphere

Figure 13 In a temperature inversion, warm air is located on top of cooler air.

Figure 14 A temperature inversion in Los Angeles traps air pollution above the city.
Describe *the effect of a temperature inversion on air quality in a metropolitan area.*

Temperature inversions that remain over an industrial area for a long time usually result in episodes of severe smog—a type of air pollution—and can cause respiratory problems.

Wind

Imagine you are entering a large, air-conditioned building on a hot summer day. As you open the door, you feel cool air rushing past you out of the building. This sudden rush of cool air occurs because the warm air outside the building is less dense and at a lower pressure than the cooler air inside the building. When the door opens, the difference in pressure causes the cool, dense air to rush out of the building. The movement of air is commonly known as wind.

Wind and pressure differences In the example above, the air in the building moves from a region of higher density to a region of lower density. In the lower atmosphere, air also generally moves from regions of higher density to regions of lower density.

These density differences are produced by the unequal heating and cooling of different regions of Earth's surface. In the atmosphere, air pressure generally increases as density increases, so regions of high and low density are also regions of high and low air pressures respectively. As a result, air moves from a region of high pressure to a region of low pressure, resulting in wind.

Wind speed and altitude Wind speed and direction change with height in the atmosphere. Near Earth's surface, wind is constantly slowed by the friction that results from contact with forms such as trees, buildings, and hills, as shown **Figure 15.** Even the surface of water affects air motion. Higher up from Earth's surface, air encounters less friction and wind speeds increase.

Wind speed is usually measured in miles per hour (mph) or kilometers per hour (km/h). Ships at sea usually measure wind in knots. One knot is equal to 1.85 km/h.

Figure 15 When wind blows over a forested area by a coast, it encounters more friction than when it blows over flatter terrain. This occurs because the wind encounters friction from the mountains, trees, and water, slowing the wind's speed.

Humidity

The distribution and movement of water vapor in the atmosphere play an important role in determining the weather of any region. **Humidity** is the amount of water vapor in the atmosphere at a given location on Earth's surface. Two ways of expressing the water vapor content of the atmosphere are relative humidity and dew point.

Relative humidity

Imagine a flask containing water. Some water molecules evaporate, leaving the liquid and becoming part of the water vapor in the flask. At the same time, other water molecules condense, returning from the water vapor to become part of the liquid. Just as the amount of water vapor in the flask might vary, so does the amount of water vapor in the atmosphere. Water on Earth's surface evaporates and enters the atmosphere and condenses to form clouds and precipitation.

In the example of the flask, if the rate of evaporation is greater than the rate of condensation, the amount of water vapor in the flask increases. **Saturation** occurs when the amount of water vapor in a volume of air has reached the maximum amount. A saturated solution cannot hold any more of the substance that is being added to it. When a volume of air is saturated, it cannot hold any more water.

The amount of water vapor in a volume of air relative to the amount of water vapor needed for that volume of air to reach saturation is called **relative humidity.** Relative humidity is expressed as a percentage. When a certain volume of air is saturated, its relative humidity is 100 percent. The volume of air cannot contain any more water vapor. If you hear a weather forecaster say that the relative humidity is 50 percent, it means that the air contains 50 percent of the water vapor needed for the air to be saturated.

 Get It?

Predict Rates of evaporation are higher than rates of condensation over a body of warm water. Would you expect the relative humidity of the air over the water to be high or low? Explain.

Dew point Another common way of describing the moisture content of air is the dew point. The **dew point** is the temperature to which air must be cooled at constant pressure to reach saturation. The term *dew point* comes from the fact that when the temperature falls to this level, water vapor condenses and begins to form dew. Dew forms when moist air near the ground cools and the water vapor in the air condenses into water droplets. If the dew point is nearly the same as the air temperature, then the relative humidity is high. If the values for dew point and air temperature are far apart, then relative humidity is low. Most meteorologists refer to dew point when describing the moisture content of air.

Latent heat As water vapor in the air condenses, thermal energy is released. Where does this energy come from? To change liquid water to water vapor, thermal energy is added to the water by heating it. The water vapor then contains more thermal energy than the liquid water. This energy is then released when the water vapor condenses. The extra thermal energy contained in water vapor compared to liquid water is called **latent heat.** As **Figure 16** shows, evaporation and condensation are opposite processes that happen at the same time. During evaporation, water molecules escape from the surface of the liquid and enter the air as water vapor. During condensation, water molecules return to a liquid state. In a closed system, these processes eventually reach equilibrium.

Get It?

Infer what happens to thermal energy when water droplets in the air evaporate.

When condensation occurs in the atmosphere, latent heat is released and warms the air. At any given time, the amount of water vapor present in the atmosphere is a significant source of energy because it contains latent heat. When water vapor condenses, the latent heat released can provide energy to a weather system, such as a hurricane, increasing its intensity.

Condensation level

An air mass can change temperature without being heated or cooled. A process in which temperature changes without the addition or removal of thermal energy from a system is called an adiabatic process. An example of an adiabatic process is the heating of air in a bicycle pump as the air is compressed. In a similar way, an air mass heats as it sinks and cools as it rises. Adiabatic heating occurs when air is compressed, and adiabatic cooling occurs when air expands.

A rising air mass cools because the air pressure around it decreases as it rises, causing the air mass to expand. A rising air mass that does not exchange thermal energy with its surroundings will cool by about 10°C for every 1000 m it rises.

Evaporation-Condensation Equilibrium

Time 1
25°C

Water molecules
begin to
evaporate.

Time 2
25°C

Evaporation continues,
and condensation
begins.

Time 3
25°C

Rate of evaporation
equals rate of
condensation or saturation.

Figure 16 At equilibrium, evaporation and condensation continue, but the amount of water in the air and amount of water in liquid form remain constant.

ACADEMIC VOCABULARY

equilibrium
a state of balance between opposing forces or processes
When the net force on an object is zero, the object is in equilibrium.

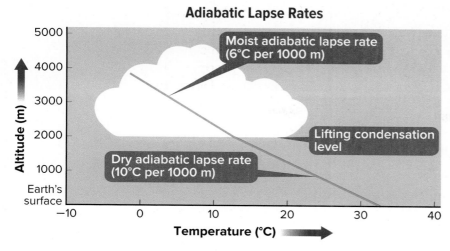

Adiabatic Lapse Rates

Moist adiabatic lapse rate (6°C per 1000 m)

Lifting condensation level

Dry adiabatic lapse rate (10°C per 1000 m)

Figure 17 Condensation occurs at the lifting condensation level (LCL). Air above the LCL is saturated and thus cools more slowly than air below the LCL.

This is called the dry adiabatic lapse rate—the rate at which unsaturated air will cool as it rises if no thermal energy is added or removed. If the air mass continues to rise, eventually it will reach saturation, and condensation will occur. The height at which condensation occurs is called the lifting condensation level (LCL).

The rate at which saturated air cools is called the moist adiabatic lapse rate. This rate ranges from about 4°C/1000 m in very warm air to almost 9°C/1000 m in very cold air. The moist adiabatic rate is slower than the dry adiabatic rate, as shown in **Figure 17,** because water vapor in the air is condensing as the air rises and is releasing latent heat.

Get It?

Explain why air above the LCL cools more slowly than air below the LCL.

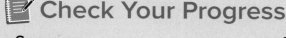

Check Your Progress

Summary

- At the same pressure, warmer air is less dense than cooler air.

- Air moves from regions of high pressure to regions of low pressure.

- The dew point of air depends on the amount of water vapor the air contains.

- Latent heat is released when water vapor condenses and when water freezes.

Demonstrate Understanding

1. **Identify** three properties of the atmosphere and describe how they vary with height in the atmosphere.

2. **Explain** what occurs during a temperature inversion.

3. **Describe** how the motion of particles in a material changes when the temperature of the material increases.

Explain Your Thinking

4. **Predict** how the relative humidity and dew point change in a rising mass of air in the troposphere.

5. **Design** an experiment that shows how average wind speeds change over different types of surfaces.

6. **MATH Connection** If the average thickness of the troposphere is 11 km, what would be the temperature difference between the top and bottom of the troposphere if the temperature decrease is the same as the dry adiabatic lapse rate?

LEARNSMART Go online to follow your personalized learning path to review, practice, and reinforce your understanding.

CLOUDS AND PRECIPITATION

What is needed to form clouds?

Cloud Formation

A cloud can form when a rising air mass cools. Recall that Earth's surface heats and cools by different amounts in different places. This uneven heating and cooling of the surface causes air masses near the surface to warm and cool. As an air mass is heated, it becomes less dense than the cooler air around it. This causes the warmer air mass to be pushed upward by the denser, cooler air.

However, as the warm air mass rises, it expands and cools adiabatically. The cooling of an air mass as it rises can cause water vapor in the air mass to condense. Recall that the lifted condensation level (LCL) is the height at which condensation of water vapor occurs in an air mass. When a rising air mass reaches the LCL, water vapor condenses around condensation nuclei, as shown in **Figure 18**. A **condensation nucleus** is a small particle in the atmosphere around which water droplets can form. These particles are usually less than about 0.001 mm in diameter and can be made of ice, salt, dust, and other materials. The droplets that form can be liquid water or ice, depending on the surrounding temperature. When the number of these droplets is large enough, a cloud is visible.

Ryan McGinnis/Flickr RF/Getty Images

Figure 18 These clouds over Nebraska formed when rising air became saturated and water collected on condensation nuclei.

3D THINKING   **DCI** Disciplinary Core Ideas   **CCC** Crosscutting Concepts   **SEP** Science & Engineering Practices

COLLECT EVIDENCE

Use your Science Journal to record the evidence you collect as you complete the readings and activities in this lesson.

INVESTIGATE

GO ONLINE to find these activities and more resources.

((•)) **Review the News**
Obtain information from a current news story about cloud formation. Evaluate your source and communicate your findings to your class.

? **Revisit the Encounter the Phenomenon Question**
What information from this lesson can help you answer the Unit and Module questions?

Atmospheric stability

As an air mass rises, it cools. However, the air mass will continue to rise as long as it is warmer than the surrounding air. Under some conditions, an air mass that has started to rise sinks back to its original position. When this happens, the air is considered stable because it resists rising. The stability of air masses determines the type of clouds that form and the associated weather patterns.

Stable air The stability of an air mass depends on how the temperature of the air mass changes relative to the atmosphere. The air temperature near Earth's surface decreases with altitude. As a result, the atmosphere becomes cooler as the air mass rises. At the same time, the rising air mass is also becoming cooler. Suppose that the temperature of the atmosphere decreases more slowly with increasing altitude than does the temperature of the rising air mass. Then the rising air mass will cool more quickly than the atmosphere. The air mass will finally reach an altitude at which it is colder than the atmosphere. It will then sink back to the altitude at which its density is the same as the atmosphere, as shown in **Figure 19**. Because the air mass stops rising and sinks downward, it is stable. Fair-weather clouds form under stable conditions.

Get It?
Describe the factors that affect the stability of air.

Unstable air Suppose that the temperature of the surrounding air cools faster than the temperature of the rising air mass. In these conditions, the air mass will always be less dense than the surrounding air. As a result, the air mass will continue to rise, as shown in **Figure 19**. The atmosphere is then considered to be unstable. Unstable conditions can produce the type of clouds associated with thunderstorms.

Note that it is common for the temperature of an air mass to rise well above the temperature of the surrounding air. Other processes, such as the release of latent heat, cause the air mass to continue rising.

Figure 19 Stable air has a tendency to resist motion. Unstable air does not resist vertical displacement. When the temperature of a mass of air is greater than the temperature of the surrounding air, the air mass rises. When the temperature of the surrounding air is greater than that of the air mass, the air mass sinks.

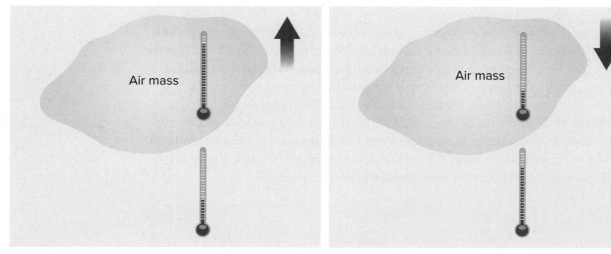

Figure 20 Orographic lifting occurs when warm, moist air is cooled because it is forced to rise over a topographic barrier, such as a mountain or cliff as shown here.

Atmospheric lifting

Clouds can form when moist air rises, expands, and cools. Air rises when it is heated and becomes warmer than the surrounding air. This process is known as convective lifting. Clouds can also form when air is forced upward or lifted by mechanical processes. Two of these processes are orographic lifting and convergence.

Orographic lifting Clouds can form when air is forced to rise over elevated land or other topographic barriers. This can happen, for example, when an air mass approaches a mountain range. **Orographic lifting** occurs when an air mass is forced to rise over a topographic barrier, as shown in **Figure 20.** The rising air mass expands and cools, with water droplets condensing when the temperature reaches the dew point. Many of the rainiest places on Earth are located on the windward side of mountain slopes, such as the coastal side of the Sierra Nevadas. The formation of clouds and the resulting heavy precipitation along the west coast of Canada are also primarily due to orographic lifting.

 Get It?

Sequence the steps involved in cloud formation through orographic lifting.

Convergence Air can be lifted by convergence, which occurs when air masses move into the same area from different directions, forcing some of the air upward. This process is even more pronounced when air masses at different temperatures collide. When a warm air mass and a cooler air mass collide, the warmer, less-dense air is forced upward and over the denser, cooler air. As the warm air rises, it cools adiabatically. If the rising air cools to the dew-point temperature, then water vapor can condense on condensation nuclei and form a cloud. This cloud formation mechanism is common at middle latitudes, where severe storm systems form as cold polar air collides with warmer air. Convergence also occurs near the equator, where the trade winds meet at the intertropical convergence zone.

 Get It?

Predict Which mechanism of cloud formation is most likely to occur in the Great Plains, located in the interior of the United States?

Types of Clouds

You have probably noticed that clouds have different shapes. Some clouds look like puffy cotton balls, while others have a thin, feathery appearance. These differences in cloud shape are due to differences in the processes that cause clouds to form. Cloud formation can also take place at different altitudes—sometimes even right at Earth's surface, in which case the cloud is known as fog.

Clouds are generally classified according to a system developed in 1803, and only minor changes have been made since it was first introduced. This system classifies clouds by the altitudes at which they form and by their shapes. There are three classes of clouds based on the altitudes at which they form: low, middle, and high. Low clouds typically form below 2000 m. Middle clouds form mainly between 2000 m and 6000 m. High clouds form above 6000 m. One type of low cloud—cumulonimbus—can develop upward vertically through the middle and high cloud levels. These are clouds with vertical development, and they are often associated with thunderstorms. **Figure 21** shows the most common types of clouds.

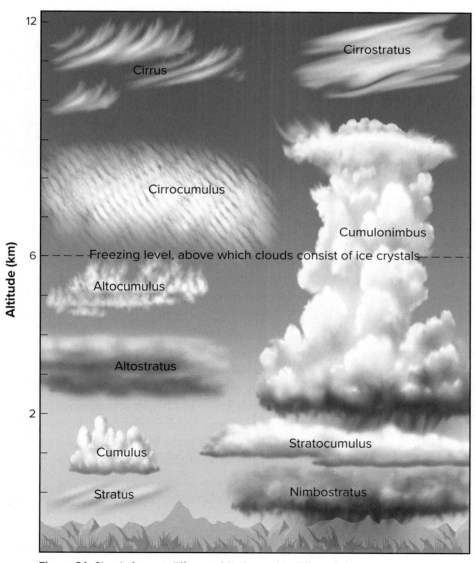

Figure 21 Clouds form at different altitudes and in different shapes.
Compare and contrast *cirrus and stratus clouds.*

Low clouds

Clouds can form when warm, moist air rises, expands, and cools. If conditions are stable, the air mass stops rising at the altitude where its temperature is the same as that of the surrounding air. If a cloud has formed, it will flatten out, and winds will spread it horizontally into stratocumulus or layered cumulus clouds, as shown in **Figure 21. Cumulus** (KYEW myuh lus) clouds are puffy, lumpy-looking clouds that usually occur below 2000 m. Also below 2000 m, layered, sheetlike **stratus** (STRAY tus) clouds can cover much or all of the sky in a given area. Stratus clouds often form when fog lifts away from Earth's surface. A nimbostratus cloud produces precipitation.

Middle clouds

Altocumulus and altostratus clouds form at altitudes between 2000 m and 6000 m. They are made up of ice crystals and water droplets due to the colder temperatures generally present at these altitudes. Middle clouds are usually layered. Altocumulus clouds are white or gray in color and form large, round masses or wavy rows. Altostratus clouds have a gray appearance, and they form thin sheets of clouds. Middle clouds sometimes produce mild precipitation.

High clouds

High clouds, made up of ice crystals, form at heights above 6000 m where temperatures are below freezing. Some, such as **cirrus** (SIHR us) clouds, often have a wispy, indistinct appearance. A second type of high cloud, cirrostratus, forms as a continuous layer in the sky, varying from almost transparent to dense enough to block out the Sun or the Moon. A third type of high cloud, cirrocumulus, has a rippled appearance.

Vertical development clouds

If the air that makes up a cumulus cloud is unstable, the cloud will be warmer than the surrounding air and will continue to grow upward. As it rises, water

Figure 22 Cumulonimbus clouds, such as the large, puffy cloud here, are associated with thunderstorms.

Describe *how a cumulonimbus cloud can form.*

vapor condenses, and the air continues to increase in temperature due to the release of latent heat. The cloud can grow through middle altitudes as a towering cumulonimbus, as shown in **Figure 22,** and, if conditions are right, it can reach the tropopause. Its top is then composed entirely of ice crystals. Strong winds can spread the top of the cloud into an anvil shape. What began as a small mass of unstable moist air is now an atmospheric giant, capable of producing the torrential rains, strong winds, and hail characteristic of some thunderstorms.

ACADEMIC VOCABULARY

continuous
uninterrupted in space or time
The stream flowed in a continuous path to the sea.

WeatherVideoHD.TV

Precipitation

All forms of water that fall from clouds to the ground are **precipitation.** Rain, snow, sleet, and hail are the four main types of precipitation. Clouds contain water droplets that are so small that the upward movement of air in the cloud can keep the droplets from falling. For these droplets to become heavy enough to fall, their size must increase by 50 to 100 times.

Coalescence

One way that cloud droplets can increase in size is by coalescence. In a warm cloud, coalescence is the primary process responsible for the formation of precipitation. **Coalescence** (koh uh LEH sunts) occurs when cloud droplets collide and join together to form a larger droplet. These collisions occur as larger droplets fall and collide with smaller droplets. As the process continues, the droplets eventually become too heavy to remain suspended in the cloud and fall to Earth as precipitation. Rain is precipitation that reaches Earth's surface as a liquid. Raindrops typically have diameters between 0.5 mm and 5 mm.

Snow, sleet, and hail

The type of precipitation that reaches Earth depends on the vertical variation of temperature in the atmosphere. In cold clouds where the air temperature is far below freezing, ice crystals can form and fall to the ground as snow. Ice crystals can reach the ground as rain if they fall through air warmer than 0°C and melt. In some cases, air currents in a cloud can cause cloud droplets to move up and down through freezing and nonfreezing air, forming ice pellets that fall to the ground as sleet. Sleet can also occur when raindrops freeze as they fall through freezing air near the surface.

If the up-and-down motion in a cloud is especially strong and occurs over large stretches of the atmosphere, large ice pellets known as hail can form. **Figure 23** shows samples of hail. Most hailstones are smaller in diameter than a dime, but some stones have been found to weigh more than 0.5 kg. Larger stones are often produced during severe thunderstorms.

Get It?

Compare the formation of different types of precipitation.

Figure 23 Hail—precipitation in the form of balls or lumps of ice—is produced by intense thunderstorms.

JosephZahnlePhotography/iStock/Getty Images

CCC CROSSCUTTING CONCEPTS

Energy and Matter Make your own diagram of the water cycle. Add captions and labels that describe how energy drives the cycling of matter within and between systems in the water cycle.

The water cycle

At any one time, only a small percentage of Earth's total water is present in the atmosphere. Still, this water is vitally important because, as it continually moves between the atmosphere and Earth's surface, it nourishes living things. The constant movement of water between the atmosphere and Earth's surface is known as the water cycle. In the water cycle, as shown in **Figure 24,** radiation from the Sun causes liquid water to evaporate. Water evaporates from lakes, streams, and oceans and rises into the atmosphere. As water vapor rises, it cools and condenses to form clouds. Water droplets combine to form larger drops that fall to Earth as precipitation. This water soaks into the ground and enters bodies of water, or it falls directly into bodies of water and eventually evaporates, continuing the water cycle.

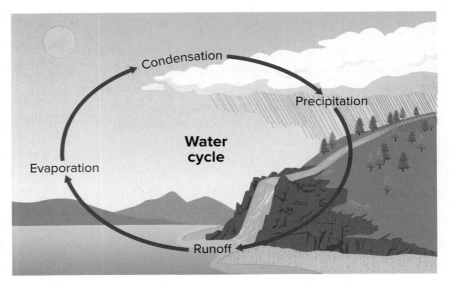

Figure 24 Water moves from Earth's surface to the atmosphere and back to the surface in the water cycle.

Check Your Progress

Summary

- Clouds form as warm, moist air is forced upward, expands, and cools.
- An air mass is stable if it tends to return to its original height after it starts rising.
- Cloud droplets form when water vapor is cooled to the dew point and condenses on condensation nuclei.
- Clouds are classified by their shape and the altitude at which they form.
- Cloud droplets collide and coalesce into larger droplets that can fall to Earth as rain, snow, sleet, or hail.

Demonstrate Understanding

1. **Summarize** the differences among low clouds, middle clouds, and high clouds.
2. **Describe** how precipitation forms.
3. **Determine** the reason precipitation will fall as snow rather than rain.
4. **Compare** stable and unstable air.
5. **Identify** Take a photo of or observe clouds in the sky. Identify the cloud type.

Explain Your Thinking

6. **Evaluate** how a reduction in the number of condensation nuclei in the troposphere would affect precipitation. Explain your reasoning.
7. **WRITING** **Connection** Suppose the amount of water vapor in the atmosphere increased. How could this change cause feedback effects that could influence the original change?

LEARNSMART Go online to follow your personalized learning path to review, practice, and reinforce your understanding.

A New Threat to the Ozone Layer

The ozone layer protects Earth from ultraviolet rays, but harmful emissions from industrial processes can harm this layer. Certain atoms in the stratosphere, such as chlorine and bromine, can destroy ozone molecules. The discovery of the ozone hole was published in 1985 in the journal *Nature*.

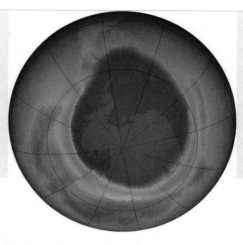

This satellite image shows the size and position of the ozone hole over Antarctica.

The Original Ozone Hole

In the journal *Nature*, researchers described a seasonal ozone depletion that covered the entire Antarctic region. The depletion was linked to chlorofluorocarbons (CFCs) used to manufacture various products. Scientists predicted that if it continued unchecked, ozone depletion would cause a global environmental crisis. Swift action and collaboration among countries resulted in the signing of the Montreal Protocol, which banned CFCs a mere two years after news of the ozone hole was published. Since 1987, CFCs and many related compounds have been banned from industrial processes. As a result, levels of chlorine and other ozone-destroying chemicals in the stratosphere have been declining since the late 1990s.

On the Mend—or Is It?

Scientists measure the area of ozone depletion through satellite images. For the past decade or so, the ozone hole has been recovering because of the efforts of both policy makers and scientists.

But now, a new threat may slow the progress. Industrial emissions of dichloromethane (CH_2CL_2), a chemical commonly used in solvents and the production of pharmaceuticals, have doubled in recent years.

Like CFCs, dichloromethane breaks apart when exposed to sunlight in the upper atmosphere. The liberated chlorine molecules destroy any ozone molecules they contact. These chemicals were not included in the Montreal Protocol because scientists at the time did not know that the chemicals damaged the ozone layer.

These emissions could slow the healing of the ozone layer by 5 to 30 years—longer if emissions continue to increase. For now, scientists suggest closely monitoring concentrations of all known ozone-depleting substances.

COMMUNICATE SCIENTIFIC INFORMATION

Make a poster that shows the chemical reactions associated with ozone depletion. Compare and contrast the reactions involved with CFCs and dichloromethane.

NASA Ozone Watch

MODULE 8
STUDY GUIDE

 GO ONLINE to study with your Science Notebook.

Lesson 1 ATMOSPHERIC BASICS

- Earth's atmosphere is composed of several gases, primarily nitrogen and oxygen, and also contains small particles.
- Earth's atmosphere consists of five layers that differ in their composition and temperatures.
- Solar energy reaches Earth's surface in the form of visible light and infrared waves.
- Solar energy absorbed by Earth's surface is transferred as thermal energy throughout the atmosphere.
- The greenhouse effect causes the atmosphere to retain heat.

- troposphere
- stratosphere
- mesosphere
- thermosphere
- exosphere
- radiation
- conduction
- convection
- greenhouse effect

Lesson 2 PROPERTIES OF THE ATMOSPHERE

- At the same pressure, warmer air is less dense than cooler air.
- Air moves from regions of high pressure to regions of low pressure.
- The dew point of air depends on the amount of water vapor the air contains.
- Latent heat is released when water vapor condenses and when water freezes.

- temperature inversion
- humidity
- saturation
- relative humidity
- dew point
- latent heat

Lesson 3 CLOUDS AND PRECIPITATION

- Clouds form as warm, moist air is forced upward, expands, and cools.
- An air mass is stable if it tends to return to its original height after it starts rising.
- Cloud droplets form when water vapor is cooled to the dew point and condenses on condensation nuclei.
- Clouds are classified by their shape and the altitude at which they form.
- Cloud droplets collide and coalesce into larger droplets that can fall to Earth as rain, snow, sleet, or hail.

- condensation nucleus
- orographic lifting
- cumulus
- stratus
- cirrus
- precipitation
- coalescence

Module Wrap-Up

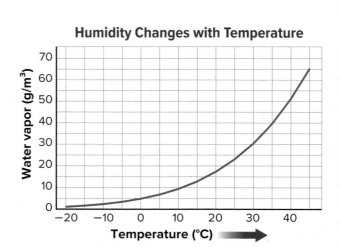

REVISIT THE PHENOMENON

How do clouds and mist form?

CER Claim, Evidence, Reasoning

Explain Your Reasoning Revisit the claim you made when you encountered the phenomenon. Summarize the evidence you gathered from your investigations and research and finalize your Summary Table. Does your evidence support your claim? If not, revise your claim. Explain why your evidence supports your claim.

STEM UNIT PROJECT
Now that you've completed the module, revisit your STEM unit project. You will summarize your evidence and apply it to the project.

GO FURTHER

SEP Data Analysis Lab

How do you calculate relative humidity?

Relative humidity is the ratio of the actual amount of water vapor in a volume of air relative to the maximum amount of water vapor needed for that volume of air to reach saturation.

Data and Observations Use the graph shown here to answer the following questions.

CER Analyze and Interpret Data

1. **Claim** What happens to the amount of water vapor in 1 m³ of air when the temperature increases from 15°C to 25°C?
2. **Evidence** Calculate the relative humidity of 1 m³ of air containing 10 g/m³ of water vapor at 20°C. Show your work.
3. **Reasoning** Can relative humidity be more than 100 percent? Explain your answer.

Humidity Changes with Temperature

METEOROLOGY

ENCOUNTER THE PHENOMENON

How do satellites help us track and predict weather?

GO ONLINE to play a video about how satellites changed meteorology.

SEP Ask Questions

Do you have other questions about the phenomenon? If so, add them to the driving question board.

CER Claim, Evidence, Reasoning

Make Your Claim Use your CER chart to make a claim about how we use satellites to help us track and predict weather. Explain your reasoning.

Collect Evidence Use the lessons in this module to collect evidence to support your claim. Record your evidence as you move through the module.

Explain Your Reasoning You will revisit your claim and explain your reasoning at the end of the module.

GO ONLINE to access your CER chart and explore resources that can help you collect evidence.

LESSON 3: Explore and Explain: Data from the Earth's Surface

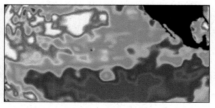

LESSON 4: Explore and Explain: Long-term Forecasts

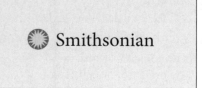

Additional Resources

(t)Video Supplied by BBC Worldwide Learning, (bl) Image Source/Getty Images, (br)JPL/NASA

FOCUS QUESTION
How does heat drive weather?

What is meteorology?

Clouds, breezes, and the warmth of sunlight are examples of atmospheric phenomena. They are often classified as types of meteors. Cloud droplets and precipitation are hydrometeors (hi droh MEE tee urz). Smoke, haze, dust, and other particles in the atmosphere are lithometeors (lih thuh MEE tee urz). Examples of electrometeors include thunder and lightning—atmospheric phenomena that you can hear and see. Meteorology is the study of the physics, chemistry, and dynamics of atmospheric phenomena. The root word of *meteorology* is the Greek word *meteoros*, which means "high in the air".

Weather v. climate

Short-term variations in atmospheric phenomena that interact and affect the environment and life on Earth are called **weather.** These variations can take place over minutes, hours, days, weeks, months, or years. **Climate** is the long-term averages and variations in weather for a particular area. Meteorologists use weather-data averages over 30 years to define an area's climate, such as that of the desert shown in **Figure 1.**

 Get It?
Differentiate between weather and climate.

Figure 1 A desert climate is dry, with extreme variations in day and night temperatures. Only organisms adapted to these conditions, such as this ocotillo, can survive in a desert.

 **3D THINKING** **DCI** Disciplinary Core Ideas **CCC** Crosscutting Concepts **SEP** Science & Engineering Practices

COLLECT EVIDENCE
Use your Science Journal to record the evidence you collect as you complete the readings and activities in this lesson.

INVESTIGATE

GO ONLINE to find these activities and more resources.

Quick Investigation: Compare the Angles of Sunlight to Earth
Use a model to describe the relationship between the angle of sunlight and the distribution of heat and energy on Earth's surface.

? Revisit the Encounter the Phenomenon Question
What information from this lesson can help you answer the Unit and Module questions?

tonda/iStockphoto/Getty Images

Heating Earth's Surface

Over the course of a year, the amount of thermal energy that Earth receives from the Sun is the same as the amount that Earth radiates back to space. A central issue in meteorology is the distribution of solar radiation around Earth.

Imbalanced heating

Why are average January temperatures warmer in Miami, Florida, than in Detroit, Michigan? Part of the explanation is that Earth's axis of rotation is tilted relative to the plane of Earth's orbit. Therefore, the number of hours of daylight and amount of solar radiation during January is greater in Miami than in Detroit.

Another factor is that Earth is a sphere, and different places on Earth are at different angles to the Sun, as shown in **Figure 2.** For most of the year, the amount of solar radiation that reaches a given region at the equator covers a larger area at latitudes nearer the poles. The greater the area covered, the smaller the amount of heat per unit of area. Because Detroit is farther from the equator than Miami is, the same amount of solar radiation that heats Miami will heat Detroit less.

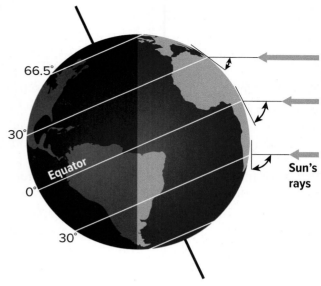

Figure 2 Solar radiation is unequal partly due to the changing angle of incidence of sunlight. In this example, it is perpendicular to Earth's surface south of the equator, 60° at the equator, and 40° north of the equator.

Explain *why average temperatures decline from the equator to the poles.*

 Get It?

Describe two factors that affect temperatures on Earth at different latitudes.

Thermal energy redistribution

Areas around Earth maintain about the same average temperatures over time due to the constant movement of air and water among Earth's surfaces, oceans, and atmosphere. The constant movement of air on Earth's surface, along with currents in Earth's oceans, redistributes thermal energy around the world. The foundation for Earth's global climate systems is electromagnetic radiation from the Sun. This thermal energy is reflected, absorbed, stored, and redistributed among the atmosphere, ocean, and land systems. Weather—from thunderstorms to large-scale weather systems—is part of the constant redistribution of Earth's thermal energy.

 Get It?

Describe the source, or foundation, for Earth's thermal energy.

Air Masses

You have learned that air over a warm surface can be heated by conduction. This heated air rises because it is less dense than the surrounding air. On Earth, this process can take place over thousands of square kilometers for days or weeks. The result is the formation of an air mass. An **air mass** is a large volume of air that has the same characteristics, such as humidity and temperature, as its **source region**—the area over which the air mass forms. Most air masses form over tropical regions or polar regions.

Types of air masses

The five types of air masses, listed in **Table 1,** influence weather in the United States. These air masses are all common in North America because there is a source region nearby.

Tropical air masses The origins of maritime tropical air are tropical bodies of water, listed in **Table 1.** In the summer, they bring hot, humid weather to the eastern two-thirds of North America. The southwestern United States and Mexico are source regions of continental tropical air, which is hot and dry, especially in summer.

Polar air masses Maritime polar air masses form over the cold waters of the North Atlantic and North Pacific. The one that forms over the North Pacific primarily affects the West Coast of the United States, occasionally bringing heavy rains in winter. Continental polar air masses form over the interior of Canada and Alaska. In winter, these air masses can carry frigid air southward. In the summer, however, cool and relatively dry continental polar air masses bring relief from hot, humid weather.

 Get It?

Compare and contrast tropical and polar air masses.

Arctic air masses Earth's ice- and snow-covered surfaces above 60°N latitude, in Siberia and the Arctic Basin, are the source regions of arctic air masses. During part of the winter, these areas receive almost no solar radiation but continue to radiate thermal energy. As a result, they can become some of the most frigid areas on Earth.

 Get It?

Explain why Siberia and the Arctic Basin are the source of arctic air masses.

Table 1 Air Mass Characteristics

Air Mass Type	Weather Map Symbol	Source Region	Characteristics	
			Winter	Summer
Arctic	A	Siberia, Arctic Basin	bitter cold, dry	cold, dry
Continental polar	cP	interiors of Canada and Alaska	very cold, dry	cool, dry
Continental tropical	cT	southwest United States, Mexico	warm, dry	hot, dry
Maritime polar	mP	North Pacific Ocean	mild, humid	mild, humid
		North Atlantic Ocean	cold, humid	cold, humid
Maritime tropical	mT	Gulf of Mexico, Caribbean Sea, tropical and subtropical Atlantic Ocean and Pacific Ocean	warm, humid	hot, humid

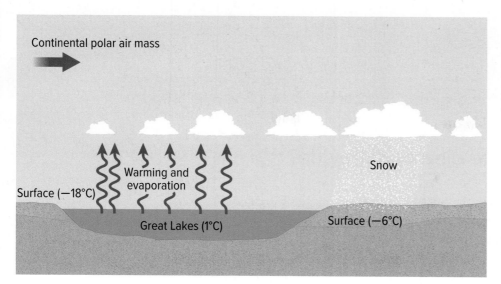

Figure 3 As cold, continental polar air moves over the warmer Great Lakes, the air gains thermal energy and moisture. This modified air cools as it is uplifted by convection and topographic features, producing lake-effect snows.

Air mass modification

Air masses do not stay in one place indefinitely. Eventually, they move, transferring thermal energy from one area to another. When an air mass travels over land or water that has characteristics different from those of its source region, the air mass can acquire some of the characteristics of that land or water. For example, when a polar air mass travels over a relatively warm body of water, as shown in **Figure 3,** the modified air mass might produce lake-effect snows. When an air mass undergoes modification, it exchanges thermal energy, moisture, or both with the surface over which it travels.

 **Check Your Progress**

Summary

- Meteorology is the study of atmospheric phenomena.

- Solar radiation is unequally distributed between Earth's equator and its poles.

- An air mass is a large body of air that takes on the moisture and temperature characteristics of the area over which it forms.

- Each type of air mass is classified by its source region.

Demonstrate Understanding

1. **Summarize** how an air mass forms.
2. **Explain** the process that prevents the poles from steadily cooling off and the tropics from heating up over time.
3. **Distinguish** between the causes of weather and climate.
4. **Differentiate** among the types of air masses.

Explain Your Thinking

5. **Predict** which type of air mass would become modified more quickly: an arctic air mass moving over the Gulf of Mexico in winter or a maritime tropical air mass moving into the southwestern United States in summer. Explain your answer.
6. **WRITING** **Connection** Describe in a paragraph how a maritime polar air mass that formed over the North Pacific is modified as it moves west over North America.

LEARNSMART Go online to follow your personalized learning path to review, practice, and reinforce your understanding.

FOCUS QUESTION

Why do winds blow in the direction they do?

Global Wind Systems

If Earth did not rotate on its axis, two large air convection currents would cover Earth, as shown in **Figure 4.** The colder and denser air at the poles would sink to the surface and flow toward the tropics. There, the cold air would force warm, equatorial air to rise. This air would cool as it gained altitude and flowed back toward the poles. However, Earth rotates from west to east, which prevents this situation.

The directions of Earth's winds are influenced by Earth's rotation. This **Coriolis effect** results in fluids and objects moving in an apparent curved path rather than a straight line. Thus, as illustrated in **Figure 5,** moving air curves to the right in the northern hemisphere and curves to the left in the southern hemisphere. Together, the Coriolis effect and the heat imbalance on Earth create distinct global wind systems. They transport colder air to warmer areas near the equator and warmer air to colder areas near the poles. Global wind systems help to equalize the thermal energy on Earth.

There are three basic zones, or wind systems, at Earth's surface in each hemisphere. They are polar easterlies, prevailing westerlies, and trade winds.

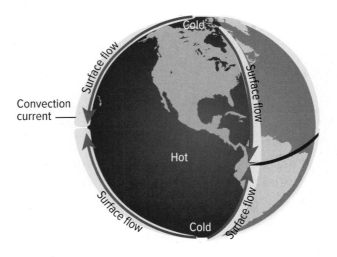

Figure 4 If Earth did not rotate, two large convection currents would form as denser polar air moved toward the equator. These currents would warm and rise as they approached the equator, and cool as they moved toward each pole.

 **3D THINKING** **DCI** Disciplinary Core Ideas **CCC** Crosscutting Concepts **SEP** Science & Engineering Practices

COLLECT EVIDENCE

 Use your Science Journal to record the evidence you collect as you complete the readings and activities in this lesson.

INVESTIGATE

 GO ONLINE to find these activities and more resources.

Investigation Lab: Modeling the Coriolis Effect
Use a model to visualize the patterns in the atmosphere due to the Coriolis effect.

Review the News
Obtain information from a current news story about the jet stream's effect on weather. Evaluate your source and communicate your findings to your class.

Figure 5 Visualizing the Coriolis Effect

The Coriolis effect results in fluids and objects moving in an apparent curved path rather than a straight line.

Recall that distance divided by time equals speed. The equator has a length of about 40,000 km—Earth's circumference—and Earth rotates west to east once about every 24 hours. This means that things on the equator, including the air above it, move eastward at a speed of about 1670 km/h.

However, not every location on Earth moves eastward at this speed. Latitudes north and south of the equator have smaller circumferences than the equator. Those objects not on the equator move less distance during the same amount of time. Therefore, their eastward speeds are slower than objects on the equator.

The island of Martinique is located at approximately 15°N latitude. Suppose that rising equatorial air is on the same line of longitude as Martinique. When this air arrives at 15°N latitude a day later, it will be east of Martinique because the air was moving to the east faster than the island was moving to the east.

The result is that air moving toward the poles appears to curve to the right, or east. The opposite is true for air moving from the poles to the equator because the eastward speed of polar air is slower than the eastward speed of the land over which it is moving.

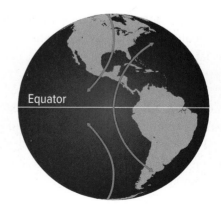

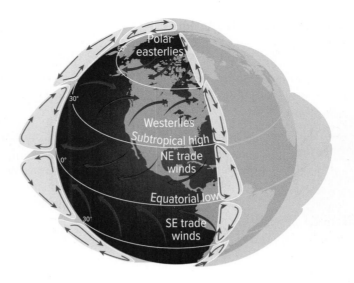

Figure 6 The directions of Earth's wind systems, such as the polar easterlies and the trade winds, vary with the latitudes in which they occur. Note that a wind is named for the direction from which it blows. A north wind blows from the north.

Polar easterlies

The wind zones between 60°N latitude and the North Pole and between 60°S latitude and the South Pole are called the **polar easterlies,** as shown in **Figure 6.** Polar easterlies begin as dense polar air that sinks. As Earth spins, this cold, descending air is deflected in an easterly direction away from each pole. The polar easterlies are typically cold winds. Unlike the prevailing westerlies, the polar easterlies are often weak and sporadic.

Between polar easterlies and prevailing westerlies is an area called a polar front. Earth has two polar fronts located near latitudes 60°N and 60°S. Polar fronts are areas of stormy weather.

Prevailing westerlies

The wind systems located between latitudes 30°N and 60°N, and between 30°S and 60°S are called the **prevailing westerlies.** In the midlatitudes, surface winds move in a westerly direction toward each pole, as shown in **Figure 6.** Because these winds originate from the west, they are called westerlies. Prevailing westerlies are steady winds that move much of the weather across the United States and Canada.

Trade winds

Between latitudes 30°N and 30°S are two circulation belts of wind known as the **trade winds,** which are shown in **Figure 6.** Air in these regions sinks and moves toward the equator in an easterly direction. When the air reaches the equator, it warms, rises, and moves back toward latitudes 30°N and 30°S, where it sinks and the process begins again.

Horse latitudes Near latitudes 30°N and 30°S, the sinking air associated with the trade winds creates an area of high pressure. This results in a belt of weak surface winds called the horse latitudes. Earth's major deserts, such as the Sahara, are under these high-pressure areas.

SCIENCE USAGE v. COMMON USAGE

circulation

Science usage: movement in a circle or circuit

Common usage: condition of being passed about and widely known; distribution

ACADEMIC VOCABULARY

generate

to bring into existence
Wind is generated as air moves from an area of high pressure to an area of low pressure.

Intertropical convergence zone Near the equator, trade winds from the north and the south meet and join, as shown in **Figure 6.** The air is forced upward, which creates an area of low pressure. This process, called convergence, can occur on a small or large scale. Near the equator, it occurs over a large area called the intertropical convergence zone (ITCZ). The ITCZ drifts south and north of the equator as seasons change. In general, it follows the positions of the Sun in relation to the equator. In March and September, it is directly over the equator. Because the ITCZ is a region of rising air, it has bands of cloudiness and thunderstorms, which deliver moisture to many of the world's tropical rain forests.

Jet Streams

Atmospheric conditions and events that occur at the boundaries between wind zones strongly influence Earth's weather. On either side of these boundaries, both surface air and upper-level air differ greatly in temperature and pressure. A large temperature gradient in upper-level air combined with the Coriolis effect results in strong westerly winds called jet streams. A **jet stream,** shown in **Figure 7,** is a narrow band of fast wind. Its speed varies with the temperature differences between the air masses at the wind zone boundaries. A jet stream can have a speed up to 400 km/h at altitudes of 10.7 km to 12.2 km.

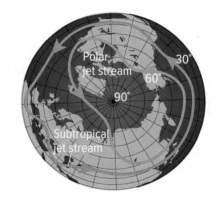

The position of a jet stream varies with the season. It generally is located along a line of strong temperature differences from the equator to a pole. A jet stream can move almost due south or north instead of its normal westerly direction. It can also split into branches and re-form later. Whatever form or position it takes, a jet stream represents the strongest core of winds.

Types of jet streams

The polar jet streams separate the polar easterlies from the prevailing westerlies. These major jet streams, which occur around latitudes 40°N to 60°N and 40°S to 60°S, move west to east. The subtropical jet streams are minor jet streams that occur where the trade winds meet the prevailing westerlies, at about 20°N to 30°N and 20°S to 30°S.

Get It?

Describe the types of jet streams at different latitudes.

Figure 7 Weather in the middle latitudes is strongly influenced by fast-moving, high-altitude jet streams.

Jet streams and weather systems

Storms form along jet streams and generate large weather systems. These systems transport cold surface air toward the tropics and warm surface air toward the poles. Weather systems generally follow the path of jet streams. Jet streams also affect the intensity of weather systems by moving air of different temperatures from place to place.

Get It?

Explain the relationship between jet streams and weather systems.

NASA

Fronts

Air masses with different characteristics can collide and result in dramatic weather changes. A collision of two air masses forms a **front**—a narrow region between two air masses of different densities. Recall that the density of an air mass results from its temperature, pressure, and humidity. Fronts can form across thousands of kilometers of Earth's surface.

Cold front

When cold, dense air displaces warm air, it forces the warm air, which is less dense, up along a steep slope, as shown in **Figure 8.** This type of collision is called a cold front. As the warm air rises, it cools, and water vapor condenses. Intense precipitation and thunderstorms are common with cold fronts. A blue line with evenly spaced blue triangles represents a cold front on a weather map. The triangles point in the direction of the front's movement.

Get It?
Define *cold front* and draw a weather map showing a cold front.

Warm front

Advancing warm air displaces cold air along a warm front. A warm front develops a gradual boundary slope, as illustrated in **Figure 8.** A warm front can cause widespread light precipitation. On a weather map, a red line with evenly spaced, red semicircles pointing in the direction of the front's movement indicates a warm front.

Get It?
Define *warm front* and draw a weather map showing a warm front.

Stationary front

When two air masses meet but neither advances, the boundary between them stalls and a stationary front forms, as shown in **Figure 8.** This front frequently occurs between two modified air masses that have small temperature and pressure gradients between them.

Get It?
Define *stationary front* and draw a weather map showing a stationary front.

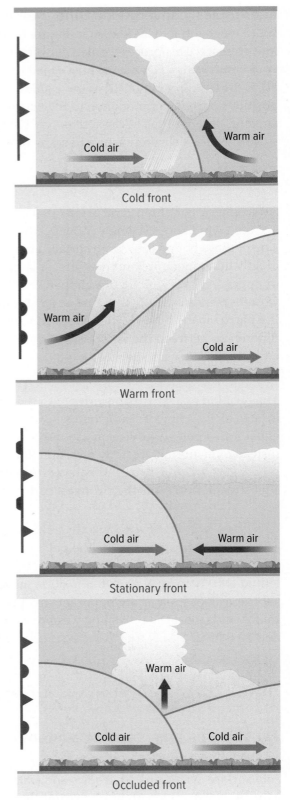

Figure 8 The type of front formed depends on the types of air masses that collide.

Identify *the front associated with high cirrus clouds.*

The air masses can continue moving parallel to the front. Stationary fronts sometimes have light winds and precipitation. A line of evenly spaced, alternating cold- and warm-front symbols pointing in opposite directions represents a stationary front on a weather map.

Occluded front

Sometimes a cold air mass moves so rapidly that it overtakes a warm front and forces the warm air upward, as shown in **Figure 8.** As the warm air is lifted, the advancing cold air mass collides with another cold air mass that was in front of the warm air. This is called an occluded front. Strong winds and heavy precipitation are common along an occluded front. An occluded front is shown on a weather map as a line of evenly spaced, alternating purple triangles and semicircles pointing in the direction of the occluded front's movement.

 Get It?

Define *occluded front* and draw a weather map showing an occluded front.

Pressure Systems

Recall that sinking air is associated with high pressure, and rising air is associated with low pressure. Air always flows from an area of high pressure to an area of low pressure. Sinking or rising air, combined with the Coriolis effect, results in the formation of rotating high- and low-pressure systems. Air in these systems moves in a circular motion around the pressure center, as shown in **Figure 9.**

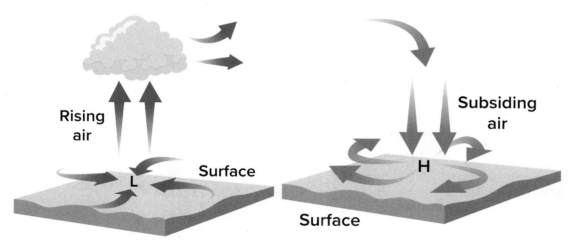

Figure 9 In the northern hemisphere, winds move counterclockwise around a low-pressure center and clockwise around a high-pressure center.

CCC CROSSCUTTING CONCEPTS

Patterns Review the information about fronts and describe the weather patterns that result from each type of front.

Low-pressure systems

In low-pressure systems, air rises. When air from outside the system replaces the rising air, this air spirals inward toward the center and then upward. Air in a low-pressure system in the northern hemisphere moves in a counterclockwise direction, as shown in **Figure 9.** The opposite occurs in the southern hemisphere; air in a low-pressure system moves in a clockwise direction. As air rises, it cools and often condenses into clouds and precipitation. Therefore, a low-pressure system is associated with cloudy weather and precipitation.

Get It?

Describe a low-pressure system and the type of weather that results from this type of system.

High-pressure systems

In a high-pressure system, sinking air moves away from the system's center when it reaches Earth's surface. The Coriolis effect causes the sinking air to move in the westerly direction. The air circulates clockwise in the northern hemisphere and counterclockwise in the southern hemisphere. High-pressure systems dominate most of Earth's subtropical oceans and usually bring fair weather.

Get It?

Describe a high-pressure system and the type of weather that results from this type of system.

Check Your Progress

Summary

- The three major wind systems are the polar easterlies, the prevailing westerlies, and the trade winds.

- Fast moving, high-altitude jet streams greatly influence weather in the middle latitudes.

- The four types of fronts are cold fronts, warm fronts, stationary fronts, and occluded fronts.

- Air moves in a generally circular motion around either a high- or low-pressure center.

Demonstrate Understanding

1. **Summarize** information about the four types of fronts. Explain how they form and lead to changes in weather.

2. **Distinguish** among the three main wind systems.

3. **Describe** the Coriolis effect.

4. **Explain** why most tropical rain forests are located near the equator.

5. **Describe** how a jet stream affects the movement of air masses.

6. **Compare and contrast** high-pressure and low-pressure systems.

Explain Your Thinking

7. **Analyze** why most of the world's deserts are located between 10° and 30° north and south latitudes.

8. **WRITING Connection** Write a summary about how the major wind systems form.

LEARNSMART® Go online to follow your personalized learning path to review, practice, and reinforce your understanding.

GATHERING WEATHER DATA

FOCUS QUESTION

How do we gather information about the weather?

Data from Earth's Surface

Meteorologists measure atmospheric conditions, such as temperature, air pressure, wind speed, and relative humidity. The quality of the data is critical for complete weather analysis and precise predictions. Two important factors in weather forecasting are the accuracy of the data and the amount of available data.

Temperature and pressure

A **thermometer,** shown in **Figure 10,** measures temperature using either the Fahrenheit or Celsius scale. There are several types of thermometers, including liquid-in-glass, bimetallic-strip, and digital. Liquid-in-glass thermometers usually contain a column of alcohol sealed in a glass tube. The liquid expands when heated, causing the column to rise, and contracts when it cools, causing the column to fall. A bimetallic-strip thermometer has a dial with a pointer. It contains a strip of metal made from two different metals that expand at different rates when heated. The strip is long and coiled into a spiral, making it more sensitive to temperature changes.

Liquid-in-glass thermometer   Bimetallic strip thermometer

Figure 10 Thermometers are common weather instruments.

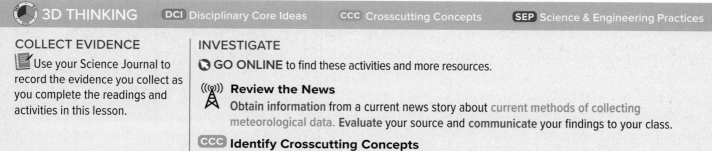

3D THINKING   **DCI** Disciplinary Core Ideas   **CCC** Crosscutting Concepts   **SEP** Science & Engineering Practices

COLLECT EVIDENCE

Use your Science Journal to record the evidence you collect as you complete the readings and activities in this lesson.

INVESTIGATE

GO ONLINE to find these activities and more resources.

((•)) **Review the News**
Obtain information from a current news story about current methods of collecting meteorological data. Evaluate your source and communicate your findings to your class.

CCC **Identify Crosscutting Concepts**
Create a table of the crosscutting concepts and fill in examples you find as you read.

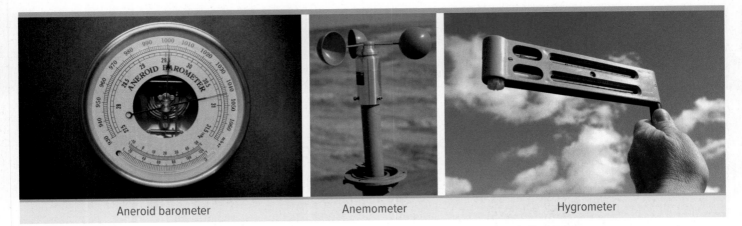

| Aneroid barometer | Anemometer | Hygrometer |

Figure 11 Barometers measure barometric pressure. Anemometers are used to measure wind speed. Hygrometers measure humidity.

A **barometer** measures air pressure. Some barometers have a column of mercury in a glass tube. One end of the tube is submerged in an open container of mercury. Changes in air pressure change the height of the column. Another type of barometer is an aneroid barometer, shown in **Figure 11.** It has a sealed, metal chamber with flexible sides. Most of the air is removed, so the chamber contracts or expands with changes in air pressure. A system of levers connects the chamber to a pointer on a dial.

Wind speed and relative humidity

An **anemometer** (a nuh MAH muh tur), shown in **Figure 11,** measures wind speed. The simplest type of anemometer has three or four cupped arms, positioned at equal angles from each other, that rotate as the wind blows. The wind's speed can be calculated using the number of revolutions of the cups over a given time. Some anemometers also have a wind vane that shows the direction of the wind.

A **hygrometer** (hi GRAH muh tur), such as the one in **Figure 11,** measures humidity. This type of hygrometer has wet-bulb and dry-bulb thermometers and requires a conversion table to determine relative humidity. When water evaporates from the wet bulb, the bulb cools. The temperatures of the two thermometers are read at the same time, and the difference between them is calculated. The relative humidity table lists the specific relative humidity for the difference between the thermometers.

Automated Surface Observing System

Meteorologists need a true "snapshot" of the atmosphere at one given moment to develop a forecast. To obtain this, meteorologists analyze and interpret data gathered at the same time, and at many different locations by weather instruments such as the one shown in **Figure 12.** Advances in sensors and computers allow instantaneous data collection.

The United States uses a surface-weather observation network known as the Automated Surface Observing System (ASOS). It gathers data 24 hours a day, every day. ASOS provides essential weather data for aviation, weather forecasting, and weather-related research.

Figure 12 This weather station consists of several instruments that measure atmospheric conditions.

Data from the Upper Atmosphere

While surface-weather data are important, weather is largely the result of changes that take place high in the troposphere. To make accurate forecasts, meteorologists must gather data at high altitudes, up to 30,000 m. This task is more difficult than gathering surface data, and it requires sophisticated technology.

An instrument used for gathering upper-atmospheric data is a **radiosonde** (RAY dee oh sahnd), shown in **Figure 13.** It consists of sensors and a radio transmitter suspended from a balloon that is about 2 m in diameter and filled with helium or hydrogen. A radiosonde's sensors measure air temperature, pressure, and humidity. Radio signals transmit these data to a ground station that tracks the radiosonde's movement. If a radiosonde also measures wind direction and speed, it is called a rawinsonde (RAY wuhn sahnd), **ra**dar + **win**d + radio**sonde.**

Tracking is a crucial component of upper-level observations. The system used since the 1980s has been replaced with one that uses Global Positioning System (GPS) and the latest computer technology. Meteorologists can determine wind speed and direction by tracking how fast and in what direction a rawinsonde moves. The various data are plotted on a chart that gives meteorologists a profile of the temperature, pressure, humidity, wind speed, and wind direction of a particular part of the atmosphere. Such charts are used to forecast atmospheric changes that affect surface weather.

Figure 13 Radiosondes gather upper-level weather data such as air temperature, pressure, and humidity.

Weather Observation Systems

There are many surface and upper-level observation sites across the United States. However, data from these sites cannot be used to locate exactly where precipitation falls without the additional help of data from weather radars and satellites.

Weather radar

A weather radar system detects specific locations of precipitation. The term *radar* stands for **ra**dio **d**etection **an**d **r**anging. A radar system generates radio waves and transmits them through an antenna at the speed of light. The transmitter is programmed to generate waves that reflect only from particles larger than a specific size. For example, when the radio waves encounter raindrops, some of the waves scatter. Because an antenna cannot send and receive signals at the same time, radars send a pulse and wait for the return before another pulse is sent. An amplifier increases the received wave signals, and then a computer processes and displays them on a monitor. From these data, the distance to precipitation and its location relative to the receiving antenna are calculated.

Doppler weather radar The pitch produced by the horn of an approaching car gets higher as it comes closer to you and lower as it passes and moves away from you. This phenomenon is called the Doppler effect. The **Doppler effect** is the change in pitch or frequency that occurs due to the relative motion of a wave, such as sound or light, as it comes toward or goes away from an observer.

Figure 14 shows the Doppler radar system that the National Weather Service (NWS) uses. Doppler radar data can be analyzed to determine the speed at which precipitation moves toward or away from a radar station. Because the movement of precipitation is caused by wind, Doppler radar can also estimate wind speeds associated with precipitation areas, including thunderstorms and tornadoes. The ability to measure wind speeds gives Doppler radar an advantage over conventional weather radar systems.

Figure 14 Norman, Oklahoma, was the site of the first Doppler radar installation.
Relate *the importance of this location to severe weather conditions.*

Weather satellites

Satellites orbiting Earth are used to observe weather. Cameras mounted aboard a weather satellite take images of Earth at regular intervals. A weather satellite can use infrared, visible-light, or water-vapor imagery to observe the atmosphere.

 Get It?

Identify the types of imagery that weather satellites can use.

Infrared imagery Infrared imagery can make observations at night. Objects radiate thermal energy at slightly different frequencies. Infrared imagery detects these different frequencies, which enables meteorologists to map either cloud cover or surface temperatures. Different frequencies are distinguishable in an infrared image, as shown in **Figure 15.**

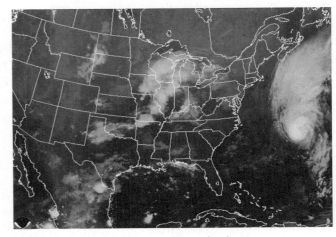

Figure 15 This infrared image shows cloud cover across most of the midwestern United States.

Clouds form at different altitudes and have different temperatures. Using infrared imagery, meteorologists can determine the cloud's temperature, its type, and its altitude. Infrared imagery is especially useful in detecting strong thunderstorms that develop and reach high altitudes. Since the troposphere cools with increasing altitude, storms appear as very cold areas on an infrared image. Because the strength of a thunderstorm is related to the altitude that it reaches, infrared imagery can be used to estimate the severity of a storm.

Visible light imagery Some satellites use cameras that require visible light to photograph Earth. These digital photos are sent back to ground stations, and their data are plotted on maps. Unlike weather radar, which tracks precipitation but not clouds, these satellites track clouds but not necessarily precipitation. They also show differences in the shading of clouds, which can relate to cloud thickness. By combining radar and visible imagery data, shown in **Figure 16** on the next page, meteorologists can determine where both clouds and precipitation are occurring.

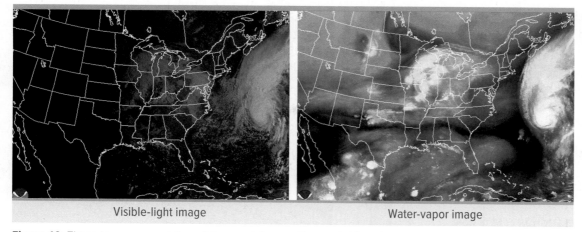

Visible-light image | Water-vapor image

Figure 16 These images were taken at the same time as the one in **Figure 15.** Each image shows different atmospheric characteristics. Together, they help meteorologists analyze and predict weather.

Water-vapor imagery Water vapor is an invisible gas and cannot be photographed, but it absorbs and emits infrared radiation at certain wavelengths. Many weather satellites have sensors that are able to provide a measure of the amount of water vapor present in the atmosphere. This is water-vapor imagery, also shown in **Figure 16.**

Water-vapor imagery is a valuable tool for weather analysis and prediction because it shows moisture in the atmosphere, not just cloud patterns. Because air currents that guide weather systems are often well defined by trails of water vapor, meteorologists can closely monitor storm systems even when clouds are not present.

✏️ Check Your Progress

Summary

- To make accurate weather forecasts, meteorologists analyze and interpret data gathered from Earth's surface by weather instruments.

- A radiosonde collects upper-atmospheric data.

- Doppler radar locates where precipitation occurs.

- Weather satellites use infrared, visible-light, or water-vapor imagery to observe and monitor changing weather conditions on Earth.

Demonstrate Understanding

1. **Identify** two important factors in collecting and analyzing weather data in the United States.

2. **Compare and contrast** methods for obtaining data from Earth's surface and Earth's upper atmosphere.

3. **State** the main advantage of Doppler radar over conventional weather radar.

4. **Summarize** the three kinds of weather satellite imagery. Use a graphic organizer.

Explain Your Thinking

5. **Analyze** the relationship between the amount of moisture in the air and the temperature of the wet bulb in a hygrometer.

6. **WRITING ▶ Connection** Write a newspaper article about the use of water-vapor imagery to detect water on the planet Mars.

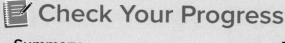

Go online to follow your personalized learning path to review, practice, and reinforce your understanding.

FOCUS QUESTION

Why is predicting the weather challenging?

Surface Weather Analysis

Newspapers, radio and television stations, and Web sites often give weather reports. These data are plotted on weather charts and maps and are often accompanied by radar and satellite imagery.

Station models

After weather data are gathered, meteorologists plot the data on a map using station models for individual cities or towns. A **station model** is a record of weather data for a particular site at a particular time. Meteorological symbols, such as the ones shown in **Figure 17,** are used to represent weather data in a station model. A station model allows meteorologists to fit a large amount of data into a small space. It also gives meteorologists a uniform way of communicating weather data.

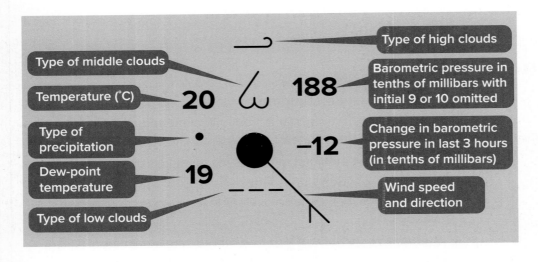

Figure 17 A station model shows temperature, wind direction and speed, and other weather data for a particular location at a particular time.

Explain *the advantage of using meteorological symbols.*

 3D THINKING   **DCI** Disciplinary Core Ideas   **CCC** Crosscutting Concepts   **SEP** Science & Engineering Practices

COLLECT EVIDENCE

 Use your Science Journal to record the evidence you collect as you complete the readings and activities in this lesson.

INVESTIGATE

GO ONLINE to find these activities and more resources.

Design Your Own: Predicting the Weather
Develop and use a model to forecast changes in weather.

GeoLAB: Interpret a Weather Map
Analyze and interpret the data on a weather map to describe the patterns in the current weather and make predictions about the future.

Figure 18 The weather map shows isobars and air pressure data for the continental United States. Air pressure is measured in millibars (mb).

Determine where on the weather map you would expect the strongest winds.

Plotting station model data

Station models provide information for individual sites. To plot data nationwide and globally, meteorologists use lines, called isopleths, that connect points of equal or constant values. The values represent different weather variables, such as pressure or temperature. Lines of equal pressure, for example, are called **isobars,** while lines of equal temperature are called **isotherms.** The lines themselves are similar to the contour lines—lines of equal elevation—on a topographic map.

Interpreting station model data

Recall that inferences about elevation can be made by studying contour intervals on a map. Inferences about weather, such as wind speed, can be made by studying isobars and isotherms on a map. Isobars that are close together indicate a large pressure difference over a small area, which means strong winds. Isobars that are far apart indicate a small difference in pressure and light winds. As shown in **Figure 18,** isobars also indicate the locations of high- and low-pressure systems. Combining this information with that of isotherms helps meteorologists to identify fronts.

Using isobars, isotherms, and station-model data, meteorologists can detect and analyze current weather conditions for a particular location. This is important because meteorologists must understand current weather conditions before they can forecast the weather.

Types of Forecasts

A meteorologist must analyze data from different levels in the atmosphere, based on current and past weather conditions, to produce a reliable forecast. Two types of forecasts are digital forecasts and analog forecasts.

Digital forecasts The atmosphere behaves like a fluid. Physical principles that apply to a fluid, such as temperature, pressure, and density, can be applied to the atmosphere and its variables. In addition, atmospheric variables can be expressed as mathematical equations to determine how they change over time.

A **digital forecast** is created by applying physical principles and mathematics to atmospheric variables and then making a prediction about how these variables will change over time. Digital forecasting relies on numerical data. Its accuracy is related directly to the amount of available data. It would take a long time for meteorologists to solve atmospheric equations on a global or national scale. Fortunately, computers can do the job quickly. Digital forecasting is the main method used by meteorologists today.

Analog forecasts

Another type of forecast, an **analog forecast,** is based on a comparison of current weather patterns with similar weather patterns from the past. Meteorologists coined the term *analog forecasting* because they look for a pattern from the past that is similar, or analogous, to a current pattern. To ensure the accuracy of an analog forecast, meteorologists, such as the one in **Figure 19,** must find a past event that had similar atmosphere, at all levels and over a large area, to a current event.

The main disadvantage of analog forecasting is the difficulty in finding the same weather pattern in the past. Still, analog forecasting is useful for conducting monthly or seasonal forecasts, which are based mainly on the past behavior of cyclic weather patterns.

Figure 19 This meteorologist is analyzing data from various sources to prepare a weather forecast.

 Get It?

Describe the advantages and disadvantages of analog forecasts.

Short-Term Forecasts

The most accurate and detailed forecasts are short term because weather systems change direction, speed, and intensity over time. For hourly forecasts, extrapolation is a reliable forecasting method because small-scale weather features that are readily observable by radar and satellites dominate current weather.

One- to three-day forecasts are no longer based on the movement of observed clouds and precipitation, which change by the hour. Instead, these forecasts are based on the behavior of larger surface and upper-level features, such as low-pressure systems. A one- to three-day forecast is usually accurate for expected temperatures as well as when and how much precipitation will occur. For this time span, however, the forecast will not be able to pinpoint an exact temperature or sky condition at a specific time.

 Get It?

Identify the type of weather forecast that is the most accurate and detailed.

ACADEMIC VOCABULARY

extrapolation
(ihk stra puh LAY shun)
the act of inferring a probable value from an existing set of values
Short-term weather forecasts can be extrapolated from data collected by radar and satellites.

M Stock/Alamy Stock Photo

Long-Term Forecasts

Because it is impossible for computers to model all of the variables that affect the weather at a given time and place, all long-term forecasts are less reliable than short-term forecasts. Features on Earth's surface, such as lakes, affect the amount of thermal energy absorbed at any location. This affects air pressure, which, in turn, affects the wind. Wind influences cloud formation and all other aspects of weather. Over time, these factors interact and create more complicated weather scenarios.

Meteorologists use changes in surface weather systems, based on circulation patterns throughout the lower atmosphere, for four- to seven-day forecasts. They can estimate each day's weather, but they cannot predict exact weather conditions. One- to two-week forecasts are based on changes in large-scale circulation patterns. These forecasts are vague and based mainly on similar conditions that have occurred in the past.

Forecasts for months and seasons are based mostly on weather cycles or patterns. These cycles, such as the one shown in **Figure 20,** can involve changes in the atmosphere, ocean currents, and solar activity that might occur at the same time. Improvements in weather forecasting depend on identifying the influences, understanding how they interact, and determining their ultimate effect on weather over time.

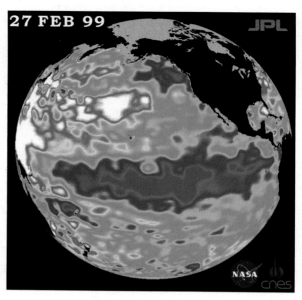

Figure 20 La Niña occurs when stronger-than-normal trade winds carry the colder water (blue) from the coast of South America to the equatorial Pacific Ocean. This happens about every three to five years and can affect global weather patterns.

Check Your Progress

Summary

- A station model is used to plot different weather variables.
- Meteorologists plot lines on a map, connecting variables of equal values to represent national and global trends.
- Two kinds of forecasts are digital and analog.
- The longer the prediction period, the less reliable the weather forecast.

Demonstrate Understanding

1. **Describe** the methods used for illustrating weather forecasts.
2. **Identify** some of the symbols used in a station model.
3. **Model** how temperature and pressure are shown in a weather map.
4. **Compare and contrast** analog and digital forecasts.
5. **Explain** why long-term forecasts are not as accurate as short-term forecasts.

Explain Your Thinking

6. **Assess** which forecast type—digital or analog—would be more accurate for three days or less.
7. **MATH** **Connection** Using a newspaper or other source, find and record the high and low temperatures in your area for five days. Calculate the average high and low temperatures for the five-day period.

The Future Looks Bright

New and improved technology is helping meteorologists forecast the weather with more precision than ever before. Cutting-edge satellites, supercomputers, and modeling systems can provide everyday weather forecasts as well as vital information for people in the path of dangerous storms.

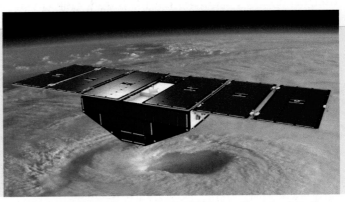

In this artist's conception, one of CYGNSS's satellites predicts the wind speed of a hurricane on Earth.

Precision forecasting

Meteorologists have not had much success in making accurate predictions more than three to ten days into the future because of unforeseen small changes in the atmosphere and sudden events that affect weather, such as forest fires. So after years of trying to improve long-range forecasts, meteorologists have changed their focus. Now they are concentrating on "precision forecasting," making predictions for extremely small areas, such as a group of city blocks.

To accomplish this "hyperlocal" forecasting, meteorologists need fast computers. They have begun using computers with chips called graphic processing units (GPUs), which are designed for parallel processing. GPUs run computations more quickly than traditional central processing unit (CPU) chips. The National Oceanic and Atmospheric Administration's (NOAA) National Weather Service is now running two supercomputers with GPUs to boost its precision forecasting abilities.

Improvements to NOAA's United States Global Forecast System (GFS), a numerical weather prediction system, have made hourly forecasts possible. The GFS also has higher-resolution models of the atmosphere and has added a fourth dimension—time—to its modeling.

Satellites

Meteorologists can better predict the wind speed of hurricanes over ocean waters with the help of the National Aeronautics and Space Administration's (NASA) Cyclone Global Navigation Satellite System (CYGNSS). The eight small satellites in the system measure the roughness of the ocean (which indicates wind speeds) by the way Global Positioning System (GPS) satellite signals bounce off the water. The GFS can also use data from two types of NASA satellites—the Geostationary Operational Environmental Satellites (GOES) and the Joint Polar Satellite System (JPSS).

 COMMUNICATE TECHNICAL INFORMATION

Work with a partner to research one of the technologies discussed in the feature. Create a presentation that explains how the technology works, and share it with your class.

NASA

MODULE 9
STUDY GUIDE

 GO ONLINE to study with your Science Notebook.

Lesson 1 THE CAUSES OF WEATHER

- Meteorology is the study of atmospheric phenomena.
- Solar radiation is unequally distributed between Earth's equator and its poles.
- An air mass is a large body of air that takes on the moisture and temperature characteristics of the area over which it forms.
- Each type of air mass is classified by its source region.

- weather
- climate
- air mass
- source region

Lesson 2 WEATHER SYSTEMS

- The three major wind systems are the polar easterlies, the prevailing westerlies, and the trade winds.
- Fast-moving, high-altitude jet streams greatly influence weather in the middle latitudes.
- The four types of fronts are cold fronts, warm fronts, stationary fronts, and occluded fronts.
- Air moves in a generally circular motion around either a high- or low-pressure center.

- Coriolis effect
- polar easterlies
- prevailing westerlies
- trade winds
- jet stream
- front

Lesson 3 GATHERING WEATHER DATA

- To make accurate weather forecasts, meteorologists analyze and interpret data gathered from Earth's surface by weather instruments.
- A radiosonde collects upper-atmosphere data.
- Doppler radar locates where precipitation occurs.
- Weather satellites use infrared, visible-light, and water-vapor imagery to observe and monitor changing weather conditions on Earth.

- thermometer
- barometer
- anemometer
- hygrometer
- radiosonde
- Doppler effect

Lesson 4 WEATHER ANALYSIS AND PREDICTION

- A station model is used to plot different weather variables.
- Meteorologists plot lines on a map, connecting variables of equal value to represent national and global trends.
- Two kinds of forecasts are digital and analog.
- The longer the prediction period, the less reliable the weather forecast.

- station model
- isobar
- isotherm
- digital forecast
- analog forecast

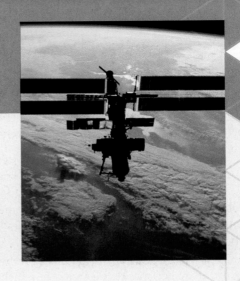

REVISIT THE PHENOMENON

How do satellites help us track and predict weather?

CER Claim, Evidence, Reasoning

Explain your Reasoning Revisit the claim you made when you encountered the phenomenon. Summarize the evidence you gathered from your investigations and research and finalize your Summary Table. Does your evidence support your claim? If not, revise your claim. Explain why your evidence supports your claim.

STEM UNIT PROJECT
Now that you've completed the module, revisit your STEM unit project. You will summarize your evidence and apply it to the project.

GO FURTHER

SEP Data Analysis Lab

How do you analyze a weather map?

Areas of high and low pressure are shown on a weather map by isobars.

CER Analyze and Interpret Data

1. Trace the diagram shown to the right on a blank piece of paper. Add the pressure values in millibars (mb) at the various locations.
2. A 1004-mb isobar has been drawn. Complete the 1000-mb isobar. Draw a 996-mb isobar and a 992-mb isobar.
3. **Claim, Evidence** Identify the contour interval of the isobars on this map and label the center of the closed 1004-mb isobar with a blue *H* for high pressure or a red *L* for low pressure.
4. **Reasoning** Determine the type of weather commonly associated with this pressure system.

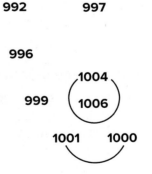

991   994   992

992   997

996

996

1004
999   1006

1001   1000

Oliver Henze/EyeEm/Getty Images

ENCOUNTER THE PHENOMENON

Why do only some thunderstorms produce tornadoes?

GO ONLINE to play a video about the formation of severe weather.

SEP Ask Questions

Do you have other questions about the phenomenon? If so, add them to the driving question board.

CER Claim, Evidence, Reasoning

Make Your Claim Use your CER chart to make a claim about why only some thunderstorms produce tornadoes. Explain your reasoning.

Collect Evidence Use the lessons in this module to collect evidence to support your claim. Record your evidence as you move through the module.

Explain Your Reasoning You will revisit your claim and explain your reasoning at the end of the module.

GO ONLINE to access your CER chart and explore resources that can help you collect evidence.

LESSON 1: Explore & Explain: Types of Thunderstorms

LESSON 2: Explore and Explain: Strong Winds

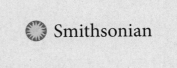

Additional Resources

THUNDERSTORMS

FOCUS QUESTION
What does a thunderstorm need to grow?

Overview of Thunderstorms

At any given moment, nearly 2000 thunderstorms are in progress around the world. Florida is one of the stormiest places on Earth. Some Florida cities get more than 100 thunderstorms per year! Some storms grow into atmospheric monsters capable of producing baseball-sized hail, tornadoes, and surface winds of more than 160 km/h. These thunderstorms can also provide the energy for nature's most destructive storms—hurricanes. Severe thunderstorms, regardless of intensity, have certain characteristics in common. **Figure 1** shows which areas of the United States experience the most thunderstorms annually.

How thunderstorms form

Recall that the stability of the air is determined by whether or not an air mass can rise up, or lift. Cool air masses are stable and, as long as they remain cooler than the surrounding air, they will resist uplift. Air masses that are warmed from the land or water below them are not stable. As long as these air masses remain warmer than the surrounding air, they will rise. Additionally, the air mass will continue to rise until its temperature becomes cool enough to resist uplift. Under the right conditions, convection causes a cumulus cloud to grow into a cumulonimbus cloud. These conditions are the same conditions that produce thunderstorms. For a thunderstorm to form, three conditions must exist: a source of moisture, lifting of the air mass, and an unstable atmosphere.

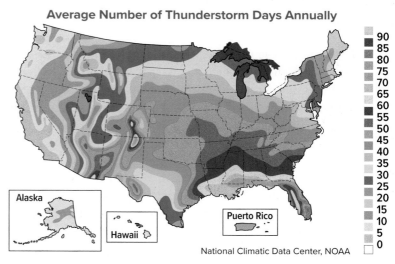

Average Number of Thunderstorm Days Annually

Alaska

Hawaii

Puerto Rico

National Climatic Data Center, NOAA

90
85
80
75
70
65
60
55
50
45
40
35
30
25
20
15
10
5
0

Figure 1 Both geography and air mass movements make thunderstorms most common in the southeastern United States.

Predict *why the Pacific Coast has so few thunderstorms and Florida has so many.*

3D THINKING **DCI** Disciplinary Core Ideas   **CCC** Crosscutting Concepts   **SEP** Science & Engineering Practices

COLLECT EVIDENCE
Use your Science Journal to record the evidence you collect as you complete the readings and activities in this lesson.

INVESTIGATE
GO ONLINE to find these activities and more resources.

? **Revisit the Encounter the Phenomenon Question**
What information from this lesson can help you answer the Unit and Module questions?

CCC **Identify Crosscutting Concepts**
Create a table of the crosscutting concepts and fill in examples you find as you read.

Figure 2 This cumulus cloud is growing as a result of unstable conditions. As the cloud continues to develop into a cumulonimbus cloud, a thunderstorm might develop.

Moisture First, for a thunderstorm to form, there must be an abundant source of moisture in the lower levels of the atmosphere. Air masses that form over tropical oceans or large lakes become more humid from water evaporating from the surface below. This humid air is less dense than the surrounding dry air and is lifted. The water vapor it contains condenses into droplets, forming clouds. Latent heat, which is released from the water vapor during the process of condensation, warms the air, causing it to rise further, cool further, and condense more of its water vapor.

Lifting Second, there must be some mechanism for condensing moisture to release its latent heat. This occurs when a warm air mass is lifted into a cooler region of the atmosphere. Dense, cold air along a cold front can push warmer air upward, just like an air mass does when moving up a mountainside. Warm land areas, heat islands such as cities, and bodies of water can also provide heat for lifting an air mass. Only when the water vapor condenses can it release latent heat and keep the cloud rising.

Stability Third, if the surrounding air remains cooler than the rising air mass, the unstable conditions can produce clouds that grow upward. This releases more latent heat and allows continued lifting. However, when the density of the rising air mass and the surrounding air are nearly the same, the cloud stops growing. **Figure 2** shows a cumulus cloud that is on its way to becoming a cumulonimbus cloud that can produce thunderstorms.

 Get It?

Describe the three conditions for thunderstorm growth.

Limits to thunderstorm growth

The conditions that limit thunderstorm growth are the same ones that form the storm. Conditions that create lift, condense water vapor, and release latent heat keep the air mass warmer than the surrounding air. The air mass will continue to rise until it reaches a layer of equal density that it cannot overcome. Because atmospheric stability increases with height, most cumulonimbus clouds are limited to about 12,000 m. Thunderstorms are also limited by duration and size.

Sea breeze

During the day, the temperature of land increases faster than the temperature of water. The warm air over land expands and rises, and the colder air over the sea moves inland and replaces the warm air. These conditions can produce strong updrafts that result in thunderstorms.

Land breeze

At night, conditions are reversed. The land cools faster than water, so the warmer sea air rises, and cooler air from above land moves over the water and replaces it. Nighttime conditions are considered stable.

Figure 3 Temperature differences exist over land and water and vary with the time of day.

Infer *why water is warmer than land at night.*

Types of Thunderstorms

Thunderstorms are classified according to the mechanism that causes the air mass that formed them to rise. There are two main types of thunderstorms: air-mass and frontal.

Air-mass thunderstorms

When air rises because of unequal heating of Earth's surface within one air mass, the thunderstorm is called an **air-mass thunderstorm.** Because the unequal heating of Earth's surface reaches its maximum during mid-afternoon, it is common for air-mass thunderstorms to occur at this time.

There are two kinds of air-mass thunderstorms. **Mountain thunderstorms** occur when an air mass rises by orographic lifting, which involves air moving up the side of a mountain. **Sea-breeze thunderstorms** are local air-mass thunderstorms that occur because land and water store and release thermal energy differently. Sea-breeze thunderstorms are common along coastal areas during the summer, especially in the tropics and subtropics. Because land heats and cools faster than water, temperature differences can develop between the air over coastal land and the air over water, as shown in **Figure 3.**

Frontal thunderstorms

Frontal thunderstorms are produced by advancing cold fronts and, more rarely, warm fronts. In a cold-front storm, dense, cold air pushes under warm air, which is less dense, rapidly lifting it up a steep cold-front boundary. This rapid upward motion can produce a thin line of thunderstorms, sometimes hundreds of kilometers long, along the leading edge of the cold front. Cold-front thunderstorms get their initial lift from the push of the cold air. Because they are not dependent on daytime heating for their initial lift, cold-front thunderstorms can persist long into the night. Flooding from soil saturation is common with these storms.

Less frequently, thunderstorms can develop along the advancing edge of a warm front. In a warm-front storm, a warm air mass slides up and over a gently sloping cold air mass. If the warm air behind the warm front is unstable and moisture levels are sufficiently high, a relatively mild thunderstorm can develop.

Thunderstorm Development

A thunderstorm usually has three stages, as shown in **Figure 4:** the cumulus stage, the mature stage, and the dissipation stage. The stages are classified according to the direction the air is moving.

Cumulus stage

During the cumulus stage, air starts to rise vertically, as shown in **Figure 4.** The water vapor condenses into visible cloud droplets and releases latent heat. As the cloud droplets coalesce, they become larger and heavier until the updrafts can no longer sustain them, and they fall to Earth as precipitation. This begins the mature stage of a thunderstorm.

Mature stage

During the mature stage, updrafts and downdrafts exist side by side in the cumulonimbus cloud. Precipitation, composed of water and ice droplets that formed at high, cool levels of the atmosphere, cools the air as it falls. The newly cooled air is more dense than the surrounding air, so it sinks rapidly to the ground along with the precipitation. This creates downdrafts. As **Figure 4** shows, the updrafts and downdrafts form a convection cell, which produces the surface winds associated with thunderstorms. The average area covered by a thunderstorm in its mature stage is 24 km.

Dissipation stage

The convection cell can exist only if there is a steady supply of warm, moist air at Earth's surface. Once that supply is depleted, the updrafts slow down and eventually stop. During a thunderstorm, the cool downdrafts spread in all directions when they reach Earth's surface. This cools the areas from which the storm draws its energy, the updrafts cease, and clouds can no longer form. The storm is then in the dissipation stage shown in **Figure 4.** This stage will last until all of the previously formed precipitation has fallen.

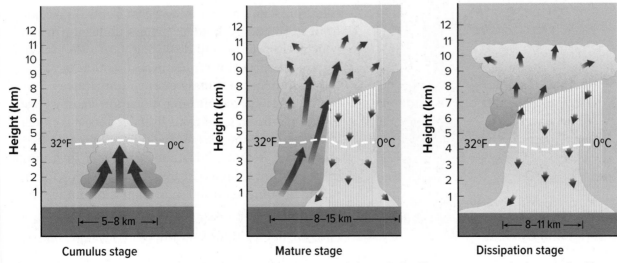

Figure 4 The cumulus stage of a thunderstorm is characterized mainly by updrafts. The mature stage is characterized by strong updrafts and downdrafts. The storm loses energy in the dissipation stage.

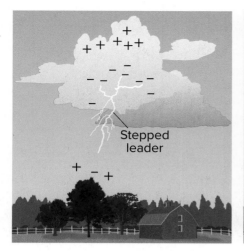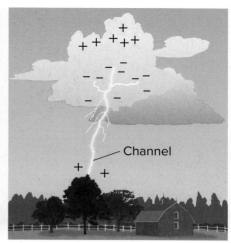

Figure 5 When a stepped leader nears an object on the ground, a powerful surge of electricity from the ground moves upward to the cloud, and lightning is produced.

Lightning

Lightning is the transfer of electricity generated by the rapid rushes of air in a cumulonimbus cloud. Clouds become charged when friction between the updrafts and downdrafts removes electrons from some of the atoms in the cloud. The atoms that lose electrons become positively charged ions. Other atoms receive the extra electrons and become negatively charged ions. As **Figure 5** shows, this creates regions of air with opposite charges. Eventually, the differences in charges break down, and a branched channel of partially charged air is formed between the positive and negative regions. The channel of partially charged air is called a **stepped leader,** and it generally moves from the center of the cloud toward the ground. When the stepped leader nears the ground, a branched channel of positively charged particles, called the **upward streamer,** rushes upward to meet it. The upward streamer surges from the ground to the cloud, illuminating the connecting return stroke channel with about 100 million volts of electricity. That illumination, the **return stroke,** is the brightest part of lightning.

Thunder

A lightning bolt heats the surrounding air to about 30,000°C. That is about five times hotter than the surface of the Sun. Thunder is the sound produced as this superheated air rapidly expands. Because sound waves travel more slowly than light waves, you might see lightning before you hear thunder, even though they are generated at the same time.

Lightning variations

There are several names given to lightning effects. Sheet lightning is reflected by clouds. Heat lightning is sheet lightning near the horizon. Spider lightning crawls across the sky for up to 150 km. Blue jets and red sprites originate in clouds and rise rapidly toward the stratosphere as cones or bursts.

Figure 6 Five times hotter than the surface of the Sun, a lightning bolt can be spectacular. But when an object such as this pine tree is struck, it can be explosive.

Thunderstorm and lightning safety

Each year in the United States, lightning causes about 12,000 wildfires, which burns thousands of square kilometers of land, and causes an average of 300 injuries and 58 deaths to humans. **Figure 6** indicates how destructive a lightning strike can be.

Avoid putting yourself in danger of being struck by lightning. If you are outdoors and feel your hair stand on end, the safest thing to do is get inside quickly. If that is not possible, squat low, on the balls of your feet. Duck your head and make yourself the smallest target possible. Small sheds, isolated trees, and convertible automobiles are hazardous as shelters. Using electrical appliances and telephones during a lightning storm can lead to electric shock. Stay out of boats and away from water during a thunderstorm.

 Check Your Progress

Summary

- The cumulus stage, the mature stage, and the dissipation stage comprise the life cycle of a thunderstorm.
- Clouds form as water is condensed and latent heat is released.
- Thunderstorms can be produced either within air masses or along fronts.
- From formation to dissipation, all thunderstorms go through the same stages.
- Lightning is a natural result of thunderstorm development.

Demonstrate Understanding

1. **List** what happens during a thunderstorm's cumulus stage.
2. **Relate** the formation of a thunderstorm to physical factors along a front.
3. **Differentiate** between a sea-breeze thunderstorm and a mountain thunderstorm.
4. **Identify** what causes a thunderstorm to dissipate.
5. **Compare and contrast** how a cold front and a warm front can create thunderstorms.
6. **Describe** two different types of lightning.

Explain Your Thinking

7. **Infer** Lightning occurs mainly during which stage of thunderstorm formation?
8. **Determine** the conditions that create lightning.
9. **WRITING Connection** Write about a setting for a movie in which a storm is part of the opening scene.

LEARNSMART Go online to follow your personalized learning path to review, practice, and reinforce your understanding.

FOCUS QUESTION
How does heat make storms more powerful?

Severe Thunderstorms

All thunderstorms are not created equal. Some die out within minutes, while others flash and thunder throughout the night. What makes one thunderstorm more severe than another? The increasing instability of the air intensifies the strength of a storm's updrafts and downdrafts, which makes a storm severe.

Supercells

Severe thunderstorms can produce some of the most violent weather conditions on Earth. They can develop into self-sustaining, extremely powerful storms called **supercells.** Supercells are characterized by intense, rotating updrafts taking 10 to 20 minutes to reach the top of the cloud. These furious storms can last for several hours and can have updrafts as strong as 240 km/h. It is not uncommon for a supercell to spawn long-lived tornadoes. **Figure 7** shows an illustration of a supercell. Notice the anvil-shaped cumulonimbus clouds associated with these severe storms. The top of a supercell is chopped off by wind shear. Of the estimated 100,000 thunderstorms that occur each year in the United States, only about 10 percent are considered to be severe, and fewer still reach supercell proportions.

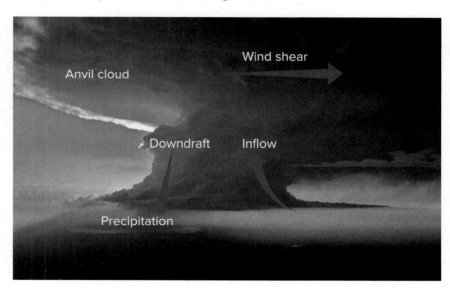

Figure 7 An anvil-shaped cumulonimbus cloud is characteristic of many severe thunderstorms. The most severe thunderstorms are called supercells.

 **3D THINKING**   **DCI** Disciplinary Core Ideas   **CCC** Crosscutting Concepts   **SEP** Science & Engineering Practices

COLLECT EVIDENCE
Use your Science Journal to record the evidence you collect as you complete the readings and activities in this lesson.

INVESTIGATE
GO ONLINE to find these activities and more resources.

 Review the News
Obtain information from a current news story about severe weather research. Evaluate your source and communicate your findings to your class.

? **Revisit the Encounter the Phenomenon Question**
What information from this lesson can help you answer the Unit and Module questions?

Strong Winds

Recall that rain-cooled downdrafts descend to Earth's surface during a thunderstorm and spread out as they reach the ground. Sometimes, instead of dispersing that downward energy over a large area underneath the storm, the energy becomes concentrated in a local area. The resulting winds are exceptionally strong, with speeds of more than 160 km/h. Violent downdrafts that are concentrated in a local area are called **downbursts.**

Based on the size of the area they affect, downbursts are classified as either macrobursts or microbursts. Macrobursts can cause a path of destruction up to 5 km wide. They have wind speeds of more than 200 km/h and can last up to 30 minutes. Smaller in size, though deadlier in force, microbursts affect areas of less than 4 km but can have winds exceeding 250 km/h. Despite lasting fewer than 10 minutes on average, a microburst is especially deadly because its small size makes it extremely difficult to predict and detect. **Figure 8** shows a microburst.

Figure 8 A microburst, such as this one, can be as destructive as a tornado.

Figure 9 This hailstorm caused slippery traffic conditions as well as property damage.

Hail

Each year in the United States, almost one billion dollars in damage is caused by hail—precipitation in the form of balls or lumps of ice. Hail can do tremendous damage to crops, vehicles, and rooftops, particularly in the central United States, where hail occurs most frequently. Hail is most common during the spring growing season. **Figure 9** shows some conditions associated with hail.

Hail forms in cumulonimbus clouds. First, water droplets enter the parts of a cloud where the temperature is below freezing. When these water droplets encounter ice pellets, the water droplets freeze on contact and cause the ice pellets to grow larger. The second characteristic that allows hail to form is an abundance of strong updrafts and downdrafts existing side by side within a cloud. The growing ice pellets are caught alternately in the updrafts and downdrafts, so that they constantly encounter more water droplets. The ice pellets keep growing until they are too heavy to stay aloft, and they finally fall to Earth as hail.

Tornadoes

In some parts of the world, the most dangerous form of severe weather is a tornado. A **tornado** is a violent, whirling column of air in contact with the ground. When a tornado does not reach the ground, it is called a funnel cloud. Tornadoes are often associated with supercells—the most severe thunderstorms. The air in a tornado is made visible by dust and debris drawn into the swirling column, sometimes called the vortex, or by the condensation of water vapor into a visible cloud.

Development of tornadoes

A tornado forms when wind speed and direction change suddenly with height, a phenomenon called wind shear. Current thinking suggests that tornadoes form when pockets of rising air are rotated horizontally, like a rolling pin, between layers of air with varying wind speeds or directions, as shown in **Figure 10.** If this rotation occurs close enough to the thunderstorm's updrafts, the twisting column of wind can be tilted from a horizontal to a vertical position. As updrafts stretch the column, the rotation is accelerated. Air is removed from the center of the column, which in turn lowers the air pressure in the center. The extreme pressure gradient between the center and the outer portion of the tornado produces the violent winds associated with tornadoes. Although tornadoes rarely exceed 200 m in diameter and usually last only a few minutes, they can be extremely destructive. A tornado is classified according to its destructive force.

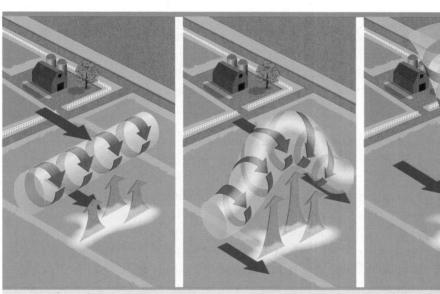

A change in wind direction and speed creates a horizontal rotation in the lower atmosphere.

Strong updrafts tilt the rotating air from a horizontal to a vertical position.

A tornado forms within the rotating winds.

Figure 10 Tornado formation is associated with changes in wind speed and direction.

Infer *what causes the updrafts that create a tornado.*

Table 1 Enhanced Fujita Tornado Damage Scale

Enhanced Fujita scale tornadoes	Weak (EF0 and EF1) 69 percent of all tornadoes Wind Speed: 105–177 km/h	Strong (EF2 and EF3) 29 percent of all tornadoes Wind speed: 178–266 km/h	Violent (EF4 and EF5) 2 percent of all tornadoes Wind speed: 267–322 km/h
Photo of tornado			

Tornado classification Tornadoes vary greatly in size and intensity. The **Enhanced Fujita Tornado Damage Scale,** which ranks tornadoes according to their destructiveness and estimated wind speed, is used to classify tornadoes. The Enhanced Fujita Scale is an update to the original Fujita Scale that was named for Japanese tornado researcher Dr. Theodore Fujita in the 1970s. The Enhanced Fujita Scale ranges from EF0, which is characterized by winds up to 137 km/h, to the incredibly violent EF5, which can pack winds of more than 322 km/h. Most tornadoes do not exceed the EF1 category. In fact, only about 2 percent reach EF4 or EF5. Those that do, however, can lift entire buildings from their foundations and toss automobiles and trucks around like toys. The Enhanced Fujita Tornado Damage Scale is shown in **Table 1.**

Tornado distribution While tornadoes can occur at any time and at any place, there are some times and locations at which they are more likely to form. Most tornadoes—especially violent ones—form in the spring during the late afternoon and evening, when the temperature contrasts between polar air and tropical air are the greatest. Large temperature contrasts occur most frequently in the central United States, where cold continental polar air collides with maritime tropical air moving northward from the Gulf of Mexico. These large temperature contrasts often spark the development of supercells, which are each capable of producing several strong tornadoes. More than 1000 tornadoes touch down each year in the United States. Many of these occur in a region called "Tornado Alley," which extends from northern Texas through Oklahoma, Kansas, and Missouri.

GEOGRAPHY Connection If you were to look at a map showing tornado occurrence on Earth, you would see that North America experiences more tornadoes than any other continent in the world. This is due to two important physical characteristics. The first is the air masses that arrive from all directions. Recall that when cold, dry air moving south from Canada encounters warm, moist air moving north from the Gulf of Mexico, storms form where they meet. The addition of dry, eastward-moving air masses from the Rocky Mountains and moist, westward-moving air masses from the Appalachian region contributes to the storms' formation. The second characteristic is the wide, flat, open land that exists between the Rocky Mountains and the Appalachian Mountains. This large expanse provides plenty of room for air masses to meet, rise, exchange energy, and develop into storms.

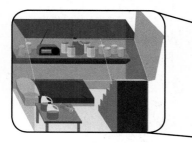

Figure 11 In some areas, tornado shelters are common. If you are caught in a tornado, take shelter in the southwest corner of a basement, a small downstairs room or closet, or a tornado shelter like this one.

Tornado safety

In the United States, an average of 80 deaths and 1500 injuries result from tornadoes each year. To help reduce these numbers, the National Weather Service issues tornado watches and warnings when conditions are conducive to the development of tornadoes or when tornadoes are indicated on weather radar or spotted in the region. These advisories are broadcast on radio and television stations. During a severe thunderstorm, the presence of dark, greenish skies; a towering wall of clouds; and a loud, roaring freight train-like noise are signs of an approaching or developing tornado.

The National Weather Service stresses that despite advanced tracking systems, some tornadoes develop very quickly, and advance warnings might not be possible. However, the number of tornado-related injuries can be substantially decreased when people seek shelter, such as the one shown in **Figure 11,** at the first sign of threatening skies.

Check Your Progress

Summary

- Intense, rotating updrafts are associated with supercells.

- Downbursts are strong winds that result in damage associated with thunderstorms.

- Hail is precipitation in the form of balls or lumps of ice that accompany severe storms.

- The worst storm damage comes from a vortex of high winds that moves along the ground as a tornado.

Demonstrate Understanding

1. **Identify** the characteristics of a severe storm.
2. **Describe** the formation of hailstones.
3. **Explain** how some hail can become baseball-sized.
4. **Compare and contrast** a macroburst and a microburst.
5. **Relate** physical factors such as wind shear to tornado formation.
6. **List** the conditions that lead to high winds, hail, and lightning.

Explain Your Thinking

7. **Explain** Why are there more tornado-producing storms in flat plains than in mountainous areas?
8. **Analyze** the data of the Enhanced Fujita Tornado Damage Scale, and determine why EF5 tornadoes often have a longer path than EF1 tornadoes.
9. **WRITING Connection** Create a pamphlet about tornado safety.

LEARNSMART Go online to follow your personalized learning path to review, practice, and reinforce your understanding.

TROPICAL STORMS

FOCUS QUESTION
Why are hurricanes so powerful?

Overview of Tropical Cyclones

During the summer and fall, the tropics experience ideal conditions for the formation of large, rotating, low-pressure tropical storms called **tropical cyclones.** In different parts of the world, the largest of these storms are known as hurricanes, typhoons, or cyclones.

Cyclone location

Favorable conditions for tropical cyclone formation exist in all tropical oceans except the South Atlantic Ocean and the Pacific Ocean off the west coast of South America. The water in these areas is somewhat cooler, and these areas contain regions of nearly permanently stable air. As a consequence, tropical cyclones do not normally occur in these areas.

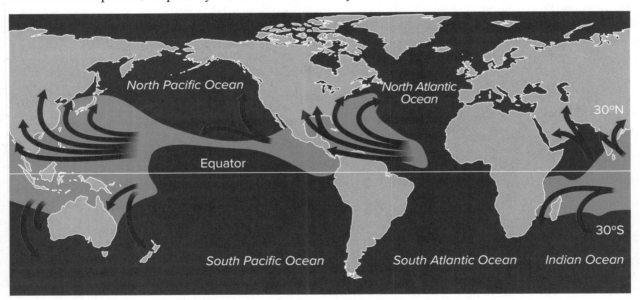

Figure 12 Tropical cyclones are common in all of Earth's tropical oceans except in the relatively cool waters of both the South Pacific and South Atlantic Oceans.

3D THINKING   **DCI** Disciplinary Core Ideas   **CCC** Crosscutting Concepts   **SEP** Science & Engineering Practices

COLLECT EVIDENCE

Use your Science Journal to record the evidence you collect as you complete the readings and activities in this lesson.

INVESTIGATE

GO ONLINE to find these activities and more resources.

Design Your Own: Building Hurricane-Proof Homes
Plan and carry out an investigation into how the structure of a building determines if the building will withstand a hurricane.

GeoLAB: Track a Tropical Cyclone
Obtain, evaluate, and communicate information about tropical cyclone patterns, paths, and life-cycles.

These storms do occur in the large expanse of warm water in the western Pacific Ocean, where they are known as typhoons. To people living near the Indian Ocean, they are known as cyclones. In the North Atlantic Ocean, the Caribbean Sea, the Gulf of Mexico, and along the western coast of Mexico, the strongest of these storms are called hurricanes. **Figure 12,** on the previous page, shows where tropical cyclones generally form.

Cyclone formation

Tropical cyclones require two basic conditions to form: an abundant supply of warm ocean water and some sort of mechanism to lift warm air and keep it rising. Tropical cyclones thrive on the tremendous amount of energy in warm, tropical oceans. As water evaporates from the ocean surface, latent heat is stored in water vapor. This latent heat is later released when the air rises and water vapor condenses.

Figure 13 The characteristic rotating nature and clearly defined eye of cyclonic storms is evident in this hurricane that formed over the Atlantic Ocean.

The air usually rises because of some sort of existing weather disturbance moving across the tropics. Many disturbances originate along the equator. Others are the result of weak, low-pressure systems called tropical waves. Tropical disturbances are common during the summer and early fall. Regardless of their origin, only a small percentage of tropical disturbances develop into cyclones. There are three stages in the development of a full tropical cyclone.

Formative stage The first indication of a building tropical cyclone is a slow-moving tropical disturbance. Less-dense, moist air is lifted, triggering rainfall and air circulation. As these disturbances produce more precipitation, more latent heat is released. In addition, the rising air creates an area of low pressure at the ocean surface. As more warm, dense air moves toward the low-pressure center to replace the air that has risen, the Coriolis effect causes the moving air to turn counterclockwise in the northern hemisphere. This produces the cyclonic (counterclockwise) rotation of a tropical cyclone, as shown in **Figure 13.** When a disturbance over a tropical ocean acquires a cyclonic circulation around a center of low pressure, it has reached the developmental stage and is known as a tropical depression, illustrated in **Figure 16.**

 Get It?

Infer what is produced when water vapor condenses.

SCIENCE USAGE v. COMMON USAGE

depression

Science usage: a pressing down or lowering; the low spot on a curved line

Common usage: a state of feeling sad

Image produced by Hal Pierce, Laboratory for Atmospheres, NASA Goddard Space Flight Center

Mature stage As the moving air approaches the center of the growing storm, it rises, rotates, and increases in speed as more energy is released through condensation. In the process, air pressure in the center of the system continues to decrease. As long as warm air is fed into the system at the surface and removed in the upper atmosphere, the storm will continue to build, and the winds of rotation will increase as the air pressure drops.

When wind speeds around the low-pressure center of a tropical depression exceed 62 km/h, the system is called a tropical storm. If air pressure continues to fall and winds around the center reach at least 119 km/h, the storm is officially classified as a tropical cyclone. Once winds reach these speeds, another phenomenon occurs—the development of a calm center of the storm called the **eye,** shown in **Figure 13.** The eye of the cyclone, which measures from 30 to 60 km in diameter, is an area of calm weather and blue sky. The strongest winds in a hurricane are usually concentrated in the **eyewall**—a tall band of strong winds and dense clouds that surrounds the eye. The heaviest rainfall is concentrated in rainbands that surround the eyewall. The eyewall is visible because of the clouds that form there and mark the outward edge of the eye.

Dissipation stage A cyclone will last until it can no longer produce enough energy to sustain itself. This usually happens when the storm has moved either over land or over colder water. During its life cycle, a cyclone can undergo several fluctuations in intensity as it interacts with other atmospheric systems. For example, examine the **Saffir-Simpson Hurricane Wind Scale,** shown in **Figure 14,** which classifies hurricanes according to wind speed, and gives an idea of the potential for property damage. In August 2005, Hurricane Katrina made landfall on the south Florida coast as a relatively weak Category 1 hurricane. But it continued across south Florida and entered the warm Gulf of Mexico waters, where it gained strength and was quickly upgraded to a Category 5 hurricane. By the time it made landfall in southeast Louisiana, it was downgraded to a Category 3 hurricane.

Tropical cyclone movement

Like all large-scale storms, tropical cyclones move according to the wind currents that steer them. Recall that many of the world's oceans are home to subtropical high-pressure systems that are present to some extent throughout the year. Tropical cyclones are often caught up in the circulation of these high-pressure systems. They move steadily west, then eventually turn poleward when they reach the far edges of the high-pressure systems. There, they are guided by prevailing westerlies and begin to interact with midlatitude systems. At this point, the movement of the storms becomes unpredictable.

Get It?
Describe the feedback that causes a tropical storm to increase in size and strength.

Saffir-Simpson Hurricane Wind Scale

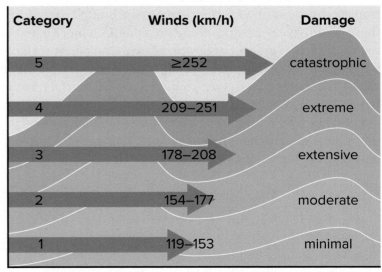

Category	Winds (km/h)	Damage
5	≥252	catastrophic
4	209–251	extreme
3	178–208	extensive
2	154–177	moderate
1	119–153	minimal

Figure 14 The Saffir-Simpson Hurricane Wind Scale classifies hurricanes according to wind speed. Note that even the winds in Category 1 or 2 hurricanes are strong enough to be considered damaging.

Figure 16 Visualizing Cyclone Formation

Moving air starts to spin as a result of the Coriolis effect.

5. As the lighter air rises, moist air from the ocean takes it place, creating a wind current.

4. Condensation releases latent heat into the atmosphere, making the air less dense.

3. As the water vapor rises, the cooler upper air condenses it into liquid droplets.

2. Water vapor is lifted into the atmosphere.

1. Warm air absorbs moisture from the ocean.

Tropical Depression The first indications of a building storm are a tropical depression with notable circulation, thunderstorms, and sustained winds of up to 62 km/h.

Tropical Storm As winds increase to speeds of 63–118 km/h, strong thunderstorms develop and become well defined.

Tropical Cyclone With sustained winds of more than 119 km/h, an intense tropical weather system with well-defined circulation becomes a cyclone, also called a typhoon or hurricane.

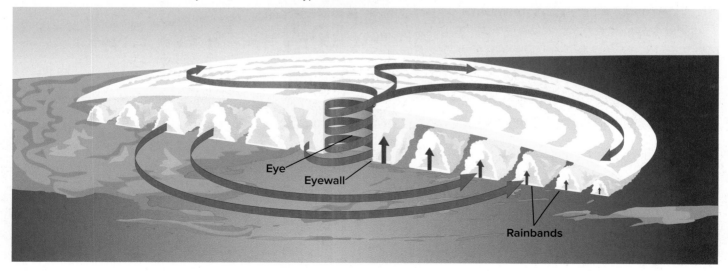

Eye

Eyewall

Rainbands

Hurricane Hazards

Recall that hurricanes are strong, regionally specific tropical cyclones. The amount of flooding and damage are dependent upon shore depth and the density of population and structures in the affected area.

Damage

Hurricanes can cause extensive damage, particularly along coastal areas, which tend to be where human populations are the most dense. Florida, in particular, has experienced hurricane damage. Out of the nine most violent hurricanes in U.S. history, six have struck Florida. **Figure 17** documents damage from hurricanes and other storms.

Winds

Much of the damage caused by hurricanes is associated with violent winds. The strongest winds in a hurricane are usually located at the eyewall. Outside of the eyewall, winds taper off as distance from the center increases, although winds of more than 60 km/h can extend as far as 400 km from the center of a hurricane.

Storm surge

Strong winds moving onshore in coastal areas are partly responsible for the largest hurricane threat—storm surges. A **storm surge** occurs when hurricane-force winds drive a mound of ocean water toward coastal areas, where it washes over the land, as shown in **Figure 15**. Storm surges can sometimes reach 6 m above normal sea level. When this occurs during high tide, the surge can cause enormous damage. In the northern hemisphere, a storm surge occurs primarily on the right side of a storm relative to the direction of its forward motion. The strongest onshore winds occur there due to the counterclockwise rotation of the storm.

Hurricanes produce heavy rain because of their continuous uptake of warm, moist ocean water. Thus, floods are an additional hurricane hazard, particularly if the storm moves over mountainous areas, where orographic lifting enhances the upward motion of air and the resulting condensation of water vapor.

Get It?

Compare Why would a storm surge be more dangerous during high tide than at low tide?

Figure 15 Storm surge has nearly flooded the first floor of this building. The wave is almost 5 stories tall.

STR/Stringer/AFP/Getty Images

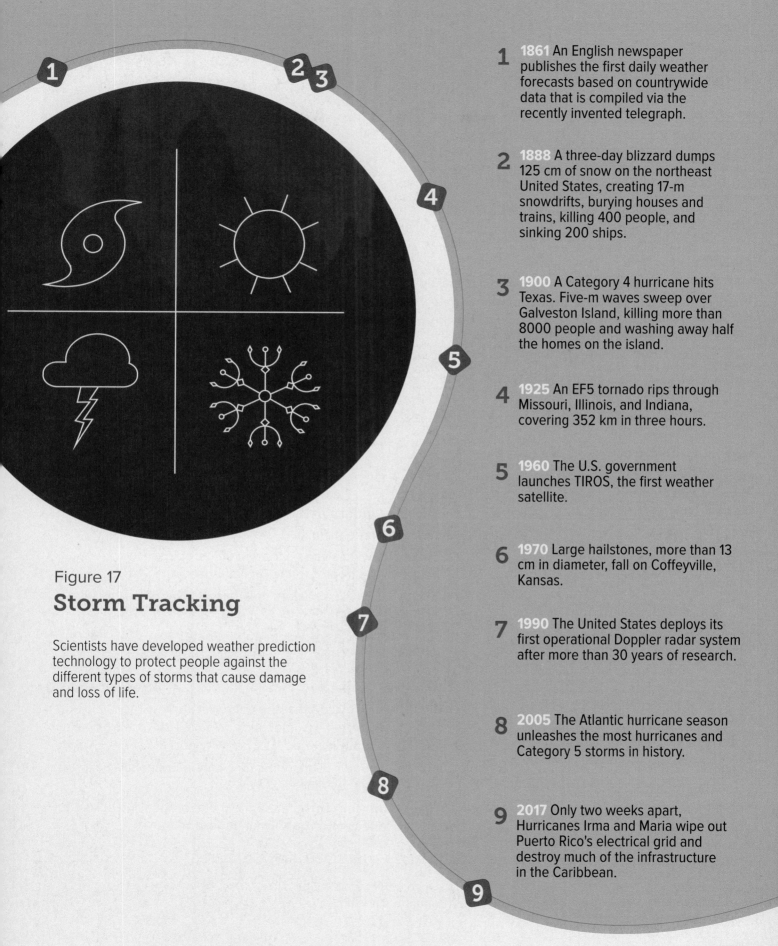

Figure 17

Storm Tracking

Scientists have developed weather prediction technology to protect people against the different types of storms that cause damage and loss of life.

1 **1861** An English newspaper publishes the first daily weather forecasts based on countrywide data that is compiled via the recently invented telegraph.

2 **1888** A three-day blizzard dumps 125 cm of snow on the northeast United States, creating 17-m snowdrifts, burying houses and trains, killing 400 people, and sinking 200 ships.

3 **1900** A Category 4 hurricane hits Texas. Five-m waves sweep over Galveston Island, killing more than 8000 people and washing away half the homes on the island.

4 **1925** An EF5 tornado rips through Missouri, Illinois, and Indiana, covering 352 km in three hours.

5 **1960** The U.S. government launches TIROS, the first weather satellite.

6 **1970** Large hailstones, more than 13 cm in diameter, fall on Coffeyville, Kansas.

7 **1990** The United States deploys its first operational Doppler radar system after more than 30 years of research.

8 **2005** The Atlantic hurricane season unleashes the most hurricanes and Category 5 storms in history.

9 **2017** Only two weeks apart, Hurricanes Irma and Maria wipe out Puerto Rico's electrical grid and destroy much of the infrastructure in the Caribbean.

Hurricane advisories and safety

The National Hurricane Center was established on July 1, 1956. Today it is located on the campus of Florida International University in Miami, Florida. The National Hurricane Center is responsible for tracking and forecasting the intensity and motion of tropical cyclones in the western hemisphere. The center issues a hurricane warning at least 36 hours before a hurricane is predicted to strike. The center also issues regular advisories that indicate a storm's position, strength, and movement. Using this information, people can then track a storm on a hurricane-tracking chart, such as the one you will use in the GeoLab associated with this module. Awareness, combined with proper safety precautions, has greatly reduced death tolls associated with hurricanes in recent years. **Figure 18** shows debris and destruction left by hurricane flooding.

Figure 18 This residential area has been engulfed in debris left behind by the floodwaters of Hurricane Katrina. Most of the deaths associated with a hurricane come from flooding, not high winds.

 Get It?

List and describe the hazards associated with hurricanes.

Check Your Progress

Summary

- Tropical cyclones rotate counter-clockwise in the northern hemisphere.
- Tropical cyclones are also known as hurricanes or typhoons.
- Tropical cyclones go through the same stages of formation and dissipation as other storms.
- Tropical cyclones are moved by various wind systems after they form.
- The most dangerous part of a tropical cyclone is the storm surge.
- Hurricane alerts are given at least 36 hours before the hurricane arrives.

Demonstrate Understanding

1. **Identify** the three main stages of a tropical cyclone.
2. **Describe** the wind systems that guide a tropical cyclone as it moves from the tropics to the midlatitudes.
3. **Relate** What physical factors must exist for a tropical cyclone to form?
4. **Explain** what causes a cyclone to dissipate.

Explain Your Thinking

5. **Analyze** Imagine that you live on the eastern coast of the United States and are advised that the center of a hurricane is moving inland 70 km north of your location. Would a storm surge be a major problem in your area? Why or why not?
6. **Compare** the Saffir-Simpson Scale with the Enhanced Fujita Scale. How are they different? Why?
7. **MATH Connection** Determine the average wind speed for each hurricane category shown in **Figure 14.**

LEARNSMART Go online to follow your personalized learning path to review, practice, and reinforce your understanding.

Paul J. Richards/AFP/Getty Images

RECURRENT WEATHER

FOCUS QUESTION

How does heat make droughts more extreme?

Floods

We all appreciate a good rainstorm if it has been particularly dry and hot. But even that welcome rain can become problematic if it lasts for too long. In fact, too much of any specific type of weather—cold, wet, warm, or dry—can be unwelcome because of the serious consequences that can result from it.

An individual thunderstorm can unleash enough rain to produce floods. Hurricanes can also cause torrential downpours, which result in extensive flooding. Floods can also occur, however, when weather patterns cause even mild storms to persist over the same area for an extended period of time. For example, a storm with a rainfall rate of 1.5 cm/h is not much of a problem if it lasts only an hour or two. If this same storm were to remain over one area for 18 hours, however, the total rainfall would be 27 cm, which is enough to create flooding in most areas. In the spring of 2010, a two-day storm caused flooding throughout much of Tennessee, Kentucky, and Mississippi. Over 10,000 properties were damaged in Nashville, Tennessee, which experienced extensive flooding, as shown in **Figure 19.**

Low-lying areas are most susceptible to flooding, making coastlines particularly vulnerable to storm surges during hurricanes. Rivers in narrow valleys can rise rapidly, creating high-powered and destructive walls of water. Building in the floodplain of a river or stream can be inconvenient and potentially dangerous.

Figure 19 Streets, cars, and buildings were flooded after heavy rains in Nashville, Tennessee.
Infer *What areas are most affected by flooding?*

3D THINKING　　**DCI** Disciplinary Core Ideas　　**CCC** Crosscutting Concepts　　**SEP** Science & Engineering Practices

COLLECT EVIDENCE

 Use your Science Journal to record the evidence you collect as you complete the readings and activities in this lesson.

INVESTIGATE

GO ONLINE to find these activities and more resources.

Investigation Lab: Observing Flood Damage
Develop and use a model to show the effect of flooding on structures, soil, and environments.

Quick Investigation: Model Flood Conditions
Use a model to visualize the effect of repeated, slow-moving storms on humans in flood-prone areas.

David Fine/FEMA

Heat Waves

A **heat wave** is an extended period of above-average temperatures. Heat waves can be formed by high-pressure systems. As the air under a large high-pressure system sinks, it warms by compression and causes above-average temperatures. The high-pressure system also blocks cooler air masses from moving into the area, so there is little relief from the heat.

Because it is difficult for condensation to occur under the sinking air of the high-pressure system, there are few, if any, clouds to block the blazing sunshine. The jet stream, or "atmospheric railway," that weather systems normally follow is farther poleward and weaker during the summer. Thus, any upper-air currents that might guide the high-pressure system are so weak that the system barely moves.

Heat index

Increasing humidity can add to the discomfort and potential danger of a heat wave. Human bodies cool by evaporating moisture from the surface of the skin. In the process, thermal energy is removed from the body. In humid air, the rate of evaporation is reduced, which lessens the body's ability to regulate internal temperature. During heat waves, this can lead to serious health problems such as heatstroke, sunstroke, and even death.

Because of the dangers posed by the combination of heat and humidity, the National Weather Service (NWS) routinely reports the heat index, shown in **Table 2.** Note that the NWS uses the Fahrenheit scale in the heat index because most United States citizens are more familiar with this scale.

The heat index considers the effect of a body's increasing difficulty in regulating its internal temperature as relative humidity rises. This index estimates how warm the air feels to the human body based on the actual air temperature and relative humidity. For example, an air temperature of 80°F (26.6°C) and a relative humidity of 70 percent would require the body to cool itself at the same rate as if the air temperature were 85°F (29.4°C).

Table 2 The Heat Index

Relative Humidity (%)	Air Temperature (°F)					
	70	80	90	100	110	120
	Apparent Temperature (°F)					
0	64	73	83	91	99	107
20	66	77	87	99	112	130
40	68	79	93	110	137	
60	70	82	100	132		
80	71	86	113			
100	72	91				

 Get It?

Identify the cause of serious health problems associated with heat waves.

CCC CROSSCUTTING CONCEPTS

Cause and Effect The interaction of many environmental factors can result in the formation of a heat wave. Make a list of these factors and explain how each contributes to the development of a heat wave. What ultimate effects do these interactions have on people and animals? Can these effects be avoided? Explain your thoughts.

Figure 20 Sunflower plants struggle to survive in dried, cracked mud during a drought.

Droughts

The same conditions that cause heat waves can also cause abnormally dry weather. Too much dry weather can cause nearly as much damage as too much rainfall. **Droughts** are extended periods of well-below-average rainfall. One of the most extreme droughts in American history occurred during the 1930s in the central United States. This extended drought put countless farmers out of business, as rainfall was inadequate to grow crops.

Droughts are usually the result of shifts in global wind patterns that allow large high-pressure systems to persist for weeks or months over continental areas. Under a dome of high pressure, air sinks on a large scale. Because the sinking air blocks moisture from rising through it, condensation cannot occur, and drought sets in until global patterns shift enough to move the high-pressure system. **Figure 20** shows one of the impacts of long-term drought.

 Get It?

Infer how the drought of the 1930s affected businesses other than farmers.

Cold Waves

The opposite of a heat wave is a **cold wave,** an extended period of below-average temperatures. Like heat waves, cold waves are also brought on by large high-pressure systems. However, cold waves are caused by systems of continental polar or arctic origin. During the winter, little sunlight is available to provide warmth. At the same time, the snow-covered surface is constantly reflecting sunlight back to space. The combined effect of these two factors is the development of large parcels of extremely cold air over polar continental areas. Because cold air sinks, the pressure near the surface increases, creating a strong high-pressure system.

Because of the location and the time of year in which they occur, winter high-pressure systems are much more influenced by the jet stream than are summer high-pressure systems. Pushed along by the fast-moving jet stream, these high-pressure systems rarely linger in any area. However, the winter location of the jet stream can remain essentially unchanged for days, or even weeks. This means that several polar high-pressure systems can follow the same path and subject the same areas to persistent numbing cold until the position of the jet stream changes. One effect of such prolonged cold weather is shown in **Figure 21.**

Figure 21 Prolonged cold or recurrent cold waves can create conditions in which a waterfall freezes.

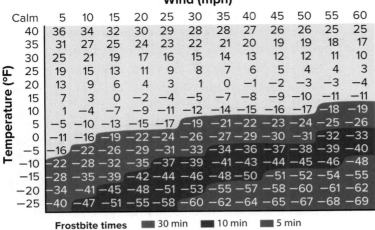

Windchill Chart

Wind (mph)

Temperature (°F)	Calm	5	10	15	20	25	30	35	40	45	50	55	60
40		36	34	32	30	29	28	28	27	26	26	25	25
35		31	27	25	24	23	22	21	20	19	19	18	17
30		25	21	19	17	16	15	14	13	12	12	11	10
25		19	15	13	11	9	8	7	6	5	4	4	3
20		13	9	6	4	3	1	0	−1	−2	−3	−3	−4
15		7	3	0	−2	−4	−5	−7	−8	−9	−10	−11	−11
10		1	−4	−7	−9	−11	−12	−14	−15	−16	−17	−18	−19
5		−5	−10	−13	−15	−17	−19	−21	−22	−23	−24	−25	−26
0		−11	−16	−19	−22	−24	−26	−27	−29	−30	−31	−32	−33
−5		−16	−22	−26	−29	−31	−33	−34	−36	−37	−38	−39	−40
−10		−22	−28	−32	−35	−37	−39	−41	−43	−44	−45	−46	−48
−15		−28	−35	−39	−42	−44	−46	−48	−50	−51	−52	−54	−55
−20		−34	−41	−45	−48	−51	−53	−55	−57	−58	−60	−61	−62
−25		−40	−47	−51	−55	−58	−60	−62	−64	−65	−67	−68	−69

Frostbite times ■ 30 min ■ 10 min ■ 5 min

Figure 22 The windchill chart was designed to show the dangers of cold air and wind.

Identify *which wind speed and temperature is the same as 10°F on a calm day?*

Windchill index

The effects of cold air on the human body are magnified by wind. Known as the windchill factor, this phenomenon is measured by the **windchill index** in **Figure 22.** The index estimates how cold the air feels to the human body based on the actual air temperature and wind speed. While the windchill index is helpful, it does not account for individual variations in sensitivity to cold, the effects of physical activity, or relative humidity. In 2001, the National Weather Service revised the calculations to utilize advances in science, technology, and computer modeling. These revisions provide a more accurate, understandable, and useful index for estimating the dangers caused by winter winds and freezing temperatures.

✎ Check Your Progress

Summary

- Too much rain in any area can cause flooding, but it is especially problematic in low-lying areas.

- The heat index estimates the effect on the human body when air is hot and humidity is high.

- Too much heat and too little precipitation causes drought.

- Too little heat and a stalled jet stream can cause a cold wave, or an extended period of cold weather.

- The windchill index tells how wind and temperature affect your body in winter.

Demonstrate Understanding

1. **Explain** how everyday weather can become recurrent and dangerous.

2. **Describe** how relatively light rain could cause flooding.

3. **Compare and contrast** a cold wave and a heat wave.

4. **Explain** why one type of front would be more closely associated with flooding than another.

Explain Your Thinking

5. **Explain** why air in a winter high-pressure system is very cold despite compressional warming.

6. **Compare** the data of the heat-index scale and the windchill scale. What variables influence each scale?

7. **MATH ⟩ Connection** A storm stalls over Virginia, dropping 0.75 cm of rain per hour. If the storm lingers for 17 hours, how much rain will accumulate?

LEARNSMART Go online to follow your personalized learning path to review, practice, and reinforce your understanding.

Weathering the Storm

It is not an easy decision to evacuate a large city in the face of an oncoming storm or natural disaster, such as a hurricane or a tsunami. If residents do not evacuate and the storm hits as predicted, they could be in danger. If they do evacuate and the storm misses the city, the government will have spent potentially millions of dollars on a threat that did not come to pass. How do officials make the decision?

Recent evacuation plans seek to avoid dangerous traffic jams caused by a failure to use both sides of a highway for evacuation traffic.

Should I stay, or should I go?

Evacuating a city "just in case" might seem to make sense, but evacuation has its own dangers. People fleeing a storm or disaster face potential car accidents, heatstroke, being stranded on the road when bad weather hits, and other issues. In addition, the areas to which residents flee may become overwhelmed. There might not be enough shelter, food, water, and medical care for all evacuees.

Evacuation planning

Cities spend years and large amounts of money creating evacuation plans. The plans take into account the needs of different populations and different parts of the city, the severity of the storm or disaster, and alternate escape routes. Plans also include ideas about where officials think residents will go, the length of time an area needs to evacuate, whether the evacuation is voluntary or mandatory, and even the effect of social media on whether people decide to stay or to go.

Cities learn from other cities' evacuation experiences. During past hurricanes, for example, traffic jams have caused many deaths. Now, many cities include in their plans "contra-flow traffic routing," which uses all lanes of a highway to evacuate people in one direction—out of the city. Another new method directs the people in the area of the city closest to the approaching storm to evacuate first.

To further improve evacuations, researchers are focusing on using computer modeling for every step of the process. Some researchers are investigating whether self-driving cars could speed up evacuations and rescue people who do not own vehicles. Others are making plans to focus on the most vulnerable populations of a city, such as older adults and people who have mobility issues.

OBTAIN, EVALUATE, AND COMMUNICATE INFORMATION

Use print and online resources to research what the federal government recommends families do to prepare for a possible evacuation. Create a public service announcement (PSA) that outlines the information most relevant to your area.

GO ONLINE to study with your Science Notebook.

Lesson 1 THUNDERSTORMS

- The cumulus stage, the mature stage, and the dissipation stage comprise the life cycle of a thunderstorm.
- Clouds form as water is condensed and latent heat is released.
- Thunderstorms can be produced either within air masses or along fronts.
- From formation to dissipation, all thunderstorms go through the same stages.
- Lightning is a natural result of thunderstorm development.

- air-mass thunderstorm
- mountain thunderstorm
- sea-breeze thunderstorm
- frontal thunderstorm
- stepped leader
- upward streamer
- return stroke

Lesson 2 SEVERE WEATHER

- Intense, rotating updrafts are associated with supercells.
- Downbursts are strong winds that result in damage associated with thunderstorms.
- Hail is precipitation in the form of balls or lumps of ice that accompany severe storms.
- The worst storm damage comes from a vortex of high winds that moves along the ground as a tornado.

- supercell
- downburst
- tornado
- Enhanced Fujita Tornado Damage Scale

Lesson 3 TROPICAL STORMS

- Tropical cyclones rotate counterclockwise in the northern hemisphere.
- Tropical cyclones are also known as hurricanes and typhoons.
- Tropical cyclones go through the same stages of formation and dissipation as other storms.
- Tropical cyclones are moved by various wind systems after they form.
- The most dangerous part of a tropical cyclone is the storm surge.
- Hurricane alerts are given at least 36 hours before the hurricane arrives.

- tropical cyclone
- eye
- eyewall
- Saffir-Simpson Hurricane Wind Scale
- storm surge

Lesson 4 RECURRENT WEATHER

- Too much rain in any area can cause flooding, but it is especially problematic in low-lying areas.
- The heat index estimates the effect on the human body when air is hot and humidity is high.
- Too much heat and too little precipitation causes drought.
- Too little heat and a stalled jet stream can cause a cold wave, or an extended period of cold weather.
- The windchill index tells how wind and temperature affect your body in winter.

- heat wave
- drought
- cold wave
- windchill index

REVISIT THE PHENOMENON

Why do only some thunderstorms produce tornadoes?

CER Claim, Evidence, Reasoning

Explain Your Reasoning Revisit the claim you made when you encountered the phenomenon. Summarize the evidence you gathered from your investigations and research and finalize your Summary Table. Does your evidence support your claim? If not, revise your claim. Explain why your evidence supports your claim.

STEM UNIT PROJECT

Now that you've completed the module, revisit your STEM unit project. You will summarize your evidence and apply it to the project.

GO FURTHER

SEP Data Analysis Lab

How can you calculate a heat wave?

The following data represent the daily maximum and minimum temperatures for seven consecutive summer days in Chicago in 2002. A heat wave in this area is defined as two or more days with an average temperature of 29.4°C or higher.

CER Analyze and Interpret Data

1. **Evidence** Calculate the average temperature for each day and add it to the table.
2. **Evidence** Plot the daily maximum and minimum temperatures on a graph, with the days on the *x*-axis and the maximum temperatures on the *y*-axis. Using the data points, draw a curve to show how the temperatures changed over the seven-day period. Add the average temperatures.
3. **Claim, Evidence** What day did the city heat wave begin? How long did it last?
4. **Evidence** Compare the average temperature for the days of the heat wave to the average temperature of the remaining days.
5. **Reasoning** Based on your results from question 4, do you think that the entire week should have been classified as a heat wave? Explain.

Data and Observations

Day	Daily Temperatures		
	Maximum (°C)	Minimum (°C)	Average (°C)
1	32	23	
2	37	24	
3	41	27	
4	39	29	
5	37	25	
6	34	24	
7	32	23	

*Data obtained from: Klinenberg, E. 2002.
Heat Wave: A social autopsy of disaster in Chicago, IL. Chicago: University of Chicago Press.

Oliver Henze/EyeEm/Getty Images

ENCOUNTER THE PHENOMENON

What is climate?

▶ **GO ONLINE** to play a video about how the movement of Earth affects climate.

SEP Ask Questions

Do you have other questions about the phenomenon? If so, add them to the driving question board.

CER Claim, Evidence, Reasoning

Make Your Claim Use your CER chart to make a claim about what climate is. Explain your reasoning.

Collect Evidence Use the lessons in this module to collect evidence to support your claim. Record your evidence as you move through the module.

Explain Your Reasoning You will revisit your claim and explain your reasoning at the end of the module.

▶ **GO ONLINE** to access your CER chart and explore resources that can help you collect evidence.

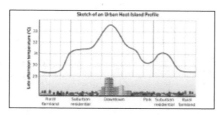

LESSON 2: Explore and Explain: Microclimates

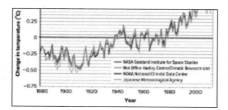

LESSON 4: Explore and Explain: Global Climate Change

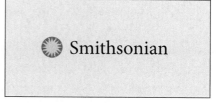

Additional Resources

Video Supplied by BBC Worldwide Learning

FOCUS QUESTION
What is the difference between climate and weather?

Annual Averages and Variations

Fifty thousand years ago, the United States had much different weather patterns than those that exist today. The average temperature was several degrees cooler, and the jet stream was probably farther south. Understanding and predicting such climatic changes are the basic goals of climatology. **Climatology** is the study of Earth's climate and the factors that cause past, present, and future climatic changes.

Climate describes the long-term weather patterns of an area. These patterns include much more than average weather conditions. Climate also describes annual variations of temperature, precipitation, wind, and other weather variables. Climate data show extreme fluctuations of these variables over time, such as the warmest and coldest temperatures for a place. **Figure 1** shows how temperature varies between summer and winter in Chicago, Illinois.

Chicago, IL, in the summer

Chicago, IL, in the winter

Figure 1 The highest temperature on record for Chicago, IL, is 41°C, which occurred in July 1934. The lowest temperature on record for Chicago, IL, is −33°C, which occurred in January 1985.

(l)marchello74/Shutterstock, (r)Henryk Sadura/Shutterstock

3D THINKING   **DCI** Disciplinary Core Ideas   **CCC** Crosscutting Concepts   **SEP** Science & Engineering Practices

COLLECT EVIDENCE
 Use your Science Journal to record the evidence you collect as you complete the readings and activities in this lesson.

INVESTIGATE
GO ONLINE to find these activities and more resources.

 GeoLAB: Identify a Microclimate
Plan and carry out an investigation to determine what weather patterns define climates.

 Investigation Lab: Heat Absorption Over Land and Water
Use a model to visualize how different surfaces affect how much heat is absorbed by the environment.

This type of information, combined with comparisons between recent conditions and long-term averages, can be used by businesses to decide where to build new facilities and by people who have medical conditions that require them to live in certain climates.

Normals

The data used to describe an area's climate are compiled from meteorological records, which are continuously gathered at thousands of locations around the world. These data include daily high and low temperatures, amounts of rainfall, wind speed and direction, humidity, and air pressure. The data are averaged on a monthly or annual basis for a period of at least 30 years to determine the **normals,** which are the standard values for a location.

Get It?

Identify data that can be used to calculate normals.

Contrasting changes in weather While normals offer valuable information, they must be used with caution. Weather conditions on any given day might differ widely from normals. For instance, the normal high temperature in January for a city might be 0°C. However, it is possible that no single day in January had a high of exactly 0°C. Normals are not intended to describe usual weather conditions; they are the average values over a long period of time.

While climate describes the average weather conditions for a region, normals apply only to the specific place where the meteorological data were collected. Most meteorological data are gathered at airports, which cannot operate without up-to-date, accurate weather information. However, many airports are located outside city limits. When climatic normals are based on airport data, they might differ from actual weather conditions in nearby cities. Changes in elevation and other factors, such as proximity to large bodies of water, can cause climates to vary.

Causes of Climate

You probably know from watching televised weather reports that climates around the country vary greatly. For example, on average, daily temperatures are much warmer in Dallas, Texas, than in Minneapolis, Minnesota. There are several reasons for such climatic variations, including differences in latitude, topography, proximity to lakes and oceans, availability of moisture, global wind patterns, ocean currents, and air masses. As you read about the factors that influence climate, recall what you have learned about incoming solar radiation. The foundation for Earth's global climate systems is this electromagnetic radiation from the Sun as well as its reflection, absorption, storage, and redistribution among the atmosphere, ocean, and land systems.

Get It?

Describe the foundation for Earth's global climate system.

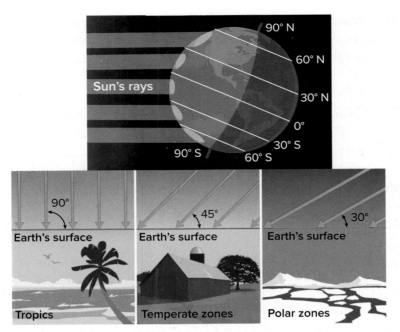

Figure 2 Latitude has a great effect on climate. The amount of solar radiation received on Earth decreases from the equator to the poles.

Describe *what happens to the angle at which the Sun's rays hit Earth's surface as one moves from the equator to the poles.*

Latitude

Recall that different parts of Earth receive different amounts of solar radiation. The amount of solar radiation received by any one place varies because Earth is tilted on its axis, and this affects how the Sun's rays strike Earth's surface. The area between 23.5°S and 23.5°N of the equator is known as the **tropics.** As **Figure 2** shows, tropical areas receive the most solar radiation because the Sun's rays are nearly perpendicular to Earth's surface. As you might expect, temperatures in the tropics are generally warm year-round. For example, Caracas, Venezuela, located at about 10°N, enjoys average maximum temperatures between 24°C and 27°C year-round. The **temperate zones** lie between 23.5° and 66.5° north and south of the equator. As their name implies, these regions experience moderate temperatures. The **polar zones** are located from 66.5° north and south of the equator to the poles. Solar radiation strikes the polar zones at a low angle, so polar temperatures tend to be cold. Thule, Greenland, located at 77°N, has average maximum temperatures between –20°C and 8°C year-round.

Topographic effects

Water heats up and cools down more slowly than land. Thus, large bodies of water such as oceans and lakes affect the climates of nearby areas. Many coastal regions are warmer in the winter and cooler in the summer than inland areas at similar latitudes.

ACADEMIC VOCABULARY

imply
to indicate by association rather than by direct statement
The title of the movie implied that it was a love story.

Also, temperatures in the lower atmosphere generally decrease with altitude, so mountain climates are usually cooler than those at sea level. In addition, climates often differ on either side of a mountain. Air rises up one side of a mountain as a result of orographic lifting. The rising air cools, condenses, and drops its moisture, as shown in **Figure 3.** The climate on this side of the mountain—the windward side—is usually wet and cool. On the opposite side of the mountain—the leeward side—the air is drier, and it warms as it descends, resulting in a desert climate.

Get It?

Explain how large bodies of water affect the climate of coastal areas.

Air masses

Two of the main causes of weather are the movement and interaction of air masses. Air masses also affect climate. Recall that air masses have distinct regions of origin, caused primarily by differences in the intensity of solar radiation. The properties of air masses also depend on whether they formed over land or water.

Figure 3 Orographic lifting leads to rain on the windward side of a mountain. The leeward side is usually dry and warm.

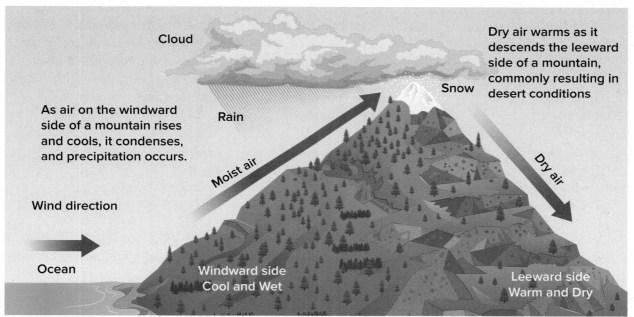

Windward side of mountains

Leeward side of mountains

Major Air Masses over North America

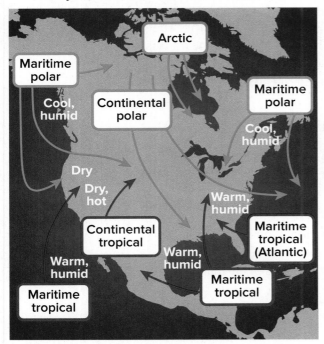

Figure 4 Air masses affect regional climates by transporting the temperature and humidity of their source regions. The warm and humid maritime tropical air mass supports the lush vegetation on the island of Dominica.

Lush vegetation on the Caribbean island of Dominica

The air masses commonly found over North America are shown in **Figure 4.** Those that form over oceans are called maritime air masses, and those that form over land are called continental air masses. Average weather conditions in and near regions of air-mass formation are similar to those of the air masses themselves. For example, the island of Dominica, shown above in **Figure 4,** is in an area where maritime tropical (mT) air masses dominate. Because of this, the island has a maritime tropical climate, with warm temperatures, high humidity, and high amounts of precipitation.

Check Your Progress

Summary

- Climate describes the long-term weather patterns of an area.
- Normals are the standard climatic values for a location.
- Temperatures vary among tropical, temperate, and polar zones.
- Climate is influenced by latitude, topographic effects, and air masses.
- Air masses have distinct regions of origin.

Demonstrate Understanding

1. **Summarize** conditions that contribute to the climate of an area.
2. **Identify** What are some limits associated with the use of normals?
3. **Compare and contrast** temperatures in the tropics, temperate zones, and polar zones.
4. **Infer** how climate data can be used by farmers.

Explain Your Thinking

5. **Assess** Average daily temperatures for City A, located at 15°S, are 5°C cooler than average daily temperatures for City B, located at 30°S. What might account for the cooler temperatures in City A, even though it is closer to the equator?
6. **WRITING** **Connection** Hypothesize why meteorological data gathered at an airport would differ from data gathered near a large lake. Assume all other factors are constant.

LEARNSMART Go online to follow your personalized learning path to review, practice, and reinforce your understanding.

CLIMATE CLASSIFICATION

FOCUS QUESTION

How does climate affect what grows in a region?

Köppen Classification System

The top graph on the right in **Figure 5** shows climate data for a desert in Reno, Nevada. The bottom graph on the right shows climate data for a tropical rain forest in New Guinea. What criteria are used to classify the climates described in the graphs? Temperature is an obvious choice, as is amount of precipitation. The **Köppen classification system** is a classification system for climates that is based on the average monthly values of temperature and precipitation. Developed by German climatologist Wladimir Köppen, the system also takes into account the distinct vegetation found in different climates.

Köppen decided that a good way to distinguish among different climatic zones was by natural vegetation. Palm trees, for instance, are not located in polar regions, but instead are largely limited to tropical and subtropical regions. Köppen later realized that quantitative values would make his system more objective and, therefore, more scientific. He revised his system to include the numerical values of temperature and precipitation. A map of global climates according to a modified version of Köppen's classification system is shown in **Figure 6.** Notice that there are five main climate types, with most variations occurring in midlatitudes.

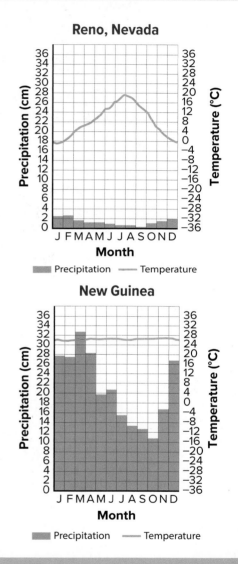

Figure 5 These graphs show temperature and precipitation for two different climates—a desert in Reno, Nevada, and a tropical rain forest in New Guinea.

Describe *the difference in temperature between these two climates.*

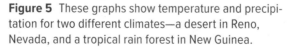

3D THINKING   **DCI** Disciplinary Core Ideas   **CCC** Crosscutting Concepts   **SEP** Science & Engineering Practices

COLLECT EVIDENCE

Use your Science Journal to record the evidence you collect as you complete the readings and activities in this lesson.

INVESTIGATE

GO ONLINE to find these activities and more resources.

Mapping Lab: Classifying Climates
Analyze and interpret data to determine the effects of precipitation and temperature on climate.

Virtual Investigation: Climate
Analyze and interpret data to visualize the patterns that define climates.

Figure 6 Visualizing Worldwide Climates

Köppen's classification system, shown here in a modified version, is made up of five main divisions based on temperature and precipitation.

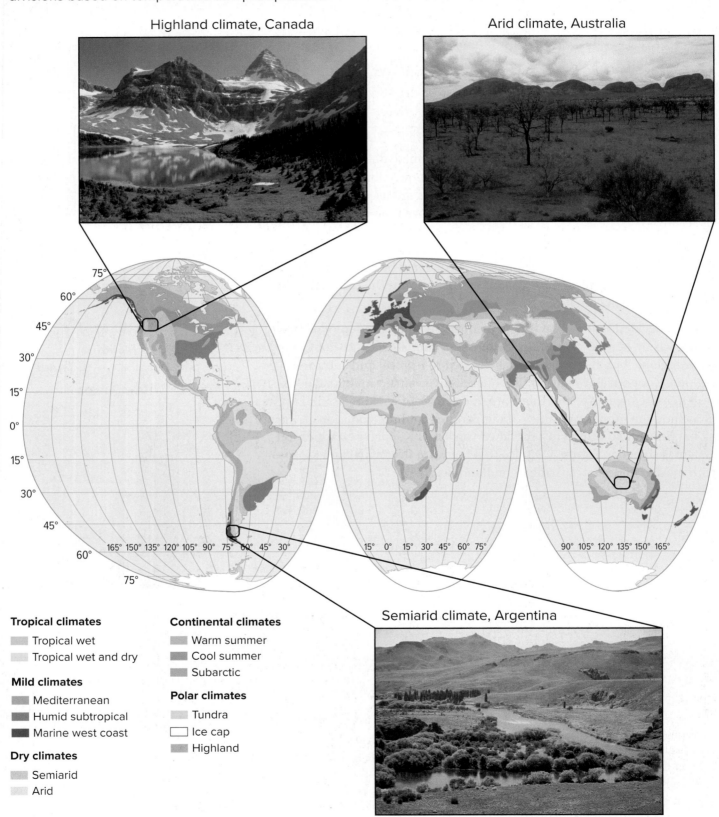

Highland climate, Canada

Arid climate, Australia

Semiarid climate, Argentina

Tropical climates
- Tropical wet
- Tropical wet and dry

Mild climates
- Mediterranean
- Humid subtropical
- Marine west coast

Dry climates
- Semiarid
- Arid

Continental climates
- Warm summer
- Cool summer
- Subarctic

Polar climates
- Tundra
- Ice cap
- Highland

(tl)Joseph J Priola/Moment/Getty Images, (tr)GeoStock/Photodisc/Getty Images, (br)Javier Perini CM/Image Source

Tropical climates

Year-round high temperatures characterize tropical climates. In tropical, wet climates, the locations of which are shown in **Figure 6,** high temperatures are accompanied by up to 600 cm of rain each year. The combination of warmth and rain produces tropical rain forests, which contain some of the most dramatic vegetation on Earth. Tropical regions are almost continually under the influence of maritime tropical air.

The areas that border the rainy tropics to the north and south of the equator are transition zones, known as the tropical wet and dry zones. Tropical wet and dry zones include savannas. These tropical grasslands are found in Africa, among other places. These areas have distinct dry winter seasons as a result of the seasonal influx of dry continental air masses. **Figure 7** shows the average monthly temperature and precipitation readings for Normanton, Australia—a savanna in northeast Australia.

Dry climates

Dry climates, which cover about 30 percent of Earth's land area, make up the largest climatic zone. Most of the world's deserts, such as the Sahara, the Gobi, and the Australian, are classified as dry climates. In these climates, continental tropical (cT) air dominates, precipitation is low, and vegetation is scarce. Many of these areas are located near the tropics. The intense solar radiation results in high rates of evaporation and few clouds. Overall, evaporation rates exceed precipitation rates. The resulting moisture deficit gives this zone its name. Within this classification, there are two subtypes: arid regions, called deserts, and semiarid regions, called semideserts. Semideserts, like the one shown in **Figure 8,** are usually more humid than deserts. They generally separate arid regions from bordering wet climates.

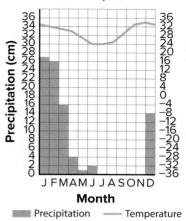

Normanton, Australia

Precipitation   Temperature

Figure 7 The graph shows the temperature and precipitation readings for a tropical savanna in Australia.

Analyze *How does the rainfall in this area differ from that of a tropical rain forest?*

Figure 8 This semidesert in South Africa is an example of a transition zone. It separates the desert from bordering climates that are more humid.

Mild climates

Mild climates can be classified into three subtypes: humid subtropical climates, marine west-coast climates, and Mediterranean climates. Humid subtropical climates are influenced by the subtropical high-pressure systems that are normally found over oceans in the summer. Florida and the southeastern United States have this type of climate. There, warm, muggy weather prevails during the warmer months, and dry, cool conditions predominate during the winter. Marine west-coast climates are dominated by the constant inland flow of air off the ocean, which creates mild winters and cool summers, with abundant precipitation throughout the year. Mediterranean climates,

Figure 9 These olive trees thrive in the Mediterranean climate of southern Spain.

named for the climate that characterizes much of the land around the Mediterranean Sea, are also found in California and parts of South America. An example of this climate is shown in **Figure 9.** Summers in Mediterranean climates are generally warm and dry because of their proximity to dry midlatitude climates from the south. Winters are cool and rainy due to midlatitude weather systems that bring storm systems from the north.

 Get It?

Identify the factors that influence mild climates.

Continental climates

Continental climates are also classified into three subtypes: warm summer climates, cool summer climates, and subarctic climates. Tropical and polar air masses often form fronts as they meet in continental climates. Thus, these zones experience rapid and sometimes violent changes in weather, including severe thunderstorms or tornadoes, like the one shown in **Figure 10.** Both summer and winter temperatures can be extreme because the influence of polar air masses is strong in winter, while warm, tropical air dominates in summer. The presence of warm, moist air causes summers to be generally wetter than winters, especially in latitudes that are relatively close to the tropics.

Figure 10 Tornadoes, such as this one in Colorado, occur in continental climates.

Polar climates

To the north of subarctic climates lies one of the polar climates—the tundra. Just as the tropics are known for their year-round warmth, tundra is known for its low temperatures. The mean temperature of the warmest month is usually less than 10°C. There are no trees in the tundra, and precipitation is generally low because cold air contains less moisture than warm air. Also, the amount of heat radiated by Earth's surface is too low to produce the strong convection currents needed to release heavy precipitation. The ice-cap polar climate, found at the highest latitudes in both hemispheres, does not have a single month in which average temperatures rise above 0°C. No vegetation grows in an ice-cap climate, shown in **Figure 11,** and the land is permanently covered by ice and snow.

Figure 11 Icebergs float in the sea in the ice-cap polar climate of Greenland.

Microclimates

Sometimes the climate of a small area can be much different from that of the larger area surrounding them. A localized climate that differs from the main regional climate is called a **microclimate.** If you climb to the top of a mountain, you can experience a type of microclimate; the climate becomes cooler with increasing elevation. **Figure 12** shows a profile of a microclimate created by the buildings and concrete in a city.

Heat islands

The presence of buildings can often create a microclimate in the area immediately surrounding them. Many concrete buildings and large expanses of asphalt can create a **heat island,** where the climate is warmer than in surrounding rural areas, as shown in **Figure 12.** This effect was first recognized in the early nineteenth century when Londoners noted that the temperature in the city was noticeably warmer than in the surrounding countryside.

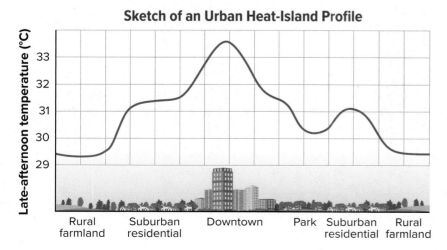

Sketch of an Urban Heat-Island Profile

Figure 12 This diagram shows the difference in temperature between the downtown area of a city and the surrounding suburban and rural areas.

Analyze *How much warmer is it in the city compared to the rural areas?*

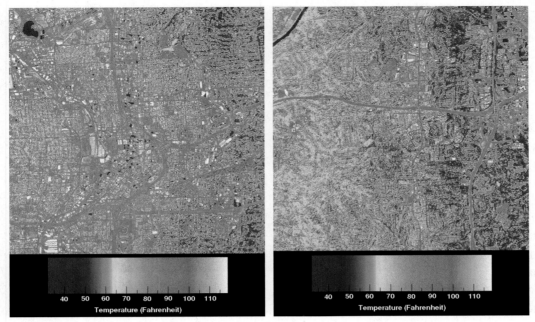

Figure 13 These thermal images compare daytime temperatures in an urban area (left) and a suburban area (right). The coolest temperatures are shown as blue; the warmest temperatures are shown as red.

Pavement, buildings, and roofs that are made of dark materials, such as asphalt, absorb more solar energy than the surrounding vegetation does. This causes the temperature of these objects to increase, heating the air around them. This also causes mean temperatures in large cities to be significantly warmer than in surrounding areas, as shown in **Figure 13.** The heat-island effect also causes greater changes in temperature with altitude, which sparks strong convection currents. This, in turn, produces increased cloudiness and up to 15 percent more total precipitation in cities.

Check Your Progress

Summary

- German scientist Wladimir Köppen developed a climate classification system.

- There are five main climate types: tropical, dry, mild, continental, and polar.

- Microclimates can occur within cities.

Demonstrate Understanding

1. **Describe** On what criteria is the Köppen climate classification system based?

2. **Explain** What are microclimates? Identify and describe one example of a microclimate.

3. **Compare and contrast** the five main climate types.

4. **Categorize** the climate of your area. In which zone do you live? Which air masses generally affect your climate?

Explain Your Thinking

5. **Construct** Make a table of the Köppen climate classification system. Include major zones, subzones, and characteristics of each.

6. **WRITING** **Connection** Write a short paragraph that explains which of the different climate types you think would be most strongly influenced by the polar jet stream.

LEARNSMART Go online to follow your personalized learning path to review, practice, and reinforce your understanding.

CLIMATIC CHANGES AND PATTERNS

FOCUS QUESTION

How does climate change naturally?

Long-Term Climatic Changes

Some years might be warmer or cooler than others, but climates change over time frames longer than human lifetimes. However, a study of Earth's history over hundreds of thousands of years shows that climates have always been, and currently are, in a constant state of change. These changes usually take place over long time periods, but human impacts are causing climate to change faster today than it has at any observed period in the past.

Ice ages

A good example of climatic change involves glaciers, which have alternately advanced and retreated over the past 2 million years. At times, much of Earth's surface was covered by vast sheets of ice. During these periods of extensive glacial coverage, called **ice ages,** average global temperatures decreased by an estimated 5°C. Global climates became generally colder, and snowfall increased, which sparked the advance of existing ice sheets. Ice ages alternate with warm periods—called interglacial intervals—and Earth is currently experiencing such an interval. The most recent ice age, shown in **Figure 14,** ended only about 10,000 years ago. In North America, glaciers spread from the east to the west and as far south as Ohio.

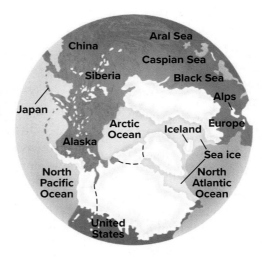

Figure 14 The last ice age covered large portions of North America, Europe, and Asia. The average sea level was approximately 130 m lower than at present.

Explain *how decreased global temperatures can destabilize Earth's systems and lead to an ice age.*

3D THINKING   **DCI** Disciplinary Core Ideas   **CCC** Crosscutting Concepts   **SEP** Science & Engineering Practices

COLLECT EVIDENCE

 Use your Science Journal to record the evidence you collect as you complete the readings and activities in this lesson.

INVESTIGATE

 GO ONLINE to find these activities and more resources.

Applying Practices: Variations in Albedo
HS-ESS2-4 Use a model to describe how variations in the flow of energy into and out of Earth's systems result in changes in climate.

 Review the News
Obtain information from a current news story about current climate research. **Evaluate** your source and **communicate** your findings to your class.

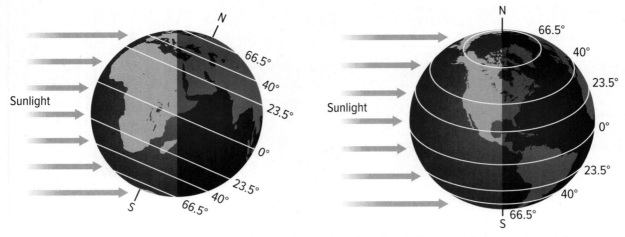

Figure 15 When the North Pole is pointed away from the Sun, the northern hemisphere experiences winter and the southern hemisphere experiences summer. During spring and fall, neither pole points toward the Sun.

Climatic Patterns

Regular changes that occur over the course of months and years temporarily alter the temperature, precipitation, and weather within a climate. These relatively short-term climatic patterns include seasons, El Niño, and La Niña.

Seasons

Seasons are short-term periods with specific weather conditions caused by regular variations in daylight, temperature, and weather patterns. The variations that occur with seasons are the result of changes in the amount of solar radiation an area receives. As **Figure 15** shows, the tilt of Earth on its axis as it revolves around the Sun causes different areas of Earth to receive different amounts of solar radiation. During winter in the northern hemisphere, the North Pole is tilted away from the Sun, and this hemisphere experiences long hours of darkness and cold temperatures. At the same time, it is summer in the southern hemisphere. The South Pole is tilted toward the Sun, and the southern hemisphere experiences long hours

of daylight and warm temperatures. Throughout the year, the seasons are reversed in the northern and southern hemispheres. During the spring and fall, neither pole points toward the Sun.

 Get It?

Compare the positions of the North Pole and the South Pole during summer in the northern hemisphere.

El Niño

Other short-term climatic patterns include those caused by **El Niño,** a band of anomalously warm ocean temperatures that periodically develops off the western coast of South America. Under normal conditions in the southeastern Pacific Ocean, atmospheric and ocean currents along the coast of South America move north, transporting cold water from the Antarctic region.

Meanwhile, the trade winds and ocean currents move westward across the tropics, keeping warm water in the western Pacific, as shown in **Figure 16.**

SCIENCE USAGE v. COMMON USAGE

pressure
Science usage: the force that a column of air exerts on the air above it
Common usage: the burden of physical or mental distress

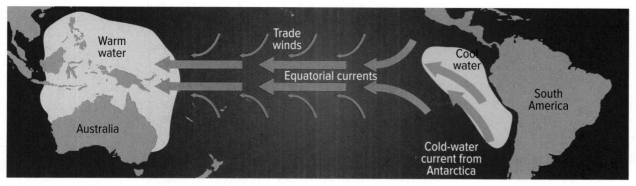

Figure 16 Under normal conditions, trade winds and ocean currents move warm water west across the Pacific Ocean.

This circulation, driven by a semi-permanent, high-pressure system, creates a cool, dry climate along much of the northwestern coast of South America.

Occasionally, however, for reasons that are not fully understood, this high-pressure system and its associated trade winds weaken drastically, which allows the warm water from the western Pacific to surge eastward toward the South American coast, as shown in **Figure 17.** These conditions are referred to as an El Niño event.

The sudden presence of this warm water heats the air near the surface of the water. Convection currents strengthen, and the normally cool and dry northwestern coast of South America becomes warmer and wetter. The increased convection pumps large amounts of heat and moisture into the upper atmosphere, where upper-level winds transport the hot, moist air eastward across the tropics. This hot, moist air in the upper atmosphere is responsible for dramatic changes in weather, including violent storms in California and the Gulf Coast, stormy weather in areas farther east that are normally dry, and drought conditions in areas that are normally wet. Eventually, the South Pacific high-pressure system becomes reestablished, and El Niño weakens.

Sometimes the trade winds blow stronger than normal, and warm water is pulled across the Pacific toward Australia. The coast of South America becomes unusually cold and chilly. These conditions are called La Niña.

 Get It?

Compare how the trade winds blow during El Niño and La Niña.

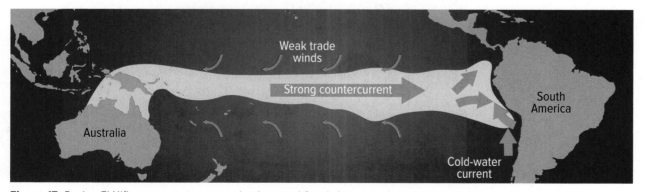

Figure 17 During El Niño, warm water surges back toward South America, changing weather patterns.

Natural Causes of Climate Changes

In order to understand how humans have affected Earth's climate, we must first understand how the climate naturally changes. The geological record shows that changes to global and regional climates can be caused by natural events that occur on a variety of time scales. These natural events, which include changes in Earth's tilt and axis, variations in solar activity, and volcanic eruptions, have short- and long-term effects on climate. Studies of tree rings, ice-core samples, fossils, and radiocarbon samples provide evidence of these past climatic changes. With this evidence, scientists have been able to distinguish between the expected natural change in climate and the anomalous anthropogenic change in climate.

Earth's orbit

Changes in Earth's orbit can trigger changes in Earth's climate. The shape of Earth's orbit becomes more elliptical, then more circular, over a 100,000-year cycle. As **Figure 18** shows, when the orbit elongates, one side of Earth's orbit is closer to the Sun, and the opposite side is farther from the Sun. As a result, Earth warms more when closer to the Sun and cools more when farther from the Sun. The increased heating and cooling can cause changes in seasonal precipitation and temperature as well as contribute to increased glaciation. When Earth's orbit is less elliptical, or more circular, Earth's temperature is more consistent. Less elliptical orbits are often correlated with interglacial periods. However, this effect is less influential on climate than the greenhouse effect. Recall that the greenhouse effect is the natural heating of Earth's surface caused by certain atmospheric gases.

 Get It?

Describe how cyclical changes in the shape of Earth's orbit cause gradual climate change and affect the intensity and distribution of sunlight falling on Earth.

Earth's tilt

As you know, seasons are caused by the angle of the tilt of Earth's axis as Earth orbits the Sun. At present, the angle of the tilt is 23.5°. However, the angle of tilt varies from a minimum of 22.1° to a maximum of 24.5° every 41,000 years. Scientists have found that these changes in angle affect the differences in seasons.

Figure 18 Scientists have found that a more elliptical orbit around the Sun could produce significant changes in Earth's climate.

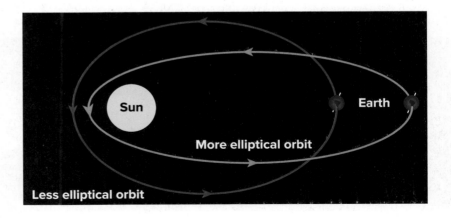

Decreased tilt
Axis with reduced angle
Existing axis
Equator
Earth
Sunlight
Sun

Figure 19 If the angle of the tilt of Earth's axis decreased, there would be less temperature contrast between summer and winter.

For example, a decrease in the angle of the tilted axis might cause a decrease in the temperature difference between winter and summer. Winters would be warmer and wetter, and summers would be cooler. The wetter winters would have more snow. The additional snow in latitudes near the poles would not melt in summer because temperatures would be cooler than average. This can result in increased glacial formation and coverage. In fact, scientists have found that changes in the angle of Earth's tilted axis, shown in **Figure 19,** have contributed to the growth and formation of ice sheets near the poles.

Get It?

Describe how cyclical changes in the tilt of Earth's axis cause gradual climate change and affect the intensity and distribution of sunlight falling on Earth.

Earth's wobble

Another movement of Earth is responsible for long-term climatic changes. Over a period of about 26,000 years, Earth wobbles as it spins around on its axis. Currently, the axis points toward the North Star, Polaris, as shown in **Figure 20.** Because of Earth's wobbling, however, the axis will eventually rotate away from Polaris and toward another star, Vega, in about 13,000 years. This rotation causes the axis to move with respect to Earth's orbit around the Sun.

Currently, the northern hemisphere tilts away from the Sun during winter, when Earth is closest to the Sun in its orbit; it tilts toward the Sun during summer, when Earth is farthest from the Sun in its orbit.

However, in 13,000 years, after Earth's axis has precessed toward Vega, the northern hemisphere will tilt away from the Sun during winter, when Earth is farthest from the Sun in its orbit; the intensity of sunlight in that hemisphere will decrease. The northern hemisphere will tilt toward the Sun during summer, when Earth is closest to the Sun in its orbit; the intensity of sunlight in that hemisphere will increase. This will result in warmer summers and cooler winters in the northern hemisphere. Axial tilt and precession have a measurable impact on Earth's climate, but, like orbit, the changes are small compared to the effects of greenhouse gases.

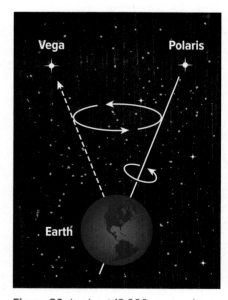

Figure 20 In about 13,000 years, when the northern hemisphere points toward the star Vega, the northern hemisphere will experience warmer summers and cooler winters.

Axial tilt and precession have a measurable impact on Earth's climate, but, like orbit, the influence is minimal compared to the effects of greenhouse gases, such as carbon dioxide, methane, and water vapor.

Get It?

Decide Besides the greenhouse effect, which factor do you think causes the greatest change in Earth's climate? Explain your answer.

Feedbacks

Earth systems are complex and interconnected. Changes in one system can cause changes in other systems. The reaction of one system to the changes in another is called **feedback.** Feedback can either amplify or dampen the changes that caused it.

Feedback that dampens changes to a system creates a negative feedback loop that returns the system to its previous state. When mine runoff turns a river acidic, the water can dissolve any limestone it flows over. The limestone reacts with the acid, neutralizes it, and returns the river to its original pH.

A positive feedback loop is created when feedback amplifies the changes to a system. As global temperatures decrease, more of Earth becomes covered in snow and ice. Snow and ice have a higher **albedo,** are more reflective, than the ocean and land that they cover, as seen in **Figure 21.** Surfaces that have a higher albedo reflect more sunlight back into space. Because less sunlight is absorbed by Earth, the global temperatures decrease. The process repeats, creating more snow and ice which lowers global temperatures and creates more snow and ice.

The reverse process also creates a positive feedback loop. As snow and ice melt, more ocean and land are uncovered. The uncovered surface has a lower albedo and absorbs more sunlight. The absorbed sunlight heats up the surface, warming Earth. The extra heat melts more of the surface, amplifying global warming.

Volcanic activity

Major volcanic eruptions release immense quantities of dust-sized particles, called aerosols. Volcanic particles can remain suspended in the atmosphere for several years, blocking incoming solar radiation and lowering global temperatures. Scientists have concluded that periods of high volcanic activity cause cool climatic periods. Climatic records from the past century show that several large eruptions have been followed by below-normal global temperatures.

High Albedo

Low Albedo

Figure 21 Light colored surfaces, like snow and ice, have high albedos. Darker surfaces, like basalt and asphalt, have low albedos.

CCC CROSSCUTTING CONCEPTS

Stability and Change Design a way to model a short-term and a long-term climatic change. Cite evidence about whether the changes are irreversible.

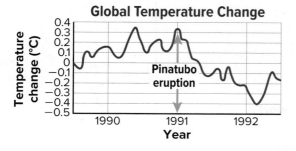

Global Temperature Change

Pinatubo eruption

Figure 22 When Mount Pinatubo erupted, some 20 million tons of sulfur dioxide gas was ejected into the atmosphere, causing a global decrease in average annual temperatures of 0.5°C for two years.

For example, the ash and gases released during the 1991 eruption of Mount Pinatubo, shown in **Figure 22,** resulted in slightly cooler average global temperatures for two years following the eruption.

The most powerful eruptions can have larger, longer-term impacts on climate. The Deccan Traps eruption, 66 million years ago, was so large that it cooled Earth by 2°C and may have had a part in the extinction of the nonavian dinosaurs.

Generally, volcanic eruptions appear to have only short-term effects on climate. These effects, as well as the other short-term and long-term effects you have read about thus far, are a result of natural causes.

Check Your Progress

Summary

- Climate change can occur on a long-term or short-term scale.
- Changes in solar activity have been correlated with periods of climate change.
- Changes in Earth's orbit, tilt, and wobble are all associated with changes in climate.

Demonstrate Understanding

1. **Identify** and explain an example of long-term climatic change.
2. **Describe** What are seasons? What causes them?
3. **Illustrate** how El Niño might affect weather in California and along the Gulf Coast.
4. **Analyze** How does volcanic activity affect climate?
5. **Compare** the time scales of climatic changes caused by a volcanic eruption and by Earth's wobble.

Explain Your Thinking

6. **Assess** What might be the effect on seasons if Earth's orbit became more elliptical and, at the same time, the angle of the tilt of Earth's axis increased?
7. **MATH ›Connection** Study **Figure 21.** During which period were sunspot numbers lowest? During which period were sunspot numbers highest?

IMPACT OF HUMAN ACTIVITIES

How is modern climate change different from natural climate change?

Our Influence on the Atmosphere

You learned how gases like methane, water vapor, and carbon dioxide help keep Earth warm enough to sustain life. These greenhouse gases absorb long-wavelength radiation and reradiate it back to Earth's surface, as shown in **Figure 23.** After measuring ice cores, sediment cores, and isotopes in rock samples, scientists have found that the concentrations of methane and carbon dioxide have fluctuated over time. As the concentrations of these gases have risen, the global climate has warmed. As the concentrations of these gases have fallen, the global climate has cooled. These concentration fluctuations and climatic changes took tens of thousands to millions of years to happen. Over the last 150 years, however, humans have increased the greenhouse effect by changing the amount of atmospheric greenhouse gases, particularly carbon dioxide and methane. Over that time, the amount of carbon dioxide in the atmosphere increased from 280 ppm to over 400 ppm. The increase in these gases has resulted in increased absorption of energy in the atmosphere.

Figure 23 Solar radiation reaches Earth's surface, where it is reradiated as long-wavelength radiation. This radiation does not easily escape through the atmosphere and is mostly absorbed and rereleased by atmospheric gases. This process is called the greenhouse effect.

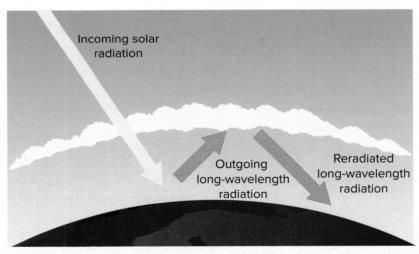

Incoming solar radiation

Outgoing long-wavelength radiation

Reradiated long-wavelength radiation

3D THINKING   **DCI** Disciplinary Core Ideas   **CCC** Crosscutting Concepts   **SEP** Science & Engineering Practices

COLLECT EVIDENCE

Use your Science Journal to record the evidence you collect as you complete the readings and activities in this lesson.

INVESTIGATE

GO ONLINE to find these activities and more resources.

Applying Practices: Exploring Relationships: Climate Change and Human Activity
HS-ESS3-6. Use a computational representation to illustrate the relationships among Earth systems and how those relationships are being modified due to human activity.

Applying Practices: Applying Practices: Forecasting Climate Change
HS-ESS3-5. Analyze geoscience data and the results from global climate models to make an evidence-based forecast of the current rate of global or regional climate change and associated future impacts to Earth systems.

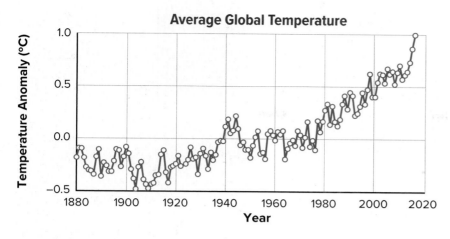

Average Global Temperature

Figure 24 The warmest years of the last century all happened within the last 20 years. This graph shows the change in (or anomalous) global surface temperature compared to average temperature between 1951 and 1980.

This has led to a rise in global temperatures, known as global warming. Since increasing the global temperature changes the long-term averages and variations in weather around the world, scientists refer to these effects as **global climate change.** Global climate change causes changes in temperature, precipitation, glaciation, frequency and strength of severe weather, and acidification of oceans across the globe.

Global climate change

Temperatures worldwide have shown an upward trend over the past 200 years, with the ten warmest years on record having occurred within the last two decades. This trend is shown in **Figure 24.** Current computer models predict that average global temperatures will continue to rise. The models strongly depend on the amount of human-generated greenhouse gases added to the atmosphere each year and the ways in which these gases are absorbed by the ocean and the biosphere. As this trend continues, polar ice caps and mountain glaciers melt at increasing rates. This will lead to a rise in sea level and the flooding of coastal cities. Other possible consequences include the spread of deserts into fertile regions, an increase in sea surface temperature, ocean acidification, and an increase in the frequency and severity of storms and droughts.

Based on empirical evidence—including temperature data from multiple U.S. government agencies, other world government agencies, and university climate researchers—scientists have concluded that global warming is occurring. An overwhelming majority of scientists from a variety of fields—including climatology, meteorology, geology, astronomy, and physics—have found that natural cycles alone cannot explain the current anomalous rise in global temperature. More than 97 percent of published climate articles from 1991 through 2011 support the idea that humans are causing the rise in global temperature that began 150 years ago.

 Get It?

Describe evidence from computer models and scientific studies about the relationship between human activities and global warming.

CCC CROSSCUTTING CONCEPTS

Cause and Effect You are an environmental scientist. Write a letter to an academic journal, citing empirical evidence used to make claims about global climate change. In your letter, be sure to differentiate between cause and correlation.

Burning fossil fuels One of the main sources of atmospheric carbon dioxide from humans is from the burning of fossil fuels, including coal, oil, and natural gas. Ninety-eight percent of the carbon dioxide emissions in the United States come from burning fossil fuels to run automobiles, heat homes and businesses, and power factories. Almost any process that involves the burning of fossil fuels results in the release of carbon dioxide. Burning fossil fuels also releases other greenhouse gases, such as methane and nitrous oxide, into the atmosphere.

Deforestation Deforestation—the mass removal of trees—also plays a role in increasing levels of atmospheric carbon dioxide. During photosynthesis, vegetation removes carbon dioxide from the atmosphere. When trees, such as the ones shown in **Figure 25,** are cut down, photosynthesis is reduced, and more carbon dioxide remains in the atmosphere. Scientists have found that deforestation intensifies global warming and other effects of climate change.

Environmental efforts Though the magnitude of human impact is greater than it has ever been, so too is our ability to manage current and future environmental impacts. As an individual, you can help reduce the amount of carbon dioxide emitted to the atmosphere by conserving energy. This, in turn, reduces fossil fuel consumption. Some easy ways to conserve energy include turning off electrical appliances and lights when not in use, turning down thermostats in the winter, recycling, and reducing the use of combustion engines, such as those in cars and lawn mowers.

Figure 25 Deforestation, the mass removal of trees, has occurred in this forest.

Explain *how deforestation leads to global warming.*

✍ Check Your Progress

Summary

- The greenhouse effect influences Earth's climate.
- Worldwide temperatures have shown an upward trend over the past 200 years.
- Human activities can influence changes in weather and climate.
- Individuals can reduce their environmental impact on climate change.

Demonstrate Understanding

1. **Describe** some human activities that impact Earth's climate.
2. **Explain** the foundation for global climates in terms of electromagnetic radiation from the Sun and the greenhouse effect.
3. **Apply** What is global warming? What are some possible consequences of global warming?

Explain Your Thinking

4. **Reason** Some computer models indicate that global warming might lead to increased cloud cover. How might this affect atmospheric processes?
5. **WRITING ⟩ Connection** Write a pamphlet that explains how changes in the atmosphere due to human activity have affected CO_2 concentrations and climate. Include tips on how individuals can reduce CO_2 emissions into the atmosphere to help manage current and future environmental impacts.

LEARNSMART Go online to follow your personalized learning path to review, practice, and reinforce your understanding.

James_Gabbert/iStock/Getty Images

On the Move: Human Migration and Climate Change

Analyze geologic data from ice cores and the ocean floor. Mix in studies of ancient archaeological remains. The result? New scientific hypotheses indicate that human migrations are linked to changes in climate that occur in long-term cycles. These hypotheses help explain the past and give scientists a window into the future.

Lake Chad in Africa is a freshwater lake that has been drying up for the last five decades. As a result, millions of people who depended on the lake for survival have had to move.

The past

Most scientists agree that humans evolved in Africa and then migrated to other regions. Many scientists also agree that this migration happened in one mass wave about 60,000 years ago.

However, researchers from the University of Hawaii have proposed a new scenario. Using a computer model that analyzes ancient glacial data, sea level changes, and fossil evidence, they theorize that humans migrated from Africa in four waves that began as far back as 100,000 years ago. These waves correspond to changes in Earth's wobble that occur about every 21,000 years. The orbital changes altered the climate of northern Africa. Dry areas became lush and green. People moved into these new ecosystems and eventually into Asia and Europe.

The data are supported by other recent studies of human migration that involve DNA analyses. However, fossil evidence in Europe does not correspond to the computer model, and will need further investigation.

The present

Studies indicate that Earth's global climate is changing once again, due to human-induced and natural factors. How will a changing climate affect human migration?

When comparing maps of modern human migration and areas adversely affected by climate change, a pattern emerges: many people are moving because of climate-related factors such as prolonged heat waves, droughts, desertification, and rising sea level. One example is Syria, where a three-year drought has led to severe crop failures and mass migration to cities.

Globally, an estimated 500 million people may migrate because of climate change. One challenge society faces is providing for the needs of immigrants without straining the resources of the places to which they move.

USE COMPUTATIONAL REPRESENTATIONS

Work with a partner to analyze digital maps of an area that is experiencing high rates of migration. Research climate data for the area. Discuss any patterns you see, and then share your findings with the class.

MODULE 11
STUDY GUIDE

 GO ONLINE to study with your Science Notebook.

Lesson 1 DEFINING CLIMATE

- Climate describes the long-term weather patterns of an area.
- Normals are the standard climatic values for a location.
- Temperatures vary among tropical, temperate, and polar zones.
- Climate is influenced by latitude, topographic effects, and air masses.
- Air masses have distinct regions of origin.

- climatology
- normal
- tropics
- temperate zones
- polar zones

Lesson 2 CLIMATE CLASSIFICATION

- German scientist Wladimir Köppen developed a climate classification system.
- There are five main climate types: tropical, dry, mild, continental, and polar.
- Microclimates can occur within cities.

- Köppen classification system
- microclimate
- heat island

Lesson 3 CLIMATIC CHANGES AND PATTERNS

- Climate change can occur on a long-term or short-term scale.
- Changes in solar activity have been correlated with periods of climate change.
- Changes in Earth's orbit, tilt, and wobble are all associated with changes in climate.

- ice age
- season
- El Niño
- Feedback
- Albedo

Lesson 4 IMPACT OF HUMAN ACTIVITIES

- The greenhouse effect influences Earth's climate.
- Worldwide temperatures have shown an upward trend over the past 200 years.
- Human activities can influence changes in weather and climate.
- Individuals can reduce their environmental impact on climate change.

- global climate change

REVISIT THE PHENOMENON

What is climate?

CER Claim, Evidence, Reasoning

Explain Your Reasoning Revisit the claim you made when you encountered the phenomenon. Summarize the evidence you gathered from your investigations and research and finalize your Summary Table. Does your evidence support your claim? If not, revise your claim. Explain why your evidence supports your claim.

STEM UNIT PROJECT

Now that you've completed the module, revisit your STEM unit project. You will summarize your evidence and apply it to the project.

GO FURTHER

SEP Data Analysis Lab

What is the temperature in Phoenix, Arizona?

The table contains temperature data for Phoenix, Arizona, based on data collected from June 1, 1933, through January 20, 2015.

Data and Observations Graph the data in the table. Plot the monthly values for average maximum temperatures. Place the month on the x-axis and temperature on the y-axis. Do the same for the monthly values for average minimum temperatures.

CER Analyze and Interpret Data

1. **Claim** Which months were warmer than the average maximum temperature?
2. **Claim** Which months were colder than the average maximum temperature?
3. **Evidence, Reasoning** What is the climate of Phoenix, Arizona, based on average temperatures?

Monthly Temperature Summary for Phoenix, AZ

Temperature (°C)	Average maximum	Average minimum
Jan	19	5
Feb	21	7
Mar	24	9
Apr	29	13
May	34	18
Jun	39	23
Jul	41	27
Aug	40	26
Sep	37	23
Oct	31	16
Nov	24	9
Dec	19	6
Average	29.8	15.1

*Data obtained from: Western Regional Climate Center. 2015.

Dmitry Ilyshev/123RF

randy andy/Shutterstock

ENCOUNTER THE PHENOMENON

How can water carve a shoreline?

GO ONLINE to play a video about how coastal features are formed.

SEP Ask Questions

Do you have other questions about the phenomenon? If so, add them to the driving question board.

CER Claim, Evidence, Reasoning

Make Your Claim Use your CER chart to make a claim about how water shapes shorelines. Explain your reasoning.

Collect Evidence Use the lessons in this module to collect evidence to support your claim. Record your evidence as you move through the module.

Explain Your Reasoning You will revisit your claim and explain your reasoning at the end of the module.

GO ONLINE to access your CER chart and explore resources that can help you collect evidence.

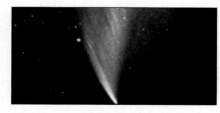

LESSON 1: Explore and Explain: Origin of the Oceans

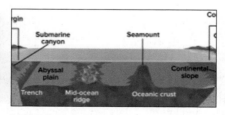

LESSON 3: Explore and Explain: The Continental Margin

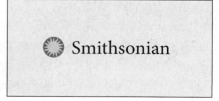

Additional Resources

AN OVERVIEW OF OCEANS

FOCUS QUESTION

How did Earth gain its oceans?

Data Collection and Analysis

Oceanography is the scientific study of Earth's oceans. In the late 1800s, the British ship *Challenger* became the first research ship to use relatively sophisticated measuring devices to study the oceans. Since then, oceanographers have been collecting data with instruments both at the surface and from the depths of the ocean floor. Technologies such as sonar, floats, satellites, submersibles, and computers have become central to the continuing exploration of the ocean. **Figure 1** chronicles some of the major discoveries that have been made about oceans.

At the surface

Sonar, which stands for **so**und **na**vigation and **r**anging, is used by oceanographers to learn more about the topography of the ocean floor. To determine ocean depth, scientists send a sonar signal to the ocean floor and measure how long it takes for the sound to reach the bottom and return to the surface as an echo. Knowing that sound travels at a constant velocity of 1500 m/s through water, scientists can determine the depth by multiplying the total time by 1500 m/s, and then dividing the answer by 2.

Get It?
Calculate If a sonar signal takes 5 seconds to return to a ship, what is the depth of the ocean at that point?

Large portions of the seafloor have been mapped using **side-scan sonar,** a technique that directs sound waves to the seafloor at an angle so that the sides of underwater hills and other topographic features can be mapped.

Oceanographers use floats that contain sensors to learn more about water temperature, salinity, and the concentration of gases and nutrients in surface water. Floats can also be used to record wave motion and the speed at which currents are moving. Satellites such as the *OSTM/Jason-3,* which you have read about, continually monitor the ocean's surface temperatures, currents, and wave conditions.

3D THINKING   **DCI** Disciplinary Core Ideas   **CCC** Crosscutting Concepts   **SEP** Science & Engineering Practices

COLLECT EVIDENCE
Use your Science Journal to record the evidence you collect as you complete the readings and activities in this lesson.

INVESTIGATE
GO ONLINE to find these activities and more resources.

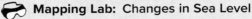
Mapping Lab: Changes in Sea Level
Analyze and interpret data to determine the effect of glaciation on sea level.

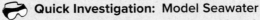
Quick Investigation: Model Seawater
Use a model to visualize the composition of sea water.

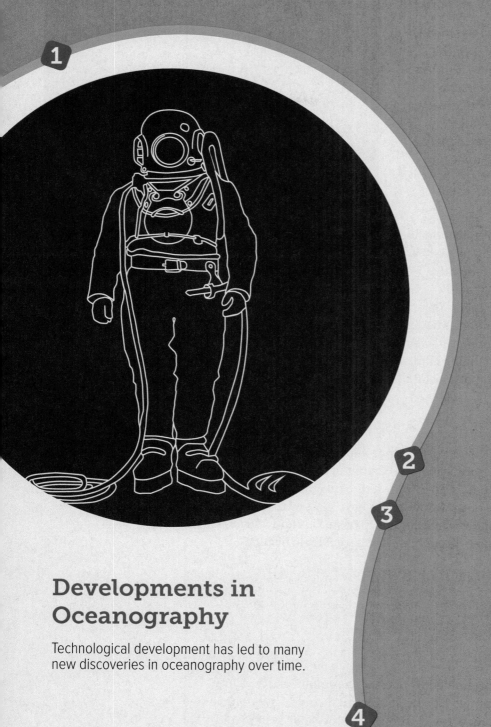

Developments in Oceanography

Technological development has led to many new discoveries in oceanography over time.

1 **1872** The Challenger expedition marks the beginning of oceanography. Scientists measure sea depth, study the composition of the seafloor, and collect a variety of oceanic data.

2 **1925** The German Meteor expedition surveys the South Atlantic floor with sonar and discovers the Mid-Atlantic Ridge.

3 **1932–1934** The first deep-ocean dives use a tethered bathysphere. The dives uncover luminescent creatures and provide sediment samples.

4 **1943** In France, the first diving equipment is invented from hoses, mouthpieces, air tanks, and a redesigned car regulator that supplies compressed air to divers.

5 **1955** A survey ship detects linear magnetic stripes along the ocean floor. These magnetic patterns lead to the formulation of the theory of plate tectonics.

6 **1977** The submersible Alvin discovers hydrothermal vents and a deep-sea ecosystem, including giant worms and clams, that can survive without energy from the Sun.

7 **1984** An observation system in the Pacific Ocean helps scientists predict El Niño and begin to understand the connection between oceanic events and weather.

8 **1995** Scientists map the entire seafloor using satellite data.

9 **2010** High-definition video of new submarine volcanoes, hydrothermal vent systems, and exotic animal ecosystems discovered off the coast of Indonesia are transmitted to scientists ashore in real time.

In the deep sea

Submersibles, underwater vessels that can be remotely operated or carry people to the deepest areas of the ocean, have allowed scientists to explore new frontiers. *Alvin,* shown in **Figure 2,** is a modern submersible that can take two scientists and a pilot to depths as great as 4500 m. *Alvin* has been used to discover geologic features, such as hydrothermal vents, and previously unknown sea creatures. It can also be used to bring sediments and water samples to the surface.

Computers

Computers are an integral tool in both the collection and analysis of data from the ocean. Information from satellites and float sensors can be transmitted and downloaded directly to computers. Sophisticated programs use mathematical equations to analyze data and produce models. When combined with observations, ocean models provide information about subsurface currents that are not observed directly. Operating in a fashion similar to weather forecasting models, global ocean models play a role in simulating Earth's changing climate. Ocean models are also used to simulate tides, tsunamis, and the dispersion of coastal pollution.

Figure 2 *Alvin* is a deep-sea submersible that can hold two scientists and a pilot.

Origin of the Oceans

Several geologic clues indicate that oceans have existed since almost the beginning of geologic history. Studies of radioactive isotopes indicate that Earth is about 4.6 billion years old. Scientists have found rocks nearly as old that formed from sediments deposited in water. Ancient lava flows are another clue. Some of these lava flows have glassy crusts that form only when molten lava is chilled rapidly underwater. Radioactive studies and lava flows offer evidence that there has been abundant water throughout Earth's geologic history.

 Get It?

Explain the evidence that suggests that oceans have existed since almost the beginning of Earth's geologic history.

Where did the water come from?

Scientists hypothesize that Earth's water originated from either a remote source, a local source, or both. Comets and meteorites are two remote sources that could have contributed to the accumulation of water on Earth. Comets travel throughout the solar system and occasionally collide with Earth. These impacts release enough water that they could have contributed to filling the ocean basins over geologic time.

Meteorites are composed of the same material that might have formed the early planets. Studies indicate that meteorites contain up to 0.5 percent water. Meteorite bombardment releases water into Earth's systems.

If early Earth contained the same percentage of water as meteorites, it would have been sufficient to form early oceans. However, some mechanism must have existed to allow the water to rise from Earth's interior to its surface. Scientists theorize that this mechanism was volcanism.

Volcanism

During volcanic eruptions, significant quantities of gases are emitted. These volcanic gases consist mostly of water vapor and carbon dioxide. Shortly after the formation of Earth, when the young planet was much hotter than it is today, an episode of massive, violent volcanism took place over the course of perhaps several hundred million years. As shown in **Figure 3,** this volcanism released huge amounts of water vapor, carbon dioxide, and other gases, which combined to form Earth's early atmosphere. As Earth's crust cooled, the water vapor gradually condensed, fell to Earth's surface as precipitation, and accumulated to form oceans. By the time the oldest known crustal rock formed billions of years ago (bya), Earth's oceans might have been close to their present size. Water is still being added to the hydrosphere by volcanism, but some water molecules in the atmosphere are continually being destroyed by ultraviolet radiation from the Sun. These two processes balance each other.

Get It?
Explain how water accumulated on Earth's surface to form oceans.

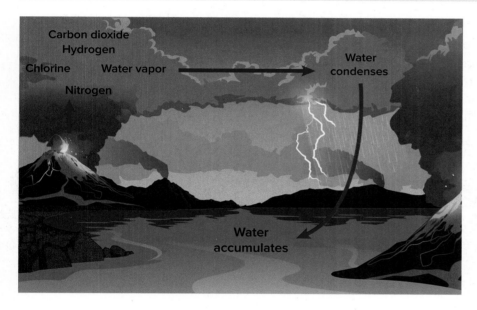

Figure 3 In addition to comets, volcanic eruptions might have been a source of water for Earth's early oceans. An intense period of volcanism occurred shortly after the planet formed. This volcanism released large quantities of water vapor and other gases into the atmosphere. The water vapor eventually condensed into oceans.

Distribution of Earth's Water

As you learned in Module 7, the oceans contain 97 percent of the water found on Earth. Another 3 percent is freshwater located in the frozen ice caps of Greenland and Antarctica and in rivers, lakes, and underground sources. The percentage of ice on Earth has varied over geologic time from near zero to perhaps as much as 10 percent of the hydrosphere. As you read further in this lesson, you will learn more about how these changes affect sea level.

The blue planet

Earth is known as the blue planet for good reason—approximately 71 percent of its surface is covered by oceans. The average depth of these oceans is 3800 m. Earth's landmasses are like huge islands, almost entirely surrounded by water. Because most landmasses are in the northern hemisphere, oceans cover only 61 percent of the surface there. However, 81 percent of the southern hemisphere is covered by water.

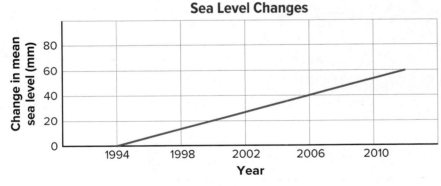

Northern Hemisphere
61% Ocean

North Pole

×

Equator

Southern Hemisphere
81% Ocean

South Pole

×

Equator

Figure 4 The northern hemisphere is covered by slightly more water than land. The southern hemisphere, however, is almost completely covered by water.

Figure 4 shows the distribution of water in the northern and southern hemispheres. Note that all the oceans are one vast, interconnected body of water. They have been divided into specific oceans and seas largely because of historic and geographic considerations.

Sea level

Global **sea level,** which is the level of the oceans' surfaces, has risen and fallen by hundreds of meters in response to melting ice during interglacial periods and expanding glaciers during ice ages. Other processes that affect sea level are tectonic forces that lift or lower portions of Earth's crust. A rising seafloor causes a rise in sea level, while a sinking seafloor causes sea level to drop. **Figure 5** shows that the sea level rose at a rate of about 3 mm per year between 1994 and 2012. Scientists hypothesize that this rise in sea level is related to water that has been released by the melting of glaciers and thermal expansion of the ocean due to warming.

Sea Level Changes

Change in mean sea level (mm)

80
60
40
20
0

1994   1998   2002   2006   2010

Year

Figure 5 Scientists at NASA used floats and satellites to collect data on sea level changes from 1994 to 2012.

Explain *What is a possible cause for rising sea level during this period?*

Major oceans

There are three major oceans: the Pacific, the Atlantic, and the Indian. The Pacific Ocean is the largest. Containing roughly half of Earth's seawater, it is larger than all of Earth's land-masses combined. The second-largest ocean, the Atlantic, extends for more than 20,000 km, from Antarctica to the Arctic Circle. North of the Arctic Circle, the Atlantic Ocean is often referred to as the Arctic Ocean. The third-largest ocean, the Indian, is located mainly in the southern hemisphere. The storm-lashed region surrounding Antarctica, south of about 50° south latitude, is known as the Southern Ocean. **Figure 6,** on the next page, can help you visualize this information about Earth's major oceans.

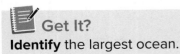 **Get It?**

Identify the largest ocean.

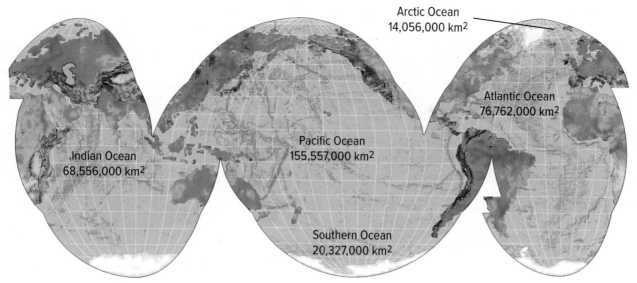

Arctic Ocean
14,056,000 km²

Atlantic Ocean
76,762,000 km²

Pacific Ocean
155,557,000 km²

Indian Ocean
68,556,000 km²

Southern Ocean
20,327,000 km²

Figure 6 The Pacific, Atlantic, and Indian oceans stretch from Antarctica to the north. The smaller Arctic Ocean and Southern Ocean are located near the North and South Poles, respectively.

Ocean and atmospheric interaction

Oceans provide moisture and heat to the atmosphere and influence large-scale circulation patterns. You have learned that warm ocean water energizes tropical cyclones, influences the position and strength of jet streams, and plays a role in El Niño events.

Oceans are also a vast reservoir of carbon dioxide. Dissolved carbon dioxide in surface waters sinks in water masses into the deep ocean, returning to the surface hundreds of years later. Without this natural uptake by the ocean, the accumulation of carbon dioxide in the atmosphere would be much larger than currently observed. There is also an uptake of carbon dioxide by phytoplankton during photosynthesis in the sunlit areas of the ocean. In the process, carbon is stored in the ocean, and excess oxygen is released to the atmosphere to make Earth habitable. **Table 1** summarizes some of the interactions between oceans and the atmosphere.

Table 1 Ocean-Atmospheric Interactions

Example	Description
Oceans are a source of atmospheric oxygen.	Fifty percent of oxygen in the atmosphere comes from marine phytoplankton, which release oxygen into surface waters as a product of photosynthesis.
Oceans are a reservoir for carbon dioxide.	Atmospheric carbon dioxide dissolves in cold water better than in warm water. When cold, dense surface water in polar oceans sinks, dissolved carbon dioxide moves to the bottom of the ocean.
Oceans are a source of heat and moisture.	Warm ocean water in equatorial regions heats the air above it. This warmer air holds more water vapor formed at the ocean's surface. These conditions can fuel hurricanes.

Chemical Composition of Seawater

Ocean water contains dissolved gases, including oxygen and carbon dioxide, and dissolved nutrients such as nitrates and phosphates. Chemical profiles of seawater vary based on both location and depth, as shown in **Figure 7**. Factors that influence the amount of a substance in an area of ocean water include wave action, vertical movements of water, and biological activity.

Figure 7 shows that oxygen levels are high at the surface in both the Atlantic and Pacific Oceans. This occurs in part because oxygen is released by surface-dwelling photosynthetic organisms.

Silica levels for both oceans are also shown in **Figure 7.** Because many organisms remove silica from ocean water and use it to make shells, silica levels near the surface are usually low. Silica levels usually increase with depth because decaying organisms sink to the ocean bottom, returning silica to the water.

Get It?

Explain why the concentration of oxygen in oceans is a result of interactions among the hydrosphere, the atmosphere, and the biosphere.

Salinity

The measure of the amount of dissolved salts in seawater is **salinity.** Oceanographers express salinity as grams of salt per kilogram of water, or parts per thousand (ppt). The total salt content of seawater averages 35 ppt, or 3.5 percent. The most abundant salt in seawater is sodium chloride. Other salts in seawater are chlorides and sulfates of magnesium, potassium, and calcium.

Get It?

Determine If a sample of seawater is 2.7 percent salt, what is the salinity of the water in ppt?

Variations in salinity

Although the average salinity of the oceans is 35 ppt, actual salinity varies from place to place. In subtropical regions where rates of evaporation exceed those of precipitation, salt left behind by the evaporation of water molecules accumulates in the surface layers of the ocean. There, salinity can be as high as 37 ppt. In equatorial regions where precipitation is abundant, salinity is lower. Even lower salinities, of 32 or 33 ppt, occur in polar regions where seawater is diluted by melting sea ice.

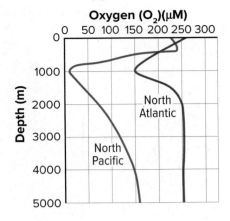

Oxygen in Seawater

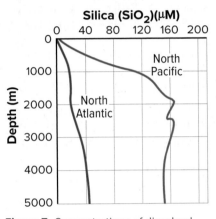

Silica in Seawater

Figure 7 Concentrations of dissolved gases and nutrients in seawater, measured in micromolars (μM), vary by location and depth.

Examine *How do oxygen levels differ between the North Atlantic and North Pacific Oceans?*

STEM CAREER Connection
Oceanographer
Being an oceanographer is an exciting job. Imagine spending days at sea collecting data and further investigating the mysteries of the deep sea! Studying the interaction between the ocean and the atmosphere is more important than ever as climate change continues to impact Earth's systems.

Ocean Salinity

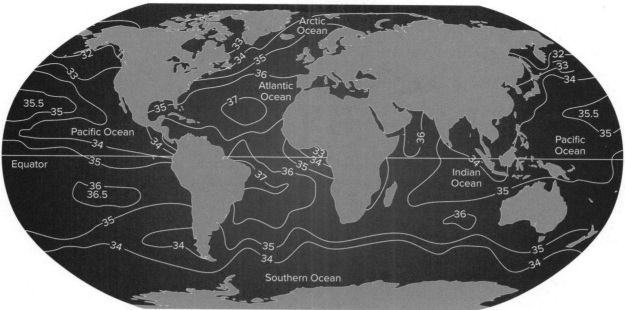

*All values are given in parts per thousand (ppt).

Figure 8 Ocean salinity varies from place to place. High salinity is common in areas with high rates of evaporation. Low salinity often occurs in estuaries where seawater mixes with fresh water.

Figure 8 shows the variation in salinity in different areas of oceans. The lowest salinity often occurs where large rivers empty into the oceans, creating areas of water called **estuaries.** Even though salinity varies, the relative proportion of major types of sea salts is constant because all ocean water continually intermingles throughout Earth's oceans.

Sources of sea salt

Geologic evidence indicates that the salinity of ancient seas was not much different from that of today's oceans. One line of evidence is based on the proportion of magnesium in the calcium-carbonate shells of some marine organisms. Present-day shells contain about the same proportion of magnesium as similar shells from throughout geologic time.

Sources of sea salts have also stayed the same over time. Sulfur dioxide and chlorine, gases released by volcanoes, dissolve in water, forming sulfate and chlorine ions. Most of the other ions in seawater, including sodium, calcium, and magnesium, come from the weathering of crustal rocks. These ions enter rivers and are transported to oceans, as shown in **Figure 9.**

Gases from volcanic eruptions

River runoff

Figure 9 Salts are added to seawater by volcanic eruptions and by the weathering and erosion of rocks.

Removal of sea salts

Although salt ions are continuously added to seawater, salinity does not increase because salts are also continuously removed. Winds pick up salty droplets from breaking waves and deposit the salt farther inland. Recall that evaporites form when water evaporates from concentrated solutions. In arid coastal regions, water evaporates from seawater and leaves solid salt behind, as shown in **Figure 10.** Marine organisms remove ions from seawater to build shells, bones, and teeth, also shown in **Figure 10.** As organisms die, their solid parts accumulate on the seafloor and become part of bottom sediments. The salt contained in these solid parts does not return to the seawater. The existing salinity of seawater represents a balance between the processes that remove salts and those that add them.

Evaporate formation

Biological activity

Figure 10 Salts are removed from seawater by the formation of evaporites and biological processes. Also, wind carries salty droplets inland.

Ocean pH

As minerals are dissolved by water and flow into the oceans, they alter pH. The global average pH of the oceans is about 8.1, meaning the water is basic. The alkalinity varies slightly throughout the year because erosion rates change with the seasons. Gases in the atmosphere also affect the pH of the oceans. As you have learned, carbon dioxide from the atmosphere dissolves into water to make carbonic acid. When the concentration of CO_2 in the atmosphere increases, more of it dissolves into the oceans. The increase in carbonic acid decreases the pH of the oceans. This process is called ocean acidification. Humans have been adding CO_2 to the atmosphere at increasing rates since the eighteenth century. Before the industrial revolution, the average pH of the oceans was 8.2, 0.1 higher than today. This does not seem like a significant difference, but that change represents a 30 percent increase in acidity because pH is a logarithmic scale.

Physical Properties of Seawater

The presence of various salts causes the physical properties of seawater to be different from those of freshwater.

Density

Freshwater has a maximum density of 1.00 g/cm³. Because salt ions add to the overall mass of the water in which they are dissolved, they increase the density of water. Seawater is therefore more dense than freshwater, and its density increases with salinity. Temperature also affects density—cold water is more dense than warm water. Because of salinity and temperature variations, the density of seawater ranges from about 1.02 g/cm³ to 1.03 g/cm³. These variations might seem small, but they are significant. They affect many oceanic processes.

Freezing point

Variations in salinity also cause the freezing point of seawater to be somewhat lower than that of freshwater. Freshwater freezes at 0°C. Because salt ions interfere with the formation of the crystal structure of ice, the freezing point of seawater is –2°C.

Absorption of light

If you have ever swum in a lake, you might have noticed that the intensity of light decreases with depth. The water might be clear, but if the lake is deep, the bottom waters will be dark. Water absorbs light, which gives rise to another physical property of oceans—darkness. In general, light penetrates only the upper 100 m of seawater. Below that depth, all is darkness.

Figure 11 illustrates how light penetrates ocean water. Notice that red light does not penetrate as far as blue light. Red objects, such as the giant red shrimp shown in **Figure 11,** appear black below a certain depth, and other reflecting objects in the water appear green or blue. Although some fading blue light can reach depths of a few hundred meters, light sufficient for photosynthesis exists only in the top 100 m of the ocean. In the darkness of the deep ocean, some organisms, including some fishes and shrimps, are blind. Other organisms attract prey by producing light through a chemical reaction called bioluminescence.

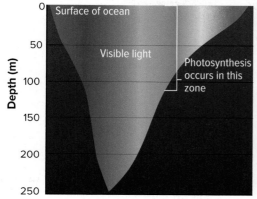

Light Penetration in Open Ocean

Figure 11 Red light does not penetrate as far as blue light in the ocean. Marine organisms that are some shades of red, such as deep-sea shrimp, appear black below a depth of 10 m. This helps them escape predators.

Identify *To what depth does blue light penetrate ocean water?*

Ocean Layering

Ocean surface temperatures range from –2°C in polar waters to 38°C in equatorial regions, with the average surface temperature being 15°C. Ocean water temperatures, however, decrease significantly with depth. Deep ocean water is always cold, even in tropical oceans.

Temperature profiles

A **temperature profile** plots water temperature against depth. Such profiles vary depending on location and season. However, beneath roughly 100 m, temperatures decrease continuously with depth to around 4°C at 1000 m. The dark waters below 1000 m have fairly uniform temperatures of less than 4°C.

ACADEMIC VOCABULARY

variation

the range in which a factor changes

The variation in temperature in New York was a shock for the tourist from California.

Pixtal/age fotostock

Based on these temperature variations, the ocean can be divided into the three layers shown in **Figure 12.** The first is a relatively warm, sunlit surface layer approximately 100 m thick. Notice that tropical areas have warmer surface temperatures than temperate or polar areas. Under the surface layer is a transitional layer known as the **thermocline,** which is characterized by rapidly decreasing temperatures with depth. The bottom layer is cold and dark with temperatures near freezing. Both the thermocline and the warm surface layer are absent in polar seas, where water temperatures are cold from top to bottom. In general, ocean layering is caused by density differences. Because cold water is more dense than warm water, cold water sinks to the bottom, while less-dense, warm water is found near the ocean's surface.

 Get It?

Describe the three main layers of water in oceans.

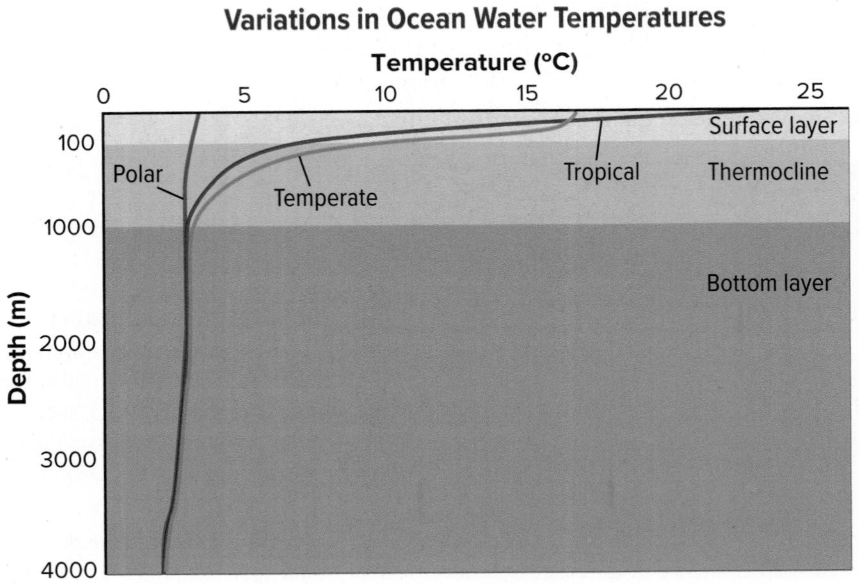

Variations in Ocean Water Temperatures

Figure 12 Ocean water temperatures decrease with depth. Tropical areas have warmer ocean surface temperatures than do temperate or polar areas.

Water Masses

The temperature of the bottom layer of ocean water is near freezing. This is true even in tropical oceans, where surface temperatures are warm. Where does all this cold water come from?

Deepwater masses

Cold water comes from Earth's polar seas. Recall that high salinity and cold temperatures cause seawater to become more dense. When seawater freezes during the arctic or antarctic winter, sea ice forms. Because salt ions are not incorporated into the growing ice crystals, they accumulate beneath the ice. Consequently, the cold water beneath the ice becomes saltier and more dense than the surrounding seawater, and this saltier water sinks. This salty water then migrates toward the equator as a cold, deepwater mass along the ocean floor. Other cold, deepwater masses form when surface currents in the ocean bring relatively salty midlatitude or subtropical waters into polar regions. In winter, these waters become colder and denser than the surrounding polar surface waters, and thus they sink. **Figure 13,** on the next page, illustrates how deepwater masses are formed.

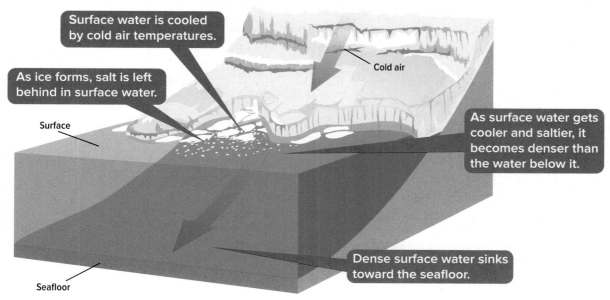

Figure 13 Dense polar water sinks, producing a deepwater mass.

Explain *the relationship between the density of water and the formation of deepwater masses.*

Three water masses account for most of the deepwater masses in the oceans—Antarctic Bottom Water, North Atlantic Deep Water, and Antarctic Intermediate Water. Antarctic Bottom Water forms when antarctic seas freeze during the winter. With temperatures below 0°C, this deepwater mass is the coldest and densest in all the oceans, as shown in **Figure 14.** North Atlantic Deep Water forms in a similar manner offshore from Greenland. Antarctic Bottom Water is colder and denser than North Atlantic Deep Water, so it sinks below it.

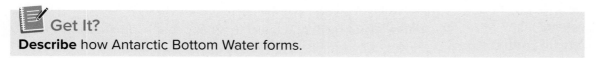

Get It?

Describe how Antarctic Bottom Water forms.

Deepwater Masses and Temperature Distribution (°C)

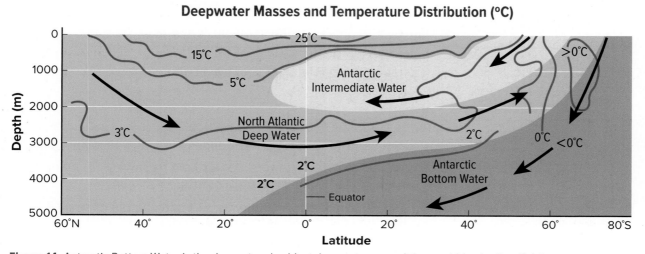

Figure 14 Antarctic Bottom Water is the densest and coldest deepwater mass. It is overridden by the slightly warmer and less dense North Atlantic Deep Water. Antarctic Intermediate Water is still warmer and less dense, and thus it overrides the other two deepwater masses.

Intermediate water masses

Antarctic Intermediate Water, shown in **Figure 14,** forms when the relatively salty waters near Antarctica decrease in temperature and sink during winter. Because Antarctic Intermediate Water is slightly warmer and less dense than North Atlantic Deep Water, it does not sink as deep as the other two deepwater masses. Antarctic Intermediate Water sinks to only about 2000 m.

While the Atlantic Ocean contains all three major deepwater masses—Antarctic Bottom Water, North Atlantic Deep Water, and Antarctic Intermediate Water—the Indian and Pacific Oceans contain only the two Antarctic deepwater masses. In Lesson 2, you will learn about other water movements in the ocean.

 Get It?

Compare and contrast Antarctic Intermediate Water and Antarctic Bottom Water.

 # Check Your Progress

Summary

- Scientists use many different instruments to collect and analyze data from oceans.

- Scientists have several ideas about where the water in Earth's oceans originated.

- Earth's oceans are the Pacific, the Atlantic, the Indian, the Arctic, and the Southern.

- Ocean water contains dissolved gases, nutrients, and salts.

- Salts are added to and removed from oceans through natural processes.

- Properties of ocean water, including temperature and salinity, vary with location and depth.

- Many of the oceans' deepwater masses sink from the surface of polar oceans.

Demonstrate Understanding

1. **State** how much of Earth is covered by oceans. How is ocean water distributed over Earth's surface?

2. **Describe** two tools scientists use to collect data about oceans.

3. **Relate** What evidence indicates that oceans formed early in Earth's geologic history?

4. **Specify** Where did the water in Earth's early oceans originate?

5. **Identify** What factors affect the chemical properties of seawater?

6. **Illustrate** the three layers into which ocean water is divided based on temperature.

7. **Sequence** the steps involved in the formation of deepwater masses.

Explain Your Thinking

8. **Predict** some possible consequences of rising sea level.

9. **Hypothesize** Which is more dense, cold freshwater or warm seawater?

10. **MATH** **Connection** If the density of a sample of seawater is 1.02716 g/mL, calculate the mass of 4.0 mL of the sample.

LEARNSMART Go online to follow your personalized learning path to review, practice, and reinforce your understanding.

FOCUS QUESTION
How does the land influence the ocean's movement?

Waves

Oceans are in constant motion. Their most obvious movement is that of waves. A **wave** is a rhythmic movement that carries energy through space or matter—in this case, ocean water. Ocean waves are generated mainly by wind blowing over the water's surface. In the open ocean, a typical wave has the characteristics shown in **Figure 15.** The highest point of a wave is the **crest,** and the lowest point is the **trough.** The vertical distance between crest and trough is the wave height, and the horizontal crest-to-crest distance is the wavelength. As energy is added, both the wavelength and speed increase. Thus, longer waves travel faster than shorter waves.

As an ocean wave passes, the water moves up and down in a circular pattern and returns to its original position, as shown in **Figure 15.** Only the energy moves steadily forward. The water itself moves in circles until the energy passes, but it does not move forward. The wavelength also determines the depth to which the wave disturbs the water. That depth, called the wave base, is equal to half the wavelength.

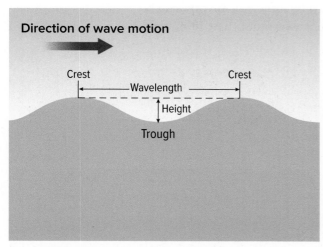

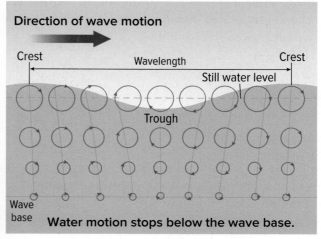

Figure 15 Wave characteristics include wave height, wavelength, crest, and trough. In an ocean wave, water moves in circles that decrease in size with depth. At a depth equal to half the wavelength, water movement essentially stops.

3D THINKING   **DCI** Disciplinary Core Ideas   **CCC** Crosscutting Concepts   **SEP** Science & Engineering Practices

COLLECT EVIDENCE

Use your Science Journal to record the evidence you collect as you complete the readings and activities in this lesson.

INVESTIGATE

GO ONLINE to find these activities and more resources.

 Investigation Lab: Making Waves
Use a model to visualize the structure of water waves.

 Mapping Lab: Ocean Surface Temperatures
Analyze and interpret data to determine how location affects ocean temperature.

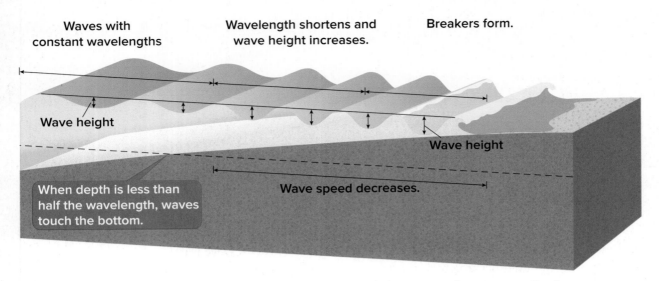

Waves with constant wavelengths

Wavelength shortens and wave height increases.

Breakers form.

Wave height

Wave height

When depth is less than half the wavelength, waves touch the bottom.

Wave speed decreases.

Figure 16 A breaker forms when wavelength decreases and wave height increases as the wave nears the shore.

Explain *what happens when waves become too steep and unstable.*

Wave height

Wave height depends on three factors: fetch, wind duration, and wind speed. Fetch refers to the expanse of water that the wind blows across. The longer the wind can blow without being interrupted (wind duration) over a large area of water (fetch), the larger the waves will be. Also, the faster the wind blows (wind speed) for a longer period of time over the ocean, the larger the waves will be. The highest waves are usually found in the Southern Ocean, an area over which strong winds blow almost continuously. Waves created by large storms can also be much higher than average. For instance, hurricanes can generate waves more than 10 m high, which is taller than a three-story building.

 Get It?

Describe the measurable properties of a wave.

Breaking waves

Study **Figure 16.** It shows that as ocean waves reach the shallow water near shorelines, the water depth eventually becomes less than one-half of the wavelength. The shallow depth causes changes to the movement of water particles at the base of the wave. This causes the waves to slow down. As the water becomes shallow, incoming wave crests gradually catch up with the slower wave crests ahead. As a result, the crest-to-crest wavelength decreases. The incoming waves become higher, steeper, and unstable, and their crests collapse forward. Collapsing waves are called **breakers.** The formation of breakers is also influenced by the motion of wave crests, which overrun the troughs. The collapsing crests of breakers, like the one shown in **Figure 17,** move at high speeds toward shore and play a major role in shaping shorelines.

Figure 17 As waves move into shallow water, breakers form.

Westend61/Brand X Pictures/Getty Images

Tides

Tides are the periodic rise and fall of sea level. The highest level to which water regularly rises is known as high tide, and the lowest level is called low tide. Because of differences in topography and latitude, the tidal range—the difference in height between high tide and low tide—varies from place to place. For example, in the Gulf of Mexico, the tidal range is less than 1 meter, whereas in New England, it can be as high as 6 meters. The greatest tidal range occurs in the Bay of Fundy between New Brunswick and Nova Scotia, Canada, where it is as much as 16.8 meters. Generally, a daily cycle of high and low tides takes 24 hours and 50 minutes. Differences in topography and latitude cause three different daily tide cycles, as illustrated in **Figure 18.** Areas with semidiurnal cycles experience two high tides in about a 24-hour period. Areas with mixed cycles have one pronounced and one smaller high tide in about a 24-hour period. Areas with diurnal cycles have one high tide in about a 24-hour period.

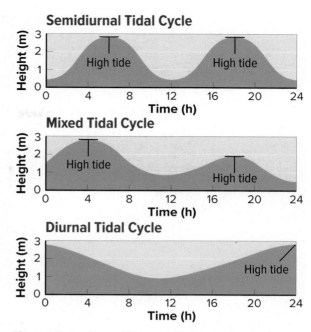

Figure 18 The three different daily tide cycles are semidiurnal, mixed, and diurnal.

The Moon's influence

The basic causes of tides are the gravitational attraction among Earth, the Moon, and the Sun, as well as the effect of Earth's rotation. Consider the Earth-Moon system. Both Earth and the Moon orbit a common center of gravity, shown as a red plus sign in **Figure 19.** As a result of their motions, both Earth and the Moon experience differing gravitational forces. These unbalanced forces generate tidal bulges on opposite sides of Earth. The gravitational effect of the Moon on Earth's oceans is similar to what happens to the liquid in a coffee cup inside a car as the car goes around a curve. The liquid sloshes toward the outside of the curve.

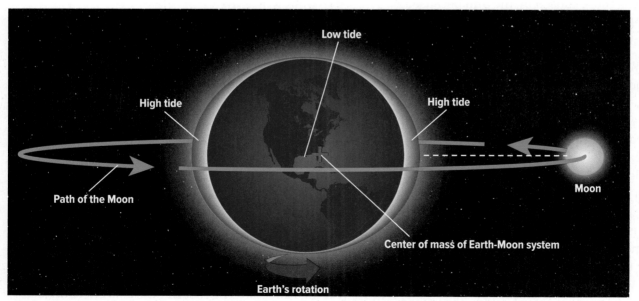

Figure 19 The Moon and Earth revolve around a common center of gravity and experience unbalanced gravitational forces. These forces cause tidal bulges on opposite sides of Earth. (Note: *Diagram is not to scale.*)

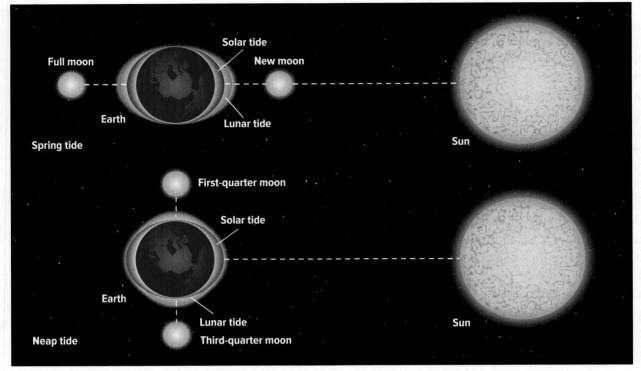

Figure 20 Spring tides occur when the Sun, the Moon, and Earth are aligned. Neap tides occur when the Sun, the Moon, and Earth form a right angle. (Note: *Diagram is not to scale.*)

The Sun's influence

The gravitational attraction between Earth and the Sun, and Earth's orbital motion around the Sun, influences tides. However, even though the Moon is much smaller than the Sun, lunar tides are more than twice as high as those caused by the Sun because the Moon is much closer to Earth. Consequently, Earth's tidal bulges are aligned with the Moon.

Depending on the phases of the Moon, solar tides can either enhance or diminish lunar tides, as illustrated in **Figure 20.** Notice in **Figure 20** that during both a full and a new moon, the Sun, the Moon, and Earth are all aligned. When this occurs, solar tides enhance lunar tides, causing high tides to be higher than normal and low tides to be lower than normal. The tidal range is highest during these times. These types of tides are called **spring tides.** Study **Figure 20** again. When there is a first- or third-quarter moon, the Sun, the Moon, and Earth form a right angle. When this occurs, solar tides diminish lunar tides, causing high tides to be lower and low tides to be higher than normal. The tidal range is lowest during these times. These types of tides are called **neap tides.** Spring and neap tides alternate every two weeks.

Currents

Currents in the ocean can move horizontally or vertically. They can also move at the surface or deep in the ocean.

Surface currents

Mainly the top 100 to 200 m of the ocean experience **surface currents,** which can move at a velocity of about 100 km per day. Surface currents follow predictable patterns and are driven by Earth's global wind systems. Recall that in the northern hemisphere, tropical trade winds blow from east to west. The resulting tropical ocean surface currents also flow from east to west. In northern midlatitudes, the prevailing westerlies and resulting ocean surface currents move from west to east. In northern polar regions, polar easterly winds push surface waters from east to west.

The direction of surface currents can also be affected by landforms, such as continents, as well as the Coriolis effect. Recall that the Coriolis effect deflects moving particles to the right in the northern hemisphere and to the left in the southern hemisphere.

Major Ocean Gyres

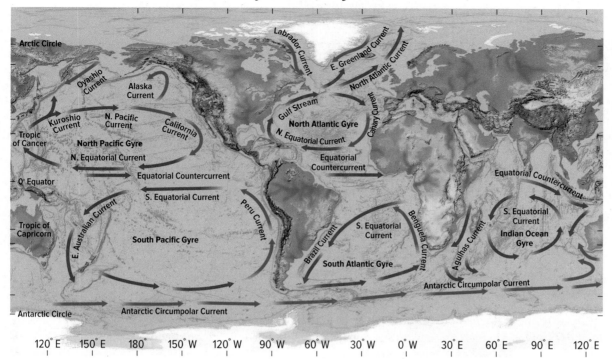

Figure 21 Large gyres in each ocean are formed by surface currents. Red arrows represent the movement of warm water, and blue arrows represent the movement of cold water.

Gyres

If Earth had no landmasses, the global ocean would have simple belts of easterly and westerly surface currents. Instead, the continents deflect ocean currents to the north and the south so that closed circular current systems, called gyres (JI urz), develop.

As shown in **Figure 21,** there are five major gyres—the North Pacific, the North Atlantic, the South Pacific, the South Atlantic, and the Indian Ocean. Due to the Coriolis effect, the gyres in the northern hemisphere move clockwise, and those in the southern hemisphere move counterclockwise. The parts of all gyres closest to the equator move west as equatorial currents. When these currents encounter a landmass, they are deflected toward the poles. These poleward-flowing currents, such as the Gulf Stream, carry warm, tropical water into higher, colder latitudes. When warm water enters polar regions, it cools and, when deflected by landmasses, moves back toward the equator. The resulting currents, such as the California Current, bring cold water from higher latitudes into tropical regions.

Upwelling

Ocean water also moves vertically. The upward motion of ocean water is called **upwelling.** Upwelling waters originate in deeper waters, below the thermocline, and thus are usually cold and nutrient-rich. Areas of upwelling exist mainly off the western coasts of continents in the trade-wind belts, such as off the coast of California. Study **Figure 22** to learn more about the steps involved in upwelling.

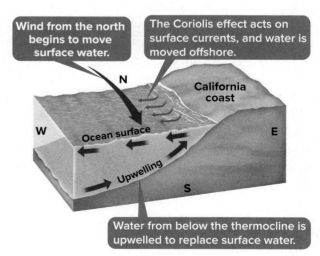

Wind from the north begins to move surface water.

The Coriolis effect acts on surface currents, and water is moved offshore.

Water from below the thermocline is upwelled to replace surface water.

Figure 22 Upwelling occurs when surface water is moved offshore and deep, colder water rises to the surface to replace it.

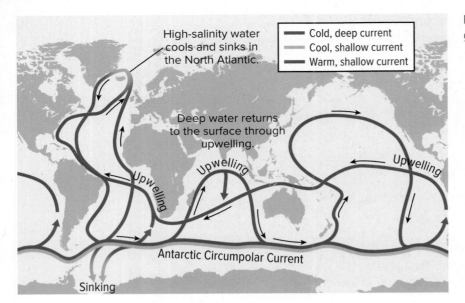

High-salinity water cools and sinks in the North Atlantic.

Legend:
- Cold, deep current
- Cool, shallow current
- Warm, shallow current

Deep water returns to the surface through upwelling.

Upwelling

Upwelling

Upwelling

Antarctic Circumpolar Current

Sinking

Figure 23 Differences in salinity and temperature generate density currents in the deep ocean.

Density currents

Another type of vertical current is a **density current,** which is caused by differences in the temperature and salinity of ocean water, which, in turn, affect density. Density currents move slowly in deep ocean waters, following a general path that is sometimes called the global conveyer belt.

The conveyor belt, a model of which is shown in **Figure 23,** begins when cold, dense water, including North Atlantic Deep Water and Antarctic Bottom Water, sinks at the poles. After sinking, these water masses slowly move away from the poles and circulate through the major ocean basins. After hundreds of years, the deep water eventually returns to the surface through upwelling. Once at the surface, the deep water is warmed by solar radiation. This water continues along the global conveyor belt until it reaches the poles, where it cools, sinks, and begins its journey again.

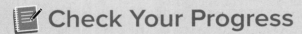

Check Your Progress

Summary

- Energy moves through ocean water in the form of waves.
- Tides are influenced by both the Moon and the Sun.
- Surface currents circulate in gyres in the major ocean basins.
- Vertical currents in the ocean include density currents and upwelling.

Demonstrate Understanding

1. **Analyze** the impacts of surface currents and density currents on the transfer of energy between the equator and the poles.
2. **Illustrate** a wave. Label the following: *crest, trough, wavelength, wave height,* and *wave base.*
3. **Explain** how tides form.
4. **Identify** the causes of surface and density currents.

Explain Your Thinking

5. **Predict** the effects on marine ecosystems if upwelling stopped.
6. **Assess** the difference between spring tides and neap tides.
7. **WRITING** **Connection** Write a step-by-step explanation of how upwelling occurs.

LESSON 3
SHORELINE AND SEAFLOOR FEATURES

FOCUS QUESTION

Why are all coasts not sandy beaches?

The Shore

Shown in **Figure 24,** the shore is the area of land between the lowest water level at low tide and the highest area of land that is affected by storm waves. Shores are places of continuous, often dramatic geologic activity—places where you can see geologic changes occurring almost daily. The shoreline is the place where the ocean meets the land. Shorelines are shaped by the action of waves, tides, and currents. The location of the shoreline constantly changes as the tide moves in and out. As waves erode some shorelines, they create some of the most impressive rock formations on Earth. In other areas, waves deposit loose material and build wide, sandy beaches.

Beaches

Long stretches of U.S. coasts are lined with wide, sandy beaches. A **beach,** shown in **Figure 24,** is the area in which sediment is deposited along the shore. Beaches are composed of loose sediments deposited and moved about by waves along the shoreline. The size of sediment particles depends on the energy of the waves striking the coast and on the source of the sediment. Beaches pounded by large waves or formed on rocky coasts usually consist of coarse materials such as pebbles and cobbles.

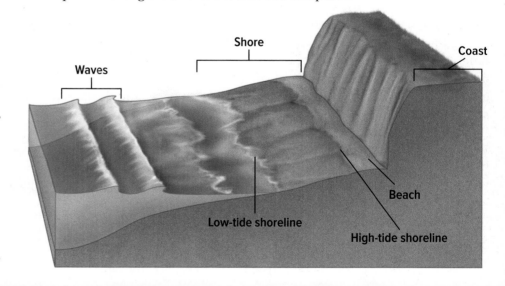

Figure 24 The location of the shoreline changes as the tide moves in and out.

Identify *How is a beach related to a shore?*

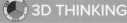 **3D THINKING**   DCI Disciplinary Core Ideas   CCC Crosscutting Concepts   SEP Science & Engineering Practices

COLLECT EVIDENCE

 Use your Science Journal to record the evidence you collect as you complete the readings and activities in this lesson.

INVESTIGATE

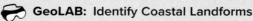 GO ONLINE to find these activities and more resources.

🥽 **GeoLAB: Identify Coastal Landforms**
Analyze and interpret data to visualize **the structure** of coastal features.

🥽 **Quick Investigation: Measure Sediment Settling Rates**
Use a model to simulate **the rate** of sedimentation in still water.

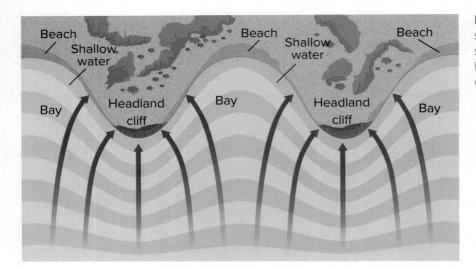

Figure 25 Wave crests advance toward the shoreline and slow down when they encounter shallow water. This causes the wave crests to bend toward the headlands and move in the direction of the arrows.

Formation of Shoreline Features

Large, breaking waves can hurl thousands of metric tons of water, along with suspended rock fragments, against a shore with such force that they are capable of eroding solid rock.

Erosional features

Waves move faster in deep water than in shallow water. This difference in wave speed causes initially straight wave crests to bend when part of the crest moves into shallow water, a process known as **wave refraction,** illustrated in **Figure 25.** Along an irregular coast with headlands and bays, the wave crests bend toward the headlands. As a result, most of the breaker energy is concentrated along the relatively short section of the shore around the tips of the rocky headlands, while the remaining wave energy is spread out along the much longer shoreline of the bays. The headlands thus undergo severe erosion. The material eroded from the headlands is swept into the bays, where it is deposited to form crescent-shaped beaches. The headlands are worn back, and the bays are filled in until the shoreline straightens.

Wave-cut platforms Many headlands have spectacular rock formations. Generally, as a headland is worn away, a flat erosional surface called a wave-cut platform is formed. The wave-cut platform terminates against a steep wave-cut cliff, as shown in **Figure 26.**

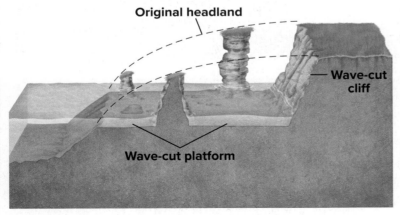

Figure 26 A headland can be modified by wave erosion. The dotted lines indicate the original shape of the headland.

Figure 27 The sea stack and sea arch shown here were formed by wave refraction at a rocky headland along the coast of France.

Sea stacks Differential erosion, the removal of weaker rocks or rocks near sea level, produces many of the other characteristic landforms of rocky headlands. As shown in **Figure 27,** a sea stack is an isolated rock tower or similar erosional remnant left on a wave-cut platform. A sea arch, also shown in **Figure 27,** is formed as stronger rocks are undercut by wave erosion. Sea caves are tubelike passages blasted into the headlands at sea level by the constant assault of the breakers.

Get It?

Describe What are sea stacks, and how are they formed?

Longshore currents

Suppose you stood on a beach at the edge of the water and began to walk out into the ocean. As you walked, the water might get deeper for a while, but then it would become shallow again. The shallow water offshore lies above a sandbar, called a **longshore bar,** that forms in front of most beaches, as illustrated in **Figure 28.** Waves break on the longshore bar in the area known as the surf zone. The deeper water between the shore and longshore bar is called the longshore trough.

The waves striking the beach are almost parallel to the shoreline, although the waves seaward of the longshore bar are generally not parallel to the shore. This is another case of wave refraction. The slowing of the waves in shallow water causes the wave crests to bend toward the shore.

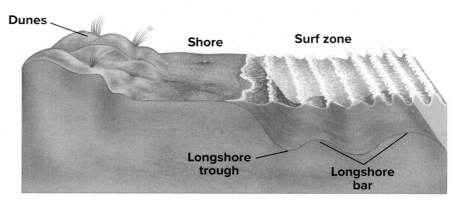

Figure 28 A typical beach profile includes longshore troughs and bars within the surf zone.

Explain *What is the difference between a longshore trough and a longshore bar?*

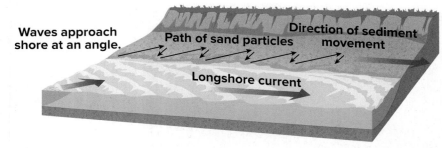

Figure 29 Longshore currents are driven by incoming waves.

As water from incoming breakers spills over the longshore bar, a current flowing parallel to the shore, the **longshore current,** is produced. This current varies in strength and direction daily. Over time, due to prevailing winds and wave patterns, one direction dominates.

Movement of sediments

Longshore currents, like the one illustrated in **Figure 29,** move large amounts of sediment along the shore. Fine-grained material, such as sand, is suspended in the turbulent, moving water, and larger particles are pushed along the bottom by the current. Incoming waves also move sediment at an angle to the shoreline in the direction of wave motion. Overall, the transport of sediment is in the direction of the longshore current.

Rip currents

Wave action also produces rip currents, which flow out to sea through gaps in the longshore bar. Rip currents, like the one shown in **Figure 30,** return the water spilled into the longshore trough to the ocean. These dangerous currents can reach speeds of several kilometers per hour.

 Get It?

Explain why rip currents are considered dangerous for swimmers at the beach.

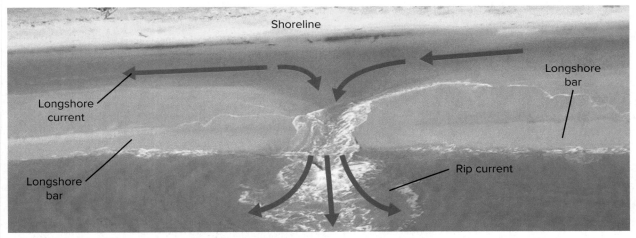

Figure 30 Rip currents return water out to sea through gaps in the longshore bar. Rip currents spread out and weaken beyond the longshore bar.

Courtesy of Dennis Decker, WCM, NWS Melbourne, FL

STEM CAREER Connection
Beach Lifeguard
Beach lifeguards have to understand currents and wave action in order to rescue people from strong currents and rip currents. They also have first aid, CPR, and water safety training.

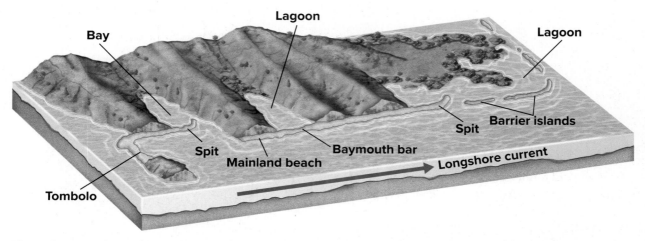

Figure 31 Depositional features of coastlines include tombolos, spits, baymouth bars, lagoons, and barrier islands.

Depositional features

As a result of wave erosion, longshore transport, and sediment deposition, most seashores are in a constant state of change. Sediments are eroded by large storm waves and deposited wherever waves and currents slow down. Sediments that are moved and deposited by longshore currents build various characteristic coastal landforms, such as spits, barrier islands, baymouth bars, and tombolos, illustrated in **Figure 31.**

Spit A narrow bank of sand that projects into the water from a bend in the coastline is called a spit. A spit, which forms where a shoreline changes direction, is protected from wave action. When a growing spit crosses a bay, a baymouth bar forms.

Barrier islands Long ridges of sand or other sediment that are deposited or shaped by the longshore current and are separated from the mainland are called **barrier islands.** Barrier islands, like the one shown in **Figure 32,** can be several kilometers wide and tens of kilometers long. Most of the Gulf Coast, including Florida, and the eastern coast south of New England are lined with an almost continuous chain of barrier islands.

Baymouth bars A baymouth bar, shown in **Figure 33,** forms when a spit closes off a bay. The shallow, protected bodies of water behind baymouth bars and barrier islands are called lagoons. Lagoons are saltwater coastal lakes that are connected to the open sea by shallow, restricted outlets.

Tombolo Another coastal landform is a tombolo, shown in **Figure 33.** A tombolo is a ridge of sand that forms between the mainland and an island and connects the island to the mainland. When this happens, the island is no longer an island, but is the tip of a peninsula.

Figure 32 This barrier island off the coast of Florida is one of many that line the Gulf of Mexico.

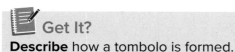

Get It?

Describe how a tombolo is formed.

USGS

Baymouth bar   Tombolo

Figure 33 Baymouth bars and tombolos are examples of features formed by the deposition of sediments.

Natural and human effects on the coast

All of these depositional coastal landforms, including large barrier islands, are unstable and temporary. Occasionally, major storms sweep away entire sections of barrier islands and redeposit the material elsewhere. **Figure 34** shows the existence of South Gosier Island, a barrier island off the coast of Louisiana, in August of 2004, a month before Hurricane Ivan passed over it. The island was completely destroyed by the strong waves generated by Hurricane Ivan. Even in the absence of storms, however, changing wave conditions can slowly erode beaches and rearrange entire shorelines. For example, the shoreline of Cape Cod, Massachusetts, is retreating by as much as 1 m per year.

People are drawn to coastal areas for their rich natural resources, mild climate, and recreation opportunities. Coastal areas in the United States make up only 17 percent of contiguous land areas but are currently home to over half of the nation's population. Increasing population has caused substantial coastal environmental changes, including pollution, shoreline erosion, and wetland and wildlife-habitat loss. These coastal changes, in turn, increase the susceptibility of the communities to natural hazards caused by hurricanes and tsunamis.

Before   After

Figure 34 The island of South Gosier, a barrier island off the coast of Louisiana, was completely washed away by Hurricane Ivan in September 2004.

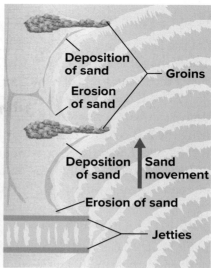

Figure 35 Seawalls deflect the energy of waves on the beach. Groins and jetties deprive downshore beaches of sand.

Protective Structures

In many coastal areas, including Florida, protective structures such as seawalls, groins, jetties, and breakwaters are built in an attempt to prevent beach erosion and destruction of oceanfront properties. However, these artificial structures interfere with natural shoreline processes and can have unexpected negative effects.

Seawalls

Structures called seawalls, shown in **Figure 35,** are built parallel to shore, often to protect beachfront properties from powerful storm waves. Seawalls reflect the energy of such waves back toward the beach, where they worsen beach erosion. Eventually, seawalls are undercut and have to be rebuilt larger and stronger than before.

Groins and jetties

Groins, shown in **Figure 35,** are wall-like structures built into the water perpendicular to the shoreline for the purpose of trapping beach sand. Groins interrupt natural longshore transport and deprive beaches down the coast of sand. The result is aggravated beach erosion down the coast from groins. Similar effects are caused by jetties, which are walls of concrete built to protect a harbor entrance from drifting sand. Jetties are also shown in **Figure 35.**

Breakwaters

Breakwaters, like the one shown in **Figure 36,** are structures that are built to provide anchorages for small boats or a calm beach area. Breakwaters are built parallel to the shoreline. When the current slows down behind a breakwater and is no longer able to move sediment, it deposits sediment behind the breakwater. If the accumulating sediment is left alone, it can eventually fill an anchorage. To prevent this, anchorages have to be dredged regularly.

Figure 36 A breakwater slows the movement of water against the coast, which results in sediment being deposited behind the structure.

Changes in Sea Level

At the height of the last ice age, approximately 20,000 years ago, the global sea level was about 130 m lower than it is at present. Since that time, the melting of most of the ice-age glaciers has raised the ocean to its present level. In the last 100 years, the global sea level has risen about 17 cm. It continues to rise slowly; estimates suggest a rise in sea level of 3 mm/year. Many scientists contend that this continuing rise in sea level is the result of global warming. During the last century, Earth's average surface temperature has increased by approximately 0.6°C. As Earth's surface temperature rises, seawater warms up and expands, which adds to the total volume of the seas. In addition, increased surface temperatures cause glaciers to melt, and the meltwater flows into the oceans, increasing their volume.

Effects of sea level changes

Computer models show that if Earth's remaining polar ice sheets in Greenland and Antarctica melted completely, their meltwater would raise the sea level by another 70 m. This rise would completely flood some countries, such as the Netherlands, along with some coastal cities in the United States. Measurements from satellites have indicated that the Greenland ice sheet is melting at an accelerating rate. Satellite measurements show that the West Antarctic Ice Sheet has also been thinning.

Many barrier islands might be former coastal dunes that were drowned by rising sea levels. Other features produced by rising sea levels are the fjords of Norway, shown in **Figure 37.** Fjords (fee ORDZ) are deep coastal valleys that were scooped out by glaciers during the ice age and later flooded when the sea level rose.

Effects of tectonic forces

Other processes that affect local sea levels are tectonic uplift and sinking. If a coastline sinks, there is a relative rise in sea level along that coast. A rising coastline, however, produces a relative drop in the local sea level. As a result of tectonic forces in the western United States, much of the West Coast is being pushed up much more quickly than the sea level is rising. Because much of the West Coast was formerly under water, it is called an emergent coast.

Figure 37 Fjords are flooded U-shaped valleys that can be up to 1200 m deep.

Ian White

ACADEMIC VOCABULARY

estimate
a rough or approximate calculation
The estimate for the new roof was $3000.

Emergent coasts tend to be relatively straight because the exposed seafloor topography is smoother than typical land surfaces with hills and valleys. Other signs of an emergent coast are former shoreline features, such as sandy beach ridges located far inland, or fossils of marine organisms within the uplifted rocks. Among the most interesting of these features are elevated marine terraces—former wave-cut platforms that are now dry and well above current sea level. **Figure 38** shows a striking example of such a platform. Some old wave-cut platforms are hundreds of meters above current sea level.

Figure 38 Elevated marine terraces are former wave-cut platforms that are now well above the current sea level. This elevated marine terrace is in New Zealand.

The Continental Margin

Until recently, most people had little knowledge of the features of the ocean floor. However, modern oceanographic techniques, including sonar and satellite data, reveal that the topography of the ocean bottom is as varied as that of the continents.

The topography of the seafloor is surprisingly rough and irregular, with numerous high mountains and deep depressions. The deepest place on the seafloor, the Mariana Trench in the Pacific Ocean, is about 11 km deep.

Study **Figure 39.** Notice that the **continental margin** is the area where edges of continents meet the ocean. It consists of continental crust, covered with sediments, that eventually meets oceanic crust. Continental margins represent the shallowest parts of the ocean. As shown in **Figure 39,** a continental margin includes the continental shelf, the continental slope, and the continental rise.

Continental shelf

The shallowest part of a continental margin extending seaward from the shore is the **continental shelf.** Continental shelves vary greatly in width, averaging 60 km wide. On the Pacific coast of the United States, the continental shelf is only a few kilometers wide, whereas the continental shelf of the Atlantic coast is hundreds of kilometers wide.

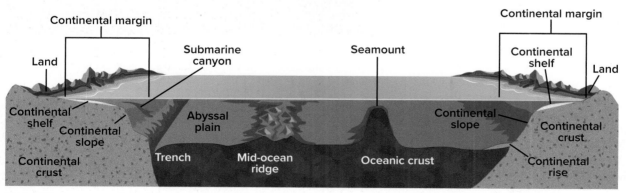

Figure 39 A cross section of the ocean reveals a diverse topography with many features that are similar to those on land.

John A. Karachewski

The average depth of the water above continental shelves is about 130 m. Recall that sea level during the last ice age was approximately 130 m lower than at present; therefore, portions of the world's continental shelves were above sea level at that time. As a result, present-day coastlines are radically different from the way they were during the last ice age. At that time, Siberia was attached to North America by the Bering land bridge, Great Britain was attached to Europe, and a large landmass existed where today there are only the widely scattered islands of the Bahamas.

When Earth's surface began to warm after the last ice age, and the continental ice sheets began to melt, the sea gradually covered up the continental shelves. Beaches and other coastal landforms from that time are now submerged and located far beyond the present shoreline. Commercially valuable fishes now inhabit the shallow waters of the continental shelves. In addition, the thick sedimentary deposits on the shelves are significant sources of oil and natural gas.

 Get It?

List the resources that can be found on the continental shelf.

Continental slope

Beyond the continental shelves, the seafloor drops away quickly to depths of several kilometers, with slopes averaging nearly 100 m/km. These sloping regions are the **continental slopes.** To marine geologists, the continental slope is the true edge of a continent because it generally marks the edge of the continental crust. In many places, this slope is cut by deep submarine canyons, which are shown in **Figure 40.** Submarine canyons are similar to canyons on land, and some are comparable in size to the Grand Canyon in Arizona.

These submarine canyons were cut by **turbidity currents,** which are rapidly flowing water currents along the bottom of the sea. These currents carry heavy loads of sediments, similar to mudflows on land. Turbidity currents might originate as underwater landslides triggered by earthquakes on the continental slope, or they might originate from sediment stirred up by large storm waves on the continental shelf. Turbidity currents can reach speeds exceeding 30 km/h and effectively erode bottom sediments and bedrock.

Figure 40 Submarine canyons are similar to canyons on land. These deep cuts in the continental slope vary in size and can be as deep as the Grand Canyon.

Explain *how submarine canyons are formed.*

Continental rise

The gently sloping accumulation of deposits that forms at the base of the continental slope is called a **continental rise.** A continental rise can be several kilometers thick. The rise gradually becomes thinner and eventually merges with the sediments of the seafloor beyond the continental margin. In some places, especially around the Pacific Ocean, the continental slope ends in deeper depressions, known as deep-sea trenches, in the seafloor. In such places, there is no continental rise at the foot of the continental margin.

 Get It?

Differentiate between the continental slope and the continental rise.

Deep-Ocean Basins

Beyond the continental margin are ocean basins, which represent about 60 percent of Earth's surface and contain some of Earth's most interesting topography. **Figure 41** shows the topography of the ocean basin beneath the Pacific Ocean.

Abyssal plains

The flattest parts of the ocean floor, 5 or 6 km below sea level, are called **abyssal plains.** Abyssal plains, shown in **Figure 41,** are plains covered with hundreds of meters of fine-grained, muddy sediments and sedimentary rocks that were deposited on top of basaltic volcanic rocks.

Deep-sea trenches

The deepest parts of the ocean basins are the **deep-sea trenches,** which are elongated, sometimes arc-shaped depressions in the seafloor several kilometers deeper than the adjacent abyssal plains. Many deep-sea trenches lie next to chains of volcanic islands, such as the Aleutian Islands of Alaska, and most of them are located around the margins of the Pacific Ocean, as shown in **Figure 41.** Deep-sea trenches are relatively narrow—about 100 km wide—but they can extend for thousands of kilometers.

Mid-ocean ridges

The most prominent features of the ocean basins are the **mid-ocean ridges,** which run through all the ocean basins and have a total length of more than 65,000 km—a distance greater than Earth's circumference. Mid-ocean ridges have an average height of 1500 m, but they can be thousands of kilometers wide. Mid-ocean ridges are sites of frequent volcanic eruptions and earthquake activity. The crests of these ridges often have valleys called rifts running through their centers. Rifts can be up to 2 km deep.

Mid-ocean ridges do not form continuous lines. They are broken into a series of shorter, stepped sections that run at right angles across each mid-ocean ridge. The areas where these breaks occur are called fracture zones, shown in **Figure 42.** Fracture zones are about 60 km wide and can be thousands of kilometers long.

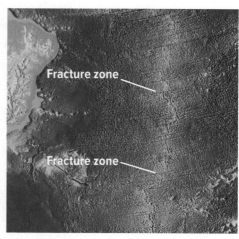

Figure 42 There are many fracture zones along the Mid-Atlantic Ridge. The fracture zones run perpendicular to the ridge.

NGDC/NOAA

Figure 41 Visualizing the Ocean Floor

The ocean floor has topographic features, including mid-ocean ridges, trenches, abyssal plains, and seamounts.

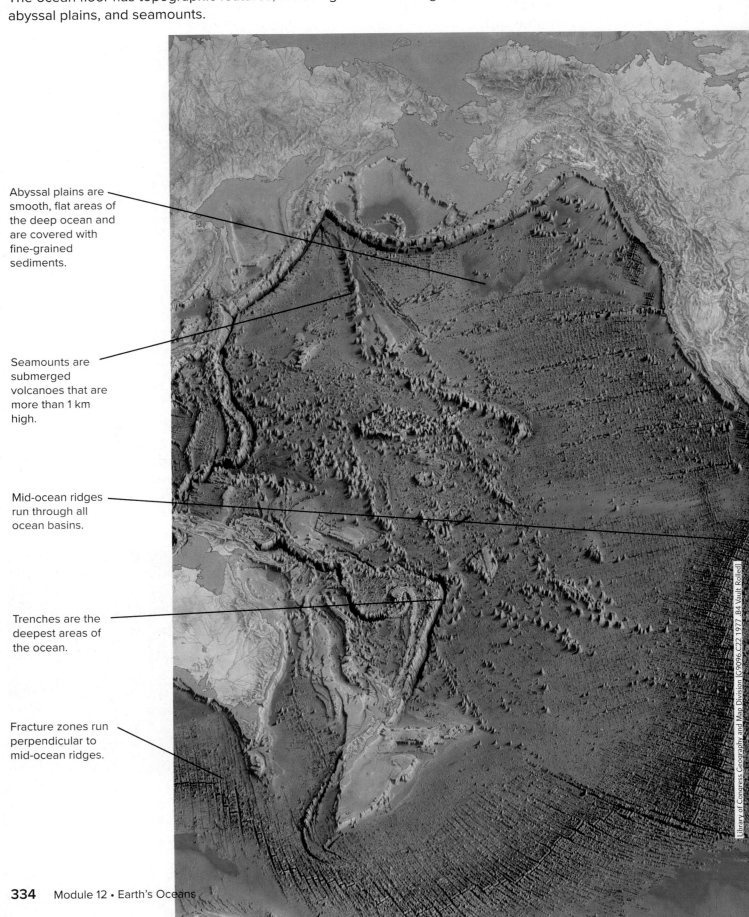

Abyssal plains are smooth, flat areas of the deep ocean and are covered with fine-grained sediments.

Seamounts are submerged volcanoes that are more than 1 km high.

Mid-ocean ridges run through all ocean basins.

Trenches are the deepest areas of the ocean.

Fracture zones run perpendicular to mid-ocean ridges.

Black Smoker   White Smoker

Figure 43 Black smokers form when metal oxides and sulfides precipitate out of fluid heated by magma. White smokers form when elements such as calcium and barium precipitate out of warm water ejected from rifts in mid-ocean ridges.

Hydrothermal vents A hydrothermal vent is a hole in the seafloor through which fluid heated by magma erupts. Most hydrothermal vents are located along the bottom of the rifts in mid-ocean ridges. When the heated fluid that erupts from these vents contains metal oxides and sulfides, they immediately precipitate out of the fluid and produce thick, black, smokelike plumes. This type of hydrothermal vent, known as a black smoker, ejects superheated water with temperatures of up to 350°C. **Figure 43** illustrates the black smokers found in the rift valley of a mid-ocean ridge. A second type of vent, known as a white smoker, is also shown in **Figure 43.** White smokers, which eject cooler water than black smokers, also get their color from the types of minerals that precipitate out of the hydrothermal fluids. Both types of hydrothermal vents are caused by seawater circulating through the hot crustal rocks in the centers of mid-ocean ridges.

 Get It?

Identify where most hydrothermal vents are located.

Seamounts and guyots

Satellite data have revealed that the ocean floor is dotted with tens of thousands of solitary mountains that are not located near areas of active volcanism. How, then, did they form? You have learned that the ocean basins are volcanically active at mid-ocean ridges and fracture zones. The almost total absence of earthquakes in most other areas of the seafloor suggests that volcanism in those areas must have ceased a long time ago. Thus, most of the mountains on the seafloor are probably extinct volcanoes.

Investigations of individual volcanoes on the seafloor have revealed that there are two types: seamounts and guyots (GEE ohz). **Seamounts** are submerged basaltic volcanoes more than 1 km high. **Guyots** are large, extinct, basaltic volcanoes with flat, submerged tops.

Marine Sediments

The sediments that cover the ocean floor come from a variety of sources, including from land, from biological activity, and from the precipitation of elements from seawater. Three types of marine sediments are described in **Table 2.**

()NOAA PMEL Vents Program, (r)Image courtesy of Submarine Ring of Fire 2004 Exploration, NOAA Vents Program

Table 2 Types of Marine Sediments

Sediment Type	Description
Terrigenous	Ocean currents disperse terrigenous sediments—fine silt, clay, and volcanic ash from land—throughout the ocean basins. The dominant type of sediment on the deep ocean floor is fine-grained, deep-sea mud. Closer to land, the sediments become mixed with coarser materials such as sand, but some sandy sediments occasionally reach the abyssal plains in particularly strong turbidity currents.
Biogenous	Deep-sea sediments that come from biological activity are called biogenous sediments. When marine organisms such as diatoms die, their shells settle on the ocean floor. Sediments containing a large percentage of particles derived from once-living organisms are called oozes. Most of these particles are small and consist of either calcium carbonate or silica.
Hydrogenous	Hydrogenous sediments are derived from elements in seawater. For example, salts that precipitate out of supersaturated lagoons are considered hydrogenous sediments, as are sulfides that form at hydrothermal vents. Manganese nodules, found on the deep-sea floor, consist of oxides of manganese, iron, copper, and other valuable metals that have precipitated directly from seawater.

 Get It?

Compare and contrast terrigenous and biogenous sediments.

Check Your Progress

Summary

- Wave erosion of headlands produces wave-cut platforms, sea stacks, sea arches, and sea caves.
- Wave action and longshore currents move sediment and build depositional features.
- A continental margin consists of the continental shelf, the continental slope, and the continental rise.
- Deep-ocean basins consist of abyssal plains, trenches, mid-ocean ridges, seamounts, and guyots.

Demonstrate Understanding

1. **Explain** Which experiences more severe erosion by breakers—headlands or bays? Why?
2. **Explain** how an increase in Earth's surface temperature can lead to changes in sea level.
3. **Identify** reasons why people are attracted to living in coastal areas.

Explain Your Thinking

4. **Cite Evidence** A video of the ocean floor shows an area with thick, black smokelike plumes erupting from the seafloor. Which feature of the ocean floor is shown? Cite evidence to support your conclusion.
5. **WRITING Connection** Write a paragraph explaining why it is important for scientists to continue to collect and study data and make models and predictions related to the interactions among Earth's systems and the impact of human activities on those systems.

LEARNSMART Go online to follow your personalized learning path to review, practice, and reinforce your understanding.

SCIENCE & SOCIETY

A Slow Flow

Ocean currents play a key role in determining Earth's climates. Scientists are concerned about a slowdown in the flow of the density current known as the global conveyor belt and how that change could affect Earth.

Melting ice can affect the salinity and temperature of ocean water, which can lead to changes in density currents.

Causes of change

Because it is a density current, the global conveyor belt is affected by both salinity and temperature. Density increases as salinity increases and as temperature decreases. Anything that affects the temperature and salinity of ocean water causes density changes that affect ocean water movement.

Scientists are concerned that an increase in global temperatures will increase the temperature and decrease the salinity of the North Atlantic Ocean, where cold, salty water sinks as part of the global conveyor belt. In turn, this might affect the transfer of thermal energy in the Earth system, altering regional climates.

Evidence of past changes

Scientists know that past changes in the global conveyor belt were associated with rapid changes in climate. Evidence shows that such a change occurred about 11,000 years ago, when a massive amount of freshwater from melting glaciers emptied into the North Atlantic. The salinity of surface water in the North Atlantic decreased, the global conveyor belt stopped, and an ice age-like period resulted.

Predicting future changes

Patterns of water movement withing the global conveyor belt are complex. Since 2004, scientists have used submerged instruments to measure the rate of the global conveyor belt. The data suggest that the global conveyor belt has weakened and slowed by about 30 percent since 2004—a big change in a very short time.

To make predictions about the possible effects of this change, scientists use evidence such as current and historical data about changes in the global conveyor belt and climate. They also use scientific reasoning. Many scientists predict that rising global temperatures will impact the global conveyor belt, leading to further changes in climate. Possible impacts of a slowed global conveyor belt include decreased temperatures in the Northern Hemisphere, increased intensity of North Atlantic storms, and changes in the patterns of monsoons in India and Asia.

DEVELOP A MODEL TO ILLUSTRATE

Make a diagram of the global conveyor belt. Then add labels and captions to provide information about how changes in ocean temperature or salinity could affect the conveyor belt, and how that could affect climates in specific places on Earth.

MODULE 12
STUDY GUIDE

 GO ONLINE to study with your Science Notebook.

Lesson 1 AN OVERVIEW OF OCEANS

- Scientists use many different instruments to collect and analyze data from oceans.
- Scientists have several ideas about where the water in Earth's oceans originated.
- Earth's oceans are the Pacific, the Atlantic, the Indian, the Arctic, and the Southern.
- Ocean water contains dissolved gases, nutrients, and salts.
- Salts are added to and removed from oceans through natural processes.
- Properties of ocean water, including temperature and salinity, vary with location and depth.
- Many of the oceans' deepwater masses sink from the surface of polar oceans.

- side-scan sonar
- sea level
- salinity
- estuary
- temperature profile
- thermocline

Lesson 2 OCEAN MOVEMENTS

- Energy moves through ocean water in the form of waves.
- Tides are influenced by both the Moon and the Sun.
- Surface currents circulate in gyres in the major ocean basins.
- Vertical currents in the ocean include density currents and upwelling.

- wave
- crest
- trough
- breaker
- tide
- spring tide
- neap tide
- surface current
- upwelling
- density current

Lesson 3 SHORELINE AND SEAFLOOR FEATURES

- Wave erosion of headlands produces wave-cut platforms, sea stacks, sea arches, and sea caves.
- Wave action and longshore currents move sediment and build depositional features.
- A continental margin consists of the continental shelf, the continental slope, and the continental rise.
- Deep-ocean basins consist of abyssal plains, trenches, mid-ocean ridges, seamounts, and guyots.

- beach
- wave refraction
- longshore bar
- longshore current
- barrier island
- continental margin
- continental shelf
- continental slope
- turbidity current
- continental rise
- abyssal plain
- deep-sea trench
- mid-ocean ridge
- seamount
- guyot

REVISIT THE PHENOMENON

How can water carve a shoreline?

CER Claim, Evidence, Reasoning

Explain Your Reasoning Revisit the claim you made when you encountered the phenomenon. Summarize the evidence you gathered from your investigations and research and finalize your Summary Table. Does your evidence support your claim? If not, revise your claim. Explain why your evidence supports your claim.

STEM UNIT PROJECT

Now that you've completed the module, revisit your STEM unit project. You will apply your evidence from this module and complete your project.

GO FURTHER

SEP Data Analysis Lab
When does the tide come in?

Tidal data is usually measured in hourly increments. The water levels shown in the data table were measured over a 24-hour period. Plot these water levels on a graph, with time on the *x*-axis and water level on the *y*-axis.

CER Analyze and Interpret Data

1. **Claim** Estimate the approximate times and water levels of high tides and low tides.
2. **Claim, Evidence** Identify the type of daily tidal cycle this area experiences. Use data to support your claim.
3. **Reasoning** Predict the water level at the next high tide, and estimate when it will occur. Explain your answer.

Data and Observations

Tidal Record			
Time (h)	**Water Level (m)**	**Time (h)**	**Water Level (m)**
00:00	2.11	**13:00**	1.70
01:00	1.79	**14:00**	1.37
02:00	1.33	**15:00**	1.02
03:00	0.80	**16:00**	0.68
04:00	0.36	**17:00**	0.48
05:00	0.10	**18:00**	0.50
06:00	0.03	**19:00**	0.69
07:00	0.20	**20:00**	1.11
08:00	0.55	**21:00**	1.58
09:00	0.99	**22:00**	2.02
10:00	1.45	**23:00**	2.27
11:00	1.74	**24:00**	2.30
12:00	1.80		

*Data obtained from: The National Oceanic and Atmospheric Administration, Center for Operational Oceanographic Products and Services.

ENCOUNTER THE PHENOMENON

Why are the rock layers sideways?

SEP Ask Questions

What questions do you have about the phenomenon? Write your questions on sticky notes and add them to the driving question board for this unit.

What are tectonic plates?

Look for Evidence

As you go through this unit, use the information and your experiences to help you answer the phenomenon question as well as your own questions. For each activity, record your observations in a Summary Table, add an explanation, and identify how it connects to the unit and module phenomenon questions.

Solve a Problem
STEM UNIT PROJECT

The Dynamic Earth Investigate and research more about how Earth is changing. Use the results of these investigations and the evidence you collected during the unit to complete your unit project.

GO ONLINE In addition to reading the information in your Student Edition, you can find the STEM Unit Project and other useful resources online.

Shawn Mahoney

PLATE TECTONICS

ENCOUNTER THE PHENOMENON

How do we know that this landmass is moving?

GO ONLINE to play a video about how Alfred Wegener developed the idea of continental drift.

SEP Ask Questions

Do you have other questions about the phenomenon? If so, add them to the driving question board.

CER Claim, Evidence, Reasoning

Make Your Claim Use your CER chart to make a claim about how we know that this landmass is moving. Explain your reasoning.

Collect Evidence Use the lessons in this module to collect evidence to support your claim. Record your evidence as you move through the module.

Explain Your Reasoning You will revisit your claim and explain your reasoning at the end of the module.

GO ONLINE to access your CER chart and explore resources that can help you collect evidence.

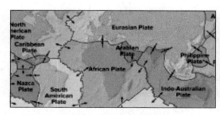

LESSON 3: Explore & Explain: Theory of Plate Tectonics

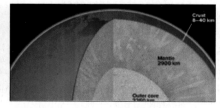

LESSON 4: Explore & Explain: Causes of Plate Motions

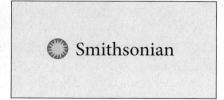

Additional Resources

Video Supplied by BBC Worldwide Learning

DRIFTING CONTINENTS

FOCUS QUESTION

How do we know the continents have moved?

Early Observations

With the exception of events such as earthquakes and landslides, most of Earth's surface appears to remain relatively unchanged during the course of a human lifetime. Over geologic time, however, Earth's surface has changed dramatically. Some of the first people to notice these changes were early cartographers. In the late 1500s, Abraham Ortelius (or TEE lee us), a Dutch cartographer, noticed the apparent fit of continents on either side of the Atlantic Ocean. He proposed that North America and South America had been separated from Europe and Africa by earthquakes and floods. During the next 300 years, many scientists noticed and commented on the matching coastlines. **Figure 1,** a proposed map by a nineteenth-century cartographer, showing the continents before they separated.

The idea of moving continents was first proposed as a scientific hypothesis in the early 1900s. In 1912, German meteorologist Alfred Wegener (VAY guh nur) presented his ideas about continental movement to the scientific community. Wegener's education and experience were primarily in meteorology, but he also studied geology, geophysics, and geography. In 1915, he published the first edition of *The Origins of Continents and Oceans.* The book was rewritten four times between 1915 and 1929; each edition contained new data and materials to address critiques and to support Wegener's hypothesis.

 Get It?

Infer why cartographers were among the first to suggest that the continents were once joined together.

Figure 1 Many early cartographers, such as Antonio Snider-Pellegrini, the author of this 1858 map, noticed the apparent fit of the continents.

3D THINKING   **DCI** Disciplinary Core Ideas   **CCC** Crosscutting Concepts   **SEP** Science & Engineering Practices

COLLECT EVIDENCE

 Use your Science Journal to record the evidence you collect as you complete the readings and activities in this lesson.

INVESTIGATE

GO ONLINE to find these activities and more resources.

? **Revisit the Encounter the Phenomenon Question**
What information from this lesson can help you answer the Unit and Module questions?

CCC **Identify Crosscutting Concepts**
Create a table of the crosscutting concepts and fill in examples you find as you read.

Science Source

Continental Drift

Wegener developed a hypothesis that he called **continental drift,** which proposed that Earth's continents had once been joined as a single landmass that broke apart and sent the continents adrift. He called this supercontinent **Pangaea** (pan JEE uh), a Greek word that means "all the earth," and suggested that Pangaea began to break apart about 200 mya. Since that time, he reasoned, the continents have continued to slowly move to their present positions, as shown in **Figure 2.**

Of the many people who had suggested continental movement, Wegener was the first to base his ideas on more than just the puzzlelike fit of continental coastlines on either side of the Atlantic Ocean. For Wegener, these gigantic puzzle pieces were just the beginning. He also collected and organized rock, climatic, and fossil data to support his hypothesis.

Figure 2 Wegener hypothesized that all the continents were once joined together. He proposed that it took 200 million years of continental drift for the continents to move to their present positions.

Locate *the parts of Pangaea that became North and South America. When were they joined? When were they separated?*

200 mya: All of the continents are assembled in a single landmass that Wegener named Pangaea.

180 mya: Continental rifting breaks Pangaea into several landmasses. The North Atlantic Ocean starts to form.

135 mya: Africa and South America begin to separate.

Present: India has collided with Asia to form the Himalayas, and Australia has separated from Antarctica. A rift valley is forming in East Africa. Continents continue to move over Earth's surface.

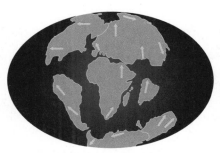

65 mya: India moves north, toward Asia.

CCC CROSSCUTTING CONCEPTS

Patterns Make a graphic organizer about the hypothesis of continental drift. Cite the empirical evidence Wegener used to develop his hypothesis.

CCC CROSSCUTTING CONCEPTS

Stability and Change Write a magazine article describing how scientists construct explanations about how things change. How does the hypothesis of continental drift provide evidence to support your ideas? Add to your article as you read other lessons in this module.

Evidence from rock formations

Wegener reasoned that when Pangaea began to break apart, large geologic structures, such as mountain ranges, became separated as the continents drifted apart. Using this reasoning, Wegener thought that there should be areas of similar rock types on opposite sides of the Atlantic Ocean. He observed that many layers of rocks in the Appalachian Mountains in the United States were identical to layers of rocks in similar mountains in Greenland and Europe.

These similar groups of rocks, older than 200 million years, supported Wegener's idea that the continents had once been joined. Some of the locations where matching groups of rock have been identified are shown in **Figure 3.**

Evidence from fossils

Wegener also gathered evidence of the existence of Pangaea from fossils. Similar fossils of several different animals and plants that once lived on or near land had been found on widely separated continents, as shown in **Figure 3.** Wegener reasoned that the land-dwelling animals, such as *Cynognathus* (sin ug NATH us) and *Lystrosaurus* (lihs truh SORE us), could not have swum the great distances that now exist between continents. Wegener also argued that because fossils of *Mesosaurus* (meh zoh SORE us), an aquatic reptile, had been found in only freshwater rocks, it was unlikely that this species could have crossed the oceans. The ages of these different fossils also predated Wegener's time frame for the breakup of Pangaea, and thus supported his hypothesis.

Figure 3 Alfred Wegener used the similarity of rock layers and fossils on opposite sides of the Atlantic Ocean as evidence that Earth's continents were once joined.

Identify *groupings that suggest that there was once a single landmass.*

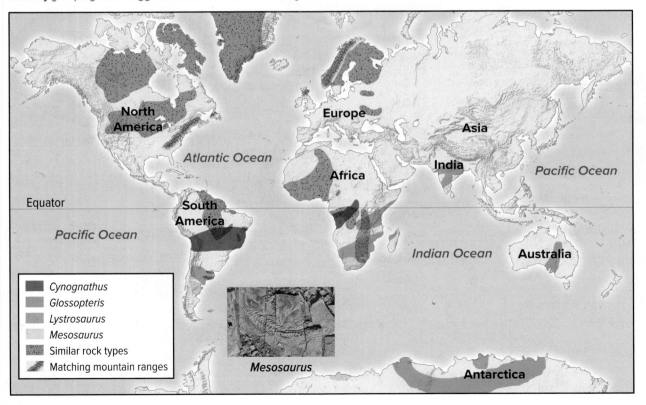

©Fotofeeling/Westend6l/Corbis

Climatic evidence

Because he had a strong background in meteorology, Wegener recognized clues about ancient climates from the fossils he studied. One fossil that Wegener used to support continental drift was *Glossopteris* (glahs AHP tur us), a seed fern that resembled low shrubs. Fossils of this plant, shown in **Figure 4,** had been found on many parts of Earth, including South America, Antarctica, and India. Wegener reasoned that the area separating these fossils was too large to have had a single climate. Wegener also argued that because *Glossopteris* grew in temperate climates, the places where these fossils had been found were once closer to the equator. This led him to conclude that the rocks containing these fossils had once been joined.

Figure 4 Wegener used the fact that fossils of *Glossopteris* were found in many parts of Earth to support his hypothesis of continental drift.

 Get It?

Infer how Wegener's background in meteorology helped him to support his ideas of continental drift.

Coal deposits Recall that sedimentary rocks provide clues to past environments and climates. In these rocks, Wegener found evidence that the climates of some continents had changed markedly. For example, coal deposits are found in Antarctica and other high-latitude locations. Coal forms from the compaction and decomposition of accumulations of ancient swamp plants that grew in warm, wet regions. The existence of coal beds in Antarctica indicated that this frozen land once had a tropical climate. Wegener used this evidence to conclude that Antarctica must have been much closer to the equator sometime in the geologic past.

Glacial deposits Another piece of climatic evidence came from glacial deposits found in parts of Africa, India, Australia, and South America. The presence of these 290-million-year-old deposits suggested to Wegener that these areas were once covered by a thick ice cap similar to the one that covers Antarctica today. Because traces of the ancient ice cap were found in regions where it is too warm for them to develop, Wegener proposed that they were once located near the South Pole, as shown in **Figure 5.** Wegener suggested two possibilities to explain the deposits. Either the South Pole had shifted its position, or these landmasses had once been closer to the South Pole. Wegener argued that it was more likely that the landmasses had drifted apart rather than that Earth had changed its axis.

Figure 5 Glacial deposits nearly 300 million years old on several continents led Wegener to propose that these landmasses might have once been joined and covered with ice. The extent of the ice is shown in white.

A Rejected Notion

In the early 1900s, many people in the scientific community considered the continents and ocean basins to be fixed features on Earth's surface. For the rest of his life, Wegener continued travelling to remote regions to gather evidence in support of continental drift. **Figure 6** shows him in Greenland on his last expedition. Although he had compiled an impressive collection of data, the continental drift hypothesis was not accepted by the scientific community during Wegener's lifetime.

Continental drift had two major flaws that prevented it from being widely accepted. First, it did not explain what force could be strong enough to push such large masses over such great distances. Wegener thought that the rotation of Earth might be responsible, but physicists were able to show that this force was not nearly great enough to move continents.

Second, scientists questioned how the continents were moving. Wegener proposed that the continents were plowing through a stationary ocean floor, but it was known that Earth's mantle below the crust was solid. So, how could continents move through something solid? These two unanswered questions were the main reasons why continental drift was rejected at the time. It was not until the early 1960s, when new technology revealed more evidence about how continents move, that scientists began to reconsider Wegener's ideas.

Figure 6 Wegener collected further evidence for his theory on a 1930 expedition to Greenland. He died during this expedition, years before his data became the basis for the theory of plate tectonics.

Check Your Progress

Summary

- The matching coastlines of continents on opposite sides of the Atlantic Ocean suggest that the continents were once joined.

- Continental drift was the idea that continents move around on Earth's surface.

- Wegener collected evidence from rocks, fossils, and glacial deposits to support his theory.

- The theory of continental drift was not accepted because there was no satisfactory explanation for how the continents moved or what caused their motion.

Demonstrate Understanding

1. **Draw** how the continents were once adjoined as Pangaea.
2. **Explain** how ancient glacial deposits in Africa, India, Australia, and South America support the idea of continental drift.
3. **Summarize** how rocks, fossils, and climate provided evidence of continental drift.
4. **Infer** what the climate in ancient North America must have been like as a part of Pangaea.

Explain Your Thinking

5. **Interpret** Examine **Figure 5.** Oil deposits that are approximately 200 million years old have been discovered in Brazil. Where might geologists find oil deposits of a similar age?
6. **WRITING › Connection** Compose a letter to the editor from a scientist in the early 1900s arguing against continental drift.

SEAFLOOR SPREADING

FOCUS QUESTION

Why does the seafloor spread?

Mapping the Ocean Floor

Until the mid-1900s, most people, including many scientists, thought that the ocean floor was essentially flat. Many people also had misconceptions that oceanic crust was unchanging and was much older than continental crust. However, advances in technology during the 1940s and 1950s showed that all of these widely accepted ideas were incorrect.

One technological advance that was used to study the ocean floor was the magnetometer. A **magnetometer** (mag nuh TAH muh tur), such as the one shown in **Figure 7**, is a device that can detect small changes in magnetic fields. Towed behind a ship, it can record the magnetic field generated by ocean-floor rocks.

Another advancement that allowed scientists to study the ocean floor was the development of echo-sounding methods, such as sonar. Sonar uses sound waves to measure distance by measuring the time it takes for sound waves sent from the ship to bounce off the seafloor and return to the ship. Developments in sonar technology enabled scientists to measure water depth and map the topography of the ocean floor. These advancements in seafloor mapping, along with advancements in scientific understanding of Earth's magnetic field, provided the necessary evidence to show not only that Earth's continents had indeed moved over time, but that the ocean floor had moved and both the continents and the ocean floor are still moving today.

 Get It?

Model Design a model to show how a magnetometer works.

Figure 7 Magnetometers are devices that can detect small changes in magnetic fields. The data collected using magnetometers lowered into the ocean furthered scientists' understanding of rocks underlying the ocean floor.

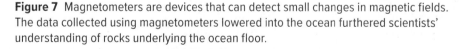

3D THINKING **DCI** Disciplinary Core Ideas   **CCC** Crosscutting Concepts   **SEP** Science & Engineering Practices

COLLECT EVIDENCE
 Use your Science Journal to record the evidence you collect as you complete the readings and activities in this lesson.

INVESTIGATE

GO ONLINE to find these activities and more resources.

Investigation Lab: Earthquakes and Subduction Zones
Use mathematical representations to visualize the result of two subducting plates.

Design Your Own: Magnetism and Ocean Ridges
Develop and use a model to illustrate how cyclic changes in Earth's magnetic field are recorded in oceanic crust.

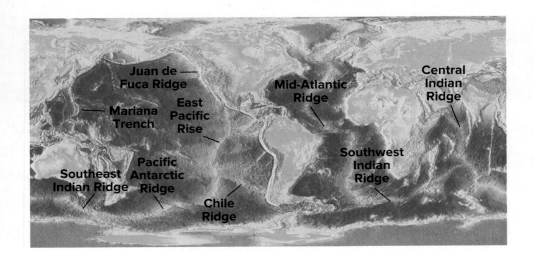

Figure 8 Sonar data revealed ocean ridges and deep-sea trenches. Earthquakes and volcanoes are common along ridges and trenches.

Ocean-Floor Topography

The maps made from data collected by sonar and magnetometers surprised many scientists. They discovered that vast, underwater mountain chains called ocean ridges run along the ocean floors around Earth, much like seams on a baseball. These ocean-floor features, shown in **Figure 8,** form the longest continuous mountain range on Earth. When they were first discovered, ocean ridges generated much discussion because of their enormous length and height—they can be more than 65,000 km long and up to 3 km above the ocean floor. Later, scientists discovered that earthquakes and volcanoes are common along the ridges.

> **Get It?**
>
> **Describe** Where is the longest continuous mountain range on Earth?

Maps generated with sonar data also revealed that underwater mountain chains had counterparts called deep-sea trenches, which are also shown on the map in **Figure 8.** Recall that a deep-sea trench is a narrow, elongated depression in the seafloor. Trenches can be thousands of kilometers long and many kilometers deep. The deepest trench, called the Mariana Trench, is in the Pacific Ocean and is more than 11 km deep. Mount Everest, the world's tallest mountain, stands at 9 km above sea level and could fit inside the Mariana Trench with six Empire State buildings stacked on top.

These two topographic features of the ocean floor—ocean ridges and deep-sea trenches— puzzled geologists for more than a decade after their discovery. What could have formed an underwater mountain range that extended around Earth? What is the source of the volcanism associated with these mountains? What forces could depress Earth's crust enough to create trenches nearly six times as deep as the Grand Canyon? You will find out the answers to these questions later in this module.

CCC CROSSCUTTING CONCEPTS
Patterns Add to the graphic organizer you made in Lesson 1 about continental drift. Cite the empirical evidence used to develop the theory of seafloor spreading.

SCIENCE USAGE v. COMMON USAGE
depress
Science usage: to cause to sink to a lower position
Common usage: to sadden or discourage

Ocean Rocks and Sediments

In addition to making maps, scientists collected samples of deep-sea sediments and the underlying oceanic crust. Analysis of these materials led to two important discoveries. First, the ages of the rocks that make up the seafloor vary across the ocean floor, and these variations are predictable. Rock samples taken from areas near ocean ridges are younger than samples taken from areas near deep-sea trenches. The samples showed that the age of oceanic crust consistently increases with distance from a ridge, as shown in **Figure 9.** This trend was symmetric across and parallel to the ocean ridges. Scientists also discovered from the rock samples that even the oldest parts of the seafloor are geologically young—about 180 million years old. Why are ocean-floor rocks so young compared to continental rocks, which can be 4 billion years old? Geologists knew that oceans had existed for more than 180 million years, so they questioned why there was no trace of older oceanic crust.

Get It?

Compare the ages of continental rocks and oceanic rocks.

The second discovery involved the sediments on the ocean floor. Measurements showed that ocean-floor sediments are typically a few hundred meters thick. Large areas of continents, on the other hand, are blanketed with sedimentary rocks that are as much as 20 km thick. Scientists knew that erosion and deposition occur in Earth's oceans but did not understand why seafloor sediments were not as thick as their continental counterparts. Scientists hypothesized that the relatively thin layer of ocean sediments was related to the age of the ocean crust. Observations of ocean-floor sediments revealed that the thickness of the sediments increases with distance from an ocean ridge, as shown in **Figure 9.** The pattern of thickness across the ocean floor was symmetrical across the ocean ridges.

Get It?

Describe how the age and thickness of oceanic crust varies with distance from the mid-ocean ridge.

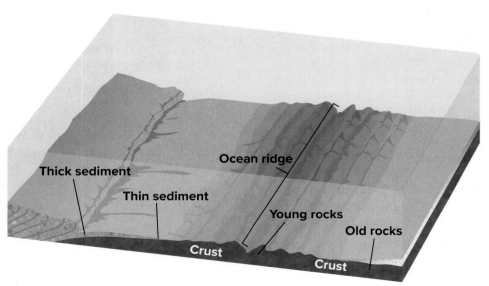

Figure 9 The age of ocean crust and the thickness of ocean-floor sediments increase with distance from the ridge.

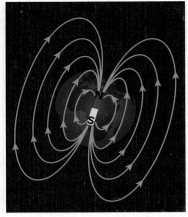

Normal magnetic field   **Reversed magnetic field**

Figure 10 Earth's magnetic field is generated by the flow of molten iron in the liquid outer core. The polarity of the field changes over time from normal to reversed.

Magnetism

Earth has a magnetic field generated by the flow of molten iron in the outer core. This field is what causes a compass needle to point to the north. A **magnetic reversal** happens when the flow in the outer core changes, and Earth's magnetic field changes direction. This would cause compasses to point to the south. Magnetic reversals have occurred many times in Earth's history. As shown in **Figure 10,** a magnetic field that has the same orientation as Earth's present field is said to have normal polarity. A magnetic field that is opposite to the present field has reversed polarity.

Magnetic polarity time scale

Paleomagnetism is the study of the history of Earth's magnetic field. When lava solidifies, iron-bearing minerals such as magnetite crystallize. As they crystallize, these minerals behave like tiny compasses and align with Earth's magnetic field. Data from paleomagnetic studies of continental lava flows allowed scientists to construct a magnetic polarity time scale, as shown in **Figure 11.**

 Get It?

Identify Study **Figure 11.** What is the current epoch and when did it begin?

Magnetic symmetry

Scientists knew that oceanic crust is mostly basaltic rock, which contains large amounts of iron-bearing minerals of volcanic origin. They hypothesized that the rocks on the ocean floor would show a record of magnetic reversals.

When scientists used magnetometers to measure the magnetic orientation of rocks on the ocean floor, a surprising pattern emerged. The regions with normal and reversed polarity formed a series of stripes across the floor parallel to the ocean ridges. Scientists were doubly surprised to discover that the ages and widths of the stripes matched from one side of the ridges to the other.

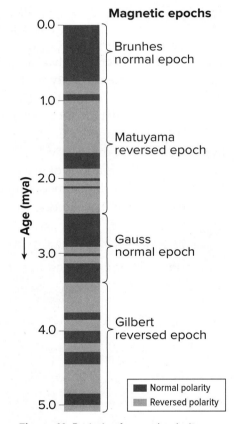

Magnetic epochs

Age (mya)

Brunhes normal epoch

Matuyama reversed epoch

Gauss normal epoch

Gilbert reversed epoch

■ Normal polarity
■ Reversed polarity

Figure 11 Periods of normal polarity alternate with periods of reversed polarity. Long-term changes in Earth's magnetic field, called epochs, are named as shown here. Short-term changes are called *events.*

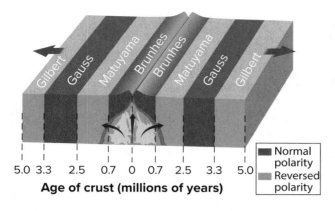

Figure 12 Reversals in the polarity of Earth's magnetic field are recorded in the rocks that make up the ocean floor.

Identify *the polarity of the most recently produced basalt at the ocean ridge.*

Gilbert Gauss Matuyama Brunhes Brunhes Matuyama Gauss Gilbert

| 5.0 | 3.3 | 2.5 | 0.7 | 0 | 0.7 | 2.5 | 3.3 | 5.0 |

Age of crust (millions of years)

■ Normal polarity
■ Reversed polarity

Compare the magnetic pattern on opposite sides of the ocean ridge shown in **Figure 12.** By matching the patterns on the seafloor with the known pattern of reversals on land, scientists were able to determine the age of the ocean floor from magnetic recording. This method enabled scientists to quickly create isochron (I suh krahn) maps of the ocean floor. An **isochron** is an imaginary line on a map that shows points that have the same age—that is, they formed at the same time. In the isochron map shown in **Figure 13,** note that relatively young ocean-floor crust is near ocean ridges, while older ocean crust is found along deep-sea trenches.

> ✏️ **Get It?**
>
> **Apply** How are isochrons similar to contour lines on a topographic map?

Figure 13 Each colored band on this isochron map of the ocean floor represents the age of that strip of the crust.
Observe *What pattern do you observe?*

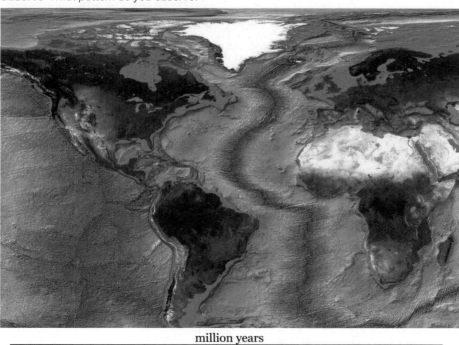

million years

| 0 | 20 | 40 | 60 | 80 | 100 | 120 | 140 | 160 | 180 | 200 | 220 | 240 | 260 | 280 |

Figure 14 Visualizing Seafloor Spreading

Data from topographic, sedimentary, and paleomagnetic research led scientists to propose seafloor spreading. Seafloor spreading is the process by which new oceanic crust forms at ocean ridges and slowly moves away from the spreading center until it is subducted and recycled at deep-sea trenches.

Magma intrudes into the ocean floor along a ridge and fills the gap that is created. When the molten material solidifies, it becomes new oceanic crust.

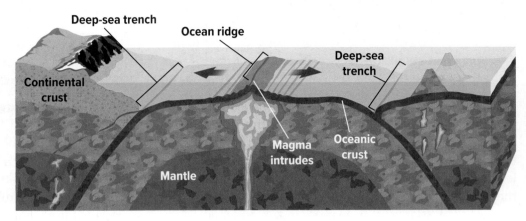

The continuous spreading and intrusion of magma result in the addition of new oceanic crust. Two halves of the oceanic crust separate slowly and move apart like a conveyor belt.

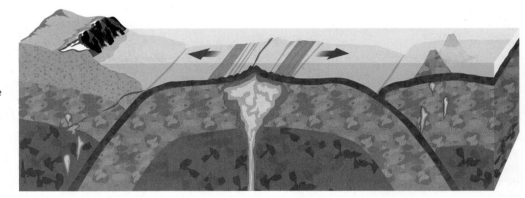

The far edges of the oceanic crust sink beneath continental crust. As it descends, water in the minerals is released, which helps melt the overlying mantle, forming magma. The magma rises and forms volcanoes on the continental crust.

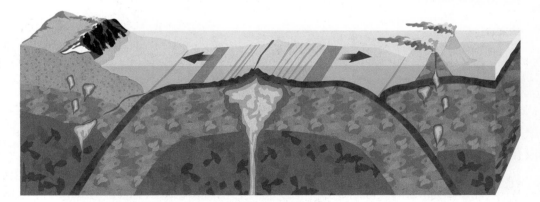

Seafloor Spreading

Using all the topographic, sedimentary, and paleomagnetic data from the seafloor, the theory of seafloor spreading was proposed. **Seafloor spreading** is the theory that explains how new oceanic crust forms at ocean ridges, slowly moves away from ocean ridges, and is destroyed at deep-sea trenches.

Figure 14 on the previous page illustrates how seafloor spreading occurs. During seafloor spreading, magma, which is hotter and less dense than surrounding mantle material, is forced up toward the crust along an ocean ridge. As the two sides of the ridge spread apart, the rising magma fills the resulting gap. When the magma solidifies, a small amount of new ocean floor is added to Earth's surface. As spreading along a ridge continues, more magma rises upward and solidifies. This cycle of spreading and the intrusion of magma forms ocean floor, which slowly moves away from the ridge. As you might guess, seafloor spreading mostly happens under the sea, but in Iceland, a portion of the Mid-Atlantic Ridge rises above sea level. **Figure 15** shows lava erupting along the ridge.

Recall that, while Wegener collected much data to support the idea that continents drift across Earth's surface, he could not explain what caused the landmasses to move or how they moved. Seafloor spreading was the missing link that completed the model of continental drift. Continents do not push through ocean crust, as Wegener proposed. Instead, they are like passengers that ride along while ocean crust slowly moves away from ocean ridges. The theory of seafloor spreading led to a new understanding of how Earth's crust and rigid upper mantle move, as you'll learn in the following lessons.

Figure 15 The island of Iceland lies on the Mid-Atlantic Ridge. Because the seafloor is spreading, Iceland is growing larger. In 1783, more than 12 km³ of lava erupted—enough to pave the entire U.S. interstate freeway system to a depth of 10 m.

Check Your Progress

Summary

- Studies of the seafloor provided evidence that the ocean floor is not flat and unchanging.
- Oceanic crust is geologically young.
- New oceanic crust forms as magma rises at ridges and solidifies.
- As new oceanic crust forms, the older crust moves away from the ridges.

Demonstrate Understanding

1. **Describe** why seafloor spreading is like a moving conveyor belt.
2. **Explain** how ocean-floor rocks and sediments provided evidence of seafloor spreading.
3. **Identify** commonly found features on the ocean floor.

Explain Your Thinking

4. **Explain** how an isochron map of the ocean floor supports the theory of seafloor spreading.
5. **Analyze** Why are magnetic bands in the eastern Pacific Ocean so far apart compared to the magnetic bands along the Mid-Atlantic Ridge?
6. **MATH Connection** Analyze **Figure 11.** What percentage of the last 5 million years has been spent in reversed polarity?

LEARNSMART Go online to follow your personalized learning path to review, practice, and reinforce your understanding.

Why do earthquakes only happen in some places?

Theory of Plate Tectonics

The evidence for seafloor spreading suggested that continental and oceanic crust move as enormous slabs, which geologists describe as tectonic plates. **Tectonic plates** are huge pieces of crust and rigid upper mantle that fit together at their edges to cover Earth's surface. As illustrated in **Figure 16,** there are about eight major plates and several smaller ones. These plates move very slowly—only a few centimeters each year—which is similar to the rate at which fingernails grow. Plate tectonics is the theory that describes how tectonic plates move and shape Earth's surface. This unifying theory explains the past and current movements of tectonic plates and provides a framework for understanding Earth's geologic history.

Figure 16 Earth's crust and rigid upper mantle are broken into enormous slabs, called tectonic plates, that interact at their boundaries.

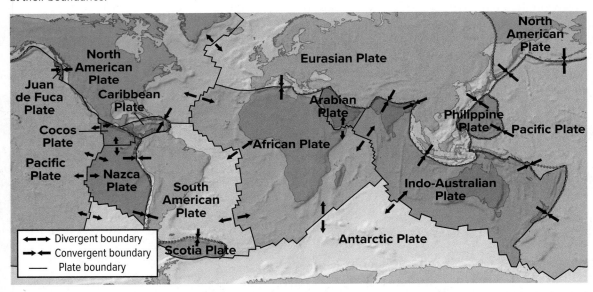

3D THINKING
DCI Disciplinary Core Ideas   **CCC** Crosscutting Concepts   **SEP** Science & Engineering Practices

COLLECT EVIDENCE
Use your Science Journal to record the evidence you collect as you complete the readings and activities in this lesson.

INVESTIGATE
GO ONLINE to find these activities and more resources.

Applying Practices: How old are crustal rocks?
HS-ESS1-5. Evaluate evidence of the past and current movements of continental and oceanic crust and the theory of plate tectonics to explain the ages of crustal rocks.

GeoLAB: Model Plate Boundaries and Isochrons
Use a model to visualize the past and current movements of continental and oceanic crust.

Plate movements are responsible for most continental and ocean-floor features and for the distribution of most rocks and minerals within Earth's crust.

Tectonic plates move in different directions and at different rates relative to one another, and they interact with one another at their boundaries. Each type of boundary has certain geologic features and processes associated with it. A divergent boundary occurs where tectonic plates move away from each other. A convergent boundary occurs where tectonic plates move toward each other. A transform boundary occurs where tectonic plates move horizontally past each other.

 Get It?

Discuss Why is plate tectonics often called a unifying theory?

Divergent boundaries

Regions where two tectonic plates are moving apart are called **divergent boundaries.** Most divergent boundaries are found along the seafloor in rift valleys. It is in this central rift that the process of seafloor spreading begins. Magma rising through the center of the rift forms a mid-ocean ridge. The mid-ocean ridge appears as a continuous mountain chain on the ocean floor. The formation of new ocean crust at most divergent boundaries accounts for the high heat flow, volcanism, and earthquakes associated with these boundaries.

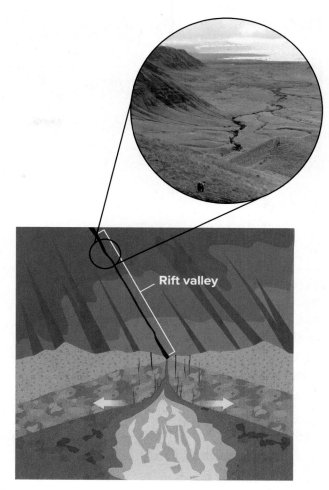

Figure 17 Divergent boundaries are places where plates separate. An ocean ridge is a divergent boundary on the ocean floor. In East Africa, a divergent boundary has also created a rift valley on land.

 Get It?

Identify the cause of volcanism and earthquakes associated with mid-ocean ridges.

Throughout millions of years, the process of seafloor spreading along a divergent boundary can cause an ocean basin to expand. Although most divergent boundaries form ridges on the ocean floor, some divergent boundaries form on continents. When continental crust begins to separate, the stretched crust forms a long, narrow depression called a **rift valley. Figure 17** shows the rift valley that is currently forming in East Africa.

Convergent boundaries

At **convergent boundaries,** two tectonic plates are moving toward each other. When two plates collide, the denser plate eventually descends below the other, less-dense plate in a process called **subduction.**

Guenter Guni/iStockphoto/Getty Images

Figure 18 Oceanic plates are mostly basalt, shown in the top photo. Continental plates are mostly granite, shown in the bottom photo, along with a thin cover of sedimentary rock, both of which are less dense than basalt.

There are three types of convergent boundaries, classified according to the type of crust involved. Recall that oceanic crust is made mostly of minerals that are high in iron and magnesium, which form dense, dark-colored basaltic rocks, such as the basalt shown in **Figure 18.** Continental crust is composed mostly of minerals such as feldspar and quartz, which form less-dense, lighter-colored granitic rocks, also shown in **Figure 18.** The differences in density of the crustal material affects how they converge. The three types of convergent boundaries and their associated landforms are shown in **Table 1.**

Oceanic-oceanic In the oceanic-oceanic convergent boundary shown in **Table 1,** a subduction zone is formed when one oceanic plate, which is denser as a result of cooling, descends below another oceanic plate. The process of subduction creates an ocean trench. The subducted plate descends into the mantle, thereby recycling oceanic crust formed at the ridge. The increased heat and pressure causes water in the minerals of the subducting plate to be released, lowering the melting temperature of the overlying mantle, causing it to melt. The molten material, called magma, is less dense, so it rises back to the surface, where it often erupts and forms an arc of volcanic islands that parallels the trench. Some examples of trenches and island arcs are the Mariana Trench and Mariana Islands in the West Pacific Ocean and the Aleutian Trench and Aleutian Islands in the North Pacific Ocean. A volcanic peak in the Aleutian Island arc is shown in **Table 1.**

Oceanic-continental Subduction zones are also found where an oceanic plate converges with a continental plate, as shown in **Table 1.** Note that the denser oceanic plate is subducted. Oceanic-continental convergence also produces trenches and volcanoes. However, instead of forming an arc of volcanic islands, oceanic-continental convergence results in a chain of volcanoes along the edge of the continental plate, creating a trench next to a mountain range with many volcanoes. The Peru-Chile Trench and the Andes mountain range, located along the western coast of South America, formed in this way.

STEM CAREER Connection

Marine Geologist

Rocks and features on the ocean floor are like a record of Earth's past. They can help scientists understand seafloor spreading and other processes related to plate tectonics. Marine geologists are scientists who specialize in studying the ocean floor.

CCC CROSSCUTTING CONCEPTS

Stability and Change Research rates of tectonic plate movement. Design a model that shows the positions of landmasses 50 million years from now. What evidence supports your model?

Table 1 Summary of Convergent Boundaries

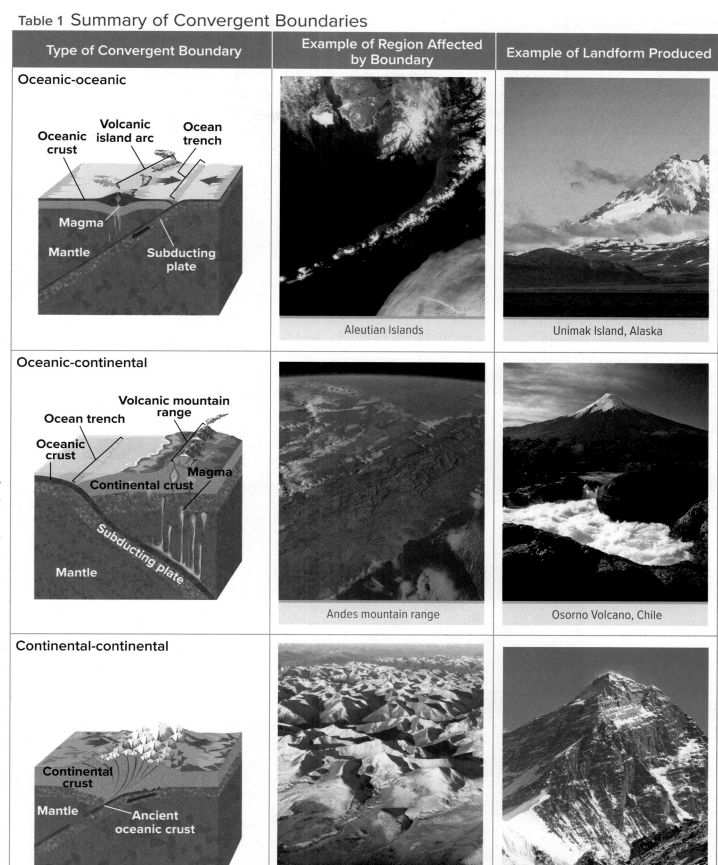

Type of Convergent Boundary	Example of Region Affected by Boundary	Example of Landform Produced
Oceanic-oceanic Volcanic island arc Ocean trench Oceanic crust Magma Mantle Subducting plate	 Aleutian Islands	 Unimak Island, Alaska
Oceanic-continental Volcanic mountain range Ocean trench Oceanic crust Magma Continental crust Subducting plate Mantle	 Andes mountain range	 Osorno Volcano, Chile
Continental-continental Continental crust Mantle Ancient oceanic crust	 Himalayas	 Mount Everest, Nepal

(left to right, top to bottom)Jeff Schmaltz/NASA, Scott Dickerson/Design Pics/Getty Images, Image courtesy of the Earth Science and Remote Sensing Unit, NASA Johnson Space Center, Philippe Thiers/Moment/Getty Images, Pixtal/age fotostock, DanielPrudek/iStockphoto/Getty Images

Continental-continental The third type of convergent boundary forms when two continental plates collide. Continental-continental boundaries form long after an oceanic plate has converged with a continental plate. Recall that continents are often carried along attached to oceanic crust. Over time, an oceanic plate can be completely subducted, dragging an attached continent behind it toward the subduction zone. Because of its denser composition, oceanic crust descends beneath the continental crust at the subduction zone. The continental crust that it pulls behind it cannot descend because continental rocks are less dense and will not sink into the mantle. As a result, the edges of both continents collide and become crumpled, folded, and uplifted. This forms a vast mountain range, such as the Himalayas, as illustrated in **Table 1.**

 Get It?

Explain why continental crust does not sink into the mantle, yet oceanic crust does.

Transform boundaries

A region where two plates slide horizontally past each other is a **transform boundary,** as shown in **Figure 19.** Transform boundaries are characterized by long faults, sometimes hundreds of kilometers in length, and by shallow earthquakes. Transform boundaries were named for the way Earth's crust changes, or transforms, its relative direction and velocity from one side of the boundary to the other. Recall that new crust is formed at divergent boundaries and destroyed at convergent boundaries. Crust can be deformed or fractured along transform boundaries.

 Get It?

Compare What happens to crust at divergent boundaries, convergent boundaries, and transform boundaries?

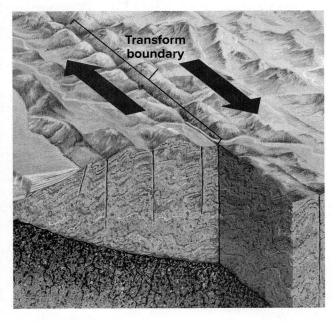

Figure 19 Plates move horizontally past each other along a transform plate boundary.

Figure 20 These twisted train tracks resulted from an earthquake-generated tsunami that struck Sri Lanka in 2004.

Most transform boundaries cause sections of ocean ridges to become offset. Sometimes transform boundaries occur on continents. The San Andreas Fault is probably the best-known example. The San Andreas Fault system is part of a transform boundary that separates southwestern California from the rest of the state. Offset train tracks, roads, fences, and creeks are telltale signs of a transform fault. Movements along the San Andreas Fault system are responsible for many of the earthquakes that strike California every year.

Get It?

Apply How does the damage caused by a transform fault reflect the movement of the tectonic plates associated with the fault?

Earthquakes can also cause tsunamis, or large ocean waves generated by vertical motions of the seafloor. Tsunamis are mainly associated with underwater earthquakes. The earthquake displaces water over a fault. The disturbance spreads out in the form of extremely long waves that can break onto shore with heights exceeding 30 m. Tsunamis can devastate coastal regions. The 2004 tsunami that struck the coasts of Indonesia, Sri Lanka, and other countries killed an estimated 225,000 people and caused widespread damage to structures, such as the train tracks shown in **Figure 20.**

Check Your Progress

Summary

- Earth's crust and rigid upper mantle are broken into large slabs of rock called tectonic plates.
- Plates move in different directions and at different rates over Earth's surface.
- At divergent plate boundaries, plates move apart. At convergent boundaries, plates come together. At transform boundaries, plates slide horizontally past each other.
- Each type of boundary is characterized by certain geologic features.

Demonstrate Understanding

1. **Explain** how plate movements are responsible for most continental and ocean-floor geologic features.
2. **Summarize** the processes of convergence that formed the Himalayan mountains.
3. **List** the geologic features associated with each type of convergent boundary.
4. **Identify** the type of location where transform boundaries most commonly occur.

Explain Your Thinking

5. **Infer** how plate movements affect the distribution of most rocks and minerals in Earth's crust.
6. **Describe** how two portions of newly formed crust move between parts of a ridge that are offset by a transform boundary.
7. **WRITING** **Connection** Write a news report on the tectonic activity that is occurring at the Aleutian Islands in Alaska.

LEARNSMART Go online to follow your personalized learning path to review, practice, and reinforce your understanding.

LESSON 4
CAUSES OF PLATE MOTIONS

FOCUS QUESTION

What makes the tectonic plates move?

Convection

One question about plate tectonics has remained unanswered since Alfred Wegener first proposed continental drift. What causes tectonic plates to move? Many scientists now think that large-scale motion in the mantle, located between the crust and the core, drives the movement of tectonic plates.

Convection currents

Recall that convection is the transfer of thermal energy by the movement of heated material from one place to another. The cooling of matter causes it to contract slightly and increase in density. The cooled matter then sinks as a result of gravity. Warmed matter is then displaced and forced to rise. This up-and-down flow produces a pattern called a convection current. Convection currents aid in the transfer of thermal energy from warmer regions of matter to cooler regions. A convection current is shown in **Figure 21.**

Beaker with H₂O · Ice cube · Convection current · Drops of blue food coloring · Burner

Matt Meadows/McGraw-Hill Education

Figure 21 Water cooled by the ice cube sinks to the bottom, where it is warmed by the burner and then rises. The process continues as the ice cube cools the water again.

Infer *What will happen to the ice cube due to convection currents?*

 3D THINKING   **DCI** Disciplinary Core Ideas   **CCC** Crosscutting Concepts   **SEP** Science & Engineering Practices

COLLECT EVIDENCE

Use your Science Journal to record the evidence you collect as you complete the readings and activities in this lesson.

INVESTIGATE

 GO ONLINE to find these activities and more resources.

Applying Practices: The Cycling of Matter Through Thermal Convection
HS-ESS2-3. Develop a model based on evidence of Earth's interior to describe the cycling of matter by thermal convection.

Revisit the Encounter the Phenomenon Question
What information from this lesson can help you answer the Unit and Module questions?

Consider that Earth's mantle is composed of primarily solid material that is heated unevenly by radioactive decay. The heat is generated from both the mantle itself and the core beneath it. The radioactive decay of unstable isotopes heats up the mantle, causing solid rock to slowly flow and creating enormous convection currents that move material throughout the mantle.

Convection in the mantle

Convection currents in the mantle, illustrated in **Figure 22,** are thought to be the driving mechanism of plate movements. Recall that even though the mantle is a solid, much of it moves like a soft, pliable plastic. The part of the mantle that is too cold and stiff to flow lies beneath the crust and is attached to it, moving as a part of tectonic plates. In the convection currents of the mantle, cooler mantle material is denser than hot mantle material. Mantle that has cooled at the base of tectonic plates slowly sinks downward toward the center of Earth. Heated mantle material is then displaced, and, like the wax warmed in a lava lamp, it rises. Convection currents in the mantle are sustained by this rise and fall of material, which results in a transfer of energy between Earth's hot interior and its cooler exterior. Although convection currents can be thousands of kilometers across, they flow at rates of only a few centimeters per year. Scientists think that these convection currents are set in motion by subducting slabs.

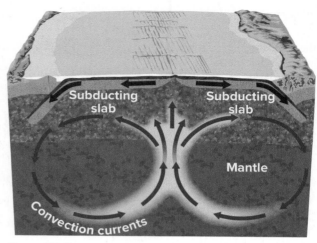

Figure 22 Convection currents develop in the mantle, moving the crust and outermost part of the mantle, and transferring thermal energy from Earth's interior to its exterior.

Plate movement

How are convergent and divergent movements of tectonic plates related to mantle convection? Plate tectonics is the surface expression of mantle convection. The rising material in the convection current spreads out as it reaches the upper mantle and causes both upward and sideways forces. These forces lift and split the lithosphere at divergent plate boundaries. As the plates separate, material rising from the mantle supplies the magma that hardens to form new ocean crust. The downward part of a convection current occurs where a sinking force pulls tectonic plates downward at convergent boundaries. So, both mantle motion and plate movements occur primarily through thermal convection, which involves the cycling of matter due to the outward flow of energy from Earth's interior and the gravitational movement of denser materials toward the interior.

 Get It?

Discuss why plate tectonics can be viewed as the surface expression of mantle convection.

Push and Pull

Scientists hypothesize that there are several processes that determine how mantle convection affects tectonic plate motion. As oceanic crust cools and moves away from a divergent boundary, it becomes denser and sinks compared to the newer, less-dense oceanic crust.

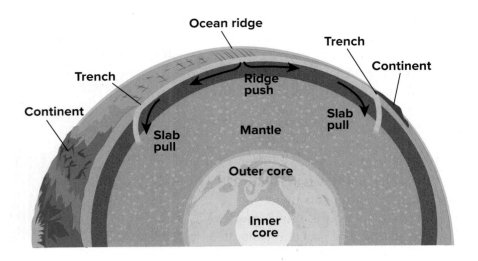

Figure 23 Ridge push and slab pull are two of the processes that move tectonic plates over the surface of Earth.

As the older portion of the seafloor sinks, the weight of the uplifted ridge is thought to push the oceanic plate toward the trench formed at the subduction zone in a process called **ridge push.** A second and possibly more significant process that determines the movement of tectonic plates is called slab pull. In **slab pull,** the weight of the relatively cool, dense subducting plate pulls the trailing slab into the subduction zone much like a tablecloth slipping off the table can pull articles off with it. Slab pull is thought to be at least twice as important as ridge push in moving an oceanic plate away from an ocean ridge. It is likely that a combination of mechanisms such as ridge push and slab pull, shown in **Figure 23,** are involved in plate motions at subduction zones.

Check Your Progress

Summary

- Convection is the transfer of energy via the movement of heated matter.
- Convection currents in the mantle result in an energy transfer between Earth's hot interior and cooler exterior.
- Plate movement results from the processes of ridge push and slab pull.

Demonstrate Understanding

1. **Draw** a diagram comparing convection in a pot of water with convection in Earth's mantle.
2. **Explain** how thermal convection drives the cycling of matter, and relate the process to mantle motion and plate movement.
3. **Restate** the relationships among mantle convection, ocean ridges, and subduction zones.
4. **Identify** the primary source of heat that drives mantle convection.
5. **Design** a model that illustrates the tectonic processes of ridge push and slab pull.

Explain Your Thinking

6. **Evaluate** this statement: Oceanic crust is moved only by convection currents.
7. **Summarize** how convection is responsible for the arrangement of continents on Earth's surface.
8. **WRITING ▸ Connection** Write dictionary definitions for *ridge push* and *slab pull* without using those terms.

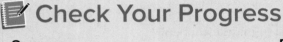

Go online to follow your personalized learning path to review, practice, and reinforce your understanding.

Solving the Biggest Puzzle on Earth

Tectonic plates are like huge puzzle pieces that make up Earth's lithosphere. Some small, but important, missing pieces are still being discovered. These discoveries are helping scientists better understand the geologic forces that raise mountains and rattle Earth's surface.

Scientists have a more precise age of the Himalayas thanks to the discovery of a microplate in the Indian Ocean.

Microplates

Imagine you are putting together a complex jigsaw puzzle. The pieces are different sizes and shapes. In the end, you notice that you are missing some pieces, and the puzzle does not quite make sense. Microplates are like those missing pieces of the tectonic jigsaw puzzle.

Major tectonic plates and microplates are both made of the lithosphere. The types of plates are distinguished by their size—most microplates are less than 1 million km² in area. They often form when major plates collide and a piece breaks off.

New discoveries

Scientists have discovered almost 60 microplates over the past several decades. Each new find adds to the knowledge of Earth's past and present. For example, scientists have long debated when the Himalayas began to form. Estimates ranged from 34 million to 59 million years ago. In 2015, scientists announced the discovery of the first microplate in the Indian Ocean. Using magnetic surveys, they determined that the microplate formed 47 million years ago when the Indian plate collided with the Eurasian plate.

This discovery offers strong evidence that the Himalayas are about 47 million years old.

Finding microplates

Finding microplates is a challenge. They can be spotted using ship-based echosounder radar, but only about 15 percent of the ocean has been mapped in this expensive, time-consuming way. Today, satellite imagery is used to make seafloor maps that are carefully analyzed to help locate underwater features, including microplates.

Analyses of plate movements are also helpful for finding microplates. Plates move at a rate of only millimeters to centimeters per year, and if the calculations do not add up, then a microplate is likely throwing off the numbers.

Plate movements can cause earthquakes and volcanism. Knowing the speed and direction of moving tectonic plates enables scientists to better forecast these major events.

USE A MODEL TO ILLUSTRATE

Conduct research on how microplates form. Then create a model to illustrate the formation of a microplate. Share your model with the class.

GO ONLINE to study with your Science Notebook.

Lesson 1 DRIFTING CONTINENTS

- The matching coastlines of continents on opposite sides of the Atlantic Ocean suggest that the continents were once joined.
- Continental drift is the hypothesis that continents move around on Earth's surface.
- Wegener collected evidence from rocks, fossils, and glacial deposits to support his theory.
- The theory of continental drift was not accepted because there was no satisfactory explanation for how the continents moved or what caused their motion.

- continental drift
- Pangaea

Lesson 2 SEAFLOOR SPREADING

- Studies of the seafloor provided evidence that the ocean floor is not flat and unchanging.
- Oceanic crust is geologically young.
- New oceanic crust forms as magma rises at ridges and solidifies.
- As new oceanic crust forms, the older crust moves away from the ridges.

- magnetometer
- magnetic reversal
- paleomagnetism
- isochron
- seafloor spreading

Lesson 3 PLATE BOUNDARIES

- Earth's crust and rigid upper mantle are broken into large slabs of rock called tectonic plates.
- Plates move in different directions and at different rates over Earth's surface.
- At divergent plate boundaries, plates move apart. At convergent boundaries, plates come together. At transform boundaries, plates slide horizontally past each other.
- Each type of boundary is characterized by certain geologic features.

- tectonic plate
- divergent boundary
- rift valley
- convergent boundary
- subduction
- transform boundary

Lesson 4 CAUSES OF PLATE MOTIONS

- Convection is the transfer of energy via the movement of heated matter.
- Convection currents in the mantle result in an energy transfer between Earth's hot interior and cooler exterior.
- Plate movement results from the processes of ridge push and slab pull.

- ridge push
- slab pull

REVISIT THE PHENOMENON

How do we know that this landmass is moving?

CER Claim, Evidence, Reasoning

Explain Your Reasoning Revisit the claim you made when you encountered the phenomenon. Summarize the evidence you gathered from your investigations and research and finalize your Summary Table. Does your evidence support your claim? If not, revise your claim. Explain why your evidence supports your claim.

STEM UNIT PROJECT
Now that you've completed the module, revisit your STEM unit project. You will summarize your evidence and apply it to the project.

GO FURTHER

SEP Data Analysis Lab
How does plate motion change along a transform boundary?

The figure at right shows the Gibbs Fracture Zone, which is a segment of the Mid-Atlantic Ridge located south of Iceland and west of the British Isles. Copy this figure.

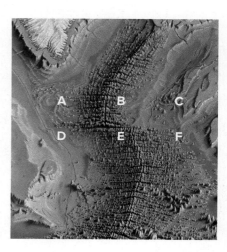

Data and Observations Draw arrows on your copy to indicate the direction of seafloor movement at locations A, B, C, D, E, and F. Compare the direction of motion for the following pairs of locations: A and D, B and E, and C and F.

CER Analyze and Interpret Data

1. **Claim, Evidence** Which three locations are on the North American Plate? How do you know?
2. **Claim, Evidence** Which portion of the fracture zone is the boundary between North America and Europe? How do you know?
3. **Reasoning** How can you determine which two locations represent the oldest crust? Use examples to explain your answer.

ENCOUNTER THE PHENOMENON
Where does lava come from?

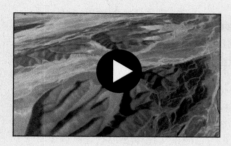

▶ **GO ONLINE** to play a video about a continuously erupting volcano called Erta Ale in Ethiopia.

SEP Ask Questions

Do you have other questions about the phenomenon? If so, add them to the driving question board.

CER Claim, Evidence, Reasoning

Make Your Claim Use your CER chart to make a claim about where lava comes from. Explain your reasoning.

Collect Evidence Use the lessons in this module to collect evidence to support your claim. Record your evidence as you move through the module.

Explain Your Reasoning You will revisit your claim and explain your reasoning at the end of the module.

▶ **GO ONLINE** to access your CER chart and explore resources that can help you collect evidence.

LESSON 1: Explore & Explain: Anatomy of a Volcano

LESSON 2: Explore & Explain: Types of Magma

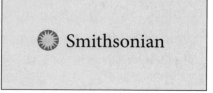

Additional Resources

FOCUS QUESTION

Where do we find most of the volcanoes?

Zones of Volcanism

Volcanoes are fueled by magma. Recall that magma is a slushy mixture of molten rock, mineral crystals, and gases. Once magma forms, it rises toward Earth's surface because it is less dense than the surrounding mantle and crust. Magma that reaches Earth's surface is called lava. **Volcanism** describes all the processes associated with the discharge of magma, hot fluids, ash, and gases.

The distribution of volcanoes on Earth's surface is not random. A map of active volcanoes, shown in **Figure 1,** reveals striking patterns on Earth's surface. Most volcanoes form at plate boundaries. The majority form at convergent boundaries and divergent boundaries. Only about 5 percent of magma erupts far from plate boundaries.

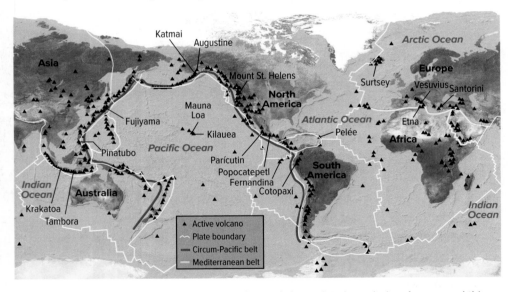

Figure 1 Most of Earth's active volcanoes are located along plate boundaries. As you read this, approximately 20 volcanoes are erupting.

 3D THINKING   **DCI** Disciplinary Core Ideas   **CCC** Crosscutting Concepts   **SEP** Science & Engineering Practices

COLLECT EVIDENCE

 Use your Science Journal to record the evidence you collect as you complete the readings and activities in this lesson.

INVESTIGATE

🌐 **GO ONLINE** to find these activities and more resources.

🥽 **Quick Investigation: Model a Caldera**
Develop and use a model to visualize the properties of a caldera.

((•)) **Review the News**
Obtain information from a current news story about volcanoes. Evaluate your source and communicate your findings to your class.

Convergent volcanism

Recall that tectonic plates collide at convergent boundaries, which can form subduction zones—places where slabs of crust descend into the mantle. As shown in **Figure 2,** an oceanic plate descends below another plate into the mantle. As the oceanic plate descends, water flows with it, which helps to melt the overlying mantle, forming magma. The magma moves upward because it is less dense than the surrounding material. As it rises, it mixes with rock, minerals, and sediment from the overlying plate. Most volcanoes located on land result from oceanic-continental subduction. These volcanoes are characterized by explosive eruptions.

Get It?

Define What is convergent volcanism?

Two major belts

The volcanoes associated with convergent plate boundaries form a major belt, shown in **Figure 1.** The Circum-Pacific Belt is also called the Pacific Ring of Fire. The name *Circum-Pacific* gives a hint about the location of the belt. *Circum* means "around" (as in *circumference*). The outline of the belt corresponds to the outline of the Pacific Plate. The belt stretches along the western coasts of North and South America, across the Aleutian Islands, and down the eastern coast of Asia. Volcanoes in the Cascade Range of the western United States and Mount Pinatubo in the Philippines are some of the volcanoes in the Circum-Pacific Belt. There is also a smaller belt called the Mediterranean Belt. It includes Mount Etna and Mount Vesuvius, two volcanoes in Italy. Its general outline corresponds to the boundaries between the Eurasian, African, and Arabian plates.

Divergent volcanism

Recall that at divergent plate boundaries, tectonic plates move apart, and new ocean floor is produced as magma rises to fill the gap. When lava is erupted under water, it often takes the form of pillows, like those in **Figure 3,** and is called pillow lava. Unlike the explosive volcanoes detailed in **Figure 4,** volcanism at divergent boundaries tends to be nonexplosive, with effusions of large amounts of lava. About two-thirds of Earth's volcanism occurs under water, along divergent plate boundaries at ocean ridges.

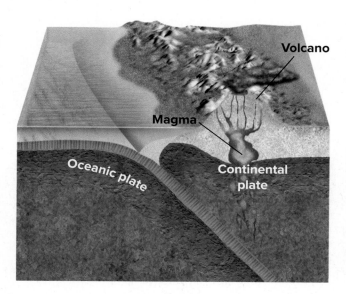

Figure 2 In an oceanic-continental subduction zone, the denser oceanic plate slides under the continental plate into the hot mantle. Parts of the overlying mantle melt, and magma rises, eventually leading to the formation of a volcano.

Identify *a volcano from **Figure 1** that is associated with oceanic-continental convergence.*

Figure 3 Eruptions at divergent boundaries tend to be nonexplosive. At the divergent boundary on the ocean floor, eruptions often form huge piles of lava called pillow lava.

OAR, National Underseas Research Program (NURP), NOAA

Figure 4

Volcanoes in Focus

Volcanoes constantly shape Earth's surface.

1 **4845 B.C.** Mount Mazama erupts in Oregon. The mountain collapses into a 9-km-wide depression known today as Crater Lake.

2 **1630 B.C.** In Greece, Santorini explodes, causing tsunamis 200 m high. Minoan civilization on the nearby Isle of Crete disappears.

3 **A.D. 79** Mount Vesuvius in Italy erupts, burying two cities in ash.

4 **1883** In Indonesia, Krakatoa erupts, destroying two-thirds of the island and generating a tsunami that kills more than 36,000 people.

5 **1912** Katmai erupts in Alaska with ten times more force than Mount St. Helens. This eruption is one of the most powerful in recorded history.

6 **1980** In Washington, Mount St. Helens' eruption blasts through the side of the volcano. Most of the 57 fatalities are from ash inhalation.

7 **1991** Mount Pinatubo erupts in the Philippines, releasing 10 km³ of ash, reducing global temperatures by 0.5°C.

8 **2010** Ash erupting from Iceland's Eyjafjallajökull volcano causes Europe's air traffic to shut down, affecting at least 10 million passengers worldwide.

Get It?

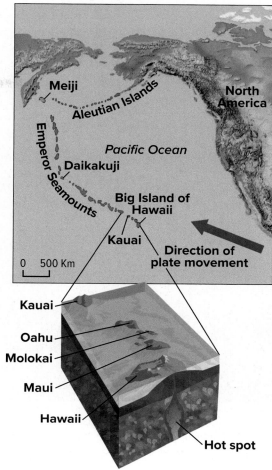

Hawaiian-Emperor Seamount Chain

Figure 5 The Hawaiian islands have been forming for millions of years as the Pacific Plate moves slowly over a stationary hot spot that is currently located under the Big Island of Hawaii.

Convert the fraction of volcanism that happens under water to a percentage.

Hot spots

Some volcanoes form far from plate boundaries, over geologic features called hot spots. Scientists hypothesize that **hot spots** are unusually hot regions of Earth's mantle where high-temperature plumes of magma rise to the surface.

Hot spot volcanoes Some of Earth's best-known volcanoes formed as a result of hot spots on the ocean floor. For example, the Hawaiian islands, shown on the map in **Figure 5**, are located over a plume of magma. As the rising magma melts through the crust, it forms volcanoes. The hot spot formed by the magma plume remains stationary, while the Pacific Plate slowly moves northwest. Over time, the hot spot has left a trail of volcanic islands on the floor of the Pacific Ocean. The volcanoes on the oldest Hawaiian island, Kauai, are inactive because the island no longer sits above the stationary hot spot. Even older volcanoes to the northwest are no longer above sea level. The world's most active volcano, Kilauea, on the Big Island of Hawaii, is currently located over the hot spot. Another volcano, Loihi, has formed on the seafloor southeast of the Big Island of Hawaii and could eventually rise above the ocean surface to form a new island.

Hot spots and plate motion Chains of volcanoes that form over stationary hot spots provide information about plate motions. The rate and direction of plate motion can be calculated from the positions of these volcanoes. The map in **Figure 5** shows that the Hawaiian islands are at one end of the Hawaiian-Emperor seamount chain. The oldest seamount, Meiji, is at the other end of the chain and is about 80 million years old, which indicates that this hot spot has existed for at least that many years. The bend in the chain at Daikakuji Seamount records a change in the direction of the Pacific Plate movement that occurred about 43 mya.

Figure 6 Huge amounts of lava erupting from fissures accumulate on the surface, often forming layers up to 1 km thick. Over time, rivers and other geologic forces erode the layers of basalt, leaving plateaus like this one in Palouse Canyon, Washington.

Flood basalts

When hot spots occur beneath continental crust, they can lead to the formation of flood basalts. **Flood basalts** form when lava flows out of long cracks in Earth's crust. These cracks are called **fissures.** Over hundreds, or even thousands, of years, these fissure eruptions can form flat plains called plateaus, as shown in **Figure 6.** As in other quiet eruptions, when the thin lava flows across Earth's surface, water vapor and other gases escape.

Columbia River basalts The volume of basalt erupted by fissure eruptions can be tremendous. For example, the Columbia River basalts, located in the northwestern United States and shown on the map in **Figure 7,** contain 170,000 km³ of basalt. This volume of basalt could fill Lake Superior, the largest of the Great Lakes, 15 times. However, the Columbia River basalts are small in comparison to the Deccan Traps in India. This area is described on the following page.

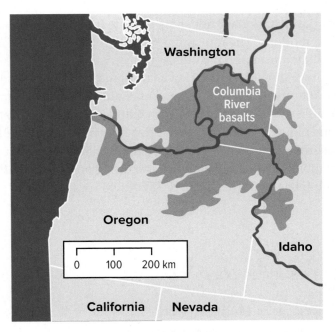

Figure 7 More than 17 mya, enormous amounts of lava poured out of large fissures, producing a basaltic plateau more than 1 km thick in the northwestern part of the United States.

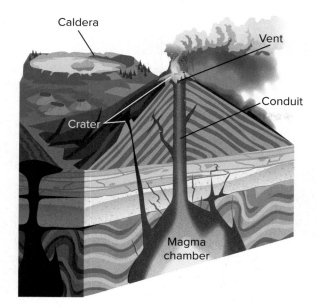

Caldera

Vent

Conduit

Crater

Magma chamber

Figure 8 Magma moves upward from deep within Earth through a conduit and erupts at Earth's surface through a vent. The area around the vent is called a crater. A caldera can form when the crust collapses into an empty magma chamber.

Deccan Traps About 65 million years ago in India, a huge flood basalt eruption created an enormous plateau called the Deccan Traps. The volume of basalt in the Deccan Traps is estimated to be about 512,000 km³. That volume would cover the island of Manhattan with a layer 10,000 km thick or the entire state of New York with a layer 4 km thick. Some geologists hypothesize that the eruption of the Deccan Traps caused a global change in climate that might have contributed to the extinction of the dinosaurs.

Anatomy of a Volcano

Recall that when magma reaches Earth's surface it is called lava. Lava reaches the surface by traveling through a tubelike structure called a **conduit,** and it emerges through an opening called a **vent.** As lava flows through the vent and out onto the surface, it cools and solidifies around the vent. Over time, layers of solidified lava can accumulate to form a mountain known as a volcano. At the top of a volcano, around the vent, is a bowl-shaped depression called a **crater.** The crater is connected to the magma chamber by the conduit. Locate the crater, conduit, and vent of the volcano shown in **Figure 8.**

Volcanic craters are usually less than 1 km in diameter. Larger depressions, called **calderas,** can be up to 100 km in diameter. Calderas often form after the magma chamber beneath a volcano empties from a major eruption. The summit or the side of a volcano collapses into the emptied magma chamber, leaving an expansive, circular depression. After the surface material collapses, water sometimes fills the caldera, forming a scenic lake. The caldera known as Crater Lake in southern Oregon formed when Mount Mazama collapsed.

STEM CAREER Connection

Volcanologist

From working to understand how and why volcanoes behave as they do to trying to predict future eruptions to keep the population safe, scientists are continually adding to the large data base of Earth's volcanoes. Volcanologists earn a bachelor's degree in geology. Most continue their education in order to specialize in a sub-area of interest.

Table 1 Types of Volcanoes

Description	Example of Volcanoes
Shield Volcanoes • Largest of the three types of volcanoes • Long, gentle slopes • Composed of layers of solidified basaltic lava • Quiet eruptions	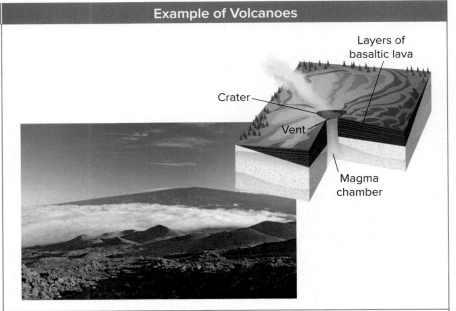
Cinder Cones • Smallest of the three types of volcanoes • Steep-sloped, cone-shaped • Usually composed of fragments of basaltic lava • Explosive eruptions • Usually form at edges of larger volcanoes	
Composite Volcanoes • Considerably larger than cinder cones • Tall, majestic mountains • Composed of layers of rock and ash from explosive eruptions and lava flows • Cycle through periods of quiet and explosive eruptions	

Types of Volcanoes

The appearance of a volcano depends on two factors: the type of material that forms the volcano and the type of eruptions that occur. Based on these two criteria, there are three major types of volcanoes, shown in **Table 1.** Each differs in size, shape, and composition.

Shield volcanoes A **shield volcano** is a mountain with broad, gently sloping sides and a nearly circular base. Shield volcanoes form when layers of thin, fluid lava accumulate during nonexplosive eruptions. They are the largest type of volcano. Mauna Loa, which is shown in **Table 1,** is a shield volcano.

Cinder cones When eruptions eject small pieces of lava into the air, **cinder cones** form as this material—often called cinders, scoria, or tephra—falls back to Earth and piles up around the vent. Cinder cones have steep sides and are generally small; most are less than 500 m high. The Lassen Volcanic Park cinder cone is 700 m high. Cinder cones commonly form on or near larger volcanoes. For instance, volcanologists have identified nearly 100 cinder cones on the flanks of Mauna Kea, a large shield volcano on Hawaii.

Composite volcanoes **Composite volcanoes** are formed of layers of ash and hardened chunks of lava from violent eruptions alternating with layers of thick, nonfluid lava that oozed downslope before solidifying. Composite volcanoes are generally cone-shaped, with concave slopes, and are much larger than cinder cones. Because of their explosive nature, they are potentially dangerous to humans and the environment. Some examples of these are Arenal Volcano in Costa Rica, shown in **Table 1,** and several in the Cascade Range of the western United States, such as Mount St. Helens.

Check Your Progress

Summary

- Volcanism includes all the processes in which magma and gases rise to Earth's surface.

- Most volcanoes on land are part of two major volcanic chains—the Circum-Pacific Belt and the Mediterranean Belt.

- The parts of a volcano include a vent, magma chamber, crater, and caldera.

- Flood basalts form when lava flows from fissures to form flat plains or plateaus.

- There are three major types of volcanoes—shield, composite, and cinder cone.

Demonstrate Understanding

1. **Explain** how the location of volcanoes is related to the theory of plate tectonics.
2. **Identify** two volcanoes in the Mediterranean Belt.
3. **Draw** a volcano and label the parts.
4. **Propose** Yellowstone National Park is an area of previous large-scale volcanism. Using a map of the United States, suggest the type(s) of tectonic processes associated with this area.

Explain Your Thinking

5. **Evaluate** the following statement: Volcanoes are present only along coastlines.
6. **Decide** whether a flood basalt is or is not a volcano.
7. **MATH** **Connection** If the Pacific Plate has moved 500 km in the last 4.7 million years, calculate its average velocity in centimeters per year. Refer to the *Skillbuilder Handbook* for more information.

LEARNSMART Go online to follow your personalized learning path to review, practice, and reinforce your understanding.

FOCUS QUESTION

Why do only some volcanoes explode?

Making Magma

What makes the eruption of one volcano quiet and the eruption of another explosively violent? A volcano's explosivity depends on the composition of the magma. As shown in **Figure 9,** lava from an eruption can be thin and runny or thick and lumpy. In order to understand why volcanic eruptions are not all the same, you first need to understand how rocks melt to make magma.

Temperature

Depending on their composition, most rocks begin to melt at temperatures between 800°C and 1200°C. Such temperatures are present in the crust and upper mantle. Recall that temperature increases with depth beneath Earth's surface.

Pressure

Pressure and the presence of water and dissolved gases also affect the formation of magma. Pressure increases with depth due to the weight of overlying rocks. As pressure increases, the temperature at which a substance melts also increases. **Figure 10** shows two melting curves for albite, a type of feldspar. Note that at Earth's surface, dry albite melts at about 1100°C, but at a depth of about 12 km, its melting point is about 1200°C. At a depth of about 100 km, the melting point of dry albite increases to 1440°C. The effect of pressure explains why most of the rocks in Earth's lower crust and upper mantle do not melt.

Figure 9 The way in which lava flows depends on the composition of the magma. The lava in the top photo is thin and runny compared to the thick and lumpy lava in the bottom photo.

(t)StockTrek/Photodisc/Getty Images, (b)moodboard/Getty Images

---

 **3D THINKING**   **DCI** Disciplinary Core Ideas   **CCC** Crosscutting Concepts   **SEP** Science & Engineering Practices

COLLECT EVIDENCE

 Use your Science Journal to record the evidence you collect as you complete the readings and activities in this lesson.

INVESTIGATE

 GO ONLINE to find these activities and more resources.

Design Your Own: Modeling a Lava Flow
Develop and use a model to visualize the factors that **affect** the geological process of lava flow.

Investigation Lab: Analyzing Volcanic-Disaster Risk
Analyze and interpret data to predict the probability of a volcanic disaster.

Composition of Magma

The composition of magma determines a volcano's explosivity, which is how it erupts and how its lava flows. What are the factors that determine the composition of magma? Scientists now know that the factors include magma's interaction with overlying crust, its temperature, pressure, amounts of dissolved gas, and—very significantly—the amount of silica a magma contains. Understanding the factors that determine the behavior of magma can aid scientists in predicting the eruptive style of volcanoes.

Dissolved gases

In general, as the amount of gases in magma increases, the magma's explosivity also increases. In the same way that gas dissolved in soda gives the soda its fizz, the gases dissolved in magma give a volcano its "bang." Important gases in magma include water vapor, carbon dioxide, sulfur dioxide, and hydrogen sulfide. Water vapor is the most common dissolved gas in magma. The presence of water vapor determines the depth and temperature at which magma forms. As shown in **Figure 10,** minerals in the mantle, such as albite, melt at high temperatures. The presence of dissolved water vapor lowers the melting temperature of minerals, causing mantle material to melt into magma. This eventually forms volcanoes and fuels their eruptions.

Viscosity

The physical property that describes a material's resistance to flow is called **viscosity.** Temperature and silica content affect the viscosity of a magma. In general, cooler magma has a higher viscosity. In other words, cool magma, much like chilled honey, tends to resist flowing.

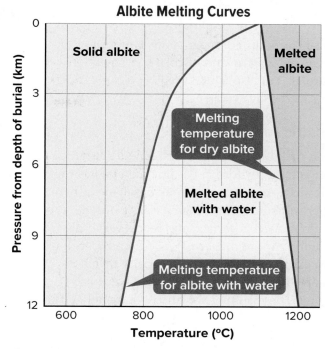

Albite Melting Curves

Solid albite

Melted albite

Melting temperature for dry albite

Melted albite with water

Melting temperature for albite with water

Pressure from depth of burial (km)

Temperature (°C)

Figure 10 Both the pressure and water content of the mineral albite affect how the mineral melts.

Locate *the melting curve of wet albite. How does the melting point of wet albite compare to that of dry albite at a depth of 3 km? At a depth of 12 km?*

 Get It?

Infer Which has a higher viscosity at room temperature—syrup or water?

Magma with high silica content tends to be thick and sticky. Because it is thick, magma with high silica content tends to trap gases, producing explosive eruptions. In general, magma with low silica content has low viscosity—it tends to be thin and runny, like warm syrup. Magma with low silica content tends to flow easily and produce quiet, nonexplosive eruptions.

ACADEMIC VOCABULARY

aid

to provide with what is useful or necessary in achieving an end

Glasses aid Omar in seeing clearly.

Types of Magma

The silica content of magma determines not only its explosivity and viscosity, as shown in **Figure 11,** but also which type of volcanic rock it forms as the lava cools.

Basaltic magma

When rock in the upper mantle melts, basaltic magma typically forms. Basaltic magma has the same silica content as the rock basalt—less than 50 percent silica. This magma rises from the upper mantle to Earth's surface and reacts very little with overlying continental crust or sediments. Its low silica content produces low-viscosity magma. Dissolved gases escape easily from basaltic magma. The resulting volcano is characterized by quiet eruptions. **Figure 12** shows how properties of magma affect the types of eruptions that occur. Volcanoes such as Kilauea and Mauna Loa produce basaltic magma. Surtsey, a volcano that was formed south of Iceland in 1963, is another volcano that produces basaltic magma.

Andesitic magma

Andesitic (an duh SIH tihk) magma has the same silica content as the rock andesite—50 to 60 percent silica. Andesitic magma forms along oceanic-continental subduction zones. Magma produced at subduction zones can change into an intermediate magma by fractional crystallization, mixing with other magma bodies, or by assimilating continental crust. The higher silica content results in a magma that has intermediate viscosity. Thus, the volcanoes it fuels are said to have intermediate explosivity. Colima Volcano in Mexico and Mt. Tambora in Indonesia are two examples of andesitic volcanoes. Both volcanoes have produced massive explosions that sent huge volumes of ash and debris into the atmosphere.

HISTORY ▶ Connection The April 1815 eruption of Mt. Tambora is considered by some volcanologists to have been the most destructive volcanic eruption in recorded history. The amount of ash, aerosols, and gases that entered the atmosphere caused the average global temperature to decrease by as much as 3°C. Crops failed, and thousands died from disease and famine. The effect on the weather was so great that the year 1816 was called the "year without a summer."

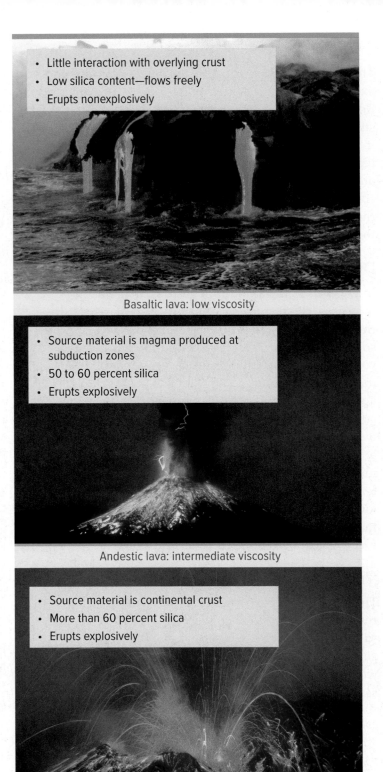

- Little interaction with overlying crust
- Low silica content—flows freely
- Erupts nonexplosively

Basaltic lava: low viscosity

- Source material is magma produced at subduction zones
- 50 to 60 percent silica
- Erupts explosively

Andestic lava: intermediate viscosity

- Source material is continental crust
- More than 60 percent silica
- Erupts explosively

Rhyolitic lava: high viscosity

Figure 11 Generally, magma and lava with a low percentage of silica have low viscosity, and those with a higher percentage of silica have high viscosity.

(t)Benny Marty/Shutterstock.com, (c)Sergio Velasco Garcia/AFP/Stringer/Getty Images, (b)Richard Roscoe/Stocktrek Images/Alamy Stock Photo

Figure 12 Visualizing Eruptions

As magma rises due to plate tectonics and hot spots, it mixes with Earth's crust. This mixing causes differences in the temperature, silica content, and gas content of magma as it reaches Earth's surface. These properties of magma determine how volcanoes erupt.

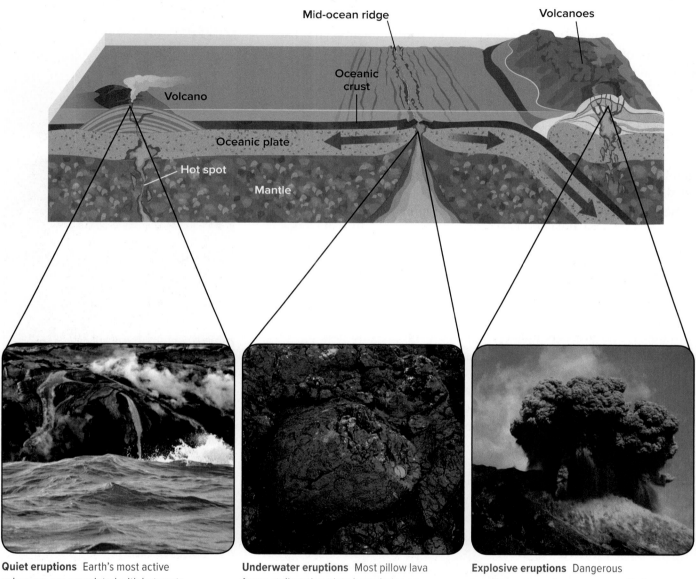

Quiet eruptions Earth's most active volcanoes are associated with hot spots under oceanic crust. Magma that upwells through oceanic crust maintains high temperature and low silica and gas contents. Lava oozes freely out of these volcanoes in eruptions that are relatively gentle.

Underwater eruptions Most pillow lava forms at diverging plate boundaries on the ocean floor. Lava oozes out of fissures in the ocean floor and forms bubble-shaped lumps as it cools.

Explosive eruptions Dangerous eruptions occur where magma melts the silica-rich rocks of the continental crust and then mixes with this material. This magma traps gases, causing tremendous pressure to build. The release of pressure drives violent eruptions.

Rhyolitic magma

When molten material rises and mixes with water and the silica-rich overlying continental crust, it forms rhyolitic (ri uh LIH tihk) magma. Rhyolitic magma has the same composition as the rock granite—more than 60 percent silica. The high viscosity of rhyolitic magma slows down its movement. High viscosity, along with the large volume of gas trapped within this magma, makes the volcanoes fueled by rhyolitic magma very explosive. The dormant volcanoes in Yellowstone National Park in the western United States were fueled by rhyolitic magma. The most recent of these eruptions, which occurred about 640,000 years ago, was so powerful that it released 1000 km³ of volcanic material into the air.

Explosive Eruptions

When lava is too viscous to flow freely from the vent, pressure builds up in the lava until the volcano explodes, throwing lava and rock into the air. The erupted materials are called **tephra.** Tephra can be pieces of lava that solidified during the eruption or pieces of the crust carried by the magma before the eruption. Tephra are classified by size. The smallest fragments, with diameters less than 2 mm, are called ash, as shown in **Figure 13.** The largest tephra thrown from a volcano are called blocks. The one shown in **Figure 13** is more than 5 m high. Large, explosive eruptions can disperse tephra over much of the planet. Ash can rise 40 km into the atmosphere during explosive eruptions and pose a threat to aircraft; it can even change the weather. The 1991 eruption of Mount Pinatubo in the Philippines, shown in **Figure 14,** sent up a plume of ash 24 km high. Tiny sulfuric acid droplets and particles remained in the stratosphere for about two years, blocking the Sun's rays and lowering global temperatures by about 0.5°C.

Ash

Block

Figure 13 Ash (top) (shown actual size) is the smallest type of tephra. This ash came from Mount St. Helens in Washington State. The volcanic block (bottom) came from the Montaña Colorada volcano in Lanzarote, Canary Islands.

Compare *the two types of tephra. What do they have in common?*

Figure 14 In 1991, the eruption of Mount Pinatubo in the Philippines sent so much ash into the stratosphere that it lowered global temperatures for two years.

Pyroclastic flow

1902 Eruption of Mount Pelée

Figure 15 This pyroclastic flow (left), from the Soufriere Hills volcano on the Caribbean island of Montserrat, has made more than half of the island unlivable. A pyroclastic flow from Mount Pelée in 1902 (right) was so powerful that it destroyed the entire town of St. Pierre in only a few minutes.

Pyroclastic Flows

Some tephra can cause tremendous damage and are deadly. Rapidly moving clouds of tephra mixed with hot, suffocating gases are called **pyroclastic flows.** Such flows can move down a slope at speeds of more than 80 km/h, as shown in **Figure 15.** They can have internal temperatures of more than 700°C. The photo on the right in **Figure 15** shows the result of another widely known and deadly pyroclastic flow that occurred on the island of Martinique in the Caribbean Sea. More than 29,000 people suffocated or were burned to death.

✏️ Check Your Progress

Summary

- There are three major types of magma—basaltic, andesitic, and rhyolitic.

- Because of their relative silica contents, basaltic magma is the least explosive magma, and rhyolitic magma is the most explosive.

- Temperature, pressure, and the presence of water are factors that affect the formation of magma.

- Rock fragments ejected during eruptions are called tephra.

Demonstrate Understanding

1. **Discuss** how the composition of magma determines an eruption's characteristics.

2. **Restate** how the viscosity of magma is related to its explosivity.

3. **Predict** the explosivity of a volcano having magma with high silica content and high gas content.

4. **Differentiate** between sizes of tephra.

Explain Your Thinking

5. **Compare and contrast** the tectonic processes that made Kilauea and Mount Etna.

6. **Infer** the composition of magma that fueled the 1991 eruption of Mount Pinatubo.

7. WRITING ▶ Connection Write a news report covering the 1902 eruption of Mount Pelée.

LEARNSMART Go online to follow your personalized learning path to review, practice, and reinforce your understanding.

INTRUSIVE ACTIVITY

FOCUS QUESTION

What happens to the magma that doesn't make it to the surface?

Plutons

Most of Earth's volcanism happens below the surface because not all magma emerges at the surface. Before it gets to the surface, rising magma can interact with the crust in several ways, as illustrated in **Figure 16.** Magma can force the overlying rock apart and enter the newly formed fissures. Magma can also cause blocks of rock to break off and sink into the magma, where the rocks eventually melt and become part of the magma. Finally, magma can melt its way through the rock into which it intrudes. What happens deep in Earth as magma slowly cools? Recall that when magma cools, minerals begin to crystallize.

Over a long period of time, minerals in the magma solidify, forming intrusive igneous rock bodies. Some of these rock bodies are ribbonlike features only a few centimeters thick and several hundred meters long. Others are massive, and range in volume from about 1 km³ to hundreds of cubic kilometers. These intrusive igneous rock bodies, called **plutons** (PLOO tahns), can be exposed at Earth's surface as a result of uplift and erosion and are classified based on their size, shape, and relationship to surrounding rocks.

Batholiths and stocks

The largest plutons are called batholiths. **Batholiths** (BATH uh lihths) are irregularly shaped masses of coarse-grained igneous rocks that cover at least 100 km² and take millions of years to form. Batholiths are common in the interior of major mountain chains. Many batholiths in North America are composed primarily of granite—the most common rock type in plutons. However, gabbro and diorite, the intrusive equivalents of basalt and andesite, are also present in batholiths.

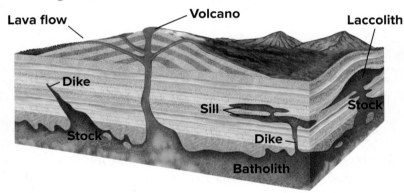

Figure 16 Magma moving upward solidifies and forms bodies of rock both at the surface and deep within Earth.

 **3D THINKING**   **DCI** Disciplinary Core Ideas   **CCC** Crosscutting Concepts   **SEP** Science & Engineering Practices

COLLECT EVIDENCE

Use your Science Journal to record the evidence you collect as you complete the readings and activities in this lesson.

INVESTIGATE

GO ONLINE to find these activities and more resources.

? **Revisit the Encounter the Phenomenon Question**
What information from this lesson can help you answer the Unit and Module questions?

CCC **Identify Crosscutting Concepts**
Create a table of the crosscutting concepts and fill in examples you find as you read.

Figure 17 shows a small portion of the largest batholith in North America, the Coast Range Batholith in British Columbia. Irregularly shaped plutons that are similar to batholiths but are smaller in size are called **stocks.** Both batholiths and stocks, as shown in **Figure 16,** cut across older rocks and generally form 5 to 30 km beneath Earth's surface.

Laccoliths

Sometimes when magma intrudes into parallel rock layers close to Earth's surface, some of the rocks bow upward as a result of the intense pressure from the magma body below. When the magma solidifies, a laccolith forms, as shown in **Figure 16.** A **laccolith** (LA kuh lihth) is a lens-shaped pluton with a round top and flat bottom. Compared to batholiths and stocks, laccoliths are relatively small; at most, they are 16 km wide. **Figure 17** shows the laccolith at Bear (Ayu-Dag) mountain near Gurzuf resort on the Crimean peninsula. North American laccoliths are present in the Black Hills of South Dakota and the Judith Mountains of Montana, among other places.

Sills

A **sill** forms when magma intrudes parallel to layers of rock, as shown in **Figure 16.** A sill can range in thickness from only a few centimeters to hundreds of meters. **Figure 17** shows the Palisades Sill, which is exposed in the cliffs above the Hudson River as it flows through New York and New Jersey. The rock that was originally above the sill has eroded. Sills affect the sedimentary rocks into which they intrude. One effect is to lift the rock above it. Because it takes great amounts of force to lift entire layers of rock, most sills form relatively close to the surface. Another effect of sills is to metamorphose the surrounding rocks.

Get It?

Infer Suppose there were still layers of rock above the Palisades Sill. How could you tell whether this was a flood basalt or a sill?

The Coast Range Batholith in British Columbia formed 5 to 30 km below Earth's surface. It is more than 1500 km long.

Laccoliths push Earth's surface up, creating a rounded top and flat bottom.

The Palisades Sill in New Jersey formed more than 200 mya. It is about 300 m thick.

Figure 17 Batholiths, laccoliths, and sills form when magma intrudes into the crust and solidifies.

Dike

Volcanic neck

Figure 18 Unlike sills, dikes cut across the rock into which they intrude. Sometimes dikes extend from the conduit of a volcano.

Infer *the size of the volcano that once surrounded this volcanic neck in New Mexico.*

Dikes

Unlike a sill, a **dike** is a pluton that cuts across preexisting rocks. Dikes often form when magma invades cracks in surrounding rock bodies. Dikes range in size from a few centimeters to several meters wide and can be tens of kilometers long. The Great Dike in Zimbabwe, Africa, is an exception—it is about 8 km wide and 500 km long.

A volcanic neck forms when the magma in a volcano conduit solidifies. When the layers of rock that make up the volcano erode, the more erosion-resistant conduit and dikes are left standing. Ship Rock in New Mexico, shown in **Figure 18,** formed in this way. Dikes are often associated with the conduit but do not always form the neck. Ship Rock has dikes extending from the neck.

Textures Recall that grain size is related to the rate of cooling. The textures of most sills and dikes are coarse-grained. This suggests that they formed deep in Earth's crust, where magma cooled slowly enough to allow large mineral crystals to develop. The fine-grained texture of dikes and sills implies that they formed both closer to the surface and more quickly. Both grain-size scenarios are shown in the intrusions in **Figure 19.**

 Get It?

Infer Suppose you find a dike that has large crystals along its center and smaller crystals along its edges. How can you explain this difference in crystal size?

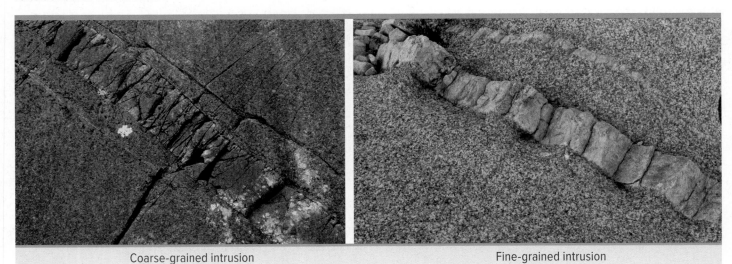
Coarse-grained intrusion   Fine-grained intrusion

Figure 19 Plutons forming deep in Earth cool slowly, producing large crystals and coarse-grained rocks. Intrusive rocks that form closer to Earth's surface cool more quickly and produce small crystals and fine-grained rocks.

Plutons and Tectonics

Many plutons form as the result of mountain-building processes. In fact, batholiths make up the cores of many of Earth's mountain ranges. From where did the enormous volume of magma that formed batholiths come? The processes that form batholiths are complex.

Recall that many major mountain chains formed along continental-continental convergent plate boundaries. Scientists think that some of these collisions might have forced continental crust down into the upper mantle, where it melted, intruded into the overlying rocks, and eventually cooled to form batholiths.

Plutons are also thought to form as a result of oceanic plate convergence and subsequent subduction. The cooler, denser oceanic plate subducts beneath the other plate and is heated, releasing water. The water expelled from the oceanic plate lowers the melting point of the surrounding mantle rocks, causing them to melt. Plutons often form when the melted material rises but does not erupt at the surface.

The Sierra Nevada Batholith formed from at least five episodes of this type of igneous activity beneath what is now California. The famous granite cliffs in Yosemite National Park, one of which is shown in **Figure 20,** are part of this vast batholith. Although they were once far below Earth's surface, uplift and erosion have brought them to their present positions. You might recognize from your study of glaciers the large U-shaped valley on the left side of Half Dome. Glaciers carved away the other half of Half Dome as they eroded their way through the valley.

Figure 20 The granite cliffs that tower over Yosemite National Park in California are part of the Sierra Nevada Batholith. The batholith on the right is named Half Dome.

Check Your Progress

Summary

- Intrusive igneous rocks are classified according to their size, shape, and relationship to the surrounding rocks.
- Most of Earth's volcanism happens below Earth's surface.
- Magma can intrude into rock in different ways, taking different forms when it cools.
- Batholiths form the core of many mountain ranges.

Demonstrate Understanding

1. **Compare and contrast** volcanic eruptions at Earth's surface with intrusive volcanic activity.
2. **Describe** the different types of plutons.
3. **Relate** the size of plutons to the locations where they form.
4. **Identify** processes that expose plutons at Earth's surface.

Explain Your Thinking

5. **Predict** why textures in the same sill might vary, with finer grains along the margin and coarser grains toward the middle.
6. **Infer** what type of pluton might be found at the base of an extinct volcano.
7. **WRITING > Connection** Write a defense or rebuttal for this statement: Sills form only deep beneath Earth's surface.

LEARNSMART Go online to follow your personalized learning path to review, practice, and reinforce your understanding.

Chris Clor/Blend Images

Some Like It Hot

A volcano erupts, and the ground trembles. Streams of lava flow like fiery rivers down the volcano's flanks. This blazing-hot landscape would send most people running for cover, but for volcanologists, an eruption is merely the signal that it is time to go to work.

Myths versus reality

In the popular imagination, volcanologists spend all their time gathering samples of lava from the summits of erupting volcanoes. Many volcanologists do travel and work in the field extensively, but others spend much of their time in offices and labs, analyzing data. Most do both. In addition to studying active volcanoes, volcanologists study extinct and dormant volcanoes. Most volcanologists work for colleges or universities, government agencies, volcano observatories, or international research organizations.

Fields of study

Volcanologists study volcanoes to understand how and why they erupt, what effects they have on people and on Earth as a whole, and the best ways to predict them. Volcanologists usually work in one of four fields. Physical volcanologists are scientists who use a variety of tools and technologies to monitor volcanic activity. They gather samples of volcanic rocks and other materials. They study all the data they have collected to learn about a

A volcanologist makes observations of an eruption at Mount Yasur on Tanna Island, Vanuatu, in the south Pacific Ocean.

volcano's past eruptions in order to better predict future ones.

Geophysicists study the earthquakes that often precede volcanic eruptions. The earthquakes, which are caused by magma moving under the crust, are usually fairly small tremors. They help scientists predict future volcanic eruptions.

Geodesic volcanologists use cutting-edge technologies to study how the underground movement of magma changes the shape of Earth, a process called ground deformation. Geochemists study the chemical composition of materials—such as lava, rocks, and gases—produced by volcanoes.

Other scientists use satellites with thermal sensors to detect changes in temperature on the surface of Earth's crust, which could point to possible volcanic activity.

COMMUNICATE SCIENTIFIC IDEAS

Use print or online sources to research one field in which volcanologists work. Create a brochure that explains the career to students who might be interested in pursuing it.

MODULE 14
STUDY GUIDE

 GO ONLINE to study with your Science Notebook.

Lesson 1 **VOLCANOES**

- Volcanism includes all the processes in which magma and gases rise to Earth's surface.
- Most volcanoes on land are part of two major volcanic chains—the Circum-Pacific Belt and the Mediterranean Belt.
- The parts of a volcano include a vent, magma chamber, crater, and caldera.
- Flood basalts form when lava flows from fissures to form flat plains or plateaus.
- There are three major types of volcanoes—shield, composite, and cinder cone.

- volcanism
- hot spot
- flood basalt
- fissure
- conduit
- vent
- crater
- caldera
- shield volcano
- cinder cone
- composite volcano

Lesson 2 **ERUPTIONS**

- There are three major types of magma—basaltic, andesitic, and rhyolitic.
- Because of their relative silica contents, basaltic magma is the least explosive magma, and rhyolitic magma is the most explosive.
- Temperature, pressure, and the presence of water are factors that affect the formation of magma.
- Rock fragments ejected during eruptions are called tephra.

- viscosity
- tephra
- pyroclastic flow

Lesson 3 **INTRUSIVE ACTIVITY**

- Intrusive igneous rocks are classified according to their size, shape, and relationship to the surrounding rocks.
- Most of Earth's volcanism happens below Earth's surface.
- Magma can intrude into rock in different ways, taking different forms when it cools.
- Batholiths form the core of many mountain ranges.

- pluton
- batholith
- stock
- laccolith
- sill
- dike

REVISIT THE PHENOMENON

Where does lava come from?

CER Claim, Evidence, Reasoning

Explain your Reasoning Revisit the claim you made when you encountered the phenomenon. Summarize the evidence you gathered from your investigations and research and finalize your Summary Table. Does your evidence support your claim? If not, revise your claim. Explain why your evidence supports your claim.

STEM UNIT PROJECT
Now that you've completed the module, revisit your STEM unit project. You will summarize your evidence and apply it to the project.

GO FURTHER

SEP Data Analysis Lab

How do zones of volcanism relate to lava production?

Researchers classify types of volcanic eruptions and study how much lava each type of volcano emits during an average year.

Data and Observations The circle graphs show data from 5337 eruptions and annual lava production for each zone.

Number of Eruptions in Average Year

Lava Production

Convergent

Hot spot

Rift

CER Analyze and Interpret Data

1. **Claim, Evidence** Describe the relationship between the type of volcanism and annual lava production.
2. **Reasoning** Consider why it is important for scientists to study this relationship.
3. **Reasoning** Evaluate what could be the next step in the researchers' investigation.

Haje Jan Kamps/EyeEm/Getty Images

ENCOUNTER THE PHENOMENON

How can we prevent people from getting hurt during an earthquake?

▶ **GO ONLINE** to play a video about how scientists can give an early warning before an earthquake strikes.

SEP Ask Questions

Do you have other questions about the phenomenon? If so, add them to the driving question board.

CER Claim, Evidence, Reasoning

Make Your Claim Use your CER chart to make a claim about how we can prevent people from getting hurt during an earthquake. Explain your reasoning.

Collect Evidence Use the lessons in this module to collect evidence to support your claim. Record your evidence as you move through the module.

Explain Your Reasoning You will revisit your claim and explain your reasoning at the end of the module.

▶ **GO ONLINE** to access your CER chart and explore resources that can help you collect evidence.

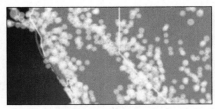

LESSON 3: Explore & Explain: Earthquake Magnitude and Intensity

LESSON 4: Explore & Explain: Earthquake Forecasting

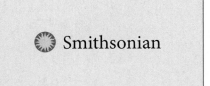

Additional Resources

FORCES WITHIN EARTH

FOCUS QUESTION
What causes earthquakes?

Stress and Strain

Most earthquakes are the result of movement of Earth's crust produced by plate tectonics. As a whole, tectonic plates tend to move very slowly. Along the boundaries between two plates, rocks in the crust often resist movement. Over time, stress builds up. **Stress** is the total force acting on crustal rocks per unit of area. When stress overcomes the strength of the rocks involved, movement occurs along fractures in the rocks. The vibrations caused by this sudden movement are felt as an earthquake. The characteristics of earthquakes are determined by the orientation and magnitude of stress applied to rocks and by the strength of the rocks involved.

There are three kinds of stress that act on Earth's rocks: compression, tension, and shear. Compression is stress that decreases the volume of a material, tension is stress that pulls a material apart, and shear is stress that causes a material to twist. The deformation of materials in response to stress is called **strain. Figure 1** illustrates the strain caused by compression, tension, and shear.

Even though rocks can be twisted, squeezed, and stretched, they fracture when stress and strain reach a critical point. At the critical point, rock can move, releasing the energy built up as a result of stress. Earthquakes are the result of this movement and release of energy. For example, the 2010 earthquake in Haiti was caused by a release of built-up compression stress. When that energy was released as an earthquake, more than 230,000 people were killed, and even more were made homeless.

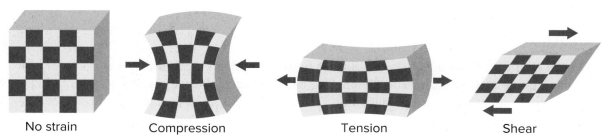

No strain   Compression   Tension   Shear

Figure 1 Compression causes a material to shorten. Tension causes a material to lengthen. Shear causes distortion of a material.

3D THINKING   **DCI** Disciplinary Core Ideas   **CCC** Crosscutting Concepts   **SEP** Science & Engineering Practices

COLLECT EVIDENCE
Use your Science Journal to record the evidence you collect as you complete the readings and activities in this lesson.

INVESTIGATE
GO ONLINE to find these activities and more resources.

? **Revisit the Encounter the Phenomenon Question**
What information from this lesson can help you answer the Unit and Module questions?

CCC **Identify Crosscutting Concepts**
Create a table of the crosscutting concepts and fill in examples you find as you read.

Laboratory experiments on rock samples show a distinct relationship between stress and strain. When the stress applied to a rock is plotted against strain, a stress-strain curve, like the one shown in **Figure 2,** is produced. A stress-strain curve usually has two segments—a straight segment and a curved segment. Each segment represents a different type of response to stress.

Elastic deformation

The first segment of a stress-strain curve shows what happens under conditions in which stress is low. Under low stress, a material shows elastic deformation. **Elastic deformation** is caused when a material is compressed, bent, or stretched. This is the same type of deformation that happens from gently pulling on the ends of a rubber band. When the stress on the rubber band is released, it returns to its original size and shape. **Figure 2** illustrates that elastic deformation is the result of stress and strain. If the stress is reduced to zero, as the graph shows, the deformation of the rock disappears.

Plastic deformation

When stress builds up past a certain point, called the elastic limit, rocks undergo **plastic deformation,** shown by the second segment of the graph in **Figure 2.** Unlike elastic deformation, this type of strain produces permanent deformation, which means that the material stays deformed even when stress is reduced to zero. Even a rubber band undergoes plastic deformation when it is stretched beyond its elastic limit. At first the rubber band stretches, then it tears slightly, and, finally, it snaps apart into two pieces. The tear in the rubber band is an example of permanent deformation. When stress increases to be greater than the strength of a rock, the rock ruptures. The point of rupture, called failure, is designated by the "X" on the graph in **Figure 2.**

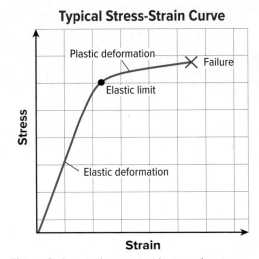

Typical Stress-Strain Curve

Figure 2 A typical stress-strain curve has two parts. Elastic deformation occurs as a result of low stress. Plastic deformation occurs under high stress. When plastic deformation is exceeded, an earthquake occurs.

Describe *what happens to a material at the point on the graph at which elastic deformation changes into plastic deformation.*

 Get It?

Differentiate between elastic deformation and plastic deformation.

Most materials exhibit both elastic and plastic behavior, although to different degrees. Brittle materials, such as dry wood, glass, and certain plastics, fail before much plastic deformation occurs. Other materials, such as metals, rubber, and silicon putty, can undergo a great deal of deformation before failure occurs, or they might not fail at all. Temperature and pressure also influence deformation. As pressure increases, rocks require greater stress to reach the elastic limit. At high enough temperatures, solid rock can also deform, causing it to flow in a fluidlike manner. This flow reduces stress.

SCIENCE USAGE v. COMMON USAGE

failure

Science usage: a collapsing, fracturing, or giving way under stress

Common usage: lack of satisfactory performance or effect

Faults

Crustal rocks fail when stresses exceed the strength of the rocks. The resulting movement occurs along a weak region in the crustal rock called a fault. A **fault** is any fracture or system of fractures along which Earth moves. **Figure 3** shows a fault. The surface along which the movement takes places is called the fault plane. The orientation of the fault plane can vary from nearly horizontal to almost vertical. The movement along a fault results in earthquakes. Several historic earthquakes are described in the time line in **Figure 4.**

Reverse and normal faults

Reverse faults form as a result of horizontal and vertical compression that squeezes rock and creates a shortening of the crust. This causes rock on one side of a reverse fault to be pushed up relative to the other side, as shown in **Table 1.** Reverse faulting can be seen near convergent plate boundaries.

Movement along a normal fault is partly horizontal and partly vertical. The horizontal movement pulls rock apart and stretches the crust. Vertical movement occurs as the stretching causes rock on one side of the fault to move down relative to the other side.

Strike-slip faults

Strike-slip faults are caused by horizontal shear. As shown in **Table 1**, the movement at a strike-slip fault is mainly horizontal and in opposite directions, similar to the way cars move in opposite directions on either side of a freeway. The San Andreas Fault, which runs through California, is a strike-slip fault.

Figure 3 A major fault passes through this section of land.

Identify *the direction of movement that occurred along this fault.*

Table 1 Types of Faults

Type of Fault	Type of Movement	Example
Reverse	Compression causes vertical movement upward along a fault plane.	
Normal	Tension causes vertical movement downward along a fault plane.	
Strike-slip	Shear causes horizontal movement along a fault plane.	

Robert E. Wallace/USGS

Figure 4

Major Earthquakes and Advances in R & D

As earthquakes cause casualties and damage around the world, scientists work to find better ways to warn and protect people.

Summarize *how earthquakes have affected human populations throughout history.*

1 **1811 - 1812** Several strong earthquakes occur along the Mississippi River Valley over three months, destroying the entire town of New Madrid, Missouri.

2 **1880** Following an earthquake in Japan, scientists invent the first modern seismograph to record the intensity of earthquakes.

3 **1906** An earthquake in San Francisco kills between 3000 and 5000 people and causes a fire that rages for three days, destroying most of the city.

4 **1923** Approximately 140,000 people die in an earthquake and subsequent fires that destroy the homes of over a million people in Tokyo and Yokohama, Japan.

5 **1960** In Chile, a 9.5 earthquake generates tsunamis that hit Hawaii, Japan, New Zealand, and Samoa. This is the largest earthquake recorded.

6 **1965** The United States, Japan, Chile, and Russia form the International Pacific Tsunami Warning System.

7 **1982** New Zealand constructs the first building with seismic isolation, using lead-rubber bearings to prevent the building from swaying during an earthquake.

8 **2004** A 9.0 earthquake in the Indian Ocean triggers the deadliest tsunami in history. The tsunami travels as far as the East African Coast.

9 **2011** The eastern coast of Japan is struck by a magnitude-9 earthquake and subsequent tsunami, which breaks off icebergs in Antarctica, 13,600 km away.

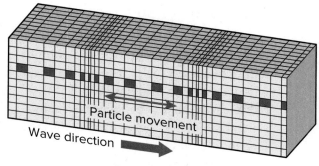

P-wave movement

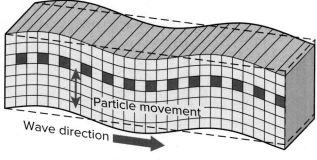

S-wave movement

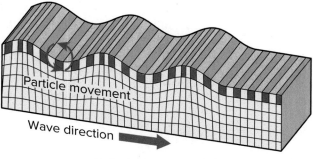

Surface wave movement

Figure 5 Seismic waves are characterized by the types of movement they cause. Rock particles move back and forth as a P-wave passes. Rock particles move at right angles to the direction of the S-wave. A surface wave causes rock particles to move both up and down and from side to side.

Earthquake Waves

Most earthquakes are caused by movements along faults. As stress continues to build in these rocks, they undergo elastic deformation. Beyond the elastic limit, they are permanently deformed. At some point after that, the rocks slip or crumble, and an earthquake occurs.

Types of seismic waves

Seismic waves are ground vibrations that are produced during an earthquake. Every earthquake generates three types of seismic waves: primary, secondary, and surface waves.

Primary waves Also referred to as P-waves, **primary waves** squeeze and push rocks in the direction along which the waves are traveling, as shown in **Figure 5.** Note how a volume of rock, which is represented by small red squares, changes length as a P-wave passes through it. The compressional movement of P-waves is similar to the movement along a loosely coiled wire. If the coil is tugged and released quickly, the vibration passes through the length of the coil parallel to the direction of the initial tug.

Secondary waves **Secondary waves,** called S-waves, are named with respect to their arrival times. They are slower than P-waves, so they are the second set of waves to be felt. S-waves have a motion that causes rocks to move perpendicular to the direction of the waves, as illustrated in **Figure 5.** The movement of S-waves is similar to the movement of a jump rope that is jerked up and down or side to side at one end. The waves travel from one end of the rope to the other end, as the rope itself moves up and down or side to side. Both P-waves and S-waves pass through Earth's interior. Collectively, P-waves and S-waves are called body waves because they travel through the interior of Earth's body.

Surface waves

The third and slowest type of waves are surface waves, which travel only along Earth's surface. Surface waves cause the ground to move sideways and up and down, like ocean waves, as shown in **Figure 5.** Surface waves usually cause the most destruction because they cause the most movement of the ground and take the longest time to pass.

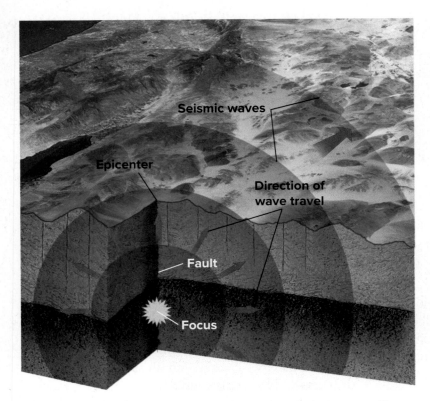

Seismic waves

Epicenter

Direction of
wave travel

Fault

Focus

Figure 6 The focus of an earthquake is the point of initial fault rupture. The surface point directly above the focus is the epicenter.

Infer *the point at which surface waves will have the potential to cause the most damage.*

Generation of seismic waves

The first body waves generated by an earthquake spread out from the point of failure of crustal rocks. The point of failure where the waves originate is called the **focus** of the earthquake. The focus is usually several kilometers below Earth's surface. Depending on the depth of the focus, an earthquake is classified as either a shallow-focus, mid-focus, or deep-focus earthquake. The point on Earth's surface directly above the focus is the **epicenter** (EH pih sen tur), shown in **Figure 6.** Surface waves originate from the epicenter and spread out.

 Get It?

Explain the relationship between the focus and the epicenter.

 Check Your Progress

Summary

- Stress is force per unit of area that acts on a material, and strain is the deformation of a material in response to stress.

- Reverse, normal, and strike-slip are the major types of faults.

- The three types of seismic waves are P-waves, S-waves, and surface waves.

Demonstrate Understanding

1. **Describe** how the formation of a fault can result in an earthquake.

2. **Explain** why a stress-strain curve usually has two segments.

3. **Compare and contrast** the movement produced by each of the three types of faults.

4. **Draw** three diagrams to show how each type of seismic wave moves through rock. How do they differ?

Explain Your Thinking

5. **Relate** the movement produced by seismic waves to the observations a person would make of the waves as they traveled across Earth's surface or through Earth's body.

6. **WRITING ▶ Connection** Relate the movement of seismic waves to the movement of something you might see every day. Make a list and share it with your classmates.

LEARNSMART Go online to follow your personalized learning path to review, practice, and reinforce your understanding.

SEISMIC WAVES AND EARTH'S INTERIOR

FOCUS QUESTION

What tools do we have to study earthquakes?

Seismometers and Seismograms

Most of the vibrations caused by seismic waves cannot be felt at great distances from an earthquake's epicenter; however, they can be detected by sensitive instruments called **seismometers** (size MAH muh turz). Some seismometers consist of a rotating drum covered with a sheet of paper, a pen or other such recording tool, and a mass, such as a pendulum. Seismometers vary in design, but all include a frame that is anchored to the ground and a mass that is suspended from a spring or wire, as shown in **Figure 7.** During an earthquake, the mass and the pen attached to it tend to stay at rest due to inertia, while the ground beneath shakes. The motion of the mass in relation to the frame is then registered on the paper with the recording tool or is directly recorded by a computer. The record produced by a seismometer is called a **seismogram** (SIZE muh gram). A portion of one is shown in **Figure 8,** on the next page.

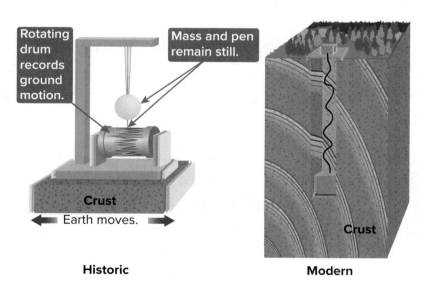

Historic

Modern

Figure 7 The frame of a historic seismometer is anchored to the ground. When an earthquake occurs, the frame moves, but the hanging mass and attached pen do not. The mass and pen record the relative movement as the rotating drum moves under them. Compare this to the modern sensor and transmitter on the right.

3D THINKING   **DCI** Disciplinary Core Ideas   **CCC** Crosscutting Concepts   **SEP** Science & Engineering Practices

COLLECT EVIDENCE

 Use your Science Journal to record the evidence you collect as you complete the readings and activities in this lesson.

INVESTIGATE

 GO ONLINE to find these activities and more resources.

 Virtual Investigation: Earthquake
Obtain, evaluate, and communicate information resulting from an earthquake.

 Review the News
Obtain information from a current news story about earthquakes. Evaluate your source and communicate your findings to your class.

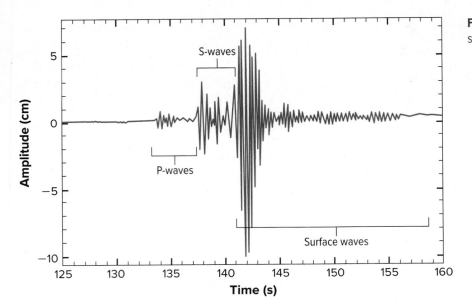

Figure 8 Seismograms provide a record of the seismic waves produced by an earthquake.

Travel-time curves

Seismic waves that travel from the focus of an earthquake are recorded by seismometers housed in seismic stations around the world. Over many years, the arrival times of seismic waves from countless earthquakes have been collected. Using these data, seismologists—earthquake scientists—have been able to construct global travel-time curves for the arrival of P-waves and S-waves of earthquakes, as shown in **Figure 9.** These curves provide the average travel times of all P- and S-waves from wherever an earthquake occurs on Earth.

 Get It?

Summarize how seismograms are used to construct global travel-time curves.

Distance from the epicenter

Note that in **Figure 9,** as in **Figure 8,** the P-waves arrive first, then the S-waves. The surface waves arrive last. With increasing travel distance from the epicenter, the time separation between the curves for the P-waves and S-waves increases. This means that waves recorded on seismograms from more distant stations are farther apart than waves recorded on seismograms at stations closer to the epicenter. This separation of seismic waves on seismograms can be used to determine the distance from the epicenter of an earthquake to the seismic station that recorded the seismogram. This method of precisely locating an earthquake's epicenter will be discussed in Lesson 3.

 Get It?

Describe how scientists use seismograms to determine the distance of a seismic station form the epicenter.

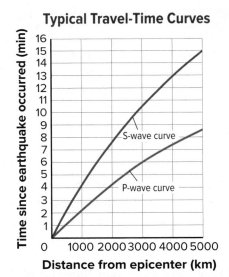

Typical Travel-Time Curves

(graph: x-axis "Distance from epicenter (km)" with gridlines at 1000 2000 3000 4000 5000; y-axis "Time since earthquake occurred (min)" from 0 to 16; labeled "S-wave curve" and "P-wave curve")

Figure 9 Travel-time curves show how long it takes for P-waves and S-waves to reach seismic stations located at different distances from an earthquake's epicenter.

Determine *how long it takes P-waves to travel to a seismogram 2000 km away. How long does it take for S-waves to travel the same distance?*

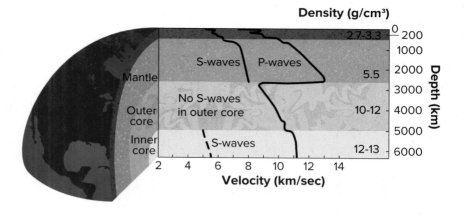

Figure 10 Earth's layers are each composed of different materials. By examining the behavior of seismic waves moving through different kinds of rock, scientists have determined the composition of layers all the way to Earth's inner core.

Clues to Earth's Interior

The seismic waves that shake the ground during an earthquake also travel through Earth's interior. This provides information that has enabled scientists to construct models of Earth's internal structure. Therefore, even though seismic waves can wreak havoc on the surface, they are invaluable for their contribution to scientists' understanding of Earth's interior.

Earth's internal structure

Seismic waves change speed and direction at the boundaries between different materials. Note in **Figure 10** that as P-waves and S-waves initially travel through the mantle, they follow fairly direct paths. When P-waves strike the core, they are refracted, which means they bend. Seismic waves also reflect off of major boundaries inside Earth. By recording the travel-time curves and path of each wave, seismologists learn about differences in density and composition within Earth.

What happens to the S-waves generated by an earthquake? To answer this question, seismologists first determined that the right-angle motion of S-waves will not travel through liquid. Then, seismologists noticed that S-waves do not travel through Earth's center. This observation led to the discovery that Earth's core must be at least partly liquid. The data collected for the paths and travel times of the waves inside Earth led to the current understanding that Earth's core has an outer region that is liquid and an inner region that is solid.

Earth's composition

Figure 11 shows that seismic waves change their paths as they encounter boundaries between zones of different materials. They also change their speed. By comparing the speed of seismic waves with measurements made on different rock types, scientists have determined the thickness and composition of Earth's different regions. As a result, scientists have determined that the upper mantle is peridotite, which is made mostly of the mineral olivine. The outer core is mostly liquid iron and nickel. The inner core is mostly solid iron and nickel.

ACADEMIC VOCABULARY

encounter

to come upon or experience, especially unexpectedly
We had never encountered such a violent storm.

Figure 11 Visualizing Seismic Waves

The travel times and behavior of seismic waves provide a detailed picture of Earth's internal structure. These waves also provide clues about the composition of Earth's layers.

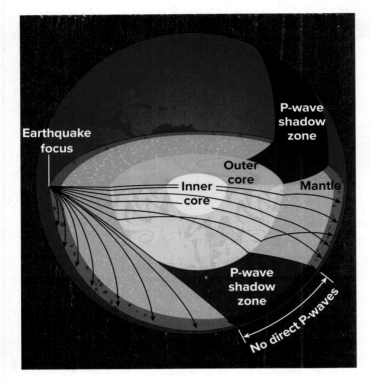

P-waves in the outer core are refracted. This generates a P-wave shadow zone on Earth's surface where no direct P-waves appear on seismograms. Other P-waves are reflected and refracted by the inner core. These can be detected by seismometers on the other side of the shadow zone.

S-waves cannot travel through the liquid outer core and thus do not reappear beyond the S-wave shadow zone.

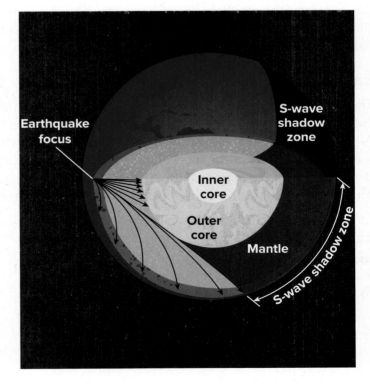

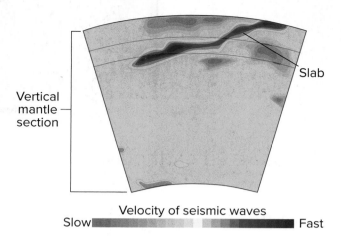

Figure 12 Images like this one from Japan are generated by capturing the path of seismic waves through Earth's interior. Areas of red indicate seismic waves that are traveling more slowly than average and areas of blue indicate seismic waves that are traveling faster than average. The blue area is a subducted plate.

Vertical mantle section

Slab

Velocity of seismic waves
Slow ▬▬▬▬▬▬▬ ▬▬▬▬▬▬ Fast

Imaging Earth's interior

Seismic wave speed and Earth's density vary with factors other than depth. Recall that the leading edge of cold, dense, subducting plates—called slabs—sink back into Earth's mantle at subduction zones. Also recall that mantle plumes are regions where hot mantle material is rising. Because the speed of seismic waves depends on temperature and composition, it is possible to use seismic waves to create images of structures such as slabs and plumes. In general, the speed of seismic waves decreases as temperature increases. Thus, waves travel more slowly in hotter areas and more quickly in cooler regions. Using measurements made at seismometers around the world and waves recorded from many thousands of earthquakes, Earth's internal structure can be constructed and visualized, and features such as slabs can be located in images like the one in **Figure 12.** These images are similar to computed tomography (CT) scans, except that the images are made using seismic waves instead of X rays.

Check Your Progress

Summary

- Seismometers are devices that record seismic wave activity on a seismogram.

- Travel times for P-waves and S-waves enable scientists to pinpoint the epicenters of earthquakes.

- P-waves and S-waves change speed and direction when they encounter different materials.

- Analysis of seismic waves provides a detailed picture of the composition of Earth's interior.

Demonstrate Understanding

1. **Explain** how P-waves and S-waves are used to determine the properties of Earth's core.

2. **Draw** a diagram of a seismometer, showing how the movement of Earth is translated into a seismogram.

3. **Describe** how seismic travel-time curves are used to study earthquakes.

4. **Differentiate** between the speed of waves through hot and cold material.

Explain Your Thinking

5. **Infer** Using the seismogram in **Figure 8,** suggest why surface waves cause so much damage even though they are the last to arrive at a seismic station.

6. **WRITING ▶ Connection** Write a newspaper article reporting on the ways scientists have determined the composition of Earth.

MEASURING AND LOCATING EARTHQUAKES

FOCUS QUESTION

How do we find where an earthquake happened?

Earthquake Magnitude and Intensity

More than one million earthquakes occur each year, but news accounts report on only the largest ones. Scientists have developed several methods for describing the size of an earthquake.

Richter scale

The **Richter scale,** devised by a geologist named Charles Richter, is a numerical rating system that measures the energy, called the **magnitude,** of the largest seismic waves that are produced during an earthquake. The numbers in the Richter scale are determined by the height, called the **amplitude,** of the largest seismic wave on a seismogram and the distance from the epicenter. Each successive number represents an increase in amplitude of a factor of 10. For example, the seismic waves of a magnitude-8 earthquake on the Richter scale are ten times larger than those of a magnitude-7 earthquake. The differences in the amounts of energy released by earthquakes are even greater than the differences between the amplitudes of their waves. Each increase in magnitude corresponds to about a 32-fold increase in seismic energy. Thus, an earthquake of magnitude-8 releases about 32 times the energy of a magnitude-7 earthquake. The damage shown in **Figure 13** was caused by an earthquake measuring 7 on the Richter scale.

Figure 13 The damage shown here was caused by a magnitude-7 earthquake that struck Haiti in January 2010.

⬤ **3D THINKING**   **DCI** Disciplinary Core Ideas   **CCC** Crosscutting Concepts   **SEP** Science & Engineering Practices

COLLECT EVIDENCE

 Use your Science Journal to record the evidence you collect as you complete the readings and activities in this lesson.

INVESTIGATE

 GO ONLINE to find these activities and more resources.

🥽 **GeoLAB:** Relate Epicenters and Plate Tectonics
Obtain, evaluate, and communicate information resulting from an earthquake using seismograms.

🥽 **Quick Investigation:** Make a Map
Develop and use a model to visualize the patterns created by seismic-intensity data.

U.S. Air Force photo by Tech. Sgt. James L. Harper Jr.

Figure 14 The modified Mercalli scale measures damage done by an earthquake. An earthquake strong enough to knock groceries off the store's shelves would probably be rated V using the modified Mercalli scale.

Moment magnitude scale

The Richter scale is often used to describe the magnitude of an earthquake. However, most seismologists use the **moment magnitude scale**—a rating scale that measures the energy released by an earthquake, taking into account the size of the fault rupture, the amount of movement along the fault, and the rocks' stiffness. The number from the moment magnitude scale is the one most often reported on the news.

Modified Mercalli scale

Another way to describe earthquakes is with respect to the amount of damage they cause, such as the damage shown in **Figure 14.** This measure, called the intensity of an earthquake, is determined using the **modified Mercalli scale,** which rates the types of damage and other effects of an earthquake as noted by observers during and after its occurrence. This scale uses Roman numerals to designate the degree of intensity. A simplified version of the modified Mercalli scale is shown in **Table 2.**

Table 2 Modified Mercalli Scale

I	Not felt except under unusual conditions.
II	Felt by only a few persons; suspended objects might swing.
III	Quite noticeable indoors; vibrations are like the passing of a truck.
IV	Felt indoors by many, outdoors by few; dishes and windows rattle; standing cars rock noticeably.
V	Felt by nearly everyone; some dishes and windows break, and some plaster cracks.
VI	Felt by all; furniture moves; some plaster falls, and some chimneys are damaged.
VII	Difficult to stand; some chimneys break; damage is slight in well-built structures but considerable in weak structures.
VIII	Chimneys, smokestacks, and walls fall; heavy furniture is overturned; partial collapse of ordinary buildings occurs.
IX	Great general damage occurs; buildings shift off foundations; ground cracks; underground pipes break.
X	Most ordinary structures are destroyed; rails are bent; landslides are common.
XI	Few structures remain standing; bridges are destroyed; railroad ties are greatly bent; broad fissures form in the ground.
XII	Damage is total; objects are thrown upward into the air.

Earthquake intensity The intensity of an earthquake depends primarily on the amplitude of the surface waves generated. Like body waves, surface waves gradually decrease in size with increasing distance from the focus of an earthquake. Because of this, the intensity also decreases as the distance from an earthquake's epicenter increases. Maximum intensity values are observed in the region near the epicenter; Mercalli values decrease to I at distances far from the epicenter.

Scientists use the modified Mercalli scale values to make seismic-intensity maps. These maps are a visual representation of an earthquake's intensity. Contour lines join points that experienced the same intensity. The contour lines show how the maximum intensity is usually found near the earthquake's epicenter.

Depth of focus

As you learned earlier in this section, earthquake intensity and magnitude reflect the size of the seismic waves generated by the earthquake. Another factor that determines the intensity of an earthquake is the depth of its focus. Recall that an earthquake can be classified as a shallow-focus, intermediate- or mid-focus, or deep-focus earthquake, depending on the location of the focus, as shown in **Figure 15.** Catastrophic earthquakes with high intensity values are almost always shallow-focus events.

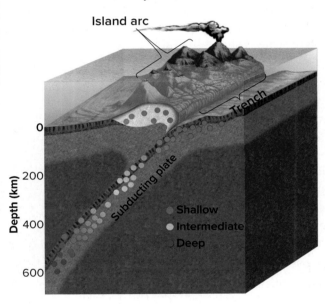

Figure 15 Earthquakes are classified as shallow, intermediate, or deep, depending on the location of the focus. Shallow-focus earthquakes are the most damaging.

Deep-focus earthquakes generally produce smaller vibrations at the epicenter than those produced by shallow-focus earthquakes. For example, a shallow-focus, moderate earthquake that measures a magnitude-6 on the Richter scale can generate a greater maximum intensity than a deep-focus earthquake of magnitude-8. Because the modified Mercalli scale is based on intensity rather than magnitude, it is a better measure of an earthquake's effect on people.

ACADEMIC VOCABULARY

intensity
the measurable amount of a property, such as force, brightness, or magnetic field
The intensity of light as perceived by the human eye is measured in a unit called a lux.

Locating an Earthquake

The location of an earthquake's epicenter, as well as the time of occurrence, can be determined using seismograms and travel-time curves.

Distance to an earthquake

P-waves reach a seismograph station before the S-waves. Consider the effect of the distance traveled on the time it takes for both waves to arrive. The gap in their arrival times will be greater when the distance traveled is longer. **Figure 16** shows the same travel-time curve graph shown in **Figure 9** of Lesson 2, but this time it is joined with the seismogram from a specific earthquake. The seismometer recorded the time that elapsed between the arrival of the first P-waves and first S-waves.

Seismologists determine the distance to an earthquake's epicenter by measuring the separation on any seismogram and identifying that same separation time on the travel-time graph. The separation time for the earthquake shown in **Figure 16** is 6 min. Based on travel times of seismic waves, the distance between the earthquake's epicenter and the seismic station that recorded the waves can only be 4500 km. This is because the known travel time over that distance is 8 min for P-waves and 14 min for S-waves. Farther from the epicenter, the gap between the travel times for both waves increases.

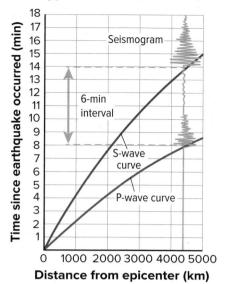

Typical Travel-Time Curves

Figure 16 This travel-time curve also shows seismographic data for an earthquake event.

 Get It?

Apply If the gap between P- and S-waves is 2 min, what can you infer about the distance from the epicenter to the seismometer?

Seismologists analyze data from many seismograms to locate the epicenter. Calculating the distance between an earthquake's epicenter and a seismic station provides enough information to determine that the epicenter was a certain distance in any direction from the seismic station.

This can be represented by a circle around the seismic station with a radius equal to the distance to the epicenter. Consider the effect of adding data from a second seismic station. The two circles will overlap at two points. When data from a third seismic station is added, the rings will overlap only at one point—the epicenter—as shown in **Figure 17.**

 Get It?

Explain how the epicenter of an earthquake can be located using information from three seismic stations.

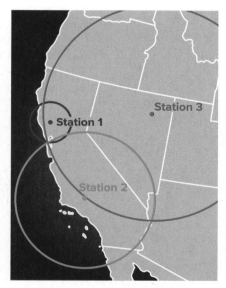

Figure 17 To locate the epicenter of an earthquake, scientists identify the seismic stations on a map and draw a circle with the radius of distance to the epicenter from each station. The point where all the circles intersect is the epicenter.

Identify *the epicenter of this earthquake.*

Time of an earthquake

The gap in the arrival times of different seismic waves on a seismogram provides information about the distance to the epicenter. Seismologists can also use the seismogram to gain information about the exact time that the earthquake occurred at the focus. The time can be determined by using a table similar to the travel-time graph shown in **Figure 9.** The exact arrival times of the P-waves and S-waves at a seismic station are recorded on the seismogram. Seismologists read the travel time of either wave to the epicenter from that station using graphs similar to the one shown in **Figure 9.** For example, consider a seismogram that registered the arrival of P-waves at exactly 10:00 A.M. If the P-waves traveled 4500 km and took 8 min according to the appropriate travel-time curve, then it can be determined that the earthquake occurred at the focus at 9:52 A.M.

 Get It?

List the information contained in a seismogram.

Seismic Belts

Over the years, seismologists have collected and plotted the locations of numerous earthquake epicenters. The global distribution of these epicenters reveals a noteworthy pattern. Earthquake locations are not randomly distributed. The majority of the world's earthquakes occur along narrow seismic belts that separate large regions with little or no seismic activity.

As with the locations of most volcanoes on Earth, most earthquakes correspond closely with tectonic plate boundaries, as shown in **Figure 18.**

Global Earthquake Epicenter Locations

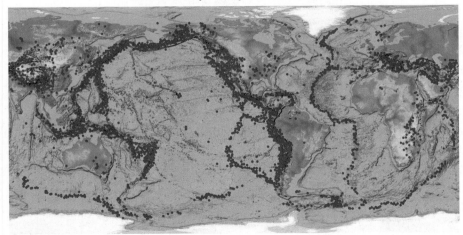

Figure 18 Notice the pattern of global epicenter locations on the map.

Identify *Based on this map, do you live near an epicenter?*

STEM CAREER Connection

Seismologist

Some seismologists study earthquakes and other hazards that result from them, such as landslides and tsunamis. Others have jobs that involve finding petroleum and natural gas underground. Seismologists may work in the field, work in a lab, or teach at a university.

In fact, almost 80 percent of all earthquakes, like the one whose effects are shown in **Figure 19,** occur on the Circum-Pacific Belt; about 15 percent occur on the Mediterranean-Asian Belt, across southern Europe and Asia. Recall that these belts are subduction zones, where tectonic plates are colliding and one plate is forced to sink beneath another. Most of the remaining earthquakes occur in narrow bands along the crests of ocean ridges, where tectonic plates are diverging.

 Get It?

Explain the relationship between the distribution of earthquakes and tectonic plate boundaries.

Figure 19 The damage shown here resulted from an earthquake in Chile. The entire coastline of Chile lies along a portion of the Circum-Pacific Belt.

Check Your Progress

Summary

- Earthquake magnitude is a measure of the energy released during an earthquake and can be measured on the Richter scale.

- Intensity is a measure of the damage caused by an earthquake and is measured with the modified Mercalli scale.

- Data from at least three seismic stations are needed to locate an earthquake's epicenter.

- Most earthquakes occur in seismic belts, which are areas associated with plate boundaries.

Demonstrate Understanding

1. **Summarize** the ways that scientists can use seismic waves to measure and locate earthquakes.

2. **Compare and contrast** earthquake magnitude and intensity and the scales used to measure each.

3. **Explain** why data from at least three seismic stations makes it possible to locate an earthquake's epicenter.

4. **Describe** how the boundaries between Earth's tectonic plates compare with the location of most of the earthquakes shown in the map in **Figure 18.**

Explain Your Thinking

5. **Formulate** a reason why a magnitude-3 earthquake can possibly cause more damage than a magnitude-6 earthquake.

6. **MATH Connection** Calculate how much more energy a magnitude-9 earthquake releases compared to that of a magnitude-7 earthquake.

LEARNSMART Go online to follow your personalized learning path to review, practice, and reinforce your understanding.

erlucho/Shutterstock

FOCUS QUESTION

Can we predict when an earthquake will happen?

Earthquake Hazards

Earthquakes are known to occur frequently along plate boundaries. An earthquake of magnitude-5 can be catastrophic in one region, but relatively harmless in another. There are many factors that determine the severity of damage produced by an earthquake. These factors are called earthquake hazards. Identifying earthquake hazards in an area can sometimes help to prevent some of the damage and loss of life. For example, the design of certain buildings can affect earthquake damage. As you can see in **Figure 20,** the most severe damage occurs to unreinforced buildings made of brittle materials such as concrete. Wooden structures, on the other hand, are more resilient and generally sustain less damage.

Figure 20 Concrete buildings are often brittle and can be easily damaged in an earthquake. The building on the right shifted and crumbled after a magnitude-7.8 earthquake in Nepal and is being held up by the building next to it.

 **3D THINKING**   **DCI** Disciplinary Core Ideas   **CCC** Crosscutting Concepts   **SEP** Science & Engineering Practices

COLLECT EVIDENCE

 Use your Science Journal to record the evidence you collect as you complete the readings and activities in this lesson.

INVESTIGATE

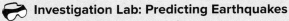 **GO ONLINE** to find these activities and more resources.

Investigation Lab: Predicting Earthquakes
Analyze geoscience data **to** predict where earthquakes are most likely to occur.

Design Your Own: Earthquake News Report
Obtain, evaluate, and communicate information on how earthquakes have influenced human activities and natural systems.

(t)Step Haiselden/Alamy Stock Photo, (b)Photo by M. Celebi, U.S. Geological Survey

Figure 21 One type of damage caused by earthquakes is called pancaking because shaking causes a building's supporting walls to collapse and the upper floors to fall one on top of the other, like a stack of pancakes.

Structural failure

In many earthquake-prone areas, buildings are destroyed as the ground beneath them shakes. In some cases, the supporting walls of the ground floor fail and cause the upper floors, which initially remain intact, to fall and collapse as they hit the ground or lower floors. The resulting debris resembles a stack of pancakes; thus, the process is called pancaking. This type of structural failure, shown in **Figure 21,** was a tragic consequence of the earthquake in Haiti in 2010.

 Get It?

Explain what happens when a building pancakes.

Another type of structural failure is related to the height of a building. During the 1985 Mexico City earthquake, for example, most buildings between 5 and 15 stories tall collapsed or were otherwise completely destroyed, as shown in **Figure 22.** Similar structures that were either shorter or taller, however, sustained only minor damage. The shaking caused by the earthquake had the same frequency of vibration as the natural sway of the intermediate buildings. This caused those buildings to sway the most violently during the earthquake. The ground vibrations, however, were too rapid to affect taller buildings, whose frequency of vibration was longer than those of the earthquake, and too slow to affect shorter buildings, whose frequency of vibration was shorter.

Figure 22 Many medium-sized buildings were damaged or destroyed during the 1985 Mexico City earthquake because they vibrated with the same frequency as the seismic waves.

Figure 23 Soil liquefaction happens when seismic vibrations cause poorly consolidated soil to liquefy and behave like quicksand. The buildings pictured here were built on this type of soil, and an earthquake caused the buildings to fall over.

Land and soil failure

In addition to their effects on structures made by humans, earthquakes can wreak havoc on Earth's landscape. In sloping areas, earthquakes can trigger massive landslides. For example, most of the estimated 30,000 deaths caused by the magnitude-7.9 earthquake that struck in Peru in 1970 resulted from a landslide that buried several towns. In areas with sand that is nearly saturated with water, seismic vibrations can cause the ground to behave like a liquid in a phenomenon called **soil liquefaction** (lih kwuh FAK shun). Soil liquefaction can generate landslides even in areas of low relief. It can cause trees and houses to fall over or to sink into the ground and underground pipes and tanks to rise to the surface. **Figure 23** shows tilted buildings that collapsed when the soil under them liquefied during an earthquake in San Francisco.

> ### Get It?
> **Summarize** how solid ground can take the properties of a liquid.

In addition to determining landslide and liquefaction risks, the type of ground material can also affect the severity of an earthquake in an area. Ground motion is amplified in some soft materials, such as unconsolidated sediments. Motion is muted in more resistant materials, such as granite. The severe damage to structures in Mexico City during the 1985 earthquake is attributed to the soft sediments on which the city is built. The thickness of the sediments caused them to resonate with the same frequency as that of the surface waves generated by the earthquake. This produced reverberations that greatly enhanced the ground motion and the resulting damage.

STEM CAREER Connection

Earthquake Engineer
How would you like to work with cutting-edge technology that could help save lives? Earthquake engineers help design buildings and other structures to make them more resistant to damage from seismic waves.

C.E. Meyer/USGS

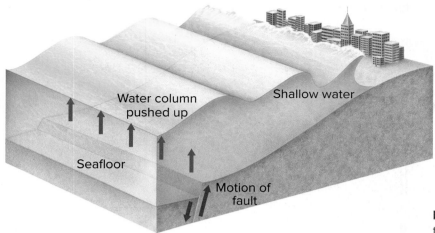

Water column
pushed up

Shallow water

Seafloor

Motion of
fault

Figure 24 A tsunami is generated when an underwater fault or landslide displaces a column of water.

Tsunami

Another type of earthquake hazard is a **tsunami** (soo NAH mee)—a large ocean wave generated by vertical motions of the seafloor during an earthquake. These motions displace the entire column of water overlying the fault, creating bulges and depressions in the water, as shown in **Figure 24.** The disturbance then spreads out from the epicenter in the form of waves with extremely long wavelengths. While these waves are in the open ocean, their height is generally less than 1 m. When the waves enter shallow water, however, they can form huge breakers with heights occasionally exceeding 30 m. These enormous wave heights, together with open-ocean speeds between 500 and 800 km/h, make tsunamis dangerous threats to coastal areas both near to and far from an earthquake's epicenter. The Indian Ocean tsunami of December 26, 2004,

originated with a magnitude-9.1 earthquake in the ocean about 250 km west of Sumatra. The 10-m-tall tsunami radiated across the Indian Ocean and struck the coasts of Indonesia, Sri Lanka, India, Thailand, Somalia, and several other nations. In 2011, a tsunami that originated with a magnitude-9.0 earthquake in the ocean struck the town of Sukuiso, Japan. The aftermath of that catastrophic event is shown in **Figure 25.**

Figure 25 The destruction from the March 11, 2011, tsunami was massive. Tsunami waves up to 39 m high began hitting the coast of Japan within an hour of the earthquake.

SCIENCE USAGE v. COMMON USAGE

column

Science usage: a hypothetical cylinder of water that goes from the surface to the bottom of a body of water

Common usage: a vertical arrangement of items

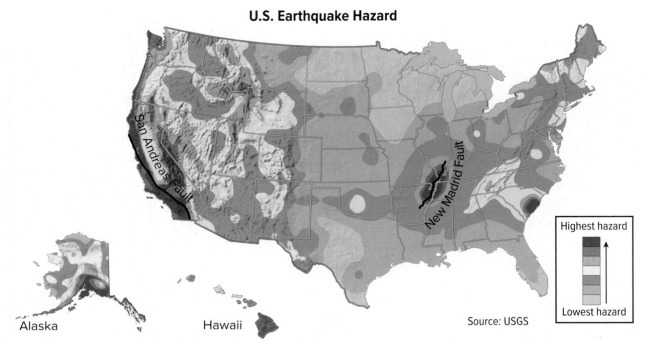

U.S. Earthquake Hazard

Highest hazard

Lowest hazard

Alaska

Hawaii

Source: USGS

Figure 26 Areas of high seismic risk in the United States include Alaska, Hawaii, and some of the western states.
Locate *the areas of highest seismic risk on the map. Locate your own state. What is the seismic risk of your area?*

Earthquake Forecasting

To minimize the damage and deaths caused by earthquakes, seismologists are searching for ways to forecast these events. There is currently no completely reliable way to forecast the exact time and location of the next earthquake. Instead, earthquake forecasting is based on calculating the probability of an earthquake. The probability of an earthquake's occurrence is based on two factors: the history of earthquakes in an area and the rate at which stress builds up in the rocks.

 Get It?

Identify the two factors seismologists use to determine the probability of an earthquake occurring in a certain area.

Seismic risk

Recall that most earthquakes occur in long, narrow bands called seismic belts. The probability of future earthquakes is much greater in these belts than elsewhere on Earth. The pattern of earthquakes in the past is usually a reliable indicator of future earthquakes in a given area. Seismometers and sedimentary rocks can be used to determine the frequency of large earthquakes. The history of an area's seismic activity can be used to generate seismic-risk maps. A seismic-risk map of the United States is shown in **Figure 26.** In addition to Alaska, Hawaii, and some western states, there are several regions of relatively high seismic risk in the central and eastern United States. These regions have experienced some of the most intense earthquakes in the past and probably will experience significant seismic activity in the future.

Recurrence rates

Earthquake-recurrence rates along a fault can indicate whether the fault ruptures at regular intervals to generate similar earthquakes. The earthquake-recurrence rate along a section of the San Andreas fault in California, for example, shows that a sequence of earthquakes of approximately magnitude 6 shook the area about every 22 years from 1857 until 1966. In 1987, seismologists forecasted a 90-percent probability that a major earthquake would rock the area within the next few decades. Several kinds of instruments at the drilling site, shown in **Figure 27,** were installed in an attempt to measure the earthquake as it occurred. In September 2004, a magnitude-6 earthquake struck. Extensive data were collected before and after the 2004 earthquake. The information obtained will be invaluable for predicting and preparing for future recurrent earthquakes around the world.

Seismic gaps

Probability forecasts are also based on the location of seismic gaps. **Seismic gaps** are sections located along faults that are known to be active, but which have not experienced significant earthquakes for a long period of time. A seismic gap in the San Andreas Fault cuts through San Francisco. This section of the fault has not ruptured since the devastating earthquake that struck the city in 1906. Because of this inactivity, seismologists currently forecast that there is a 72-percent probability that the San Francisco area will experience a magnitude-6.7 or higher earthquake within the next 30 years. **Figure 28** shows the seismic-gap map for a fault that passes through an area of Turkey.

Figure 27 This drilling rig was used to drill a hole 2.3 km deep in Parkfield, California. Once completed, the hole was rigged with instruments to record data during major and minor tremors. The goal of the project was to better understand how earthquakes work and what triggers them. This information could help scientists predict when earthquakes will occur.

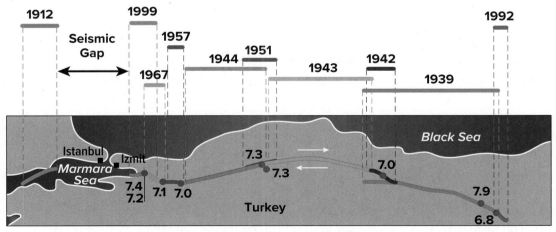

Figure 28 Earthquakes in 1912 and 1999 happened on either side of Istanbul, a city of 18 million people. The earthquakes around the city leave a seismic gap that indicates that an earthquake is likely to occur in that area.

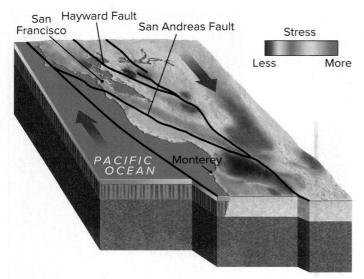

San Francisco, Hayward Fault, San Andreas Fault

PACIFIC OCEAN, Monterey

Stress — Less / More

Figure 29 Stress-accumulation maps help scientists determine the probability of an earthquake in any particular place.

Explain Why does stress build up in the areas indicated?

Stress accumulation

The rate at which stress builds up in rocks is another factor seismologists use to determine the earthquake probability along a section of a fault. Eventually this stress is released, generating an earthquake. Scientists use satellite-based technology such as GPS to measure the stress that accumulates along a fault. The stress accumulated in a particular part of a fault, together with the amount of stress released during the last earthquake in a particular part of the fault, can be used to develop images like **Figure 29.** It shows that this section of the San Andreas Fault is experiencing a wide distribution of tectonically induced stress.

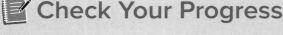

Check Your Progress

Summary

- Earthquake forecasting is based on seismic history and measurements of accumulated stress.

- Earthquakes cause damage by creating vibrations that can shake Earth.

- Earthquakes can cause structural collapse, landslides, soil liquefaction, and tsunamis.

- Seismic gaps are sections along an active fault that have not experienced significant earthquakes for a long period of time.

Demonstrate Understanding

1. **List** some examples of how scientists determine the probability of an earthquake occurring.

2. **Summarize** the effects of the different types of hazards caused by earthquakes.

3. **Draw** before-and-after pictures of what can happen when the ground ruptures along a fault.

4. **Summarize** the events that lead to a tsunami.

Explain Your Thinking

5. **Assess** where an earthquake is most likely to occur—in the same place that a magnitude-7.5 earthquake occurred 20 years ago or at a location between areas that had earthquakes 20 and 60 years ago, respectively.

6. **WRITING Connection** Imagine you are on an international aid committee. Write a report suggesting ways to identify earthquake-vulnerable areas.

LEARNSMART® Go online to follow your personalized learning path to review, practice, and reinforce your understanding.

All Shook Up

Science has not yet developed a way to accurately predict earthquakes. However, in recent years, countries that often experience earthquakes have developed and put into use early detection systems that can warn people before an earthquake hits. The United States is following their lead, but it does not yet have a fully operating warning system for the entire West Coast, where earthquakes are most likely to occur in the U.S.

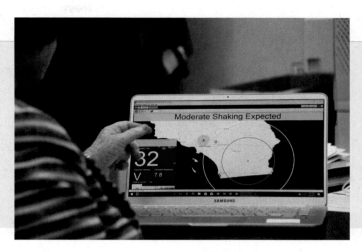

Once the ShakeAlert system is fully operational, West Coast residents will receive earthquake warnings on phones and computers.

How warning systems work

When an earthquake strikes, it generates P-waves (primary waves) and S-waves (secondary waves). P-waves travel at 4–8 km per second, while S-waves travel at 2.5–4 km per second, depending on the composition of the crust. P-waves are the less destructive of the waves. They travel faster and are detected first by warning systems. The system then puts out an alert that gives people from 15 seconds to a minute or more of warning.

People can use these crucial seconds to stop trains and elevators, secure dangerous equipment, and move to safer locations. Then, when the slower, more destructive S-waves hit, they will do less damage than they would have otherwise.

Existing systems

Mexico was the first country in the world to construct a seismic early warning system, which was put into operation in Mexico City in 1991. Scientists worked hard to develop it after a deadly earthquake hit in 1985, killing thousands. Japan has had a seismic early warning system in operation since 2007.

The ShakeAlert system

Since 2006, the U.S. Geological Survey (USGS) and several partners have been developing ShakeAlert, an earthquake early warning system for California, Oregon, and Washington. The USGS unveiled ShakeAlert updates in 2016 and 2017. While the system cannot yet send out public alerts, it has been sending alerts to a few test users. In the next few years, ShakeAlert will provide alerts on smartphones and through other public means of notification. The alerts will include the earthquake's estimated intensity and estimated time of arrival.

ASK QUESTIONS TO CLARIFY

Brainstorm several questions you have about earthquake early detection systems, and use print or online sources to find answers. Create a presentation to share your research with others.

Frederic J. Brown/AFP/Getty Images

MODULE 15
STUDY GUIDE

 GO ONLINE to study with your Science Notebook.

Lesson 1 FORCES WITHIN EARTH

- Stress is force per unit of area that acts on a material, and strain is the deformation of a material in response to stress.
- Reverse, normal, and strike-slip are the major types of faults.
- The three types of seismic waves are P-waves, S-waves, and surface waves.

- stress
- strain
- elastic deformation
- plastic deformation
- fault
- seismic wave
- primary wave
- secondary wave
- focus
- epicenter

Lesson 2 SEISMIC WAVES AND EARTH'S INTERIOR

- Seismometers are devices that record seismic wave activity on a seismogram.
- Travel times for P-waves and S-waves enable scientists to pinpoint the epicenters of earthquakes.
- P-waves and S-waves change speed and direction when they encounter different materials.
- Analysis of seismic waves provides a detailed picture of the composition of Earth's interior.

- seismometer
- seismogram

Lesson 3 MEASURING AND LOCATING EARTHQUAKES

- Earthquake magnitude is a measure of the energy released during an earthquake and can be measured on the Richter scale.
- Intensity is a measure of the damage caused by an earthquake and is measured with the modified Mercalli scale.
- Data from at least three seismic stations are needed to locate an earthquake's epicenter.
- Most earthquakes occur in seismic belts, which are areas associated with plate boundaries.

- Richter scale
- magnitude
- amplitude
- moment magnitude scale
- modified Mercalli scale

Lesson 4 EARTHQUAKES AND SOCIETY

- Earthquake forecasting is based on seismic history and measurements of accumulated strain.
- Earthquakes cause damage by creating vibrations that can shake Earth.
- Earthquakes can cause structural collapse, landslides, soil liquefaction, and tsunamis.
- Seismic gaps are sections along an active fault that have not experienced significant earthquakes for a long period of time.

- soil liquefaction
- tsunami
- seismic gap

Module Wrap-Up

REVISIT THE PHENOMENON

How can we prevent people from getting hurt during an earthquake?

CER Claim, Evidence, Reasoning

Explain Your Reasoning Revisit the claim you made when you encountered the phenomenon. Summarize the evidence you gathered from your investigations and research and finalize your Summary Table. Does your evidence support your claim? If not, revise your claim. Explain why your evidence supports your claim.

STEM UNIT PROJECT
Now that you've completed the module, revisit your STEM unit project. You will summarize your evidence and apply it to the project.

GO FURTHER

SEP Data Analysis Lab

How can you find an earthquake's epicenter?

To pinpoint the epicenter, analyze the P-wave and S-wave data recorded at seismic stations. Obtain a map of the western hemisphere from your teacher, and mark the seismic stations listed in the table.

CER Analyze and Interpret Data

1. For each station, calculate and record the arrival time differences by subtracting the P-wave arrival time from the S-wave arrival time.
2. Use the arrival time differences and the travel-time curve **(Figure 9)** to find the distance between the epicenter and each seismic station. Record the distances.
3. Draw a circle around each station. Use the distance from the epicenter as the radius for each circle. Repeat for each seismic station.
4. **Claim, Evidence, Reasoning** Identify the epicenter of the earthquake. Explain how you arrived at your answer.
5. **Reasoning** Explain why data from more seismic stations would be useful for finding the epicenter.

Data and Observations

Seismic Station	P-wave Arrival Time (PST)	S-wave Arrival Time (PST)	Arrival Time Difference (min)	Distance from Epicenter (km)
Newcomb, NY	8:39:02	8:44:02		
Idaho Springs, CO	8:35:22	8:37:57		
Darwin, CA	8:35:38	8:38:17		

*Data obtained from: Significant earthquakes of the world. 2006.
USGS earthquake center.

ENCOUNTER THE PHENOMENON

How do mountains grow so large?

GO ONLINE to play a video about how deeply thrust rock layers affect the surface of Earth.

SEP Ask Questions

Do you have other questions about the phenomenon? If so, add them to the driving question board.

CER Claim, Evidence, Reasoning

Make Your Claim Use your CER chart to make a claim about how mountains grow so large. Explain your reasoning.

Collect Evidence Use the lessons in this module to collect evidence to support your claim. Record your evidence as you move through the module.

Explain Your Reasoning You will revisit your claim and explain your reasoning at the end of the module.

GO ONLINE to access your CER chart and explore resources that can help you collect evidence.

LESSON 2: Explore & Explain: Mountain Building at Convergent Boundaries

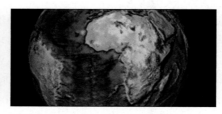

LESSON 3: Explore & Explain: Other Mountain Types

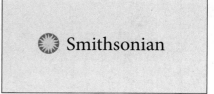

Additional Resources

CRUST-MANTLE RELATIONSHIPS

FOCUS QUESTION

Why does continental crust sit higher on Earth's surface than oceanic crust?

Earth's Topography

On a globe or a map of Earth's surface, the oceans and continents are easily distinguished. From these representations of Earth, you can estimate that about 71 percent of Earth's surface is below sea level, and about 29 percent lies above sea level. What is not obvious from most maps and globes, however, is the variation in elevations of the crust, which is referred to as its **topography.** Recall that topographic maps show an area's hills and valleys. When a very large map scale is used, such as the one in **Figure 1,** the topography of Earth's entire crust can be shown.

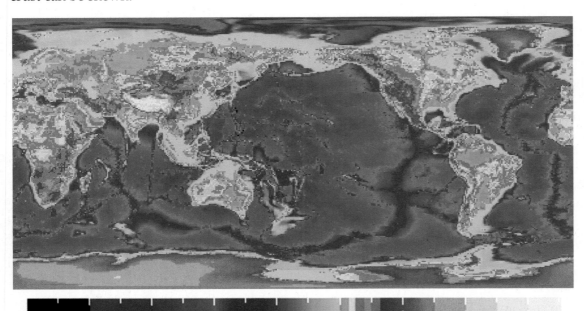

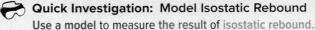

Figure 1 Topographic maps show differences in elevation on Earth's surface.

Interpret *the map to determine Earth's highest and lowest elevations. Where are they?*

3D THINKING   **DCI** Disciplinary Core Ideas   **CCC** Crosscutting Concepts   **SEP** Science & Engineering Practices

COLLECT EVIDENCE

Use your Science Journal to record the evidence you collect as you complete the readings and activities in this lesson.

INVESTIGATE

GO ONLINE to find these activities and more resources.

Quick Investigation: Model Isostatic Rebound
Use a model to measure the result of isostatic rebound.

? **Revisit the Encounter the Phenomenon Question**
What information from this lesson can help you answer the Unit and Module questions?

When Earth's topography is plotted on a graph such as **Figure 2,** a pattern in the distribution of elevations emerges. Note that most of Earth's elevations cluster around two main ranges of elevation. Above sea level, elevation averages around 0 to 1 km. Below sea level, elevations range between −4 and −5 km. These two ranges dominate Earth's topography and reflect the basic differences in density and thickness between continental and oceanic crust.

Continental crust

Imagine this experiment: blocks of wood with different densities are placed in water. The blocks displace different amounts of water, and thus float at different heights above the surface of the water. Given the same thickness, blocks of higher density displace more water than blocks of lower density. Recall that oceanic crust is composed mainly of basalt, which has an average density of about 2.9 g/cm³. Continental crust, which can be billions of years older than oceanic crust, is composed of more granitic rock, which has an average density of about 2.8 g/cm³. The slightly higher density of oceanic crust causes it to displace more of the mantle—which has a density of about 3.3 g/cm³—than the same thickness of continental crust.

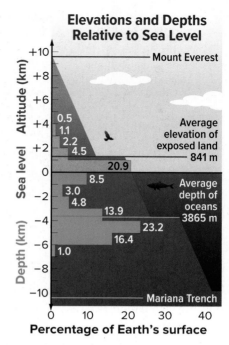

Figure 2 About 29 percent of Earth is land and 71 percent is water.

Interpret *At what elevation does most of Earth's surface lie? At what depth?*

 Get It?

Compare the composition and average density of continental crust and oceanic crust.

Differences in elevation, however, are not caused by density differences alone. A thick wood block will displace more water than a thin, denser wood block. However, the thicker, less dense block will float higher in the water than the denser block. Continental crust, which is thicker and less dense than oceanic crust, behaves similarly. It extends deeper into the mantle because of its thickness, and it rises higher above Earth's surface than oceanic crust because of its lower density, as shown in **Figure 3.**

Figure 3 Continental crust is thicker and less dense than oceanic crust, so it extends higher above Earth's surface and deeper into the mantle than oceanic crust.

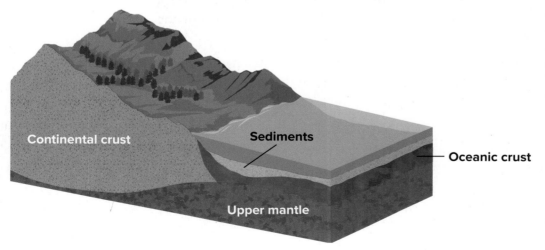

 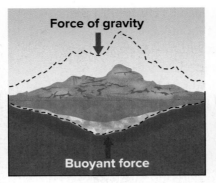 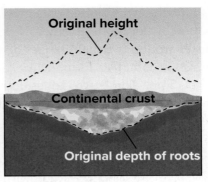

Figure 4 According to the principle of isostasy, parts of Earth's crust rise or subside until they are buoyantly supported by their roots.

Isostasy

The displacement of the mantle by Earth's continental and oceanic crust is a condition of equilibrium called **isostasy** (i SAHS tuh see). The crust and mantle are in equilibrium when the downward force of gravity on the mass of crust is balanced by the upward force of buoyancy that results from displacement of the mantle by the crust. This balance might be familiar to you if you have ever watched people get in and out of a small boat. As the people boarded the boat, it sank deeper into the water. Conversely, as the people got out of the boat, it displaced less water and floated higher in the water. A similar sinking and rising that results from the addition and removal of mass occurs within Earth's crust. Gravitational and seismic studies have detected thickened areas of continental material, called **roots,** that extend into the mantle below Earth's mountain ranges.

 Get It?

Describe What conditions are present when the crust and mantle are in equilibrium?

Mountain roots

A mountain range requires large roots to counterbalance the enormous mass of the range above Earth's surface. **Figure 4** illustrates how, according to the principle of isostasy, parts of the crust rise or subside until these parts are buoyantly supported by their roots. Continents and mountains are said to float on the mantle because they are less dense than the underlying mantle. They project into the mantle to provide the necessary buoyant support.

What do you think happens when erosion causes mass to be removed from a mountain or mountain range? If mass is removed from a mountain, the roots will rise in response. If erosion continues, the mountain will eventually disappear, exposing the roots.

ACADEMIC VOCABULARY

displace
to physically move out of position or space
The block displaced the water in the container.

Figure 5 Before erosion, the Appalachian Mountains were thousands of meters taller than they are now. Because of isostatic rebound, as the mountains eroded, the deep roots also rose thousands of meters closer to the surface. The mountains visible today are only the roots of an ancient mountain range. They, too, are being eroded and will someday resemble the low relief of the craton in northern Canada.

Isostasy and Erosion

The Appalachian Mountains, shown in **Figure 5,** in the eastern United States formed hundreds of millions of years ago when the North American continent collided with Europe and Africa. Rates of erosion on land are such that these mountains should have been completely eroded millions of years ago. Why, then, do these mountains still exist? As the mountains rose above Earth's surface, deep roots formed until isostatic equilibrium was achieved and the mountains were buoyantly supported. As peaks eroded, the mass decreased. This allowed the roots themselves to rise and erode.

A balance between erosion and decrease in the size of the roots will continue for hundreds of millions of years, until the mountains disappear and the roots are exposed at the surface. This slow process of the crust's rising as the result of the removal of overlying material is called **isostatic rebound.** Erosion and rebound allows metamorphic rocks formed at great depths to rise to the top of mountain ranges such as the Appalachians.

 Get It?

Predict what will eventually happen to the Appalachian Mountains.

Seamounts

Crustal movements resulting from isostasy are not restricted to Earth's continents. They can also occur in oceanic crust. For example, recall that hot spots under the ocean floor can produce a chain of individual volcanic mountains. When these mountains are under water, they are called seamounts. On the geologic time scale, these mountains form very quickly. So, what happens to the seafloor after these seamounts form? The seamounts are added mass. As a result of isostasy, the oceanic crust around these peaks displaces the underlying mantle until equilibrium is achieved.

Deep roots

You have learned that the elevation of Earth's crust depends on the thickness of the crust as well as its density. You also learned that a mountain peak is countered by a root. Mountain roots can be many times as deep as a mountain is high. Mount Everest, shown in **Figure 6,** towers nearly 9 km above sea level and is the tallest peak in the Himalayas. Some parts of the Himalayas are underlain by roots that are nearly 70 km deep. Recall that these mountains are still growing at a rate of more than 1 cm/yr. As India continues to push northward into Asia, the Himalayas, including Mount Everest, will continue to grow in height. Currently, the combined thickness of the Himalayas is approximately equal to 868 football fields lined up end to end. Where do the immense forces required to produce such crustal thickening originate? You will read about these forces in Lesson 2 of this module.

Get It?

Contrast the depth of the roots and the height of the peaks of the Himalayas.

Figure 6 Mount Everest, a peak in Asia, is currently the highest mountain on Earth. A deep root supports its mass. Scientists have determined that Mount Everest has a root that is nearly 70 km deep.

Check Your Progress

Summary

- The majority of Earth's elevations are either 0 to 1 km above sea level or 4 to 5 km below sea level.

- The mass of a mountain above Earth's surface is supported by a root that projects into the mantle.

- The addition of mass to Earth's crust depresses the crust, while the removal of mass from the crust causes the crust to rebound in a process called isostatic rebound.

Demonstrate Understanding

1. **Relate** density and crustal thickness to mountain building.
2. **Describe** the pattern in Earth's elevations, and explain what causes the pattern in distribution.
3. **Explain** why isostatic rebound slows over time.
4. **Infer** why the crust is thicker beneath continental mountain ranges than it is under flat-lying stretches of landscape.

Explain Your Thinking

5. **Apply** the principle of isostasy to explain how the melting of the ice sheets that once covered the Great Lakes has affected the land around the lakes.
6. **Consider** how the term *root* applies differently to mountains than it does to plants.
7. **MATH** **Connection** Suppose a mountain is being uplifted at a rate of 1 m/1000 y. It is also being eroded at a rate of 1 cm/y. Is this mountain getting larger or smaller? Explain.

FOCUS QUESTION
How were the Appalachian Mountains formed?

Mountain Building at Convergent Boundaries

Orogeny (oh RAH jun nee) refers to all processes that form mountain ranges. You have learned about many of these processes. Recall that metamorphism can cause rocks to be squeezed and folded and that rising magma can form igneous intrusions or erupt at Earth's surface. You have also learned about movement along faults. The result of all these processes can be broad, linear regions of deformation known as mountain ranges; in geology, they are also known as orogenic belts. Look at **Figure 7** and recall what you have read about the interaction of converging tectonic plates at their boundaries. Most orogenic belts are associated with convergent plate boundaries. Here, **compressive forces** squeeze the crust and cause intense deformation in the form of folding, faulting, metamorphism, and volcanism.

In general, the tallest and most varied orogenic belts form at convergent boundaries. Recall that convergence can occur between two oceanic plates, between an oceanic plate and a continental plate, or between two continental plates. Interactions at each of these convergent boundaries create different types of mountain ranges.

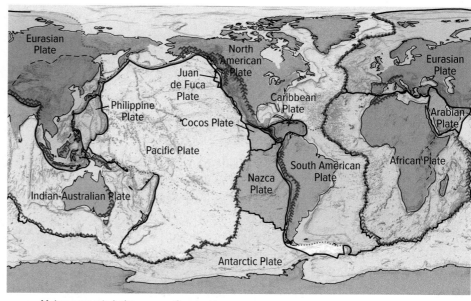

Figure 7 Most of Earth's mountain ranges (blue and red peaks on the map) formed along plate boundaries.

Identify *the mountain ranges that lie along the South American Plate by comparing a world map with the one shown here.*

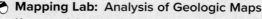 Major mountain belts   Ocean ridges

---

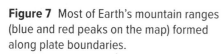 **3D THINKING**   **DCI** Disciplinary Core Ideas   **CCC** Crosscutting Concepts   **SEP** Science & Engineering Practices

COLLECT EVIDENCE
 Use your Science Journal to record the evidence you collect as you complete the readings and activities in this lesson.

INVESTIGATE
🧭 GO ONLINE to find these activities and more resources.

🥽 **Investigation Lab: Plate Tectonics of North America**
Conduct an investigation to identify and describe the tectonic plates of North America and how they affect the mountains of North America.

🥽 **Mapping Lab: Analysis of Geologic Maps**
Use a model to visualize the result of tectonic forces on North America.

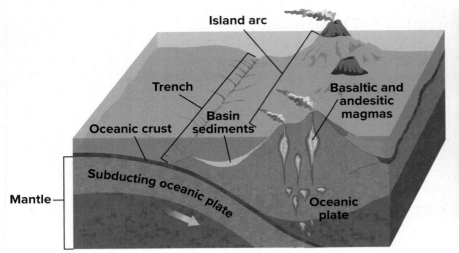

Lesser Antilles island arc

Inactive volcano on Lesser Antilles

Figure 8 Convergence between two oceanic plates results in the formation of individual volcanic peaks that make up an island arc complex. Several volcanic peaks make up the island arc complex known as the Lesser Antilles, in the southern Caribbean.

Oceanic-oceanic convergence

Recall that when an oceanic plate converges with another oceanic plate, one plate descends into the mantle to create a subduction zone. As parts of the mantle above the subducting plate melt, magma is forced upward, where it can form a series of individual volcanic peaks that, together, are called an island arc complex. The Aleutian Islands, off the coast of Alaska, and the Lesser Antilles, in the Caribbean, are examples of island arc complexes. The tectonic relationships and processes associated with oceanic-oceanic convergence are detailed in **Figure 8.**

What kinds of rocks make up island arc complexes? Often, they are a jumbled mixture of rock types. They are partly composed of basaltic and andesitic magmas. In addition to these volcanic rocks, some large island arcs contain sedimentary rocks. How do these sedimentary rocks eventually become part of a mountain? Recall that between an island arc and a trench is a depression, called a basin. This basin fills with sediments that have been eroded from the island arc. If subduction continues for tens of millions of years, some of these sediments can be uplifted, folded, faulted, and thrust against the existing island arc. This ultimately forms complex new masses of sedimentary and volcanic rocks. Parts of Japan formed in this way.

Get It?

Sequence how sedimentary rocks can become part of an island arc complex.

SCIENCE USAGE v. COMMON USAGE

uplift
Science usage: to cause a portion of Earth's surface to rise above adjacent areas

Common usage: to improve the spiritual, social, or intellectual condition

(t)Jacques Descloitres, MODIS Rapid Response Team, NASA/GSFC, (b)Westend61 Premium/Shutterstock

Oceanic-continental convergence

Oceanic-continental boundaries are similar to oceanic-oceanic boundaries in that convergence along both creates subduction zones and trenches. Unlike convergence at oceanic-oceanic boundaries, convergence between oceanic and continental plates produces mountain belts that are much bigger and more complicated than island arc complexes. When an oceanic plate converges with a continental plate, the descending oceanic plate forces the edge of the continental plate upward. This uplift marks the beginning of orogeny. In addition to uplift, compressive forces can cause the continental crust to fold and thicken. As the crust thickens, higher mountains form. Deep roots develop to support these enormous masses of rocks.

Recall that volcanic mountains can form over the subducting plate. As illustrated in **Figure 9,** sediments eroded from such volcanic mountains can fill the low areas between the trench and the coast. These sediments, along with ocean sediments and material scraped off the descending plate, are shoved against the edge of the continent to form a jumble of highly folded, faulted, and metamorphosed rocks. The metamorphosed rocks shown in **Figure 9** are near Dorset, in the United Kingdom. They formed when an oceanic plate subducted under the Eurasian Plate millions of years ago.

Continental-continental convergence

Earth's tallest mountain ranges, including the Himalayas, are formed at continental-continental plate boundaries. Because of its relatively low density, continental crust cannot be subducted into the mantle when two continental plates converge. Instead, the low-density continental crust becomes highly folded, faulted, and thickened.

Figure 9 At an oceanic-continental boundary, compression causes continental crust to fold and thicken. Igneous activity and metamorphism are also common along such boundaries. This uplifted outcrop of metamorphosed rock formed as the result of convergence of an oceanic plate with a continental plate.

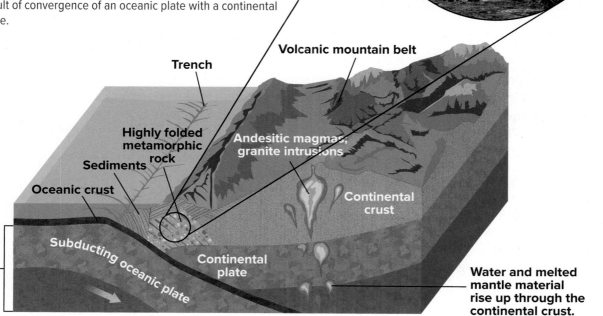

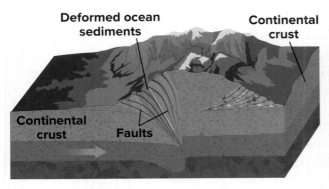

Figure 10 Intense folding and faulting along continental-continental boundaries produce some of the highest mountain ranges on Earth. The Himalayas, shown at right, are the result of convergence between the Indian and Eurasian plates.

Compressional forces break the crust into thick slabs that are thrust onto each other along low-angle faults. This process, shown in **Figure 10,** can double the thickness of the deformed crust. Deformation can also extend laterally for hundreds of kilometers into the continents involved. For example, studies of rocks in southern Tibet suggest that the original edge of Asia has been pushed approximately 2000 km eastward since the collision of the Indian and Eurasian plates. The magma that forms as a result of continental-continental mountain building solidifies beneath Earth's surface to form granite batholiths.

 Get It?

Explain why continental crust does not subduct.

Marine sedimentary rock Another common characteristic of the mountains that form when two continents collide is the presence of marine sedimentary rock near the mountains' summits. Such rock forms from the sediments deposited in the ocean basin that existed between the continents before their collision. For example, Mount Godwin Austen, in the western Himalayas, is composed of thousands of meters of marine limestone that sit upon a granite base. The limestone represents the northern portions of the old continental margin of India that were pushed up and over the rest of the continent when India began to collide with Asia about 50 mya.

The Appalachian Mountains—A Case Study

Recall that Alfred Wegener used the matching rocks and geologic structures in the Appalachians and mountains in Greenland and northern Europe to support his hypothesis of continental drift. In addition to Wegener, many other scientists have studied the Appalachian Mountains. Based on these studies, geologists have divided the Appalachians into several distinct regions.

Each region, illustrated in **Figure 11,** is characterized by rocks that show different degrees of deformation. For example, rocks of the Valley and Ridge Province are highly folded sedimentary rocks. In contrast, the rocks of the Piedmont Province consist of older, deformed metamorphic and igneous rocks that are overlain by relatively undeformed sedimentary layers. These regions, two of which are shown in **Figure 12,** are different because they formed in different ways.

The early Appalachians

The tectonic history of the Appalachians is illustrated in **Figure 13.** It began about 800 to 700 mya when ancestral North America separated from ancestral Africa along two divergent boundaries to form two oceans. The ancestral Atlantic Ocean was located off the western coast of ancestral Africa. A shallow, marginal sea formed along the eastern coast of ancestral North America. A continental fragment was located between the two divergent boundaries. About 700 to 600 mya, the directions of plate motions reversed. The ancestral Atlantic Ocean began to close as the plates converged. This convergence resulted in the formation of a volcanic island arc east of ancestral North America. About 200 million years passed before the continental fragment became attached to ancestral North America, as illustrated in **Figure 13.**

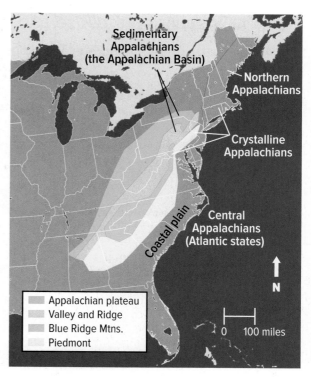

Figure 11 The Appalachian Mountain Range is made up of more than one type of mountain. It has several distinct regions, each with its own orogenic history.

Folded rock from the Valley and Ridge Province

Undeformed rock from the Piedmont Province

Figure 12 The Valley and Ridge Province of the Appalachians has highly folded rocks. Rocks from the Piedmont Province are relatively undeformed.

Figure 13 Visualizing the Rise and Fall of the Appalachians

The Appalachians formed hundreds of millions of years ago as a result of convergence.

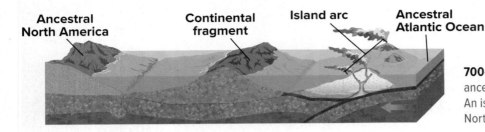

700–600 mya Convergence causes the ancestral Atlantic Ocean to begin to close. An island arc develops east of ancestral North America.

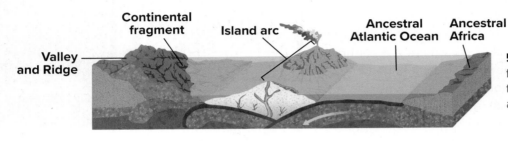

500–400 mya The continental fragment, which eventually becomes the Blue Ridge Province, becomes attached to ancestral North America.

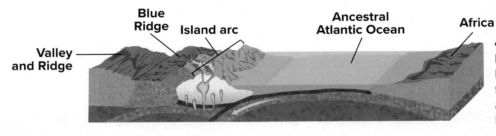

400–300 mya The island arc becomes attached to ancestral North America, and the continental fragment is thrust farther onto ancestral North America. The arc becomes the Piedmont Province.

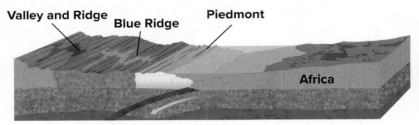

300–260 mya Pangaea forms. Ancestral Africa collides with ancestral North America to close the ancestral Atlantic Ocean. Compression forces the Blue Ridge and Piedmont rocks farther west, and the folded Valley and Ridge Province forms.

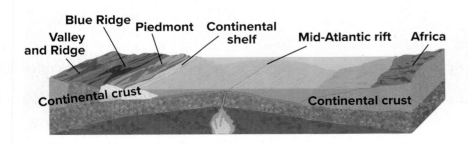

Present After the breakup of Pangaea, tension forces open the modern Atlantic Ocean and separates the continents. North America and Africa continue to move apart as the Atlantic Ocean widens.

These highly metamorphosed rocks were thrust over younger rocks to become the Blue Ridge Province, parts of which are shown in **Figure 14.**

Final stages of formation

Between about 400 and 300 mya, the island arc became attached to North America, as illustrated in **Figure 13.** Evidence of this event is preserved in the Piedmont Province as a group of metamorphic and igneous rocks. These rocks were also faulted over the continent, pushing the Blue Ridge rocks farther west.

Between about 300 and 260 mya, the ancestral Atlantic Ocean closed as ancestral Africa, Europe, and South America collided with ancestral North America. Around this time, all the ancestral continents were joined together in one large supercontinent—Pangaea. This collision of landmasses resulted in extensive folding and faulting to form the Valley and Ridge Province, as illustrated in **Figure 13.** When rifting caused Pangaea to break apart about 200 mya, the modern Atlantic Ocean formed, and the continents moved to their present positions, as illustrated in **Figure 13.**

The Appalachian Mountains are only one example of the many mountain ranges that have formed along convergent boundaries. In Lesson 3, you will read about the type of orogeny that takes place along divergent plate boundaries.

Figure 14 Outcrops, such as this one, are common in the Blue Ridge Province.

Check Your Progress

Summary

- Orogeny refers to all of the processes that form mountain belts.

- Most mountain belts are associated with plate boundaries.

- Island arc complexes, highly deformed mountains, and very tall mountains form as a result of the convergence of tectonic plates.

- The Appalachian Mountains are geologically ancient; they began to form 700 to 800 mya.

Demonstrate Understanding

1. **Describe** how plate movements are responsible for orogenic belts.
2. **Identify** the tectonic plate that includes the Lesser Antilles.
3. **Explain** why you might find fossil shells at the top of a mountain.
4. **Differentiate** the types of mountains that form at convergent plate boundaries.

Explain Your Thinking

5. **Infer** how the Aleutian Islands in Alaska formed.
6. **Evaluate** this statement: The Appalachian Mountains are younger than the Himalayas.
7. **WRITING** **Connection** Write and illustrate the story of the formation of the Appalachian Mountains. A middle-school student is the audience for your written piece.

LEARNSMART Go online to follow your personalized learning path to review, practice, and reinforce your understanding.

OTHER TYPES OF MOUNTAIN BUILDING

FOCUS QUESTION

How do mountains that aren't on plate boundaries form?

Divergent-Boundary Mountains

When ocean ridges were first discovered, people in the scientific community were stunned by their length. These underwater volcanic mountains form a continuous chain that snakes along Earth's ocean floor for over 65,000 km. In addition to being much longer than most of their continental counterparts, these mountains formed as a result of different orogenic processes. Recall that ocean ridges are regions of broad uplift that form when new oceanic crust is created by seafloor spreading. The newly formed crust and underlying mantle at the ocean ridge are hot. When rocks are heated, they expand, which results in decreased density. This decrease allows the ridge to bulge upward, as illustrated in **Figure 15.** As the oceanic plates move away from the ridge, the newly formed crust and mantle cool and contract, and the surface of the crust subsides. As a result, the crust stands highest where the ocean crust is youngest, and the underwater mountain chains have gently sloping sides.

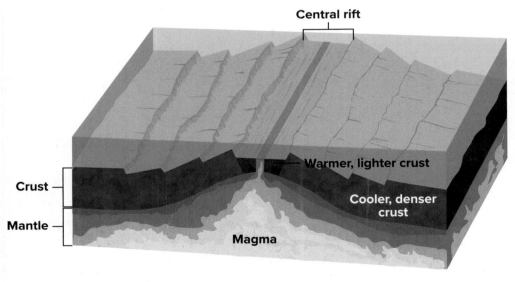

Figure 15 An ocean ridge is a broad, topographic high that forms as the lithosphere bulges upward due to an increase in temperature along a divergent boundary.

Determine *where the ocean is deeper—near the ridge or far from it.*

 3D THINKING　　**DCI** Disciplinary Core Ideas　　**CCC** Crosscutting Concepts　　**SEP** Science & Engineering Practices

COLLECT EVIDENCE

📝 Use your Science Journal to record the evidence you collect as you complete the readings and activities in this lesson.

INVESTIGATE

🔍 **GO ONLINE** to find these activities and more resources.

⚙️ **Applying Practices: Modeling Earth's Internal and Surface Processes**
HS-ESS2-1. Develop a model to illustrate how Earth's internal and surface processes operate at different spatial and **temporal scales** to form continental and ocean-floor features.

🥽 **GeoLAB: Make a Map Profile**
Develop and use a model to visualize the structure of Earth's surface features.

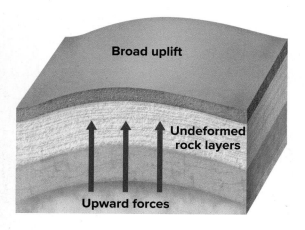

Broad uplift

Undeformed rock layers

Upward forces

Figure 16 The Adirondack Mountains of New York State are uplifted mountains. Uplifted mountains form when large sections of Earth's crust are forced upward without much structural deformation.

Uplifted Mountains

As illustrated in **Figure 16,** some mountains form when large regions of Earth have been slowly forced upward as a unit. These mountains are called **uplifted mountains.** The Adirondack Mountains in New York State, shown in **Figure 16,** are uplifted mountains. Generally, the rocks that compose uplifted mountains undergo less deformation than rocks associated with plate-boundary orogeny, which, as you have just read, are highly folded, faulted, and metamorphosed. The cause of large-scale regional uplift is not well understood. One popular hypothesis is that the part of the lithosphere made of mantle rocks becomes cold and dense enough that it sinks into the underlying mantle. The mantle lithosphere is replaced by hotter and less-dense mantle. The lower density of the new mantle provides buoyancy, which vertically lifts the overlying crust. This process has been used to explain the uplift of the Sierra Nevadas, in California, which are shown in **Figure 17.** When a whole region is uplifted, a relatively flat-topped area called a **plateau** can form. An example is the Colorado Plateau, which extends through Colorado, Utah, Arizona, and New Mexico. Erosion eventually carves these relatively undeformed, uplifted masses to form peaks, valleys, and canyons.

Figure 17 The Sierra Nevadas are the result of regional uplift.

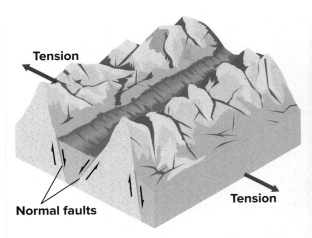

Tension

Normal faults

Tension

Figure 18 Fault-block mountains are areas of Earth's crust that are higher than the surrounding landscape as the result of faulting. The Basin and Range Province consists of hundreds of mountains separated by normal faults.

Fault-Block Mountains

Another type of mountain that is not necessarily associated with plate boundaries is a fault-block mountain. Recall how movement at faults can lift land on one side of a fault or drop it on the other. When Earth's crust is stretched, a series of normal faults can form. **Figure 18** illustrates how **fault-block mountains** form between these large faults when pieces of crust are dropped downward due to tensional forces. The Basin and Range Province of the southwestern United States and northern Mexico, a part of which is shown in **Figure 18,** consists of hundreds of nearly parallel mountains separated by normal faults. The Grand Tetons in Wyoming are also fault-block mountains.

Check Your Progress

Summary

- Divergent boundaries, uplift, and faulting produce some of Earth's mountains.

- Underwater volcanic mountains at divergent boundaries form Earth's longest mountain chain.

- Regional uplift can result in the formation of uplifted mountains that are made of nearly unde-formed layers of rock.

- Fault-block mountains form when large pieces of crust are dropped downward between normal faults.

Demonstrate Understanding

1. **Explain** why all of Earth's mountains do not form at convergent plate boundaries.
2. **Identify** the kinds of rocks associated with ocean ridges.
3. **Explain** why an ocean ridge is higher than the surrounding crust.
4. **Compare** ocean ridges and fault-block mountains.
5. **Compare and contrast** the formation of uplifted and fault-block mountains.

Explain Your Thinking

6. **Formulate** criteria for identifying an uplifted mountain.
7. **WRITING > Connection** Write three practice test questions for your classmates to assess their knowledge of mountain building.

©William Perry/age fotostock

LEARNSMART® Go online to follow your personalized learning path to review, practice, and reinforce your understanding.

Triple Crown of Hiking

The "Triple Crown" of hiking goes to hikers who trek the entire lengths of the Appalachian Trail, the Continental Divide Trail, and the Pacific Crest Trail. In total, these trails cover more than 12,000 km, with a vertical gain of 300 km. They traverse iconic mountain ranges in the continental United States.

Appalachian Trail

The Appalachian Trail is the oldest of the three trails, as is the mountain range that gives the trail its name. The trail stretches from Mount Katahdin, Maine, in the north to Springer Mountain, Georgia, in the south. Thru-hikers—those who hike the entire trail from one end to the other—pass through forests, mountains, meadows, streams, and deep ravines. On their trek, which takes about six months to complete, hikers see evidence of past geologic activity. More than 25,000 years ago, glaciers in the northern Appalachians sculpted two large cirque basins—the Great Basin and the North Basin—just east of the Katahdin trailhead.

Continental Divide Trail

Moving west, the Continental Divide Trail (CDT) follows the U.S. portion of the Continental Divide, from the Canadian border to the Mexican border. The U.S. Continental Divide runs through the Rocky Mountains and separates the main waterways of the continent. Formed by ancient tectonic activity, the Rocky Mountains are made of three formidable ranges with distinct geological origins.

The CDT is a challenging trail, some 1,500 m above sea level. Altitude sickness is a risk for hikers, who

Millions of hikers hit trails across the United States each year, but only a fraction are thru-hikers. Even fewer will achieve the "Triple Crown" of hiking.

travel from rolling deserts to icy glaciers high on mountainsides. The hike must be carefully timed in order to avoid hazards such as heat waves in the deserts and avalanches in the mountains.

Pacific Crest Trail

Farther west still is the Pacific Crest Trail, which stretches from northern Mexico to Canada. The five regions of the trail—Southern California, Central California, Northern California, Oregon, and Washington—each have unique geologic features. At the southern tip of the trail, hikers tackle deserts before climbing up into mountains surrounded by active faults. Heading north, hikers cross the Sierra Nevada range, uplifted by geologic forces, and enter the Cascades. In Oregon, the trail winds through old-growth forests dotted with volcanoes. At the northernmost section in Washington, hikers trek through the rugged northern Cascades, including Columbia Gorge.

COMMUNICATE SCIENTIFIC INFORMATION

Create a poster highlighting the mountains on one of the trails. Include information about the geologic processes that formed and shaped the mountain range.

MODULE 16
STUDY GUIDE

 GO ONLINE to study with your Science Notebook.

Lesson 1 CRUST-MANTLE RELATIONSHIPS

- The majority of Earth's elevations are either 0 to 1 km above sea level or 4 to 5 km below sea level.
- The mass of a mountain above Earth's surface is supported by a root that projects into the mantle.
- The addition of mass to Earth's crust depresses the crust, while the removal of mass from the crust causes the crust to rebound in a process called isostatic rebound.

- topography
- isostasy
- root
- isostatic rebound

Lesson 2 OROGENY

- Orogeny refers to all of the processes that form mountain belts.
- Most mountain belts are associated with plate boundaries.
- Island arc complexes, highly deformed mountains, and very tall mountains form as a result of the convergence of tectonic plates.
- The Appalachian Mountains are geologically ancient; they began to form 700 to 800 mya.

- orogeny
- compressive force

Lesson 3 OTHER TYPES OF MOUNTAIN BUILDING

- Divergent boundaries, uplift, and faulting produce some of Earth's mountains.
- Underwater volcanic mountains at divergent boundaries form Earth's longest mountain chain.
- Regional uplift can result in the formation of uplifted mountains that are made of nearly undeformed layers of rock.
- Fault-block mountains form when large pieces of the crust are dropped downward between normal faults.

- uplifted mountain
- plateau
- fault-block mountain

Module Wrap-Up

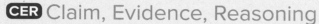

REVISIT THE PHENOMENON

How do mountains grow so large?

CER Claim, Evidence, Reasoning

Explain Your Reasoning Revisit the claim you made when you encountered the phenomenon. Summarize the evidence you gathered from your investigations and research and finalize your Summary Table. Does your evidence support your claim? If not, revise your claim. Explain why your evidence supports your claim.

STEM UNIT PROJECT
Now that you've completed the module, revisit your STEM unit project. You will apply your evidence from this module and complete your project.

GO FURTHER

SEP Data Analysis Lab
Can you get a rebound?

The rate of isostatic rebound changes over time. An initially rapid rate often declines to a very slow rate. The data shown in the table indicate rebound after the North American ice sheet melted 10,000 years ago.

Isostatic Rebound Data					
Years before present	8,000	6,000	4,000	2,000	0
Total amount of rebound (m)	50	75	88	94	97

Data and Observations Plot a graph with *Years before present* on the x-axis and *Total amount of rebound* on the y-axis. Study the graph. Note how the rate of isostatic rebound decreases with time.

CER Analyze and Interpret Data

1. **Claim, Evidence** What percentage of the total rebound occurred during the first 2,000 years? Show your work.
2. **Claim** Predict how much rebound will still occur and approximately how long this will take.
3. **Reasoning** What is similar about mountain erosion and glaciation in terms of isostatic rebound? What is different?

Daniel Prudek/Shutterstock

ENCOUNTER THE PHENOMENON

What can these fossils tell us about Earth millions of years ago?

SEP Ask Questions

What questions do you have about the phenomenon? Write your questions on sticky notes and add them to the driving question board for this unit.

How long is four billion years?

Look for Evidence

As you go through this unit, use the information and your experiences to help you answer the phenomenon question as well as your own questions. For each activity, record your observations in a Summary Table, add an explanation, and identify how it connects to the unit and module phenomenon questions.

Solve a Problem
STEM UNIT PROJECT

Geologic Time Investigate and research more about geologic time. Use the results of these investigations and the evidence you collected during the unit to complete your unit project.

GO ONLINE In addition to reading the information in your Student Edition, you can find the STEM Unit Project and other useful resources online.

ENCOUNTER THE PHENOMENON

How has information about Earth's history been preserved for so long?

GO ONLINE to play a video about how paleontology can be done in local communities.

SEP Ask Questions

Do you have other questions about the phenomenon? If so, add them to the driving question board.

CER Claim, Evidence, Reasoning

Make Your Claim Use your CER chart to make a claim about how information about Earth's history has been preserved. Explain your reasoning.

Collect Evidence Use the lessons in this module to collect evidence to support your claim. Record your evidence as you move through the module.

Explain Your Reasoning You will revisit your claim and explain your reasoning at the end of the module.

GO ONLINE to access your CER chart and explore resources that can help you collect evidence.

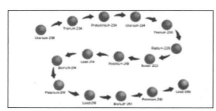

LESSON 3: Explore & Explain: Radioactive Isotopes

LESSON 4: Explore & Explain: The Fossil Record

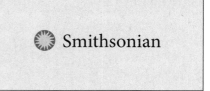

Additional Resources

FOCUS QUESTION

Why do we need a geologic time scale?

Organizing Time

A hike down the Grand Canyon reveals the multicolored layers of rock, called strata, that make up the canyon walls, as shown in **Figure 1.** Some of the layers contain fossils, which are the remains, traces, or imprints of ancient organisms. By studying rock layers and the fossils within them, geologists can reconstruct aspects of Earth's history and interpret ancient environments.

To help in the analysis of Earth's rocks, geologists have divided the history of Earth into time units. These time units are based largely on the fossils contained within the rocks. The time units are part of the **geologic time scale,** a record of Earth's history from its origin 4.6 billion years ago (bya) to the present. Since the naming of the Jurassic period (juh RA sihk) in 1795, additions and revisions to the time scale have continued to the present day. Some of the units have remained unchanged for centuries, while others have been reorganized as scientists have gained new knowledge. The geologic time scale is shown in **Figure 2.**

Figure 1 The rock layers of the Grand Canyon represent geologic events spanning nearly 2 billion years. Geologists study the rocks and fossils in each layer to learn about Earth's history during different units of time.

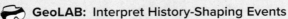

🔺 **3D THINKING**   **DCI** Disciplinary Core Ideas   **CCC** Crosscutting Concepts   **SEP** Science & Engineering Practices

COLLECT EVIDENCE

📝 Use your Science Journal to record the evidence you collect as you complete the readings and activities in this lesson.

INVESTIGATE

🔵 **GO ONLINE** to find these activities and more resources.

🥽 **GeoLAB: Interpret History-Shaping Events**
Communicate information about why some changes in Earth's development are important in the scheme of Earth's history.

❓ **Revisit the Encounter the Phenomenon Question**
What information from this lesson can help you answer the Unit and Module questions?

Figure 2 Visualizing the Geologic Time Scale

The geologic time scale begins with Earth's formation 4.6 billion years ago (bya). Geologists organize Earth's history according to groupings called eons. Each eon contains eras, which, in turn contain periods. Each period in the geologic time scale contains epochs. The current geologic epoch is called the Holocene epoch. Each unit on the scale is labeled with its range of time in millions of years ago (mya).

Identify *the period, era, and eon representing the most modern unit of time.*

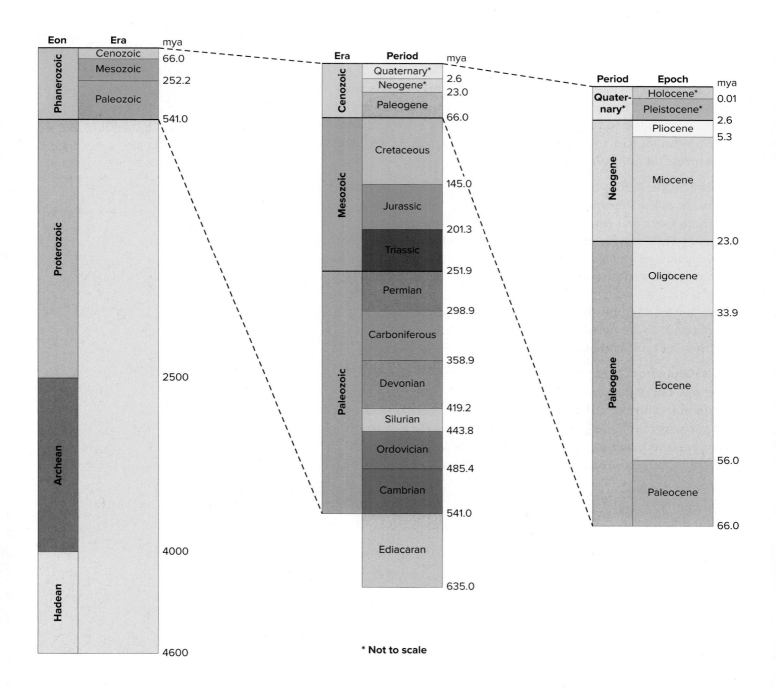

* Not to scale

The Geologic Time Scale

The geologic time scale enables scientists to find relationships among the geological events, environmental conditions, and fossilized life-forms that are preserved in the rock record. The oldest division of time is at the bottom of the scale, shown in **Figure 2.** Moving upward, each division is more recent, just as the rock layers in the rock record are generally younger toward the surface.

Get It?

Explain why scientists need a geologic time scale.

Eons

The time scale is divided into units called eons, eras, periods, and epochs. An **eon** is the largest of these time units and includes all the others. From oldest to youngest, they consist of the Hadean (HAY dee un), Archean (ar KEE un), Proterozoic (pro tuh ruh ZOH ihk), and Phanerozoic (fa nuh ruh ZOH ihk) eons.

The three earliest eons make up 90 percent of geologic time, known informally as the **Precambrian** (pree KAM bree un). During the Precambrian, Earth was formed and became hospitable to life. Fossil evidence, while scarce, suggests that simple life-forms evolved during the Archean eon and that by the end of the Proterozoic eon, life had evolved to the point that some organisms might have been able to move in complex ways. Most of these fossils, such as the one shown in **Figure 3,** formed from soft-bodied organisms, many of which resemble modern animals. Others had bodies with rigid parts. All life-forms until then had soft bodies without shells or skeletons.

Fossils dating from the most recent eon, the Phanerozoic, are the best preserved, not only because they are younger than those from the Precambrian, but because many of them represent organisms with hard parts such as bones, teeth, shells, and scales. Hard parts are more easily preserved than soft tissues. The time line in **Figure 4** shows some important fossil and age-dating discoveries.

Get It?

Explain why a fish is more likely to become fossilized than a snake is.

Figure 3 This is a well-preserved fossil found in a sedimentary rock of the Precambrian. During the latest span of that time, the first complex life-forms evolved on Earth.

Infer *Was this organism able to move on its own?*

SCIENCE USAGE v. COMMON USAGE

eon
Science usage: the largest unit of time in the geologic time scale
Common usage: implying a very long period of time

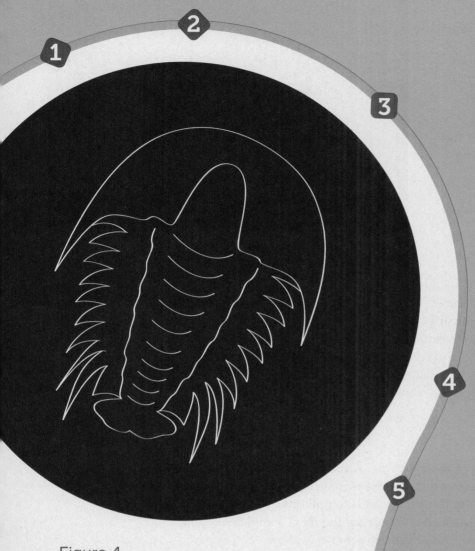

Figure 4

Fossil Discoveries and Technology

Fossil discoveries and dating technology have changed our understanding of life on Earth.

1 **1796** William Smith, a canal surveyor, creates the first geologic map based on distinct fossil layers.

2 **1820s** Mary Anning, the daughter of a cabinet maker, finds and identifies fossils of many ancient creatures, sparking great interest in paleontology.

3 **1857** Quarry workers uncover a skeleton identified as Neanderthal, a species similar to modern humans.

4 **1909** The discovery of the Burgess Shale fossils in the Canadian Rocky Mountains reveals the diversity of invertebrate life that thrived during the Cambrian period.

5 **1929** An Anasazi ruin becomes the first prehistoric site to be dated using tree-ring chronology.

6 **1946** University of Chicago scientists show that the age of relatively recent organic objects and artifacts can be determined with radiocarbon dating.

7 **1976** The most complete adult female skeleton of *Australopithecus afarensis*, named Lucy, is discovered in Northern Ethiopia.

8 **1987** Jenny Clack leads an expedition to Greenland and unearths fossils of animals that lived 360 mya, showing that animals developed legs prior to moving onto land.

9 **1993** Fossils found in western Australia provide evidence that bacteria existed 3.5 bya.

10 **2010** Scientists discover the oldest known animal fossils—spongelike creatures that lived about 650 mya—in South Australia.

Eras

All eons are made up of eras, the next-largest unit of time. **Eras** are usually tens to hundreds of millions of years in duration. Like all other time units, they are defined by the different fossils of life-forms present in the rocks. The names of the three eras of the Phanerozoic eon are named for the relative ages of the life-forms that lived during those times. For example, in Greek, *paleo* means "old," *meso* means "middle," and *ceno* means "recent." *Zoic* means "of life" in Greek. Thus, *Paleozoic* means "old life," *Mesozoic* means "middle life," and *Cenozoic* means "recent life."

Periods

All eras are divided into periods. **Periods** are generally tens of millions of years in duration, though some periods of the Precambrian are considerably longer. Some periods are named for the geographic region in which the rocks or fossils characterizing the age were first observed and described. Consider, for example, the Ediacaran (ee dee A kuh run) period at the end of the Proterozoic era. It is named for the Ediacara Hills in South Australia, shown in **Figure 5.** It was here that fossils typical of the period were first found. The Ediacaran period was added to the geologic time scale in 2004.

Figure 5 The Ediacara Hills of South Australia yielded the first fossils typical of the Ediacaran period. Fossils from that time, found anywhere in the world, are called Ediacaran fossils.

Epochs

Epochs (EH puhks) are even smaller divisions of geologic time. Although the time scale in **Figure 2** shows epochs only for periods of the Cenozoic era, all periods of geologic time are divided into epochs. **Epochs** are generally hundreds of thousands to millions of years in duration. The rock record from the epochs of the Cenozoic era are the most complete because there has been less time for weathering and erosion to remove evidence of this part of Earth's history. This also means that the fossil record for these periods of time is more complete. Because the time units are defined based on the presence of fossils, geologists are able to more finely subdivide the rock units of the Cenozoic. For this reason, the epochs of the Cenozoic are relatively short in duration. For example, the Holocene (HOH luh seen) epoch, which includes modern time, began only about 12,000 years ago.

The years that mark the divisions of the geologic time scale are constantly updated. This is due to a technique called radiometric dating, which allows scientists to use radioactive elements to refine the boundaries of the time periods. You will learn about this technique in Lesson 3 of this module.

Ray Warren/Shutterstock

Succession of Life-Forms

During the Phanerozoic eon, multicellular life diversified and left abundant fossils. *Phanerozoic* means "visible life" in Greek. During the first era of the Phanerozoic, the Paleozoic (pay lee uh ZOH ihk), the oceans were home to a wide variety of organisms. Trilobites, such as those shown in **Figure 6,** were among the first life-forms with hard parts. They dominated the oceans during the early Paleozoic era. Land plants evolved later, followed by land animals. Swamps of the Carboniferous (kar buh NIH fuh rus) period provided the plant material that developed into the coal deposits of today. The Paleozoic ended with the largest mass extinction event in Earth's history. In a **mass extinction,** many groups of organisms die out over a relatively short period of geologic time. At the end of the Paleozoic era, 95 percent of all marine organisms became extinct.

The age of dinosaurs

The second era, the Mesozoic (mez uh ZOH ihk), is known for the emergence of dinosaurs, but many other organisms also appeared during this time. Large predatory reptiles ruled the oceans, and corals closely related to today's corals built huge reef systems. Water-dwelling amphibians began adapting to terrestrial environments. Insects, some as large as birds, thrived. Mammals evolved and diversified. Flowering plants and trees emerged. The Mesozoic ended with another large extinction event. Notable extinctions include the nonavian dinosaurs and large marine reptiles.

The rise of mammals

The last era is the Cenozoic (sen uh ZOH ihk). During this time, mammals began to dominate the land, increasing both in number and diversity. Human ancestors, the first primates, emerged during the Paleocene epoch, and modern humans appeared during the Pleistocene (PLYS tuh seen) epoch.

Figure 6 Trilobites are Paleozoic arthropod fossils found all over the world. The last trilobites became extinct at the end of the Paleozoic era.

✏️ Check Your Progress

Summary

- Scientists organize geologic time into eons, eras, periods, and epochs.
- Scientists divide time into units based largely on fossils of plants and animals.
- The Precambrian makes up nearly 90 percent of geologic time.
- The geologic time scale changes as scientists learn more about Earth.

Demonstrate Understanding

1. **Explain** the purpose of the geologic time scale.
2. **Distinguish** among eons, eras, periods, and epochs, using specific examples.
3. **Describe** the importance of extinction events to geologists.
4. **Explain** why scientists know more about the Cenozoic than they do about other eras.

Explain Your Thinking

5. **Discuss** why scientists know so little about the Precambrian.
6. **MATH Connection** Make a bar graph showing the relative percentage of time spanned by each era of the Phanerozoic eon. For more help, refer to the *Skillbuilder Handbook*.

LEARNSMART Go online to follow your personalized learning path to review, practice, and reinforce your understanding.

Alan Morgan

RELATIVE-AGE DATING

How do we know one rock layer is older than another?

Interpreting Geology

Recall that Earth's history stretches back billions of years. Scientists have not always thought that Earth was this old. Early ideas about Earth's age were generally placed in the context of time spans that a person could understand relative to his or her own life. This changed as people began to explore Earth and Earth processes in scientific ways. James Hutton, a Scottish geologist who lived during the late 1700s, was one of the first scientists to think of Earth as very old. He attempted to explain Earth's history in terms of geologic forces, such as erosion and sea-level changes, that operate over long stretches of time. His work helped set the stage for the development of the geologic time scale.

Uniformitarianism

Hutton's work lies at the foundation of **uniformitarianism,** a concept which states that geologic processes occurring today have been occurring since Earth formed. The only exception is that the rate, scale, and intensity of the processes may have changed. For example, Earth's glaciers today cause erosion of the landscape just like the much larger ice sheets that formed during the ice ages did. Another example of uniformitarianism is shown in **Figure 7.**

Figure 7 The waves crashing on an ancient Jurassic beach probably looked much like those crashing on this modern beach in Oregon. The geologic processes that formed it are unchanged.

Bogdan Bratosin/Moment/Getty Images

3D THINKING **DCI** Disciplinary Core Ideas **CCC** Crosscutting Concepts **SEP** Science & Engineering Practices

COLLECT EVIDENCE

 Use your Science Journal to record the evidence you collect as you complete the readings and activities in this lesson.

INVESTIGATE

 GO ONLINE to find these activities and more resources.

Quick Investigation: Determine Relative Age
Create and use a model to visualize the patterns used to determine the relative ages of rock layers.

? **Revisit the Encounter the Phenomenon Question**
What information from this lesson can help you answer the Unit and Module questions?

Principles for Determining Relative Age

By applying the concept of uniformitarianism, scientists are able to learn about the past by studying the present. One way to do this is by studying the order in which geologic events occurred using a process called **relative-age dating.** This does not allow scientists to determine exactly how many years ago an event occurred, but it gives scientists a clearer understanding about the relative order of geologic events in Earth's history.

In 1669, a Danish physician named Nicolaus Steno set forth four principles that enabled scientists to interpret the relative ages of rock layers and events in Earth's history. These principles were so profound that they are still in use today. The principles are original horizontality, superposition, cross-cutting relationships, and lateral continuity.

Original horizontality

The principle of **original horizontality** states that sediments that make up sedimentary rocks are deposited in horizontal or nearly horizontal layers. This can be seen in the walls of the Grand Canyon, illustrated in **Figure 8.** Sediment is deposited in horizontal layers for the same reason that layers of sand on a beach are mostly flat; that is, gravity combined with wind and water spreads the sediment and sand evenly.

Superposition

Geologists cannot determine the numeric ages of the rock layers in the Grand Canyon using relative-age dating methods. However, they can assume that the oldest rocks are at the bottom and that each successive layer above is younger. Thus, they can infer that the Kaibab Limestone at the top of the canyon is much younger than the Vishnu Schist, which is at the bottom. This is an application of the principle of **superposition,** which states that in an undisturbed rock sequence, the oldest rocks are at the bottom, and each consecutive layer is younger than the layer beneath it.

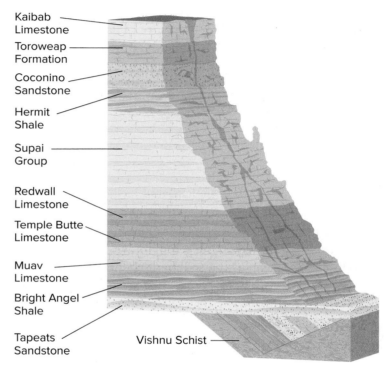

Kaibab Limestone
Toroweap Formation
Coconino Sandstone
Hermit Shale
Supai Group
Redwall Limestone
Temple Butte Limestone
Muav Limestone
Bright Angel Shale
Tapeats Sandstone
Vishnu Schist

Figure 8 The horizontal layers of the Grand Canyon were formed by deposition of sediment over millions of years. The principle of original horizontality states that the tilted strata at the bottom were deposited horizontally and sometime later were tilted by geologic forces.

ACADEMIC VOCABULARY

principle
a general hypothesis that has been tested repeatedly; sometimes also called a law
The geologic principle was illustrated in the rock layers the students observed.

Cross-cutting relationships

Rocks exposed in the deepest part of the Grand Canyon are mostly igneous and metamorphic. Within the metamorphic schist of the Vishnu Group in the bottom sequence are intrusions—also called dikes—of granite, as shown in **Figure 9.** You have learned that intrusions are rocks that form when magma intrudes and solidifies in existing rock. The principle of **cross-cutting relationships** states that an intrusion is younger than the rock it cuts across. Therefore, the granite intrusion in the Grand Canyon is younger than the schist because the granite cuts across the schist.

The principle of cross-cutting relationships also applies to faults. Recall that a fault is a fracture in Earth's crust along which movement takes place. Many faults exist in earthquake-prone areas, such as California, and in ancient, mountainous regions, such as the Adirondacks of New York. A fault is younger than the strata and surrounding geologic features because the fault cuts across them.

Figure 9 According to the principle of cross-cutting relationships, this igneous intrusion is younger than the Precambrian-aged red beds it cuts across.

Infer *how the igneous intrusion was formed.*

Lateral continuity

Recall that all sediment is deposited in horizontal or nearly horizontal layers. It is possible to trace such layers over very large distances and in all directions. Steno called this relationship the principle of **lateral continuity.** He noticed that layers of rock extend in all directions until the environment in which the sediment was deposited changes. Picture, for example, the shoreline of an ocean as the sediment changes from that which forms limestone to that which forms sandstone. This principle is useful when matching rock layers across a large river valley or canyon. The rock layers have been cut through. But knowing that they were once laterally continuous allows us to determine which widely separated rock layers are the same age. This matching technique, called correlation, is discussed later in this lesson.

Inclusions

Relative age can also be determined where one rock layer contains pieces of rock from the layer next to it. This might occur after an exposed layer has eroded and the loose material on the surface has become incorporated into the layer deposited on top of it. The **principle of inclusions** states that the fragments, called inclusions, in a rock layer must be older than the rock layer that contains them. As you have learned, once a rock has eroded, the resulting sediment might be transported and redeposited many kilometers away. In this way, a rock formed during the Triassic period might contain inclusions from a Cambrian-aged rock. Inclusions can also form from pieces of rock that are trapped within a lava flow.

 Get It?

Sketch a hypothetical area that illustrates all five geologic principles. Label the section of the sketch that shows each principle.

Figure 10 An unconformity is any erosional surface separating two layers of rock that have been deposited at different times. The three types of unconformities are shown below.

Disconformity Horizontal sedimentary layer overlies horizontal sedimentary layer

Nonconformity Horizontal sedimentary layer overlies nonsedimentary layer

Angular unconformity Horizontal sedimentary layer overlies tilted sedimentary layers

Unconformities

Earth's surface is constantly changing as a result of weathering, erosion, earthquakes, volcanism, and other processes. This makes it difficult to find a sequence of rock layers in which a layer has not been disturbed. Sometimes the record of a past event or time period is missing entirely. For example, if rocks from a volcanic eruption erode, the record of that eruption is lost. If an eroded area is covered at a later time by a new layer of sediment, the eroded surface represents a gap in the rock record. Buried surfaces of erosion are called **unconformities.** The rock immediately above an unconformity is sometimes considerably younger than the rock immediately below it. Three different types of unconformities are recognized. They are shown in **Figure 10.**

Disconformity When a horizontal layer of sedimentary rock overlies another horizontal layer of sedimentary rock that has been eroded, a disconformity forms. This type of unconformity often forms when there is a change in sea level. As the water raises or lowers, the action of the waves at the shoreline causes erosion of the rock layers or sediment beneath them. Once the change in sea level stops, sedimentation begins again. The eroded surface gets buried and the disconformity is formed. Disconformities are easiest to identify when the eroded surface is uneven.

Nonconformity When a layer of sedimentary rock overlies an eroded layer of igneous or metamorphic rock, such as granite or marble, a nonconformity is formed. The eroded surface in this situation is easy to identify. Both granite and marble form deep in Earth. A nonconformity records a gap in the rock record during which rock layers were uplifted and eroded at Earth's surface. Then new layers of sedimentary rock formed on top of them. If the igneous or metamorphic rock were younger than the overlying sedimentary rock, the bottom of the sedimentary rock layer would have been metamorphosed rather than eroded.

 Get It?

Distinguish between a disconformity and a nonconformity.

Angular unconformity When horizontal layers of sedimentary rock are deformed during mountain building or other geologic events involving compressional forces, they are usually uplifted and tilted. During this process, the layers are exposed to weathering and erosion. If horizontal layers of sedimentary rock are later laid down on top of the tilted, eroded layers, the resulting unconformity is called an angular unconformity. Angular unconformities indicate the complex history of uplift and erosion.

 Get It?

Apply the principle of superposition to interpret the relative ages of the rock layers in the bottom photo of **Figure 10.**

Correlation

The Kaibab Limestone layer rims the top of the Grand Canyon in Arizona, but it is also found more than 100 km away at the bottom of the rock layers in Zion National Park in Utah. How do geologists know that these layers, which are far apart from each other, formed at the same time? One method is by correlation (kor uh LAY shun). **Correlation** is the matching of rock outcrops or fossils exposed in one geographic region to similar outcrops or fossils exposed in other geographic regions. Through correlation of many different layers of rocks, geologists have determined that Zion National Park, Bryce Canyon, and the Grand Canyon are all part of one layered sequence called the Grand Staircase, as illustrated in **Figure 11.**

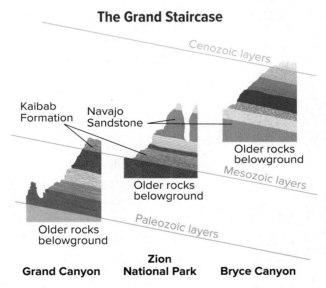

The Grand Staircase

Cenozoic layers

Kaibab Formation

Navajo Sandstone

Older rocks belowground

Mesozoic layers

Older rocks belowground

Older rocks belowground

Paleozoic layers

Grand Canyon

Zion National Park

Bryce Canyon

Figure 11 The top layers of rocks at the Grand Canyon are identical to the bottom layers at Zion National Park, and the top layers at Zion are the same as the bottom layers at Bryce Canyon.

Infer *the makeup of the buried layer below Zion's Kaibab layer.*

CCC CROSSCUTTING CONCEPTS

Patterns The ability of geologists to correlate rock layers across large distances is due to the fact that patterns can be recognized in the rock layers. Make a list of the types of patterns geologists can use in correlation.

Key beds Distinctive rock layers are sometimes deposited over wide geographic areas as a result of a large meteorite strike, volcanic eruption, or other brief event. For example, the key-bed ash layer that marks the 1980 eruption of Mount St. Helens can be found in many states in the U.S. and in parts of Canada. Because these types of layers are easy to recognize, they help geologists correlate rock formations in different geographic areas where the layers are exposed. A rock or sediment layer used as a marker in this way is called a **key bed.** Using the principle of superposition, geologists know that the layers above a key bed are younger than the layers below it.

Get It?

Infer What other types of geologic events or processes could produce key beds?

Fossil correlation Geologists also use fossils to correlate rock formations in locations that are geographically distant. As shown in **Figure 12,** fossils of organisms that lived at the same time can be correlated across large regions. Fossils can indicate similar times of deposition even though the sediments in which they were deposited, and resulting rocks, might be entirely different.

The correlation of fossils and rock layers aids in the relative dating of rock sequences and helps geologists understand the history of larger geographic regions. Petroleum geologists also use correlation to help them locate reserves of oil and gas. For example, if a sandstone layer in one area contains oil, it is possible that the same layer in other areas also contains oil. It is largely through correlation that geologists have constructed the geologic time scale.

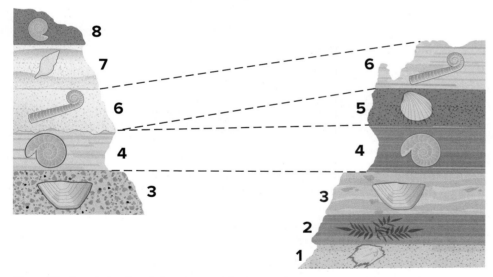

Figure 12 Correlating fossils from rock layers in one location to fossils from rock layers in another location shows that the layers were deposited during roughly the same time period.

Index Fossils Some fossils are more useful than others for relative-age dating. **Index fossils** are fossils that are easily recognized, abundant, and widely distributed geographically. They also represent species that existed for relatively short periods of geologic time. The different species of trilobites shown in **Figure 13** make excellent index fossils for periods within the Paleozoic era because each was distinct, abundant, and existed for a certain range of time. Ammonites, extinct marine organisms related to nautiloids and squids, are excellent index fossils for the Mesozoic. If a geologist finds an index fossil in a rock layer, he or she can immediately determine an approximate age of the layer.

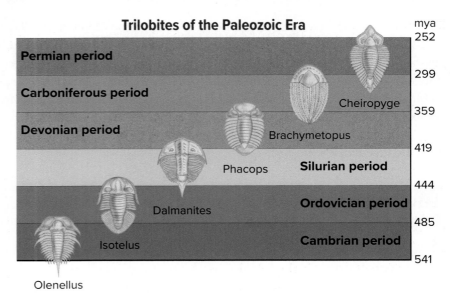

Trilobites of the Paleozoic Era

mya
252

Permian period

299

Carboniferous period

Cheiropyge

359

Devonian period

Brachymetopus

419

Phacops

Silurian period

444

Dalmanites

Ordovician period

485

Isotelus

Cambrian period

541

Olenellus

Figure 13 These trilobite species make excellent index fossils because each species lived for a relatively short period of time before becoming extinct, and they were abundant and widespread.

Check Your Progress

Summary

- The principle of uniformitarianism states that processes occurring today have been occurring since Earth formed.
- Scientists use geologic principles to determine the relative ages of rock sequences.
- An unconformity represents a gap of time in the rock record.
- Geologists use correlation to compare rock layers in different geographic areas.
- Index fossils help scientists correlate rock layers in the geologic record.

Demonstrate Understanding

1. **Summarize** the principles that geologists use to determine relative ages of rocks.
2. **Make a diagram** to compare and contrast the three types of unconformities.
3. **Explain** how geologists use fossils to determine the relative ages of rock layers within a large region.
4. **Discuss** how a coal seam might be used as a key bed.
5. **Apply** Explain how the principle of uniformitarianism would help geologists determine the source of a layer of particular igneous rock.

Explain Your Thinking

6. **Propose** how a scientist might support a hypothesis that rocks from one quarry were formed at the same time as rocks from another quarry 50 km away.
7. **WRITING** **Connection** Write a paragraph that explains how an event, such as a large hurricane, might result in a key bed. Use a specific example in your paragraph.

ABSOLUTE-AGE DATING

FOCUS QUESTION

How can we measure how old a rock is?

Radioactive Isotopes

As you have learned, relative-age dating is a method of comparing past geologic events based on the order of strata in the rock record. In contrast, **absolute-age dating** enables scientists to determine the numerical age of rocks and other objects. In one method of absolute-age dating, scientists measure the decay of the radioactive isotopes in igneous and metamorphic rocks and in the remains of some organisms preserved in sediments.

Radioactive decay

Radioactive isotopes emit nuclear particles at a constant rate. Recall that an element is defined by the number of protons it contains. As the number of protons changes with each emission, the original radioactive isotope, called the parent, is gradually converted to a different element, called the daughter. For example, a radioactive isotope of uranium, U-238, will decay into the daughter isotope lead-206 (Pb-206) over a specific span of time, as illustrated in **Figure 14.** Eventually, enough of the parent decays that traces of it are undetectable, and only the daughter product is measurable. The emission of radioactive particles and the resulting change into other isotopes over time is called **radioactive decay.** Because the rate of radioactive decay is constant regardless of pressure, temperature, or any other physical changes, scientists use it to determine the absolute age of a rock or a geologic event.

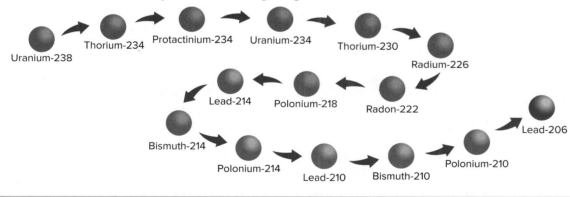

Uranium-238 → Thorium-234 → Protactinium-234 → Uranium-234 → Thorium-230 → Radium-226 → Radon-222 → Polonium-218 → Lead-214 → Bismuth-214 → Polonium-214 → Lead-210 → Bismuth-210 → Polonium-210 → Lead-206

Figure 14 The decay of U-238 to Pb-206 follows a specific and unchanging path.

![3D THINKING icon] **3D THINKING**　　**DCI** Disciplinary Core Ideas　　**CCC** Crosscutting Concepts　　**SEP** Science & Engineering Practices

COLLECT EVIDENCE

 Use your Science Journal to record the evidence you collect as you complete the readings and activities in this lesson.

INVESTIGATE

GO ONLINE to find these activities and more resources.

 Review the News
Obtain information from a current news story about absolute-age dating. Evaluate your source and communicate your findings to your class.

 Revisit the Encounter the Phenomenon Question
What information from this lesson can help you answer the Unit and Module questions?

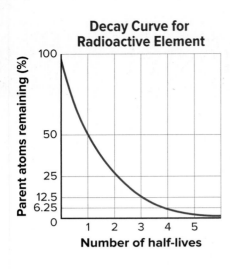

Decay Curve for Radioactive Element

Parent atoms remaining (%)

Number of half-lives

Growth Curve for Daughter Product

Daughter atoms forming (%)

Number of half-lives

Figure 15 During radioactive decay, the number of parent atoms decreases at the same rate that the number of daughter atoms increases.

Interpret *What percentage of daughter isotope would exist in a sample containing 50 percent parent isotope?*

Radiometric Dating

As the number of parent atoms decreases during radioactive decay, the number of daughter atoms increases, as shown in **Figure 15.** The ratio of parent isotope to daughter isotope in a mineral can be used to determine the amount of time that has passed since the object formed. For example, by measuring this ratio in the minerals of an igneous rock, geologists pinpoint when the minerals first crystallized from magma. When scientists date an object using radioactive isotopes, they are using a method called **radiometric dating.**

Half-life

Scientists measure the length of time it takes for one-half of the original parent isotope to decay, called its **half-life.** After one half-life, 50 percent of the parent remains, resulting in a 1:1 ratio of parent-to-daughter product. After two half-lives, one-half of the remaining 50 percent of the parent decays. The result is a 25:75 percent ratio of the original parent to the daughter product—a 1:3 ratio. This process is shown in **Figure 16.**

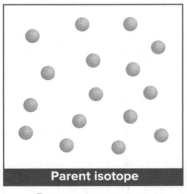

Parent isotope

- 100 percent parent

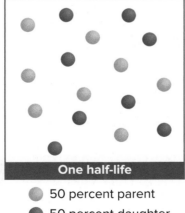

One half-life

- 50 percent parent
- 50 percent daughter

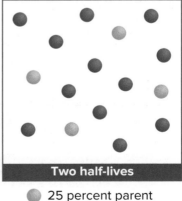

Two half-lives

- 25 percent parent
- 75 percent daughter

Figure 16 After one half-life, a sample contains 50 percent parent and 50 percent daughter. After two half-lives, the sample contains 25 percent parent and 75 percent daughter.

Table 1 Half-Lives of Selected Radioactive Isotopes

Radioactive Parent Isotope	Approximate Half-Life	Daughter Product
Rubidium-87 (Rb-87)	48.8 billion years	Strontium-87 (Sr-87)
Thorium-232 (Th-232)	14.0 billion years	Lead-208 (Pb-2080)
Uranium-238 (U-238)	4.5 billion years	Lead-206 (Pb-206)
Potassium-40 (K-40)	1.3 billion years	Argon-40 (Ar-40)
Uranium-235 (U-235)	0.7 billion years	Lead-207 (Pb-207)
Carbon-14 (C-14)	5730 years	Nitrogen-14 (N-14)

Dating rocks

To date an igneous or metamorphic rock using radiometric dating, scientists examine the parent-daughter ratios of the radioactive isotopes in the minerals that comprise the rock. **Table 1** lists some of the radioactive isotopes they might use. The best isotope to use for dating depends on the approximate age of the rock being dated. For example, scientists might use uranium-235 (U-235), which has a half-life of 700 million years, to date a rock that is a few tens of millions of years old. Conversely, to date a rock that is hundreds of millions of years old, scientists might use U-238, which has a longer half-life. If an isotope with a shorter half-life is used for an ancient rock, there might be a point when the parent-daughter ratio becomes too small to measure. This technique is also used to determine the age of meteorites that have been found on Earth's surface. Although they range in age from about 4.5 billion years old to about 200 million years old, scientists can confirm that the oldest meteorites date back to the formation of Earth and the solar system. Radiometric dating is not useful for dating sedimentary rocks because, as you have learned, the minerals in most sedimentary rocks were formed from pre-existing rocks. **Figure 17** shows how geologists can learn the approximate age of sedimentary layers by dating layers of igneous rock that lie between them.

> ### Get It?
> **Describe** The production of nuclear energy results in radioactive waste materials. Refer to the second column in **Table 1** to describe why it is important to dispose of these materials responsibly.

Radiocarbon dating

Notice in **Table 1** that the half-life of carbon-14 (C-14) is much shorter than the half-lives of other isotopes. Scientists use C-14 to determine the age of organic materials, which contain abundant carbon, in a process called **radiocarbon dating.** Organic materials used in radiocarbon dating include plant and animal material such as bones, charcoal, and amber.

The tissues of all living organisms, including humans, contain small amounts of C-14. During an organism's life, the C-14 decays but is continually replenished by the process of respiration. When the organism dies, it no longer takes in C-14, so over time, the amount of C-14 decreases. Scientists can measure the amount of C-14 in organic material to determine how much time has passed since the organism's death. This method is used for dating only recent geologic events—within the last 60,000 years.

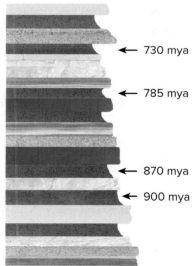

← 730 mya

← 785 mya

← 870 mya

← 900 mya

Radiometric Dating of Volcanic Ash

Figure 17 To help them determine the age of sedimentary rocks, scientists date layers of igneous rock or volcanic ash above and below the sedimentary rock layers.

Other Ways to Determine Absolute Age

In addition to radiometric dating, geologists also use tree rings, ice cores, and lake-bottom and ocean-bottom sediments to help determine the ages of objects or events.

Tree rings

Many trees contain a record of time in the growth rings of their trunks. These rings are called annual tree rings. Each annual tree ring consists of a pair of early-season and late-season growth rings. The width of the rings depends on certain conditions in the environment. For example, when rain is plentiful, trees grow fast and rings are wide. The harsh conditions of drought result in narrow rings. Trees from the same geographic region tend to have the same patterns of ring widths for a given time span. By matching the rings in these trees, as shown in **Figure 18,** scientists have established tree-ring chronologies that can span time periods up to 10,000 years.

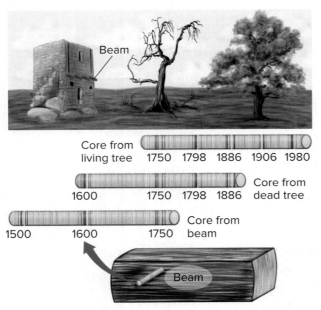

Figure 18 Tree-ring chronologies can be established by matching tree rings from different wood samples, both living and dead. The science of using tree rings to determine absolute age is called dendrochronology.

Calculate *the number of years represented in this tree-ring chronology.*

The science of using tree rings to determine absolute age is called **dendrochronology** and has helped geologists date relatively recent geologic events—such as volcanic eruptions, earthquakes, and glaciation—that have toppled tress. It is also useful in archaeological studies. In Mesa Verde National Park in Colorado, archaeologists used dendrochronology to determine the age of the wooden rafters in the pueblos of the Anasazi, an ancient group of Native Americans. This allowed scientists to approximate a time during which the Anasazi people lived. Also, dendrochronology provides a reliable way for geologists to confirm the results from radiocarbon dating.

Ice cores

Ice cores are analogous to tree rings. Like tree rings, they contain a record of past environmental conditions such as temperature and atmospheric composition, but in annual layers of snow deposition. Summer ice tends to have more bubbles and larger crystals than winter ice. Geologists use ice-core chronologies to study glacial cycles through geologic history. Several facilities around the world store thousands of meters of ice cores from ice sheets, such as the core shown in **Figure 19.** Because ice cores contain information about past environmental conditions, scientists also use them to study climate change.

Figure 19 Ice cores are stored in facilities such as the National Ice Core Facility in Denver, Colorado. Scientists use ice cores to date glacier deposits and to learn about ancient climates.

Figure 20 The alternating bands of sediment in varves help scientists date the cycles of deposition in glacial lakes.

Varves

Bands of alternating light- and dark-colored sediments of sand, clay, and silt are called **varves.** Varves represent the seasonal deposition of sediments, usually in lakes. Summer deposits are generally sand-sized particles with traces of organic matter. These bands are usually lighter and thicker than the dark, fine-grained sediments that represent the winter. Varves, shown in **Figure 20,** are typical of lake deposits near glaciers, where summer meltwaters actively carry sand into the lake, and little to no sedimentation occurs in the winter. Using varve cores, scientists can date cycles of glacial sedimentation over periods as long as 120,000 years.

Check Your Progress

Summary

- Techniques of absolute-age dating help identify numeric dates of geologic events.

- The decay rate of certain radioactive elements can be used as a kind of geologic clock.

- The concept of half-life is used to calculate the numeric ages of igneous and metamorphic rocks.

- Annual tree rings, ice cores, and sediment deposits can be used to date recent geologic events.

Demonstrate Understanding

1. **Point out** the differences between relative-age dating and absolute-age dating.

2. **Explain** how the process of radioactive decay can provide more accurate measurements of age compared to relative-age dating methods.

3. **Compare and contrast** the use of U-238 and C-14 in absolute-age dating.

4. **Describe** the usefulness of varves to geologists who study glacial lake deposits.

5. **Discuss** the link between uniformitarianism and absolute-age dating.

Explain Your Thinking

6. **Infer** why scientists might choose to use two different methods to date a tree felled by an advancing glacier. What methods might the scientists use?

7. **MATH** **Connection** A rock sample contains 25 percent K-40 and 75 percent daughter product Ar-40. If K-40 has a half-life of 1.3 billion years, how old is the rock?

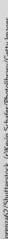

LEARNSMART Go online to follow your personalized learning path to review, practice, and reinforce your understanding.

FOSSIL REMAINS

FOCUS QUESTION
How do living things become fossils?

The Fossil Record

Fossils are the preserved remains or traces of once-living organisms. They provide evidence of the past existence of a wide variety of life-forms, most of which are now extinct. The diverse fossil record also provides evidence that species—groups of closely related organisms—have evolved. **Evolution** (eh vuh LEW shun) is the change in species over time.

When geologists find fossils in rocks, they know that the rocks are about the same age as the fossils, and they can infer that the same fossils found elsewhere are also of the same age. This is another application of correlation, as you learned about in Lesson 3. Some fossils, such as the radiolarian microfossils shown in **Figure 21**, also provide information about past climates and environments. Radiolarians are unicellular organisms with hard shells that have populated the oceans since the Cambrian period. When they die, their shells can be deposited in large quantities and can form an ocean sediment called radiolarian ooze.

Petroleum geologists use radiolarians and other microfossils to determine the ages and types of rocks where oil might be found. Microfossils can also indicate whether the rocks had ever been subjected to the temperatures and pressures necessary to form oil or gas.

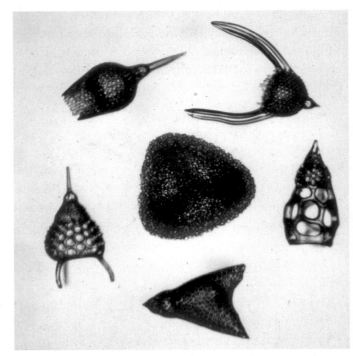

Figure 21 These tiny radiolarian microfossils—each no bigger than 1 mm in diameter—provide geologists clues about ancient marine environments. This photograph is a color-enhanced SEM magnification.

Get It?
Infer The fossils in the photo are called microfossils. What type of organisms might be preserved as macrofossils?

 **3D THINKING** **DCI** Disciplinary Core Ideas   **CCC** Crosscutting Concepts   **SEP** Science & Engineering Practices

COLLECT EVIDENCE
Use your Science Journal to record the evidence you collect as you complete the readings and activities in this lesson.

INVESTIGATE
GO ONLINE to find these activities and more resources.

Design Your Own: Analysis of a Climate Change Timeline Using Planktonic Foraminifera
Conduct an investigation into the effect of climate change on an indicator fossil.

Investigation Lab: Fossilization and Earth's History
Create and use models to visualize the result of fossilization.

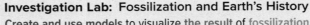

Original preservation

Fossils with **original preservation** are the remains of plants and animals that have been altered very little since the organisms' deaths. Such fossils are rare because their preservation requires extraordinary circumstances, such as freezing, arid, or oxygen-free environments. For example, soft parts of mammoths are preserved in the sticky tar of California's La Brea Tar Pits. Original woody parts of plants are embedded in the permafrost of 10,000-year-old Alaskan bogs. Tree sap from prehistoric trees that has hardened into amber contains the remains of entrapped insects, as illustrated in **Figure 22.** Soft parts may also be preserved when plants or animals are dried and their remains are mummified.

Original preservation fossils can be surprisingly old. In 2005, a dinosaur bone with preserved soft tissue was excavated in Montana. Carbon dating of this tissue revealed the bone to be 70 million years old. Scientists have since found preserved tissue with other dinosaur bones.

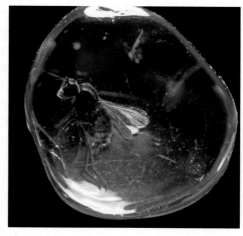

Figure 22 This insect was trapped in tree sap millions of years ago.

Explain *why fossils with original preservation are rare.*

Altered hard parts

Under most circumstances, the soft organic material of plants and animals decays quickly. However, over time, the remaining hard parts, such as shells, bones, or cell walls, can become fossils as **altered hard parts.** These fossils are the most common type of fossil and can form from two processes.

Mineral replacement In the process of **mineral replacement,** the pore spaces of an organism's buried hard parts are filled in with minerals from groundwater. The groundwater comes in contact with the hard part and gradually replaces the hard part's original mineral material with a different mineral. A shell's calcite ($CaCO_3$), for example, might be replaced by silica (SiO_2). Mineral replacement can occur in trees that are buried by volcanic ash. Over time, minerals dissolved from the ash fill the microscopic spaces within the wood. The result is a fossil called petrified wood, shown in **Figure 23.**

Figure 23 Petrified wood is an example of mineral replacement in fossils. The blowout shows that tree rings and cell walls are still evident at 100× magnification with a light microscope.

Describe *from where the minerals in the petrified wood came.*

Recrystallization Another way in which hard parts can be altered and preserved is the process of recrystallization (ree krihs tuh luh ZAY shun). This occurs when a buried hard part is subjected to changes in temperature and pressure. This process is similar to that of mineral replacement. However, in mineral replacement the original mineral is replaced by a different mineral from the water, whereas in recrystallization the original mineral is transformed into a new mineral with the same chemical composition. The original shape of the fossil is preserved, but the fine details are often destroyed. A snail shell, for example, is composed of the mineral aragonite ($CaCO_3$). Through recrystallization, the aragonite undergoes a change in internal structure to become calcite, the basic material of limestone or chalk. Though calcite has the same composition ($CaCO_3$) as aragonite, its crystal structure is more stable than that of aragonite over long periods of time. In fact, it is rare to find fossil shells older than a few hundred thousand years that are still composed of aragonite. **Figure 24** shows how mineral replacement and recrystallization differ.

Minerals in water replace original materials.

Shell mineral replaced by different form of same material

Mineral replacement   **Recrystallization**

Figure 24 During mineral replacement, the minerals in a buried hard part are replaced by other minerals in groundwater. During recrystallization, temperature and pressure change the crystal structure of the hard part's original material.

Molds and casts Some fossils do not contain any original or altered material of the original organism. These fossils might instead be molds or casts. A **mold** forms when sediments cover the original hard part of an organism, such as a shell, and the hard part is later removed by erosion, weathering, or dissolution. As shown in **Figure 25** on the next page, a hollowed-out impression of the shell, called the mold, is left in its place. A mold might later become filled with material to create a **cast** of the shell.

HISTORY ▶ Connection Some very well-known molds and casts can be seen in the ancient Italian city of Pompeii. In A.D. 79, the volcano Mt. Vesuvius, thought at the time to be simply a mountain, erupted and buried the area around it in pyroclastic material. The inhabitants of the region were caught by surprise, and very few survived the eruption. The city of Pompeii was completely buried in volcanic ash, and the nearby city of Herculaneum was covered in ash and lava. Moisture from the decomposing bodies in Pompeii mixed with the volcanic ash to create a seal around the bodies. Details as fine as facial expressions were preserved as the wet ash hardened into a cementlike material. Nearly 2000 years later, an archaeologist discovered these preserved bodies. By carefully pouring plaster into holes drilled in the cement coverings and then removing the cement, he was able to reveal nearly perfect mold and cast fossils of the residents of Pompeii. Because so much of the cities were preserved under the ash and lava, excavations in the area have revealed a near-perfect snapshot of life during that time.

CCC CROSSCUTTING CONCEPTS
Cause and Effect Conditions within a depositional environment determine what type of fossil will form. Review the types of fossilization presented in this lesson. Draw a flow chart that describes how a clam goes from living on the ocean floor to becoming a mold and cast fossil.

Trace fossils Sometimes the only fossil evidence of an organism is indirect. Indirect fossils, called **trace fossils,** include worm trails, footprints, and tunneling burrows. Trace fossils can provide information about how an organism lived, moved, and obtained food. For example, dinosaur tracks provide clues about dinosaur size and walking characteristics. Groupings of dinosaur tracks have shown scientists that at least some species of dinosaurs traveled in herds, while others traveled alone.

Other trace fossils include gastroliths (GAS truh lihths) and coprolites (KAH pruh lites). Gastroliths are smooth, rounded rocks once present in the stomachs of dinosaurs to help them grind and digest food. We know that descendents of one group of dinosaurs are today's modern birds. Gravel and grit are often found in the stomachs of birds. This characteristic helps confirm the link to their dinosaur ancestors. Coprolites are the fossilized solid waste materials of animals. How do scientists know that this is what these rocks are? Recall that igneous and metamorphic rocks are crystalline, with orderly internal structures. Coprolites do not show this arrangement. Rather, the pieces are more of a randomly arranged mixture with no crystalline structure. Also, they are not made of individual grains cemented together, as are sedimentary rocks. By analyzing coprolites, scientists learn about animals' eating habits.

Figure 25 These molds formed from ancient ammonites. In some cases, the cavities were later filled in with minerals to create casts.

Check Your Progress

Summary

- Fossils provide evidence that species have evolved.
- Fossils help scientists date rocks and locate reserves of oil and gas.
- Fossils can be preserved in several different ways.

Demonstrate Understanding

1. **Describe** how the fossil record helps scientists understand Earth's history.
2. **List** ways in which fossils can form, and give an example of each.
3. **Explain** how scientists can use a trilobite to determine the relative age of a sedimentary rock layer in which it is found.
4. **Compare and contrast** a mold and a cast.

Explain Your Thinking

5. **Evaluate** Why are the best index fossils widespread?
6. **WRITING Connection** Imagine that you have just visited a petrified forest. Write a letter to a friend describing the forest. Explain what the forest looks like and how it was fossilized.

LEARNSMART Go online to follow your personalized learning path to review, practice, and reinforce your understanding.

Greg Dale/National Geographic RF/Getty Images

How to Find Fossils

Over the years, many great discoveries of fossils have been serendipitous. Large crews head out to remote areas where scientists think fossils may be found. They work for days, maybe weeks hoping that between the hard work and some amount of good fortune, they will find fossils.

The Great Divide Basin in Wyoming is one area that has been studied extensively using Landsat images and other technologies.

Now, scientists are looking to take more of the guesswork out of finding fossils by are using satellite images from NASA's series of Landsat missions. While satellites cannot detect fossils, they can distinguish between different types of rock. This information helps scientists narrow down where fossils might be.

One area in the United States that scientists have mapped using this technology is the Great Divide Basin in Wyoming. This area, shown in the photograph, has been a treasure trove of fossil finds from across the Paleocene-Eocene boundary (approximately 55 mya).

The equipment onboard the satellites can measure electromagnetic radiation that is not directly observable to humans. For example, Landsat 7 has a tool called Enhanced Thematic Mapper Plus, which detects infrared radiation as well as blue, green, and red wavelengths.

Using the detailed satellite images, scientists distinguish between areas of different types of ground cover based on the wavelengths of light reflected by different rock and mineral types.

Scientists use computers to analyze the data, but not just any computers. They are building networks of computer analysis systems inspired by human neurons.

Scientists designed the computers to analyze each pixel of the satellite image. Using information about sites that are known to be fossil-rich, the computer system was taught to separate land cover types to identify promising fossil beds.

 EVALUATE A DESIGN SOLUTION

Consider the process described in the feature. Suggest ways to refine and improve scientists' methods for finding fossils.

MODULE 17
STUDY GUIDE

 GO ONLINE to study with your Science Notebook.

Lesson 1 THE ROCK RECORD

- Scientists organize geologic time into eons, eras, periods, and epochs.
- Scientists divide time into units based largely on fossils of plants and animals.
- The Precambrian makes up nearly 90 percent of geologic time.
- The geologic time scale changes as scientists learn more about Earth.

- geologic time scale
- eon
- Precambrian
- era
- period
- epoch
- mass extinction

Lesson 2 RELATIVE-AGE DATING

- The principle of uniformitarianism states that processes occurring today have been occurring since Earth formed.
- Scientists use geologic principles to determine the relative ages of rock sequences.
- An unconformity represents a gap of time in the rock record.
- Geologists use correlation to compare rock layers in different geographic areas.
- Index fossils help scientists correlate rock layers in the geologic record.

- uniformitarianism
- relative-age dating
- original horizontality
- superposition
- cross-cutting relationship
- lateral continuity
- principle of inclusions
- unconformity
- correlation
- key bed
- index fossil

Lesson 3 ABSOLUTE-AGE DATING

- Techniques of absolute-age dating help identify numeric dates of geologic events.
- The decay rate of certain radioactive elements can be used as a kind of geologic clock.
- The concept of half-life is used to calculate the numeric ages of igneous and metamorphic rocks.
- Annual tree rings, ice cores, and sediment deposits can be used to date recent geologic events.

- absolute-age dating
- radioactive decay
- radiometric dating
- half-life
- radiocarbon dating
- dendrochronology
- varve

Lesson 4 FOSSIL REMAINS

- Fossils provide evidence that species have evolved.
- Fossils help scientists date rocks and locate reserves of oil and gas.
- Fossils can be preserved in several different ways.

- evolution
- original preservation
- altered hard part
- mineral replacement
- mold
- cast
- trace fossil

REVISIT THE PHENOMENON

How has information about Earth's history been preserved for so long?

CER Claim, Evidence, Reasoning

Explain your Reasoning Revisit the claim you made when you encountered the phenomenon. Summarize the evidence you gathered from your investigations and research and finalize your Summary Table. Does your evidence support your claim? If not, revise your claim. Explain why your evidence supports your claim.

STEM UNIT PROJECT
Now that you've completed the module, revisit your STEM unit project. You will summarize your evidence and apply it to the project.

GO FURTHER

SEP Data Analysis Lab
How do you interpret the relative ages of rock layers?

The diagram illustrates a sequence of rock layers. Geologists use the principles of relative-age dating to determine the order in which layers such as these were formed.

CER Analyze and Interpret Data
1. **Claim, Evidence** Identify a type of unconformity between any two layers of rock. Justify your answer.
2. **Evidence** Interpret which rock layer is oldest.
3. **Evidence, Reasoning** Infer where inclusions might be found. Explain.
4. **Evidence, Reasoning** Compare and contrast the rock layers on the right and left sides of the diagram. Why do they not match?
5. **Claim, Evidence, Reasoning** Which feature is younger, the dike or the folded strata? What geologic principle did you use to determine your answer?
6. **Reasoning** Propose why there is no layer labeled E on the right side of the diagram.

MarcelC/iStock/Getty Images

ENCOUNTER THE PHENOMENON

How was this pre-historic dragonfly different from a modern-day dragonfly?

GO ONLINE to play a video about the first trees and the effects those trees had on Earth.

SEP Ask Questions

Do you have other questions about the phenomenon? If so, add them to the driving question board.

CER Claim, Evidence, Reasoning

Make Your Claim Use your CER chart to make a claim about how pre-historic dragonflies are different from modern-day dragonflies. Explain your reasoning.

Collect Evidence Use the lessons in this module to collect evidence to support your claim. Record your evidence as you move through the module.

Explain Your Reasoning You will revisit your claim and explain your reasoning at the end of the module.

GO ONLINE to access your CER chart and explore resources that can help you collect evidence.

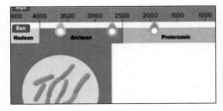

LESSON 1: Explore & Explain: The Age of Earth

LESSON 2: Explore & Explain: Formation of the Atmosphere

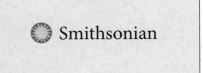

Additional Resources

(t)Video Supplied by BBC Worldwide Learning, (b)Stephen M. Wessells/USGS

EARLY EARTH

FOCUS QUESTION
What did Earth look like in the beginning?

The Age of Earth

The Precambrian, which includes the Hadean, Archean, and Proterozoic eons, is an informal time unit that spans nearly 90 percent of Earth's history. When Earth first formed it was hot and volcanically active, and no continents existed on its surface. Rocks of Earth's earliest eon—the Hadean—are extremely rare, so scientists know very little about Earth's first 600 million years. The earliest signs of life, shown in **Figure 1,** are simple, unicellular organisms from the Archean.

Crustal rock evidence

Absolute-age dating has revealed that the oldest known rocks are 4.28 billion years in age. Evidence that Earth is older than 4.28 billion years exists in small grains of the mineral zircon ($ZrSiO_4$) found in certain metamorphosed Precambrian rocks in Australia. Because **zircon** is a stable and common mineral that can survive erosion and metamorphism, scientists often use it to age-date ancient rocks. Geologists theorize that the zircon in the Australian rocks is residue from crustal rocks that no longer exist.

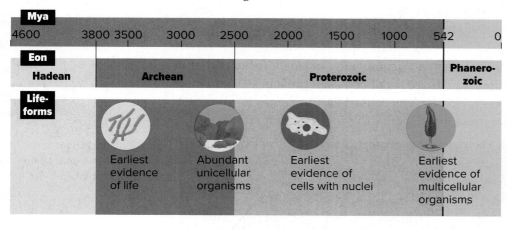

Figure 1 The Precambrian lasted for nearly 4 billion years. Multicellular organisms did not appear until the end of the Proterozoic.

3D THINKING   **DCI** Disciplinary Core Ideas   **CCC** Crosscutting Concepts   **SEP** Science & Engineering Practices

COLLECT EVIDENCE
Use your Science Journal to record the evidence you collect as you complete the readings and activities in this lesson.

INVESTIGATE
GO ONLINE to find these activities and more resources.

Applying Practices: Earth's Formation and Early History
HS-ESS1-6. Apply scientific reasoning and evidence from ancient Earth materials, meteorites, and other planetary surfaces to construct an account of Earth's formation and early history.

 Investigation Lab: Sequencing Time
Create a model to compare the time scale of life on Earth to the time scale of our individual lives.

Based on radiometric dating, which shows that the zircon is at least 4.4 billion years old, Earth must also be at least this old.

Solar system evidence

Plate tectonics, erosion, and other geologic processes have destroyed or altered most of the very early rock record on Earth. But certain objects from space have changed little over billions of years. Studying these objects can provide information about Earth's formation and early history. Evidence from meteorites (MEE tee uh rites) suggests that Earth is more than 4.4 billion years old. **Meteorites** are small fragments of orbiting bodies that have fallen on Earth's surface. They have fallen to Earth throughout Earth's history, but most have been dated at between 4.5 and 4.7 billion years old. Many scientists agree that all parts of the solar system formed at the same time, so Earth and meteorites are about the same age.

In addition, lunar rocks collected during the *Apollo* missions in the 1970s have been dated at 4.4 to 4.5 billion years old. Scientists think that the Moon formed very early in Earth's history when a massive solar system body the size of Mars collided with Earth. Considering all the evidence, scientists agree that Earth is about 4.6 billion years old.

 Get It?

Explain why certain objects in space can provide information about Earth's formation and early history.

Early Earth's Heat Sources

Earth was extremely hot after it formed. There were three likely sources of this heat: Earth's gravitational contraction; radioactivity; and bombardment by asteroids, meteorites, and other solar system bodies, as shown in **Figure 2.**

Gravitational contraction

Scientists think that Earth formed by the gradual accumulation of small, rocky bodies in orbit around the Sun. As Earth accumulated these small bodies, it grew in size and mass. With increased mass came increased gravity. Gravity caused Earth's center to squeeze together with so much force that the pressure raised Earth's internal temperature.

Figure 2 Impacts from asteroids and meteorites were a source of heat for early Earth.

Radioactivity

A second source of Earth's heat was the decay of radioactive isotopes. Scientists know that certain radioactive isotopes were more abundant in Earth's past than they are today. While some of these isotopes, such as uranium-238, are long-lasting and continue to decay today, others were short-lived and have nearly disappeared. Radioactive decay releases energy in the form of heat. Because there were more radioactive isotopes in early Earth, more heat was generated, making early Earth hotter than it is today.

Asteroid and meteorite bombardment

A third source of heat in early Earth came from the impacts of meteors, asteroids (AS tuh roydz), and other objects in the solar system. **Asteroids** are carbon or mineral-rich objects between 1 m and 950 km in diameter. Today, most asteroids orbit the Sun between the orbits of Mars and Jupiter. Large asteroids seldom collide with Earth. Planetary geologists estimate that only about 60 objects with diameters of 5 km or more have struck Earth during the last 600 million years. Most objects that hit Earth today are meteorites—fragments of asteroids.

However, evidence from the surfaces of the Moon and other planets suggests that for the first 500 to 700 million years of Earth's history, many more asteroids were distributed throughout the solar system than there are today, and collisions were much more frequent. The impacts of these bodies on Earth's surface generated a tremendous amount of thermal energy. For example, scientists think that the massive collision that likely formed the Moon generated so much heat that parts of Earth melted. The debris (duh BREE) from these impacts also caused a blanketing effect around Earth, which prevented the newly generated heat from escaping to space.

Cooling

The combined effects of gravitational contraction, radioactivity, and bombardment by other objects in the solar system made Earth's beginning very hot. Eventually, Earth's surface cooled enough for an atmosphere and oceans to form. Scientists do not know exactly how long it took for this to happen, but evidence suggests that Earth cooled enough for liquid water to form within its first 200 million years. The cooling process continues even today. As much as half of Earth's internal heat remains from Earth's formation.

 Get It?

Identify the factors that made early Earth a hot planet.

Formation of the Crust

Because of the intense heat in early Earth, many scientists think that much of the planet consisted of hot, molten magma. As Earth cooled, the minerals and elements in this magma differentiated and became concentrated in specific density zones.

SCIENCE USAGE v. COMMON USAGE

differentiate
Science usage: to layer into distinct zones
Common usage: to distinguish; to mark as different

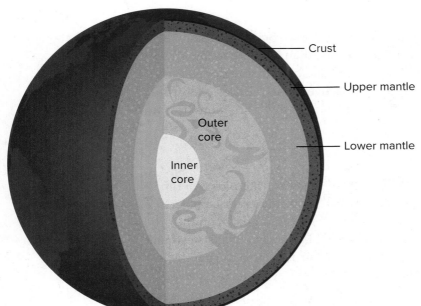

Figure 3 Earth differentiated into layers shortly after it formed.

Analyze *What is the densest part of Earth?*

Differentiation

Scientists know that less-dense materials float on top of denser materials. As you may know, oil floats on water because oil is less dense than water. This same general principle operated on the early molten Earth. The element with the highest density—iron—sank toward the center. In contrast, the light elements, such as silicon and oxygen, remained closer to the surface. The process by which a planet becomes internally zoned when heavy materials sink toward its center and lighter materials accumulate near its surface is called **differentiation** (dih fuh ren shee AY shun). The differentiated zones of Earth are illustrated in **Figure 3.**

Relative densities The process of differentiation explains the relative densities of parts of Earth today. **Figure 4** compares the proportions of elements in Earth's crust and in Earth as a whole. Notice that iron, a dense element, is much less abundant in the crust than it is in the entire Earth, while the crust has a higher proportion of less-dense elements, such as silicon and oxygen. This also explains why granite occurs on Earth's surface. Granite is composed mainly of feldspar, mica, and quartz, which, as you have learned, are minerals with low densities.

 Get It?

Explain why there is more iron in Earth's core than in its crust.

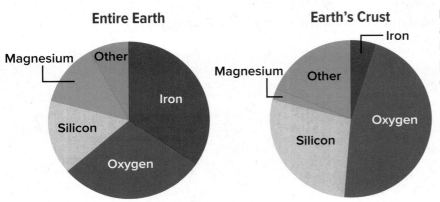

Figure 4 Larger amounts of dense elements are found in Earth as a whole than are found in Earth's crust.

Estimate *the percentage of iron in Earth's crust and in the entire Earth.*

Earliest crust

Some type of early crust formed as soon as Earth's upper layer began to cool. This crust was probably similar to the basaltic crust that underlies Earth's oceans today. Recall that present-day oceanic crust is recycled at subduction zones. Pieces of Earth's early crust were also recycled, though scientists do not know how the recycling occurred. Some suggest that it occurred by a process that does not occur on Earth today. Most agree that the recycling was vigorous—so vigorous that none of Earth's earliest crust exists today.

Continental crust

As the early crustal pieces were returned to the mantle, they carried water. The introduction of water into the mantle was essential for the formation of the first continental crust. The water reacted with the mantle material to produce new material that was less dense than the original crustal pieces. As this material crystallized and reemerged on Earth's surface, small fragments of granite-containing crust were formed. Granite makes up much of the crust that forms Earth's continents today. As volcanic activity continued during the Archean, small fragments of granite-rich crust continued to form. These crustal fragments are called **microcontinents.** They are called this because they were not large enough to be considered continents.

Cratons Most of the microcontinents that formed during the Archean and Early Proterozoic still exist as the cores of today's continents. A **craton** (KRAY tahn) is the oldest and most stable part of a continent. It is made up of the crust and a part of the upper mantle and can extend to a depth of 200 km. Cratons are composed of granitic rocks, such as granite and gneiss, with alternating bands of metamorphosed basaltic rocks, which represent ancient continental collisions. As shown in **Figure 5,** the Archean cratons represent about 10 percent of Earth's total landmass.

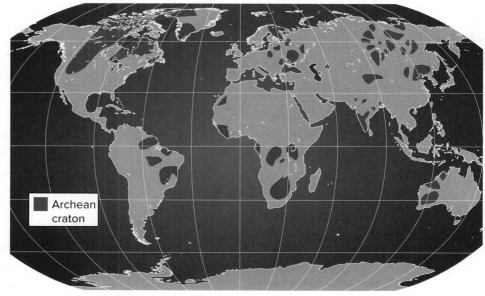

Figure 5 Archean cratons make up about 10 percent of Earth's continents. These granite-rich cores extend into the mantle as deep as 200 km.

Precambrian shields Most of the cratons are buried beneath sedimentary rocks. However, in some places deep erosion has exposed the rocks of the craton. This exposed area is called a **Precambrian shield.**

In North America, the Precambrian shield is called the **Canadian Shield** because much of it is exposed in Canada. The Canadian Shield, shown in **Figure 6,** also occupies a large part of Greenland as well as the northern parts of Minnesota, Wisconsin, and Michigan. Valuable minerals such as nickel, silver, and gold are found in the rocks of the Canadian Shield. Some of the oldest known crustal rocks on Earth that date back 4.28 billion years are from the Canadian Shield. In contrast, North America's platform rocks are generally younger than about 600 million years.

 Get It?

Explain why the Precambrian shield is exposed.

Growth of the Continents

Recall that all of Earth's continents were once consolidated into a single landmass called Pangaea. Pangaea formed relatively recently in Earth's history—only about 250 million years ago. The plate tectonic forces that formed Pangaea have been at work since at least the end of the Archean.

Mountain building

During the Proterozoic, the microcontinents that formed during the Archean collided with each other, becoming larger but fewer in number. As they collided, they formed massive mountains. Recall that mountain-building episodes are called orogenies. Orogenies form long belts of deformed rocks called orogens, or orogenic belts. The mountain-building events that formed North America are illustrated in **Figure 7.**

Laurentia One of Earth's largest Proterozoic landmasses was Laurentia (law REN shuh). **Laurentia** was the ancient continent of North America. As shown in **Figure 7,** the growth of Laurentia involved many different mountain-building events. For example, near the end of the early Proterozoic, between 1.8 and 1.6 bya, thousands of square kilometers were added to Laurentia when Laurentia collided with a volcanic island arc. This collision is called the Yavapai-Mazatzal Orogeny.

Figure 7 Visualizing Continent Formation

North America was formed by a succession of mountain-building episodes over billions of years. This map shows mountain-building events that occurred during the Precambrian. By the end of the Precambrian, about 75 percent of North America had formed.

The **Grenville Orogeny** occurred when Laurentia collided with Amazonia, the ancient continent of South America. A huge mountain range rose from Newfoundland in Canada to western North Carolina.

The **Trans-Hudson Orogeny** occurred when the Superior province collided with the Wyoming and Hearne-Rae provinces. Remnants of this collision exist in the Black Hills of South Dakota.

The **Yavapai-Mazatzal Orogeny** added what is now New Mexico and Arizona, as well as parts of Utah and California. The oldest rocks of the Grand Canyon formed in this event.

A **midcontinent rift** began to split the continent about 1 bya, but it stopped a few million years later. Scientists do not know why.

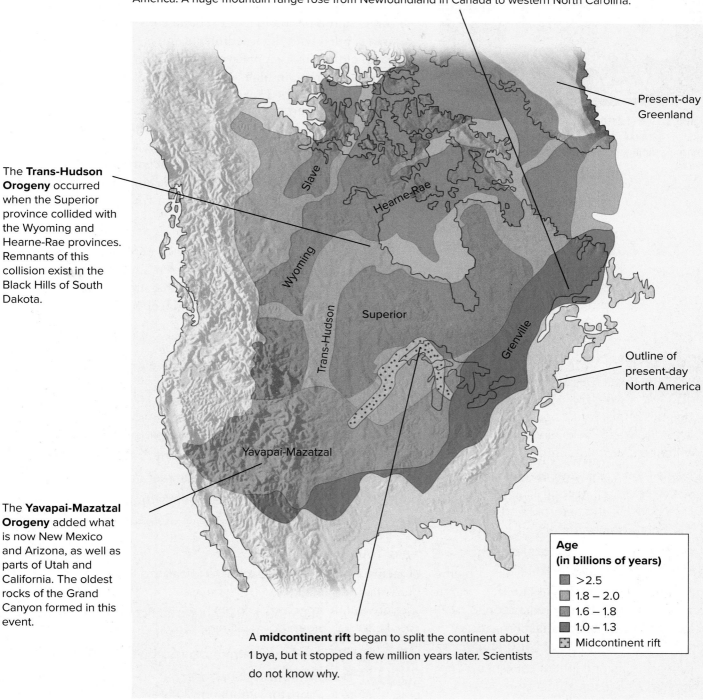

Present-day Greenland

Outline of present-day North America

Age (in billions of years)
- >2.5
- 1.8 – 2.0
- 1.6 – 1.8
- 1.0 – 1.3
- Midcontinent rift

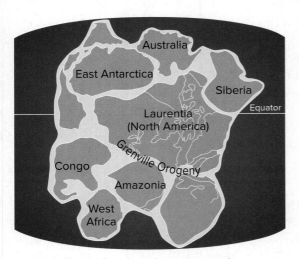

Figure 8 One of Earth's earliest supercontinents—Rodinia—formed when Laurentia collided with Amazonia during the Grenville Orogeny.

Notice the midcontinent rift in **Figure 7.** When rifting stops and a continent is no longer being split apart, the rift is called a failed rift. The failed midcontinent rift in North America runs between present-day Arkansas, Louisiana, western Tennessee, and Mississippi. Though this rift—called the Reelfoot Rift—failed, the area remains weakly tectonically active. Three of the largest North American earthquakes in recorded history, known as the New Madrid sequence, occurred here in a three-month period in late 1811 and early 1812. Today, the rift area is filled with thick sediments and is the southern drainage basin of the Mississippi River. Chances for a future large earthquake remain high.

An early supercontinent The collision of Laurentia with Amazonia, the ancestral continent of South America, occurred during the mid–Proterozoic, about 1.2 bya. This collision coincided with the formation of one of Earth's earliest super-continents, called Rodinia (roh DIN ee ah), shown in **Figure 8.** Rodinia was positioned on the equator with Laurentia near its center.

By the time Rodinia formed, nearly 75 percent of Earth's continental crust was in place. The remaining 25 percent was added during the three eras of the Phanerozoic eon. The breakup of this supercontinent began about 750 mya.

Check Your Progress

Summary

- Scientists use Earth rocks, zircon crystals, Moon rocks, and mete-orites to determine Earth's age.
- Likely heat sources of early Earth were gravitational contraction, radioactivity, and asteroid and meteorite bombardment.
- Earth differentiated into specific density zones early in its formation.
- Plate tectonics caused microcon-tinents to collide and fuse throughout the Proterozoic.
- The ancient continent of Laurentia formed as a result of many mountain-building episodes.

Demonstrate Understanding

1. **Summarize** the data that scientists use to determine Earth's age.
2. **Desribe** how gravitational contraction, radioactivity, and asteroid and meteorite bombardment heated early Earth.
3. **Explain** why pieces of Earth's earliest crust do not exist today.
4. **Deduce** how a craton is like a continent's root.
5. **Describe** the importance of zircon as an age-dating tool.

Explain Your Thinking

6. **Evaluate** whether it is reasonable to call the Proterozoic the age of continent building.
7. **Assess** Which of Earth's early sources of heat are not major contributors to Earth's present-day internal heat?
8. **Infer** why little evidence of Proterozoic orogenies exists today.
9. **WRITING ▸ Connection** Suppose you are the North American craton. Write a short story about how Laurentia formed around you.

LEARNSMART Go online to follow your personalized learning path to review, practice, and reinforce your understanding.

FOCUS QUESTION

What was the first life on Earth?

Formation of the Atmosphere

Earth's atmosphere likely began to form soon after the planet itself formed. Asteroids, meteorites, and other objects that collided with Earth probably contained water. The water vaporized on impact, forming a haze around Earth. Hydrogen and helium were probably also present, with lesser amounts of ammonia and methane. However, hydrogen and helium have small atomic masses, and Earth's gravity was, and still is, too weak to keep them from escaping to space. Much of the ammonia and methane surrounding Earth likely broke apart due to the Sun's intense ultraviolet radiation, releasing more hydrogen into space. Once Earth formed, its volcanic gases changed the atmosphere. Volcanic eruptions such as the one in **Figure 9** release large quantities of gases, and the Precambrian had considerable volcanic activity.

Figure 9 The eruption of Mount St. Helens in 1980 released a large amount of carbon dioxide, water vapor, and other gases.

 3D THINKING   **DCI** Disciplinary Core Ideas   **CCC** Crosscutting Concepts   **SEP** Science & Engineering Practices

COLLECT EVIDENCE

 Use your Science Journal to record the evidence you collect as you complete the readings and activities in this lesson.

INVESTIGATE

 GO ONLINE to find these activities and more resources.

Applying Practices: WebQuest: The Coevolution of Living Things and the Atmosphere HS-ESS2-7. Construct an argument based on evidence about the simultaneous coevolution of Earth's systems and life on Earth.

Mapping Lab: What came first?
Create a model to visualize the temporal scale of the appearance and disappearance of important lifeforms.

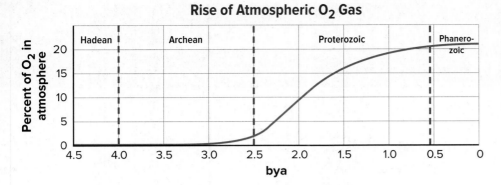

Rise of Atmospheric O₂ Gas

Figure 10 There were only negligible amounts of free oxygen in Earth's atmosphere until the early Proterozoic.

Analyze *How old was Earth when oxygen began to accumulate in its atmosphere?*

Outgassing

Recall that present-day volcanoes release large amounts of water vapor and carbon dioxide and trace amounts of nitrogen and other gases in a process called outgassing. While scientists do not know the exact concentration of gases in Earth's early atmosphere, it probably contained the same gases that vent from volcanoes today.

Oxygen in the Atmosphere

One gas that volcanoes do not generally produce is oxygen. There was little oxygen in the Hadean and Archean atmospheres that was not bonded with carbon or other elements. As illustrated in **Figure 10,** atmospheric oxygen did not begin to accumulate until the early Proterozoic. Where did the oxygen gas come from?

First oxygen producers

The answer to this question is found in Australian and South African fossils that are about 3.5 billion years old. These fossils appear to be traces of tiny, threadlike organisms called **cyanobacteria.** Like present-day plants and other producers, ancient cyanobacteria used photosynthesis to produce the nutrients they needed to survive. In the process of photosynthesis, organisms use light energy and convert carbon dioxide and water into sugar. Oxygen gas is given off as a waste product. This process is an example of a biosphere-atmosphere interaction that causes feedback effects that increase the original change.

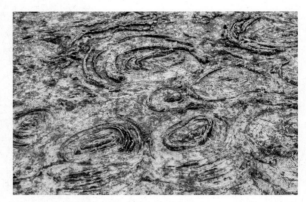

Figure 11 These well-preserved fossil stromatolites are evidence that cyanobacteria existed during the Precambrian.

Stromatolites Most scientists think that cyanobacteria captured carbon dioxide and released enough oxygen to gradually change the composition of Earth's atmosphere during the Archean. By the Early Proterozoic, large, coral reeflike mounds of cyanobacteria called **stromatolites** (stroh MA tuh lites) dominated the shallow seas that covered most of the continents. Stromatolites are made by billions of cyanobacteria colonies that trap and bind sediments together. These structures are similar in size and shape to the Precambrian fossil stromatolites, shown in **Figure 11.**

 Get It?

Explain how organisms gradually changed Earth's atmosphere.

Figure 12 This iron mine contains banded-iron formations that date from the Proterozic.

Explain *how banded-iron formations are evidence of atmospheric oxygen gas.*

Evidence in rocks

Scientists can determine whether there was oxygen in Earth's Archean atmosphere by looking for oxidized iron in Archean rocks. Scientists know that iron reacts with oxygen in the atmosphere to form iron oxides, more commonly called rust. Iron oxides are identified by their red color and provide evidence of oxygen in the atmosphere. The absence of iron oxides in rocks of the late Archean indicates that there was no oxygen gas in the atmosphere at that time. Had atmospheric oxygen gas been present, it would have reacted with the iron ions in the water or with the iron contained in sediments.

Banded iron By the beginning of the Proterozoic, however, cyanobacteria had increased oxygen gas levels enough that iron oxides began to form in localized areas. These locally high concentrations of iron oxides are called **banded-iron formations.** Banded-iron formations consist of alternating bands of iron oxide and chert, an iron-poor sedimentary rock. The iron oxides appear to have been deposited cyclically, perhaps in response to seasonal variations. Today, these formations are mined for iron ore. An iron mine and a banded-iron rock are shown in **Figure 12.**

Red beds Many sedimentary rocks that date from the mid-Proterozoic, beginning about 1.8 bya, are rusty red in color. These rocks are called **red beds** because they contain so much iron oxide. The presence of red beds in mid-Proterozoic and younger rocks is strong evidence that the atmosphere by the mid-Proterozoic contained oxygen gas.

Importance of oxygen

Oxygen is important not only because most animals require it for respiration, but also because it provides protection from harmful ultraviolet radiation (UV) from the Sun. Today, only a small fraction of the Sun's UV radiation reaches Earth's surface. This is because Earth is protected by ozone in Earth's upper atmosphere.

Recall that an ozone molecule consists of three oxygen atoms bonded together. As oxygen accumulated in Earth's atmosphere, an ozone layer began to develop. Ozone filtered out much of the UV radiation, providing an environment in which new life-forms could develop.

Get It?
Describe the importance of oxygen for the evolution of life.

Formation of the Ocean

As you have learned, some scientists think that the oceans reached their current volume of water very early in Earth's history. The water that filled the oceans probably originated from the two major sources that provided water in Earth's atmosphere: volcanic outgassing and asteroids, meteorites, and other objects that bombarded Earth's surface. Earth's Early Precambrian atmosphere was rich with water vapor from these sources. As Earth cooled, the water vapor condensed to form liquid water. Recall that condensation occurs when matter changes state from a gas to a liquid.

Rain

As liquid water formed, a tremendous amount of rain fell. The rain filled low-lying basins and eventually formed the oceans. Rainwater dissolved the soluble minerals exposed at Earth's surface and—just as they do today—rivers, runoff, and groundwater transported these minerals to the oceans. Dissolved minerals made the oceans of the Precambrian salty, just as dissolved minerals make today's oceans salty.

 Get It?

Describe the geologic development of Earth's oceans.

Water and life

The Precambrian began with an environment that was inhospitable to life. When it ended, much of Earth was covered with oceans that were teeming with tiny cyanobacteria and other life-forms. Life as it exists on Earth today cannot survive without liquid water.

Scientists think that Earth is not the only object in the solar system that contains or has contained water. Some scientists estimate that the asteroid Ceres contains more freshwater than does Earth. Scientists also think that some surface features on Mars, such as the gullies shown in **Figure 13,** were carved by liquid water. They recently found strong evidence that water still flows in brief spurts on Mars. Some moons of Saturn and Jupiter might also contain water in their interiors.

Today, the search for life elsewhere in the solar system and universe is typically centered on the search for water. Life on Earth has been found in almost every environment that contains water, from Antarctic ice to hot, deep-water ocean vents. Scientists think that simple life-forms might exist in similar environments on other objects in the solar system.

 Get It?

Explain Why is the search for life elsewhere in the solar system often centered on the search for water?

Figure 13 This photograph taken by the *Mars Reconnaissance Orbiter* reveals evidence suggesting that liquid water once flowed on the Martian surface.

NASA/JPL/University of Arizona

Origin of Life

You have learned that fossil evidence suggests that cyanobacteria existed on Earth as early as 3.5 bya. Though cyanobacteria are simple organisms, photosynthesis—the process by which they produce oxygen—is complex, and it is likely that cyanobacteria evolved from simpler life-forms. Most scientists think that intense asteroid and meteorite bombardment prevented life from developing on Earth until at least 3.9 bya. Where and how the first life-form developed, however, remains an active area of research.

Primordial soup

During the first half of the twentieth century, scientists thought that Earth's earliest atmosphere contained hydrogen, methane, and ammonia. Some biologists suggested that such an atmosphere, with energy supplied by lightning, would give rise to an organic "primordial soup" in Earth's shallow oceans. *Primordial* (pry MOR dee al) means "earliest" or "original". In 1953, Stanley Miller and Harold Urey devised an apparatus, shown in **Figure 14,** to test this hypothesis. They connected an upper chamber containing hydrogen, methane, and ammonia to a lower chamber designed to catch any particles that condensed in the upper chamber. They added sparks from tungsten electrodes to model lightning. Within a week, organic molecules had formed in the lower chamber—the primordial soup!

Uncertainties The organic molecules that formed in Miller and Urey's experiment included **amino acids,** the building blocks of proteins. Miller and Urey were the first to show experimentally that amino acids and other molecules necessary for the origin of life could have formed in conditions thought present on early Earth. However, Earth's early atmosphere contained gases like those that vent from volcanoes—carbon dioxide, water vapor, and traces of ammonia, methane, and hydrogen. When combinations of these gases are used in simulations, amino acids do not form in high quantities. This led scientists to question whether those processes were sufficient for the origin of life. Some scientists continue to explore the possibility that amino acids (and, therefore, life) arose in Earth's oceans under localized conditions similar to those in the Miller-Urey experiment.

Figure 14 In 1953, Stanley Miller, shown here, and Harold Urey performed experiments to test whether organic molecules could form on early Earth.

CCC CROSSCUTTING CONCEPTS
Stability and Change Make a time line that shows the evidence scientists use to construct explanations about how early Earth changed over geologic time.

ACADEMIC VOCABULARY
simulate
to create a representation or model of something
The video game simulated the airplane's flight with impressive realism.

Other scenarios

Because of uncertainties with the conditions in the Miller-Urey experiment, scientists continue to propose different scenarios and conduct new research into sources and conditions for the origin of life. Some of these are shown in **Table 1.** Some scientists think that amino acids organized elsewhere in the universe and were transported to Earth in asteroids or comets. Their experiments show that chemical synthesis of organic molecules is possible in interstellar clouds, and amino acids have been found in meteorites. Other scientists hypothesize that amino acids originated deep in Earth or its oceans. Experiments show that conditions there are favorable for chemical synthesis, and organisms have been found at depths exceeding 3 km.

One current area of research explores the possibility that life emerged deep in the ocean at hydrothermal vents. The energy and nutrients necessary for the origin of life are present in this environment. As shown in **Figure 15,** a variety of unique organisms called extremophiles (from the Latin *extremus*, meaning "extreme," and Greek *philía*, meaning "love"), live near hydrothermal vents.

 Get It?

Decide Which hypothesis about the origin of life on Earth do you find most valid? Give evidence to support your answer.

No single theory needs to be exclusive; it is possible that all of these contributed to the origin of life. Regardless of how life arose, it is known that conditions during that time were not hospitable, and life probably had many starts and restarts on early Earth. Asteroid impacts were probably still common between 3.9 and 3.5 bya when life arose. Large impacts during this time could have vaporized many early life-forms.

Table 1 How Life Might Have Begun on Earth: Three Hypotheses

	Earth's Surface	Deep Earth	Space
Hypothesis	Life originated on Earth's surface in warm, shallow oceans.	Life originated in hydrothermal vents deep in the oceans.	Organic molecules were brought to Earth in asteroids or comets.
Requirement	Hydrogen, methane, and ammonia must be present in the atmosphere.	Life must survive at high temperatures and pressures.	Organic molecules must be present in extraterrestrial bodies.
Evidence	Simulations produce amino acids.	Simulations of deep-sea vents produce amino acids.	Some meteorites contain amino acids that survived impact.
Drawback	The composition of the early atmosphere likely did not have large amounts of the required gases.	It might have been too hot for organic molecules to survive.	It is difficult to test at this time due to technical limitations.

An RNA world While experiments have shown the likelihood that amino acids existed on early Earth, scientists are still learning how the amino acids were organized into complex proteins and other molecules of life. One essential characteristic of life is the ability to reproduce. All cells require RNA and DNA to reproduce. In modern organisms, RNA carries and translates the instructions necessary for cells to function. Both RNA and DNA use proteins called enzymes to replicate.

Recent experiments have shown that RNA molecules called ribozymes can act as enzymes. They can replicate without the aid of enzymes. This suggests that RNA molecules might have been the first replicating molecules on Earth. An RNA-based world might have been intermediate between an inorganic world and today's DNA-based organic world.

Proterozoic Life

Fossil evidence indicates that unicellular organisms dominated Earth until the end of the Precambrian. These organisms are **prokaryotes** (proh KE ree ohts)—organisms that do not contain nuclei. Nuclei are separate compartments in cells that contain DNA and RNA. Organisms whose cells contain RNA and DNA in nuclei are called **eukaryotes** (yew KE ree ohts). **Figure 16** illustrates how prokaryotes and eukaryotes differ in the packaging of their DNA and RNA.

Simple eukaryotes

Eukaryotes can be unicellular or multicellular, but because they contain nuclei and other internal structures, they tend to be larger than prokaryotes. This general observation is useful in determining whether a fossil represents a prokaryote or a eukaryote because it is rare for a fossil to be preserved in enough detail to determine whether its cells had nuclei. The oldest-known eukaryote fossil is unicellular. It was found in a banded-iron formation, about 2.1 billion years old, in Michigan.

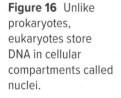 **Get It?**

Explain how the relative sizes of eukaryotes and prokaryotes are useful to paleontologists.

Figure 15 These tubeworms tolerate extreme pressures and temperatures near hydrothermal vents 2 km below the ocean's surface.

Deduce *why pressure is high in a hydrothermal vent environment.*

Figure 16 Unlike prokaryotes, eukaryotes store DNA in cellular compartments called nuclei.

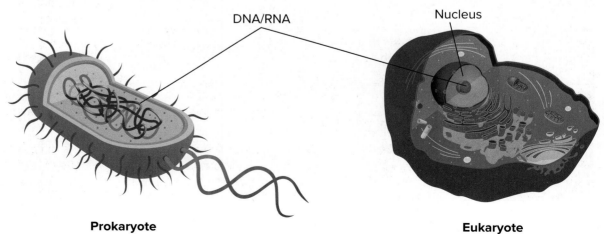

DNA/RNA

Nucleus

Prokaryote

Eukaryote

Snowball Earth Some scientists think that glaciation events 850–550 mya played a critical role in the extinction of many early unicellular eukaryotes. These glaciation events were so widespread that some geologists compare Earth at that time to a giant snowball. Evidence from ancient glacial deposits around the world suggests that glaciers might have advanced as far as the equator and that even the oceans might have been frozen. Though many organisms went extinct during this time, some life-forms survived, perhaps near hydrothermal vents or in pockets of sunlight streaming through openings in ice, as illustrated in **Figure 17.**

Figure 17 Sunbeams streaming through ice might have provided a refuge for some life-forms 750 mya, when ice covered Earth.

Multicellular organisms

Shortly after the ice retreated toward the poles, the climate warmed dramatically. Although probably not Earth's first multicellular life, many multicellular marine organisms appear in the rock record. Certain fossils of this time period were discovered in 1947 in Australia's Ediacara Hills. Collectively called the **Ediacaran biota** (ee dee A kuh ruhn by OH tuh), these fossils show the impressions of large, soft-bodied eukaryotes. **Figure 18** shows what these organisms might have looked like.

Ediacaran biota The discovery of the Ediacaran biota at first seemed to solve one of the great mysteries in geology: why there are no fossils of the ancestors of the complex and diverse animals that existed during the Cambrian period—the first period of the Paleozoic era. The Ediacaran biota seemed to provide fossil evidence of an ancestral stock of complex organisms.

Figure 18 This reconstruction of an ocean during the Ediacaran period shows how Earth's early multicellular organisms might have looked. They ranged from several centimeters to two meters in length.

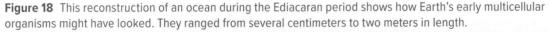

Claudio Gazzaroli/WaterFrame/Getty Images

Figure 19 One type of Ediacaran organism resembles a present-day sea pen. Some scientists think that the two are related.

As shown in **Figure 19,** one type of Ediacaran organism seemed similar to modern sea pens. Others appeared similar to jellyfish, segmented worms, and arthropods—just the type of ancestral stock that geologists had hoped to find.

However, upon closer examination, some scientists have questioned that conclusion and suggest that Ediacaran organisms are not relatives of present-day animal groups, but instead represent unique organisms. These scientists point out that Ediacaran organisms do not show evidence of a mouth, anus, or gut, and there is little evidence that they could move. As a result, there is ongoing debate in the scientific community about the precise nature of many of these fossils.

Mass extinction

In recent years, geologists have found Ediacaran fossils in all parts of the world. This suggests that these organisms were widely distributed throughout the shallow seas of the Late Proterozoic. They seem to have flourished between 600 mya and 540 mya. Then, in an apparent mass extinction, most of them disappeared, and organisms more likely related to modern organisms began to inhabit the oceans.

✏ Check Your Progress

Summary

- The atmosphere and oceans formed early in Earth's history.

- Oxygen gas began to accumulate in the Proterozoic due to photosynthesizing cyanobacteria.

- The water that filled Earth's oceans most likely came from two major sources.

- Life on Earth likely began between 3.9 and 3.5 bya.

- Stanley Miller and Harold Urey were the first to show experimentally that organic molecules could have formed on early Earth.

- Earth's multicellular organisms evolved at the end of the Precambrian.

Demonstrate Understanding

1. **Explain** why an atmosphere rich in oxygen was important for the evolution of life.
2. **Summarize** how scientists conclude that ancient cyanobacteria produced oxygen.
3. **Relate** What is the relationship between banded-iron formations and oxygen gas?
4. **Describe** three scientific explanations about the origin of life on Earth. Include evidence for each.
5. **Explain** why scientists think that life on Earth began after 3.9 bya.
6. **Compare and contrast** eukaryotes and prokaryotes.

Explain Your Thinking

7. **Hypothesize** why Ediacaran organisms went extinct.
8. **Discuss** why some scientists think that Ediacaran organisms do not represent present-day animal groups.
9. **WRITING >Connection** Write a newspaper article about the discovery of a new fossil outcrop that dates to the end of the Precambrian. Describe the fossil organisms found in this outcrop.

LEARNSMART®   Go online to follow your personalized learning path to review, practice, and reinforce your understanding.

FOCUS QUESTION

Why were there so many mass extinctions in the Paleozoic era?

Paleozoic Paleogeography

The geologic activity of the three eras of the Phanerozoic eon are well represented in the rock record. By studying this record, geologists can reconstruct estimates of landscapes that have long since disappeared. The ancient geographic setting of an area is called its **paleogeography** (pay lee oh jee AH gruh fee). The paleogeography of the Paleozoic era—the first era of the Phanerozoic—is defined by the breakup of the supercontinent Rodinia. As this breakup proceeded, multicellular life evolved with increasing complexity, as illustrated in **Figure 20.**

Passive margins

Recall that the ancient North American continent of Laurentia split off from Rodinia by the early Paleozoic. Laurentia was located near the equator and was surrounded by ocean. In addition, it was almost completely covered by a shallow, tropical sea. Throughout the Cambrian, there was no tectonic activity on Laurentia, so no mountain ranges formed. The edge of a continent is called a margin. When there is no tectonic activity along a margin, it is called a **passive margin.** During the Cambrian, Laurentia was completely surrounded by passive margins—there was no tectonic activity along its edges.

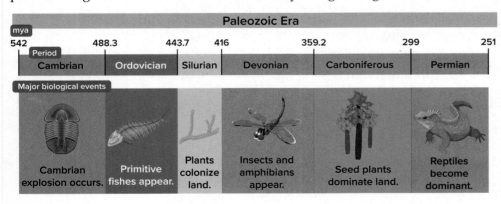

Figure 20 Life-forms became more complex during the six periods of the Paleozoic.

3D THINKING　　**DCI** Disciplinary Core Ideas　　**CCC** Crosscutting Concepts　　**SEP** Science & Engineering Practices

COLLECT EVIDENCE

Use your Science Journal to record the evidence you collect as you complete the readings and activities in this lesson.

INVESTIGATE

GO ONLINE to find these activities and more resources.

Mapping Lab: Water to Land
Obtain, evaluate, and communicate Information to discover how fossil distribution is affected by the original environment.

? Revisit the Encounter the Phenomenon Question
What information from this lesson can help you answer the Unit and Module questions?

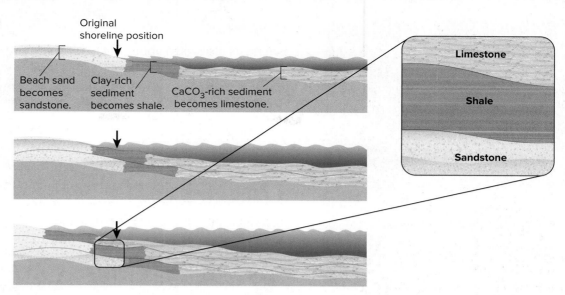

Figure 21 A vertical sequence of sandstone-shale-limestone in the rock record indicates that an ancient shoreline moved inland as sea level rose. This inland movement is called a transgression.

Sea-Level Changes in the Rock Record

Rock sequences preserved in passive margins tell paleogeographers a great deal about ancient shorelines. These sequences are useful in charting the rise and fall of sea level. To understand this, it is first necessary to understand how sediment is deposited on a shoreline.

Shoreline deposition

Ocean tides wash small grains of sand and sediment ashore to make beaches. Tides also deposit offshore sediment the size of clay particles (<0.002 mm). Calcium carbonate ($CaCO_3$) sediment accumulates farther from shore as calcium muds form from seawater and organisms containing calcium carbonate die and fall to the seafloor. The sand deposited on beaches eventually becomes sandstone, the offshore clay sediment compacts to form shale, and the calcium carbonate sediment farther offshore turns into limestone, as shown in **Figure 21.**

Transgression When sea level rises or falls, the deposition of sediment shifts. As illustrated in **Figure 21,** a rise in sea level causes the water to move inland to an area that previously had been dry. The area where clay sediment was deposited also moves shoreward, on top of the old beach. This movement is called a **transgression.** The result of the transgression is the formation of deep-water deposits overlying shallow-water deposits. This appears in the rock record as a vertical or stepwise sequence of sandstone-shale-limestone.

Regression When sea level falls, the shoreline moves seaward in a process called **regression.** This process results in shallow-water deposits overlying deep-water deposits. A stacked sequence of limestone-shale-sandstone is evidence of a regression.

SCIENCE USAGE v. COMMON USAGE

transgression
Science usage: movement of a shoreline inland as sea level rises
Common usage: violation of a law or moral duty

Evaporites

Scientists also learn about fluctuating sea level by studying evaporite deposits. Recall that evaporite deposits are rocks that have crystallized out of water that is saturated with dissolved minerals. Some evaporite deposits can be associated with fossilized reefs.

Fossilized reefs are made of the carbonate skeletons of tiny organisms called corals. Reefs form in long, linear mounds parallel to a continent or island, where they absorb the energy of the waves that crash against them on their seaward side. The area behind the reef, called a lagoon, is protected from the waves' energy. In a tropical setting, water in the lagoons evaporates in the warm sunshine, and minerals such as halite and gypsum precipitate out. Over time, cycles of evaporite deposition mark changes in water level.

Get It?

Explain how evaporites and reefs are related.

Mineral deposits Huge amounts of gypsum and halite evaporites were deposited in Paleozoic lagoons. The white sands of White Sands National Park, shown in **Figure 22,** are the remains of one such evaporite deposit. Other deposits, such as those in the Great Lakes area of North America, are mined commercially. Halite is used as road salt. Gypsum is an ingredient in plaster and drywall.

Impermeability As shown in **Figure 23,** reef limestones tend to have large pore spaces, allowing oil and other liquids to move through them. Evaporite rocks, in contrast, are impermeable.

Figure 22 The white sands of New Mexico's White Sands National Park are made of gypsum from ancient evaporite deposits.

Reef rock   Evaporite

Figure 23 The left photo shows a thin section of reef rock. The right photo shows an evaporite. Reef rocks have large pores that can contain oil or other liquids, in contrast to evaporite rock, which is impermeable to liquids. The blue areas in the reef rock are petroleum-filled pores.

Infer *why ancient evaporite deposits are important to petroleum geologists.*

This means that they contain very little connected pore space, and liquid cannot move through them. When an evaporite deposit overlies a reef rock that contains oil, it seals in the oil and prevents the oil from migrating. A good example is the Permian Basin, home to the Great Permian Reef Complex in western Texas and southeastern New Mexico. The oil in this complex rarely leaks to Earth's surface because of its tight evaporite seal.

The salt in evaporite deposits is less dense than sedimentary rock, and the thick salt beds often rise upward toward the surface, deforming the strata to form pillar-shaped structures called salt domes. Salt domes make excellent seals for upward-rising oil and gas, and they are often associated with rich deposits of these important natural resources. Because salt domes are relatively impermeable, some are used to store natural gas, oil, and radioactive waste. For example, human-made caverns in natural salt domes in Texas and Louisiana contain the U.S. Strategic Petroleum Reserve.

Factors of sea-level change

Scientists have determined that sea level transgressed and regressed as many as 50 times during the Late Paleozoic. Geologists have found a number of reasons for relative sea level change—climate and glaciation cycles, crustal subsidence and uplift, varying sedimentation rates, and plate motions. These were all factors in the transgressive and regressive cycles of the Paleozoic.

 Get It?

Infer how glaciation affects sea level.

Mountain Building

Laurentia's margins were passive during the first period of the Paleozoic, and mountains were not forming. However, changes occurred during the Ordovician (or duh VIH shun) period, as shown in **Figure 24.**

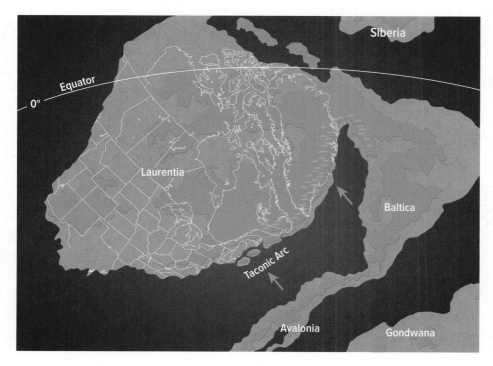

Figure 24 Baltica and Avalonia collided with the Taconic Island Arc during the Ordovician. This was one of the many Paleozoic tectonic events that transformed eastern Laurentia.

At that time, Laurentia, which was located near the equator, collided with the Taconic Island Arc, and mountains began to rise in what is now northeastern North America. This event is called the Taconic Orogeny. The Taconic Orogeny added new land and established an active volcanic zone along Laurentia's eastern margin. Remnants of this event are present in New York's Taconic Mountains.

Laurentia deformed

Laurentia was further transformed in the Silurian (si LUR ee uhn) period, about 400 million years ago, when Laurentia's eastern margin collided with Baltica and Avalonia, as shown in **Figure 24.** Baltica was a landmass that today is part of northern Europe and parts of Russia. Avalonia was an island ocean arc.

The deformation caused by these collisions—called the Acadian Orogeny—further transformed Laurentia and added folds, faults, and igneous intrusions to the already deformed Taconic rocks.

Ouachita Orogeny Another Laurentian mountain-building event—the Ouachita (WAH shuh taw) Orogeny—occurred during the Carboniferous period when southeastern Laurentia began to collide with Gondwana. Recall that Gondwana was the large landmass that eventually formed the southern continents, including Africa and South America. This collision formed the Ouachita Mountains of Arkansas and Oklahoma and was so intense that it caused the crust to uplift inland as far as present-day Colorado. Vertical faults raised rocks more than 2 km, forming a mountain range that geologists call the Ancestral Rockies.

Alleghenian Orogeny As Gondwana continued to push against Laurentia, the Appalachian Mountains began to form. This event, called the Alleghenian Orogeny, was the last of the Paleozoic mountain-building events to affect eastern North America. When it was completed at the end of the Paleozoic, the Appalachians were possibly higher than the Himalayas, and one giant supercontinent—Pangaea—had formed on Earth's surface.

 Get It?

Compare the Ouachita Orogeny and the Alleghenian Orogeny.

Paleozoic Life

The formation of Pangaea was the major geologic story of the Paleozoic, but Paleozoic rocks also tell another dramatic story. Fossils within these rocks show that multicellular animals went through extensive diversification at the beginning of this era. As you have learned, fossils help geologists in numerous ways. They are used to correlate geologic landscapes and piece together geologic time.

ACADEMIC VOCABULARY

transform
to change in a major way
The continent was transformed by a massive orogeny.

Fossils also help paleoecologists (pay lee oh ih KAH luh jists) learn about the ecology of ancient environments. Ecology refers to the relationships between organisms and their environments. Changes within the environment drove changes within the ecologies of Paleozoic life.

 Get It?

Describe what scientists learn from studying fossils.

Figure 25 The *Anomalocaris* shown in this artist's reconstruction was among the Cambrian organisms that had hard parts. Its name means "abnormal shrimp."

Cambrian explosion

Nearly every major marine group living today appeared during the first period of the Paleozoic era. The geologically rapid diversification of such a large collection of organisms in the Cambrian fossil record is known as the **Cambrian explosion.**

Some of the best-preserved Cambrian organisms occur in the Burgess Shale in the Canadian Rocky Mountains and in southern China. A spectacular array of fossil organisms with hard parts has been found in these locations. A reconstruction of one such organism is shown in **Figure 25.**

Ordovician extinction

At the end of the Ordovician, more than half of the marine groups that appeared in the Cambrian became extinct. Those that survived suffered large losses in their numbers. What caused this extinction?

Geologists have found evidence of glacial deposits in rocks of northern Africa, which at the time was situated at the South Pole. As you have learned, when water freezes in glaciers, sea level drops. Then, as now, most marine organisms lived in the relatively shallow waters of the continental shelves. When sea level is high, the shelves are flooded, and marine animals have many places to live. During regression, however, sea level decreases, and continental shelves can become too narrow to support diverse marine habitats.

Devonian extinction

Following the Late Ordovician extinction, marine life recovered, and new species evolved. There was a tremendous diversification of vertebrates, including fish, and the first appearance of tetrapods on land. Tetrapods are organisms with four feet. In the Late Devonian (dih VOH nee un), another extinction event eliminated approximately 50 percent of the marine groups. Some scientists think that global cooling was again the cause; there is evidence that some continents had glaciers at this time.

 Get It?

Describe what may have caused the Devonian extinction.

Terrestrial plants

The Ordovician and Devonian extinction events appear to have mainly affected marine life. They had little effect on life-forms living on land. Simple land plants began to appear on Earth in the Ordovician. During the Carboniferous, the first plants with seeds, called seed ferns, diversified. Because seeds contain their own moisture and food sources, they enabled terrestrial plants to survive in a variety of environments.

Carboniferous plants Many Carboniferous plants lived in low-lying swamps, such as the one shown in **Figure 26.** As these plants died, they were buried by layers of sediment. Chemical changes and increased pressure and heat from overlying layers of sediment transformed and compacted the ancient plant remains into coal deposits.

Swamps were also breeding grounds for insects. Fossils of the largest known insects, including dragonflies with 74-cm wingspans, have been found in Carboniferous sediment deposits. Compare this to the largest known wingspan of a modern dragonfly—19 cm.

Permian changes

At the end of the Permian, the largest mass extinction in the history of Earth occurred. The Permo-Triassic Extinction Event caused the extinction of nearly 95 percent of marine life-forms. Unlike the mass extinctions at the end of the Ordovician and Devonian, this extinction affected both marine and terrestrial organisms. More than 65 percent of the amphibians and almost one-third of all insects did not survive.

What could have caused such a widespread catastrophe? It was probably a combination of causes. First, there was a dramatic drop in sea level as the coalescence of Pangaea drained and closed the shallow seas. A regression would have been particularly critical for organisms inhabiting the continental shelves when there was only one continent. Other contributing factors likely included extreme volcanism in Siberia, low atmospheric oxygen levels, and climate change.

Figure 26 This artist's reconstruction shows what a Carboniferous swamp might have looked like.

Explain *why Carboniferous swamps produced coal deposits.*

✎ Check Your Progress

Summary

- Scientists study sediment and evaporite deposits to learn how sea level fluctuated in the past.

- Eastern Laurentia was transformed by many mountain-building events during the Paleozoic.

- A great diversity of multicellular life appeared during the first period of the Paleozoic.

- The largest extinction event in Earth's history occurred at the end of the Paleozoic.

Demonstrate Understanding

1. **Explain** how the formation of Pangaea affected the evolution of life-forms.

2. **Compare** transgression and regression.

3. **Discuss** the relationship between oil deposits and evaporites.

4. **Assess** the significance of the Cambrian explosion.

Explain Your Thinking

5. **Infer** what has happened to the Ancestral Rockies since their formation.

6. **Predict** changes in the fossil and rock record that might indicate a marine extinction event.

7. **MATH Connection** If 10 million species exist today and 5.5 species become extinct every day, calculate how many years it would take for 96 percent of today's species to become extinct.

LEARNSMART Go online to follow your personalized learning path to review, practice, and reinforce your understanding.

THE MESOZOIC ERA

FOCUS QUESTION

How did dinosaurs become so prevalent in the Mesozoic era?

Mesozoic Paleogeography

The mass extinction event that ended the Paleozoic era ushered in new opportunities for animals and plants of the Mesozoic era. Earth's life-forms changed drastically as new kinds of organisms, shown in **Figure 27,** evolved to fill empty niches. While some groups of these organisms remain on Earth today, none of the giant reptiles that dominated the land, sea, and air, and typified the era, survived. The nonavian dinosaurs all became extinct at the end of the era.

Breakup of Pangaea

When the Mesozoic era began, a single global ocean and a single continent—Pangaea—defined Earth's paleogeography. During the Late Triassic period, Pangaea began to break apart. The heat beneath Pangaea caused the continent to expand, and Pangaea's brittle lithosphere began to crack. Some of the large cracks, called rifts, gradually widened, and the landmass began spreading apart. The ocean flooded the rift valleys to form seaways, and large blocks of crust collapsed to form deep valleys. The Mesozoic climate was warm and tropical; no glaciers formed.

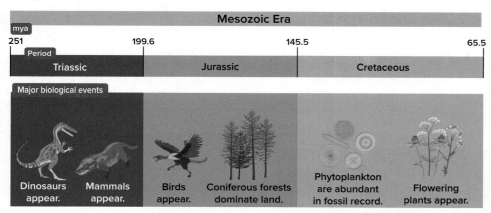

Figure 27 Although dinosaurs are the most famous of the Mesozoic life-forms, other organisms also appeared during this era.

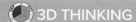 **3D THINKING** **DCI** Disciplinary Core Ideas   **CCC** Crosscutting Concepts   **SEP** Science & Engineering Practices

COLLECT EVIDENCE

Use your Science Journal to record the evidence you collect as you complete the readings and activities in this lesson.

INVESTIGATE

GO ONLINE to find these activities and more resources.

GeoLAB: Solve Dinosaur Fossil Puzzles
Engage in argument from evidence to identify the environments inhabited by different dinosaurs based on their dental structures.

Quick Investigation: Model Continental Shelf Area
Develop and use a model to visualize the result of shelf area change when continents collide.

Seaways

As the continents continued to split apart, mid-ocean rift systems developed at the junctures, and the widening seaways became oceans. The Atlantic Ocean began forming in the Triassic as North America rifted away from Europe and Africa. Some of the spreading areas at this juncture joined to form a long, continuous rift system called the Mid-Atlantic Ridge. As you have learned, this mid-ocean ridge system is still active today, erupting magma deep in the ocean as it widens. The Red Sea, shown in **Figure 28,** and the Gulf of Aden are new seaways in East Africa that are today slowly widening by a few centimeters a year as a result of continental breakup.

Changing sea level

The formation of mid-ocean rift systems was partly responsible for a rise in sea level during the Mesozoic. The magma that erupted at the ridges displaced a considerable amount of seawater onto the continents. However, sea level dropped at the end of the Triassic, and desertlike conditions developed in western North America. The climate became arid and, as evidenced in ancient sand dunes, a thick blanket of sand covered some of the land. Sea level rose again during the Jurassic, and a shallow sea formed in North America's center. The ocean continued to rise during the Cretaceous (krih TAY shus). **Figure 29** shows that nearly one-third of Earth's landmasses were covered with water.

Figure 28 The Red Sea and the Gulf of Aden (visible in the lower right-hand corner) are widening into a new seaway.

Identify *the tectonic force behind the creation of this new seaway.*

Mountain Building

Recall that the collision of continents during the Paleozoic transformed the eastern margin of Laurentia, while the continent's western margin remained passive. During the Mesozoic and Early Cenozoic, the reverse was true. During the breakup of Pangaea, multiple mountain-building episodes occurred along Laurentia's western margin, while little was happening along its eastern edge.

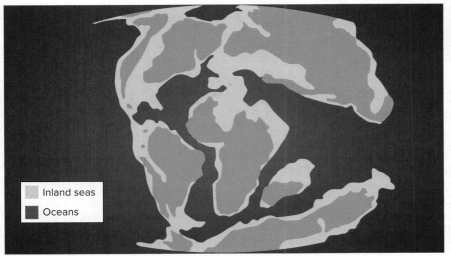

Figure 29 Nearly one-third of Earth's land surface was covered with water during the Late Cretaceous.

Inland seas
Oceans

Jeff Schmaltz, MODIS Rapid Response Team, NASA/GSFC

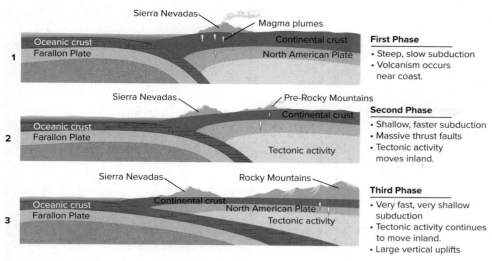

Figure 30 During the three phases of the Cordilleran Orogeny, mountains formed farther inland as the angle of subduction became more shallow and the speed of subduction increased, causing massive faulting and uplift.

First Phase
- Steep, slow subduction
- Volcanism occurs near coast.

Second Phase
- Shallow, faster subduction
- Massive thrust faults
- Tectonic activity moves inland.

Third Phase
- Very fast, very shallow subduction
- Tectonic activity continues to move inland.
- Large vertical uplifts

Cordillera

Much of the mountain building that occurred in western Laurentia was caused by the subduction of the oceanic Farallon Plate beneath Laurentia's western margin. As the plate descended, many structural features of the present-day Rocky Mountains, Sierra Nevada, and other western mountain ranges were formed. Geologists call these ranges collectively the North American Cordillera (kor dee AYR uh). *Cordillera* means "mountain range" in Spanish. The Cordilleran Orogeny consisted of three distinct phases. As shown in **Figure 30,** each phase was characterized by a different rate and angle of subduction.

First phase The first phase occurred during the Late Jurassic and Early Cretaceous. Subduction proceeded slowly, and the oceanic plate descended at a steep angle, and magma that formed from the surrounding mantle rose at the site of the emerging Sierra Nevada.

Second phase The second phase of the Cordilleran Orogeny occurred during the Cretaceous. Subduction increased in speed, but the oceanic plate descended at a shallow angle. As a result, there was less volcanism along Laurentia's margin and more tectonic activity inland, with massive thrust faults occurring in the Rocky Mountain area.

Third phase During the third phase of the Cordilleran Orogeny, which began during the Late Cretaceous and continued into the Cenozoic, subduction was even more shallow and rapid than it was during the second phase. The subduction rate was so fast that some scientists suggest the oceanic plate was pushed almost horizontally beneath the North American Plate. As a result, this phase was characterized by large, vertical uplifts, which formed the Rocky Mountains, and a decrease in volcanism. This range now extends from northern Mexico into Canada.

 Get It?

Sequence the phases in the formation of the North American Cordillera.

Mesozoic Life

As Pangaea broke apart during the Early Mesozoic, much of the habitat on the continental shelves that was lost during Pangaea's formation once again became available. New marine organisms, ranging from large predatory reptiles to tiny photosynthetic phytoplankton, evolved to fill these niches.

Figure 31 England's White Cliffs of Dover are made mostly of the fossil remains of tiny phytoplankton, shown as colored in the artist's rendition. The fossilized shells, called tests, are a translucent white, giving the cliffs their pale coloring.

Phytoplankton are microscopic organisms at the base of the marine food chain. These organisms were abundant during the Cretaceous. The remains of their shell-like hard parts make up the many chalk deposits around the world, including England's famous White Cliffs of Dover, shown in **Figure 31.**

Plant life

As the cool climate that characterized the Late Paleozoic came to an end during the Mesozoic, plant life changed sharply. The large, temperate swamps dried as the climate warmed. Tall cycad trees are seed plants without true flowers. These evolved during the Jurassic, along with ginkgos, pine trees, and other conifers. Flowering plants appeared during the Cretaceous.

Terrestrial animals

Mammals appeared during the Late Triassic, around the same time as the dinosaurs. However, the dominant Mesozoic animals were reptiles. Unlike amphibians, whose eggs need to be laid in water to prevent drying out, reptiles can lay their eggs on dry land. These eggs, called **amniotic** (am nee AH tihk) **eggs,** contain the food and water required by developing embryos inside. Amniotic eggs made it possible for reptiles, including dinosaurs, to roam widely.

Dinosaurs Archosaurs are a group of reptiles that includes dinosaurs and crocodilians. Archosaurs have a unique skeletal structure that allows for speed and flexibility of movement. While lizards and turtles walk with a sprawling posture, archosaurs have a hip structure that allows their legs to be held underneath the body. This enabled some dinosaurs to run with an upright posture, as shown in **Figure 32.**

Figure 32 Dinosaurs have a unique hip structure that enabled some, like this *Velociraptor*, to develop an erect posture and run on two legs.

 Get It?

Explain how a dinosaur's posture differed from that of other reptiles.

Table 2 Major Extinctions in the Phanerozoic

Extinction event	End Ordovician	Late Devonian	Permo-Triassic	End Cretaceous
Approximate age	400 mya	350 mya	200 mya	65 mya
Percentage groups extinct	57 percent marine	50 percent marine	95 percent marine; 70 percent land	75 percent marine; 56 percent land

Mass extinctions

At the end of the Mesozoic, an extinction event devastated terrestrial dinosaurs, most marine reptiles, plants, and many other organisms. Today, most scientists agree that the combination of massive volcanism, which stressed Earth's climate, and a large meteorite impact that occurred at the end of the Cretaceous is responsible for the extinction event. It is thought that the meteorite was at least 10 km in diameter. An impact of this size could have blown up to 25 trillion metric tons of rock into the atmosphere, causing long-lasting global warming.

Evidence for this impact includes an impact site—Chicxulub Crater—on Mexico's Yucatán Peninsula as well as a unique layer of clay that separates Cretaceous rocks from rocks of the first period of the Cenozoic. Found worldwide, this layer contains an unusually high amount of **iridium** (ih RID ee um), a rare metal in Earth's rocks but a relatively common metal in asteroids. As shown in **Table 2,** the extinction event at the end of the Mesozoic was relatively mild compared with the Permo-Triassic Extinction Event at the end of the Paleozoic.

Check Your Progress

Summary

- The breakup of Pangaea triggered a series of tectonic events that transformed western Laurentia.

- The Atlantic Ocean began to form during the Mesozoic as North America broke away from Europe.

- Dinosaurs and other new organisms evolved to fill niches left empty by the Permo-Triassic Extinction Event.

- All dinosaurs, except birds, and many other organisms became extinct during a mass extinction event at the end of the Mesozoic.

Demonstrate Understanding

1. **Describe** the significance of the Permo-Triassic Extinction Event for the animals that populated the Mesozoic.

2. **Explain** how rifts are related to the formation of oceans.

3. **Compare** the tectonic events that transformed Laurentia's western margin with the tectonic events that changed Laurentia's eastern margin.

4. **Discuss** the evidence that suggests a meteorite impact was responsible for the extinctions at the end of the Mesozoic era.

Explain Your Thinking

5. **Deduce** what happened to the oceanic plate that subducted beneath western North America during the Mesozoic.

6. **WRITING Connection** Prepare a report documenting the chain of events that might have occurred once a meteorite hit Earth in the Late Mesozoic. Include a discussion of the effect on climate, air quality, and plant and animal life.

LEARNSMART Go online to follow your personalized learning path to review, practice, and reinforce your understanding.

FOCUS QUESTION
How did the ice ages affect life in the Cenozoic era?

Cenozoic Paleogeography

Although the Cenozoic era makes up only about 1.5 percent of Earth's total history—about the last 66 million years—scientists know more about this era than any other. Humans evolved during the Cenozoic, appearing in their present-day form during the Pleistocene epoch. **Figure 33** shows that you live in the Holocene, the current epoch of the Cenozoic.

Cooling trend

The Mesozoic era was relatively warm, and Earth remained warm during the Early Cenozoic. However, as Australia was splitting apart from Antarctica during the Eocene (EE uh seen) epoch, global climate began to cool. The cooling may have been caused in part by a change in ocean currents. When Antarctica and Australia were connected, a current of warm water flowing from the Pacific, Atlantic, and Indian Oceans moderated Antarctica's temperature.

Figure 33 Mammals diversified widely during the Cenozoic, but modern humans did not appear until relatively recently.

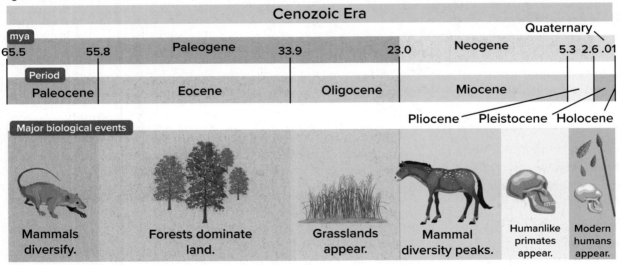

Cenozoic Era

| mya | | | | | | | | |
| 65.5 | 55.8 | Paleogene | 33.9 | | 23.0 | Neogene | 5.3 2.6 .01 | Quaternary |

Period: Paleocene | Eocene | Oligocene | Miocene | Pliocene — Pleistocene — Holocene

Major biological events:
Mammals diversify. | Forests dominate land. | Grasslands appear. | Mammal diversity peaks. | Humanlike primates appear. | Modern humans appear.

3D THINKING   DCI Disciplinary Core Ideas   CCC Crosscutting Concepts   SEP Science & Engineering Practices

COLLECT EVIDENCE
Use your Science Journal to record the evidence you collect as you complete the readings and activities in this lesson.

INVESTIGATE
GO ONLINE to find these activities and more resources.

Applying Practices: Mammoths...In Ohio?
HS-ESS2-2. Analyze geoscience data to make the claim that one change to Earth's surface can create feedbacks that cause changes to other Earth systems.

Mapping Lab: Cenozoic Ice Sheets and Plant Distribution
Create and use a model to visualize the result of the last ice age on plant distribution.

Figure 34 At the peak of Pleistocene glaciation, glaciers covered nearly one-third of Earth's land surfaces.

Infer *why patches of glaciation existed near the equator.*

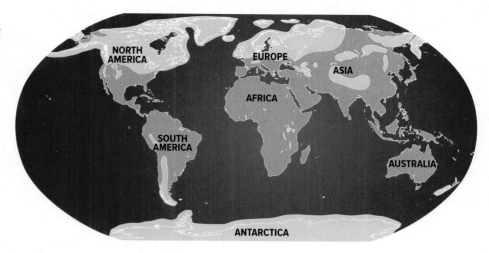

After Antarctica and Australia split, Antarctica was over the South Pole. A cold current began to flow around it, and a permanent ice cap began growing during the Oligocene.

Micocene warming

In the early Miocene epoch, the climate warmed again. The ice cap on Antarctica began to melt, and the ocean flooded the margins of North America. This trend reversed during the Middle and Late Miocene. Antarctica's ice cap stopped melting, and the Arctic Ocean began to freeze, resulting in the formation of the arctic ice cap. This set the stage for the ice ages.

Ice ages

Throughout the Pleistocene, ice covered much of the northern hemisphere. Glaciers advanced and retreated in at least four stages over North America and the northern latitudes. During the peak of these ice ages, glaciers up to 3 km thick covered nearly one-third of Earth's land surfaces, as shown in **Figure 34.** In North America, the paths of the Ohio and Missouri rivers roughly mark the southernmost point of glacial coverage. Glaciers carved out lakes and valleys, dropped huge boulders, and left behind abundant deposits of clay, sand, and gravel. At the end of the last ice age, glacial melting in northeastern Washington State caused a rush of water that created the largest waterfall recorded on Earth. The remnants are shown in **Figure 35.**

Figure 35 This photo shows Dry Falls, the remnants of Earth's largest waterfall in what is now east-central Washington State. The waterfall, more than 5 km long and 120 m high, once flowed with water from glacial melting.

Cenozoic Mountain Building

The mountain-building events of the Mesozoic uplifted massive blocks of crust to form the Rocky Mountains. During the Cenozoic, erosion wore down the Rockies, but uplift continued. Eroded sediment filled large basins adjacent to the mountains. Today, this sediment is mined for coal. It also contains well-preserved fossils of fish, insects, plants, and birds. A fossil bird from one of the most famous of these deposits—Wyoming's Green River Formation—is shown in **Figure 36** on the next page.

Subduction in the West

Volcanism returned to the western coast of North America at the end of the Eocene epoch when the oceanic Farallon Plate began a steep subduction beneath the Pacific Northwest. As a result, the Cascade Mountains began to rise. Volcanoes in the Cascade range remain active today, as shown in **Figure 37.**

While subduction continued in northwestern North America, the Farallon Plate disappeared under what is now California. The North American Plate came into contact with another oceanic plate—the Pacific Plate—that was moving in a different direction. As a result, the San Andreas Fault formed. The San Andreas Fault is a transform boundary between the two plates. The two plates slide against each other, and there is no subduction. Because there is no subduction along the fault, there is no volcanic activity there.

Figure 36 This 38-million-year-old fossil bird was found in Wyoming's Green River Formation. The fossil is about 25 cm long.

Basin and Range Province

The beginning of the interaction between the North American Plate and the Pacific Plate coincided with the formation of the Basin and Range Province in the southwestern United States and northern Mexico. The Basin and Range Province consists of hundreds of nearly parallel mountains. These mountains were formed when stresses in Earth's crust—called tension—pulled it apart. This process, illustrated in **Figure 38,** continues today.

Continental collisions

The final breakup of Pangaea during the early Cenozoic resulted in several separate continents. It also brought some continents together. During the Paleocene, Africa began to collide with Eurasia, creating the Alps and narrowing the ancient Tethys (TEE thus) Ocean, which once separated Eurasia and Gondwana. The remnants of this ocean now exist as four seas in Europe and central Asia—the Black, Caspian, Aral, and Mediterranean Seas.

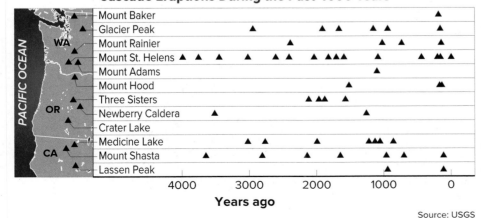

Figure 37 The Cascade Mountain Range includes active volcanoes that have erupted many times during the past 4000 years.

Source: USGS

Figure 38 Visualizing the Basin and Range Province

The Basin and Range Province is a series of mountains and basins that is bordered on the west by California's Sierra Nevada and on the east by Utah's Wasatch Mountains. During the past 25 million years, crustal stretching has increased the distance between these two points by over 250 km.

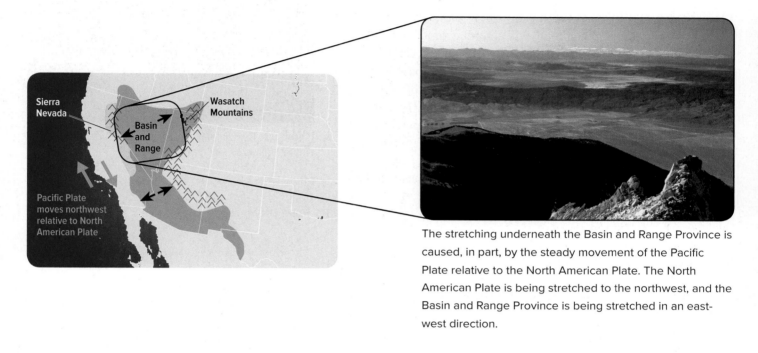

The stretching underneath the Basin and Range Province is caused, in part, by the steady movement of the Pacific Plate relative to the North American Plate. The North American Plate is being stretched to the northwest, and the Basin and Range Province is being stretched in an east-west direction.

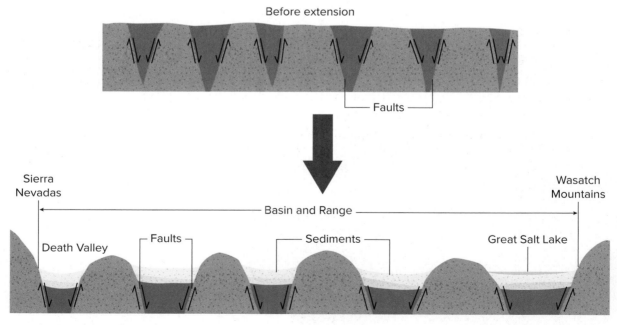

To compensate for crustal stretching, the rocks broke into hundreds of blocks along normal fault lines. Some blocks rose to form mountains, while adjacent areas dropped to form basins. The mountains are still being pushed upward, rising as quickly as they erode, and the basins are still dropping and filling with eroded debris. The crust underneath the Basin and Range Province has stretched so much that it is one of the thinnest parts of Earth's crust today.

John A. Karachewski

Also during the Paleocene, India began colliding into the southern margin of Asia to form the Himalayas, a mountain range that is still rising today. **Figure 39** shows the Himalayas as an abrupt junction where India joins Asia. The rocks on the top of Mount Everest are Ordovician marine limestone. Tectonic forces have pushed what was the Ordovician seafloor to the highest elevation on Earth.

Tectonic forces continue

Many scientists think that Earth is now in a relatively warm phase and that in the future the climate will again become cooler. No one can predict when or if this will happen. What is clear is that the tectonic forces that have shaped Earth over the past 4.6 billion years continue today. Some scientists think that in 250 million years, those forces will have largely eliminated the Atlantic Ocean and pushed the continents together into another supercontinent, as shown in **Figure 40.**

Figure 39 The Himalayas appear as an abrupt junction where India is colliding into Asia.

 Get It?

Predict what will happen to the Atlantic Ocean 250 million years from now.

Cenozoic Life

Many marine organisms, including clams, sea urchins, and sharks, survived the mass extinction at the end of the Cretaceous and populated the oceans during the Cenozoic. On land, forests dominated the Early Cenozoic landscapes. As the climate cooled during the Late Eocene, forests gave way to open land, and grasses appeared. By the Late Oligocene, grassy savannas, like those in east Africa today, were common worldwide. The rise of grasslands led to the diversification of many new mammal groups. Many scientists call the Cenozoic the Age of Mammals.

Future Continents (+250 million years)

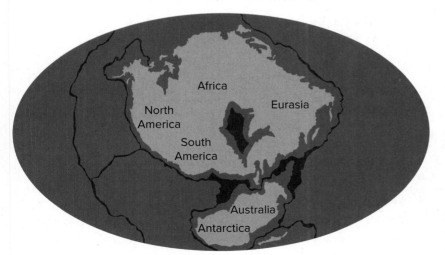

Figure 40 The Atlantic Ocean has nearly disappeared in this hypothetical map of Earth 250 million years in the future.

Ice age mammals

As the ice ages began, the climate began to cool, and new animals evolved in northern latitudes. Two of the most famous mammals of the Late Pleistocene are the woolly mammoth and the saber-toothed cat, shown in **Figure 41.** By the time these animals roamed Earth, modern humans—called *Homo sapiens*—were well established.

Humans

The defining characteristic of humans is their upright, or **bipedal,** locomotion. The fossil record, while incomplete, shows that the first bipedal human-like primates appeared about 6 mya, during the Late Miocene. The fossil remains of the earliest modern humans—found in Africa—are about 195,000 years old.

Migrations The migrations of early humans were undoubtedly influenced by the ice ages of the Late Pleistocene. For example, scientists think that the Bering Strait, which now separates Russia and Alaska, was exposed during the Late Pleistocene because much of Earth's water was frozen in glaciers. It is likely that the humans who walked across the strait were North America's first inhabitants.

Climate change and other natural hazards, such as droughts, earthquakes, floods, and volcanic eruptions, may have caused humans to migrate to new places. Geologic events such as these have shaped the course of human history and have driven human migrations.

Figure 41 The woolly mammoth and saber-toothed cat, shown in this artist's reconstruction, were adapted to the cool Pleistocene climate.

Check Your Progress

Summary

- Ice covered nearly one-third of Earth's land surface at the peak of the Cenozoic ice ages.
- The Cascade Mountains began to rise and the San Andreas Fault formed during the Cenozoic.
- The Cenozoic is known as the Age of Mammals.
- Fossil evidence suggests that modern humans appeared during the Pleistocene.

Demonstrate Understanding

1. **Describe** why the Cenozoic is called the Age of Mammals.
2. **Assess** the extent of glaciation in North America.
3. **Discuss** how the Basin and Range Province and the San Andreas Fault are tectonically related.
4. **Explain** how the positions of the continents contributed to Cenozoic climate change.
5. **Describe** how geologic events can drive human migrations.

Explain Your Thinking

6. **Propose** Why do you think early humans migrated to North America?
7. **MATH ▶ Connection** If the glacial ice on Earth were to melt, sea level would rise about 50 m. If sea level rose at an average rate of 2 mm per year, how long would it take for all the ice on Earth to melt? Use this relationship: distance = rate × time.

LEARNSMART Go online to follow your personalized learning path to review, practice, and reinforce your understanding.

Another Mass Extinction?

In Earth's history, there have been mass extinction events during which large percentages of Earth's species became extinct. Many scientists propose that Earth is facing another mass extinction event.

Passenger pigeons were hunted by humans, causing their extinction in 1914. This photograph shows the last individual of this species.

Past extinction events

Scientists recognize five major mass extinction events in Earth's history. At the end of the Ordovician and again in the Late Devonian, Earth lost more than 50 percent of its marine species. At the end of the Permian, the vast majority of species on Earth became extinct. The Triassic-Jurassic extinction event was a series of extinction events that caused about 80 percent of Earth's species to become extinct. Finally, the mass extinction at the end of the Mesozoic caused a large number of species, including dinosaurs, to become extinct.

All of these extinction events have some factors in common. For example, they were all caused by dramatic changes in conditions on Earth. Some were caused by global cooling during ice ages, others by massive amounts of volcanic activity that created inhospitable conditions. Changes in ocean water chemistry and increases in greenhouse gases were factors in some extinction events. Still others may have been caused by meteorite impacts and the resulting environmental changes.

A sixth mass extinction?

Many scientists suggest that Earth is entering, or has entered, a sixth mass extinction event.

To support their claim, they point to Earth's current extinction rate and evidence of environmental changes due to human activity. Extinction occurs when the last individual of a species dies. Extinctions do not occur only during mass extinction events. "Background extinction" is a phrase used to describe a typical rate of extinction. Earth's current rate of extinction is much greater than the background extinction rate. By some measures, the current extinction rate is 100 times greater than the background extinction rate.

Earth's elevated extinction rate can be attributed to changes caused by human activity. Hunting, pollution, habitat destruction, climate change, and sea level change are all affecting Earth's species. Scientists point out that many of these changes, such as changing global temperatures, are similar to changes that contributed to previous mass extinctions.

 EVALUATE EVIDENCE
SUPPORTING CLAIMS

Conduct additional research about the claim that Earth is currently undergoing a mass extinction event. Work with a small group to discuss and evaluate the evidence supporting this claim.

GO ONLINE to study with your Science Notebook.

Lesson 1 EARLY EARTH

- Scientists use Earth rocks, zircon crystals, Moon rocks, and meteorites to determine Earth's age.
- Likely heat sources of early Earth were gravitational contraction, radioactivity, and asteroid and meteorite bombardment.
- Earth differentiated into specific density zones early in its formation.
- Plate tectonics caused microcontinents to collide and fuse throughout the Proterozoic.
- The ancient continent of Laurentia formed as a result of many mountain-building episodes.

- zircon
- meteorite
- asteroid
- differentiation
- microcontinent
- craton
- Precambrian shield
- Canadian Shield
- Laurentia

Lesson 2 THE ATMOSPHERE, OCEANS, AND EARLY LIFE ON EARTH

- The atmosphere and oceans formed early in Earth's history.
- Oxygen gas began to accumulate in the Proterozoic due to photosynthesizing cyanobacteria.
- The water that filled Earth's oceans most likely came from two major sources.
- Life on Earth likely began between 3.9 and 3.5 bya.
- Stanley Miller and Harold Urey were the first to show experimentally that organic molecules could have formed on early Earth.
- Earth's multicellular organisms evolved at the end of the Precambrian.

- cyanobacteria
- stromatolite
- banded-iron formation
- red bed
- amino acid
- prokaryote
- eukaryote
- Ediacaran biota

Lesson 3 THE PALEOZOIC ERA

- Scientists study sediment and evaporite deposits to learn how sea level fluctuated in the past.
- Eastern Laurentia was transformed by many mountain-building events during the Paleozoic.
- A great diversity of multicellular life appeared during the first period of the Paleozoic.
- The largest extinction event in Earth's history occurred at the end of the Paleozoic.

- paleogeography
- passive margin
- transgression
- regression
- Cambrian explosion

GO ONLINE to study with your Science Notebook.

Lesson 4 THE MESOZOIC ERA

- The breakup of Pangaea triggered a series of tectonic events that transformed western Laurentia.
- The Atlantic Ocean began to form during the Mesozoic as North America broke away from Europe.
- Dinosaurs and other new organisms evolved to fill niches left empty by the Permo-Triassic Extinction Event.
- Many organisms became extinct at the end of the Mesozoic.
- All dinosaurs, except birds, and many other organisms became extinct during a mass extinction event at the end of the Mesozoic.

- phytoplankton
- amniotic egg
- iridium

Lesson 5 THE CENOZOIC ERA

- Ice covered nearly one-third of Earth's land surface at the peak of the Cenozoic ice ages.
- The Cascade Mountains began to rise and the San Andreas Fault formed during the Cenozoic.
- The Cenozoic is known as the Age of Mammals.
- Fossil evidence suggests that modern humans appeared during the Pleistocene.

- *Homo sapiens*
- bipedal

REVISIT THE PHENOMENON

How was this pre-historic dragonfly different from a modern-day dragonfly?

CER Claim, Evidence, Reasoning

Explain Your Reasoning Revisit the claim you made when you encountered the phenomenon. Summarize the evidence you gathered from your investigations and research and finalize your Summary Table. Does your evidence support your claim? If not, revise your claim. Explain why your evidence supports your claim.

STEM UNIT PROJECT

Now that you've completed the module, revisit your STEM unit project. You will apply your evidence from this module and complete your project.

GO FURTHER

SEP Data Analysis Lab
Can you find the time?

Paleoecologists study the shapes and compositions of fossil organisms to interpret how and in what types of environments they lived. Fossils are also used to interpret climatic changes and the passage of time.

Time Record Data

Geologic Era	Hours per Day	Days per Year	Geologic Time (mya)
Cenozoic	23.5–24	365–377	0–65
Mesozoic	23.5–22.4	377–392	65–248
Paleozoic	22.4–20	392–430	248–543

*Data obtained from: Prothero, D.R., and R.H. Dott, Jr. 2004. *Evolution of the Earth*. New York: McGraw-Hill

Data and Observations Graph the data in the table. Label the x-axis *Geologic time (mya)*. Label one y-axis *Hours per day,* and the second y-axis *Days per year.*

CER Analyze and Interpret Data

1. **Claim, Evidence** Determine the number of hours in a day 400 mya. Show your work.
2. **Claim, Evidence** Determine the number of hours in a day 200 mya. Show your work.
3. **Claim, Evidence** Determine the number of hours in a day 150 mya. Show your work.
4. **Reasoning** Predict when there will be 24.5 hours in a day.

ENCOUNTER THE PHENOMENON

How does coal mining affected both Earth and human communities?

SEP Ask Questions

What questions do you have about the phenomenon? Write your questions on sticky notes and add them to the driving question board for this unit.

Is it renewable or non-renewable?

Look for Evidence

As you go through this unit, use the information and your experiences to help you answer the phenomenon question as well as your own questions. For each activity, record your observations in a Summary Table, add an explanation, and identify how it connects to the unit and module phenomenon questions.

Solve a Problem
STEM UNIT PROJECT

Resources and the Environment Investigate and research more about natural resources and human effects on the environment. Use the results of these investigations and the evidence you collected during the unit to complete your unit project.

GO ONLINE In addition to reading the information in your Student Edition, you can find the STEM Unit Project and other useful resources online.

traffic_analyzer/iStock/Getty Images

ENCOUNTER THE PHENOMENON

What happened to the lake?

GO ONLINE to play a video about water shortages.

SEP Ask Questions

Do you have other questions about the phenomenon? If so, add them to the driving question board.

CER Claim, Evidence, Reasoning

Make Your Claim Use your CER chart to make a claim about what happened to the lake. Explain your reasoning.

Collect Evidence Use the lessons in this module to collect evidence to support your claim. Record your evidence as you move through the module.

Explain Your Reasoning You will revisit your claim and explain your reasoning at the end of the module.

GO ONLINE to access your CER chart and explore resources that can help you collect evidence.

LESSON 2: Explore and Explain: Managing Land

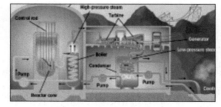

LESSON 5: Explore and Explain: Nuclear Energy

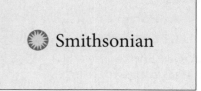

Additional Resources

(t)Video supplied by BBC Worldwide Learning

FOCUS QUESTION

What makes something a renewable resource?

Resources

Resource availability has aided the development of human society. You and every other living thing on Earth must have certain resources to grow, develop, maintain life processes, and reproduce. The resources that Earth provides are known as **natural resources.** Natural resources include Earth's organisms, nutrients, rocks, water, and minerals. Natural resources might come from the soil, air, water, or deep in Earth's crust. All items that you use every day, like those shown in **Figure 1,** come from natural resources.

Figure 1 Most of the items in this photo originated as natural resources.

Identify *three natural resources represented in this photo.*

 3D THINKING   **DCI** Disciplinary Core Ideas   **CCC** Crosscutting Concepts   **SEP** Science & Engineering Practices

COLLECT EVIDENCE

 Use your Science Journal to record the evidence you collect as you complete the readings and activities in this lesson.

INVESTIGATE

🔎 **GO ONLINE** to find these activities and more resources.

((૧)) **Review the News**
Obtain information from a current news story about current research into renewable resourses. **Evaluate** your source and **communicate** your findings to your class.

CCC **Identify Crosscutting Concepts**
Create a table of the crosscutting concepts and fill in examples you find as you read.

Figure 2 Bamboo can be grown as a sustainable-yield crop because it grows fast and needs no replanting. Bamboo can be used to produce a variety of items, including flooring, cooking utensils, and clothing.

Renewable resources

If you cut down a tree, you can replace that tree by planting a seedling. A tree is an example of a **renewable resource,** which is a natural resource that can be replaced by nature in a short period of time. Renewable resources include fresh air; fresh surface water in lakes, rivers, and streams; and most groundwater. When used properly, fertile soil is a renewable resource. However, if soil is exposed to wind and water, it can be eroded. Renewable resources also include all living things and elements that cycle through Earth's systems, such as nitrogen, carbon, and phosphorus. Resources that exist in an inexhaustible supply, such as solar energy, are also renewable resources.

Sustainable yield of organisms Humans can use natural resources responsibly by replacing resources as they are used. The replacement of renewable resources at the same rate at which they are consumed results in a **sustainable yield.**

Organisms in the biosphere are important renewable resources. Plants and animals reproduce; therefore, as long as some mature individuals of a species survive, they can be replaced. Crops can be planted every spring and harvested every fall from the same land as long as the Sun shines, rain falls, and organic matter or fertilizers provide the required nutrients. Animals that are raised for food, such as chickens and cattle, can also be replaced in short periods of time. Forests that are cut down for the production of paper products can be replanted and ready for harvest again in 10 to 20 years. Trees that are cut down for timber can be replaced after a period of up to 60 years.

Bamboo, shown in **Figure 2,** is one of Earth's most versatile renewable resources. Used by more than half the world's population for food, shelter, fuel, and clothing, bamboo is one of the fastest-growing plants. Because bamboo is a grass, it can be harvested without replanting. Bamboo grows without fertilizers or pesticides and is harvested in three to five years.

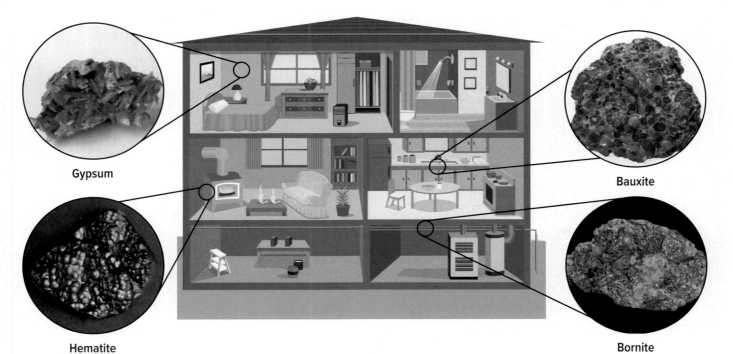

Figure 3 Nonrenewable resources are all around us. Aluminum from bauxite is used to make pots and pans, copper sulfides are used in copper plumbing, calcium sulfate is used to make drywall for houses and buildings, and iron from hematite is used to make appliances such as wood stoves.

Gypsum

Hematite

Bauxite

Bornite

Sunlight Some of Earth's renewable resources are not provided by Earth. The Sun provides an inexhaustible source of energy for all processes on Earth. Sunlight is considered a renewable resource because it will be available for at least the next five billion years.

Nonrenewable resources Many homes have copper pipes that transport water to the faucets. The price of copper fluctuates daily, but has steadily increased over the past 15 years. Copper is expensive because there are a limited number of copper mines, and demand continues to increase. When all the resources in the operating mines have been exhausted, no more copper will be mined unless new sources can be located. Copper is an example of a **nonrenewable resource**—a resource that exists in a fixed amount in various places in Earth's crust and can be replaced only by geological, physical, and chemical processes that take millions of years. Resources such as fossil fuels; diamonds and other gemstones; and elements such as gold, copper, and silver are, therefore, considered to be nonrenewable. **Figure 3** shows some materials you use every day and the nonrenewable resources used to make them.

Distribution of Resources

You have probably noticed that natural resources are not distributed evenly on Earth. Ohio, Pennsylvania, and West Virginia have an abundance of coal. California is known for its gold deposits. Georgia has large stands of trees used for paper and lumber. Some regions of the world, such as the United States, have an abundance of different types of natural resources. Other areas might have limited types of resources, but in abundant supply. For example, Saudi Arabia and Kuwait, in the Middle East, have more petroleum reserves than other areas of the world.

Global Consumption of Natural Resources

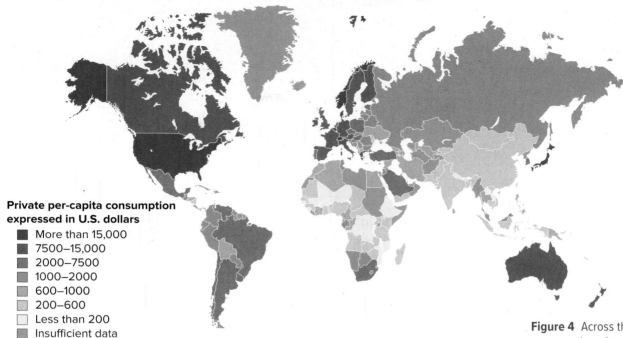

Private per-capita consumption expressed in U.S. dollars

- More than 15,000
- 7500–15,000
- 2000–7500
- 1000–2000
- 600–1000
- 200–600
- Less than 200
- Insufficient data

Figure 4 Across the globe, consumption of natural resources varies from country to country. Note that the average person in the United States consumes more than $15,000 a year in natural resources.

Determine *How does this compare with the consumption of natural resources in Canada or India?*

Consumption of resources

Natural resources are not evenly distributed on Earth. Not only are natural resources distributed unevenly, they are likewise consumed unevenly. Although people in the United States make up less than 5 percent of the world's population, they consume approximately 25 percent of Earth's mineral and energy resources each year, as shown in **Figure 4.**

 ## Check Your Progress

Summary

- Natural resources are the resources that Earth provides, including organisms, nutrients, rocks, minerals, air, and water.

- Renewable resources can be replaced within a short period of time.

- Nonrenewable resources exist in a fixed amount and take millions of years to replace.

Demonstrate Understanding

1. **Explain** how organisms, including humans, use natural resources and how resources aided the development of society.

2. **Explain** why costs of copper and other materials continue to increase.

3. **Categorize** the following as a renewable or nonrenewable resource: trees, aluminum, cotton, gemstones, and corn. Which are produced by sustainable yield?

Explain Your Thinking

4. **Propose** why consumption of natural resources is higher in the United States. Why is it important to be aware of this?

5. **MATH Connection** Aluminum production from bauxite ore costs $2000 per ton, whereas aluminum recycling costs $800 per ton. What is the percent saved by recycling?

LEARNSMART Go online to follow your personalized learning path to review, practice, and reinforce your understanding.

Why are mineral deposits only found in some areas?

Managing Land

Land provides places for humans and other organisms to live and interact. Land also provides spaces for the growth of crops, forests, grasslands, and wilderness areas.

Publicly managed land

Federal, state, and local governments manage more than 828 million acres of land in the United States. **Figure 5** shows that about 28 percent of land is managed by federal government agencies.

Federally Managed Land in the United States

Bureau of Indian Affairs
Bureau of Land Management
Bureau of Reclamation
Department of Defense
Fish and Wildlife Service
Forest Service
National Park Service
Tennessee Valley Authority
Other Agencies

Figure 5 Many government agencies—including the Bureau of Land Management, the USDA Forest Service, and the Fish and Wildlife Service—help manage land across the United States.

 3D THINKING   **DCI** Disciplinary Core Ideas   **CCC** Crosscutting Concepts   **SEP** Science & Engineering Practices

COLLECT EVIDENCE

Use your Science Journal to record the evidence you collect as you complete the readings and activities in this lesson.

INVESTIGATE

GO ONLINE to find these activities and more resources.

Review the News
Obtain information from a current news story about human caused desertification. **Evaluate** your source and **communicate** your findings to your class.

? **Revisit the Encounter the Phenomenon Question**
What information from this lesson can help you answer the Unit and Module questions?

These land areas are managed to support recreational uses, grazing, and mineral and energy resources. National forests are managed for sustainable yield and to provide recreational spaces. Wilderness areas are places that are maintained in their natural state and protected from development. The sustainablity of human societies and the biodiversity that supports them requires responsible management of natural resources, including managing the land. National forests, national parks, and national wildlife refuges are examples of places where these resources are managed.

National parks The national park system in the United States preserves scenic and unique natural landscapes, preserves and interprets the country's historic and cultural heritage, protects wildlife habitats and wilderness areas, and provides areas for various types of recreation. About 50 percent of the land in the national park system is designated as wilderness.

National wildlife refuges National wildlife refuges provide protection of habitats and breeding areas for wildlife, and some provide protection for endangered species. Other uses of the land in wildlife refuges, such as fishing, trapping, farming, and logging, are permitted as long as they are compatible with the purpose of the refuge.

Soil, Aggregates, Bedrock, and Ores

We also depend on four additional resources from the land: soil, aggregates, bedrocks, and ores.

Soil

In some parts of Earth's crust, it can take up to 1000 years to form just a few centimeters of topsoil, yet it can be lost in a matter of minutes as a result of erosion by wind or water. Plowing and leaving the ground without plant cover can increase topsoil loss.

The loss of topsoil makes soil less fertile and less able to hold water, causing crop losses. Today, topsoil is eroding more quickly than it forms on about one-third of Earth's croplands. Each decade, Earth loses about 7 percent of its topsoil, yet the eroded croplands must feed an increasing human population.

In arid and semiarid areas of the world, the loss of topsoil leads to **desertification,** which is the process whereby productive land becomes desert. Desertification can occur when too many grazing animals are kept on arid lands or when trees and shrubs are cut down for use as fuel in areas with few energy resources. Desertification can be prevented by reducing overgrazing and by planting trees and shrubs to anchor soil and retain water.

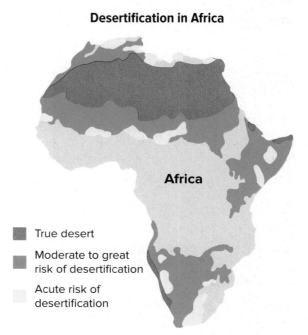

Desertification in Africa

Africa

■ True desert

■ Moderate to great risk of desertification

□ Acute risk of desertification

Figure 6 Desertification is a growing concern in many areas. Clearcutting and over-farming have led some parts of Africa to be considered in great risk of desertification.

Desertification is a growing problem in Africa, as shown in **Figure 6.** It is also a growing problem in the Middle East, in the western half of the United States, and in Australia. Desertification can be prevented by reducing overgrazing and by planting trees and shrubs to anchor soil and retain water.

 Get It?

Describe activities that can lead to erosion of topsoil.

Topsoil

Aggregate

Bedrock

Figure 7 Different layers of Earth's surface have value as resources. Topsoil provides nutrients for crop production and aggregate can be used for constructing roads and sidewalks.

Aggregates

Have you ever observed the construction of a highway? You might have seen workers place layers of materials on the ground before they began to build the highway surface. In some instances, the materials used for this first layer come from **aggregate,** which is sand and gravel and crushed stone that can naturally accumulate on or near Earth's surface.

You have learned how Earth processes transport materials. Some aggregates are transported by water and are found on floodplains in river valleys and in alluvial fans in mountainous areas. Other aggregates were deposited by glacial activity in moraines, eskers, kames, and outwash plains. Aggregates used in construction are often mixed with cement, lime, or other materials to form concrete, mortar, or asphalt.

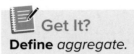 Get It?

Define *aggregate.*

Bedrock

Recall that underneath topsoil is a layer of soil consisting of inorganic matter, including weathered rock, sand, silt, clay, and gravel, as shown in **Figure 7.** This deeper soil layer lies on a base of unweathered parent rock called bedrock. **Bedrock** is solid rock, and it can consist of limestone, granite, marble, or other rocks that can be mined in quarries. Slabs of bedrock are often cut from quarry faces. Large pieces of bedrock are used in the construction of buildings, monuments, flooring, countertops, and fireplaces. Bedrock is also crushed for use as stone aggregate.

Ores

An **ore** is a natural resource that can be mined for a profit; that is, it can be mined as long as its value on the market is greater than the cost of its extraction. For example, the mineral hematite is an iron ore because it contains 70 percent iron by weight. Other minerals, such as limonite, also contain iron, but they are not considered ores because the percentage of iron contained in them is too low to make extraction profitable. Ores can be classified by the manner in which they formed. Some ores are associated with igneous rocks, and other ores are formed from processes that occur at Earth's surface.

Settling of crystals Iron, chromium, and platinum are examples of metals that are extracted from ores associated with igneous rocks. Chromium and platinum come from ores that form when minerals crystallize and settle to the bottom of a cooling body of magma. Chromite ore deposits are often found near the bases of igneous intrusions. One of the largest deposits of chromite is found in the Bushveldt Complex in South Africa. A chromite band is shown in **Figure 8.**

Hydrothermal fluids The most important sources of metallic ore deposits are hydrothermal fluids. Hot water and other fluids might be part of the magma that is injected into surrounding rock during the last stages of magma crystallization. Because atoms of metals such as copper and gold do not fit into the crystals of minerals during the cooling process, they become concentrated in the remaining magma. Eventually, a solution rich in metals and silica moves into the surrounding rocks to create ore deposits known as hydrothermal veins, shown in **Figure 8.** Hydrothermal veins commonly form along faults and joints in rock.

Get It?

Describe how deposits of gold and copper are formed.

Chemical precipitation Manganese and iron ores most commonly originate in layers formed through chemical precipitation. Iron ores in sedimentary rocks are often found in bands made up of alternating layers of iron-bearing minerals and chert, as shown in **Figure 8.** The origin of these ores, called banded iron formations, is not fully understood. Scientists think that banded iron formations resulted from an increase in atmospheric oxygen during the Precambrian.

Get It?

Explain how manganese and iron ores commonly form.

Chromite band

Hydrothermal veins

Banded iron

Figure 8 The chromite bands in the Bushveldt Complex are up to 0.5 m thick. Ores are also found in hydrothermal veins and banded formations.

CCC CROSSCUTTING CONCEPTS

Stability and Change Humans use materials from Earth's surface, including soil, aggregates, bedrock, and ores. With a partner, discuss the stability and changes of these resources. List your arguments and supporting evidence.

Placer deposits

Figure 9 Sand and gravel bars that contain heavy sediments, such as gold, are called placer deposits.

Placer deposits Some sediments, such as grains of gold and silver, are more dense than other sediments. When stream velocity decreases, as, for example, when a stream flows around a bend, heavy sediments are sometimes dropped by the water and deposited in bars of sand and gravel. Sand and gravel bars that contain heavier sediments, such as gold nuggets, gold dust, diamonds, platinum, gemstones, and rounded pebbles of tin and titanium oxides, are known as placer deposits, as shown in **Figure 9.** Some of the gold found during the Gold Rush in California during the late 1840s was located in placer deposits.

Check Your Progress

Summary

- Loss of topsoil can lead to desertification.

- Aggregates, composed of sand, gravel, and crushed stone, can be found in glacial deposits.

- An ore is a resource that can be mined at a profit. Ores can be associated with igneous rocks or formed by processes on Earth's surface.

Demonstrate Understanding

1. **Describe** three natural resources derived from Earth's crust.

2. **Explain** why topsoil loss is considered a worldwide problem.

3. **Identify** three reasons why it is important to responsibly manage Earth's land resources.

4. **Determine** where placer materials might have originated.

Explain Your Thinking

5. **Predict** what would happen if a land resource, such as aluminum, was depleted.

6. **WRITING ⟩ Connection** Create a three-fold pamphlet explaining the purposes and use of national parks and National Wildlife Refuge lands.

Diarmuid/Alamy Stock Photo

FOCUS QUESTION

How is the atmosphere a resource?

Origin of Oxygen

Most organisms on Earth require oxygen to maintain their life processes, but oxygen has not always been a part of Earth's atmosphere. As you have learned, scientists think that 4.6 to 4.5 bya Earth's atmosphere was similar to the mixture of gases released by erupting volcanoes. These gases included carbon dioxide, nitrogen, and water vapor. As Earth's crust cooled and became more solid, rains washed most of the carbon dioxide out of the atmosphere and into the oceans. Early life-forms in the seas used carbon dioxide during photosynthesis and released oxygen. Over time, gradual atmospheric changes were due to plants and other organisms that captured carbon dioxide and released oxygen, as shown in **Figure 10.**

 Get It?

Describe how the gradual atmospheric changes occurred due to plants and other organisms.

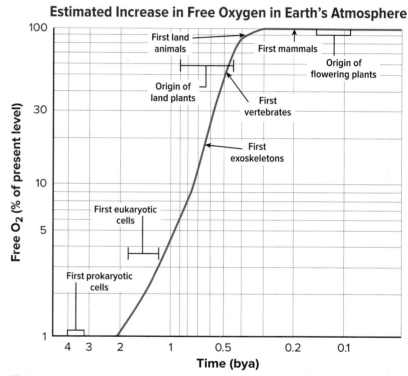

Figure 10 Scientists think that prokaryotes first appeared about 4 bya. It was not until 2 bya that eukaryotes appeared on Earth. Notice the difference in oxygen (O_2) gas levels between when prokaryotes and eukaryotes appeared.

 **3D THINKING** **DCI** Disciplinary Core Ideas   **CCC** Crosscutting Concepts   **SEP** Science & Engineering Practices

COLLECT EVIDENCE

Use your Science Journal to record the evidence you collect as you complete the readings and activities in this lesson.

INVESTIGATE

GO ONLINE to find these activities and more resources.

Investigation Lab: Assessing Wind Energy
Carry out an investigation to determine if wind energy is a viable energy resource for your area.

CCC **Identify Crosscutting Concepts**
Create a table of the crosscutting concepts and fill in examples you find as your read.

Cycles of Matter

The law of conservation of mass states that the amount of matter on Earth never changes. Earth's elements cycle among organisms and the nonliving environment. You have already learned about how water cycles on Earth. Earth's atmosphere plays a significant role in other cycles, such as the nitrogen and carbon cycles. These cycles are sometimes called biogeochemical cycles.

Earth's cycles are in delicate balance. When fossil fuels burn, the carbon that was stored in them for millions of years is released into Earth's atmosphere. Clearing forests results in fewer trees to take in carbon and release oxygen through photosynthesis.

Carbon cycle

Life on Earth would not exist without carbon because carbon is the key element in the sugars, starches, proteins, and other compounds that make up living things. The carbon cycle is illustrated in **Figure 11 on the next page.** During photosynthesis, green plants and algae convert carbon dioxide and water into carbohydrates and release oxygen back into the air. These carbohydrates are used as a source of energy for all organisms in a food web. Other organisms release carbon dioxide back into the air during respiration.

Carbon is also stored when organic matter is buried underground and, over millions of years, is converted to peat, coal, oil, or natural gas deposits. Carbon dioxide gas is released into the atmosphere when these fossil fuels are burned for energy.

 Get It?

Analyze the movement of matter and energy through the carbon cycle.

Nitrogen cycle

Nitrogen is an element that organisms need in order to produce proteins. Nitrogen makes up 78 percent of the atmosphere, but plants and animals cannot use nitrogen directly from the atmosphere. Some species of bacteria, called **nitrogen-fixing bacteria,** live in water or soil or grow on the roots of some plants and can capture nitrogen gas. Nitrogen-fixing bacteria convert the nitrogen into a form that can be used by plants to build proteins. Nitrogen continues through the food chain as one organism eats another. As organisms excrete waste and later die, the nitrogen returns to the soil and air. **Figure 11** shows the nitrogen cycle. Nitrogen moves from the atmosphere, to the soil, to living organisms, and then back to the atmosphere.

Natural Air Pollution Sources

A **pollutant** is a substance that enters Earth's geochemical cycles and can harm the well-being of living things or adversely affect their activities. Air pollution can come from natural or human sources and can affect air outside or inside buildings. Natural sources of air pollution include volcanoes, fires, and radon.

 Get It?

Examine the carbon cycle in **Figure 11** on the next page. What is a source of human-made pollution that enters the carbon cycle?

Figure 11 Visualizing Carbon and Nitrogen Cycles

All life-forms depend on carbon and nitrogen in many different ways.

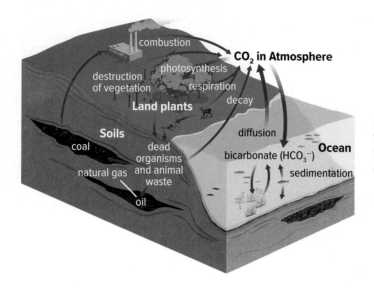

Humans have influenced the carbon cycle through the combustion of fuels. When fuels such as coal or oil are burned, one by-product of this combustion is carbon dioxide. Once released, carbon dioxide enters the atmosphere and continues in the carbon cycle.

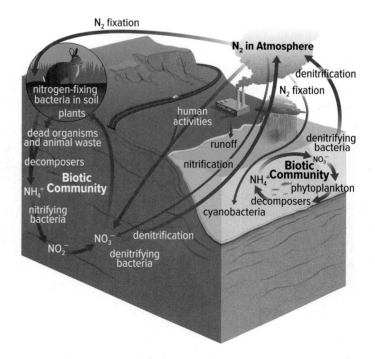

Nitrogen-fixing bacteria are an integral part of the nitrogen cycle. When animals produce waste, or when plants or animals die and begin to decompose, one by-product of this process is nitrogen. Nitrogen-fixing bacteria can break down the nitrogen, making it accessible for use by other plants and animals.

Volcanoes

Volcanoes can be significant sources of air pollution. On May 18, 1980, Mount St. Helens in Washington State shot an enormous column of ash 24 km into the sky. It continued to eject ash for about nine hours. Some of the ash reached the eastern United States within three days. Small particles entered the jet stream and circled Earth within two weeks. Mount St. Helens started erupting again between 2004 and 2008 and pumped out between 45,000 and 270,000 kg a day of sulfur dioxide. Italy's Mount Etna produces 100 times more sulfur dioxide than Mount St. Helens and is located in the middle of a heavily populated area. This sulfur contributes to acid rain and a type of bluish smog that volcanologists call vog, shown in **Figure 12,** which can cover large areas of land.

Get It?

Describe how volcanoes contribute to air pollution.

Figure 12 Vog, shown here over Kilauea, Hawaii, is formed when sulfur dioxide and other particulates emitted from a volcano mix with oxygen and moisture in the presence of sunlight.

Fires

Smoke is a mixture of gases and fine particles produced when wood and other organic matter burn. The most significant health threat from smoke comes from fine particles. These microscopic particles can get into your eyes and respiratory system, where they can cause health problems such as burning eyes, a runny nose, and illnesses such as chronic bronchitis. People with chronic lung disease can be at risk of serious injury from smoke.

Forest fires can release thousands of tons of carbon monoxide, a gas that interferes with oxygen transport in your blood. Gases from forest fires can also contribute to particulate and smog pollution hundreds of kilometers from the burning forest. In 2004, a large fire in Alaska and Canada, similar to the one shown in **Figure 13,** added about 30 billion kg of carbon monoxide to the atmosphere—about as much as was released during human activities in the United States that month.

Figure 13 Forest fires can release dangerous gases into the atmosphere. People with respiratory problems can be at risk of injury from high levels of smoke and gas.

Radon

The gas known as radon-222 (Rn-222) is colorless, odorless, tasteless, and naturally occurring. Rn-222 is produced by the radioactive decay of Uranium-238 (U-238). Small amounts of U-238 are found in most soils and rocks and in underground geologic formations, mainly in the northern third of the United States. Usually, radon gas from such deposits seeps upward through the soil and is released into the atmosphere, where it is diluted to harmless levels. However, when buildings are constructed with hollow concrete blocks, or when they have cracks in their foundations, radon gas can enter and build up to high levels indoors, as shown in **Figure 14.** Once indoors, radon gas decays into other radioactive particles that can be inhaled.

Radon is responsible for about 21,000 lung cancer deaths every year. About 2900 of these deaths occur among people who have never smoked. Because it is impossible to see or smell a buildup of radon gas in a building, the EPA suggests that people test the radon levels in their homes and offices.

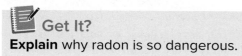
Get It?

Explain why radon is so dangerous.

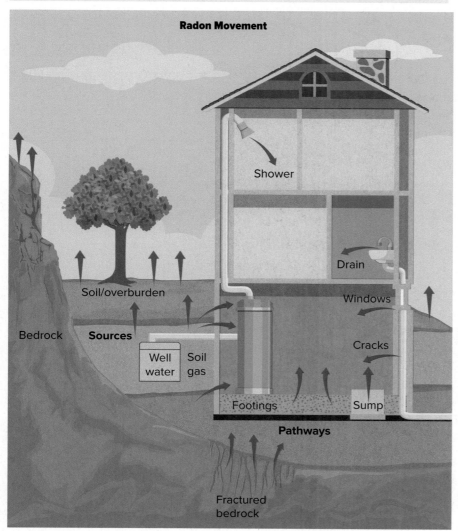

Figure 14 There are many ways radon can enter a home or other building. Once inside, radon is colorless and odorless, making it difficult to detect. Many homes are equipped with radon detectors that sound an alarm if safe levels are exceeded. Although radon often enters through cracks in the foundation or through drains or other openings in the basement, it can also enter through other pathways, such as showerheads.

Transport and Dilution

As air in the lower atmosphere moves across Earth's surface, it collects both naturally occurring and human-made pollutants. These pollutants are often transported, diluted, transformed, or removed from the atmosphere.

Some pollutants are carried downwind from their origin. Transport depends on wind direction and speed, topographical features, and the altitude of the pollutants. For example, hills, valleys, and tall buildings interrupt the flow of winds and thus influence the transport of pollutants. Some air pollutants may travel great distances. Many of the pollutants in acid precipitation that falls in the mountain ranges of North Carolina, shown in **Figure 15,** were transported from coal-burning power plants in the midwestern states. If air movement in the troposphere is turbulent, some pollutants are diluted and spread out, which reduces the damage they cause.

Some air pollutants undergo physical changes. For example, dry particles might clump together and become heavy enough to fall back to Earth's surface. These and other air pollutants are removed from the atmosphere in the form of snow, mist, fog, and rain.

Figure 15 When acid rain falls on a forest, the pH of the soil changes. As a result, tree growth can slow, and trees can become susceptible to disease.

Check Your Progress

Summary

- Earth's early atmosphere had no oxygen; it was changed over time by photosynthetic organisms.
- Oxygen, carbon, and nitrogen cycle from living organisms to the nonliving environment.
- Volcanoes, fires, and radon are natural sources of air pollution.

Demonstrate Understanding

1. **Explain** what caused changes to the early atmosphere.
2. **Compare and contrast** the carbon and nitrogen cycles.
3. **Describe** how coal-burning power plants in the Midwest can cause acid precipitation in New York.

Explain Your Thinking

4. **Predict** what might happen if there were no nitrogen-fixing bacteria on Earth.
5. **Apply** How might increasing the energy efficiency of a home lead to increased radon levels indoors?
6. **MATH ⟩Connection** Each year, about 21,000 people die from lung cancer related to radon. Of these, 2900 have never smoked. What percentage of people who die from radon-related lung cancer have never smoked?

FOCUS QUESTION

How do we get water to people who live in areas affected by shortages?

Properties of Water

About 71 percent of Earth's surface is covered by water. The world's oceans help regulate climate, provide habitats for marine organisms, dilute and degrade many pollutants, and even have a role in shaping Earth's surface. Freshwater is an important resource for agriculture, transportation, recreation, and numerous other human activities. In addition, the organisms that live on Earth are made up mostly of water. Most animals are about 50 to 65 percent water by mass, and even trees can be composed of up to 60 percent water.

Liquid water

What unique properties of water allow it to be so versatile? Water has a high boiling point, 100°C, and a low freezing point, 0°C. As a result, water remains liquid in most of the environments on Earth. Water can exist as a liquid over a wide range of temperatures because of the hydrogen bonds between water molecules. **Hydrogen bonds** form when the positive ends of some water molecules are attracted to the negative ends of other water molecules. Hydrogen bonds, shown in **Figure 16,** also cause water's surface to contract and allow water to adhere to and coat a solid. These properties enable water to rise from the roots of a plant, through its stem, to its leaves.

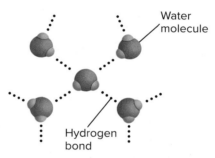

Figure 16 The attractions between the slightly positive and slightly negative ends of water molecules are called hydrogen bonds.

 Get It?

Describe some of water's unique physical and chemical properties.

 **3D THINKING**   **DCI** Disciplinary Core Ideas   **CCC** Crosscutting Concepts   **SEP** Science & Engineering Practices

COLLECT EVIDENCE
Use your Science Journal to record the evidence you collect as you complete the readings and activities in this lesson.

INVESTIGATE

GO ONLINE to find these activities and more resources.

GeoLAB: Monitor Daily Water Usage
Obtain and evaluate information to visualize the scale of your impact on water resources.

Investigation Lab: Water Usage
Analyze and interpret data to determine the effect of humans on the United States' freshwater resources.

Thermal energy storage capacity Liquid water can store a large amount of thermal energy without a significant increase in temperature. This property protects aquatic organisms from rapid temperature changes, and it also contributes to water's ability to regulate Earth's climate. Because of this same property, water is used as a coolant for automobile engines, power plants, and other thermal energy-generating processes. Have you ever perspired heavily while participating in an outdoor activity on a hot day? Evaporation of perspiration from your skin helps you cool off because large quantities of thermal energy are released as the water in the perspiration changes into water vapor.

Get It?

Describe examples of water's capacity to absorb, store, and release large amounts of energy.

Water as a solvent Liquid water can dissolve a variety of compounds. This enables water to carry nutrients into, and waste products out of, the tissues of living things. The diffusion of water across cell membranes enables cells to regulate their internal pressure.

Get It?

Explain how water dissolves and transports materials.

Solid water

Unlike most liquids, water expands when it freezes. Because ice has a lower density than liquid water, it floats on top of water. As a result, bodies of water freeze from the top down. If water did not have this property, ponds and streams would freeze solid, and aquatic organisms would die each winter. **Figure 17** shows that expansion of water as it freezes can also fracture rocks. Thus, ice formation in cracks in Earth's surface becomes part of the weathering process.

Figure 17 In a rock formation where weathering has previously occurred, water can enter cracks in the formation.

Identify *the physical property of water that causes these cracks in the rock to widen when the water freezes.*

Location of freshwater resources

Freshwater resources are not distributed evenly across Earth's landmasses. The eastern United States receives ample precipitation, and most freshwater in these states is used for cooling, energy production, and manufacturing. By contrast, southwestern states often have little precipitation. In the southwestern United States, the largest use of freshwater is for agricultural purposes such as irrigation. Water tables in these areas might drop as people continue to use groundwater faster than it can be recharged.

Water distribution is a continuing problem worldwide, even though most continents have plenty of water. Since the 1970s, scarcity of water has resulted in the deaths of more than 24,000 people each year. In areas where water is scarce, women and children often walk long distances each day to collect water for domestic uses. Millions of people also try to survive on land that is prone to drought. About 25 countries, primarily in Africa, experience chronic water shortages. **Figure 18** on the next page shows projected water stress levels across the globe for the year 2025. These stress levels are predicted in large part by projected population growth as well as other factors.

Projected World Water Stress Levels in 2025

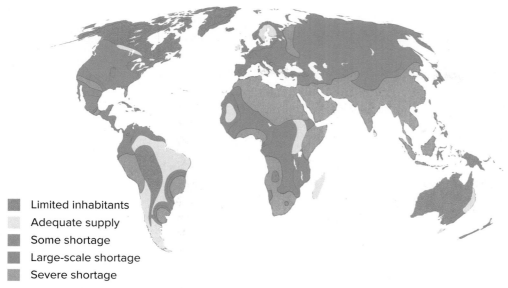

- ■ Limited inhabitants
- ☐ Adequate supply
- ■ Some shortage
- ■ Large-scale shortage
- ■ Severe shortage

Figure 18 By the year 2025, scientists predict that water stress levels will reach those shown here. The areas with projected adequate water supply will be limited. Most of the United States is projected to have some shortage, while much of Asia is predicted to have large-scale shortage.

Use of freshwater resources

Recall that the upper surface of groundwater is called the water table and that the water-saturated rock through which groundwater flows is called an aquifer. Aquifers are refilled naturally as rain percolates through soil and rock.

In the United States, about 20 percent of all freshwater used is groundwater pumped from aquifers. Water moves through aquifers at an average rate of about 15 m/day. If the withdrawal rate of an aquifer exceeds its natural recharge rate, the water table around the withdrawal point is lowered, called drawdown. If too many wells are drilled into the same aquifer in a limited area, the drawdown can lower the water table, and, as a result, wells might run dry.

Worldwide consumption Uses of freshwater vary worldwide, but about 70 percent of the water withdrawn each year is used to irrigate 18 percent of the world's croplands. About 23 percent of freshwater is used for cooling purposes in power plants, for oil and gas production, and in industrial processing. Domestic and municipal uses account for only 7 percent of the freshwater withdrawal.

Managing freshwater resources

Most countries manage their supplies of freshwater by building dams, transporting surface water, or tapping groundwater. The dam shown in **Figure 19** was built to hold back the floodwaters of the Yangtze River in China.

Figure 19 Dams are often built to contain freshwater resources in rivers. While this provides a readily available source of freshwater for human use, there are many other factors involved that make the damming of rivers controversial, including the flooding of farmland and displacement of people.

Dams and reservoirs Building dams is one of the primary ways that countries manage their freshwater resources. Large dams are built across river valleys, and the reservoirs behind dams capture the river's flow as well as rain and melting snow. Because the runoff is captured, flooding downstream is controlled. The water held in these reservoirs can be released as necessary to provide water for irrigation; to meet municipal needs, such as in homes and businesses; or to produce hydroelectric power. Reservoirs also provide opportunities for recreational activities, such as fishing and boating. Dams and reservoirs currently control between 25 and 50 percent of the total runoff on every continent.

Get It?
Explain several advantages of building dams.

Transporting surface water If you were to visit Europe or the Middle East, you would likely see many ancient aqueducts. The Romans built aqueducts 2000 years ago to bring water from other locations to their cities. Today, many countries use aqueducts, tunnels, and underground pipes to move water from areas where it is plentiful to areas that need freshwater.

The State Water Project in California, illustrated in **Figure 20,** is one example of the benefits, as well as the costs, of transporting surface water. In California, about 75 percent of the precipitation occurs north of the city of Sacramento, yet 75 percent of the state's population lives south of that city. The California Water Project uses a system of dams, pumps, and aqueducts to transport water from northern California to southern California. Eighty-two percent of this water is used for agriculture. The residents of Los Angeles and San Diego are withdrawing groundwater faster than it is being replenished. As a result, there is a demand for even more water to be diverted to the south. Conflicts over the transport of surface water could increase as human populations increase.

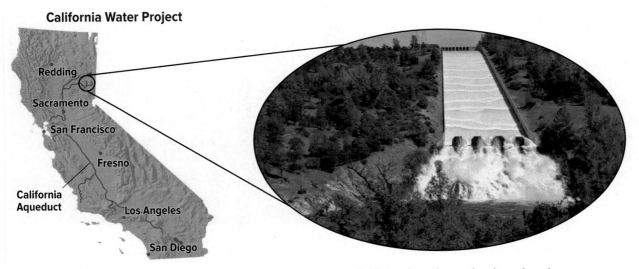

Figure 20 A system of dams, pumps, and aqueducts moves water in California from the north, where there is more rainfall, to the south, where the climate is more arid.

Desalination

With so much water available in the oceans, some countries have explored the possibility of removing salt from seawater to provide freshwater in a process called **desalination.** Several methods are available to desalinate seawater. One way is through distillation—a process in which water is heated until it evaporates, and then it is condensed and collected. This evaporation process leaves the salts behind. Most countries that use desalination to produce freshwater use solar energy to evaporate seawater. This process is slow, but relatively inexpensive. Some desalination plants, such as the one shown in **Figure 21,** use fuel to distill seawater, but because this process is expensive, it is used primarily to provide drinking water. Desalination provides about 25 percent of the freshwater supply in Florida. This state is the top user of desalination technology in the country.

Figure 21 Desalination can be accomplished using several different methods. One method, called distillation, removes salt by boiling the water. Another process involves pumping the water through a filtration system to remove the salt.

Check Your Progress

Summary

- Water has unique properties that allow life to exist on Earth.
- Water is not evenly distributed on Earth's surface.
- Water management methods distribute freshwater resources more evenly through the use of dams, aqueducts, and wells.

Demonstrate Understanding

1. **Describe** how the distribution of freshwater resources affects humans.
2. **Explain** why the thermal energy storage capacity of water is important to life on Earth.
3. **Explain** why water in a pond freezes from the top down.

Explain Your Thinking

4. **Propose** Do you think the process of desalination is a good option for areas like the southwestern United States where there is a high demand for freshwater? Explain your reasoning.
5. **Analyze** What are two things you could do to reduce your daily water usage?
6. **WRITING Connection** Imagine there is a large river near your hometown. For years, residents have used the river to fish, canoe, and swim. Recently a group has proposed damming the river to provide a clean, renewable energy source. Write two newspaper editorials—one in support of the construction of a dam and one against it.

LEARNSMART Go online to follow your personalized learning path to review, practice, and reinforce your understanding.

ENERGY RESOURCES

Why do people use different resources for their energy?

Earth's Main Energy Source

The energy that humans and all other organisms use comes mostly from the Sun. How is solar energy used by organisms? Plants are producers—they capture the Sun's light energy in the process of photosynthesis. The light energy is converted into a form that the plant can use for maintenance, growth, and reproduction. When other organisms called consumers eat producers, they use that stored energy for their own life processes. For example, when a rabbit eats grass, it consumes the energy stored by the plant. The rabbit stores energy as well, and this energy can be transferred to other organisms when the rabbit is eaten, when the rabbit produces waste, or when it dies and decomposes back into the ground. **Figure 22** shows how trapped light energy can be transferred from plants to humans.

Humans use energy to keep them warm in cold climates, to cook food, to pump water, and to provide light. There are many different fuel sources available to humans to provide this energy. Most of these fuels also store energy that originated from the Sun.

Figure 22 Humans need energy to live. When you eat a bowl of cereal, you use energy derived from the Sun. The wheat plant harnessed the Sun's light energy through photosynthesis. Some of this energy was stored in the seed of the wheat, which humans can consume to get energy they need to survive.

Image Source/Getty Images

3D THINKING   **DCI** Disciplinary Core Ideas   **CCC** Crosscutting Concepts   **SEP** Science & Engineering Practices

COLLECT EVIDENCE

 Use your Science Journal to record the evidence you collect as you complete the readings and activities in this lesson.

INVESTIGATE

🖥 **GO ONLINE** to find these activities and more resources.

 Design Your Own: Solar Water-Heater
Develop and use a model to examine the efficience and usefulness of solar heating as a source of energy.

Quick Investigation: Model Oil Migration
Use a model to visualize how the structure of rocks affects how oil resources flow through and are stored in the rocks.

Biomass Fuels and Biofuels

Fuels are materials that are consumed to produce energy. The total amount of living matter in an ecosystem is its biomass. Therefore, fuels derived from living things are called **biomass fuels.** Biomass fuels, shown in **Figure 23,** are renewable resources.

One type of fuel available for human use is derived directly from plant material. Plant materials burn readily because of the presence of **hydrocarbons**—molecules with hydrogen and carbon bonds only. Hydrocarbons are the result of the combination of carbon dioxide and water during photosynthesis. When plant materials burn, carbon dioxide is released as a waste product.

Wood

Humans have been using wood for fuel for thousands of years. Billions of people, mostly in developing countries of the world, use wood as their primary source of fuel for heating and cooking. Unfortunately, the need to use wood as a fuel has resulted in deforestation of many areas. As forests near villages are cut down for fuel, people travel farther to gather the wood they need. In some parts of the world, this demand for wood has led to the complete removal of forests, which can result in erosion and the loss of topsoil.

Field crops

Another biomass fuel commonly used in developing countries is field crops. Burning field crops, such as corn, hay, and straw, is the simplest way to use them as fuel. Crop residues left after harvest, including the stalks, hulls, pits, and shells from corn, grains, and nuts, are other sources of energy.

Fecal material

Feces are the solid wastes of animals. In many cases, dried feces contain undigested pieces of grass that help the material to burn. Feces from cows often meet the energy needs of people in developing countries with limited forest resources. Some people collect animal fecal matter for fuel and dry it outdoors, as shown in **Figure 23.**

Wood | Fecal material

Figure 23 Biomass fuel, such as wood, field crops, and fecal material, is the primary source of fuel for people in many countries. The fecal matter in the image above is being dried before it is burned.

Peat

Bogs are poorly drained areas with spongy, wet ground that is composed mainly of dead and decaying plant matter. When plants in a bog die, they fall into the water. Bog water is acidic and has low levels of oxygen; these conditions slow down or stop the growth of the bacteria that decompose dead organic matter, including plants. As a result, partially decayed plant material accumulates on the bottom of the bog. Over time, as the plant material is compressed by the weight of water and by other sediments that accumulate, it becomes a light, spongy material called **peat,** shown in **Figure 24.** Most of the peat used as fuel today is thousands of years old.

Peat has been used as a low-cost fuel for centuries because it can easily be cut out of a bog, dried in sunlight, and then burned in a stove or furnace to produce heat. Highly decomposed peat burns with greater fuel efficiency than wood. Today, peat is used to heat many homes in Ireland, England, parts of northern Europe, and the United States.

Biofuels

Biomass fuels include wood, dried field crops, and fecal materials from animals. Biomass is a renewable energy resource as long as the organisms that provide the biomass are replaced. Scientists are developing ways to produce fuels similar to gasoline from crops such as corn and soybeans. These fuels are called biofuels.

Ethanol Ethanol is a liquid produced by fermenting crops such as barley, wheat, and corn, which is shown in **Figure 25.** Ethanol can be blended with gasoline to reduce consumption of fossil fuels. Ethanol fuels burn more cleanly than pure gasoline. In 2011, the Environmental Protection Agency approved a gasoline blend with 15 percent ethanol for cars manufactured after 2001. Some vehicles, called flexible fuel vehicles, can run on mixtures containing 85 percent ethanol.

Biodiesel Biodiesel can be manufactured from vegetable oils, animal fats, recycled restaurant greases, and even algae. Biodiesel is safe and biodegradable, and it reduces air pollution. Blends of 20 percent biodiesel with 80 percent petroleum diesel (B20) can generally be used in unmodified diesel engines; however, it is currently more expensive than regular diesel.

Figure 24 This peat is being dried in the Sun before it is burned as fuel.

Figure 25 Biofuels, like biomass fuels, are derived from renewable resources. Crops like corn can be processed to create ethanol, a cleaner burning fuel than gasoline.

Fossil Fuels

Energy sources that formed over geologic time as a result of the compression and incomplete decomposition of plants and other organic matter are called **fossil fuels.** Although coal, oil, and natural gas originally formed from once-living things, these energy sources are considered nonrenewable. Nonrenewable resources are used at a faster rate than they can be replaced. Fossil fuels are nonrenewable resources because their formation occurs over millions of years, and we are using them at a much faster rate.

Fossil fuels, consisting mainly of hydrocarbons, can be transported wherever energy is needed and used on demand. This is why most industrialized countries, including the United States, depend primarily on coal, natural gas, and petroleum to fuel electric power plants and vehicles. Although fossil fuels are diverse in their appearance and composition, all of them originated from organic matter trapped in sedimentary rock.

 Get It?

Explain why fossil fuels are considered nonrenewable energy sources.

Coal

Coal forms from peat over millions of years and is the most abundant of fossil fuels. The hydrogen and oxygen in peat are lost as it is compressed; eventually, only carbon remains. The greater the carbon concentrations in coal, the hotter it burns. **Figure 26** on the next page shows how the different types of coal form.

Petroleum and natural gas

Most petroleum deposits formed from the accumulation of microscopic organisms on the seafloor. Other deposits are found in ancient lake beds. At the time of deposition, if the water was stagnant and lacked oxygen, organisms did not decay completely, and a layer of organic matter formed. Over time, layers of sediment buried the organic matter, and as more and more sediments were added, the temperature and pressure increased, forming liquid oil, also called crude oil. Crude oil that is collected on Earth's surface or pumped out of the ground is refined into a wide variety of petroleum products, such as gasoline, diesel fuel, and kerosene. Natural gas forms along with oil and is found beneath layers of solid rock. The rock prevents the gas from escaping to Earth's surface.

Natural gas can also be found in the form of solid methane hydrate. Methane hydrate is methane gas that is trapped inside a crystal structure made of water molecules. It is formed in sediments that are very deep underwater, where the temperature is low and the pressure is high.

 Get It?

Explain what a *solid methane hydrate* is and where you might find it.

SCIENCE USAGE v. COMMON USAGE

consume
Science usage: to use up completely
Common usage: to eat

ACADEMIC VOCABULARY

diverse
made up of distinct characteristics, qualities, or elements
The United States has diverse climates—the Northwest is cool and wet, while the Southwest is hot and dry.

Figure 26 Visualizing Coal

Coal forms from the compression of organic material over time.

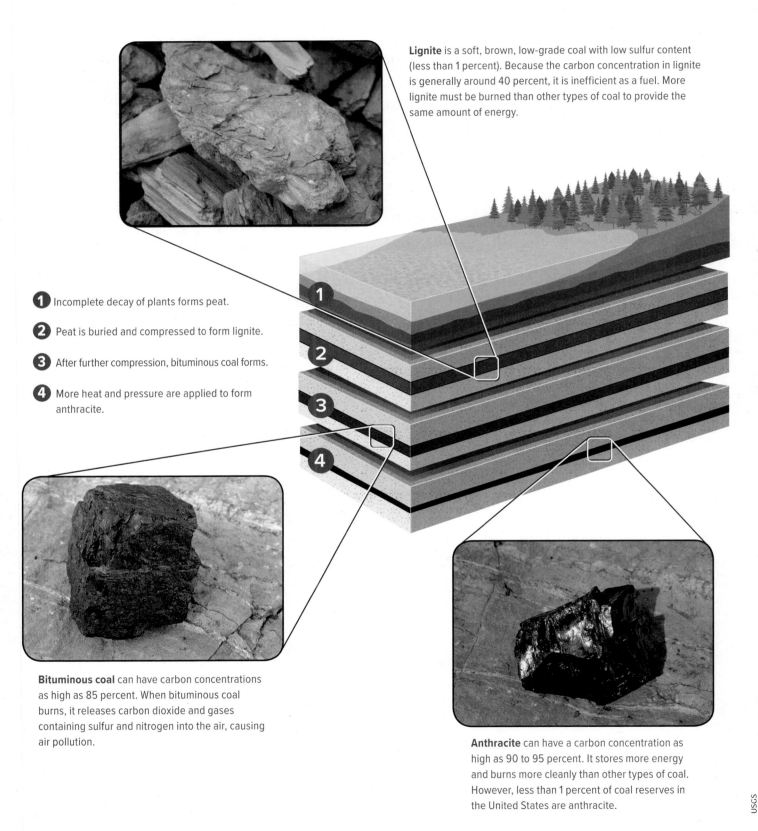

Lignite is a soft, brown, low-grade coal with low sulfur content (less than 1 percent). Because the carbon concentration in lignite is generally around 40 percent, it is inefficient as a fuel. More lignite must be burned than other types of coal to provide the same amount of energy.

1 Incomplete decay of plants forms peat.

2 Peat is buried and compressed to form lignite.

3 After further compression, bituminous coal forms.

4 More heat and pressure are applied to form anthracite.

Bituminous coal can have carbon concentrations as high as 85 percent. When bituminous coal burns, it releases carbon dioxide and gases containing sulfur and nitrogen into the air, causing air pollution.

Anthracite can have a carbon concentration as high as 90 to 95 percent. It stores more energy and burns more cleanly than other types of coal. However, less than 1 percent of coal reserves in the United States are anthracite.

USGS

Migration Rock containing pores or spaces that liquid can move through is called permeable rock. Crude oil and natural gas migrate sideways and upward from their place of formation, or source rock. As they migrate, they accumulate in permeable sedimentary rocks, called reservoir rocks, such as limestone and sandstone. Because petroleum is less dense than water, oil and gas continue to rise until they reach a barrier of impermeable rock, such as slate or shale, which prevents their continued upward movement. The barrier effectively seals the reservoir and creates a trap for the petroleum. Geologic formations such as faults and anticlines—folds of rock—can trap petroleum deposits, as shown in **Figure 27**.

 Get It?

Describe how oil migrates upward through sedimentary rock.

Oil shale Some petroleum resources are trapped in different types of rocks. For example, oil shale is a fine-grained rock that contains a solid, waxy mixture of hydrocarbon compounds called kerogen. Oil shale can be mined, then crushed and heated until the kerogen vaporizes. The kerogen vapor can then be condensed to form a heavy, slow-flowing, dark-brown oil known as shale oil. Shale oil is processed to remove nitrogen, sulfur, and other impurities before it can be sent through the pipelines to a refinery.

 Get It?

Explain how oil is extracted from oil shale.

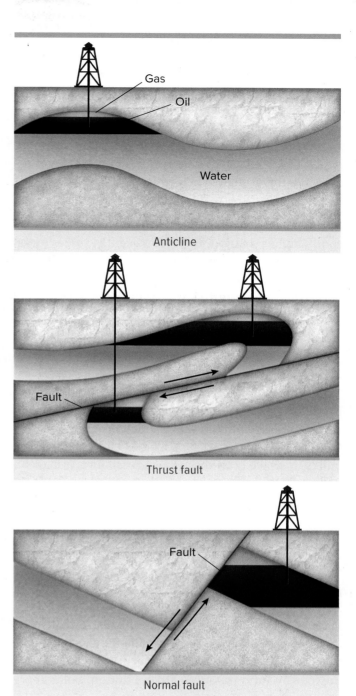

Figure 27 These diagrams show typical structural traps for oil and gas deposits.

STEM CAREER Connection

Conservation Scientist

These scientists manage forests, parks, rangelands, and other natural resources. They usually work for federal, state, or local governments or social advocacy organizations. Conservation scientists seek ways to use resources responsibly to ensure that they are available for future generations.

The largest deposits of oil shale in the world are found in the Green River Formation of Utah, Wyoming, and Colorado, shown in **Figure 28**. This geologic formation consists of lake sediments that were deposited during the Eocene epoch, around 50 mya. Within these sediments are an estimated 800 billion barrels of recoverable oil, which is three times greater than the proven oil reserves of Saudi Arabia. People in the United States use about 20 million barrels of oil per day. If oil shale could be used to meet a quarter of that demand, the estimated 800 billion barrels of recoverable oil from the Green River Formation would last for more than 400 years.

Historically, the cost of oil derived from oil shale has been significantly higher than pumped oil. Recently, prices for crude oil have again risen to levels that might make oil-shale-based oil production commercially viable, and both governments and industries are interested in pursuing the development of oil shale.

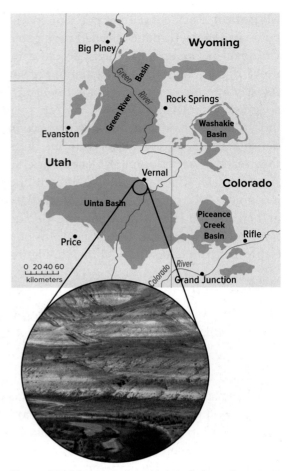

Figure 28 Oil shale is found primarily in sedimentary rocks. One of the most abundant sources of oil shale known is the Green River Formation, shown on the map as dark green regions.

 Get It?

Describe the advantages and disadvantages of obtaining oil from the Green River Formation.

Fossil fuels add CO_2 to the atmosphere because the processes that would naturally remove them do not occur over human time scales. Without a way to draw the added CO_2 out of the atmosphere, it builds up as we burn these fuels. Fossil fuels also create other pollutants harmful to the atmosphere and hydrosphere, such as sulfur dioxide and nitrogen dioxide. Scientists, private companies, and governmental agencies are all studying renewable resources, such as solar energy, as lower-pollution alternatives to traditional energy resources, including fossil fuels.

Solar Energy

Have you ever used a calculator with a solar collector? Solar-powered calculators use the Sun's energy to provide power. The Sun is the source of most of the energy on Earth. The main advantages of solar energy are that, once the generator is constructed, energy is inexpensive and produces little pollution. Solar energy can also be used directly to meet human energy needs through passive and active solar heating.

 Get It?

Explain the advantages of solar power.

CCC CROSSCUTTING CONCEPTS

Stability and Change Explain whether or not changes to fossil fuel use is causing reversible or irreversible damage to the fossil fuel system.

John A. Karachewski

Passive solar heating

If you have ever sat in a car that has been in the sunlight, you know that the Sun can heat up the inside of a car just by shining through its windows and on its surface. In the same way, the Sun's energy can be captured in homes. Thermal energy from the Sun enters through windows, as shown in **Figure 29.** Floors and walls made of concrete, adobe, brick, stone, or tile have heat-storing capacities and can help to hold the thermal energy inside the home. These materials collect solar energy during the daytime and slowly release it during the evening, as the surroundings cool.

In some warm climates, these materials alone can provide enough energy to keep a house warm. Solar energy that is trapped in materials and slowly released is called passive solar heating. Passive solar designs can provide up to 70 percent of the energy needed to heat a house. Although a passive solar house can be slightly more expensive to build than a traditional home, the cost of operating such a house is 30 to 40 percent lower.

 Get It?

Explain the process of heating a home using passive solar heating.

Active solar heating

Even in areas that do not receive consistent sunlight, the Sun's energy can still be used for heating. Active solar-heating systems include solar panels that absorb solar energy, as shown in **Figure 29,** and fans or pumps that distribute that energy throughout the house.

Solar panels mounted on a roof can have unobstructed exposure to the Sun. Energy collected by these solar panels can be used to heat a house directly, or it can be stored for later use in insulated tanks that contain rocks, water, or a heat-absorbing chemical. Solar panels on a roof can heat water up to 65°C, which is hot enough to wash clothing.

 Get It?

Identify a place where active solar heating could be or is used.

Passive solar heating

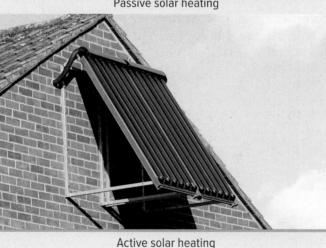

Active solar heating

Figure 29 Solar heating is considered a good alternative to conventional energy resources because it is clean and readily available in some areas. However, sunlight is available during limited hours each day, and it's energy is difficult to store for later use. More research needs to be done to make solar power a reasonable alternative for more people.

Passive and active solar heating rely on direct sunlight. Florida ranks third in solar-panel potential in the country, and solar energy use is projected to grow at a fast rate over the next five years. Using direct sunlight is relatively easy, but energy is also needed during hours of darkness, or in areas that are often overcast. Solar energy is difficult to store for later use. An economical and practical method of storing large amounts of solar energy for long periods of time has not yet been developed.

Photovoltaic cells

Solar energy can be converted into electric energy by using a **photovoltaic cell,** a structure that is made of two layers of two types of silicon. The cell absorbs energy from the sunlight that strikes it. The electricity produced by photovoltaic cells can be stored in batteries. Photovoltaic cells are reliable, quiet, and typically last more than 30 years. Large-scale groups of panels can be set up in deserts and in other land areas that are not useful for other human purposes.

One example of a large-scale group of panels is a solar power tower. The solar power tower generates electricity by harnessing the solar heating of the desert surface. A glass canopy surrounds the tower and acts as a greenhouse to heat the earth beneath it. The heat creates a self-contained wind field, driving a network of turbines, which generates electricity.

Energy from Water

Hydroelectric power is generated by converting the energy of free-falling water to electricity. When a dam is built across a large river to create a reservoir, the water stored in the reservoir can flow through pipes at controlled rates and cause turbines to spin to produce electricity. Hydroelectric power can also be generated from free-flowing water, such as the Niagara River. Today, hydroelectric power provides about 16 percent of the world's electricity and 6 percent of its total energy. Approximately 6 percent of the electricity used in the United States is generated by water, while Canada obtains about 60 percent of its electricity from this source. Many of the hydroelectric power resources of North America and Europe have been developed, but sites have not yet been developed in Africa, South America, or Asia.

Energy from the oceans

Ocean water is another potential source of energy. The energy of motion in waves, which is created primarily by wind, can be used to generate electricity. Barriers built across estuaries or inlets can capture the energy associated with the ebb and flow of tides for use in tidal power plants.

 Get It?
Describe the limitations of ocean water as a source of energy.

Nuclear Energy

Recall that atoms lose particles in the process of radioactive decay. One process by which atomic particles are emitted is called nuclear fission. **Nuclear fission** is the process in which a heavy nucleus (mass number greater than 200) divides to form smaller nuclei and one or two neutrons. This process releases a large amount of energy.

Radioactive elements consist of atoms that have a natural tendency to undergo nuclear fission. Uranium is one such radioactive element that is commonly used in the production of nuclear energy. Nuclear energy is one energy source that does not come directly from the Sun.

In the late 1950s, power companies in the United States began developing nuclear power plants similar to the one shown in **Figure 30.** Scientists suggested that nuclear power could produce electricity at a much lower cost than coal and other types of fossil fuels. Another advantage is that nuclear power plants do not produce carbon dioxide or any other green-house gases. After 60 years of development, however, 438 nuclear reactors are currently producing only 11 percent of the world's electricity. Construction of new nuclear power plants has slowed considerably in the United States and elsewhere.

What happened to using nuclear energy as a new source of power? High operating costs, poor reactor designs, and public concerns about radioactive wastes contributed to the decline of nuclear power. In addition, nuclear accidents, such as those at Three Mile Island in Pennsylvania in 1979 and at Chernobyl, Ukraine, in 1986, alerted people to the dangers of nuclear power plants. Because of its hazards, nuclear power has not been developed further in the United States as an alternative energy source.

Other advances in technology, such as those described in **Figure 31** on the next page, might make renewable energy sources more accessible for future generations.

Get It?

Explain why nuclear power plant construction has slowed in recent years.

Figure 30 Nuclear reactors rely on fission to generate heat. Heated water is converted to steam, which turns a turbine to generate electricity.

Identify *how many separate systems are in this reactor.*

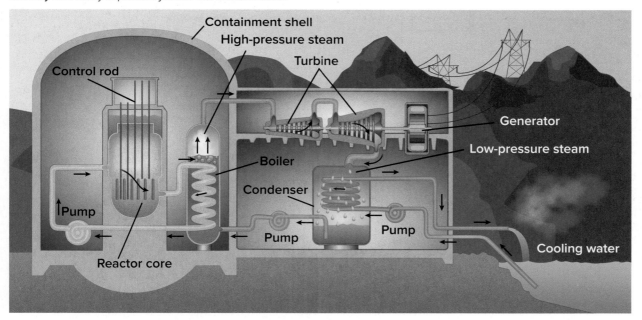

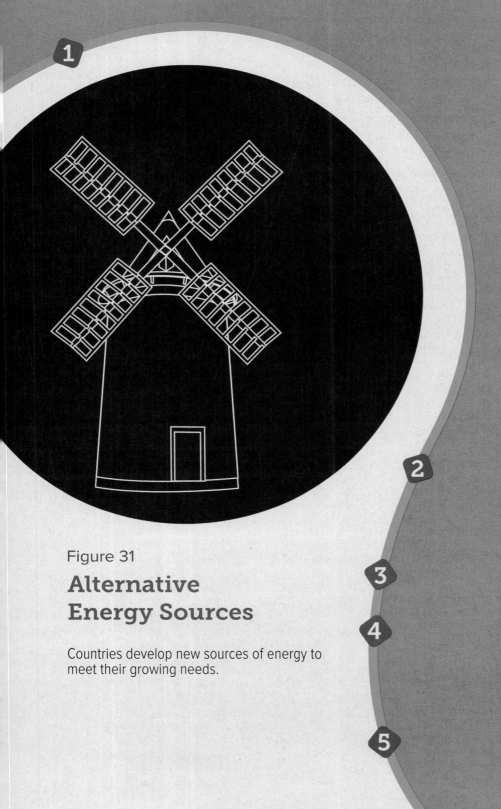

Figure 31

Alternative Energy Sources

Countries develop new sources of energy to meet their growing needs.

1 **1800** Holland boasts 9000 windmills that are used chiefly for land drainage and grinding grain.

2 **1933–1935** The United States builds the Hoover Dam and the Grand Coulee Dam to produce hydroelectric power, the country's main energy source, second to coal, until 1984.

3 **1952** Coal, which had replaced wood in much of Europe due to deforestation, causes a smog that kills 12,000 Londoners. England enacts new antipollution laws.

4 **1957** The first large-scale commercial nuclear power plant in the United States begins operating in Shippingport, Pennsylvania.

5 **1969** Iceland builds its first geothermal power plant. Today, geothermal energy heats 87 percent of the country's homes and supplies 17 percent of its energy needs.

6 **1995** A United States' program uses landfill gas to make electricity, reducing certain greenhouse gas emissions.

7 **1997** The first hybrid car to run on a gasoline engine and an electrical motor is mass-produced and released in Japan.

8 **2005** Ninety percent of all homes in Israel use solar panels to heat water. Other countries have adopted this technology in recent decades.

9 **2010** China leads the world in total renewable consumption of electricity, followed closely by the United States, Brazil, and Canada.

Geothermal Energy

Geothermal energy doesn't come from the Sun. Instead, it originates from Earth's internal heat. Steam produced when water is heated by hot magma beneath Earth's surface can be used to turn turbines and generate electricity. A geothermal power plant is shown in **Figure 32.** Energy produced by naturally occurring heat, steam, and hot water is called **geothermal energy.** While some geothermal energy escapes from Earth in small amounts that are barely noticeable, large amounts of geothermal energy are released at other surface locations. In these areas, which usually coincide with plate boundaries, geothermal energy can be used to produce electricity.

Wind Energy

Windmills in the Netherlands have been capturing wind power for human use for almost 2000 years. The windmills used today are more accurately called wind turbines because they convert energy of the wind into electrical energy. Wind turbines provide about 40 percent of the electricity used in Denmark. Experts suggest that wind power could supply more than 30 percent of the world's electricity by the year 2050.

Figure 32 Geothermal energy plants produce clean energy by harnessing the naturally occurring heat often found at plate boundaries.

Analyze *Is geothermal energy a renewable resource? Explain.*

Check Your Progress

Summary

- The Sun is the source of most energy on Earth.
- Humans have used materials derived from living things, such as wood, as renewable fuels for thousands of years.
- Fossil fuels formed from organisms that lived millions of years ago.
- Solar energy is unlimited, but technological advances are needed to find solutions to collect and store it.
- Nuclear energy is produced when atoms of radioactive elements emit particles in the process known as nuclear fission.

Demonstrate Understanding

1. **Explain** how energy stored in coal was obtained from the Sun.
2. **List** four types of biomass fuels.
3. **Illustrate** how coal forms.
4. **Compare** passive solar energy and active solar energy.

Explain Your Thinking

5. **Evaluate** this statement: Anthracite is usually found deeper in Earth's crust than lignite is.
6. **Analyze** In theory, solar energy could supply all of the world's energy needs. Why isn't it used to do so?
7. **WRITING** **Connection** Write a newspaper article that describes the associated economic, social, environmental, and geopolitical costs as well as risks and benefits of energy production.

LEARNSMART Go online to follow your personalized learning path to review, practice, and reinforce your understanding.

Flickr RF/Getty Images

Why Wildlife Refuges Make Good Neighbors

The term *refuge* means "shelter" or "safe haven." National wildlife refuges are certainly places where plants and animals can exist without threat from humans. They also provide many other benefits to a region.

Wildlife refuges do more than just provide protection for animals.

Many kilometers, many functions

The first National Wildlife Refuge was established in 1903 through an executive order by President Teddy Roosevelt. This order made it illegal to "hunt, trap, capture, willfully disturb, or kill any bird of any kind whatever" on Pelican Island, a small piece of land located just off of Florida's eastern coast. This first refuge was just the beginning. It took only ten years for the number of national refuges to reach 30. Today, there are over 600 protected areas across all 50 states covering more than 600,000 square kilometers.

Most people tend to think of wildlife refuges in the context of Teddy Roosevelt's original order. Wildlife refuges do continue to play an extremely important role in the protection of wildlife. For example, the Alligator River National Wildlife Refuge in North Carolina (ARNWR) is home to some of the last red wolves on Earth. This species of wolf is considered to be one of the five most endangered animal species in the world. Only an estimated 50 wild wolves remain in the southeastern United States. Areas like ARNWR are places where breeding pairs of endangered animals like the red wolf can be released into the wild to try to restore their numbers.

Wildlife refuges also provide a region with ecosystem services. Ecosystem services can range from the essential, like clean drinking water, to the aesthetic, like beautiful scenery. Wildlife refuges are able to provide these services because they tend to be areas of unspoiled wilderness, which supports the presence of a balanced ecosystem. For example, the Big Muddy refuge in Missouri has large, open spaces in which wildflowers grow and bloom. The flowers attract insects and other animals that are essential to the pollination of crops growing on neighboring farms.

Successful cultivation of crops is just one economic benefit of national wildlife refuges. Wildlife refuges like Chincoteague, in Virginia, and the Oregon Islands are responsible for bringing in millions of tourist dollars to the towns nearby. The refuges themselves employ thousands of people, and many people pay for permits to use the lands for fishing, camping, and other types of outdoor recreation.

CONSTRUCT AN ARGUMENT FROM EVIDENCE

Using data from your research, construct an argument that supports the statement "National wildlife refuges are good for the economy." Present your argument in the form of a letter to the editor.

MODULE 19
STUDY GUIDE

 GO ONLINE to study with your Science Notebook.

Lesson 1 NATURAL RESOURCES

- Natural resources are the resources that Earth provides, including organisms, nutrients, rocks, minerals, air, and water.
- Renewable resources can be replaced within a short period of time.
- Nonrenewable resources exist in a fixed amount and take millions of years to replace.

- natural resource
- renewable resource
- sustainable yield
- nonrenewable resource

Lesson 2 LAND RESOURCES

- Loss of topsoil can lead to desertification.
- Aggregates, composed of sand, gravel, and crushed stone, can be found in glacial deposits.
- An ore is a resource that can be mined at a profit. Ores can be associated with igneous rock or formed by processes on Earth's surface.

- desertification
- aggregate
- bedrock
- ore

Lesson 3 AIR RESOURCES

- Earth's early atmosphere had no oxygen; it was changed over time by photosynthetic organisms.
- Oxygen, carbon, and nitrogen cycle from living organisms to the nonliving environment.
- Volcanoes, fires, and radon are natural sources of air pollution.

- nitrogen-fixing bacteria
- pollutant

Lesson 4 WATER RESOURCES

- Water has unique properties that allow life to exist on Earth.
- Water is not evenly distributed on Earth's surface.
- Water management methods distribute freshwater resources more evenly through the use of dams, aqueducts, and wells.

- hydrogen bond
- desalination

Lesson 5 ENERGY RESOURCES

- The Sun is the source of most energy on Earth.
- Humans have used materials derived from living things, such as wood, as renewable fuels for thousands of years.
- Fossil fuels formed from organisms that lived millions of years ago.
- Solar energy is unlimited, but technological advances are needed to find solutions to collect and store it.
- Nuclear energy is produced when atoms of radioactive elements emit particles in the process known as nuclear fission.

- fuel
- biomass fuel
- hydrocarbon
- peat
- fossil fuel
- photovoltaic cell
- hydroelectric power
- nuclear fission
- geothermal energy

REVISIT THE PHENOMENON

What happened to the lake?

CER Claim, Evidence, Reasoning

Explain your Reasoning Revisit the claim you made when you encountered the phenomenon. Summarize the evidence you gathered from your investigations and research and finalize your Summary Table. Does your evidence support your claim? If not, revise your claim. Explain why your evidence supports your claim.

STEM UNIT PROJECT

Now that you've completed the module, revisit your STEM unit project. You will summarize your evidence and apply it to the project.

GO FURTHER

SEP Data Analysis Lab

What is the rate of deforestation in the Amazon?

Many experts are concerned about the loss of the forest cover in tropical rain forests worldwide. In the Amazon River Basin, scientists estimate that one hectare (ha, about 2.47 acres) of forest is cut down each hour.

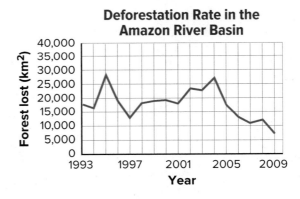

Deforestation Rate in the Amazon River Basin

CER Analyze and Interpret Data

1. How many square kilometers of the Amazon River Basin were deforested between 2002 and 2004?
2. According to the graph, what was the peak year for deforestation of the Amazon River Basin?
3. Calculate the rates of deforestation (or percent change yearly) for the periods 1993 to 2001 and 2001 to 2009.
4. Compare the rates of deforestation for the periods from 1993 to 2001 and 2001 to 2009.
5. **Claim, Evidence** Predict what will happen to the Amazon Rain Forest over the next 30 years.
6. **Reasoning** Explain how loss of rain forest could affect the carbon cycle.

traffic_analyzer/iStock/Getty Images

ENCOUNTER THE PHENOMENON

Why would we grow and use seaweed instead of standard crops like rice and corn?

GO ONLINE to play a video about growing seaweed on the ocean floor.

SEP Ask Questions

Do you have other questions about the phenomenon? If so, add them to the driving question board.

CER Claim, Evidence, Reasoning

Make Your Claim Use your CER chart to make a claim about why we would want to grow seaweed as a crop. Explain your reasoning.

Collect Evidence Use the lessons in this module to collect evidence to support your claim. Record your evidence as you move through the module.

Explain Your Reasoning You will revisit your claim and explain your reasoning at the end of the module.

GO ONLINE to access your CER chart and explore resources that can help you collect evidence.

LESSON 2: Explore and Explain: Agriculture

LESSON 4: Explore and Explain: Water Pollution

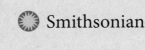

Additional Resources

Populations and the Use of Natural Resources

FOCUS QUESTION

What factors limit human population?

Resources and Organisms

Like all organisms, humans need natural resources to grow, reproduce, and maintain life. Among the resources that organisms require are air, food, water, and shelter. To meet their basic needs, most organisms are adapted to their immediate environment. They live in balance with the natural resources provided within their environment. For example, song-birds live in grassy meadows, forage for grass seeds to eat, weave nests out of dried grasses and twigs, and drink water from ponds or streams nearby.

Other organisms alter their environment to better meet their needs. For example, beavers build dams, like the one in **Figure 1,** across streams to create ponds where none previously existed. Such alteration of the environment has both positive and negative impacts. It creates a new wetland environment for other organisms, but at the same time, it kills some trees and displaces both aquatic and terrestrial organisms. Of all organisms, however, humans have an unequaled capacity to modify their environments. This capacity allows humans to live in every terrestrial environment on Earth. As a result, humans also have the greatest impact on Earth's natural resources.

Figure 1 Beavers can alter their environment to suit their needs. Notice how, by damming the stream, beavers have changed the water level.

Infer How might this affect the other organisms living in this environment?

 **3D THINKING** **DCI** Disciplinary Core Ideas   **CCC** Crosscutting Concepts   **SEP** Science & Engineering Practices

COLLECT EVIDENCE

 Use your Science Journal to record the evidence you collect as you complete the readings and activities in this lesson.

INVESTIGATE

🌐 GO ONLINE to find these activities and more resources.

⚙ **Applying Practices:** Human Activity, Natural Resources, Hazards, and Climate Change
HS-ESS3-1. Construct an explanation based on evidence for how the availability of natural resources, occurrence of natural hazards, and changes in climate have influenced human activity.

Population Growth

Population growth is defined as an increase in the size of a population over time. A graph of a growing population resembles a J-shaped curve at first. Whether the population is one of dandelions in a lawn, squirrels in a city park, or herring gulls on an island, the initial increase in population is small because the number of adults capable of reproducing is low.

As the number of reproducing adults increases, however, the rate of population growth increases rapidly. As shown in **Figure 2,** the population then experiences **exponential growth,** which is a pattern of growth in which a population grows faster as it increases in size.

 Get It?

Explain exponential growth. Why is this an important concept to understand in relation to how organisms affect their environment?

Limits to population growth

If the population graphed in **Figure 2** were studied for an extended period of time, what do you think would happen to the size of the population? Would it continue to grow exponentially? Many of Earth's natural resources are in limited supply, and therefore, most populations cannot continue to grow forever.

Eventually, one or more limiting factors, such as the availability of food, water, or shelter will cause a population to stop increasing. This leveling-off of population size results in an S-shaped curve, similar to the one in **Figure 3.**

Carrying capacity

The number of organisms that any given environment can support is its **carrying capacity.** When population size has not yet reached the carrying capacity of a particular environment, the population will continue to grow for several reasons. First, there will be more births than deaths because of adequate resources. Second, because of the availability of resources, more individuals might move to the area than die or leave.

If the population size temporarily exceeds the carrying capacity, the number of deaths will increase or the number of births will decrease until the population size returns to the carrying capacity. A population at the carrying capacity for its environment is in equilibrium. The population will continue to fluctuate around the carrying capacity as long as natural resources remain available.

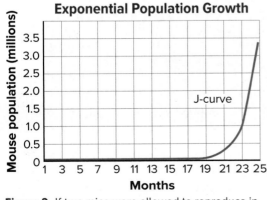

Figure 2 If two mice were allowed to reproduce in perfect conditions and all their offspring survived, the population would grow slowly at first, but then would accelerate quickly.

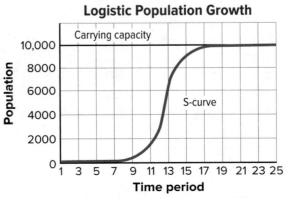

Figure 3 A population grows exponentially until it reaches its carrying capacity. Carrying capacity is limited by the resources available to the population.

Environmental limits

Environmental factors that do not depend on population size, such as storms and fires, are **density-independent factors.** Density-independent factors affect all populations that they come in contact with, regardless of population size, as **Figure 4** shows. Environmental factors that depend on population size, such as disease, predators, and competition for food, are called **density-dependent factors.** Density-dependent factors are often biotic factors that increasingly affect a population as the population's size increases. In this way, density-dependent factors affect population growth.

Figure 4 A forest fire is one example of a density-independent factor of population growth. Fires can affect trees, birds, mammals, and other populations. Fires, like the one that occurred here, can also encourage new growth.

Human Population Growth

During your lifetime, you might have seen an increase in the number of cars, houses, and roads around you. The human population on Earth is exponentially growing. The growth curve is still in the J-shaped stage. In fact, at its current rate, the human population is expected to reach approximately 9.6 billion by the year 2050.

The human population has not yet reached carrying capacity, but the current rate of growth cannot continue forever. As the population increases, demand for natural resources will also continue to increase steadily. Use of natural resources has already had global environmental implications.

Check Your Progress

Summary

- All organisms use resources to maintain their existence. The use of these resources impacts the environment.

- As populations increase, the demand for resources increases. With limited resources, populations will stop growing when they reach carrying capacity.

- Populations grow exponentially at early stages.

Demonstrate Understanding

1. **Explain** how an increasing human population places more demands on Earth's natural resources.

2. **Identify** three limiting factors that keep populations from growing indefinitely.

3. **Compare** density-dependent and density-independent factors that limit population growth.

Explain Your Thinking

4. **Predict** how a small population of bacteria placed in a petri dish with limited nutrients will change over time. Draw a graph to represent the population growth.

5. **MATH Connection** If a city has 300,000 residents and an average birth rate of 1.5 children per person, how many people will there be in the next generation?

LEARNSMART Go online to follow your personalized learning path to review, practice, and reinforce your understanding.

Barrett Hedges/Getty Images

HUMAN IMPACT ON LAND RESOURCES

FOCUS QUESTION
How does agriculture affect the land?

Mining for Resources

How much land per year do you think is necessary to provide the raw materials that you use? Each year, a typical person in the United States consumes resources equal to the renewable yield from approximately 6.8 ha (about 16.8 acres) of forest and farmland. Many of these raw materials come from under the surface of Earth. To access these resources for human use, they must be extracted through one of many mining techniques.

Effects of mining

Mining techniques can have a significant impact on Earth's surface. Modern societies require huge amounts of land resources, including iron, aluminum, copper, sand, gravel, and limestone. Unfortunately, the extraction of these resources often disturbs large areas of Earth's surface, as shown in **Figure 5.** Groundwater can become polluted, natural habitats can be disturbed or destroyed, and air quality can suffer. Finding a balance between the need for mineral resources and controlling the environmental change caused by extraction can be difficult, but scientists, in conjunction with mining companies, have created ways to reduce the impact of mining on the environment.

Figure 5 Mines, such as the one shown here, can have negative environmental impacts, such as topsoil erosion.

Determine *Where would the eroded topsoil go? What other environmental impacts might a mine like this one have?*

National Geographic Image Collection/Alamy Stock Photo

 **3D THINKING**   **DCI** Disciplinary Core Ideas   **CCC** Crosscutting Concepts   **SEP** Science & Engineering Practices

COLLECT EVIDENCE

 Use your Science Journal to record the evidence you collect as you complete the readings and activities in this lesson.

INVESTIGATE

🔎 **GO ONLINE** to find these activities and more resources.

Quick Investigation: Model Nutrient Loss
Use a model to simulate the effect of farming, strip-mining, or development on the nutrients in soil.

 Review the News
Obtain information from a current news story about the effect of urban development on land resources. **Evaluate** your source and **communicate** your findings to your class.

Figure 6 Waste rock, such as this tailings pile, is discarded after minerals are extracted.

Surface mining

Mines that are used to remove materials from Earth's surface destroy the original ground contours. Open-pit mines can leave behind waste rock, shown in **Figure 6,** which can weather over time. The extraction of mineral ores often involves grinding parent rock to separate the ore. The material left after the ore is extracted, called **tailings,** might release toxic elements such as mercury and arsenic into the groundwater or surface water. These materials can form acids as they weather and pollute the environment.

Reclamation In the United States, the Surface Mining Control and Reclamation Act of 1977 requires mining companies to restore the land to its original contours and to replant vegetation in a process called **reclamation.** However, vegetation cannot grow well without topsoil. Mining companies can scrape the topsoil off of the land surface prior to mining and stockpile it for reclamation after materials have been removed. **Figure 7** shows a strip-mined area that is being reclaimed. Although reclamation repairs much of the damage that surface mining causes, it can be extremely difficult to restore land to its original contours and vegetation.

 Get It?

Explain why it is important to have regulations requiring mining companies to restore land to its original contours.

Figure 7 This land is in the process of being reclaimed. It is being sprayed with a combination of grass seed and mulch.

Underground mining

Underground mining, also called subsurface mining, is used where mineral resources lie deep under the ground. Underground mining is less disruptive to the land surface than surface mining, but it still impacts the environment. For example, although the underground mines cannot be seen, the mountains of waste rock dug from under the ground are stockpiled on the surface. The water in **Figure 8** is orange because precipitation seeps through mine waste piles and causes a decrease in pH, dissolving many harmful metals in the waste. When runoff from the piles reaches the stream, which has a higher pH, the higher pH causes the metals to come out of solution and discolor the water. Aquatic plants and animals can also be affected by runoff.

Figure 8 Runoff from a nearby mine pollutes this river. The presence of metals, including iron, causes the orange color in the runoff.

Forestry

Although lumber is a valuable resource that provides society with many important products, extracting lumber and other resources from forests has environmental impacts. Clearing forested land is another way in which topsoil is lost. Worldwide, thousands of hectares of forests are cut down annually for firewood, charcoal, paper, and lumber. In parts of the world, the clearing of forested land results in **deforestation,** which is the removal of trees from a forested area without adequate replanting.

Fortunately, the negative environmental impacts of deforestation can be minimized through the practices of selective logging and the retention of buffer zones of trees along streambeds. In selective logging, workers remove only designated trees. This practice reduces the amount of ground left bare, and thus helps prevent erosion.

Agriculture

Crops grown in the agricultural industry provide society with much of the food it consumes. Vegetation, including agricultural crops, needs nutrients from topsoil to grow. It can take thousands of years for topsoil to form, and once it is lost, it is hard to replace. Whenever fields are plowed and the plants whose roots hold the soil in place are removed, topsoil can be eroded by wind and water, and nutrients can be lost. The use of fertilizers can help replace some of the nutrients, but there are other substances in topsoil that fertilizers cannot provide.

Gustavo Basso/NurPhoto/Getty Images

ACADEMIC VOCABULARY

adequate
sufficient to satisfy a requirement or meet a need
She filled her gas tank to ensure she had an adequate supply of gas for the trip.

Topsoil contains trace minerals as well as organisms such as earthworms and nitrogen-fixing bacteria. Earthworms burrow into soil, providing oxygen and space for plant roots to grow, and nitrogen-fixing bacteria take nitrogen out of the air and make it available to plants. Topsoil also has an abundance of organic matter, including decaying organisms and fecal material from organisms that live in the soil. Organic matter helps hold moisture, reduces erosion, and releases nutrients back into the soil. Soil erosion can be reduced and fertility can be increased by using a variety of farming practices, as shown in **Figure 9** on the next page.

Effects of pesticides

Chemicals applied to farm fields to control weeds, insects, and fungi are called **pesticides.** Pesticides have played an important role in boosting food production worldwide by eliminating or controlling organisms that destroy crops. However, some pesticides remain in, and can potentially harm, the environment for long periods of time.

Pesticides can slowly accumulate in organisms higher on the food chain, such as fishes and birds. Some pesticides also kill beneficial insect predators along with the targeted destructive insects. When pesticides kill decomposers, such as worms, the overall fertility of topsoil deteriorates. Insects can develop resistance to an insecticide, causing some farmers to use ever-increasing amounts to control pests. Further problems can be created when wind and rain carry pesticides away from farm fields and cause pollution in nearby waterways.

Safety Concerns

In addition to causing environmental problems, extracting and/or gathering resources can pose risks to society. For example, the National Safety Council has identified mining as one of the most dangerous occupations in the United States. It has one of the highest yearly death rates of all occupations. Working in the forestry and agriculture industries also carries a risk. Annual fatality rates in the forestry industry are approximately 130 fatalities per 100,000 full-time workers per year. In the agriculture industry, there are about 22 fatalities per 100,000 full-time workers per year.

Urban Development

As the human population continues to increase, more people live in cities and towns. Agricultural land located near cities is being converted to suburban housing. As people populate areas that were once agricultural or rural, stores and industry follow. Eighty percent of the population in North America lives in urban and suburban areas, and an estimated 6 billion people worldwide will be living in urban areas by the year 2045.

The development of urban areas has many environmental impacts. When towns and cities expand into rural areas, natural habitats are lost to roads, houses, and other buildings. Development leaves less land for agricultural use, which puts pressure on the remaining farmland for increased production. Other problems are created when concrete and asphalt cover large areas. Because there are fewer opportunities for rainwater to soak into the ground, groundwater supplies are not recharged, and flooding increases during heavy rains.

Figure 9 Visualizing Agricultural Practices

Using agricultural conservation practices can help protect precious nutrients in the soil as well as help reduce topsoil loss. Contour farming, crop rotation, and no-till farming are shown below.

Contour farming is often done on hillsides or other areas prone to erosion. Farmers plant crops with the contour of the earth, slowing the flow of runoff and helping to prevent erosion.

No-till farming Farmers leave the unused portion of the crops on the field instead of plowing them under each year. In this image, the crop in the previous year was wheat. After the seeds were harvested, the stalks were left on the field to prevent erosion and maintain topsoil.

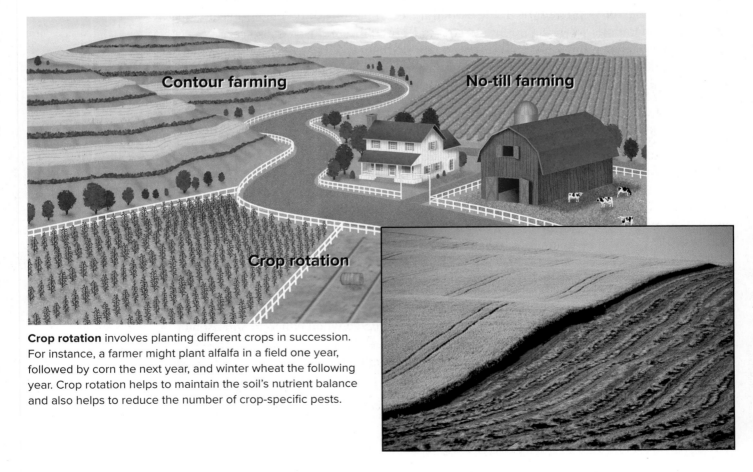

Contour farming

No-till farming

Crop rotation

Crop rotation involves planting different crops in succession. For instance, a farmer might plant alfalfa in a field one year, followed by corn the next year, and winter wheat the following year. Crop rotation helps to maintain the soil's nutrient balance and also helps to reduce the number of crop-specific pests.

Figure 10 Barriers such as these are often used on construction sites to prevent erosion and reduce loss of topsoil.

Conservation

People are becoming increasingly aware of the need to protect the environment, and communities are making increased efforts to do so as urban development continues. For example, developers are often required to place barriers, like the ones shown in **Figure 10,** around construction sites to catch sediment from increased erosion. In the United States, wetlands are now recognized as valuable ecosystems and are protected from development.

Solid Waste

Each person in the United States generates an average of 2.0 kg of solid waste per day. Where does it all go? Much of it is buried in landfills. **Figure 11** shows the percentages of various materials that were disposed in landfills in the United States in 2013.

Improper disposal of wastes can result in the contamination of land and water resources. Heavy metals, such as lead and mercury, and poisonous chemicals, such as arsenic, are by-products of many industrial processes and can pollute the soil and groundwater. Some of this type of contamination has been caused by industries that operated before the dangers of improper waste disposal were known.

Waste disposal remains a problem because of the immense volume of trash. Modern landfills are carefully designed and use new technology to minimize the environmental impacts of landfills. Impermeable clay or plastic layers are placed beneath a landfill, and trash is compacted by machines and buried under a layer of dirt to reduce volume and eliminate wind-blown trash. Vents in landfills release methane and other gases that are produced as the garbage decomposes. In some areas of California, the methane from landfills is captured and used to generate renewable energy.

Figure 11 This circle graph shows the total solid waste generated in the United States in 2013.

Determine *What material composed the highest percentage of solid waste in the United States in 2013? Why do you think this was the case? Could this material be recycled?*

2013 Total Waste Generation–254 million metric tons (before recycling)

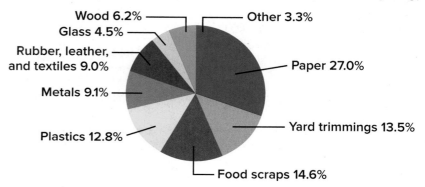

Wood 6.2%
Glass 4.5%
Rubber, leather, and textiles 9.0%
Metals 9.1%
Plastics 12.8%
Other 3.3%
Paper 27.0%
Yard trimmings 13.5%
Food scraps 14.6%

bruceman/E+/Getty Images

Cleaning Up

Several methods are available for cleaning up industrial waste sites. Contaminated soil can be removed and disposed of at hazardous waste landfills. Soil can also be incinerated to destroy toxic chemicals. The drawbacks to this method are that it can be expensive to treat large volumes of soil, and it can produce toxic ash that may pollute the air.

Bioremediation is the use of organisms to clean up or break down toxic wastes. These organisms, like the bacteria shown in **Figure 12,** actually eat pollutants for food, neutralizing their negative impacts on the environment. Bioremediation is useful for contamination caused by spilled gasoline and oil.

 Get It?

Explain how bioremediation works.

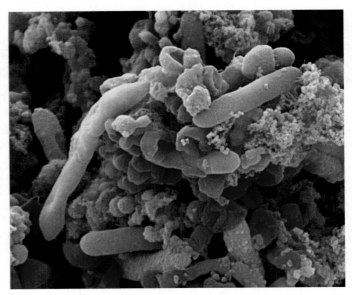

Figure 12 Bacteria such as these can be used to clean up an area contamined by oil, pesticides, or toxic chemicals.

Check Your Progress

Summary

- Humans require large amounts of land resources.

- The extraction of resources can disrupt Earth's surface.

- Growing populations increase the demand for food and result in increased urban development.

- Agriculture, poor forestry practices, and urban development can cause habitat loss, increased erosion, and soil and water pollution.

- Human impact on land resources can be minimized through the use of modern techniques.

Demonstrate Understanding

1. **Describe** the environmental and social impacts of mining, forestry, and agriculture; and how technology, regulations, and other methods can be used to reduce these impacts.

2. **Explain** the impacts of urban development on the environment and ways these impacts can be reduced.

3. **Predict** How many items will you throw away during lunch today? How much will you throw away in one week? One month? How does this relate to the impact of urban development on the environment?

Explain Your Thinking

4. **Suggest** methods of development that will reduce soil erosion and damage to streams.

5. **Infer** why responsible management of natural resources is important.

6. **WRITING ⟩Connection** For your school website, write an article suggesting ways that everyone can reduce the amount of waste they produce.

Eye of Science/Science Source

HUMAN IMPACT ON AIR RESOURCES

FOCUS QUESTION

How will climate change affect me?

Global Impacts of Air Pollution

It has become evident that human activities have been affecting Earth on a global scale. The global atmospheric effects of air pollution include ozone depletion, acid precipitation, and climate change.

Photochemical smog

On sunny days, you might notice a yellow-brown haze near densely populated areas. This haze is a type of air pollution, called **photochemical smog,** that forms mainly from automobile exhaust in the presence of sunlight. **Figure 13** shows how air pollutants from car exhaust form ground-level ozone. In the upper atmosphere, solar radiation converts oxygen gas into ozone. Ozone in the upper atmosphere is beneficial because it absorbs and filters out harmful ultraviolet (UV) radiation. However, ground-level ozone can irritate the eyes, noses, throats, and lungs of humans and other animals. It also has harmful effects on plants. When smog occurs in a city, the air becomes harmful to breathe.

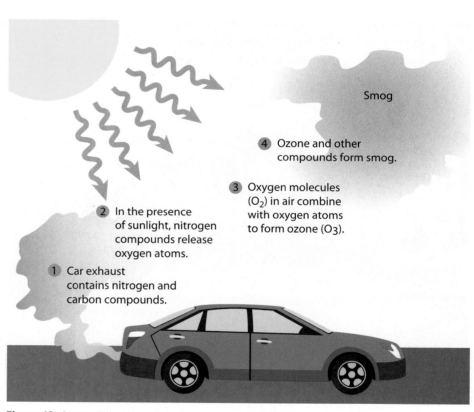

Smog

4 Ozone and other compounds form smog.

3 Oxygen molecules (O_2) in air combine with oxygen atoms to form ozone (O_3).

2 In the presence of sunlight, nitrogen compounds release oxygen atoms.

1 Car exhaust contains nitrogen and carbon compounds.

Figure 13 Automobile exhaust, in the presence of sunlight, can form a haze called photochemical smog.

3D THINKING   **DCI** Disciplinary Core Ideas   **CCC** Crosscutting Concepts   **SEP** Science & Engineering Practices

COLLECT EVIDENCE

 Use your Science Journal to record the evidence you collect as you complete the readings and activities in this lesson.

INVESTIGATE

 GO ONLINE to find these activities and more resources.

⚙ **Applying Practices: Locking Up Carbon**
HS-ESS3-4. Evaluate or refine a technological solution that reduces impacts of human activities on natural systems.

⚙ **Applying Practices: Carbon Cycling Through Earth's Spheres**
HS-ESS2-6. Develop a quantitative model to describe the cycling of carbon among the hydrosphere, atmosphere, geosphere, and biosphere.

One way to reduce this type of air pollution is to remove older, highly polluting vehicles from roadways. It is estimated that just 10 percent of the vehicles in operation produce 50 to 60 percent of the air pollution generated by gasoline-powered engines. Switching to newer cars with more efficient engines could significantly reduce air pollution throughout the world.

Ozone depletion

Recall that the ozone layer in the stratosphere serves as a protective shield as it absorbs and filters out harmful UV radiation. UV radiation has been linked to eye damage, skin cancer, and reduced crop yields. In the early 1970s, scientists found that chlorofluorocarbons (CFCs) destroy ozone molecules. All of the CFCs present in the atmosphere are a result of human activity. CFCs are released from old and leaky refrigerators and propellants in aerosol cans.

Although CFCs are stable and harmless near Earth's surface, they destroy ozone molecules, as shown in **Figure 14,** when they migrate into the upper atmosphere. Since the mid-1980s, atmospheric studies have revealed an extremely thin area of the ozone layer over Antarctica, called an **ozone hole.** Even though there is a natural, seasonal ozone hole, CFCs had thinned the ozone layer around the South Pole to dangerously low levels. Since 1987, 200 nations, have signed the Montreal Protocol, which seeks to regulate and eliminate the use of CFCs and other chemicals that deplete the ozone layer. This international effort has allowed the ozone layer to begin regenerating. Scientists project a full recovery of the ozone layer in this century.

Figure 14 One chlorine atom from one CFC molecule can destroy many ozone molecules.

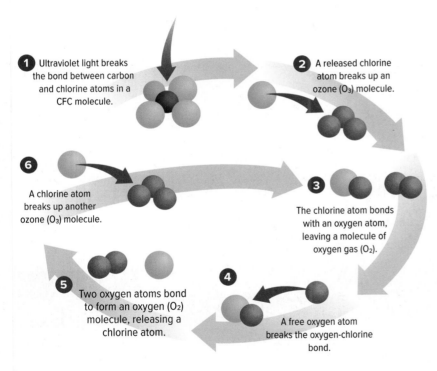

1 Ultraviolet light breaks the bond between carbon and chlorine atoms in a CFC molecule.

2 A released chlorine atom breaks up an ozone (O_3) molecule.

3 The chlorine atom bonds with an oxygen atom, leaving a molecule of oxygen gas (O_2).

4 A free oxygen atom breaks the oxygen-chlorine bond.

5 Two oxygen atoms bond to form an oxygen (O_2) molecule, releasing a chlorine atom.

6 A chlorine atom breaks up another ozone (O_3) molecule.

ACADEMIC VOCABULARY

particulate
of or relating to minute, separate, solid particles
People with asthma might have more symptoms when there is a high level of particulate matter in the air.

CCC CROSSCUTTING CONCEPTS

Stability and Change Research to find out more about the state of the ozone layer. Is it recovering? Why or why not? How do models help scientists make predictions about its recovery? Would you consider both the depletion and recovery of the ozone layer to be a short-term change or a long-term change? Present the results of your research in a format of your choice.

Precipitation pH in the Continental U.S.

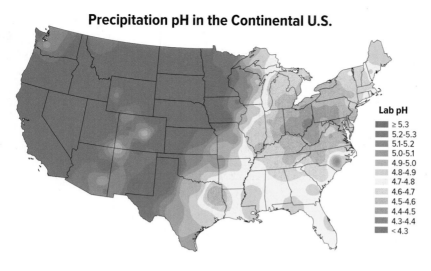

Lab pH
- ≥ 5.3
- 5.2-5.3
- 5.1-5.2
- 5.0-5.1
- 4.9-5.0
- 4.8-4.9
- 4.7-4.8
- 4.6-4.7
- 4.5-4.6
- 4.4-4.5
- 4.3-4.4
- < 4.3

Figure 15 This map shows the pH levels of precipitation across the continental United States.

Explain *why acid precipitation is an example of an environmental impact that results from energy production.*

Acid precipitation

Another major air pollution problem is acid precipitation, which is defined as rain, snow, fog, and mist with a pH of less than 5.0. Recall that pH is a measure of the acidity of a substance on a scale of 0 to 14, with 7 being neutral.

Natural precipitation has a pH of about 5.0 to 5.6, which is slightly acidic. **Acid precipitation** forms when sulfur dioxide and nitrogen oxides combine with atmospheric moisture to create sulfuric acid and nitric acid. Although volcanoes and marshes add sulfur gases to the atmosphere, 90 percent of the sulfur emissions in eastern North America are of human origin.

One cause of acid precipitation is coal-burning power plants. Coal contains significant amounts of the mineral pyrite (FeS_2) and other sulfur-bearing compounds. The sulfur dioxide generated by midwestern power plants rises high into the air and is carried by winds toward the East Coast, where they mix with precipitation and fall to the ground. The distribution of acid precipitation in shown in **Figure 15.**

 Get It?

Explain how acid precipitation forms.

To reduce emissions of particulate matter and sulfur dioxide, many coal-burning power plants have installed devices such as the ones shown in **Figure 16.** In North America and western Europe, the use of low-sulfur coal and natural gas has helped to reduce such emissions. In 1963, the United States Congress passed clean air laws that set specific reduction goals and enforcement policies for many types of air pollution. Because of this and other regulations, there have been significant reductions in air pollutants in the United States since 1970.

Figure 16 Scrubbers are often required to clean out the smokestacks on coal plants. Scrubbers help remove gases and particulate matter before they enter the air.

Mark Winfrey/Shutterstock

Climate change

Recall that the greenhouse effect is a natural phenomenon in which atmospheric gases trap thermal energy in the troposphere to warm Earth. As the concentrations of greenhouse gases increase, particularly carbon dioxide, Earth's average surface temperature increases with them.

Human activities, especially those involving the burning of fossil fuels, have increased levels of carbon dioxide. Fossil fuels contain carbon, and when they are burned, the carbon combines with oxygen to form carbon dioxide. Since the start of the industrial revolution, around 1750, humans have been burning fossil fuels at an ever-increasing rate. **Figure 17** shows how atmospheric carbon dioxide has increased since 1960.

Studies have found that Earth's mean surface temperature has risen about 1.26°C from 1917 to 2016. Current climate models predict that if concentrations of carbon dioxide and other greenhouse gases continue to increase, average global temperatures will rise another 1.1°C to 4.8°C by 2100, likely exceeding the 2.0°C mark.

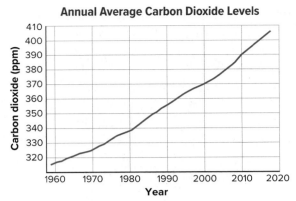

Annual Average Carbon Dioxide Levels

Figure 17 The graph shows the increased levels of carbon dioxide in the atmosphere since 1960, based on data gathered in Mauna Loa, in Hawaii.

Explain *What are some possible contributing factors to this rise in carbon dioxide levels? What impacts might this have on the environment?*

Effects of climate change on weather The added heat in the atmosphere fuels increasingly extreme weather around the globe. Heat waves, one of the most direct effects, are becoming more frequent and more severe. The increasingly warm air evaporates more water more quickly. This further dries already parched lands, providing fuel for wildfires, shown in **Figure 18,** increasing the frequency and duration of droughts, and turning once-fertile land into desert through desertification. In wetter regions, the heat has the opposite effect. Warm air holds more water vapor than cooler air. The extra heat and water in the air creates larger, stronger storms. If global temperatures continue to rise, many regions will experience more dangerous storms and more powerful tropical cyclones, greater amounts of precipitation, and more frequent and destructive floods.

Figure 18 In the western United States, wildfires have increased in frequency since the 1980s.

CCC CROSSCUTTING CONCEPTS
Systems and System Models What types of models do scientists use to learn more about climate change? How do these models help them understand more about the interactions among Earth's systems as related to climate change? Write a short report summarizing your findings.

DaveAlan/E+/Getty Images

Effects of climate change on the oceans

When excess carbon dioxide and other greenhouse gases cause the atmosphere to warm, the oceans warm too. Warm oceans produce their own host of problems. The combined warmth from oceans and the atmosphere have been melting the ice caps at the poles. Scientists project that the arctic will be free of ice by 2050.

Also, when water warms, it expands and takes up more space. With glacial meltwater flooding into the oceans and the ocean water expanding as it heats, sea level will rise. Sea level has already risen over 20 cm since 1880. It currently rises over 3.0 mm per year. Scientists project that the sea will rise another 30 to 120 cm by 2100.

Figure 19 Increased acidity of ocean water can lead to coral bleaching.

The excess carbon dioxide in our atmosphere also dissolves into the oceans, creating carbonic acid. As more carbonic acid is created, the pH of the oceans drops. Since the beginning of the industrial revolution, the acidity of the oceans has increased by 30 percent. This acidity bleaches corals, shown in **Figure 19,** and hurts other wildlife by damaging their carbonate shells or preventing them from making shells in the first place.

Get It?

Identify ways that climate change affects oceans.

Check Your Progress

Summary

- Many human activities create air pollution. Air pollution can cause human health problems.

- CFCs are a major cause of ozone depletion.

- Clean air laws have resulted in a decrease in air pollution emissions since 1970.

- Climate change affects weather and ocean pH.

Demonstrate Understanding

1. **Name** two forms of pollutants found in air. What are some of the sources of these pollutants?

2. **Describe** how CFCs cause ozone depletion.

3. **List** some of the causes of acid precipitation and its effects on ecosystems.

4. **Relate** human activities to climate change and the greenhouse effect.

Explain Your Thinking

5. **Argue** Make an argument that air is a natural resource that needs to be managed responsibly.

6. **MATH ⟩ Connection** If carbon monoxide emissions were reduced from 102 to 87 million metric tons in one year, what would be the percent decrease?

LEARNSMART Go online to follow your personalized learning path to review, practice, and reinforce your understanding.

FOCUS QUESTION

How does growing seaweed help with water conservation?

Use of Water Resources

Humans depend on water in many ways. Most people use freshwater in their homes for bathing, drinking, cooking, and washing. The irrigation of crops also requires water. Because water supplies are not distributed evenly on Earth, some areas have less water than is needed.

Water conservation

Is there a leaky faucet in your home? In the United States alone, 20 to 35 percent of the water taken from public water supplies is lost through leaky toilets, bathtubs, and faucets.

When there is not enough water to go around, people have two choices: decrease demand or develop new supplies. When new supplies are not readily available or are too expensive to develop, water conservation can help. Because large amounts of water are used for crops, efficient irrigation practices can greatly reduce water usage. Monitoring soil moisture to irrigate only when the soil is dry, using equipment that places water near plant roots to reduce evaporation, as shown in **Figure 20,** and raising water prices have all been effective in minimizing the amount of water used for irrigation. Industries can also conserve water by recycling cooling water and wastewater or by using conservation practices.

Photo by Lynn Betts, USDA Natural Resources Conservation Service

Figure 20 Farmers develop methods of water conservation such as the drip irrigation system shown here. In a drip irrigation system, the water is released slowly so that more is absorbed into the soil and less is lost to runoff and evaporation.

3D THINKING   **DCI** Disciplinary Core Ideas   **CCC** Crosscutting Concepts   **SEP** Science & Engineering Practices

COLLECT EVIDENCE

Use your Science Journal to record the evidence you collect as you complete the readings and activities in this lesson.

INVESTIGATE

GO ONLINE to find these activities and more resources.

GeoLAB: Pinpoint a Source of Pollution
Obtain, evaluate, and communicate information on the effect of pollution on a city water supply.

Investigation Lab: Algal Blooms
Develop and use a model to visualize the effect of adding fertilizers to freshwater.

Point source

Nonpoint source

Figure 21 Point-source pollution comes from a single source, while nonpoint-source pollution is generated from a widespread area.

Water Pollution

Pollution is another area in which humans have an impact on water supplies. Water-pollution sources are grouped into two main types. **Figure 21** shows that **point sources** originate from a single point of origin, such as a sewage-treatment plant or an industrial site, while **nonpoint sources** generate pollution from widespread areas.

Most water used for domestic purposes, including showering, laundry, cooking, and using the bathroom, is treated at a sewage treatment facility. Treated sewage is then released through a point source to a receiving stream. Point sources also include wastes that enter streams from illegal dumping, accidental spills, and industries that use water in manufacturing processes and then discharge waste into streams and rivers.

Precipitation can absorb air pollutants and deposit them far from their source. Runoff can wash pesticides and fertilizers into streams as it flows over farms or lawns. It can also carry oil, gasoline, and other chemicals from roads and parking lots. Each of these is an example of nonpoint-source pollution.

Pollution of groundwater

Leaking chemical-storage barrels, underground gasoline-storage tanks, landfills, road salts, nitrates from fertilizers, sewage from septic systems, and other pollutants can seep into the ground and pollute underground water supplies. Polluted groundwater can reach the drinking-water supplies of people who rely on wells. Once groundwater is contaminated, the pollutants can be difficult to remove.

Pollution in the oceans Although human activities have the greatest impact on freshwater supplies, pollution of ocean waters is also a concern. Pollutants from land often end up in estuaries and other nearshore regions. Pollution of nearshore zones can affect organisms that depend on estuaries for breeding and raising young.

Another common ocean pollutant is mercury. Mercury released into the air and water from burning coal and manufacturing is ingested by fish. The fish are then eaten by larger predators, and the mercury is passed along the food chain. Mercury has been detected in bears that do not live near polluted waters but have eaten salmon that migrated from the oceans.

Reducing Water Pollution

In the past few decades, steps have been taken to prevent and reduce water pollution. Two major laws have been passed in the United States to combat water pollution: the Safe Drinking Water Act and the Clean Water Act.

The Safe Drinking Water Act

In 1974, the Safe Drinking Water Act was passed. This act was designed to ensure that everyone in the United States has access to safe drinking water. Progress is being made, but many water supplies still do not consistently meet the standards. In 2009, 20 percent of public water supplies were in violation of the act at least once in a one-year period. The goal of the Safe Drinking Water Act is to reduce this number to less than 5 percent.

The Clean Water Act

The primary federal law that protects U.S. waters is the Clean Water Act of 1972. The two main goals of the Clean Water Act are to eliminate discharge of pollutants into rivers, streams, lakes, and wetlands and to restore water quality to levels that allow for recreational uses of waters, including fishing and swimming.

Is the Clean Water Act working? Since 1972, the number of people served by sewage-treatment plants has increased from 85 million to 190 million. During that same time period, the annual rate of wetland losses has decreased from 146,000 ha/y to about 32,000 ha/y. Two-thirds of the nation's waters are now safe for swimming and fishing, compared to only one-third in 1972. Continued monitoring and improvement are still needed. In 2000, the EPA reported that 39 percent of tested rivers were polluted and 45 percent of tested lakes were polluted.

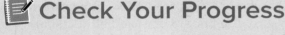 Check Your Progress

Summary

- Humans use water to irrigate crops, for industry, cooking, bathing, and drinking.
- Conserving water can stretch limited supplies.
- Water can be polluted from point sources and nonpoint sources. Groundwater and oceans can also become polluted.
- The United States has passed laws to limit water pollution.

Demonstrate Understanding

1. **Identify** ways that surface waters can be polluted.
2. **Determine** how residents of a city might reduce water consumption.
3. **Analyze** What are some of the positive impacts of the Clean Water Act?
4. **Predict** What are some ways to minimize the need for irrigation?

Explain Your Thinking

5. **Infer** which type of pollution is easier to eliminate: point sources or nonpoint sources. Give an example of each type, and explain how it might be controlled.
6. **WRITING Connection** The sustainability of human societies, in part, depends on access to clean water. Write your own Clean Water Act that would require responsible management of water as a resource. What regulations would you place on businesses or homes? How quickly would you expect change?

LEARNSMART Go online to follow your personalized learning path to review, practice, and reinforce your understanding.

HUMAN IMPACT ON ENERGY RESOURCES

FOCUS QUESTION

What can I do at home to be more energy efficient?

Global Use of Energy Resources

As you have learned, fossil fuels are nonrenewable, pollute our air and water, and are the largest source of the carbon dioxide that is warming the Earth. Yet people on Earth consume these resources at increasing rates. **Figure 22** shows national consumption of natural resources, both renewable and nonrenewable. However, consumption is not equal in all parts of the world. Costa Rica, for example, obtained 98 percent of their electrical energy from renewable resources in 2016, while renewable resources account for only about 9 percent of the energy used in industrialized countries.

Using renewable energy resources that are locally available reduces pollution from fossil fuels by being cleaner and by conserving the fuel that would be used to transport and process resources at a different location. Using a variety of energy resources rather than a single, nonrenewable energy resource, such as fossil fuels, can also help conserve resources. For example, a community that has hydroelectric energy resources might also use solar or wind energy to generate electricity during months when water levels are low.

Figure 22 Petroleum is the most widely used energy resource nationwide, followed closely by coal and natural gas.

Explain *Why do you think nonrenewable resources account for such a high percentage of the nation's energy consumption?*

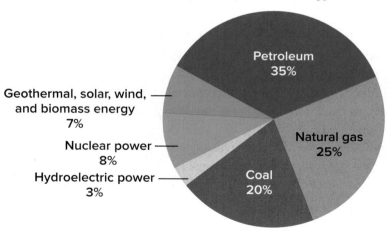

U.S. Consumption of Energy Resources

Petroleum 35%

Geothermal, solar, wind, and biomass energy 7%

Nuclear power 8%

Hydroelectric power 3%

Coal 20%

Natural gas 25%

 **3D THINKING**   **DCI** Disciplinary Core Ideas   **CCC** Crosscutting Concepts   **SEP** Science & Engineering Practices

COLLECT EVIDENCE

 Use your Science Journal to record the evidence you collect as you complete the readings and activities in this lesson.

INVESTIGATE

 GO ONLINE to find these activities and more resources.

🥽 **GeoLAB: Design an Energy-Efficient Building**
Develop a model to determine building designs that are energy efficient.

📡 **Review the News**
Obtain information from a current news story about energy efficiency research. Evaluate your source and communicate your findings to your class.

Energy Efficiency

Energy is the ability to do work. The amount of work produced compared to the amount of energy used is called **energy efficiency.** Energy resources do not produce 100 percent of the potential work that is stored in the energy source.

When a car uses gasoline, some of the energy stored in the gasoline is converted to mechanical energy that moves the car, while some of the energy is used to power accessories, like the car's air conditioner. Most of the energy in the gasoline is lost as heat. Decreasing heat loss is one way that more of the stored energy can be converted to do work. To find ways to use resources more efficiently, scientists study exactly how energy resources are used and where improvements are needed. Using resources more efficiently is a type of conservation. For example, adding insulation to a house reduces heat loss, so less energy is needed to heat the air inside.

Get It?

Explain energy efficiency.

Improving efficiency in industry

Most of the electricity in the United States is generated by burning fossil fuels to heat water, forming steam. Recall that increasing the temperature of a gas also increases pressure. Steam pressure spins the turbines that drive the generators, creating electricity. Unfortunately, this is an inefficient process, as only approximately one-third of the energy potential within the original fuel source can be converted into steam pressure. Practices for increasing energy efficiency in industry include conducting an energy assessment to learn where energy efficiency opportunities occur within an operating plant. Changing how energy is managed, rather than replacing equipment, can also help improve energy efficiency at a plant.

Improving efficiency in transportation

Transportation is necessary to move people, food, and other goods from one place to another. Although most transportation currently relies on oil, conservation practices can help reduce dependency on oil resources used for transportation. **Table 1** lists some of the advantages of public transportation, which is one way people can improve energy efficiency in transportation. New technologies have also led to increased energy efficiency in public transportation. For example, some cities have started using biodiesel, a fuel obtained from plant oils or animal fats, to power their city buses. About 8% of cities use biodiesel to run their buses. About 17% of cities have switched to hybrid or electric buses. Hybrid buses get about 50% more miles to the gallon than buses fueled by regular diesel.

Table 1 Advantages of Public Transportation in the U.S.

Using public transportation to get to work can save a person between $300 and $3000 in fuel costs per year.
Using public transportation saves more than 3 billion liters of gasoline every year—equal to all energy used by U.S. manufacturers of computers and electronic equipment.
If Americans used public transportation for roughly 10 percent of daily travel needs, the United States would reduce its dependence on foreign oil by more than 40 percent.
One person switching to public transit can reduce daily carbon emissions by 9.1 kg, or more than 3300 kg in a year.

Commuting efficiently

People who live in metropolitan areas can improve energy efficiency by using public transportation to get from place to place. Major U.S. cities, such as New York City, use subways or elevated trains to move people. In Europe, mass transportation includes long-distance rail systems as well as electric trams and trolleys. When it is necessary to drive private automobiles, carpooling can reduce the number of vehicles on the highways. Some metropolitan areas encourage carpooling by providing express lanes for cars with multiple passengers.

Automobiles

The use of fuel-efficient vehicles is another way to reduce the amount of petroleum resources consumed. Automobile manufacturers can build vehicles that achieve high rates of fuel efficiency without sacrificing performance. The future of this industry is promising, as hybrid, fuel cell, and electric technologies are reaching the consumer market. Also, less energy is needed to move something that weighs less, thus smaller cars use less gasoline. Another way to conserve gasoline is to drive slower than 100 km/h (62 mph) on the freeway and use alternate forms of transportation.

Getting more for less Increased demand for fuels requires a greater supply and results in higher costs. Electricity is costly to produce, and it is not usually used efficiently in homes or industry. In the United States, approximately 40 percent of the energy used to fuel motor vehicles and to heat homes and businesses is lost as thermal energy. If energy were used more efficiently, less energy would be needed, thus decreasing the total cost of energy.

Harnessing waste thermal energy

Generating electricity produces energy that can be recovered. This recoverable excess energy is known as waste thermal energy, and it can be harnessed for use. The simultaneous production of two usable forms of energy is called **cogeneration.** Cogeneration technology captures the waste thermal energy (steam) for domestic or industrial heating or for use in a large air conditioning unit. In an air conditioning unit, the waste thermal energy turns a turbine connected to a compressor that chills water. This chilled water is then sent to an air handler unit in a different building. Excess thermal energy can also be used to generate electricity that operates electrical devices within the power plant, such as sulfur-removing scrubbers on smokestacks. While industries use one-third of all energy produced in the United States, cogeneration has allowed some industries to increase production while reducing energy use. California has set a target of 4000 megawatts of energy produced from cogeneration by 2020. The power station shown in **Figure 23,** on the next page, utilizes cogeneration for an oil refinery and a chemical plant.

ACADEMIC VOCABULARY

efficient
productive without waste
The automobile was more efficient when the proper tune-ups had been done.

STEM CAREER Connection

Home Energy Specialist
Would you like to evaluate how people are using energy in their homes and make recommendations about how they could use energy more efficiently? Home energy specialists help people learn how to make their homes less expensive to cool and heat and how to increase energy efficiency with simple changes.

Sustainable Energy

Energy resources on Earth are interrelated, and they affect one another. **Sustainable energy** involves the global management of Earth's natural resources to meet current and future energy needs. A good management plan incorporates both conservation and energy efficiency. New technology that extends the supply of fossil fuels is a vital part of such a plan. Global cooperation can help maintain the necessary balance between protection of the environment and economic growth. The achievement of these goals will depend on the commitment made by all so that future generations have access to the energy resources required to maintain a high quality of life on Earth.

Figure 23 This cogeneration power station helps reduce energy use at an oil refinery and chemical plant in Hampshire, UK.

 Get It?

Identify the components of a good energy management plan.

Check Your Progress

Summary

- Energy resources will last longer if conservation and energy-efficiency measures are developed and used.
- Energy efficiency results in the use of fewer resources to provide more usable energy.
- Cogeneration, in which two usable forms of energy are produced at the same time from the same process, can help save resources.
- Sustainable energy can help meet current and future energy needs.

Demonstrate Understanding

1. **Summarize** why the conservation and efficient use of energy resources is important.
2. **List** three ways in which you could conserve electric energy in your home.
3. **Identify** new technology that can be used to help increase energy efficiency.
4. **Analyze** Why is it important to responsibly manage energy resources?

Explain Your Thinking

5. **Summarize** how cogeneration can be used in your home to save energy resources.
6. **MATH** **Connection** If the national consumption of coal were reduced by 25 percent, what would the percentage consumption of coal be? Refer to **Figure 22** for more information.

LEARNSMART Go online to follow your personalized learning path to review, practice, and reinforce your understanding.

A Breath of Fresh Air

In their daily weather reports, meteorologists discuss the Air Quality Index (AQI), a scale that shows an area's air quality. Urban smog impacts the AQI and affects the health of plants, animals, and humans. In response, scientists and engineers have designed various solutions to help reduce the unhealthful effects of smog.

Urban areas experience the highest levels of smog due to vehicle exhaust and industrial pollution.

Smog and the Air Quality Index

The Environmental Protection Agency (EPA) designed the AQI to inform the public about air quality and the unhealthful effects of poor air quality. AQI scores range from 0 to 500, with 500 indicating an extremely hazardous level of pollution. A score of 100 or less is considered satisfactory for the general public.

The EPA considers five air pollutants—including ground-level ozone and particulate matter, the two primary components of smog—when it calculates an area's AQI. These two air pollutants present the most risk to people with heart and lung diseases, asthma, chronic bronchitis, and emphysema.

Smog is a problem in many large cities, such as London, Los Angeles, Beijing, and Hong Kong. New Delhi, India, had an AQI of 703 on November 7, 2017, partly due to the burning of agricultural waste. The smog generated in cities does not stay there; studies have found that smog can move across oceans and affect air quality in other countries.

Battling smog

Scientists and engineers—as well as students in those fields of study—are finding ways to reduce or prevent the effects of smog. In 2014, California engineering students designed a roof tile coating of titanium dioxide that removes 88 to 97 percent of the nitrogen oxides in the surrounding atmosphere. This is an important advance in smog prevention because smog is formed when nitrogen oxides react with other substances and with sunlight.

Other recent smog-fighting inventions include a machine that filters smog and releases clean air and a high-rise building with a façade that includes many species of air-filtering trees and other plants.

PLAN AND CONDUCT AN INVESTIGATION

Work with a partner to collect information about your area's AQI scores over a week's time. Make a poster to present the results of your investigation. Participate in a class discussion about the causes of air pollution in your area.

Yenwen Lu/Getty Images

MODULE 20
STUDY GUIDE

 GO ONLINE to study with your Science Notebook.

Lesson 1 POPULATIONS AND THE USE OF NATURAL RESOURCES

- All organisms use resources to maintain their existence. The use of these resources impacts the environment.
- As populations increase, the demand for resources increases. With limited resources, populations will stop growing when they reach carrying capacity.
- Populations grow exponentially at early stages.

- exponential growth
- carrying capacity
- density-independent factor
- density-dependent factor

Lesson 2 HUMAN IMPACT ON LAND RESOURCES

- Humans require large amounts of land resources.
- The extraction of resources can disrupt Earth's surface.
- Growing populations increase the demand for food and result in increased urban development.
- Agriculture, poor forestry practices, and urban development can cause habitat loss, increased erosion, and soil and water pollution.
- Human impact on land resources can be minimized through the use of modern techniques.

- tailings
- reclamation
- deforestation
- pesticide
- bioremediation

Lesson 3 HUMAN IMPACT ON AIR RESOURCES

- Many human activities create air pollution. Air pollution can cause human health problems.
- CFCs are a major cause of ozone depletion.
- Clean air laws have resulted in a decrease in air pollution emissions since 1970.
- Climate change affects weather and ocean pH.

- photochemical smog
- ozone hole
- acid precipitation

Lesson 4 HUMAN IMPACT ON WATER RESOURCES

- Humans use water to irrigate crops, and for industry, cooking, bathing, and drinking.
- Conserving water can stretch limited supplies.
- Water can be polluted from point sources and nonpoint sources. Groundwater and oceans can also become polluted.
- The United States has passed laws to limit water pollution.

- point source
- nonpoint source

Lesson 5 HUMAN IMPACT ON ENERGY RESOURCES

- Energy resources will last longer if conservation and energy-efficiency measures are developed and used.
- Energy efficiency results in the use of fewer resources to provide more usable energy.
- Cogeneration, in which two usable forms of energy are produced at the same time from the same process, can help save resources.
- Sustainable energy can help meet current and future energy needs.

- energy efficiency
- cogeneration
- sustainable energy

REVISIT THE PHENOMENON

Why would we grow and use seaweed instead of standard crops like rice and corn?

CER Claim, Evidence, Reasoning

Explain Your Reasoning Revisit the claim you made when you encountered the phenomenon. Summarize the evidence you gathered from your investigations and research and finalize your Summary Table. Does your evidence support your claim? If not, revise your claim. Explain why your evidence supports your claim.

STEM UNIT PROJECT
Now that you've completed the module, revisit your STEM unit project. You will apply your evidence from this module and complete your project.

GO FURTHER

SEP Data Analysis Lab
Are you breathing cleaner air?

The table lists changes in emissions in the United States since the Clean Air Act of 1972. Graph the data from the table. Put years on the *x*-axis and the pollutant emissions per year on the *y*-axis. Use different colors for each pollutant.

CER Analyze and Interpret Data

1. **Reasoning** Infer why emissions of lead have declined so drastically since 1970.
2. **Claim, Evidence** Could you estimate the reductions for 2014 by looking at the graph? Explain.

Data and Observations					
Pollutant (millions of tons)	1970	1980	1990	2000	2004
Particulate matter < 10 microns	12.2	6.2	3.2	2.3	2.5
Sulfur dioxide	31.2	25.9	23.1	16.3	15.2
Nitrogen dioxides	26.9	27.1	25.1	22.3	18.8
Volatile organic compounds	33.7	30.1	23.1	16.9	15.0
Carbon monoxide	197.3	177.8	143.6	102.4	87.2
Lead	0.221	0.074	0.005	0.003	0.003

*Data obtained from: Air Emission Trends—Continued Progress Through 2004. *U.S. Environmental Protection Agency*.

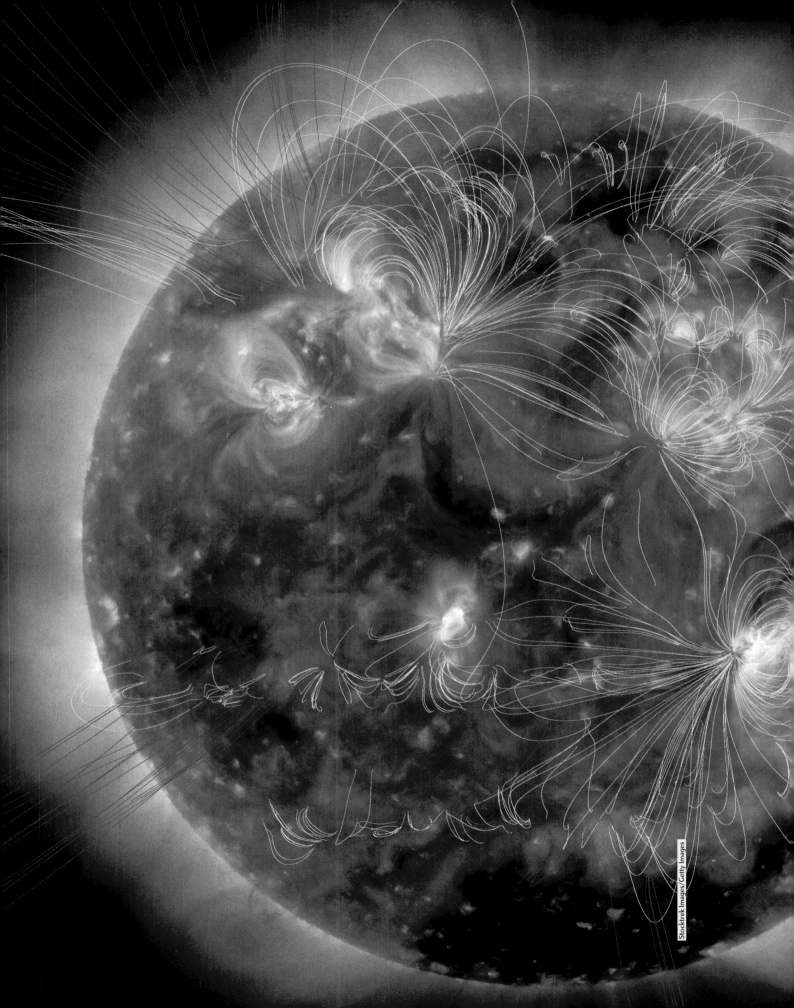

ENCOUNTER THE PHENOMENON

How is space different from the planet we call home?

SEP Ask Questions

What questions do you have about the phenomenon? Write your questions on sticky notes and add them to the driving question board for this unit.

What's beyond Earth?

Look for Evidence

As you go through this unit, use the information and your experiences to help you answer the phenomenon question as well as your own questions. For each activity, record your observations in a Summary Table, add an explanation, and identify how it connects to the unit and module phenomenon questions.

Solve a Problem
STEM UNIT PROJECT

Beyond Earth Investigate and research more about the universe beyond Earth. Use the results of these investigations and the evidence you collected during the unit to complete your unit project.

🔖 **GO ONLINE** In addition to reading the information in your Student Edition, you can find the STEM Unit Project and other useful resources online.

Jeffrey Schwartz Photography/Shutterstock

ENCOUNTER THE PHENOMENON

Why doesn't the Moon eclipse the Sun every month?

GO ONLINE to play a video about the 2017 American solar eclipse.

SEP Ask Questions

Do you have other questions about the phenomenon? If so, add them to the driving question board.

CER Claim, Evidence, Reasoning

Make Your Claim Use your CER chart to make a claim about why the Moon does not eclipse the Sun every month. Explain your reasoning.

Collect Evidence Use the lessons in this module to collect evidence to support your claim. Record your evidence as you move through the module.

Explain Your Reasoning You will revisit your claim and explain your reasoning at the end of the module.

GO ONLINE to access your CER chart and explore resources that can help you collect evidence.

LESSON 1: Explore and Explain: Space-Based Astronomy

LESSON 3: Explore and Explain: Phases of the Moon

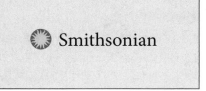

Additional Resources

(t)Video Supplied by BBC Worldwide Learning, (bl)NASA/JSC, (br)Igor Korionov/Alamy Stock Photo

FOCUS QUESTION

How do we track objects in space?

Radiation

The radiation that scientists study from distant bodies throughout the universe is called electromagnetic radiation. Electromagnetic radiation consists of electric and magnetic disturbances traveling through space as waves. Electromagnetic radiation includes visible light, infrared and ultraviolet radiation, radio waves, microwaves, X-rays, and gamma rays.

You might be familiar with some forms of electromagnetic radiation. For example, overexposure to ultraviolet waves can cause sunburn, microwaves heat your food, and X-rays help doctors make diagnoses. All types of electromagnetic radiation, arranged according to wavelength and frequency, form the **electromagnetic spectrum,** shown in **Figure 1.**

Wavelength and frequency

Electromagnetic radiation is classified by wavelength—the distance between peaks on a wave. Red light has a longer wavelength than blue light, and radio waves have a much longer wavelength than gamma rays. Electromagnetic radiation is also classified according to frequency, the number of waves or oscillations that pass a given point per second.

Figure 1 The electromagnetic spectrum identifies the different radiation frequencies and wavelengths.

Explain *What is the relationship between the wave properties of frequency and wavelength?*

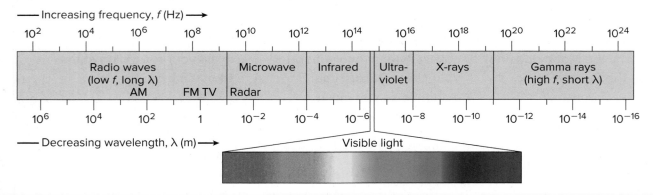

 3D THINKING   `DCI` Disciplinary Core Ideas   `CCC` Crosscutting Concepts   `SEP` Science & Engineering Practices

COLLECT EVIDENCE

Use your Science Journal to record the evidence you collect as you complete the readings and activities in this lesson.

INVESTIGATE

 GO ONLINE to find these activities and more resources.

Investigation Lab: Make Your Own Telescope
Develop and use a model telescope to visualize how its structure influences the electromagnetic radiation it collects.

((ᄋ)) **Review the News**
Obtain information from a current news story about telescopes. Evaluate your source and communicate your findings to your class.

The visible light portion of the spectrum has frequencies ranging from red to violet, or 4.3×10^{14} to 7.5×10^{14} Hertz (Hz)—a unit equal to one cycle per second. Frequency is related to wavelength by the mathematical relationship $c = \lambda f$, where c is the speed of light (3.0×10^8 m/s), λ is the wavelength, and f is the frequency. Note that all types of electromagnetic radiation travel at the speed of light in a vacuum.

Astronomers choose their tools based on the type of radiation they wish to study. For example, to see stars forming in interstellar clouds, they use special telescopes that are sensitive to infrared wavelengths, and to view remnants of supernovas, they often use telescopes that are sensitive to UV, X-ray, and radio wavelengths.

Telescopes

Objects in space emit radiation in all portions of the electromagnetic spectrum. Telescopes, such as the one shown in **Figure 2,** allow us to observe wavelengths beyond what the human eye can detect. In addition, a telescope collects more electromagnetic radiation from distant objects and focuses it so that an image of the object can be recorded. The pupil of a typical human eye has a diameter of up to 7 mm when it is adapted to darkness; the diameter of a telescope's opening, or aperture, might be as large as 10 m. Larger apertures can collect more electromagnetic radiation, making dim objects in the sky appear much brighter.

Telescopes also surpass the human eye in collecting electromagnetic radiation by using cameras and other imaging devices to create time exposures. The human eye responds to visible light within one-tenth of a second, so objects too dim to be perceived in that time cannot be seen. Telescopes can collect light over periods of minutes or hours. In this way telescopes can detect objects that are too faint for the human eye to see. Also, astronomers can add specialized equipment. A photometer, for example, measures the intensity of light, and a spectrophotometer displays the intensities of different wavelengths of radiation.

Reflecting and refracting telescopes

Two different types of telescopes are used to focus visible light. The first telescopes, invented around 1608, used lenses to bring visible light to a focus and are called **refracting telescopes,** or refractors. The largest lens on such telescopes is called the objective lens.

Figure 2 The top photo of the Iris nebula was taken by the Mayall 4-m telescope, shown here with its observatory.

STEM CAREER Connection

Telescope Operator

If you enjoy gazing at the stars and don't mind working nights, you might consider a career as a telescope operator. A telescope operator maintains the telescopes and related equipment at an observatory. He or she has at least a high-school diploma and often undergoes training to earn a certificate in the field.

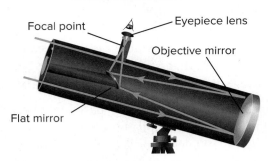

Refracting telescope

Eyepiece lens

Focal point

Convex lens

Reflecting telescope

Focal point

Eyepiece lens

Objective mirror

Flat mirror

Figure 3 Refracting telescopes use a lens to collect light. Reflecting telescopes use a mirror to collect light.

In 1668, a new telescope that used mirrors to focus light was built. Telescopes that bring visible light to a focus with mirrors are called **reflecting telescopes,** or reflectors. **Figure 3** illustrates how simple refracting and reflecting telescopes work. Although refracting telescopes are still in use today, most astronomers use reflectors because mirrors can be made larger than lenses and can, therefore, collect more light. Most telescopes used for scientific study are located in observatories far from city lights, usually at high elevations where there is less atmosphere overhead to blur images. Some of the best observatory sites in the world are located high atop mountains in the southwestern United States, along the peaks of the Andes mountain range in Chile, and on the summit of Mauna Kea, a volcano in Hawaii.

Telescopes using nonvisible wavelengths

For all telescopes, the goal is to bring as much electromagnetic radiation as possible into focus. Infrared and ultraviolet radiation can be focused by mirrors in a way similar to that used for visible light. X-rays cannot be focused by normal mirrors, and thus special designs must be used. Gamma rays cannot be focused, so telescopes designed to detect this type of radiation can determine only the direction from which the rays come.

A radio telescope collects the longer wavelengths of radio waves with a large dish antenna, which resembles a satellite TV dish. The dish plays the same role as the primary mirror in a reflecting telescope by reflecting radio waves to a point above the dish. There, a receiver converts the radio waves into electric signals that can be stored in a computer for analysis.

The data are converted into visual images by a computer. The resolution of the images produced can be improved using a process called **interferometry,** which is a technique that uses the images from several telescopes to produce a single image. By combining the images from several telescopes, astronomers can create a highly detailed image that has the same resolution of one large telescope with a dish diameter as large as the distance between the two telescopes. One example of this is the moveable telescopes shown in **Figure 4.** Both radio and optical telescopes can be linked this way. Technology used in astronomy has changed over time, as shown in **Figure 5.**

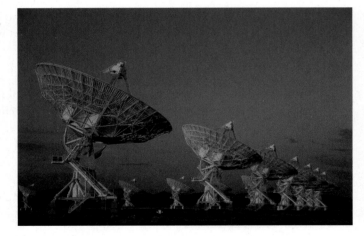

Figure 4 The Very Large Array is situated near Socorro, New Mexico. The dish antennae of these radio telescopes are mounted on tracks so they can be moved to improve resolution.

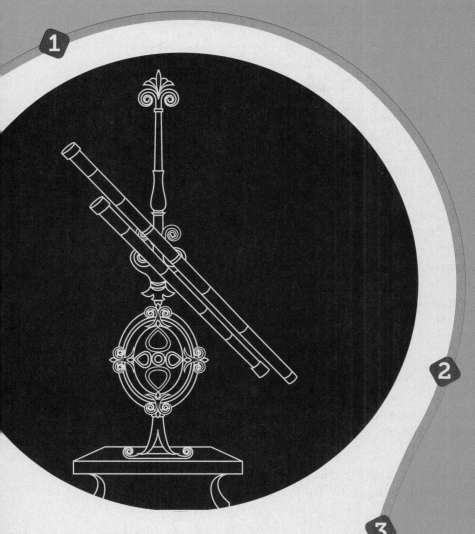

Figure 5

Development of Astronomy

Humanity's curiosity about the night sky was limited to Earth-bound explorations until the first probe was sent into space in 1957.

1 **4236 B.C.** After lunar and solar calendars predict agricultural seasons, Egyptians adopt a 365-day calendar based on the movement of the star Sirius.

2 **A.D. 90s** Arab astronomers improve the accuracy of the Greek astrolabe —a tool for celestial navigation that determines time and location.

3 **1054** Chinese astronomers document the supernova that creates the Crab nebula, believing it foretells the arrival of a wealthy visitor to the emperor.

4 **1608** The telescope is invented, allowing astronomers to discover planets such as Uranus and Neptune, moons, and stars that are invisible to the naked eye.

5 **1860s** The invention of spectroscopy suggests that the celestial bodies are composed of some of the same elements that make up Earth's atmosphere.

6 **1957** Russia launches the first two satellites into orbit around Earth, marking the beginning of space exploration.

7 **1969** U.S. astronauts become the first humans to walk on the Moon.

8 **2004** A Mars rover discovers rock formations and sulfate salts, which indicate that the planet once had flowing water.

9 **2010** Japanese probe *Hayabusa* returns the first-ever sample from the surface of an asteroid back to Earth.

Table 1 Orbiting Telescopes

Name	Launch	Wavelengths	Studies	Host
MOST	2003	visible	stars	Canada
Swift	2004	X-ray, UV, visible	black holes	NASA
Agile	2007	gamma ray	wide-ranging	ESA
Kepler	2009	visible	extrasolar planets	NASA
WISE	2009	IR	survey	NASA
Spektr-R	2011	radio	radio sources beyond Milky Way	Russia
Astrosat	2015	UV, IR, X-ray, visible	wide-ranging	India
HXMT	2017	X-ray	neutron stars, black holes	China

Space-Based Astronomy

Astronomers often send instruments into space to collect information because Earth's atmosphere interferes with most radiation. It blurs visual images and absorbs infrared and ultraviolet radiation, X-rays, and gamma rays. Space-based telescopes allow astronomers to study radiation that would be blurred by our atmosphere.

Hubble Space Telescope

One of the best-known space-based observatories, the *Hubble Space Telescope (HST)*, was launched in 1990. The *Hubble Space Telescope* was designed to obtain sharp visible-light images without atmospheric interference and also to make observations in infrared and ultraviolet wavelengths. *Hubble,* seen in **Figure 6,** has observed galaxies well over 12 billion light-years away.

The successor to the *HST* is the *James Webb Space Telescope (JWST),* scheduled to launch in 2021. JWST is an infrared telescope, with some capability in the visible-light range. Other space-based telescopes are listed above in **Table 1.**

Spacecraft

Some spacecraft can be sent directly to the bodies being observed. Robotic probes are spacecraft that can make close-up observations and sometimes land to collect information directly. Probes are practical only for objects within our solar system. *New Horizons* reached Pluto in 2015 and has now moved deeper into the solar system. It has visible, infrared, and ultraviolet cameras as well as equipment to measure magnetic fields. *OSIRIS-REx,* launched by the *Atlas V* rocket, will briefly touch the surface of an asteroid to gather rocks and other materials.

Figure 6 In 1990, the *Hubble Space Telescope* was launched into orbit around Earth.

NASA/JSC

Human spaceflight

Before humans can safely explore space, scientists must learn about the effects of space, such as weightlessness and radiation. A multicountry space station called the *International Space Station (ISS)*, shown in **Figure 7,** is the ideal environment for studying the effects of space on humans. In 2010, NASA and its international partners celebrated ten years of permanent human habitation on the *ISS.* The crew members conduct many different experiments in this weightless environment.

Spinoff technology

Space-exploration programs benefit not only astronomers, but society as well. Many technologies that were originally developed for use in space programs are now used by people throughout the world. Did you know that the technology for the space shuttle's fuel pumps led to the development of pumps used in artificial hearts? Or that NASA's quest to improve crash protection led to the memory foam found in mattresses? In fact, more than 1500 different NASA technologies have been passed on to commercial industries for common use; these technologies are called spinoffs.

Figure 7 This view of the *International Space Station* was taken from the space shuttle *Discovery.*

Review *What types of studies can be carried out in the space station?*

Check Your Progress

Summary

- Telescopes collect and focus electromagnetic radiation emitted or reflected from distant objects.

- Electromagnetic radiation is classified by wavelength and frequency.

- The two main types of optical telescopes are refractors and reflectors.

- Space-based astronomy includes the study of orbiting telescopes, satellites, and probes.

- Technology originally developed to explore space is now used by people on Earth.

Demonstrate Understanding

1. **Explain** how electromagnetic radiation helps scientists study the universe.

2. **Distinguish** between refracting and reflecting telescopes and how they work.

3. **Name** two benefits of using a telescope.

4. **Report** on how interferometry affects the images that are produced by telescopes.

5. **Examine** the reasons why astronomers send telescopes and probes into space.

Explain Your Thinking

6. **Assess** the benefits of technology spinoffs to society.

7. **Consider** the advantages and disadvantages of using robotic probes to study distant objects in space.

8. **MATH Connection** Calculate the wavelength of radiation with a frequency of 10^{12} Hz. (*Hint: Use the equation $c = \lambda f$.*)

LEARNSMART Go online to follow your personalized learning path to review, practice, and reinforce your understanding.

FOCUS QUESTION

What is the "Man in the Moon?"

Exploring the Moon

Astronomers have learned much about the Moon from observations with telescopes. However, most knowledge of the Moon comes from explorations by space probes, such as *Kaguya* and the *Lunar Reconnaissance Orbiter (LRO)*, and from landings by astronauts. The first step toward reaching the Moon was in 1957, when the Soviet Union launched the first artificial satellite, *Sputnik I.* Four years later, Soviet cosmonaut Yuri A. Gagarin became the first human in space.

That same year, the United States launched the first American, Alan B. Shepard, Jr., into space during Project Mercury. This was followed by Project Gemini, which launched two-person crews. Finally, on July 20, 1969, the Apollo program landed Neil Armstrong and Edwin "Buzz" Aldrin on the Moon during the *Apollo 11* mission. Astronauts of the Apollo program explored several areas of the Moon, often using special vehicles, such as the *Lunar Roving Vehicle* shown in **Figure 8.**

After a gap of many years, scientists hope to return to the Moon someday. Astronauts hope to remain longer on the Moon and eventually establish a permanent base there. NASA is also assisting private companies in their efforts to build piloted spacecraft.

Figure 8 *Apollo 15* astronauts used the *Lunar Roving Vehicle (LRV)* to explore the Moon's surface.

Explain *how the LRV might have resulted in improved mission performance.*

(NASA-HQ-GRIN)

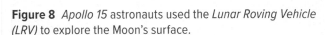 **3D THINKING**　　**DCI** Disciplinary Core Ideas　　**CCC** Crosscutting Concepts　　**SEP** Science & Engineering Practices

COLLECT EVIDENCE

 Use your Science Journal to record the evidence you collect as you complete the readings and activities in this lesson.

INVESTIGATE

 **GO ONLINE** to find these activities and more resources.

Design Your Own: Observing the Moon
Use mathematical and computational thinking to describe the patterns of the Moon's motion.

GeoLab: Determining Relative Ages of Lunar Features
Interpret lunar data to determine the relative age and history of structures on the Moon.

The Lunar Surface

Although the Moon is the brightest object in our night sky, the lunar surface is dark. The albedo of the Moon, the fraction of incoming sunlight that its surface reflects, is only 0.12, reflecting only 12 percent of sunlight. For comparison, worn asphalt has an albedo around 0.10 and Earth has an average albedo of nearly 0.30. Sunlight that is absorbed by the surface of the Moon produces extreme differences in temperature. Because the Moon has no atmosphere to absorb heat, sunlight can heat the Moon's surface to 400 K (127°C), while the temperature of its unlit surface can drop to a chilly 40 K (−233°C).

The "man in the Moon" pattern seen from Earth is produced by the Moon's surface features. Lunar **highlands** are heavily cratered regions of the Moon that are light in color and mountainous. Other regions, called **maria** (MAH ree uh) (singular, *mare* [MAH ray]), are dark, relatively smooth plains, which average 3 km lower in elevation. Although maria are mostly smooth, they do have a few scattered craters and rilles. **Rilles** are valleylike structures that might be collapsed lava tubes. In addition, there are mountain ranges near some of the maria.

Get It?

Explain what lunar features produce the "man in the Moon."

Lunar craters

The craters on the Moon, called **impact craters,** formed when objects from space crashed into the lunar surface. The material blasted out during these impacts fell back to the Moon's surface as **ejecta.** Some craters have long trails of ejecta, called **rays,** that radiate outward from the impact site much like the spokes of a bicycle tire, as shown in **Figure 9.** Rays are visible as light-colored streaks.

Figure 9 You can see some of the details of maria and highlands in the view of the full moon. Craters, ejecta, rilles, and rays are visible in close-up views of the Moon's surface.

Daedalus Crater

Ejecta

Highlands and maria on the Moon

Rilles

Rays

Lunar properties

Earth's moon is unique among moons in the solar system. First, it is the largest moon compared to the radius and mass of the planet it orbits, as shown in **Table 2.** Also, it is a solid, rocky body, in contrast with the icy compositions of most other moons of the solar system. Finally, the Moon's orbit is farther from Earth relative to the distance of many moons from the planets they orbit. **Figure 10** shows a photo mosaic of Earth and the Moon taken from space.

Composition The Moon is made up of minerals similar to those of Earth—mostly silicates. Recall that silicates are compounds containing silicon and oxygen, which make up 96 percent of the minerals in Earth's crust. The highlands, which cover most of the lunar surface, are predominately lunar breccias (BRE chee uhs), which are rocks formed by the fusion of smaller pieces of angular rock during impacts. Unlike sedimentary breccias on Earth, most of the lunar breccias are composed of plagioclase feldspar, a silicate containing high quantities of calcium and aluminum but low quantities of iron. The maria are predominately basalt, but unlike basalt on Earth, they contain no water.

Table 2 The Moon and Earth

	The Moon	Earth
Mass (kg)	7.349×10^{22}	5.974×10^{24}
Radius (km)	1737.1	6371.0
Volume (km³)	2.196×10^{10}	1.083×10^{12}
Density (kg/m³)	3350	5515

 Get It?

Describe the compositions of the lunar highlands and maria.

History of the Moon

The entire lunar surface is old—radiometric dating of rocks from the highlands indicates an age between 3.8 and 4.6 billion years—about the same age as Earth. Based on the ages of the highlands and the frequency of the impact craters that cover them, scientists theorize that the Moon was heavily bombarded during its first 800 million years. This caused the breaking and heating of surface rocks and resulted in a layer of loose, ground-up rock called **regolith** on the surface. Regolith averages several meters in thickness, but it varies greatly depending on location.

Figure 10 This photo mosaic shows images of Earth and the Moon at the relative size that each appears when viewed from a similar distance away. The images were taken by the *Mariner 10* spacecraft.

NASA/JPL/Northwestern University

Layered structure

Scientists infer from seismic data that the Moon, like Earth, has a layered structure, which consists of the crust, upper mantle, lower mantle, and core, as illustrated in **Figure 11.** The crust varies in thickness and is thickest on the far side. The far side of the Moon is the side that is always facing away from Earth. The Moon's upper mantle is solid, its lower mantle is thought to be partially molten, and its core is mostly solid iron.

Formation of maria

After the period of intense bombardment that formed the highlands, lava welled up from the Moon's interior and filled in the large impact basins. This lava fill created the dark, smooth plains of the maria. Scientists estimate the maria formed between 3.1 and 3.8 bya, making them younger than the highlands. Flowing lava in the maria scarred the surface with rilles. Rilles are much like lava tubes found on Earth, through which lava flows in underground streams. The maria have remained relatively free of craters because fewer impacts have occurred on the Moon since they formed.

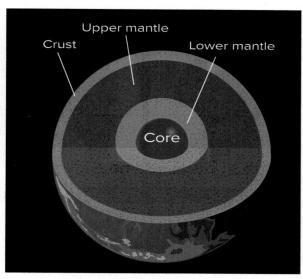

Figure 11 Scientists deduce the structure of the Moon's interior from seismic data obtained from seismometers left on the Moon's surface.

Often lava did not fill the basins completely and left the rims of the basins above the lava. This left behind the mountain ranges that now surround many maria. As shown in **Figure 12,** there are virtually no maria on the far side of the Moon, which is covered almost completely with highlands. Scientists hypothesize that this is because the crust is thicker on the far side, which made it difficult for lava to reach the lunar surface.

Tectonics

Seismometers measure the strength and frequency of moonquakes. Seismic data show that on average, the Moon experiences an annual moonquake that would be strong enough to cause dishes to fall out of a cupboard if it happened on Earth. Despite these moonquakes, scientists think that the Moon is not tectonically active. The Moon has no active volcanoes and no significant magnetic field. Scientists know from the locations and shapes of mountains on the Moon that they were not formed tectonically, as mountain ranges on Earth are formed. Lunar mountains are actually higher elevations that surround ancient impact basins filled with lava.

Figure 12 The heavily cratered far side of the Moon, shown on the left, has many fewer maria than the more familiar near side of the Moon, shown on the right.

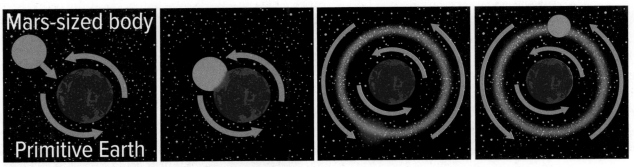

Figure 13 The impact theory of the Moon's formation states that material ejected from Earth and from the striking object eventually merged to form the Moon.

Formation

Several theories have been proposed to explain the Moon's formation. The theory accepted by most astronomers, the giant impact theory, was developed using computer models. According to this theory, illustrated in **Figure 13,** the Moon formed as the result of a collision between Earth and a Mars-sized object about 4.5 billion years ago. The object likely struck young Earth with a glancing blow. Materials from the incoming body and Earth's outer layers were ejected into space, where—trapped by Earth's gravity—they began to orbit Earth. Over time, the materials merged to form the Moon. According to this model, the Moon has iron in its small core, and its crust is mostly silicates that came from Earth's mantle and crust. This is why the chemical compositions of Earth's crust and the Moon's crust are so similar. Evidence for this theory is found in analyses of rock samples from the Moon and Earth. Because lunar rocks have changed so little over billions of years, they provide information about Earth's formation and early history.

Check Your Progress

Summary

- Astronomers have gathered information about the Moon using telescopes, space probes, and astronaut exploration.

- Like Earth's crust, the Moon's crust is composed mostly of silicates.

- Surface features on the Moon include highlands, maria, ejecta, rays, and rilles. The Moon is heavily cratered.

- The Moon probably formed about 4.5 bya in a collision between Earth and a Mars-sized object.

Demonstrate Understanding

1. **Compare** the Moon and the moons of other planets.
2. **Classify** the following according to age: maria, highlands, and rilles.
3. **Explain** how scientists determined that the Moon has no tectonics.
4. **Sequence** the steps in the giant impact theory.

Explain Your Thinking

5. **Infer** how the surface of the Moon would look if the crust on the far side were the same thickness as the crust on the near side.
6. **Outline** the major ideas in this lesson. Include the following terms: *highlands, crust, lava, maria, craters, tectonics,* and *giant impact theory.*
7. **WRITING ▸ Connection** Write the introductory paragraph to an article entitled *History of the Moon.*

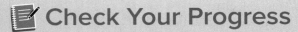 Go online to follow your personalized learning path to review, practice, and reinforce your understanding.

FOCUS QUESTION

What causes a solar eclipse?

Daily Motions

On Earth, the most obvious pattern of motion in the sky is the rising and setting of the Sun, the Moon, stars, and everything visible in the night sky. The Sun appears to rise in the east and set in the west, as do the Moon, planets, and stars. These daily motions result from Earth's rotation. The Sun, the Moon, planets, and stars do not orbit around Earth every day. It only appears that way because we observe the sky from a planet that rotates. But how do we know that Earth rotates?

Earth's rotation

There are two relatively simple ways to demonstrate that Earth is rotating. One is to use a Foucault pendulum, like the one shown in **Figure 14.** A Foucault pendulum swings in a constant direction. But as Earth turns under it, the pendulum seems to shift its orientation. The second way is to observe the way that air on Earth is diverted from a north-south direction to an east-west direction by the Coriolis effect.

Day length The time period from one noon to the next is called a solar day. Our timekeeping system is based on the solar day. But the length of a day as we observe it is roughly four minutes longer than the time it takes Earth to rotate once on its axis. As Earth rotates, it also revolves and has to turn a little farther each day to align again with the Sun.

Figure 14 This Foucault pendulum is surrounded by pegs. As Earth rotates under it, the pendulum knocks over the pegs, showing the progress of the rotation.

 3D THINKING   **DCI** Disciplinary Core Ideas   **CCC** Crosscutting Concepts   **SEP** Science & Engineering Practices

COLLECT EVIDENCE

 Use your Science Journal to record the evidence you collect as you complete the readings and activities in this lesson.

INVESTIGATE

 GO ONLINE to find these activities and more resources.

🥽 **Quick Investigation: Predict the Sun's Summer Solstice Position**
Analyze data to visualize the cyclical changes in the position of the Sun in the sky.

❓ **Revisit the Encounter the Phenomenon Question**
What information from this lesson can help you answer the Unit and Module questions?

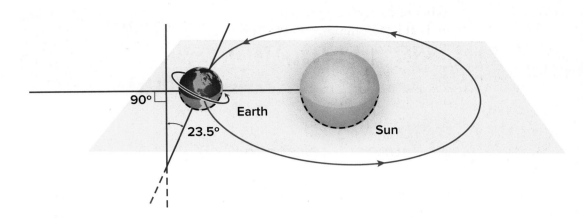

Figure 15 Earth's nearly circular orbit around the Sun lies on the ecliptic plane. When looking toward the horizon and the plane of the ecliptic, different stars are visible during the year.

Describe *How do the positions of stars vary when you look overhead?*

Annual Motions

Earth orbits the Sun in a slightly elliptical orbit, as shown in **Figure 15.** The plane of Earth's orbit is called the **ecliptic plane.** As Earth rotates, the Sun and planets appear to move across the sky in a path known as the ecliptic. As Earth moves in its orbit, different constellations are visible.

The effects of Earth's tilt

Earth's axis is tilted relative to the ecliptic at approximately 23.5°. As Earth orbits the Sun, the orientation of Earth's axis remains fixed in space so that, at a given time, the northern hemisphere of Earth is tilted toward the Sun, while at another point, six months later, the northern hemisphere is tilted away from the Sun. A cycle of the seasons is a result of this tilt and Earth's orbital motion around the Sun. Another effect is the changing angle of the Sun above the horizon from summer to winter. More hours of daylight cause the summer months to be warmer than the winter months.

 Get It?

Describe the cause of seasons on Earth.

Solstices

Earth's orbit around the Sun and the tilt of Earth's axis are illustrated in **Figure 16.** Positions 1 and 3 correspond to the solstices. At a **solstice,** the Sun is overhead at its farthest distance either north or south of the equator. The lines of latitude that correspond to these positions on Earth have been identified as the Tropic of Cancer and the Tropic of Capricorn. The area between these latitudes is commonly known as the tropics. Position 1 corresponds to the summer solstice in the northern hemisphere, when the Sun is directly overhead at the Tropic of Cancer, 23.5° north latitude. At this time, around June 21 each year, the number of daylight hours reaches its maximum, and the Sun is in the sky continuously within the region of the Arctic Circle.

ACADEMIC VOCABULARY

cycle
recurring sequence of events or phenomena
The cycle of seasons repeats every year.

On this day, the number of daylight hours in the southern hemisphere is at its minimum, and the Sun does not appear in the region within the Antarctic Circle.

 Get It?

Identify where the Sun is directly overhead at the summer solstice in the northern hemisphere.

As Earth moves past Position 2, the Sun's altitude decreases in the northern hemisphere until Earth reaches Position 3, known as the winter solstice for the northern hemisphere. Here, the Sun is directly overhead at the Tropic of Capricorn, 23.5° south latitude. This happens around December 21. On this day, the number of daylight hours in the northern hemisphere is at its minimum, and the Sun does not appear in the region within the Arctic Circle. Then, as Earth continues around its orbit past Position 4, the Sun's altitude increases again until it returns to Position 1.

Notice that the summer and winter solstices are reversed for those living in the southern hemisphere—June 21 is the winter solstice and December 21 is the summer solstice in the southern hemisphere.

Equinoxes

Positions 2 and 4, where Earth is midway between solstices, represent the equinoxes, a term meaning "equal nights." At an **equinox,** Earth's axis is perpendicular to the Sun's rays, and at noon the Sun is directly overhead at the equator. Those living in the northern hemisphere refer to Position 2 as the autumnal equinox, or, more simply, fall, and Position 4 as the vernal equinox, or spring.

People living in the southern hemisphere experience the reverse—Positions 2 and 4 are the vernal and autumnal equinoxes, respectively, or spring and fall.

Changes in altitude

The Sun's maximum height at midday, called its zenith, varies throughout the year depending on the viewer's location. For example, on the summer solstice, a person located at 23.5° north latitude sees the Sun's zenith directly overhead.

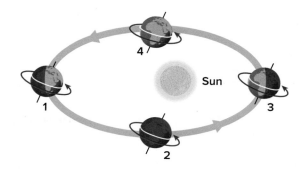

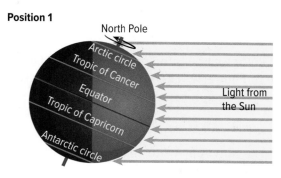

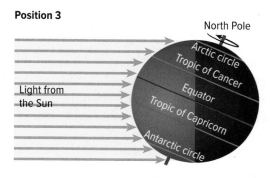

Figure 16 Earth's axis remains tilted at the same angle as it orbits the Sun. It points either toward or away from the Sun at solstices, as in Positions 1 and 3, and to the side at equinoxes, as in Positions 2 and 4.

Identify *the correct term for each position in each hemisphere.*

Figure 17 For a person standing at 23.5° north latitude, the Sun would be directly overhead on the summer solstice. It would be at its lowest position on the horizon at the winter solstice.

Draw *a diagram showing how the Sun's angle changes throughout the year at your latitude.*

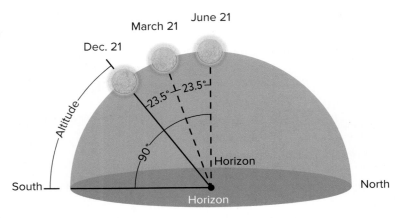

At the equinox, the Sun appears lower, and at the winter solstice, it is at its lowest position, shown in **Figure 17.** Then it starts moving higher again to complete the cycle.

Phases of the Moon

Just as the Sun appears to change its position in the sky during the year, the Moon also changes position relative to the ecliptic plane as it orbits Earth. The Moon's cycle is complex, as you will learn in this lesson. One striking effect of interactions in the Sun-Earth-Moon system are the changing views of the lit side of the Moon as it orbits Earth. The sequential changes in the appearance of the Moon are called lunar phases, shown in **Figure 18.**

The light given off by the Moon is a reflection of the Sun's light. In fact, one-half of the Moon is illuminated at all times. How much of this lighted half is visible from Earth varies as the Moon revolves around Earth. When the Moon is between Earth and the Sun, for instance, the side that is illuminated is not visible from Earth. This phase is called a new moon.

 Get It?

Explain what is meant by the term *lunar phases.*

Waxing and waning

Starting at the new moon, as the Moon orbits Earth, more of the sunlit side of the Moon becomes visible. This increase in the visible sunlit surface of the Moon is called the waxing phase. The waxing phases are waxing crescent, first quarter, and waxing gibbous. Then, as the Moon moves to the far side of Earth from the Sun, the entire sunlit side of the Moon faces Earth. This is known as a full moon. After the full moon, the portion of the Moon's sunlit side that is visible begins to decrease. This is called the waning phase. The waning phases are named similarly to the waxing phases—that is, waning gibbous and waning crescent. When exactly half of the sunlit portion is visible, it is called the third quarter.

SCIENCE USAGE v. COMMON USAGE

altitude
Science usage: angular elevation of a celestial body above the horizon
Common usage: vertical elevation of a body above a surface

Figure 18 Visualizing the Phases of the Moon

One-half of the Moon is always illuminated by the Sun's light, but the entire lighted half is visible from Earth only at full moon. The rest of the time, you see portions of the lighted half. These cyclical changes are called lunar phases.

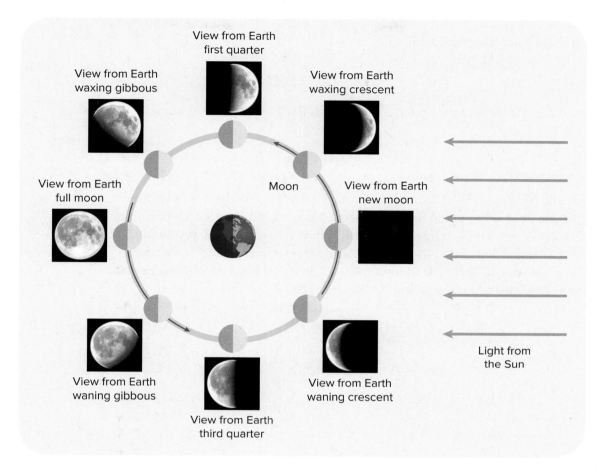

View from Earth
first quarter

View from Earth
waxing gibbous

View from Earth
waxing crescent

View from Earth
full moon

Moon

View from Earth
new moon

View from Earth
waning gibbous

View from Earth
third quarter

View from Earth
waning crescent

Light from
the Sun

Sometimes a dim image of the full moon is seen along with a crescent. This is caused by Earth's reflected light on the Moon's surface. It is often referred to as "the new moon with the old moon in its arms."

Because of variations in the plane of the Moon's orbit, the phases might appear different—either tipped or misshapen.

Synchronous rotation

You might have noticed that the surface features of the Moon always look the same. As the Moon orbits Earth, the same side faces Earth at all times. This is because the Moon rotates with a period equal to its orbital period. In other words, the Moon spins on its axis exactly once each time it goes around Earth. This is no coincidence. Scientists theorize that Earth's gravity slowed the Moon's original spin until the Moon reached **synchronous rotation,** the state at which its orbital and rotational periods are equal.

Get It?

Explain why the same side of the Moon always faces Earth.

Lunar Motions

The length of time it takes for the Moon to go through a complete cycle of phases, for example—from one new moon to the next—is called a lunar month. The length of a lunar month is about 29.5 days. This is longer than the 27.3 days it takes for one revolution, or orbit, around Earth, as illustrated in **Figure 19.** The Moon also rises and sets about 50 minutes later each day because the Moon moves 13° in its orbit over a 24-hour period, and Earth has to turn an additional 13° for the Moon to rise.

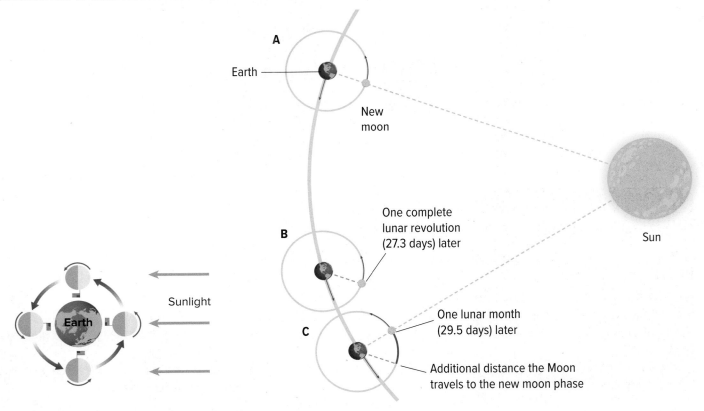

Figure 19 As the Moon moves from point A, where it is in the new moon phase as seen from Earth, to point B, it completes one revolution but is now in the waning crescent phase as seen from Earth. It must travel an additional 2.2 days to return to the new moon phase. The Moon rotates as it revolves, keeping the same side facing Earth, as shown in the smaller diagram to the left.

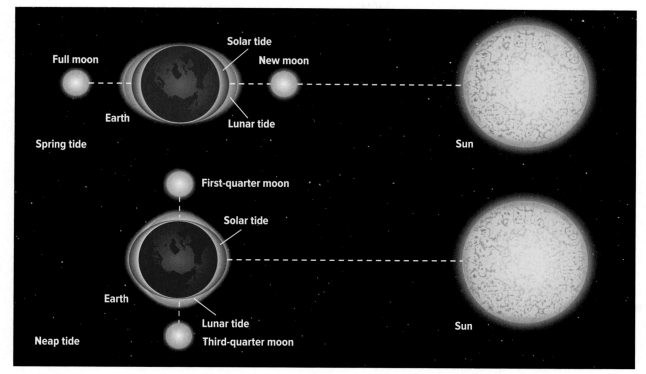

Figure 20 Alignment of the Sun and the Moon produces the spring tides shown in the top diagram. Neap tides, shown in the bottom diagram, occur when the Sun, Earth, and the Moon form a right angle.

Tides

The Moon has a direct effect on Earth in the form of ocean tides. The Moon's gravity pulls on Earth along an imaginary line connecting Earth and the Moon, and this creates bulges of ocean water on both the near and far sides of Earth. Recall that Earth's rotation also contributes to the formation of tides. As Earth rotates, these bulges remain aligned with the Moon, so that a person at a shoreline on Earth's surface would observe that the ocean level rises and falls every 12 hours. Most coastal areas on Earth experience two high tides and two low tides each day. A semidiurnal tidal pattern occurs when the two daily high tides have a similar range and the two low tides have a similar range. A mixed tidal pattern occurs when the two daily high tides have different ranges and the two low tides have different ranges.

Spring and neap tides The Sun's gravitational pull also affects tides, but the Sun's influence is half that of the Moon's because the Sun is farther away. However, when the Sun and the Moon are aligned along the same direction, their effects are combined, and tides are higher than normal. These tides, called spring tides, are especially high when the Moon is nearest Earth and Earth is nearest the Sun in their slightly elliptical orbits.

When the Moon is at a right angle to the Sun-Earth line, the result is lower-than-normal tides, called neap tides. This occurs because the Sun's and the Moon's gravitational forces are competing. The Sun and Moon alignments during spring and neap tides are shown in **Figure 20.**

 Get It?

Model Describe how you could model spring tides and neap tides.

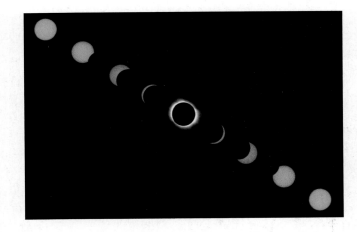

Figure 21 The stages of a total solar eclipse are seen in this multiple-exposure photograph.

Explain *why the Moon seems to cross the Sun at an angle rather than directly right to left.*

Solar Eclipses

A **solar eclipse** occurs when the Moon passes directly between the Sun and Earth and blocks the Sun from view. Although the Sun is much larger than the Moon, it is far enough away that they appear to be the same size when viewed from Earth. When the Moon perfectly blocks the Sun's disk, only the dim, outer gaseous layers of the Sun are visible. This spectacular sight, shown in **Figure 21,** is called a total solar eclipse. A partial solar eclipse is seen when the Moon blocks only a portion of the Sun's disk.

How solar eclipses occur

Each object in the solar system creates a shadow as it blocks the path of the Sun's light. This shadow is totally dark directly behind the object and has a cone shape. During a solar eclipse, the Moon casts a shadow on Earth as it passes between the Sun and Earth. This shadow consists of two regions, as illustrated in **Figure 22.** The inner, cone-shaped portion, which blocks the direct sunlight, is called the umbra, or umbral shadow. People who witness an eclipse from within the umbral shadow see a total solar eclipse. That means they see the Moon completely cover the face of the Sun. The outer portion of this shadow, where some of the Sun's light still reaches, is called the penumbra, or penumbral shadow. People in the region of the penumbral shadow see a partial solar eclipse, where only a part of the Sun's disk is blocked by the Moon. Typically, the umbral shadow is never wider than 270 km, so a total solar eclipse is visible from a very small portion of Earth, whereas a partial solar eclipse is visible from a much larger portion.

Figure 22 During a solar eclipse, the Moon passes between Earth and the Sun. Those on Earth within the darkest part of the Moon's shadow (umbra) see a total eclipse. Those within the lighter part, or penumbral shadow, see only a partial eclipse.

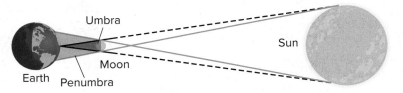

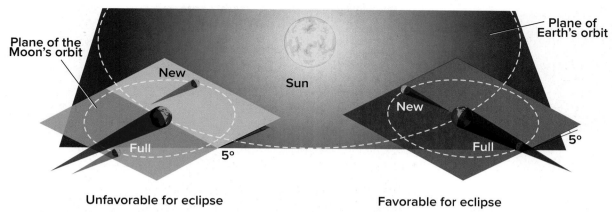

Figure 23 Eclipses can take place only when Earth, the Moon, and the Sun are perfectly aligned. This can happen only when the Moon's orbital plane and the ecliptic plane intersect along the Sun-Earth line, as shown in the diagram on the right. In the left diagram, this does not happen, and the Moon's shadow misses Earth.

Effects of tilted orbits You might wonder why a solar eclipse does not occur every month when the Moon passes between the Sun and Earth during the new moon phase. This does not happen because the Moon's orbit is tilted 5° relative to the ecliptic plane. Normally, the Moon passes above or below the Sun as seen from Earth, so no solar eclipse takes place. Only when the Moon crosses the ecliptic plane is it possible for the proper alignment for a solar eclipse to occur, but even that does not guarantee a solar eclipse. The plane of the Moon's orbit also rotates slowly around Earth, and a solar eclipse occurs only when the intersection of the Moon and the ecliptic plane is in a line with the Sun and Earth, as **Figure 23** illustrates.

 Get It?

Determine why a total solar eclipse does not occur every month.

Annular eclipses Not only does the Moon move above and below the plane of Earth and the Sun, but the Moon's distance from Earth increases and decreases as the Moon moves in its elliptical orbit around Earth. The closest point in the Moon's orbit to Earth is called **perigee,** and the farthest point is called **apogee.** When the Moon is near apogee, it appears smaller from Earth, and thus will not completely block the disk of the Sun during an eclipse. This is called an annular eclipse because, as **Figure 24** shows, a ring of the Sun, called the annulus, appears around the dark Moon.

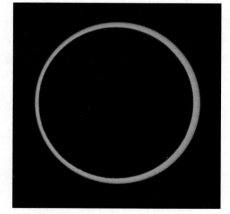

Figure 24 An annular eclipse takes place when the Moon is too far away for its umbral shadow to reach Earth. A ring, or annulus, is left uncovered.

Predict *Would annular eclipses occur if the Moon's orbit were a perfect circle?*

Earth's orbit also has a closest point in its orbit around the Sun, called perihelion, and a farthest point, called aphelion. When Earth is nearest the Sun and the Moon is at apogee, the Moon would not block the Sun entirely. The opposite is true for Earth at aphelion and the Moon at perigee.

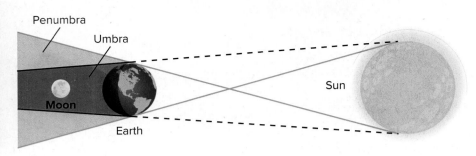

Figure 25 When the Moon is completely within Earth's umbra, a total lunar eclipse takes place, as shown in the diagram. The darkened Moon often has a reddish color, as shown in the photo, because Earth's atmosphere bends and scatters the Sun's light.

Lunar Eclipses

A **lunar eclipse** occurs when the Moon passes behind Earth in relation to the Sun, and through Earth's shadow. As illustrated in **Figure 25,** this can happen only at the time of a full moon when the Moon is on the opposite side of Earth from the Sun. The shadow of Earth has umbral and penumbral portions, just as the Moon's shadow does. A total lunar eclipse occurs when the entire Moon is within Earth's umbral shadow. This lasts for approximately two hours. During a total lunar eclipse, the Moon is faintly visible, as shown in **Figure 25,** because sunlight that has passed near Earth has been filtered and refracted by Earth's atmosphere. This light can give the eclipsed Moon a reddish color as Earth's atmosphere bends the red light into the umbra, much like a lens. Like solar eclipses, lunar eclipses do not occur every full moon because the Moon in its orbit usually passes above or below the Sun as seen from Earth.

 Check Your Progress

Summary

- Earth's rotation makes the Sun and other objects in space appear to rise and set each day.
- Seasons are caused by the tilt of Earth's axis relative to the ecliptic plane as Earth revolves around the Sun.
- The gravitational attraction of the Moon and, to a lesser extent, the Sun, causes tides.
- Lunar phases result from our view of the Moon's lighted side as it orbits Earth.
- Solar and lunar eclipses occur when the Sun's light is blocked.

Demonstrate Understanding

1. **State** one proof that Earth rotates, one proof that Earth rotates in 24 hours, and one observation that proves Earth revolves around the Sun in one year.

2. **Compare** solar and lunar eclipses, including the positions of the Sun, Earth, and the Moon.

3. **Diagram** the waxing and waning phases of the Moon.

4. **Analyze** why the Moon has a greater effect on Earth's tides than the more massive Sun.

Explain Your Thinking

5. **Relate** what you have learned about lunar phases to how Earth would appear to an observer on the Moon. To explain your answer, diagram the positions of the Sun, Earth, and the Moon, and draw how Earth would appear in several positions.

6. **MATH ❯ Connection** What would happen if Earth's axis were tilted 45°? At what latitudes would the Sun be directly overhead on the solstices and the equinoxes?

LEARNSMART® Go online to follow your personalized learning path to review, practice, and reinforce your understanding.

Countdown to Mars

The world was transfixed when the first human walked on the Moon in 1969. Imagine what it will be like to see humans exploring another planet. That goal, once the stuff of science fiction, may be a reality in a few short decades.

NASA plans to send astronauts into Mars orbit by the early 2030s. The ultimate goal is to land humans on the red planet.

NASA has designed a three-stage plan to send astronauts in orbit around Mars by the early 2030s. The space agency has already sent orbiters, landers, and rovers to study the red planet, including the *Curiosity* rover in 2012. NASA's plan for a crewed mission to Mars is broken into three stages: Earth Reliant, Proving Ground, and Earth Independent.

Earth Reliant

The first stage, Earth Reliant, is already underway. This stage is focused on research aboard the *International Space Station (ISS)*. There, astronauts are conducting studies to learn how humans can survive long-term space missions. These studies involve practical matters, such as growing crops to supply fresh food, as well as medical issues, such as how to mitigate the effects of microgravity on the human body.

Proving Ground

Stage two, Proving Ground, will involve missions near the Moon. These missions will test the ability of astronauts to live and work far from Earth, in preparation for eventually exploring Mars.

While the *ISS* is only hours away from Earth, Proving Ground missions will take place in "cislunar" space,

which is days away from our home planet. This will help NASA assess the ability of astronauts to live in deep space.

Also during stage two, robotic spacecraft will capture an asteroid and put it in orbit around the Moon. Astronauts will gather samples of the asteroid to test their spacewalking abilities and data-gathering techniques in preparation for an eventual mission to Mars.

Earth Independent

The third stage, Earth Independent, will involve sending humans into low-Mars orbit by 2033. Astronauts will travel farther and longer than ever before. As part of this stage, NASA will test methods for landing humans on the surface of Mars and identify potential resources for living on Mars. For example, scientists are developing technology for extracting oxygen from the atmosphere of Mars.

USE A MODEL TO DESCRIBE

Pat Rawlings/NASA

Find out more about how NASA will send humans into low-Mars orbit. Summarize your findings in a visual model. Create a presentation to explain your model, and share your presentation with the class.

GO ONLINE to study with your Science Notebook.

Lesson 1 **TOOLS OF ASTRONOMY**

- Telescopes collect and focus electromagnetic radiation emitted or reflected from distant objects.
- Electromagnetic radiation is classified by wavelength and frequency.
- The two main types of optical telescopes are refractors and reflectors.
- Space-based astronomy includes the study of orbiting telescopes, satellites, and probes.
- Technology originally developed to explore space is now used by people on Earth.

- electromagnetic spectrum
- refracting telescope
- reflecting telescope
- interferometry

Lesson 2 **THE MOON**

- Astronomers have gathered information about the Moon using telescopes, space probes, and astronaut exploration.
- Like Earth's crust, the Moon's crust is composed mostly of silicates.
- Surface features on the Moon include highlands, maria, ejecta, rays, and rilles. The Moon is heavily cratered.
- The Moon probably formed about 4.5 bya in a collision between Earth and a Mars-sized object.

- albedo
- highland
- maria
- rille
- impact crater
- ejecta
- ray
- regolith

Lesson 3 **THE SUN-EARTH-MOON SYSTEM**

- Earth's rotation makes the Sun and other objects in space appear to rise and set each day.
- Seasons are caused by the tilt of Earth's axis relative to the ecliptic plane as Earth revolves around the Sun.
- The gravitational attraction of the Moon and, to a lesser extent, the Sun, causes tides.
- Lunar phases result from our view of the Moon's lighted side as it orbits Earth.
- Solar and lunar eclipses occur when the Sun's light is blocked.

- ecliptic plane
- solstice
- equinox
- synchronous rotation
- solar eclipse
- perigee
- apogee
- lunar eclipse

REVISIT THE PHENOMENON

Why doesn't the Moon eclipse the Sun every month?

CER Claim, Evidence, Reasoning

Explain Your Reasoning Revisit the claim you made when you encountered the phenomenon. Summarize the evidence you gathered from your investigations and research and finalize your Summary Table. Does your evidence support your claim? If not, revise your claim. Explain why your evidence supports your claim.

STEM UNIT PROJECT
Now that you've completed the module, revisit your STEM unit project. You will summarize your evidence and apply it to the project.

GO FURTHER

SEP Data Analysis Lab

How can you predict how a solar eclipse will look to an observer at various positions?

The diagram below shows the Moon eclipsing the Sun. The Sun will appear differently to observers located at Points A through E.

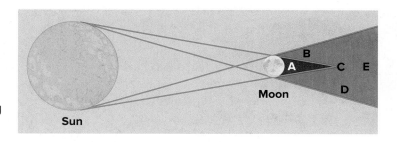

Data and Observations Observe the points in relation to the position of the Moon's umbra and penumbra.

CER Analyze and Interpret Data

1. **Claim** Draw how the solar eclipse would appear to an observer at each labeled point.
2. **Evidence** Design a data table to display your drawings.
3. **Reasoning** What type of solar eclipse is represented in each of your drawings? How do you know?

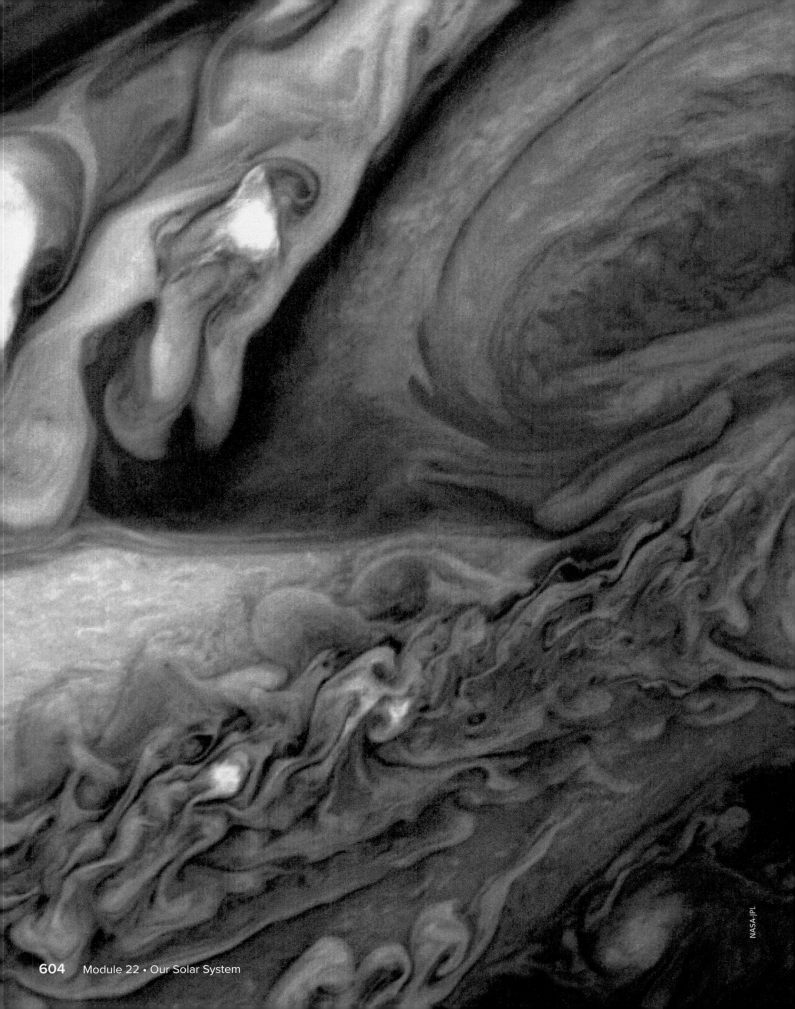

NASA-JPL

ENCOUNTER THE PHENOMENON

Why do we have to use satellites to take detailed pictures of Jupiter and the other planets?

GO ONLINE to play a video about the Voyager missions.

SEP Ask Questions

Do you have other questions about the phenomenon? If so, add them to the driving question board.

CER Claim, Evidence, Reasoning

Make Your Claim Use your CER chart to make a claim about why we have to use satellites to take detailed pictures of Jupiter and the other planets. Explain your reasoning.

Collect Evidence Use the lessons in this module to collect evidence to support your claim. Record your evidence as you move through the module.

Explain Your Reasoning You will revisit your claim and explain your reasoning at the end of the module.

GO ONLINE to access your CER chart and explore resources that can help you collect evidence.

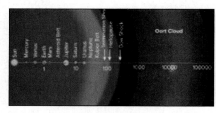

LESSON 1: Explore and Explain: Present-Day Viewpoints

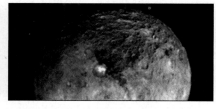

LESSON 4: Explore and Explain: Dwarf Planets

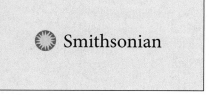

Additional Resources

FORMATION OF THE SOLAR SYSTEM

FOCUS QUESTION

Why do the planets move in the way they do?

Formation Theory

Theories of the origin of the solar system rely on direct observations and data from probes. Scientific theories must explain observed facts, such as the shape of the solar system, differences among the planets, and the nature of the oldest planetary surfaces—asteroids, meteorites, and comets.

A Collapsing Interstellar Cloud

Stars and planets form from interstellar clouds, which exist in space between the stars. These clouds consist mostly of hydrogen and helium gas, with small amounts of other elements and dust. Dust makes interstellar clouds look dark because it blocks the light from stars within or behind the clouds. Often, starlight reflects off of the dust and partially illuminates the clouds. Also, stars can heat clouds, making them glow on their own. This is why interstellar clouds often appear as blotches of light and dark, as shown in **Figure 1.** This interstellar dust can be thought of as a kind of smog that contains elements formed in older stars, which expelled their matter long ago.

Figure 1 Stars form in collapsing interstellar clouds, such as in the Eagle nebula, pictured here.

NASA

 **3D THINKING**   **DCI** Disciplinary Core Ideas   **CCC** Crosscutting Concepts   **SEP** Science & Engineering Practices

COLLECT EVIDENCE

 Use your Science Journal to record the evidence you collect as you complete the readings and activities in this lesson.

INVESTIGATE

🔎 **GO ONLINE** to find these activities and more resources.

⚙️ **Applying Practices: Planetary Orbits**
HS-ESS1-4. Use mathematical or computational representations to predict the motion of orbiting objects in the solar system.

 Design Your Own: Relating Gravitational Force and Orbits
Use mathematical and computational thinking to describe the pattern of planetary motion using Kepler's Laws.

Figure 2 The interstellar cloud that formed our solar system collapsed into a rotating disk of dust and gas. When concentrated matter in the center acquired enough mass, the Sun formed in the center, and the remaining matter gradually condensed, forming the planets.

At first, the density of interstellar gas is low. However, gravity slowly draws matter together until it is concentrated enough to form a star and, possibly, planets. Astronomers think that the solar system began this way. They have also observed planets around other stars and hope that studying such planet systems will provide clues to how our solar system formed.

Collapse accelerates

At first, the collapse of an interstellar cloud is slow, but it gradually accelerates, and the cloud becomes much denser at its center. If rotating, the cloud spins faster as it contracts for the same reason that ice skaters spin faster as they pull their arms close to their bodies—centripetal force. As the collapsing cloud spins, the rotation slows the collapse in the equatorial plane, and the cloud becomes flattened. Eventually, the cloud becomes a rotating disk with a dense concentration of matter at the center, as shown in **Figure 2.**

 Get It?
Explain why the rotating disk spins faster as it contracts.

Matter condenses

Astronomers think our solar system began in this manner. The Sun formed when the dense concentration of gas and dust at the center of a rotating disk reached a temperature and pressure high enough to fuse hydrogen into helium. The rotating disk surrounding the young Sun became our solar system. Within this disk, the temperature varied greatly with location; the area closest to the dense center was still warm, while the outer edge of the disk was cold. This temperature gradient resulted in different elements and compounds condensing, depending on their distance from the Sun. This also affected the distribution of elements in the forming planets. The inner planets are richer in the higher-melting-point elements, and the outer planets are composed mostly of the more volatile elements. That is why the outer planets and their moons consist mostly of gases and ices. Eventually, the condensation of materials into liquid and solid forms slowed.

Table 1 Physical Data of the Planets

Planet	Diameter (km)	Relative Mass (Earth = 1)	Average Density (kg/m³)	Atmosphere	Distance from the Sun (AU)	Moons
Mercury	4,880	0.055	5430	none	0.39	0
Venus	12,104	0.815	5240	CO_2, N_2	0.72	0
Earth	12,742	1.00	5520	N_2, O_2, H_2O	1.00	1
Mars	6,779	0.11	3930	CO_2, N_2, Ar	1.52	2
Jupiter	139,822	317.83	1330	H_2, He	5.20	79
Saturn	116,464	95.16	690	H_2, He	9.54	62
Uranus	50,724	14.50	1270	H_2, He, CH_4	19.20	27
Neptune	49,244	17.10	1640	H_2, He, CH_4	30.05	14

Planetesimals

Next, the tiny grains of condensed material started to accumulate and merge, forming larger particles. These particles grew as grains collided and stuck together and as gas particles collected on their surfaces. Eventually, colliding particles in the early solar system merged to form **planetesimals**—objects ranging from one kilometer to hundreds of kilometers in diameter. Growth continued as planetesimals collided and merged. Sometimes, collisions destroyed planetesimals, but the overall result was a smaller number of larger bodies—the planets. Some of their properties are given in **Table 1.**

Gas giants form

The first large planet to develop was Jupiter. Jupiter increased in size through the merging of icy planetesimals that contained mostly lighter elements. It grew larger as its gravity attracted additional gas, dust, and planetesimals. Saturn and the other gas giants formed similarly, but they could not become as large because Jupiter had collected so much of the available material. As each gas giant attracted material from its surroundings, a disk formed in its equatorial plane, much like the disk of the early solar system. In this disk, matter clumped together to form rings and satellites.

Terrestrial planets form

Planets also formed by the merging of planetesimals in the inner part of the main disk, near the young Sun. These were composed primarily of elements that resist vaporization, so the inner planets are rocky and dense, in contrast to the gaseous outer planets. Also, scientists think that solar wind swept away much of the gas in the area of the inner planets and prevented them from acquiring much of this material from their surroundings.

STEM CAREER Connection

Science Teacher

Helping students envision how the solar system formed, what the surface of Jupiter is like, or how cold it is on Neptune are all part of being a science teacher. Making sure future consumers, voters, and decision makers are scientifically literate is one of the most rewarding parts of a teacher's job.

Debris

Material that remained after the formation of the planets and satellites is called debris. Eventually, the amount of interplanetary debris diminished as it crashed into planets or was diverted out of the solar system. Some debris that was not ejected from the solar system became icy objects known as comets. Other debris formed rocky bodies known as asteroids. Most asteroids are found in the area between Jupiter and Mars known as the asteroid belt, shown in **Figure 3.** They remain there because Jupiter's gravitational force prevented them from merging to form a planet.

 Get It?

Describe why interplanetary debris diminished over time.

Modeling the Solar System

Ancient astronomers assumed that the Sun, planets, and stars orbited a stationary Earth in an Earth-centered model of the solar system. They thought this explained the most obvious daily motion of the stars and planets rising in the east and setting in the west. But as you have learned previously, this does not happen because these bodies orbit Earth, but rather that Earth spins on its axis.

This geocentric (jee oh SEN trihk), or Earth-centered, model could not readily explain some other aspects of planetary motion. For example, the planets might appear farther to the east one evening, against the background of the stars, than they had the previous night. Sometimes a planet seems to reverse direction and move back to the west. The apparent backward movement of a planet is called **retrograde motion.** For example, Mars normally appears to move from west to east from night to night. However, approximately every two years, Mars's position in the night sky seems to change and move from east to west instead. The retrograde motion of Mars is shown in the diagram in **Figure 4.** Today, we know that Mars's retrograde motion is due to the fact that Earth "catches up" to Mars and overtakes it as both planets orbit the Sun. Earth orbits the Sun faster than Mars. For every full orbit Mars completes, Earth completes two. As a result, every 26 months, Mars exhibits retrograde motion as Earth passes it in orbit.

 Get It?

Explain why early astronomers thought that Mars had retrograde motion, and explain what actually happens.

Figure 3 Hundreds of thousands of asteroids have been detected in the asteroid belt, which lies between Mars and Jupiter.

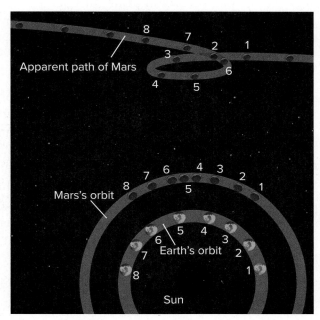

Figure 4 The diagram shows the apparent retrograde motion of Mars and how the changing angles of view from Earth create this effect.

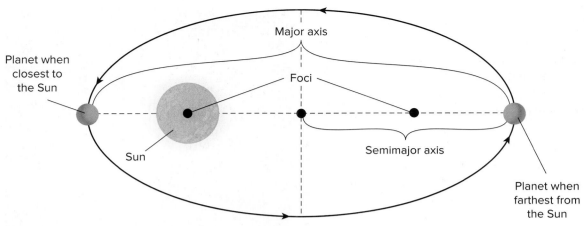

Figure 5 This diagram shows the geometry of an ellipse using an exaggerated planetary orbit. The Sun lies at one of the two foci. The minor axis of the ellipse is its shorter diameter. The major axis of the ellipse is its longer diameter, which equals the distance between a planet's closest and farthest points from the Sun. The semimajor axis represents the average distance of the planet to the Sun.

Heliocentric model

In 1543, Polish scientist Nicolaus Copernicus suggested that the Sun was the center of the solar system. In this Sun-centered, or heliocentric (hee lee oh SEN trihk) model, Earth and all the other planets orbit the Sun. In a heliocentric model, the increased gravity due to proximity to the Sun causes the inner planets to move faster in their orbits than do the outer planets. This model also provides a simple explanation for retrograde motion.

Kepler's first law

Within a century, the ideas of Copernicus were confirmed by other astronomers, who found evidence that supported the heliocentric model. For example, Tycho Brahe (TEE coh BRAH), a Danish astronomer, designed and built very accurate equipment for observing the stars. From 1576–1601, before the telescope was used in astronomy, he made accurate observations of the planets' positions. Using Brahe's data, German astronomer Johannes Kepler demonstrated that each planet orbits the Sun in a path shaped like an ellipse rather than a circle. This is known as Kepler's first law of planetary motion. An **ellipse** is an oval shape that is centered on two points instead of a single point, as in a circle. The two points are called the foci (singular, *focus*). The major axis is the line that runs through both foci at the maximum diameter of the ellipse, as illustrated in **Figure 5.**

Each planet has its own elliptical orbit, but the Sun is always at one focus. For each planet, the average distance between the Sun and the planet is its semimajor axis, which equals half the length of the major axis of its orbit, as shown in **Figure 5.** Earth's semimajor axis is of special importance because it is a unit used to measure distances within the solar system. Earth's average distance from the Sun is 1.496×10^8 km, or 1 **astronomical unit** (AU). Distance in space is often measured in AU. For example, Mars is 1.52 AU from the Sun.

SCIENCE USAGE v. COMMON USAGE

law

Science usage: a general relation proved or assumed to hold between mathematical expressions

Common usage: a rule of conduct prescribed as binding and enforced by a controlling authority

ACADEMIC VOCABULARY

collapse

to fall down, give way, or cave in

The hot-air balloon collapsed when the fabric was torn.

Eccentricity

A planet in an elliptical orbit does not orbit at a constant distance from the Sun. The shape of a planet's elliptical orbit is defined by **eccentricity,** which is the ratio of the distance between the foci to the length of the major axis. The orbits of most planets are not very eccentric; in fact, some are almost perfect circles. The eccentricity of a planet can change slightly. Earth's eccentricity today is about 0.02, but the gravitational attraction of other planets can stretch the eccentricity to 0.05 or cause it to fall to 0.01.

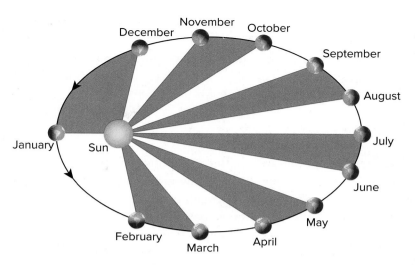

Figure 6 Kepler's second law states that planets move faster when close to the Sun and slower when farther away. This means that a planet sweeps out equal areas in equal amounts of time. (Note: *not drawn to scale*)

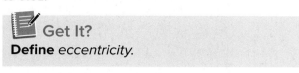 **Get It?**

Define *eccentricity.*

Kepler's second and third laws

In addition to discovering the shapes of planetary orbits, Kepler showed that planets move faster when they are closer to the Sun. He demonstrated this by proving that an imaginary line between the Sun and a planet sweeps out equal amounts of area in equal amounts of time, as shown in **Figure 6.** This is known as Kepler's second law. The length of time it takes for a planet or other body to travel a complete orbit around the Sun is called its orbital period. In Kepler's third law of planetary motion, he determined the mathematical relationship between the size of a planet's ellipse and its orbital period. This relationship is written as follows:

$$P^2 = a^3$$

P is time measured in Earth years, and *a* is length of the semimajor axis measured in astronomical units.

 Get It?

State Kepler's three laws in your own words.

Galileo

While Kepler was developing his ideas, Italian scientist Galileo Galilei became the first person to use a telescope to observe the sky. Galileo made many discoveries that supported Copernicus's ideas. The most famous of these was his discovery that four moons orbit the planet Jupiter, proving that not all celestial bodies orbit Earth, and demonstrating that Earth was not necessarily the center of the solar system.

CCC CROSSCUTTING CONCEPTS

Scale, Proportion, and Quantity Use Kepler's third law to predict the effect of an increase in the size of a planet's ellipse on its orbital period. Explain your answer.

Galileo's view of Jupiter's moons was not as clear as our present-day view of them, shown in **Figure 7.** The underlying explanation for the heliocentric model remained unknown until 1684, when English scientist Isaac Newton published his law of universal gravitation.

Gravity

Newton first developed an understanding of gravity by observing falling objects. He described falling as downward acceleration produced by gravity, an attractive force between two objects. He determined that both the masses of and the distance between two bodies determined the force between them. This relationship is expressed in his law of universal gravitation, illustrated in **Figure 8,** which is stated mathematically as follows:

$$F = \frac{Gm_1 m_2}{r^2}$$

F is the force measured in newtons, G is the universal gravitational constant (6.67×10^{-11} m^3/ kg·s^2), m_1 and m_2 are the masses of the bodies in kilograms, and r is the distance between the two bodies in meters.

Figure 7 Galileo would probably be astounded to see Jupiter's four largest moons in the composite image above. Still, his view of Jupiter and its moons proved a milestone in support of heliocentric theory.

Gravity and orbits

Newton realized that this attractive force could explain why planets move according to Kepler's laws. He observed the Moon's motion and realized that its direction changes because of the gravitational attraction of Earth. In a sense, the Moon is constantly falling toward Earth. If it were not for this attraction, the Moon would continue to move in a straight line and would not orbit Earth. The same is true of the planets and their moons, stars, and all orbiting bodies throughout the universe.

Center of mass

Newton also determined that each planet orbits a point between it and the Sun called the center of mass. For any planet and the Sun, the center of mass is just above or within the surface of the Sun, because the Sun is much more massive than any planet. **Figure 9,** on the next page, illustrates how the center of mass between objects is similar to the balance point on a seesaw.

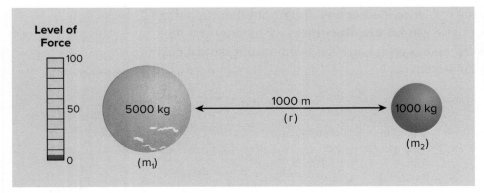

Figure 8 The gravitational attraction between these two objects is 3.3×10^{-10} N.

Predict *the effect of doubling the masses of both objects, and check your prediction using Newton's equation.*

NASA/JPL/DLR

Present-Day Viewpoints

Astronomers traditionally divided the planets into two groups: the four smaller, rocky, inner planets—Mercury, Venus, Earth, and Mars—and the four outer gas planets—Jupiter, Saturn, Uranus, and Neptune. It was not clear how to classify Pluto because it is different from the gas giants in composition and orbit. Pluto also did not fit the present-day theory of how the solar system developed. In the early 2000s, astronomers discovered a vast number of small, icy bodies inhabiting the outer reaches of the solar system, beyond the orbit of Neptune. At least one of these is larger than Pluto.

These discoveries have led many astronomers to rethink traditional views of the solar system. Some already define it in terms of three zones: the inner terrestrial planets, the outer gas giant planets, and the dwarf planets and comets. In science, views change as new data becomes available and new theories are proposed. Astronomy today is a rapidly changing field.

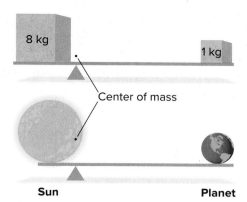

Figure 9 Just as the balance point on a seesaw is closer to the heavier box, the center of mass between two orbiting bodies is closer to the more massive body.

 Get It?

Explain why information in science is constantly changing.

Check Your Progress

Summary

- A collapsed interstellar cloud formed the Sun and planets from a rotating disk.

- The inner planets formed closer to the Sun than the outer planets, leaving debris to produce asteroids and comets.

- Copernicus created the heliocentric model, and Kepler defined its shape and mechanics.

- Newton explained the forces governing the solar system bodies and provided proof for Kepler's laws.

- Present-day astronomers divide the solar system into three zones.

Demonstrate Understanding

1. **Explain** the formation of planets based on our knowledge of the solar system.

2. **Explain** why retrograde motion is an apparent motion.

3. **Describe** how the gravitational force between two bodies is related to their masses and the distance between them.

4. **Explain** Kepler's first law and how it is related to the motion of orbiting objects.

5. **Compare** the shapes of two ellipses having eccentricities of 0.05 and 0.75.

Explain Your Thinking

6. **Infer** Based on what you have learned about Kepler's third law, which planet moves faster in its orbit, Jupiter or Neptune? Explain.

7. **MATH ▶ Connection** Use Newton's law of universal gravitation to calculate the force of gravity between two students standing 12 m apart. Their masses are 65 kg and 50 kg.

FOCUS QUESTION

How did we see through Venus's thick atmosphere?

Terrestrial Planets

The four inner planets are called **terrestrial planets** because they are similar in density to Earth and have solid, rocky surfaces. Their average densities, obtained by dividing the mass of a planet by its volume, range from about 3.9 to just over 5.5 g/cm^3. Average density is an important indicator of internal conditions, and densities in this range indicate that the interiors of these planets are compressed.

Mercury

Mercury, shown in **Figure 10,** is the planet closest to the Sun, and for this reason it is difficult to see from Earth. During the day it is lost in the Sun's light; it is more easily seen at sunset and sunrise.

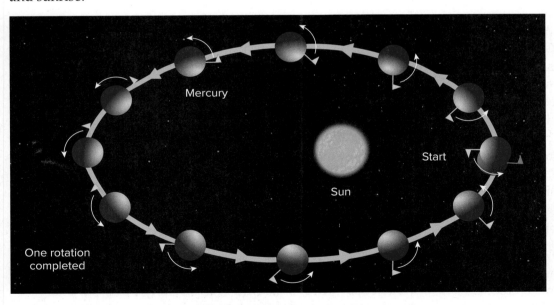

Mercury

Sun

Start

One rotation completed

Figure 10 Because of Mercury's odd rotation, its day lasts for two-thirds of its year.

Compare *Mercury's orbital motion with that of Earth's Moon.*

3D THINKING **DCI** Disciplinary Core Ideas **CCC** Crosscutting Concepts **SEP** Science & Engineering Practices

COLLECT EVIDENCE

 Use your Science Journal to record the evidence you collect as you complete the readings and activities in this lesson.

INVESTIGATE

GO ONLINE to find these activities and more resources.

 Review the News
Obtain information from a current news story about current research on Mars. Evaluate your source and **communicate** your findings to your class.

 Revisit the Encounter the Phenomenon Question
What information from this lesson can help you answer the Unit and Module questions?

Mercury is about one-third the size of Earth and has a smaller mass. Mercury has no moons. Radio observations in the 1960s revealed that Mercury has a slow spin of 1407.6 hours. In one orbit around the Sun, Mercury rotates one and one-half times. As Mercury spins, the side facing the Sun at the beginning of the orbit faces away from the Sun at the end of the orbit. This means that two complete Mercury years equal three complete Mercury rotations.

Atmosphere

Unlike Earth and the other planets, Mercury's atmosphere is constantly being replenished by the solar wind. What little atmosphere does exist is composed primarily of oxygen, sodium, and hydrogen deposited by the Sun. The daytime surface temperature on Mercury is 700 K (427°C), while temperatures at night fall to 100 K (−173°C). This is the largest day-night temperature difference among the planets.

Surface

Early knowledge about Mercury was based on radio observations from Earth and images from U.S. space probe *Mariner 10,* which passed close to Mercury three times in 1974 and 1975. The *MESSENGER (**ME**rcury **S**urface, **S**pace **EN**vironment, **GE**ochemistry and **R**anging)* space probe also made flybys and, in 2011, became the first spacecraft to orbit Mercury. Images show that Mercury's surface, like that of the Moon, is covered with craters and plains, as shown in **Figure 11.** The plains on Mercury's surface are smooth and relatively crater free. Scientists think that the plains formed from lava flows that covered cratered terrain, much like the maria formed on the Moon. The surface gravity of Mercury is much greater than that of the Moon, resulting in larger crater diameters and shorter lengths of ejecta.

Mercury has a planetwide system of cliffs called **scarps,** such as the one shown in **Figure 12.** Mercury's scarps are much higher than Earth's. Scientists hypothesize that the scarps developed as Mercury's crust shrank and fractured early in the planet's geologic history.

Interior

Without seismic data, scientists have no way to analyze the interior of Mercury. However, its high density suggests that Mercury has a large nickel-iron core. Mercury's small magnetic field indicates that some of its core is molten.

Figure 11 This mosaic of Mercury's heavily cratered surface was made by *Mariner 10.* Craters range in size from 100 to 1300 km in diameter.

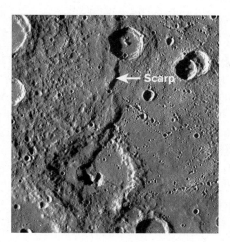

Figure 12 This image from the *MESSENGER* probe shows Victoria Rupes, a scarp on Mercury. *Rupes* is Latin for "cliff."

(t)NASA-JPL, (b)NASA/Johns Hopkins University Applied Physics Laboratory/Carnegie Institution of Washington

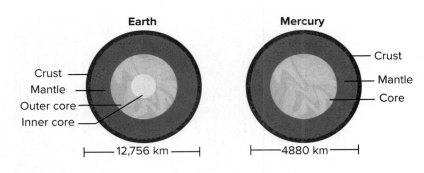

Earth

Crust
Mantle
Outer core
Inner core

├────── 12,756 km ──────┤

Mercury

Crust
Mantle
Core

├── 4880 km ──┤

Figure 13 The structure of Mercury's interior, which contains a proportionally larger core than Earth, suggests that Mercury was once much larger.

Early Mercury

Mercury's small size, high density, and probable molten interior are similar to Earth, without much of the mantle, as shown in **Figure 13.** There are three major theories to explain these observations. Mercury may have collided with another body early in its history, stripping off its crust and mantle. Or perhaps the heat of the early solar nebula vaporized the outer layers. Finally, before Mercury formed, the lighter gas near the Sun might have slowed and fallen inward, leaving higher density material behind.

Venus

Venus and Mercury are the only two planets closer to the Sun than is Earth. Like Mercury, Venus has no moons. Venus is the brightest planet in the sky because it is close to Earth and because its albedo is 0.90—the highest of any planet. Venus is the first bright "star" to be seen after sunset in the western sky or the last "star" to be seen before sunrise, depending on which side of the Sun it is on. For these reasons it is often called either the evening or morning star.

Thick clouds around Venus prevent astronomers from observing the surface directly. However, astronomers learned much about Venus from spacecraft launched by the United States and the Soviet Union. The 1978 *Pioneer-Venus* and 1989 *Magellan* missions of the United States used radar to map 98 percent of the surface of Venus.

For images of the surface like the one shown in **Figure 14,** radar data was combined with images from *Magellan* spacecraft and the radio telescope in Arecibo, Puerto Rico. This view uses false colors to outline the major landmasses. In 2006, a European space probe, called *Venus Express*, went into orbit around Venus. It found signs that Venus has been—and still may be—volcanically active.

Figure 14 On the surface of Venus, the highlands are shown in red, and valleys are shown in blue. Large highland regions are like continents on Earth.

Infer *What do green areas represent?*

NASA/JPL/USGS

Retrograde rotation

Radar measurements show that Venus rotates slowly—a day on Venus is equivalent to 243 Earth days. Also, Venus rotates clockwise, while most planets spin counterclockwise. This backward spin, called retrograde rotation, means that an observer on Venus would see the Sun rise in the west and set in the east. Astronomers theorize that this retrograde rotation might be the result of a collision between Venus and another body early in the solar system's history.

Atmosphere

Venus is the planet most similar to Earth in physical properties, such as diameter, mass, and density, but its surface conditions and atmosphere are vastly different from those on Earth. The atmospheric pressure on Venus is 92 atmospheres (atm), compared to 1 atm at sea level on Earth. If you were on Venus, the pressure of the atmosphere would make you feel like you were under 920 m of water.

The atmosphere of Venus is composed primarily of carbon dioxide and small amounts of nitrogen and water vapor. Like Earth, Venus also has clouds, as shown in **Figure 15,** an image taken of Venus by the *Pioneer Venus Orbiter*. However, instead of being composed of water vapor and ice, as on Earth, the thick bands of clouds on Venus consist of sulfuric acid and produce concentrated acid rain.

Figure 15 Clouds swirl around Venus in this image taken using ultraviolet wavelengths.

Greenhouse effect

Venus also experiences a greenhouse effect similar to Earth's, but Venus's is more efficient. As you have learned, greenhouse gases in Earth's atmosphere trap infrared radiation and keep Earth much warmer than it would be if it had no atmosphere. The concentration of carbon dioxide is so high in Venus's atmosphere that it keeps the surface extremely hot—hot enough to melt lead. In fact, Venus is the hottest planet, with an average surface temperature of about 737 K (464°C), compared with Earth's average surface temperature of 288 K (15°C). It is so hot on the surface of Venus that no liquid water can exist.

Surface

The *Magellan* orbiter used radar reflection measurements to map the surface of Venus. This revealed that Venus has a surface smoothed by volcanic lava flows and with few impact craters. Observations from *Venus Express* indicate that volcanic activity took place as recently as 2.5 mya and that Venus might still be volcanically active. There is little evidence of current tectonic activity or well-defined crustal plates on Venus.

Interior

Because the size and density of Venus are similar to Earth's, it is probable that the internal structure is similar also. Astronomers theorize that Venus has a liquid metal core that extends halfway to the surface. Despite this core, Venus has no measurable magnetic field, probably because of its slow rotation.

CCC **CROSSCUTTING CONCEPTS**

Systems and System Models Make a model of Venus's atmosphere and how it influences the greenhouse effect and surface temperature of Venus. Research typical models of the greenhouse effect on Earth. How does your model of Venus differ from these models?

Earth

Earth, shown in **Figure 16,** has many unique properties when compared with other planets. Its distance from the Sun and its nearly circular orbit allow water to exist on its surface in all three states—solid, liquid, and gas. Liquid water is required for life, and Earth's abundance of water has been important for the development and existence of life on Earth. In addition, Earth's mild greenhouse effect and moderately dense atmosphere of nitrogen and oxygen provide conditions suitable for life.

Earth is the most dense of the terrestrial planets. It is the only known planet where plate tectonics currently occurs. Unlike Venus and Mercury, Earth has a moon, likely acquired by an impact.

Get It?

Identify the importance of liquid water on Earth.

Mars

Mars is often referred to as the red planet because of its reddish surface color, shown in **Figure 16.** Mars is smaller and less dense than Earth and has two irregularly shaped moons, Phobos and Deimos. Mars has been the target of recent exploration: the *Mars Exploration Rovers* in 2004, *Mars Reconnaissance Orbiter* in 2006, the *Phoenix Mars Lander* in 2008, and the *Mars Science Laboratory* in 2012.

Atmosphere

Both Mars and Venus have atmospheres of similar composition. The density and pressure of the atmosphere on Mars are much lower; therefore, Mars does not have a strong greenhouse effect like Venus does. Although the atmosphere is thin, it is turbulent—there is constant wind, and dust storms can last for months at a time.

Earth

Mars

Figure 16 Earth's blue seas and white clouds contrast sharply with the reddish, barren Mars.

ACADEMIC VOCABULARY

composition
the qualitative and quantitative makeup of an object or substance
The composition of water is hydrogen and oxygen atoms.

Olympus Mons Volcano   Gusev Crater

Figure 17 Orbital probes and landers have provided photographic details of the Martian features and surface, such as Olympus Mons and Gusev crater.

Surface

The southern and northern hemispheres of Mars vary greatly, as shown in **Figure 17.** The southern hemisphere is a heavily cratered, highland region resembling the highlands of the Moon. The northern hemisphere has sparsely cratered plains. Scientists theorize that great lava flows covered the once-cratered terrain of the northern hemisphere. Four gigantic shield volcanoes are located close to the equator, near a region called the Tharsis Plateau. The largest volcano on Mars is Olympus Mons. The base of Olympus Mons is larger than the state of Colorado, and the volcano rises 3 times higher than Mount Everest in the Himalayas.

Tectonics

An enormous canyon, Valles Marineris, shown in **Figure 18,** lies on the Martian equator, splitting the Tharsis Plateau. This canyon is 4000 km long—almost 10 times the length of the Grand Canyon on Earth and more than 3 times its depth. It probably formed as a fracture during a period of tectonic activity 3 bya, when the Tharsis Plateau was uplifted. The gigantic volcanoes were caused during the same period by upwelling of magma at a hot spot, much like the Hawaiian Island chain was formed. However, with no plate movement on Mars, magma accumulated in one area.

Erosional features

Other Martian surface features include dried riverbeds and lake beds, gullies, outflow channels, and runoff channels. These erosional features suggest that liquid water once existed on the surface of Mars. Astronomers think that the atmosphere was once much warmer, thicker, and richer in carbon dioxide, allowing liquid water to flow on Mars. The *Mars Reconnaissance Orbiter* found water ice below the surface at mid-latitudes, and the Mars rover *Curiosity* found evidence that ancient Mars had long-standing rivers and lakes.

Figure 18 Valles Marineris is a 4000-km-long canyon on Mars.

Ice caps

Ice caps cover both poles on Mars. The caps grow and shrink with the seasons. Martian seasons are caused by a combination of a tilted axis and a slightly eccentric orbit. Both caps are made of carbon dioxide ice, sometimes called dry ice.

Water ice lies beneath the carbon dioxide ice in the northern cap, shown in **Figure 19,** and is exposed during the northern hemisphere's summer, when the north pole is tilted closer to the Sun, and the carbon dioxide ice evaporates. There is also water ice beneath the southern cap, although the carbon dioxide ice does not completely evaporate to expose it.

Interior

The internal structure of Mars remains unknown. Astronomers hypothesize that there is a core of iron, nickel, and possibly sulfur that extends somewhere between 1200 km and 2400 km from the center of the planet. Because Mars has no magnetic field, astronomers think that the core is probably solid. Above the solid core is a mantle. There is no evidence of current tectonic activity or tectonic plates on the surface of the crust.

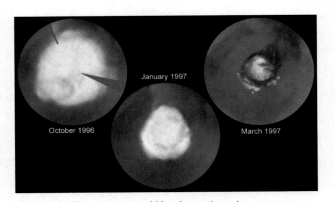

January 1997

October 1996

March 1997

Figure 19 These images of Mars's northern ice cap were taken three months apart by the *Hubble Space Telescope*.

Interpret *What do these images indicate about the orientation of Mars's axis?*

Check Your Progress

Summary

- Mercury is heavily cratered and has high cliffs. It has no real atmosphere and the largest day-night temperature difference among the planets.

- Venus has clouds containing sulfuric acid and an atmosphere of carbon dioxide that produces a strong greenhouse effect.

- Earth is the only planet that has all three forms of water on its surface.

- Mars has a thin atmosphere. Surface features include four volcanoes and channels that suggest that liquid water once existed on the surface.

Demonstrate Understanding

1. **Identify** the reason that the inner planets are called terrestrial planets.

2. **Summarize** the characteristics of each of the terrestrial planets.

3. **Compare** the average surface temperatures of Earth and Venus, and describe what causes them.

4. **Describe** the evidence that indicates there was once tectonic activity on Mercury, Venus, and Mars.

Explain Your Thinking

5. **Consider** what the inner planets would be like if impacts had not shaped their formation and evolution.

6. **MATH** **Connection** Create a graph showing the distance from the Sun for each terrestrial planet on the *x*-axis and their orbital periods in Earth days on the *y*-axis.

JPL/NASA/STScI

THE OUTER PLANETS

How did we get close enough to take pictures of Neptune?

The Gas Giant Planets

Jupiter, Saturn, Uranus, and Neptune are known as the gas giants. The **gas giant planets** are all very large, ranging from 15 to more than 300 times the mass of Earth and from about 4 to more than 10 times Earth's diameter. Their interiors are either gases or liquids, and they might have small, solid cores. They are made primarily of lightweight elements such as hydrogen, helium, carbon, nitrogen, and oxygen, and they are very cold at their surfaces. The gas giants have many satellites as well as ring systems.

Jupiter

Jupiter is the largest planet, with a diameter one-tenth that of the Sun and 11 times larger than Earth's. Jupiter's mass makes up 70 percent of all planetary matter in the solar system. Jupiter appears bright because its albedo is 0.343. Telescopic views of Jupiter show a banded appearance as a result of flow patterns in its atmosphere. Nestled among Jupiter's cloud bands is the Great Red Spot, an atmospheric storm that has raged for more than 300 years. This is shown in **Figure 20.**

Rings

The *Galileo* spacecraft observed Jupiter and its moons during a 7-year mission in the 1990s and 2000s. It revealed two faint rings around the planet in addition to a 6400-km-wide ring around Jupiter that had been discovered by *Voyager 1*.

Figure 20 Jupiter's cloud bands contain the Great Red Spot. The planet is circled by three faint rings that are probably composed of dust particles.

3D THINKING   **DCI** Disciplinary Core Ideas   **CCC** Crosscutting Concepts   **SEP** Science & Engineering Practices

COLLECT EVIDENCE
Use your Science Journal to record the evidence you collect as you complete the readings and activities in this lesson.

INVESTIGATE
GO ONLINE to find these activities and more resources.

Investigation Lab: Your Age and Weight on Other Planets
Use mathematical and computational thinking to determine the strength of gravity on a planet's surface using Newton's Law of Gravitation.

((•)) Review the News
Obtain information from a current news story about current research about Saturn. Evaluate your source and communicate your findings to your class.

Figure 21 The four largest moons of Jupiter are Io, Europa, Ganymede, and Callisto. Ganymede is larger than Mercury. Callisto's bright scars illustrate a long history of impacts. Io is the most volcanically active object in the solar system. Scientists think that Europa's subsurface ocean could possibly support life.

Io

Europa

Ganymede

Callisto

Atmosphere and interior

Jupiter has a density of 1326 kg/m³, which is low for its size, because it is composed mostly of hydrogen and helium in gaseous or liquid form. Below the liquid hydrogen is a layer of **liquid metallic hydrogen,** a form of hydrogen that has properties of both a liquid and a metal, which can exist only under conditions of very high pressure. Electric currents exist within the layer of liquid metallic hydrogen and generate Jupiter's magnetic field. Models suggest that Jupiter might have an Earth-sized solid core containing heavier elements.

Rotation

Jupiter rotates very rapidly for its size; it spins once on its axis in a little less than 10 hours, giving it the shortest day among the planets. This rapid rotation distorts the shape of the planet so that the diameter through its equatorial plane is 7 percent larger than the diameter through its poles. Jupiter's rapid rotation causes its clouds to flow rapidly as well, in bands of alternating dark and light colors called belts and zones. **Belts** are low, warm, dark-colored clouds that sink, and **zones** are high, cool, light-colored clouds that rise. These are similar to cloud patterns in Earth's atmosphere that are caused by Earth's rotation.

Get It?

Explain what the belts are on Jupiter and what causes them.

Moons

Jupiter has more than 75 moons, most of which are extremely small. Jupiter's four largest moons, Io, Europa, Ganymede, and Callisto, shown in **Figure 21,** are called Galilean satellites after their discoverer. Three of them are bigger than Earth's moon, and all four are composed of ice and rock. The ice content is lower in Io and Europa because they have been squeezed and heated by Jupiter's gravitational force more than the outer Galilean moons have been. In fact, Io is almost completely molten inside and undergoes constant volcanic eruptions. Gravitational heating has melted Europa's ice in the past, and astronomers hypothesize that it still has a subsurface ocean of liquid water. Cracks and water channels mark Europa's icy surface.

Jupiter's four, small, inner moons are thought to be the source of Jupiter's rings. Scientists think that the rings are produced as meteoroids strike these moons and release fine dust into Jupiter's orbit.

Gravity assist

A technique first used to help propel *Mariner 10* to Mercury was to use the gravity of Venus to boost the speed of the satellite. Today it is common for satellites to use a planet's gravity to help propel them deeper into space. Jupiter is the most massive planet, so any satellite passing deeper into space than Jupiter can use its gravity to give it an assist. Flybys on their way to Saturn and Pluto by the *Cassini* and *New Horizons* missions used that assist.

Saturn

Saturn, shown in **Figure 22,** is the second-largest planet in the solar system. Saturn is slightly smaller than Jupiter, and its average density is lower than that of water. Like Jupiter, Saturn rotates rapidly for its size and has a layered cloud system.

Atmosphere and interior

Saturn's atmosphere is mostly hydrogen and helium, with ammonia ice near the cloud tops. The internal structure of Saturn is probably similar to Jupiter's—fluid throughout, except for a small, solid core. Saturn's magnetic field is 1000 times stronger than Earth's and is aligned with its rotational axis. This is highly unusual among the planets.

Rings

Saturn's most striking feature is its rings, which are shown in **Figure 22.** Saturn's rings are much broader and brighter than those of the other gas giant planets. They are composed of pieces of ice that range from microscopic particles to house-sized chunks. There are seven major rings, and each ring is made up of narrower rings, called ringlets. The rings contain many open gaps.

These ringlets and gaps are caused by the gravitational effects of Saturn's many moons. The rings are thin—less than 200 m thick—because rotational forces keep the orbits of all the particles confined to Saturn's equatorial plane. The ring particles have not combined to form a large satellite because Saturn's gravity prevents particles located close to the planet from sticking together. This is why the major moons of the gas giant planets are always beyond the rings.

Origin of the rings Until recently, astronomers thought that the ring particles were left over from the formation of Saturn and its moons. Now, many astronomers think it is more likely that the ring particles are debris left over from collisions of asteroids and other objects or from moons broken apart by Saturn's gravity.

Figure 22 *Cassini-Huygens* provided detailed views of Saturn and its rings. Saturn's largest ring was discovered by NASA's *Spitzer Space Telescope* in 2009, orbiting six million kilometers away from the planet.

Explain *why the ring particles orbit Saturn in the same plane.*

Get It?

Explain the origin of Saturn's rings.

Moons

Saturn has more than 60 satellites, including the giant Titan, which is larger than the planet Mercury. Titan is unique among planetary satellites because it has a dense atmosphere made of nitrogen and methane. Methane can exist as a gas, a liquid, and a solid on Titan's surface. In 2005, *Cassini* released the *Huygens* (HOY gens) probe into Titan's atmosphere. *Cassini* detected plumes of ice and water vapor ejected from Saturn's moon Enceladus, suggesting geologic activity.

Uranus

Uranus was discovered accidentally in 1781, when a bluish object was observed moving relative to the stars. In 1986, *Voyager 2* flew by Uranus and provided detailed information about the planet, including the existence of new moons and rings. Uranus's average temperature is 58 K (–215°C).

NASA/JPL/Space Science Institute

Atmosphere

Uranus is 4 times larger and 15 times more massive than Earth. It has a blue, velvety appearance, shown in **Figure 23,** which is caused by methane gas in Uranus's atmosphere. Most of Uranus's atmosphere is composed of helium and hydrogen, which are colorless. There are few clouds, and they differ little in brightness and color from the surrounding atmosphere, contributing to Uranus's featureless appearance. The internal structure of Uranus is similar to that of Jupiter and Saturn; it is completely fluid except for a small, solid core. Uranus also has a strong magnetic field.

Moons and rings

Uranus has at least 27 moons and a faint ring system. Many of Uranus's rings are dark—almost black and nearly invisible. They were discovered only when the brightness of a star behind the rings dimmed as Uranus moved in its orbit and the rings blocked the starlight.

Rotation

The rotational axis of Uranus is tipped so far that its north pole almost lies in its orbital plane, as shown in **Figure 24.** Astronomers hypothesize that Uranus was knocked sideways by a massive collision with a passing object, such as a large asteroid, early in the solar system's history. Each pole on Uranus spends 42 Earth years in darkness and 42 Earth years in sunlight due to this tilt.

 Get It?

Explain why the poles of Uranus spend so much time in sunlight and in darkness.

Neptune

The existence of Neptune was predicted before it was discovered, based on small deviations in the motion of Uranus and the application of Newton's universal law of gravitation. In 1846, Neptune was discovered where astronomers had predicted it to be. In 1989, *Voyager* 2 flew past Neptune and took the image of its cloud-streaked atmosphere shown in **Figure 25,** on the next page. Neptune is the last of the gas giant planets and orbits the Sun almost 4.5 billion km away.

Figure 23 The blue color of Uranus is caused by methane in its atmosphere, which reflects blue light. These images, infrared composite images of the two hemispheres of Uranus, were taken with Keck Telescope adaptive optics.

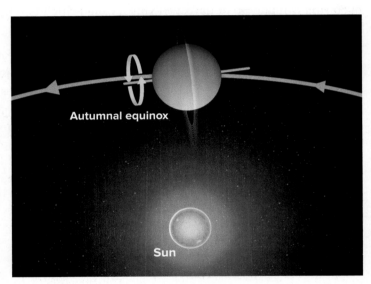

Figure 24 The axis of rotation of Uranus is tipped 98 degrees. This view shows its position at an equinox.

Draw *a diagram showing Uranus's position at the other equinox and the solstices.*

Neptune Cloud Streaks

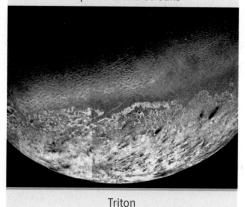

Triton

Figure 25 *Voyager 2* took these images showing Neptune's cloud streaks and a close-up view of Neptune's largest moon, Triton.

Atmosphere

Neptune is slightly smaller and denser than Uranus, but its radius is about 4 times greater than Earth's. Another similarity between Neptune and Uranus is their bluish color, caused by methane in the atmosphere. Neptune's atmospheric composition, temperature, magnetic field, interior, and particle belts or rings are also comparable with those of Uranus. Unlike Uranus, however, Neptune has distinctive clouds and atmospheric belts and zones, similar to those of Jupiter and Saturn. In fact, Neptune can have persistent storms. One such storm, called the Great Dark Spot, was similar to Jupiter's Great Red Spot, although Neptune's storm disappeared by 1994.

 Get It?

Describe the similarities between Neptune and Uranus.

Moons and rings

Neptune has 14 moons, the largest of which is Triton. Triton has a retrograde orbit, which means that it orbits backward, unlike other large satellites in the solar system. Triton, shown in **Figure 25,** has a thin atmosphere and nitrogen geysers. The geysers are caused by nitrogen gas below Triton's south polar ice, which expands and erupts when heated by the Sun.

Neptune's six rings are composed of microscopic dust particles, which do not reflect light well. Therefore, Neptune's rings are not as visible from Earth as are Saturn's rings.

Check Your Progress

Summary

- The gas giant planets are composed mostly of hydrogen and helium.
- The gas giant planets have ring systems and many moons.
- Some moons of Jupiter and Saturn have water and experience volcanic activity.
- All four gas giant planets have been visited by space probes.

Demonstrate Understanding

1. **Create** a table that lists the gas giant planets and their characteristics.
2. **Compare** the composition of the gas giant planets with that of the Sun.
3. **Compare** Earth's moon with the moons of the gas giant planets.
4. **Explain** how a possible collision with a passing object affected the rotational axis of Uranus.

Explain Your Thinking

5. **Evaluate** Where do you think are the most likely sites on which to find extraterrestrial life? Explain.
6. **WRITING** **Connection** Research and describe one of the *Voyager* missions to interstellar space.

LEARNSMART® Go online to follow your personalized learning path to review, practice, and reinforce your understanding.

NASA-JPL

OTHER SOLAR SYSTEM OBJECTS

FOCUS QUESTION

Why is Pluto no longer classified as a planet?

Dwarf Planets

In the early 2000s, astronomers began to detect large objects in the region of the then-planet Pluto, about 40 AU from the Sun, called the Kuiper belt. Then in 2003, an object now known as Eris was discovered, and it was larger than Pluto. At this time, the scientific community began to take a closer look at the planetary status of Pluto and other solar system objects.

Ceres

In 1801, Giuseppe Piazzi discovered a large object in orbit between Mars and Jupiter. Scientists had predicted that there was a planet somewhere in that region. However, Ceres was extremely small for a planet. Then hundreds of thousands of other objects were discovered in the same region. Therefore, Ceres was no longer thought of as a planet, but as the largest of the asteroids in what would be called the asteroid belt.

Pluto

After its discovery by Clyde Tombaugh in 1930, Pluto was called the ninth planet. But it was an unusual planet. It is not a terrestrial or gas planet; it is made of rock and ice. It does not have a circular orbit; its orbit is long, elliptical, and overlaps the orbit of Neptune. It has five moons that orbit at a widely odd angle from the plane of the ecliptic, and it is smaller than Earth's moon. Pluto is one of many similar objects that exist outside of the orbit of Neptune.

How many others?

Pluto, Eris, and Ceres have been placed into a new classification of objects in space called dwarf planets. The International Astronomical Union (IAU) has defined a **dwarf planet** as an object that, due to its own gravity, is spherical in shape, orbits the Sun, is not a satellite, and has not cleared the area of its orbit of smaller debris. The IAU has limited this classification to Pluto, Eris, Ceres, Makemake, and Haumea. There are at least 10 other objects whose classifications are undecided, some of which are shown in **Figure 26.**

 **3D THINKING**   **DCI** Disciplinary Core Ideas   **CCC** Crosscutting Concepts   **SEP** Science & Engineering Practices

COLLECT EVIDENCE

 Use your Science Journal to record the evidence you collect as you complete the readings and activities in this lesson.

INVESTIGATE

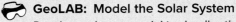 GO ONLINE to find these activities and more resources.

GeoLAB: Model the Solar System
Develop and use a model to visualize the structure of the solar system.

(((•))) **Review the News**
Obtain information from a current news story about current research from the *New Horizons* spacecraft. Evaluate your source and communicate your findings to your class.

Figure 26 Visualizing Other Solar System Objects

Recent findings of objects beyond Pluto have forced scientists to rethink what features define a planet.

(Note: *Buffy (2004 XR190) is a nickname used by its discoverer.*)

Characteristics of Objects Beyond Neptune

Characteristics	Pluto	Sedna	Eris	Haumea	Buffy	Makemake
Average distance from the Sun, AU	40	519	68	43	58	46
Relative size	1	0.67	1.05	0.33	0.25	0.63
Moons	5	?	1	2	?	?
Orbital period, years	248	10,500	557	284	436	305
Orbital tilt, degrees	17	12	44	28	47	29
Orbital eccentricity	0.25	0.85	0.44	0.19	0.11	0.16

NASA

Small Solar System Bodies

Once the IAU defined planets and dwarf planets, they had to identify what was left. In the early 1800s, a name was given to the rocky planetesimals between Mars and Jupiter—the asteroid belt. Objects beyond the orbit of Neptune have been called trans-Neptunian objects (TNOs), Kuiper belt objects (KBOs), comets, and members of the Oort cloud. But what would the collective name for these objects be? The IAU calls them small solar system bodies.

Asteroids

There are hundreds of thousands of asteroids orbiting the Sun between Mars and Jupiter. They are rocky bodies that vary in diameter and have pitted, irregular surfaces. Some asteroids have satellites of their own, such as the asteroid Ida, shown in **Figure 27.** Astronomers estimate that the total mass of all the known asteroids in the solar system is equivalent to only about 0.08 percent of Earth's mass.

As asteroids orbit, they occasionally collide and break into fragments. An asteroid fragment, or any other interplanetary material, is called a **meteoroid.** When a meteoroid passes through Earth's atmosphere, the air around it is heated by friction and compression, producing a streak of light called a **meteor.** If the meteoroid does not burn up completely and part of it strikes the ground, the part that hits the ground is called a **meteorite.** When large meteorites strike Earth, they produce impact craters. Any craters that are still visible on Earth must be young, otherwise they would have been erased by erosion.

Kuiper belt

Like the rocky asteroid belt, another group of small solar system bodies that are made mostly of rock and ice lies outside the orbit of Neptune, in the **Kuiper** (KI pur) **belt.** Most of these bodies probably formed in this region—30 to 50 AU from the Sun—from the material left over from the formation of the Sun and planets. Some, however, might have formed closer to the Sun and been knocked into this area by Jupiter and the other gas giant planets. Eris, Pluto, Pluto's moons, and an ever-growing list of objects have been detected within this band; however, none of them has been identified as a comet. Comets usually come from the farthest limits of the solar system, the Oort cloud, shown in **Figure 28.**

Figure 27 The asteroid Ida and its tiny moon, Dactyl, are shown in this image gathered by the *Galileo* spacecraft.

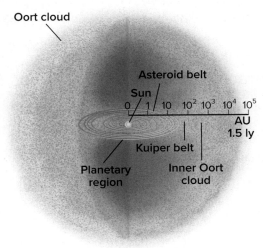

Figure 28 The Kuiper belt appears as the outermost limit of the planetary disk. The Oort cloud surrounds the Sun, echoing its solar sphere.

NASA-JPL

Comets

Comets are small, icy bodies that have highly eccentric orbits around the Sun. Ranging from 1 to 10 km in diameter, most comets orbit in a continuous distribution that extends from the Kuiper belt to 100,000 AU from the Sun. The outermost region, known as the Oort cloud, expands into a sphere surrounding the Sun. Occasionally, a comet is disturbed by the gravity of another object and is thrown into the inner solar system.

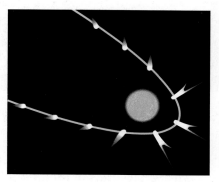

Figure 29 A comet's tail always points away from the Sun and is driven by a stream of particles and radiation.

Comet structure

When a comet comes within 3 AU of the Sun, it begins to evaporate. It forms a head and one or more tails. The head is surrounded by an envelope of glowing gas, and it has a small, solid core. The tails form as gas and dust are pushed away from the comet by particles and radiation from the Sun. This is why comets' tails always point away from the Sun, as illustrated in **Figure 29.**

Periodic comets

Comets that orbit the Sun, and thus repeatedly return to the inner solar system, are known as periodic comets. One example is Halley's comet, which has a 76-year period—it appeared last in 1985 and is expected to appear again in 2061. Each time a periodic comet comes near the Sun, it loses some of its matter, leaving behind a trail of particles. When Earth crosses the trail of a comet, particles left in the trail burn up in Earth's upper atmosphere, producing bright streaks of light called a **meteor shower.** In fact, most meteors are caused by dust particles from comets.

Check Your Progress

Summary

- Dwarf planets, asteroids, and comets formed from the debris of the solar system formation.
- Meteoroids are rocky bodies that travel through the solar system.
- Mostly rock and ice, the Kuiper belt objects are currently being detected and analyzed.
- Periodic comets are in regular, permanent orbit around the Sun, while others might pass this way only once.
- The outermost regions of the solar system house most comets in the Oort cloud.

Demonstrate Understanding

1. **Identify** the kinds of small solar system bodies and their compositions.
2. **Compare** planets and dwarf planets.
3. **Distinguish** among meteors, meteoroids, and meteorites.
4. **Explain** why a comet's tail always points away from the Sun.
5. **Compare** and contrast the asteroid belt and the Kuiper belt.
6. **Explain** how a comet's orbit can occasionally be changed.

Explain Your Thinking

7. **Infer** why comets have highly eccentric orbits.
8. **WRITING ▶ Connection** Suppose you are traveling from the outer reaches of the solar system toward the Sun. Write a scientifically accurate description of the things you see.

LEARNSMART Go online to follow your personalized learning path to review, practice, and reinforce your understanding.

Probing for Astronomical Answers

Over the past century, scientists have gained extensive knowledge about the solar system and what lies beyond. With the invention of new technologies, they continue to gather information that frequently modifies theories about the universe. Now, a space probe is hurtling through the Kuiper belt, and the data it collects may provide even more insights into the mysteries of the universe.

Even with a gravity assist from Jupiter, it took *New Horizons* **nine years to travel the 4.8 billion km (3 billion mi) to Pluto.**

New Horizons

In 2006, the National Aeronautics and Space Administration (NASA) launched *New Horizons*, a space probe with a mission of studying Pluto and the Kuiper belt, a disc-shaped area with thousands of icy bodies and a trillion or more comets. Scientists want to learn more about Pluto's geology, atmospheric composition, and temperatures. They also want to know more about other Kuiper belt objects (KBOs). Scientists hope that exploring the Kuiper belt will yield clues about the origins of our solar system.

Pluto

On its journey, *New Horizons* swung by Jupiter in 2007, using the planet's gravity to increase the probe's velocity and slingshot it farther out into space. This technique, known as a "gravity assist," allowed the probe to reach Pluto six years earlier than projected. In 2015, the probe began a six-month study of Pluto.

During a close-up flyby, the probe used technologies—including types of spectrometers and telescopic cameras—to send back data and stunning photos of the dwarf planet and its moons.

KBO MU69

Scientists put *New Horizons* into hibernation mode while it travels long distances to other KBOs. This allows scientists to complete other tasks, including planning the next flyby before it reaches its target—a KBO called (486958) 2014 MU$_{69}$ that is 1.6 billion km (1 billion mi) past Pluto.

Scientists who have observed 2014 MU$_{69}$ with the Hubble telescope think the KBO is either a binary orbiting pair of objects or two connected objects of roughly the same size. They are eagerly awaiting the 2019 flyby to help them answer this question and many others.

DEVELOP A MODEL TO ILLUSTRATE

Work with a partner to make a 2D or 3D model of *New Horizons'* journey through our solar system. Use print or online sources to find a diagram to aid you as you create the model.

GO ONLINE to study with your Science Notebook.

Lesson 1 FORMATION OF THE SOLAR SYSTEM

- A collapsed interstellar cloud formed the Sun and planets from a rotating disk.
- The inner planets formed closer to the Sun than the outer planets, leaving debris to produce asteroids and comets.
- Copernicus created the heliocentric model, and Kepler defined its shape and mechanics.
- Newton explained the forces governing the solar system bodies and provided proof for Kepler's laws.
- Present-day astronomers divide the solar system into three zones.

- planetesimal
- retrograde motion
- ellipse
- astronomical unit
- eccentricity

Lesson 2 THE INNER PLANETS

- Mercury is heavily cratered and has high cliffs. It has no real atmosphere and the largest day-night temperature difference among the planets.
- Venus has clouds containing sulfuric acid and an atmosphere of carbon dioxide that produces a strong greenhouse effect.
- Earth is the only planet that has all three forms of water on its surface.
- Mars has a thin atmosphere. Surface features include four volcanoes and channels that suggest that liquid water once existed on the surface.

- terrestrial planet
- scarp

Lesson 3 THE OUTER PLANETS

- The gas giant planets are composed mostly of hydrogen and helium.
- The gas giant planets have ring systems and many moons.
- Some moons of Jupiter and Saturn have water and experience volcanic activity.
- All four gas giant planets have been visited by space probes.

- gas giant planet
- liquid metallic hydrogen
- belt
- zone

Lesson 4 OTHER SOLAR SYSTEM OBJECTS

- Dwarf planets, asteroids, and comets formed from the debris of the solar system formation.
- Meteoroids are rocky bodies that travel through the solar system.
- Mostly rock and ice, the Kuiper belt objects are currently being detected and analyzed.
- Periodic comets are in regular, permanent orbit around the Sun, while others might pass this way only once.
- The outermost regions of the solar system house most comets in the Oort cloud.

- dwarf planet
- meteoroid
- meteor
- meteorite
- Kuiper belt
- comet
- meteor shower

Module Wrap-Up

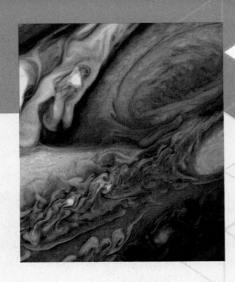

REVISIT THE PHENOMENON

Why do we have to use satellites to take detailed pictures of Jupiter and the other planets?

CER Claim, Evidence, Reasoning

Explain your Reasoning Revisit the claim you made when you encountered the phenomenon. Summarize the evidence you gathered from your investigations and research and finalize your Summary Table. Does your evidence support your claim? If not, revise your claim. Explain why your evidence supports your claim.

STEM UNIT PROJECT
Now that you've completed the module, revisit your STEM unit project. You will summarize your evidence and apply it to the project.

GO FURTHER

SEP Data Analysis Lab

How well do the orbits of the planets conform to Kepler's third law?

For the six planets closest to the Sun, Kepler observed that $P^2 = a^3$, where P is the orbital period in years and a is the semimajor axis in AU.

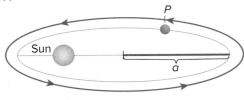

CER Analyze and Interpret Data

1. Use this typical planet orbit diagram and data from **Table 1** and the *Reference Handbook* to confirm the relationship between P_2 and a_3 for each of the planets.
2. Prepare a table showing your results and how much they deviate from predicted values.
3. **Claim** Determine which planets conform most closely to Kepler's law and which do not seem to follow it.
4. **Evidence and Reasoning** Would Kepler have formulated this law if he had been able to study Uranus and Neptune? Explain.
5. **Claim** Predict the orbital period of an asteroid orbiting the Sun at 2.5 AU.
6. **Evidence and Reasoning** Find the semimajor axis of Halley's comet, which has an orbital period of 76 years.

picturist/iStock/Getty Images

ENCOUNTER THE PHENOMENON

How do telescopes tell us what elements are in stars?

GO ONLINE to play a video about how stars fuse smaller elements into larger ones.

SEP Ask Questions

Do you have other questions about the phenomenon? If so, add them to the driving question board.

CER Claim, Evidence, Reasoning

Make Your Claim Use your CER chart to make a claim about how telescopes tell us what elements are in stars. Explain your reasoning.

Collect Evidence Use the lessons in this module to collect evidence to support your claim. Record your evidence as you move through the module.

Explain Your Reasoning You will revisit your claim and explain your reasoning at the end of the module.

GO ONLINE to access your CER chart and explore resources that can help you collect evidence.

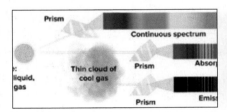

LESSON 1: Explore and Explain: Solar Spectra

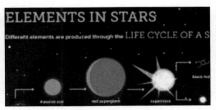

LESSON 3: Explore and Explain: Summarizing Stellar Life Cycles

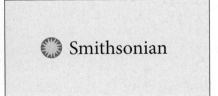

Additional Resources

FOCUS QUESTION

What is inside the Sun?

Properties of the Sun

The Sun is the largest object in the solar system, in both diameter and mass. It would take 109 Earths, or almost 10 Jupiters, lined up edge to edge to fit across the Sun. The Sun is about 330,000 times the mass of Earth and 1048 times the mass of Jupiter. In fact, the Sun contains more than 99 percent of all the mass in the solar system. It should not be surprising, then, that the Sun's mass affects the motions of the planets and other objects.

The Sun's average density is similar to the densities of the gas giant planets, represented by Jupiter in **Table 1.** Astronomers deduce densities at specific points inside the Sun, as well as other information, by using computer models that explain the observations they make. These models show that the density in the center of the Sun is about 1.50×10^5 kg/m^3, which is about 7.5 times the density of gold. A pair of dice as dense as the Sun's center would have a mass of about 1 kg.

Like many other stars, the Sun's interior is gaseous throughout because of its high temperature—about 1×10^7 K in the center. At this temperature, all of the gases are completely ionized, meaning the interior is composed only of atomic nuclei and electrons. This state of matter is known as plasma. Though partially ionized, the outer layers of the Sun are not hot enough to be plasma. The Sun produces the equivalent of 4×10^{24} 100-W lightbulbs of light each second. Only 1 billionth of this amount actually reaches Earth.

Table 1 Relative Properties of the Sun

	Sun	Earth	Jupiter
Diameter (km)	1.4×10^6	1.3×10^4	1.4×10^5
Mass (kg)	2.0×10^{30}	6.0×10^{24}	1.9×10^{27}
Density (kg/m³)	1.4×10^3	5.5×10^3	1.3×10^3

 **3D THINKING**   **DCI** Disciplinary Core Ideas   **CCC** Crosscutting Concepts   **SEP** Science & Engineering Practices

COLLECT EVIDENCE

 Use your Science Journal to record the evidence you collect as you complete the readings and activities in this lesson.

INVESTIGATE

 GO ONLINE to find these activities and more resources.

⚙ **Applying Practices: Element Production in Stars**
HS-ESS1-3. Communicate scientific ideas about the way stars, over their life cycle, produce elements.

👓 **GeoLAB: Identify Stellar Spectral Lines**
Evaluate information to determine the composition of a star using spectral lines.

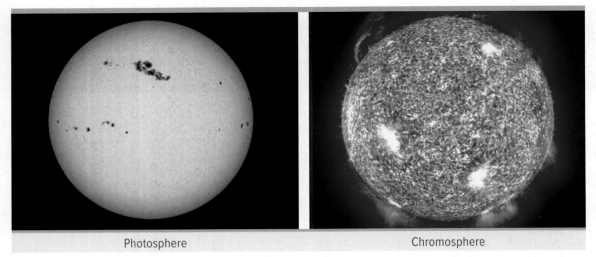

Photosphere

Chromosphere

Figure 1 Sunspots appear dark on the photosphere, the visible surface of the Sun. The white-hot areas are almost 6000 K while the darker, red areas are closer to 3000 K. The chromosphere of the Sun appears red.

Explain *why the innermost layer of the Sun's atmosphere is visible.*

The Sun's Atmosphere

You might ask how the Sun can have an atmosphere when it is already gaseous. Like many stars, the outer regions of the Sun are organized into layers. Each layer emits energy at wavelengths resulting from its temperature.

Photosphere

The **photosphere,** shown in **Figure 1,** is the visible surface of the Sun. It is approximately 400 km thick and has an effective temperature of 5800 K. It is also the innermost layer of the Sun's atmosphere. You might wonder how it is the visible surface of the Sun if it is the innermost layer. This is because most of the visible light emitted by the Sun comes from this layer. The two outermost layers are transparent at most wavelengths of visible light.

Chromosphere

The next layer is the **chromosphere,** which is approximately 2500 km thick and has an average temperature of 15,000 K. The chromosphere is visible only during a solar eclipse, when the photosphere is blocked. However, astronomers can use special filters to observe the chromosphere when the Sun is not eclipsed. The chromosphere appears red, as shown in **Figure 1,** because its strongest emissions are in a single band in the red wavelength.

Corona

The outermost layer of the Sun's atmosphere is the **corona.** It extends several million kilometers from the outside edge of the chromosphere and usually has a temperature of about 3 to 5 million K. The density of the gas in the corona is very low, which explains why the corona is so dim that it can be seen only during a solar eclipse, as shown in **Figure 2,** or by using specially equipped instruments. The temperature is so high in these outer layers of the solar atmosphere that most of the radiation emitted is of ultraviolet wavelengths for the chromosphere and X-rays for the corona.

Figure 2 The Sun's hottest and outermost layer, the corona, can be seen only seen during a solar eclipse.

Aurora from Earth

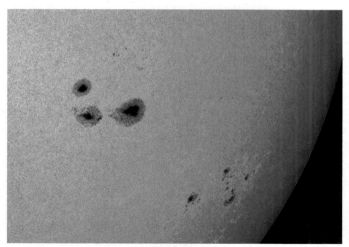

Aurora from space

Figure 3 The aurora is the result of particles from the Sun colliding with gases in Earth's atmosphere. It is best viewed from high-latitude regions.

Infer *When can you see the aurora?*

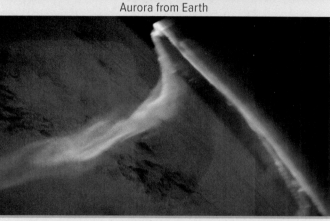

Figure 4 Sunspots are dark, relatively cool spots on the surface of the photosphere. These dark areas are associated with the Sun's magnetic field. Sunspots typically last several days, but can last for many months.

Solar wind

The corona of the Sun does not have an abrupt edge. Instead, plasma flows outward from the corona at high speeds and forms the **solar wind.** As this wind of charged particles, called ions, flows outward through the entire solar system, it bathes each planet in a flood of particles. The solar wind is not uniform. Streams of 300 km/s and 800 km/s alternately pass by Earth as the Sun rotates. The charged particles are deflected by Earth's magnetic field and are trapped in two huge rings, called the Van Allen belts. The high-energy particles in these belts collide with gases in Earth's atmosphere and cause the gases to give off light. This light, called the aurora, can be seen from Earth or from space, as shown in **Figure 3.** Aurorae are generally seen from Earth in the polar regions.

Solar Activity

While the solar wind and layers of the Sun's atmosphere are permanent features, other features on stars change over time in a process called solar activity. Some of the Sun's activity includes fountains and loops of glowing gas. Some of this gas has structure—a certain order in both time and place. This structure is driven by magnetic fields.

The Sun's magnetic field and sunspots

The Sun's magnetic field disturbs the solar atmosphere periodically and causes new features to appear. The most obvious features are **sunspots,** shown in **Figure 4,** which are dark spots on the surface of the photosphere. Sunspots are bright, but they appear darker than the surrounding areas on the Sun because they are cooler. They are located in regions where the Sun's intense magnetic fields penetrate the photosphere. Magnetic fields create pressure that counteracts the pressure from the hot, surrounding gas. This stabilizes the sunspots despite their lower temperature. Sunspots occur in pairs with opposite magnetic polarities—a north pole and a south pole, similar to a magnet.

 Get It?

Explain What makes up the solar wind?

Solar activity cycle

Astronomers have observed that the number of sunspots changes in a predictable pattern from minimum to maximum to minimum again. This is called the sunspot cycle, which takes about 11 years to complete. Then, the Sun's magnetic field reverses, so that the north magnetic pole becomes the south magnetic pole, and vice versa. Because sunspots are caused by magnetic fields, the polarities of sunspot pairs reverse when the Sun's magnetic poles reverse. Therefore, when the polarity of the Sun's magnetic field is taken into account, the length of the cycle is approximately 22 years. At this point, the magnetic field then switches back to the original polarity, and the solar activity cycle starts again.

Other solar features

Coronal holes, shown in **Figure 5,** are detectable only in X-ray photography. They are often located over sunspot groups. Coronal holes are areas of low density in the gas of the corona and are the main regions from which the particles that comprise the solar wind escape.

Solar flares, shown in **Figure 5,** are also associated with sunspots. **Solar flares** are violent eruptions of particles and radiation from the Sun's surface. Often, these particles escape the surface in the solar wind, and Earth gets bombarded with the particles a few days later. The largest recorded solar flare, which occurred in November 2003, hurled solar particles at nearly 9 million km/h.

Another active feature, sometimes associated with flares, is a **prominence,** shown in **Figure 5.** It is an arc of gas that is ejected from the chromosphere or is gas that condenses in the inner corona and rains back to the surface. A prominence can reach temperatures greater than 50,000 K and can last from a few hours to a few months. Like flares, prominences are also associated with sunspots and the magnetic field, and occurrences of both vary with the solar-activity cycle.

A coronal mass ejection (CME) is a huge discharge of highly charged particles from the corona. If these particles get caught up in Earth's magnetic field, they can form a geomagnetic storm. The effects are both positive and negative. Unusually strong auroras occur and can be visible at much lower latitudes than usual. But CMEs can interrupt radio communications and electrical power delivery and can even damage electronic components of satellites. On March 13, 1989, a CME struck Earth's magnetosphere. The resulting geomagnetic storm created auroras visible as far south as Florida. The geomagnetism interfered with radio communication across the globe and cause a 9-hour power outage in Quebec, Canada, by inducing a current in the powerlines. Since this mass ejection event, governments around the world have begun preparing themselves for when Earth is hit by the next CME.

 Get It?

Describe How might a CME affect your daily life?

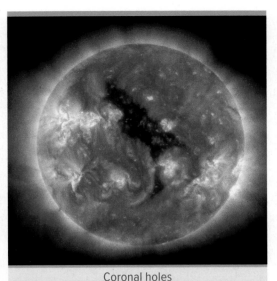
Coronal holes

Solar flares

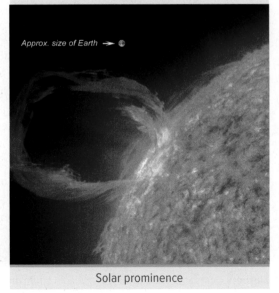
Solar prominence

Approx. size of Earth ➜

Figure 5 Features of the Sun's surface include coronal holes into the surface and solar flares and prominences that erupt from the surface.

(t)NASA/SDO/AIA, (c)NASA/GSFC/Solar Dynamics Observatory, (b)NASA/GSFC/SDO

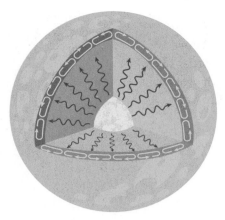

Figure 6 Energy in the Sun is transferred mostly by radiation from the core outward to about 75 percent of its radius. The outer layers transfer energy in convection currents.

The Solar Interior

You might be wondering where all the energy that causes solar activity and light comes from. Fusion occurs in the core of the Sun, where the pressure and temperature are extremely high. **Fusion** is the combination of lightweight, atomic nuclei into heavier nuclei, such as hydrogen fusing into helium. This is the opposite of the process of **fission,** which is the splitting of heavy atomic nuclei into smaller, lighter nuclei, like uranium into lead.

Energy production in the Sun

In the Sun's core, helium is a product of the process in which hydrogen nuclei fuse. The mass of the helium nucleus is less than the combined mass of the four hydrogen nuclei, which means that mass is lost during the process.

 Get It?

Describe What two elements does the Sun use to create energy? During what process does this occur?

Albert Einstein's special theory of relativity shows that mass and energy are equivalent, and that matter can be converted into energy, and vice versa. This relationship can be expressed as $E = mc^2$, where E is energy measured in joules, m is the quantity of mass that is converted to energy measured in kilograms, and c is the speed of light measured in m/s. This theory explains that the mass lost in the fusion of hydrogen to helium is converted to energy, which powers the Sun. At the Sun's rate of hydrogen fusing, it is about halfway through its lifetime, with approximately 5 billion years left. Even so, the Sun has used only about 3 percent of its hydrogen.

Energy transport

If the Sun's energy is produced in its core, how does the energy get to the surface before it travels to Earth? The answer lies in the two zones in the solar interior illustrated in **Figure 6.** In the inner portion of the Sun, energy is transferred by radiation. This is the radiation zone. Above that, in the convection zone, energy is transferred by gaseous convection currents. As energy moves outward, the temperature is reduced from a central value of about 1×10^7 K to about 5800 K at the photosphere. Leaving the Sun's outermost layer, energy moves in a variety of wavelengths in all directions. A tiny fraction of that immense amount of solar energy eventually reaches Earth.

ACADEMIC VOCABULARY

convection
the circular, up-and-down motion that occurs when less-dense, warmer fluid rises while cooler, denser fluid sinks
Convection currents enable solar energy to move outward from the Sun's interior.

Solar energy on Earth

The quantity of energy that arrives on Earth every day from the Sun is enormous. Above Earth's atmosphere, 1354 J of energy is received in 1 m²/s (1354 W/m²). In other words, 13 100-W lightbulbs could be operated with the solar energy that strikes a 1-m² area. However, not all of this energy reaches the ground because some is absorbed and scattered by the atmosphere.

Spectra

You are probably familiar with the rainbow that appears when white light is shined through a prism. This rainbow is a spectrum (plural, *spectra*), which is visible light arranged according to wavelengths. There are three types of spectra: continuous, emission, and absorption, as shown in **Figure 7.**

A spectrum that has no breaks in it, such as the one produced when light from an ordinary bulb is shined through a prism, is called a continuous spectrum. A continuous spectrum can also be produced by a glowing solid or liquid or by a highly compressed, glowing gas. The spectrum from a noncompressed gas contains bright lines at certain wavelengths. This is called an emission spectrum, and the lines are called emission lines. The wavelengths of the visible lines depend on the element being observed because each element has its own characteristic emission spectrum.

 Get It?

Describe continuous and emission spectra.

A spectrum produced from the Sun's light shows a series of dark bands. These dark spectral lines are caused by different chemical elements that absorb light at specific wavelengths. This is called an absorption spectrum, and the lines are called absorption lines. Absorption is caused by a cooler gas in front of a source that emits a continuous spectrum. The pattern of the dark absorption lines of an element is exactly the same as the bright emission lines for that same element. Thus, by comparing laboratory spectra of different gases with the dark lines in the solar spectrum, it is possible to identify the elements that make up the Sun's outer layers.

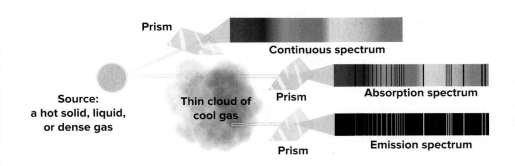

Figure 7 Energy excites the elements of a substance so that the substance emits different wavelengths of light.

Infer *what the colors of a spectrum represent.*

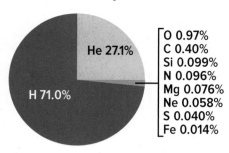

Element Composition of the Sun by Mass

He 27.1%

H 71.0%

O 0.97%
C 0.40%
Si 0.099%
N 0.096%
Mg 0.076%
Ne 0.058%
S 0.040%
Fe 0.014%

Figure 8 The Sun is composed primarily of hydrogen and helium, with small amounts of other gases.

Solar Composition

Although scientists have not been able to take samples from the Sun directly, they have learned a great deal about the Sun's composition from its spectra. Using the lines of the absorption spectra like fingerprints, astronomers have identified more than 60 elements that compose the Sun. The Sun consists primarily of hydrogen (H), at about 71.0 percent by mass; helium (He), at 27.1 percent; and a small amount of other elements, as illustrated in **Figure 8.**

This composition is similar to that of the gas giant planets. It suggests that the Sun and the gas giants represent the composition of the interstellar cloud from which the solar system formed. While the terrestrial planets have lost most of the lightweight gases, their heavier element composition probably came from a contribution to the interstellar cloud of by-products from long-extinct stars.

The Sun's composition represents that of the galaxy as a whole. The proportions of elements in most stars are similar to those in the Sun. Hydrogen and helium are the predominant gases in stars and in the rest of the universe. Even dying stars still have hydrogen and helium in their outer layers because their internal temperatures might fuse only about 10 percent of their total hydrogen into helium. All other elements are in small proportions compared to hydrogen and helium. The larger the star's mass at its inception, the more heavy elements it will produce in its lifetime. But, as you will read in this module, there are different stages and results of a star's death. As stars die, they return as much as 50 percent of their mass to interstellar space to be recycled into new generations of stars and planets.

Check Your Progress

Summary

- Most of the mass in the solar system is found in the Sun.
- The Sun's average density is approximately equal to that of the gas giant planets.
- The Sun has a layered atmosphere.
- The Sun's magnetic field causes sunspots and other solar activity.
- The fusion of hydrogen into helium provides the Sun's energy and composition.

Demonstrate Understanding

1. **Identify** which features of the Sun are typical of stars.
2. **Describe** the three outer layers of the Sun's atmosphere.
3. **Classify** the different types of spectra by how they form.
4. **Describe** the process of fusion in the Sun.
5. **Compare** the composition of the Sun and the gas giant planets.

Explain Your Thinking

6. **Infer** how the Sun would affect Earth if Earth did not have a magnetic field.
7. **Relate** the solar activity cycle with solar flares and prominences.
8. **WRITING ▸ Connection** Design a brochure explaining the physical properties of the Sun and its dynamic nature.

LEARNSMART Go online to follow your personalized learning path to review, practice, and reinforce your understanding.

MEASURING THE STARS

FOCUS QUESTION

How do we know anything about stars that are hundreds of light-years away?

Patterns of Stars

Long ago, many civilizations looked at the brightest stars and named groups of them after animals, mythological characters, or everyday objects. These groups of stars are called **constellations.** Today, astronomers group stars by the 88 constellations named by ancient peoples. Some constellations are visible throughout the year, depending on the observer's location. In the northern hemisphere, you can see constellations that appear to rotate around the North Pole. These constellations are called circumpolar constellations. Ursa Major, which contains the Big Dipper, is a circumpolar constellation for most of the northern hemisphere.

Unlike circumpolar constellations, the other constellations can be seen only at certain times of the year because of Earth's changing position in its orbit around the Sun, as illustrated in **Figure 9.** For example, the constellation Orion can be seen in the northern hemisphere's winter, and the constellation Hercules can be seen in the northern hemisphere's summer. For this reason, constellations are classified as summer, fall, winter, and spring constellations. Ancient people relied on the constellations to tell them when to prepare for planting, harvest, and ritual celebrations. The most familiar constellations are the ones that are part of the zodiac.

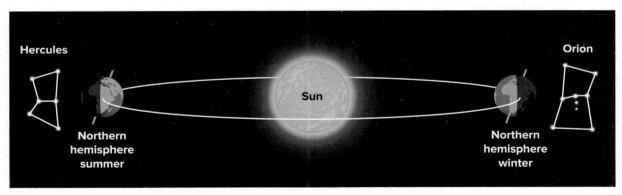

Figure 9 Different constellations are visible in the sky due to Earth's movement around the Sun.

 3D THINKING   **DCI** Disciplinary Core Ideas   **CCC** Crosscutting Concepts   **SEP** Science & Engineering Practices

COLLECT EVIDENCE

 Use your Science Journal to record the evidence you collect as you complete the readings and activities in this lesson.

INVESTIGATE

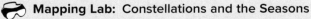 **GO ONLINE** to find these activities and more resources.

Mapping Lab: Constellations and the Seasons
Use a model to visualize the seasonal patterns of stars and constellations from your location on Earth.

Quick Investigation: Model Parallax
Use a model to visualize how parallax it used to determine the structure of the Universe from Earth's reference frame.

Pleiades   M13

Figure 10 Star clusters are groups of stars that are gravitationally bound to one another. The Pleiades is an open cluster group, and M13 is a globular cluster.

Star clusters Although the stars in constellations appear to be close to each other, few are gravitationally bound to one other. The reason that they appear to be close together is that human eyes cannot distinguish how far or near stars are. Two stars could appear to be located next to each other in the sky, but one might be 100 trillion km from Earth, and the other might be 200 trillion km from Earth. However, by measuring distances to stars and observing how their gravities interact with each other, scientists can determine which stars are gravitationally bound to each other. A group of stars that are gravitationally bound to each other is called a cluster. The Pleiades (PLEE uh deez) in the constellation Taurus, shown in **Figure 10,** is an open cluster because the stars are not densely packed. In contrast, a globular cluster is a group of stars that are densely packed into a spherical shape, such as M13 in the constellation Hercules, also shown in **Figure 10.** Different kinds of clusters are explained in **Figure 12** on the next page.

Binaries When only two stars are gravitationally bound together and orbit a common center of mass, they are called **binary stars.** More than half of the stars in the sky are either binary stars or members of multiple-star systems. The bright star Sirius is half of a binary system, shown in **Figure 11.** Most binary stars appear to be single stars to the human eye, even with the aid of a telescope. The two stars are usually too close together to appear separately, and one of the two is often much brighter than the other.

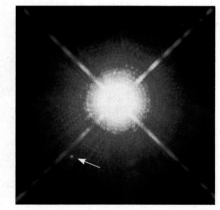

Figure 11 Sirius and its companion star, seen below and to the left, are the simplest form of stellar grouping, known as a binary.

Astronomers are able to identify binary stars through the use of several methods. For example, even if only one star is visible, accurate measurements can show that its position shifts back and forth as it orbits the center of mass between it and the unseen companion star. Also, the orbital plane of a binary system can sometimes be seen edgeways from Earth. In such cases, the two stars alternately block each other and cause the total brightness of the two-star system to dip each time one star passes in front of the other. This type of binary star is called an eclipsing binary.

Get It?

Describe the movements of an eclipsing binary star system.

(tl) T.A. Rector (University of Alaska Anchorage), Richard Cool (University of Arizona) and WIYN, (tr)NASA, The Hubble Heritage Team, STScI, AURA, (b)NASA, ESA, H. Bond (STScI), and M. Barstow (University of Leicester)

Figure 12 Visualizing Star Groupings

When you look into the night sky, the stars seem to be randomly spaced from horizon to horizon. Upon closer inspection, you begin to see groups of stars that seem to cluster in one area. Star clusters are gravitationally bound groups of stars, which means that their gravities interact to hold the stars in a group. The Milky Way galaxy, a portion of which is shown in the center photo, contains all of the groupings described below.

Galaxy Not a true cluster, a galaxy is a very large star grouping that contains a variety of different clusters of stars. This is the Andromeda Galaxy.

Globular clusters are made from densely packed groups of stars that are the same age. Their gravities hold them into a rounded cluster. Many globular clusters are found in the halos of galaxies.

Open clusters are loosely organized groups of stars that are not densely packed. This open cluster in the constellation Cygnus is young and contains a mixture of stellar types, from stars dimmer than the Sun to giants and supergiants.

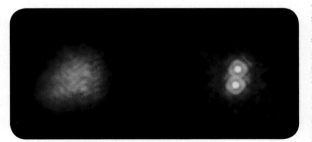

Binaries, the smallest of all star groupings, consist of only two stars orbiting around a single center of gravity. Adaptive optics technology refined the view of the IW Tau binary system from the fuzzy view on the left to the clear image of two distinct stars on the right.

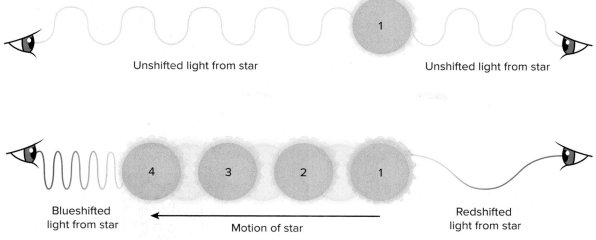

Unshifted light from star

Unshifted light from star

Blueshifted light from star

Motion of star

Redshifted light from star

Figure 13 When a star moves toward the observer, the light emitted by the star shifts toward the blue end of the electromagnetic spectrum. When a star moves away from the observer, its light shifts toward the red. Scientists use Doppler shift to determine the speed and direction of a star's motion.

Explain *how the Doppler shift causes color changes.*

Doppler shifts

The most common way to tell that a star is one of a binary pair is to find subtle wavelength shifts, called Doppler shifts. As the star moves back and forth along the line of sight, as shown in **Figure 13,** its spectral lines shift. If a star is moving toward the observer, the spectral lines are shifted toward shorter wavelengths, which is called a blueshift. However, if the star is moving away, the wavelengths become longer, which is called a redshift. The higher the speed, the greater the shift, thus careful measurements of spectral line wavelengths can be used to determine the speed of a star's motion. Because there is no Doppler shift for motion that is at a right angle to the line of sight, astronomers can learn only about the portion of a star's motion that is directed toward or away from Earth. The Doppler shift in spectral lines can be used to detect binary stars as they move about their center of mass toward and away from Earth with each revolution. It is also important to note that there is no way to distinguish whether the star, the observer, or both are moving. A star undergoing periodic Doppler shifts can be interpreted only as one of a binary. Stars identified in this way are called spectroscopic binaries. Binaries can reveal much about the individual properties of stars.

Stellar Positions and Distances

Astronomers use two units of measure for long distances. One, which you are probably familiar with, is a light-year (ly). A light-year is the distance that light travels in one year, equal to 9.461×10^{12} km. Astronomers often use a unit larger than a light-year—a parsec. A **parsec** (pc) is equal to 3.26 ly, or 3.086×10^{13} km.

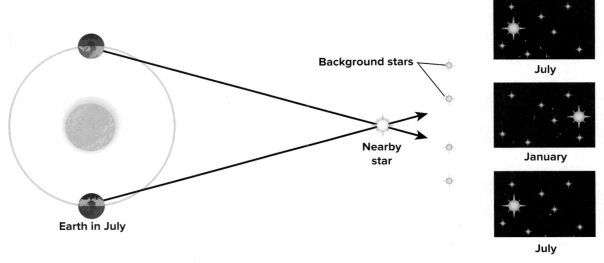

Figure 14 As Earth orbits the Sun, nearby stars appear to change position in the sky compared to faraway stars. Earth reaches its maximum change in position at six months, so the angle measured to the star from these two positions is also at the maximum. This shift in observation position is called parallax and can be used to estimate the distance to the star being observed.

Predict *the position of the star in September.*

Parallax

Precise position measurements are important for determining distances to stars. When determining the distance of stars from Earth, astronomers must account for the fact that nearby stars shift in position as observed from Earth. This apparent shift in position caused by the motion of the observer is called **parallax.** In this case, the motion of the observer is the change in position of Earth as it orbits the Sun. As Earth moves from one side of its orbit to the opposite side, a nearby star appears to be shifting back and forth, as illustrated in **Figure 14.** The closer the star, the larger the shift. The distance to a star can be estimated from its parallax shift by measuring the angle of the change. Using the parallax technique, astronomers could find accurate distances to stars up to only 50 ly, or approximately 15 pc, until recently. With advancements in technology, such as the *Hipparcos* satellite, astronomers can now find accurate distances up to 100 pc by using parallax.

Basic Properties of Stars

The basic properties of a star are mass, diameter, and luminosity, which are all related to each other. Temperature is another property and is estimated by finding the spectral type of a star. Temperature controls the nuclear reaction rate and governs the luminosity, or absolute magnitude. The absolute magnitude compared to the apparent magnitude can be used to find the distance to a star.

 Get It?

Identify the direction of motion of the observer in **Figure 14.**

ACADEMIC VOCABULARY
precise
exactly or sharply defined or stated
The builder's precise measurements ensured that all of the boards were cut to the same length.

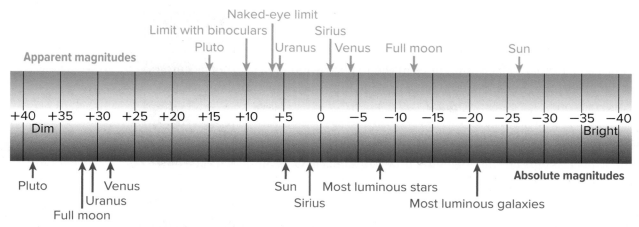

Figure 15 Apparent magnitude is how bright the stars and planets appear in the sky from Earth. Absolute magnitude takes into account the distance to that star or planet and makes adjustments.

Magnitude One of the most basic observable properties of a star is how bright it appears, or its **apparent magnitude.** The ancient Greeks established a classification system based on the brightness of stars. The brightest stars were given a ranking of +1, the next brightest +2, and so on. Today's astronomers still use this system, but they have refined it. In this system, a difference of 5 magnitudes corresponds to a factor of 100 in brightness. Thus, a magnitude +1 star is 100 times brighter than a magnitude +6 star.

Absolute magnitude Apparent magnitude does not indicate the actual brightness of a star because it does not account for distance. A faint star can appear to be very bright because it is relatively close to Earth, while a bright star can appear to be faint because it is far away. To account for this phenomenon, astronomers have developed another classification system for brightness. **Absolute magnitude** is how bright a star would appear if it were placed at a distance of 10 pc. The classification of stars by absolute magnitude allows comparisons that are based on how bright the stars would appear at equal distances from an observer. The disadvantage of absolute magnitude is that it can be difficult to determine unless the actual distance to a star is known. The apparent and absolute magnitudes for several objects are shown in **Figure 15.**

Luminosity Apparent magnitudes do not give an actual measure of energy output. To measure the energy output from the surface of a star per second, called its power or **luminosity,** an astronomer must know both the star's apparent magnitude and how far away it is. The brightness observed depends on both a star's luminosity and distance from Earth, and because brightness diminishes with the square of the distance, a correction must be made for distance. Luminosity is measured in units of energy emitted per second, or watts. The Sun's luminosity is about 3.85×10^{26} W. This is equivalent to 3.85×10^{24} 100-W light-bulbs. The values for other stars vary widely, from about 0.0001 to more than 1 million times the Sun's luminosity. No other stellar property varies as much.

Classification of Stars

You have learned that the Sun has dark absorption lines at specific wavelengths in its spectrum. Other stars also have dark absorption lines in their spectra and are classified according to their patterns of absorption lines. Spectral lines provide information about a star's temperature and composition.

Temperature

Stars are assigned spectral types in the following order: O, B, A, F, G, K, and M. Each class is subdivided into more specific divisions with numbers from 0 to 9. For example, a star can be classified as being a type A4 or A5.

The classes were originally based on only the pattern of spectral lines, but astronomers later discovered that the classes also correspond to stellar temperatures, with O stars being the hottest and M stars being the coolest. Thus, by examination of a star's spectrum, it is possible to estimate its temperature.

Surface temperatures range from about 50,000 K for the hottest O stars to as low as 2000 K for the coolest M stars. The Sun is a type G2 star, which corresponds to a surface temperature of about 5800 K. **Figure 16** shows how spectra from some different star classes appear.

Temperature is also related to luminosity and absolute magnitude. Hotter stars put out more light than do stars with lower temperatures. In most normal stars, the temperature corresponds to the luminosity. Distance can be determined by calculating a star's luminosity based on its temperature.

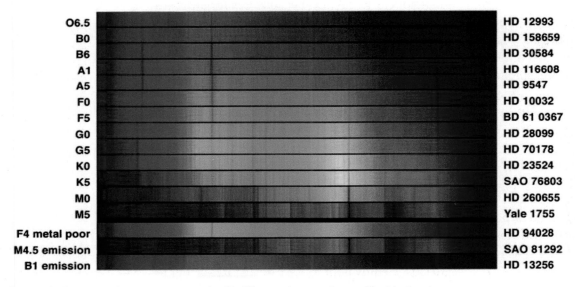

Figure 16 These are absorption spectra for 13 different classes of stars. The black stripes are absorption lines telling us each star's element composition.

KPNO 0.9-m Telescope, AURA, NOAO, NSF

CCC CROSSCUTTING CONCEPTS

Patterns Classification of stars is based on two types of data: temperature and spectra. With **Figure 16** as your guide, use the information in this lesson to list the types of patterns a scientist uses to classify a star. Think about the usefulness of a figure such as this. Do you have any suggestions for improvement? What would you change to make classifications more precise?

Table 2 Relationships of Spectral Types of Stars

Color of Star	Spectral Type	H-R diagram
	O5	
	B5	
	F5	
	G5	
	M5	

Composition

All stars, including the Sun, have nearly identical compositions, despite the differences in their spectra. The differences in the appearance of their spectra are almost entirely a result of temperature differences, shown in **Table 2** on the previous page. Hotter stars have fairly simple visible spectra, while cooler stars have spectra with more lines. The coolest stars have bands in their spectra due to molecules such as titanium oxide in their atmospheres. Typically, about 73 percent of a star's mass is hydrogen (H), about 25 percent is helium (He), and the remaining 2 percent is composed of all the other elements. While there are some variations in the composition of stars, particularly in the final 2 percent, all stars have this general composition.

H-R diagrams

The properties of mass, luminosity, temperature, and diameter are closely related. Each class of star has a specific mass, luminosity, temperature, and diameter. These relationships can be demonstrated on a graph called the **Hertzsprung-Russell diagram** (H-R diagram), on which absolute magnitude is plotted on the vertical axis and temperature or spectral type is plotted on the horizontal axis, as shown in **Table 2.** Spectroscopists first plotted this graph in the early twentieth century. An H-R diagram with luminosity plotted on the vertical axis looks similar to the one in **Table 2** and is used to calculate the evolution of stars.

Get It?

Analyze Antares is an M1 red supergiant with a radius of more than 3 AUs. It makes up the heart of the constellation Scorpius. Look at **Table 2** and determine Antares' approximate temperature.

Most stars occupy the region in the diagram called the **main sequence,** which runs diagonally from the upper-left corner, where hot, luminous stars are represented, to the lower-right corner, where cool, dim stars are represented. **Table 3** shows some properties of main-sequence stars.

Table 3 Properties of Main-Sequence Stars

Spectral Type	Mass*	Surface Temperature (K)	Luminosity*	Radius*
O5	40.0	40,000	5×10^5	18.0
B5	6.5	15,500	800	3.8
A5	2.1	8500	20	1.7
F5	1.3	6580	2.5	1.2
G5	0.9	5520	0.8	0.9
K5	0.7	4130	0.2	0.7
M5	0.2	2800	0.008	0.3

STEM CAREER Connection

Telescope Software Engineer Do you love looking at the stars through telescopes? It is easy to stand next to a telescope and adjust your field of view. But what about the telescopes in orbit around Earth? Who is adjusting them? Scientists with strong math and physics backgrounds write the software that schedules, controls, and processes data from telescopes such as Hubble and, soon, the James Webb Space Telescope. A bachelor's degree in computer science, physics, math, or systems engineering is your pathway to this career.

Main sequence About 90 percent of stars, including the Sun, fall along a broad strip of the H-R diagram called the main sequence. While stars are in the main sequence, they are fusing hydrogen in their cores. As stars evolve off the main sequence, they begin to fuse helium in their cores and burn hydrogen around the core edges.

The Sun, with its average temperature and luminosity, lies near the center of the main sequence. A star's mass determines almost all its other properties, including its lifetime on the main sequence. The more massive a star is, the higher its central temperature and the more rapidly it burns its hydrogen fuel. This is due primarily to the ratio of radiation pressure to gravitational pressure. Higher pressures cause the fuels to burn faster. The star runs out of hydrogen more rapidly, and thus evolves off the main sequence faster than a lower-mass star.

Red giants and white dwarfs The stars plotted at the upper right of the H-R diagram in **Table 2** are cool, yet luminous. Because cool surfaces emit much less radiation per square meter than hot surfaces do, these cool stars must have large surface areas to be so bright. For this reason, these larger, cool, luminous stars are called red giants. In some cases, red giants are more than 100 times the size of the Sun! The largest of these are called red supergiants. Conversely, the dim, hot stars plotted in the lower-left corner of the H-R diagram must be small, or they would be more luminous. These small, dim, hot stars are called white dwarfs. A white dwarf is about the size of Earth but has a mass about as large as the Sun's. You will learn how different stars are formed in Lesson 3.

✏️ Check Your Progress

Summary

- Most stars exist in clusters held together by their gravity.
- The simplest cluster is a binary.
- Parallax is used to measure distances to stars.
- The brightness of stars is related to their temperature.
- Stars are classified by their spectra.
- The H-R diagram relates the basic properties of stars: class, temperature, and luminosity.

Demonstrate Understanding

1. **Relate** stellar temperature to the classification of a star.
2. **Explain** the difference between apparent and absolute magnitude.
3. **Explain** how parallax is used to measure the distance to stars.
4. **Compare and contrast** luminosity and magnitude.
5. **Contrast** the apparent magnitude and the absolute magnitude of a star.
6. **Compare** a light-year and a parsec.

Explain Your Thinking

7. **Design** a model to explain parallax.
8. **Explain** the relationship between radius and mass using **Table 3.**
9. 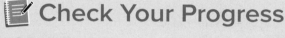 **Connection** Compare Regulus (B class), the brightest star in Leo, to Barnard's Star (M class), one of the closest stars to the Sun, using **Table 3** as a reference.

LEARNSMART Go online to follow your personalized learning path to review, practice, and reinforce your understanding.

Lesson 2 • Measuring the Stars   **651**

FOCUS QUESTION

What happens to the elements in a star when it dies?

Basic Structure of Stars

Mass governs nearly all of a star's properties, including temperature, luminosity, and diameter.

Mass effects

The more massive a star is, the greater the gravity pressing inward, and the hotter and more dense the star must be inside to balance this gravity. The temperature inside a star governs the rate of nuclear reactions, which in turn determines the star's energy output—its luminosity. The balance between gravity squeezing inward and pressure expanding outward is maintained by heat due to nuclear reactions and compression. This balance is called hydrostatic equilibrium, and it must hold for any stable star, as illustrated in **Figure 17,** otherwise the star would expand or contract.

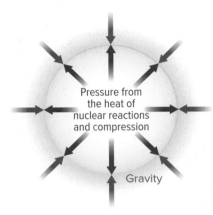

Figure 17 When the pressure from radiation and fusion is balanced by gravity, a star is stable.

Fusion

Inside a star, conditions are similar to those inside the Sun. Density and temperature increase toward the center, where energy is generated by nuclear fusion. Stars on the main sequence produce energy by fusing hydrogen into helium, as the Sun does. Stars that are not on the main sequence either fuse elements other than hydrogen in their cores or do not undergo fusion at all.

Stellar Evolution

A star begins to change from the moment of formation because its internal composition changes as nuclear-fusion reactions in the star's core convert one element into another. As the core composition changes, the star's density increases, its temperature rises, and its luminosity increases. As long as the star is stable and converting hydrogen to helium, it remains on the main sequence. When the nuclear fuel runs out, the star must change to counteract gravity.

3D THINKING   **DCI** Disciplinary Core Ideas   **CCC** Crosscutting Concepts   **SEP** Science & Engineering Practices

COLLECT EVIDENCE
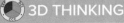 Use your Science Journal to record the evidence you collect as you complete the readings and activities in this lesson.

INVESTIGATE

 GO ONLINE to find these activities and more resources.

Applying Practices: The Sun's Energy Formation and Radiation
HS-LS2-5. Develop a model based on evidence to illustrate the life span of the sun and the role of nuclear fusion in the sun's core to release energy in the form of radiation.

 Review the News
Obtain information from a current news story about current research on stellar evolution. Evaluate your source and communicate your findings to your class.

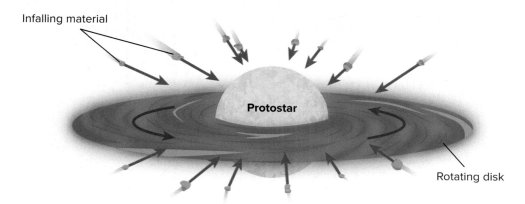

Infalling material

Protostar

Rotating disk

Figure 18 Temperatures will continue to build as gravity pulls the infalling matter to the center of the rotating disk. The center region is a protostar until fusion initiates and a star ignites.

Infer *what happens to the remaining material in the disk.*

Star formation

All stars form in much the same manner as the Sun did. The formation of a star begins with a cloud of interstellar gas and dust, called a **nebula** (plural, *nebulae*), which collapses on itself as a result of its own gravity. As the cloud contracts, its rotation forces it into a disk shape with a hot, condensed object at the center, called a **protostar,** as illustrated in **Figure 18.** Friction from gravity continues to increase the temperature of the protostar, until the condensed object reaches the ignition temperature for nuclear reactions and becomes a new star. A protostar is brightest at infrared wavelengths.

 Get It?

Infer what causes the disk shape to form.

Fusion begins

When the temperature inside a protostar becomes hot enough, nuclear fusion reactions begin. The first reaction to ignite is always the conversion of hydrogen to helium. Once this reaction begins, the star becomes stable because it then has sufficient internal heat to produce the pressure needed to balance gravity. The object is then truly a star and takes its place on the main sequence according to its mass. A new star often illuminates the gas and dust surrounding it, as shown in **Figure 19.**

Life Cycles of Stars Like the Sun

What happens next during a star's life cycle depends on its mass. For example, as a star like the Sun converts hydrogen into helium in its core, it gradually becomes more luminous because the core density and temperature rise slowly and increase the reaction rate. It takes about 10 billion years for a star with the mass of the Sun to convert all of the hydrogen in its core into helium. Thus, such a star has a main-sequence lifetime of 10 billion years. From here, the next step in the life cycle of a medium-mass star is to become a red giant.

Figure 19 Using the Spitzer telescope's infrared wavelengths, protostars are imaged inside the Elephant Trunk nebula.

Red giant

Only about the innermost 10 percent of a star's mass can undergo nuclear reactions because temperatures outside of this core never become hot enough for reactions to occur. Thus, when the hydrogen in its core is gone, a star has a helium center and outer layers made of hydrogen-dominated gas. Some hydrogen continues to react in a thin layer at the outer edge of the helium core, as illustrated in **Figure 20.** The energy produced in this layer forces the outer layers of the star to expand and cool. The star then becomes a red giant because its luminosity increases while its surface temperature decreases due to the expansion.

While the star is a red giant, it loses gas from its outer layers. The star is so large that its surface gravity is low, and thus the outer layers can be released by small expansions and contractions, or pulsations, of the star due to instability. Meanwhile, the core of the star becomes hot enough, at 100 million K, for helium to react and form carbon. The star contracts back to a more normal size, where it again becomes stable for awhile. The helium-reaction phase lasts only about one-tenth as long as the earlier hydrogen-burning phase. Afterward, when the helium in the core is depleted, the star is left with a core made of carbon.

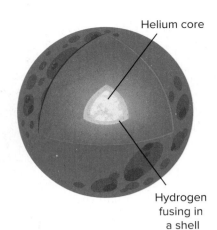

Figure 20 If the central region of a red giant becomes hot enough, helium is converted to carbon. In the spherical shell just outside, hydrogen continues to be converted to helium. The low temperature of the outer atmosphere due to expansion and cooling causes the red color.

The final stages

A star with the same mass as the Sun never becomes hot enough for carbon to undergo fusion, so its energy production ends. The outer layers expand again and are expelled by pulsations that develop in the outer layers. This shell of gas is called a planetary nebula. In the center of a planetary nebula, shown in **Figure 21,** the core of the star becomes exposed as a small, hot object about the size of Earth. The star is then a white dwarf made of carbon.

Internal pressure in white dwarfs

A white dwarf is stable despite its lack of nuclear reactions because it is supported by the resistance of electrons being squeezed together. A white dwarf is also stable because it does not require a source of heat to be maintained. This pressure counteracts gravity and can support the core as long as the mass of the remaining core is less than about 1.4 times the mass of the Sun. The main-sequence lifetime of such a star is much longer than the main-sequence lifetime of a more massive star because low-mass stars are dim and do not deplete their nuclear fuel rapidly. The electron pressure does not require ongoing reactions, so it can last indefinitely. The white dwarf gradually cools, eventually losing its luminosity, and becomes an undetectable black dwarf.

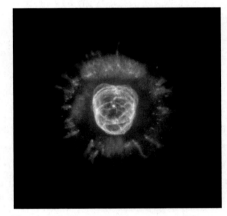

Figure 21 The star at the center of the Eskimo nebula, now a white dwarf, was the source of the remnant gases surrounding it.

 Get It?

Compare the stages in the life cycle of a sun-like star.

Table 4 Comparing Proxima Centauri to the Sun

Property	Sun	Proxima Centauri
Mass ($\times 10^{30}$ kg)	1.99	0.245
Luminosity ($\times 10^{24}$ W)	382.8	0.65
Life expectancy (billion years)	10	2000

Life Cycles of Small Stars

Stars with less than half the mass of the Sun have different life cycles than the Sun.

Brown dwarfs

Protostars with masses that are less than about 0.08 solar masses become very dim stars called brown dwarfs. They are massive enough to fuse two hydrogen nuclei together to form deuterium. Deuterium is an isotope of hydrogen having one proton and one neutron. This minimal amount of fusion gives off a very dim red light. Brown dwarfs are not massive enough, however, to fuse hydrogen into helium in a sustainable manner. For this reason, they are sometimes called failed stars. Brown dwarfs do not evolve; they remain brown dwarfs for their entire lives.

Red dwarfs

Stars that are large enough to fuse hydrogen into helium but are less than about 0.5 solar masses are known as red dwarfs. Red dwarfs have a simpler life cycle than Sun-like stars. They fuse hydrogen into helium, but do not fuse helium into carbon. Instead of swelling into red giants, they slowly collapse into white dwarfs without a planetary nebula.

Red dwarfs burn very slowly and give off very dim red light. They are so dim that they are often overlooked, despite comprising over 70% of all stars. Red dwarfs are, however, very long-lived, with lifespans in the trillions of years.

The closest star to our Sun is the red dwarf Proxima Centauri. This star, described in **Table 4** and shown in **Figure 22,** is visible only with a telescope. This star is actually the closest of a triple star system. From Earth, we see it as a single star called Alpha Centauri.

Figure 22 Proxima Centauri, photograp[hed] dwarf star about 4.25 ly from the Sun.

CCC CROSSCUTTING CONCEPTS

Energy and Matter This lesson discusses the life cycles of stars from the moment they begin to produce energy via fusion to the moment they die. Think of the types of chemical reactions taking place in the stars and the types of energy that are transferred. What happens to the matter involved? Once you have the answers to these questions, make a flow chart depicting the life and death of the smallest stars in the universe.

Life Cycles of Massive Stars

For a star that is more massive than the Sun, evolution is different. Its life begins in the same way, with hydrogen being converted to helium, but it is much higher up on the main sequence. The star's lifetime in this phase is short because the star is very luminous and uses up its fuel quickly.

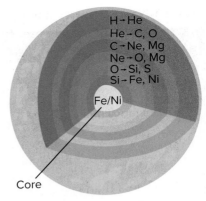

Supergiant

A massive star undergoes many more reaction phases and thus produces a rich stew of many elements in its interior. The star becomes a red giant several times as it expands following the end of each reaction stage. As more shells are formed by the fusion of different elements, illustrated in **Figure 23,** the star expands to a larger size and becomes a supergiant, such as Betelgeuse in the Orion constellation.

Figure 23 A massive star can have many shells fusing different elements. These stars are the source of heavier elements in the universe.

Supernova formation

A star that begins with a mass between about 8 and 20 times the Sun's mass will end up with a core that is too massive to be supported by electron pressure. Such a star comes to a violent end. Once reactions in the core of the star have created iron, no further energy-producing reactions can occur, and the core of the star violently collapses in on itself, as illustrated in **Figure 24.** Protons and electrons in the core merge to form neutrons. Like electrons, a neutron's resistance to being squeezed close together creates a pressure that halts the collapse of the core, and the core becomes a collapsed stellar remnant—a **neutron star.** A neutron star has a mass of 1.4 to 3 times the Sun's mass but a diameter of only about 20 km. Its density is extremely high—about 100 trillion times the density of water—and is comparable to that of an atomic nucleus.

Pu' Some neutron stars are unique in that they have a pulsating pattern of light. The m?ic fields of these stars focus the light they emit into cones. As these stars rotate on t?s, the light from each spinning neutron star is observed as a series of pulses of light af the cones sweeps out a path in Earth's direction. This pulsating star is known as a

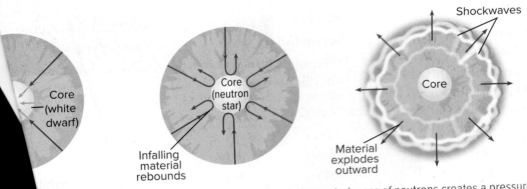

outer layers of a star collapse into the neutron core, the central mass of neutrons creates a pressure to explode outward as a supernova, leaving a neutron star.

Supernova A neutron star forms while the outer gas layers of the star are still falling inward. The gas rebounds when it strikes the hard surface of the neutron star and explodes outward. The entire outer portion of the star is blown off in a massive explosion called a **supernova** (plural, *supernovae*). This explosion enriches the universe with newly formed elements that are heavier than iron. A distant supernova explosion, such as that shown in **Figure 25,** might be brighter than the galaxy in which it formed.

Black holes

Some stars are too massive to form neutron stars. The pressure from the resistance of neutrons being squeezed together cannot support the core of a star if the star's mass is greater than about 25 times the mass of the Sun. The resistance of neutrons to being squeezed is not great enough to stop the collapse, and the core of the star continues to collapse, compacting matter into a smaller volume. The small, extremely dense object that remains is called a **black hole** because its gravity is so immense that nothing, not even light, can escape it. Astronomers cannot observe what goes on inside a black hole, but they can observe the X-ray-emitting gas that spirals into it.

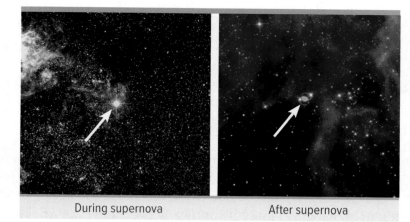

During supernova   After supernova

Figure 25 The region of sky in the Large Magellanic Cloud seemed ordinary before one of its stars underwent a supernova explosion in 1987. The remnants of the star can be seen in the right-hand photo.

Check Your Progress

Summary

- The mass of a star determines its internal structure and its other properties.
- Gravity and pressure balance each other in a stable star.
- If the temperature in the core of a star becomes high enough, elements heavier than hydrogen can fuse together.
- A supernova occurs when the outer layers of a star bounce off the neutron star core and explode outward.

Demonstrate Understanding

1. **Describe** and predict how the initial mass of a star determines its evolution.
2. **Infer** how mass determines hydrostatic equilibrium.
3. **Determine** how mass controls a star's lifetime.
4. **Determine** why only the most massive stars enrich the galaxy with heavy elements.

Explain Your Thinking

5. **Explain** how the universe would be different if massive stars did not explode at the end of their lives.
6. **Distinguish** which stars on the H-R diagram exhibit hydrostatic equilibrium.
7. **WRITING ▸ Connection** Write a description of an observation of a supernova in another galaxy.

(l)Celestial Image Co./Science Source, (r)Science & Society Picture Library/SSPL/Getty Images

LEARNSMART Go online to follow your personalized learning path to review, practice, and reinforce your understanding.

Trailblazers: A Map to the Stars

The sheer number of stars in the night sky is an overwhelming sight. Who could possibly number them? In the late nineteenth and early twentieth centuries, a group of female astronomers at the Harvard College Observatory took on this task. In the process, they catalogued hundreds of thousands of stars, created a star classification system that scientists still use today, and shaped the field of modern astrophysics.

The "computers," including astronomer Annie Jump Cannon, worked to catalogue and classify stars.

Harvard's "Computers"

At the Harvard College Observatory, scientists used telescopes equipped with cameras to transfer images onto photographic plates. The goal was to catalogue the entire night sky. With new photographic advancements, there was an overwhelming amount of data. Edward Pickering, the director, soon realized he needed more employees. He decided to hire a team of female staff, coined "computers."

Williamina Fleming was one of the first women on the staff of "computers," but she was soon joined by others, including Annie Jump Cannon, Henrietta Swan Leavitt, Antonia Maury, and Cecilia Payne-Gaposchkin. Their job was to meticulously examine photographic plates to identify properties and locations of stars. Eventually, more than 80 women worked on the project, cataloguing stars and their spectra.

Many of the astronomers did groundbreaking work. First Fleming and then Maury worked to create a stellar classification system that Cannon later refined and simplified. The system, which classifies stars by their temperature, was adopted in 1922 by International Astronomical Union as the official stellar classification system.

After analyzing Cannon's data, Payne-Gaposchkin confirmed that stars' spectra corresponded to their temperatures. She also debunked a common belief of the time—that Earth and the stars were composed of the same materials. Stars, unlike rocky planets, are made mostly of hydrogen and helium. Payne was the first female to receive a doctorate degree in astronomy from Harvard, and she went on to become the first female professor at the college.

Leavitt's major contribution to the field was her discovery that variable stars, including Cepheid variables, fluctuate in brightness at regular intervals. Because of their consistency, these stars can be used to accurately measure distances in space. In the 1920s, astronomer Edwin Hubble used Cepheid variables to measure the distances to other galaxies and determined that the universe is expanding.

COMMUNICATE SCIENTIFIC IDEAS

Research the work of one of the female astronomers discussed in the feature. Write a short paper on your findings.

Bettmann/Getty Images

GO ONLINE to study with your Science Notebook.

Lesson 1 THE SUN

- Most of the mass in the solar system is found in the Sun.
- The Sun's average density is approximately equal to that of the gas giant planets.
- The Sun has a layered atmosphere.
- The Sun's magnetic field causes sunspots and other solar activity.
- The fusion of hydrogen into helium provides the Sun's energy and composition.

- photosphere
- chromosphere
- corona
- solar wind
- sunspot
- solar flare
- prominence
- fusion
- fission

Lesson 2 MEASURING THE STARS

- Most stars exist in clusters held together by their gravity.
- The simplest cluster is a binary.
- Parallax is used to measure distances to stars.
- The brightness of stars is related to their temperature.
- Stars are classified by their spectra.
- The H-R diagram relates the basic properties of stars: class, temperature, and luminosity.

- constellation
- binary star
- parsec
- parallax
- apparent magnitude
- absolute magnitude
- luminosity
- Hertzsprung-Russell diagram
- main sequence

Lesson 3 STELLAR EVOLUTION

- The mass of a star determines its internal structure and its other properties.
- Gravity and pressure balance each other in a stable star.
- If the temperature in the core of a star becomes high enough, elements heavier than hydrogen can fuse together.
- A supernova occurs when the outer layers of a star bounce off the neutron star core and explode outward.

- nebula
- protostar
- neutron star
- pulsar
- supernova
- black hole

REVISIT THE PHENOMENON

How do telescopes tell us what elements are in stars?

CER Claim, Evidence, Reasoning

Explain your Reasoning Revisit the claim you made when you encountered the phenomenon. Summarize the evidence you gathered from your investigations and research and finalize your Summary Table. Does your evidence support your claim? If not, revise your claim. Explain why your evidence supports your claim.

STEM UNIT PROJECT
Now that you've completed the module, revisit your STEM unit project. You will summarize your evidence and apply it to the project.

GO FURTHER

SEP Data Analysis Lab
Can you identify elements in a star?

Astronomers study the composition of stars by observing their absorption spectra. Each element in a star's outer layer produces a set of lines in the star's absorption spectrum. From the pattern of lines, astronomers can determine what elements are in a star.

Hydrogen

Helium

Sodium

Calcium

Sun

Mystery star

*James B. Kaler, Professor Emeritus of Astronomy, University of Illinois. 1998.

Data and Observations

Study the spectra of the four elements shown. Examine the spectra for the Sun and the mystery star. To identify the elements of the Sun and the mystery star, use a ruler to help you line up the spectral lines with the known elements.

CER Analyze and Interpret Data
1. Identify the elements that are present in the part of the absorption spectrum shown for the Sun.
2. **Claim, Evidence** Identify the elements that are present in the absorption spectrum for the mystery star.
3. **Claim, Evidence, Reasoning** Determine which elements are common to both stars.

picturist/iStock/Getty Images

MODULE 24

GALAXIES AND THE UNIVERSE

Viktar Malyshchyts/Shutterstock

GALAXIES AND THE UNIVERSE

ENCOUNTER THE PHENOMENON

Why did it take so long to discover other galaxies?

GO ONLINE to play a video about how scientists discovered the first galaxy outside of our own.

SEP Ask Questions

Do you have other questions about the phenomenon? If so, add them to the driving question board.

CER Claim, Evidence, Reasoning

Make Your Claim Use your CER chart to make a claim about why it took so long to discover other galaxies. Explain your reasoning.

Collect Evidence Use the lessons in this module to collect evidence to support your claim. Record your evidence as you move through the module.

Explain Your Reasoning You will revisit your claim and explain your reasoning at the end of the module.

GO ONLINE to access your CER chart and explore resources that can help you collect evidence.

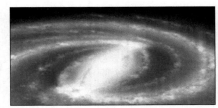

LESSON 1: Explore and Explain: Spiral Arms

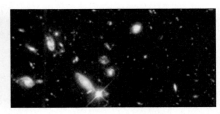

LESSON 3: Explore and Explain: Contents of the Universe

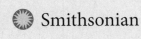

Additional Resources

FOCUS QUESTION

Where are we in our galaxy?

Discovering the Milky Way

When you observe the Milky Way galaxy, you are looking at the edge of a disk from the inside of the disk. You cannot see its size and shape because you are inside the galaxy, looking out. It is difficult to tell how big the galaxy is, where its center is, or what Earth's location is within this vast expanse of stars, gas, and dust. Though astronomers have answers to these questions, they are still refining their measurements.

Variable stars

In the 1920s, astronomers focused their attention on mapping the locations of globular clusters of stars. These huge, spherical star clusters are located above or below the plane of the galactic disk. Astronomers estimated the distances to the clusters by identifying variable stars in them. **Variable stars** are located in the giant branch of the Hertzsprung-Russell diagram and pulsate in brightness because of the expansion and contraction of their outer layers. Variable stars are brightest at their largest diameters and dimmest at their smallest diameters. **Figure 1** shows the varying brightness of a variable star. This particular variable star, called V1, was discovered by the astronomer Edwin Hubble in 1923. The discovery helped prove that hazy nebulae in the night sky were actually distant galaxies beyond our own. In late 2010, the *Hubble Space Telescope* captured images of the variable star to mark this important milestone.

Figure 1 Variable stars brighten and dim because their diameters change over a period of hours or days. These images show the varying brightness of Cepheid variable star V1 over about one month.

 3D THINKING **DCI** Disciplinary Core Ideas   **CCC** Crosscutting Concepts   **SEP** Science & Engineering Practices

COLLECT EVIDENCE

Use your Science Journal to record the evidence you collect as you complete the readings and activities in this lesson.

INVESTIGATE

GO ONLINE to find these activities and more resources.

((•)) **Review the News**
Obtain information from a current news story about current cosmological research. **Evaluate** your source and **communicate** your findings to your class.

CCC **Identify Crosscutting Concepts**
Create a table of the crosscutting concepts and fill in examples you find as your read.

Types of variables

For certain types of variable stars, there is a relationship between a star's luminosity and its pulsation period, which is the time between its brightest pulses. The longer the period of pulsation, the greater the luminosity of the star. **RR Lyrae variables** are stars that have periods of pulsation between 1.5 hours and 1.2 days, and on average, they have the same luminosity. **Cepheid variables,** however, have pulsation periods between 1 and 100 days, and the luminosity increases as much as 100 times from the dimmest star to the brightest. By measuring a star's period of pulsation, astronomers can determine the star's absolute magnitude. This allows them to compare the star's luminosity (energy) to its apparent magnitude (brightness) and calculate how far away the star must be in order to appear this dim or bright.

The galactic center

After reasoning that there were globular clusters orbiting the center of the Milky Way, astronomers then used RR Lyrae variables to determine the distances to them. They discovered that these clusters are located far from our solar system and that their distribution in space is centered on a distant point 26,000 light-years (ly) away. The galactic center is a region of high star density, shown in **Figure 2,** much of which is obscured by interstellar gas and dust. The direction of the galactic center is toward the constellation Sagittarius. **Figure 2** also shows the galaxy as you might see it on a clear night as well as a view of the galactic bulge.

 Get It?

Describe how astronomers located the galactic center of the Milky Way.

The Shape of the Milky Way

Only by mapping the galaxy with radio waves have astronomers been able to determine its shape. This is because radio waves are long enough that they can penetrate interstellar gas and dust without being scattered or absorbed. By measuring radio waves as well as infrared radiation, astronomers have discovered that the galactic center is surrounded by a nuclear bulge, which sticks out of the galactic disk much like the yolk in a fried egg. Around the nuclear bulge and disk is the **halo,** a spherical region where globular clusters are located.

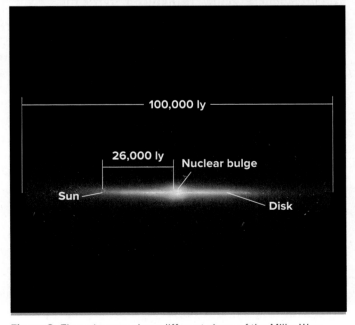

Figure 2 These images show different views of the Milky Way galaxy.

Spiral arms

Knowing that the Milky Way galaxy has a disklike shape with a central bulge, astronomers speculated that it might also have spiral arms, as do many other galaxies. This was difficult to prove. Because of the distance, astronomers have no way to get outside of the galaxy and look down on the disk. Astronomers decided to use hydrogen atoms to look for the spiral arms.

To locate the spiral arms, hydrogen emission spectra are helpful for three reasons. First, hydrogen is the most abundant element in space; second, the interstellar gas, composed mostly of hydrogen, is concentrated in the spiral arms; and third, the 21-cm wavelength of hydrogen emission can penetrate the interstellar gas and dust and be detected all the way across the galactic disk.

Using hydrogen emission and infrared images as a guide, astronomers have identified two major spiral arms, called Scutum-Centaurus and Perseus, as well as several minor arms in the Milky Way, as shown in **Figure 3.** Using these data, scientists discovered that the Sun is located in the partial Orion arm at a distance of about 26,000 ly from the galactic center. The Sun's orbital speed is about 220 km/s, and thus its orbital period is about 225 million years. In its 5-billion-year life, the Sun has orbited the galaxy approximately 20 times.

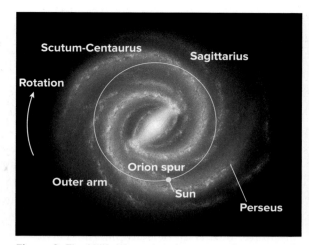

Figure 3 The Milky Way galaxy has two major arms and several minor arms. The Sun is located on the partial Orion arm or spur. (Note: *Drawing is not to scale.*)

Get It?

Explain how astronomers used the Milky Way's hydrogen emission spectrum to locate the arms.

Nuclear bulge or not?

Many spiral galaxies have a barlike shape rather than a round disk to which the arms are attached. Radio observation of interstellar gas indicates that the Milky Way has a slightly elongated shape. Recent evidence suggests that two of the arms begin at the ends of a central bar. **Figure 4** shows a satellite view of the barred galaxy NGC 1300, which has an elongated central bulge.

Using a variety of wavelengths, astronomers are discovering what the center of the Milky Way looks like. The nuclear bulge of a galaxy is typically made up of older, red stars. The bar in a galaxy center, however, is associated with younger stars and a disk that forms from neutral hydrogen gas.

Figure 4 A barred galaxy has an elongated central bulge.

Star formation does continue to occur in the bulge of the Milky Way, and most stars are about 1000 AU apart in this region, compared to the 200,000 AU separation in the locale of the Sun. Infrared measurements of 30 million stars in the Milky Way indicate a bar about 27,000 ly in length.

(t)NASA Jet Propulsion Laboratory (NASA-JPL), (b)Digital Vision/Getty Images

Mass of the Milky Way

The mass located within the circle of the Sun's orbit through the galaxy is about 100 billion times the mass of the Sun. Using this figure, astronomers have concluded that the galaxy contains about 100 billion stars within its disk.

Mass of the halo

Evidence of the movement of outer-disk stars and gas suggests that as much as 90 percent of the galaxy's mass is contained in the halo. Some of this unseen matter is probably in the form of dim stellar remnants such as white dwarfs, neutron stars, or black holes, but the nature of the remainder of this mass is unknown. As you will read in Lesson 2, the nature of unseen matter extends to other galaxies and to the universe as a whole. **Figure 5** shows the halo of the Sombrero galaxy.

A galactic black hole

Weighing in at a few million to a few billion times the mass of the Sun, supermassive black holes occupy the centers of most galaxies. When the center of the galaxy is observed at infrared and radio wavelengths, several dense star clusters and supernova remnants stand out. Among them is a complex source called Sagittarius A (Sgr A), with a sub-source called Sagittarius A* (Sgr A*), which appears to be an actual point around which the whole galaxy rotates.

Studies of the motions of the stars that orbit close to Sagittarius A* (pronounced A–star) indicate that this region has about 2.6 million times the mass of the Sun but is smaller than our solar system. Data gathered by the *Chandra X-Ray Observatory* reveal intense X-ray emissions. Astronomers think that Sagittarius A* is a supermassive black hole that glows brightly because of the hot gas surrounding it and spiraling into it. This black hole probably formed early in the history of the galaxy, at the time when the galaxy's disk was forming. Gas clouds and stars within the disk probably collided and merged to form a single, massive object that collapsed to form a black hole.

Figure 6 illustrates how a supermassive black hole develops. This kind of black hole should not be confused with the much smaller, stellar black hole, which is usually made from the collapsing core of a massive star.

Figure 5 Both the galaxy halo and central bulge are populated by older, dimmer stars. The central bulge, however, has a higher density of stars and contains some newer, brighter stars, as shown in this view of the Sombrero galaxy.

Figure 6 The formation of a supermassive black hole begins with the collapse of a dense gas cloud. The accumulation of mass releases photons of many wavelengths and perhaps even a jet of matter, as shown here.

NASA/JPL-Caltech

Stellar populations in the Milky Way

Even though the basic composition of all stars is the same, there are several distinct differences in detail. Stars differ in location, motion, and age, leading to the notion of stellar populations. The population of a star provides information about its galactic history. In fact, the galaxy could be divided into two components: the round part made up of the halo and bulge, where the stars are old and contain only traces of heavy elements, and the disk, especially the spiral arms. To astronomers, heavy elements are any elements with a mass larger than helium.

Astronomers divide stars in these two regions into two classes. **Population I stars** are in the disk and arms and have small amounts of heavy elements. **Population II stars** are found in the halo and bulge and contain even smaller traces of heavy elements. Refer to **Table 1** for more details.

Population I Most of the young stars in the galaxy are located in the spiral arms of the disk, where the interstellar gas and dust are concentrated. Most star formation takes place in the arms. Population I stars tend to follow circular orbits with low (flat) eccentricity, and their orbits lie close to the plane of the disk. Finally, Population I stars have normal compositions, meaning that approximately 2 percent of their mass is made up of elements heavier than helium. The Sun is a Population I star.

 Get It?

Describe characteristics of Population I stars such as the Sun.

Population II There are few stars and little interstellar material currently forming in the halo or the nuclear bulge of the galaxy, and this is one of the distinguishing features of Population II stars. Age is another. The halo of the Milky Way contains the oldest known objects in the galaxy—globular clusters. These clusters are estimated to be 12 to 14 billion years old. Stars in the globular clusters have extremely small amounts of elements that are heavier than hydrogen and helium. All stars contain small amounts of these heavy elements, but in globular clusters, the amounts are mere traces. Stars like the Sun are composed of about 98 percent hydrogen and helium, whereas in globular cluster stars, this composition can be as high as 99.9 percent. This indicates their extreme age. The nuclear bulge of the galaxy also contains stars with compositions like those in globular clusters. **Table 1** points out some other comparisons of Population I and II stars.

Table 1 Population I and II Stars of the Milky Way

	Location in Milky Way	Percent of H and He	Percent Heavy Elements	Age (years)	Type of Star	Example
Population I stars	disk arms and open clusters	98	2.0	<10 billion	young sequence stars	Sun, most giants, and supergiants
Population II stars	bulge and halo	99.9	0.1	>10 billion	old main-sequence stars (type K and M)	most white dwarfs and globular cluster stars

Formation and Evolution of the Milky Way

The fact that the halo is made exclusively and the nuclear bulge is made primarily of old stars suggests that these parts of the galaxy formed first, before the disk that contains only younger stars. Astronomers therefore hypothesize that the galaxy began as a spherical cloud in space. The first stars formed while this cloud was round. This explains why the halo, which contains the oldest stars, is spherical. The nuclear bulge, which is also round, represents the inner portion of the original cloud. The cloud eventually collapsed under the force of its own gravity, and rotation forced it into a disklike shape. Stars that formed after this time have orbits lying in the plane of the disk. They also contain greater quantities of heavy elements because they formed from gas that had been enriched by previous generations of massive stars. In **Figure 7,** the nuclear bulge makes up the bright glow of the Sombrero galaxy.

Spiral Arms

Most of the main features of the galaxy are understood by astronomers, except for the way in which the spiral arms are retained. The Milky Way is subject to gravitational tugs by neighboring galaxies and is periodically disturbed by supernova explosions from within, both of which can create or affect spiral arms. There are several hypotheses about why galaxies keep this spiral shape.

One hypothesis is that a kind of wave called a spiral density wave is responsible. A **spiral density wave** has spiral regions of alternating density, which rotate as a rigid pattern.

Figure 7 Easily seen through small telescopes, the Sombrero galaxy looks like a hat in visible light, but resembles a bull's-eye in this infrared image.

Predict *which type of stars would be found in the nuclear bulge.*

NASA/JPL-Caltech and The Hubble Heritage Team (STScI/AURA)

STEM CAREER Connection

Planetarium Lecturer
A planetarium lecturer presents programs about all aspects of cosmology to students and the public. Projection technology is often used to enhance presentations. He or she should have a bachelor's degree in astronomy or Earth science and excellent communication skills.

As the wave moves through gas and dust, it causes a temporary buildup of material, like a slow truck on the highway causes a buildup of cars, shown in **Figure 8.** Like cars surrounding a slow truck, the stars, gas, and dust that encounter the density wave form spiral arms.

A second hypothesis is that the spiral arms are not permanent structures but instead are continually forming as a result of disturbances such as supernova explosions. The Milky Way has a broken-spiral-arm pattern, which most astronomers think fits this second model best. However, some galaxies have a prominent two-armed pattern that was more likely created by density waves.

A third possibility is considered for faraway galaxies. It suggests that the arms are visible only because they contain hot, blue stars that stand out more brightly than dimmer, redder stars. When viewed in UV wavelengths, the arms stand out, but when viewed in infrared wavelengths, they seem to disappear.

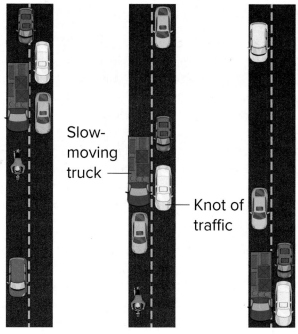

Figure 8 This buildup of cars around a slow truck on a highway illustrates one theory as to how spiral density waves maintain spiral arms in a galaxy.

 Get It?

Summarize the hypotheses about how spiral arms are maintained.

Check Your Progress

Summary

- The discovery of variable stars aided in determining the shape of the Milky Way.
- RR Lyrae and Cepheid are two types of variable stars used to measure distances.
- Globular clusters of old stars are found in the nuclear bulge and halo of the Milky Way.
- The spiral arms of the Milky Way are made of younger stars and gaseous nebulae.
- Population I stars are found in the spiral arms, while Population II stars are in the central bulge and halo.

Demonstrate Understanding

1. **Explain** How did astronomers determine where Earth is located within the Milky Way?
2. **Determine** What do measurements of the mass of the Milky Way indicate?
3. **Analyze** How are Population I stars and Population II stars different?
4. **Summarize** How can variable stars be used to determine the distance to globular clusters?

Explain Your Thinking

5. **Explain** If our solar system were slightly above the disk of the Milky Way, would astronomers still have difficulty determining the shape of the galaxy? Explain.
6. **Hypothesize** What would happen to the stellar orbits near the center of the Milky Way if there were no black hole?
7. **WRITING** **Connection** Write a description of riding a space-ship from above the Milky Way into its center. Point out all of the galaxy's parts and star types.

LEARNSMART® Go online to follow your personalized learning path to review, practice, and reinforce your understanding.

FOCUS QUESTION

How did we discover there were objects outside our galaxy?

Discovering Other Galaxies

Long before they knew what galaxies were, astronomers observed many objects scattered throughout the sky. Some astronomers hypothesized that these objects were nebulae or star clusters within the Milky Way. Others hypothesized that they were distant galaxies that were as large as the Milky Way. The question of what these objects were was answered by Edwin Hubble in 1924, when he discovered Cepheid variable stars in the Great Nebula in the Andromeda constellation. Using these stars to measure the distance to the nebula, Hubble showed that they were too far away to be located in our own galaxy. The Andromeda nebula then became known as the Andromeda Galaxy, shown in **Figure 9.**

Properties of galaxies

Masses of galaxies range from dwarf ellipticals, which have masses of approximately 1 million times the mass of the Sun; to large spirals such as the Milky Way, with masses of around 100 billion times the mass of the Sun; to the largest galaxies, called giant ellipticals, which have masses as high as 1 trillion times that of the Sun. Measurements of the masses of many galaxies indicate that they have extensive halos containing more mass than is visible, just as the Milky Way does. **Figure 9** shows a large spiral and several smaller galaxies.

Figure 9 Andromeda is a spiral galaxy like the Milky Way. The bright elliptical object and the sphere-shaped object near the center are small galaxies orbiting the Andromeda Galaxy.

3D THINKING   **DCI** Disciplinary Core Ideas   **CCC** Crosscutting Concepts   **SEP** Science & Engineering Practices

COLLECT EVIDENCE

 Use your Science Journal to record the evidence you collect as you complete the readings and activities in this lesson.

INVESTIGATE

 GO ONLINE to find these activities and more resources.

 GeoLAB: Classify Galaxies
Obtain, evaluate, and communicate information that uses **the structure** of galaxies to describe the evolution of the Universe.

Investigation Lab: Modeling Spiral Galaxies
Use a model to visualize how **the structure** of galaxies direct the life cycle of stars.

NASA/JPL-Caltech

Luminosities of galaxies also vary widely, from the dwarf spheroidals—not much larger or more brilliant than a globular cluster—to supergiant elliptical galaxies, more than 100 times more luminous than the Milky Way. All galaxies show evidence that an unknown substance called dark matter dominates their masses. **Dark matter** is thought to be made up of a form of subatomic particle that interacts only weakly with other matter.

Galaxy Classification

Hubble went on to study galaxies and categorize them according to their shapes, using a letter-number system.

Disklike galaxies

Hubble classified the disklike galaxies with spiral arms as spiral galaxies. These were subdivided into normal spirals and barred spirals. As shown in **Figures 10** and **12,** barred spirals *(SB)* have an elongated central region—a bar—from which the spiral arms extend, while normal spirals *(S)* do not have bars. Normal and barred spirals are further subdivided by how tightly the spiral arms are wound and how large and bright the nucleus is. The letter *a* represents tightly wound arms and a large, bright nucleus. The letter *c* represents loosely wound arms and a small, dim nucleus. Thus, a normal spiral with tightly wound arms and a bright nucleus is denoted *Sa,* while a barred spiral with class *a* arms and nucleus is denoted *SBa.* Galaxies with flat disks that do not have spiral arms are denoted as *S0.*

Elliptical galaxies

Some galaxies are not flattened into disks and do not have spiral arms, as shown in **Figure 12.** Called elliptical galaxies, they are divided into subclasses based on the apparent ratio of their major and minor axes. Round ellipticals are classified as *E0,* while elongated ellipticals are classified as *E7. E1* through *E6* are progressively less round and more elongated. The classification of spiral and elliptical galaxies can be summarized by Hubble's tuning-fork diagram, illustrated in **Figure 11.**

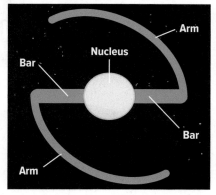

Barred Spiral Galaxy

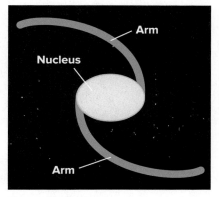

Spiral Galaxy

Figure 10 Measurements have indicated that the Milky Way's central region might be a bar, not a spiral.

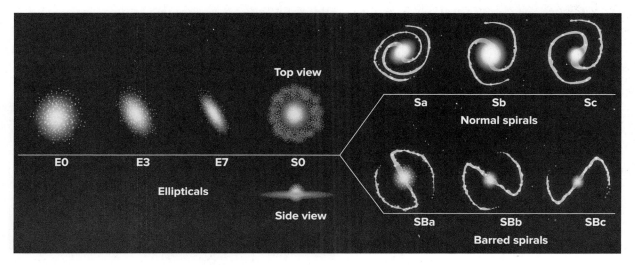

Figure 11 The Hubble tuning-fork diagram summarizes Hubble classification for normal galaxies.

Figure 12 Visualizing the Local Group

All of the stars easily visible in the night sky belong to a single galaxy, the Milky Way. Just as stars are a part of galaxies, galaxies are gravitationally drawn into galactic groups, or clusters. The 40 galaxies closest to Earth are members of the Local Group of galaxies.

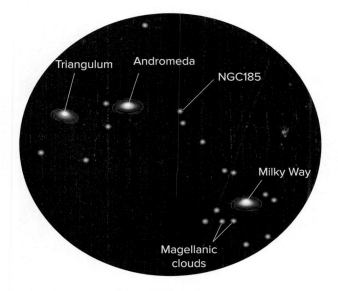

▲ **Spiral galaxies** The two largest galaxies in the Local Group, Andromeda and the Milky Way, are large, flat disks of interstellar gas and dust with arms of stars extending from the disk.

▲ **Elliptical galaxies** are nearly spherical in shape and consist of a tightly packed group of relatively old stars. Nearly half of the Local Group are ellipticals.

▲ **Barred spiral galaxies** Sometimes the flat disk that forms the center of a spiral galaxy is elongated into a bar shape. Recent evidence suggests that the Milky Way galaxy has a bar.

Irregular galaxies Some galaxies are neither spiral nor elliptical. Their shape seems to follow no set pattern, so astronomers have given them the classification of irregular. ▶

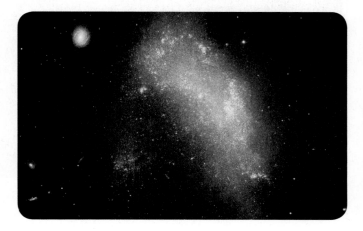

Figure 13 The Large and Small Magellanic Clouds are small galaxies that orbit the Milky Way.

Figure 14 The force of gravity of the nearby Virgo cluster of about 2000 galaxies is so strong that it is pulling the Milky Way toward it.

At first, Hubble's tuning-fork diagram was viewed as an evolutionary diagram for galaxies. It was theorized that all galaxies began as ellipticals and evolved into spirals. As astronomers were able to look farther into the universe, and hence farther back in time, they observed spirals as well as ellipticals. If astronomers were looking far back in time, then the young galaxies should be ellipticals, but this was not the case. The tuning-fork diagram is simply a convenient way to organize the different classes.

Irregular galaxies

Some galaxies do not have distinct shapes. These irregular galaxies are denoted by *Irr*. The Large and Small Magellanic Clouds, shown in **Figure 13,** two satellite galaxies of the Milky Way, are irregular galaxies.

Groups and Clusters of Galaxies

Most galaxies are located in groups, as shown in **Figure 14,** rather than being spread uniformly throughout the universe. **Figure 12** shows some of the features of the Local Group of galaxies.

Local Group

The Milky Way belongs to a small cluster of galaxies called the Local Group. The diameter of the Local Group is nearly 10 million ly. There are about 40 known members, of which the Milky Way and Andromeda galaxies are the largest. Most of the members are dwarf ellipticals that are companions to the larger galaxies.

The Large and Small Magellanic Clouds were thought to be the closest galaxies to the Milky Way until 1994, when the Sagittarius Dwarf Elliptical galaxy was discovered. However, the Canis Major dwarf galaxy, discovered in 2003, is now our closest known neighbor. This galaxy is being pulled apart by the Milky Way's gravity and is leaving streams of dust, gas, and stars in its wake. As dim galaxies continue to be found, more could be added to the Local Group.

 Get It?

Identify the kinds of galaxies in the Local Group.

Large clusters

Galaxy clusters larger than the Local Group, such as the Virgo cluster, might have thousands of members. Their diameters may be in the range of about 5 to 30 million ly. Most of the galaxies in the inner region of a large cluster are ellipticals, while there is a more even mix of ellipticals and spirals in the outer portions.

In regions where galaxies are as close together as they are in large clusters, gravitational interactions among galaxies have many important effects. Galaxies often collide and form strangely shaped galaxies, as shown in **Figure 15,** or they form galaxies with more than one nucleus.

Figure 15 This galactic merger will eventually form a single galaxy. The streams of gas and dust will form new stars.

Masses of clusters For clusters of galaxies, the mass determined by analyzing the motion of member galaxies is always much larger than the sum of the total masses of each of the galaxies, as determined by their total luminosity. This suggests that most of the mass in a cluster of galaxies is invisible, which provides astronomers with strong evidence that the universe contains a great amount of dark matter.

 Get It?

Describe the evidence used by scientists to infer the existence of dark matter.

Superclusters

Clusters of galaxies are organized into even larger groups called **superclusters.** These gigantic formations, hundreds of millions of light-years in size, can be observed only when astronomers map out the locations of many galaxies ranging over huge distances. The superclusters appear in sheetlike and threadlike shapes, giving the appearance of a gigantic bubble bath, with galaxies located on the surfaces of the bubbles and the inner air pockets void of galaxies.

The Expanding Universe

In 1929, Edwin Hubble made another dramatic discovery. It was known at the time that most galaxies have redshifts in their spectra, indicating that all but the nearest galaxies are moving away from Earth. Hubble measured the redshift and distances of many galaxies and found that the farther away a galaxy is, the faster it is moving away. In other words, the universe is expanding.

Implications of redshift

The human view of Earth's place in the cosmos has changed drastically on several occasions in history. Each time, the new perception placed Earth in a less central position. Now we realize that Earth orbits an ordinary star far from the center of a galaxy that is one of billions, occupying no special place in the universe. So, though you might infer that Earth is at the center of the universe, this is not the case.

NASA, H. Ford (JHU), G. Illingworth (UCSC/LO), M.Clampin (STScI), G. Hartig (STScI), the ACS Science Team, and ESA

An observer located in any galaxy, at any place in the universe, will observe the same thing in a medium that is uniformly expanding—all points are moving away from all other points, and no point is at the center. At greater distances, the expansion increases the rate of motion.

If the universe is expanding now, it must have been smaller and denser in the past. In fact, there must have been a time when all contents of the universe were compressed together. The Big Bang theory, which you will learn about in Lesson 3, has been proposed to explain this expansion.

Hubble's law

By making a graph comparing a galaxy's distance to the speed at which it is moving, Hubble determined that the universe is expanding. The result is a straight line, which can be expressed as a simple equation, $v = Hd$, where v is the velocity at which a galaxy is moving away measured in kilometers per second; d is the distance to the galaxy measured in megaparsecs (Mpc), where 1 Mpc = 3,260,000 ly; and H is a number called the **Hubble constant,** measured in kilometers per second per megaparsec. H represents the slope of the line.

Figure 16 In addition to radio lobes, M87 has a jet that emits visible light.

Measuring H Determining the value of H requires finding distances and speeds for many galaxies and constructing a graph to find the slope. This is a difficult task because it is hard to measure accurate distances to the most remote galaxies. Edwin Hubble could obtain only a crude value for H. Obtaining an accurate value for H was one of the key goals of astronomers who designed the *Hubble Space Telescope* (HST). It took nearly ten years after the launch of the HST to gather enough data to pinpoint the value of H. Currently, the best measurements indicate a value of approximately 70 km/s/Mpc.

New way to measure distance Once the value of H is known, it can be used to find distances to faraway galaxies. By measuring the speed at which a galaxy is moving, astronomers use the graph to determine the corresponding distance to the galaxy. This method works for the most remote galaxies that can be observed and allows astronomers to measure distances to the edge of the observable universe.

The only galaxies that do not seem to be moving apart are those within a cluster. The internal gravity of the cluster keeps them from separating.

Active Galaxies

Galaxies that emit large amounts of energy from their cores, such as the one shown in **Figure 16,** are called active galaxies. The core of an active galaxy, where highly energetic objects or activities are located, is called the **active galactic nucleus** (AGN).

NASA and The Hubble Heritage Team (STScI/AURA)

An AGN emits an equal or greater amount of energy compared with the rest of the galaxy. The output of this energy often varies over time, sometimes in as little as a few days. About 10 percent of all known galaxies are active, including radio galaxies and quasars.

Radio galaxies

Radio-telescope surveys of the sky have revealed a number of galaxies that are extremely luminous. These galaxies, called **radio galaxies,** are often giant elliptical galaxies that emit as much energy in radio wavelengths as they do in wavelengths of visible light. Radio galaxies have many unusual properties. The radio emission usually comes from two huge lobes of very hot gas located on opposite sides of the visible galaxy. These lobes are linked to the galaxy by jets of hot gas. The type of emission that comes from these regions indicates that the gas is ionized and that electrons in the gas jets are traveling near the speed of light. Many radio galaxies have jets that can be observed only at radio wavelengths. One of the brightest radio galaxies is shown in **Figure 16** on the previous page.

 Get It?

Describe the unusual properties of a radio galaxy.

Quasars

In the 1960s, astronomers discovered objects that looked like ordinary stars, but some emitted strong radio waves. Most stars do not. Also, whereas most stars have spectra with absorption lines, these new objects had mostly emission lines in their spectra. These starlike objects with emission lines in their spectra were called **quasars.** The word *quasar* is a contraction of the original designation of these objects, which was *quasi-stellar objects*. Some astronomers refer to them as QSOs.

Quasars such as the one shown in **Figure 17** are very luminous, very distant, active galaxies that often vary in brightness every few days. The emission lines of quasars are those of common elements, such as hydrogen, shifted far toward longer wavelengths. Once astronomers had identified the large spectral-line shifts of quasars, they wondered whether they could have redshifts caused by the expansion of the universe.

Figure 17 Quasars are distant celestial objects that emit several thousand times more energy than does our entire galaxy.

Recall *What other objects emit jets of matter?*

Quasar redshift The redshift of quasars was much larger than any that had been observed in galaxies up to that time, which would mean that the quasars were much farther away than any known galaxy.

At first, some astronomers doubted that quasars were far away, but in the decades since quasars were discovered, more evidence supports this hypothesis. One piece of supporting evidence indicates that those quasars associated with clusters of galaxies have the same redshift, verifying that they are the same distance away.

NASA/ESA

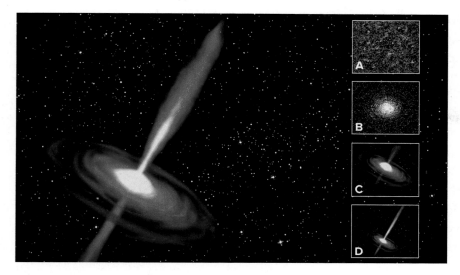

Figure 18 An interstellar gas cloud (A) collapses gravitationally (B) on its way to forming a galaxy. The nucleus (C) forms a black hole as the gas there is compressed. Magnetic fields of the rapidly rotating disk surrounding the black hole form two highly energetic jets (D) that are perpendicular to the disk's equatorial plane.

Another, more important, discovery is that most quasars are nuclei of very dim galaxies, whose formation is illustrated in **Figure 18.** The quasars appear to be extra-bright AGNs—so much brighter than their surrounding galaxies that astronomers could not initially see those galaxies.

 Get It?

Explain why quasars appear to be AGNs.

Looking back in time Because quasars are distant, it takes their light a long time to reach Earth. Therefore, observing a quasar is seeing it as it was a long time ago. For example, it takes light from the Sun approximately 8 minutes to reach Earth. When you observe the Sun, you are seeing it as it was 8 minutes earlier. When you observe the Andromeda Galaxy, you see the way it looked 2 million years earlier. The most remote quasars are several billion light-years away, which indicates that the stage you see is from billions of years ago.

If quasars are extra-bright AGNs, then the many distant ones are nuclei of galaxies as they existed when the universe was young. This suggests that many galaxies went through a quasar stage when they were young. Consequently, today's AGNs might be former quasars that are not as energetic as they were long ago.

Looking far back into time, the early universe had many quasars. Current theory suggests that they existed around supermassive black holes that pulled gas into the center, where, in a violent swirl, friction heated the gas to extreme temperatures, resulting in the bright light energy that was first detected.

SCIENCE USAGE v. COMMON USAGE

stage

Science usage: a step in a process
Common usage: a platform in a theater

Source of power AGNs and quasars emit far more energy than do ordinary galaxies, but they are as small as solar systems. This suggests that AGNs and quasars contain supermassive black holes. Recall that the black hole thought to exist in the core of our own galaxy has a mass of about 1 million Suns. The black holes thought to exist in AGNs and the cores of quasars are much more massive—up to hundreds of millions of times the mass of the Sun.

The beams of charged particles that stream out of the cores of radio galaxies and form jets are probably created by magnetic forces. As material falls into a black hole, the magnetic forces push the charged particles out into jets. There is evidence that similar beams or jets occur in other types of AGNs and in quasars. In fact, radio-lobed quasars have jets that are essentially related to radio galaxies.

Figure 19 shows evidence of a supermassive black hole in the center of the Centaurus A galaxy. In modeling a supermassive black hole of this magnitude, the mass of nearly 1 billion Suns may be needed to pull the stars in this galaxy into the center. A plasma jet, ejected from the nucleus, extends 13,000 ly into space.

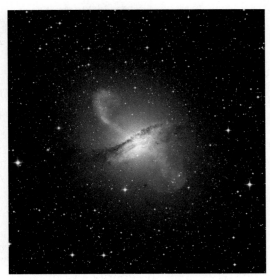

Figure 19 A jet of energetic X-ray particles is emitted from the AGN of an elliptical galaxy called Centaurus A, which probably hides a supermassive black hole.

📝 Check Your Progress

Summary

- Galaxies can be elliptical, disk-shaped, or irregular.
- Galaxies range in mass from 1 million Suns to more than a trillion Suns.
- Many galaxies seem to be organized in groups called clusters.
- Hubble's law helped astronomers discover that the universe is expanding.
- Quasars are the nuclei of faraway galaxies that are dim and seen as they were long ago, due to their great distances.

Demonstrate Understanding

1. **Explain** how astronomers discovered that there are other galaxies beyond the Milky Way.
2. **Summarize** why astronomers theorize that most of the matter in galaxies and clusters of galaxies is dark matter.
3. **Explain** why it is difficult for astronomers to accurately measure a value for the Hubble constant, H. Once a value is determined, describe how it is used.
4. **Contrast** the appearances of normal spiral, barred spiral, elliptical, and irregular galaxies.

Explain Your Thinking

5. **Deduce** how the nighttime sky would look from Earth if the Milky Way were an elliptical galaxy.
6. **Infer** how black holes cause both AGNs and quasars to be so luminous.
7. **MATH ⟩ Connection** Convert the distance across the Milky Way to Mpc if the diameter of the Milky Way is 100,000 ly. What is the distance in Mpc across a supercluster of galaxies whose diameter is 200 million ly? (1 Mpc = 3,260,000 ly)

LEARNSMART® Go online to follow your personalized learning path to review, practice, and reinforce your understanding.

ESO/WFI (Optical); MPIfR/ESO/APEX/A.Weiss et al. (Submillimetre); NASA/CXC/CfA/R.Kraft et al. (X-ray)

FOCUS QUESTION
How will the universe end?

Big Bang Model

The study of the universe—its nature, origin, and evolution—is called **cosmology.** The mathematical basis for cosmology is general relativity, from which equations were derived to describe both the energy and matter content of the universe. These equations, combined with observations of density and acceleration, led to the most accurate model so far—the Big Bang model.

The fact that the universe is expanding implies that it had a beginning. The theory that the universe began as a point and has been expanding ever since is called the **Big Bang theory.** Although the name implies an explosion into space, space itself actually expands while gravity holds matter in check.

Outward expansion

In the Big Bang model, the momentum of the outward expansion of the universe is opposed by the inward force of gravity acting on matter to slow that expansion, as shown in **Figure 20.** What ultimately will happen depends on which force is stronger. When the rate of expansion of the universe is known, it is possible to calculate the time since the expansion started and determine the age of the universe.

In astronomical terms, if the value of H, the expansion (Hubble) constant, is known, then the age of the universe can be determined. Corrections are needed because the expansion has not been constant—it has slowed since the beginning and is now accelerating.

Based on the best value for H that has been calculated from *Hubble Space Telescope* data and other sources, the age of the universe is 13.7 billion years. This fits with what astronomers know about the Milky Way galaxy, which is estimated to be 12 to 14 billion years old, based on the ages of the oldest star clusters.

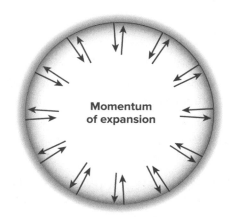

Momentum of expansion

Force of gravity

Figure 20 The universe is either open, flat, or closed, depending on whether gravity or the momentum of expansion dominates.

3D THINKING **DCI** Disciplinary Core Ideas **CCC** Crosscutting Concepts **SEP** Science & Engineering Practices

COLLECT EVIDENCE
 Use your Science Journal to record the evidence you collect as you complete the readings and activities in this lesson.

INVESTIGATE
GO ONLINE to find these activities and more resources.

 Applying Practices: The Big Bang Theory
HS-ESS1-2. Construct an explanation of the Big Bang theory based on **astronomical evidence** of light spectra, motion of distant galaxies, and composition of matter in the universe.

 Quick Investigation: Model Expansion
Use a model to visualize the changes in the electromagnetic spectrum due to the expansion of the Universe.

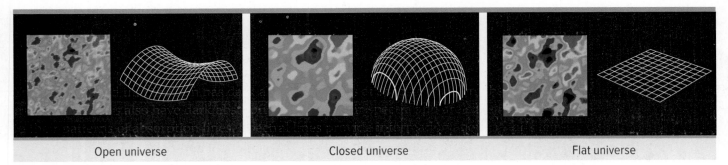

| Open universe | Closed universe | Flat universe |

Figure 21 The universe could expand forever and be open; it could snap back at the end and be closed; or it could be flat and die out like a glowing ember. The red and blue spots in the green squares show the estimated cosmic background radiation necessary for each result. In an open universe, the curvature of space makes these variations seem smaller than they are, and in a closed universe, they appear larger.

Possible outcomes

Based on the Big Bang theory, there are three possible outcomes for the universe, as shown in **Figure 21.** The average density of the universe is an observable quantity with vast implications to the outcome.

Open universe An open universe is one in which the expansion will never stop. This would happen if the density of the universe is insufficient for gravity to ever halt the expansion.

Closed universe A closed universe will result if the expansion stops and becomes a contraction. That would mean the density is high enough that eventually the gravity caused by the mass will halt the expansion and pull all of the mass back to the original point of origin.

Flat universe A flat universe results if the expansion slows to a halt in an infinite amount of time but never contracts. This means, that while the universe would continue to expand, its expansion would be so slow that it would seem to stop.

Critical density

All three outcomes are based on the premise that the rate of expansion has slowed since the beginning of the universe, but the density of the universe is unknown. At the critical density, there is a balance, so that the expansion will come to a halt in an infinite amount of time. The critical density, about 6×10^{-27} kg/m³, means that, on average, there are only two hydrogen atoms for every cubic meter of space. When astronomers attempt to count the galaxies in certain regions of space and divide by the volume, they get an even smaller value. So, they would conclude that the universe is open, except that the dark matter has not been included. But even the best estimates of dark matter density are not enough to conclude that the universe is a closed system.

Cosmic Background Radiation

If the universe began in a highly compressed state before the Big Bang, it would have been extremely hot. Then, as the universe expanded, the temperature cooled. After about 300,000 years, the universe was filled with short-wavelength radiation. With continued expansion, the wavelengths became longer. Today this radiation is in the form of microwaves.

Discovery

In 1965, scientists discovered a persistent background noise in their radio antenna. This noise was caused by weak radiation, called **cosmic background radiation,** that appeared to come from all directions in space and corresponded to an emitting object with a temperature of about 2.725 K (−270°C). This was very close to the temperature predicted by the Big Bang theory, and the radiation was interpreted to be from the Big Bang. This radiation, along with observations of distant galaxies receding from our own and the compositions of stars and nonstellar gases, provides evidence for the Big Bang theory.

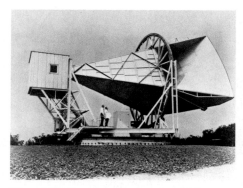

Figure 22 Cosmic background radiation was discovered by accident with this radio antenna at Bell Labs in Holmdel, New Jersey.

Mapping the radiation

Since the discovery of cosmic background radiation by the radio antenna shown in **Figure 22,** extensive observations have confirmed that it matches the properties of the predicted leftover radiation from the early, hot phase in the expansion of the universe. A space observatory called the *Wilkinson Microwave Anisotropy Probe* (*WMAP*), launched by NASA in 2001, mapped the radiation in greater detail. The peak of the radiation it measured has a wavelength of approximately 1 mm; thus, it is microwave radiation in the radio portion of the electromagnetic spectrum.

Acceleration of the expansion

The data produced by *WMAP* have provided enough detail to refine cosmological models. In particular, astronomers have found small wiggles in the radiation, representing the first major structures in the universe. This helped to pinpoint the time at which the first galaxies and clusters of galaxies formed and also the age of the universe. According to every standard model, the expansion of the universe is slowing down due to gravity. However, the debate about the future of the universe based on this model came to a halt with the surprising discovery that the expansion of the universe is now accelerating, as illustrated in **Figure 23.** Astronomers have labeled this acceleration dark energy. Although they do not know its cause, they can determine the rate of acceleration and estimate the amount of dark energy.

Figure 23 While standard models predict deceleration of the expansion of the universe, data show the expansion accelerating.

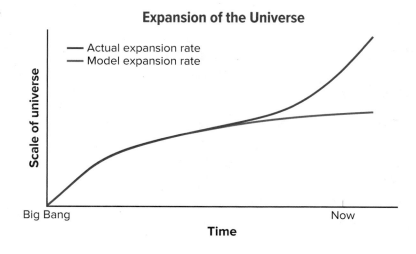

Expansion of the Universe

— Actual expansion rate
— Model expansion rate

Scale of universe

Big Bang

Now

Time

CCC **CROSSCUTTING CONCEPTS**
Scale, Proportion, and Quantity Make a graphic organizer that describes the Big Bang in terms of scale, proportion, and quantity.

SCIENCE USAGE v. COMMON USAGE
cosmic
Science usage: of or relating to the universe in contrast to Earth alone
Common usage: characterized by greatness of thought or intensity

Contents of the Universe

Given the evidence, astronomers can interpret the contents of the universe. Their best clue comes from the radiation left in space from the universe's beginning. The ripples left during the cooling phase of the universe's beginning radiation set the density at that point of time and dictated how matter and energy would separate. This, in turn, laid the groundwork for future galaxies. **Figure 24** gives one view into the universe.

Dark matter and energy

Cosmologists estimate that the universe is composed of dark matter (27 percent), dark energy (68 percent), and luminous matter. If you compare the universe to Earth's surface, dark energy is like the water covering it. That would be like saying that the majority of Earth is covered with something that is not identified.

What is unknown is the nature of dark matter and dark energy. Dark matter may consist of subatomic particles, but of the known particles, none display the right properties to explain or fully define dark matter. Although scientists recognize the effects of dark energy, they still do not know precisely what it is.

Figure 24 In this view of deep space, galaxies appear as glowing flecks. Astronomers estimate that only about 5 percent of the universe is composed of luminous matter.

✎ Check Your Progress

Summary

- The study of the universe's origin, nature, and evolution is cosmology.
- The Big Bang model came from observations of density and acceleration.
- The critical density of the universe, along with the amount of dark energy, will determine if the universe is open or closed.
- Cosmic background radiation supports the Big Bang theory.
- The universe is made mostly of dark matter and dark energy, whose natures are unknown.

Demonstrate Understanding

1. **Compare and contrast** What are the differences among the three possible outcomes of the universe?
2. **Describe** how the age of the universe can be calculated using the Big Bang model.
3. **Explain** why dark matter is important in determining the density of matter in the universe.
4. **Explain** why cosmic background radiation was an important discovery.

Explain Your Thinking

5. **Determine** What does dark matter have to do with the critical density of the universe?
6. **Analyze** All of the models tell us that the expansion of the universe should be slowing down, but instead it is speeding up. How does this affect our model of the universe?
7. **WRITING ⟩ Connection** Write one paragraph summarizing evidence for the Big Bang theory.

LEARNSMART Go online to follow your personalized learning path to review, practice, and reinforce your understanding.

Robert Williams and the Hubble Deep Field Team (STScI) and NASA

GPS Detectives: Investigating the Invisible

If you have ever used a map-based app to get directions or find a place to eat, then you have used Global Positioning System (GPS) technology. GPS consists of dozens of satellites that orbit above Earth's surface and provide positioning, navigation, and timing services. In a recent research project, scientists used GPS technology to search for something far more elusive than the closest restaurant: They were looking for dark matter.

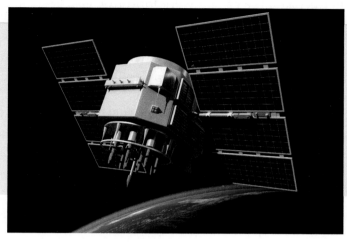

Scientists analyzed data from GPS satellites orbiting Earth to look for evidence of dark matter.

Searching for dark matter

About 27 percent of all matter in the universe is dark matter—a mysterious material that scientists cannot observe directly. How do scientists know it is there? They studied the motion of distant stars and galaxies. Scientists are determined to find direct evidence of dark matter interacting with Earth.

That is where the GPS satellites come in. GPS relies on atomic clocks, which are so accurate that they will not gain or lose more than a second over billions of years. It is hypothesized that dark matter would disrupt the clocks by causing interferences with the atomic processes.

Quantum physicist Andrei Derevianko and geophysicist Geoff Blewitt at the University of Reno in Nevada, teamed up to look for dark matter by using data from each atomic clock satellite. Together they analyzed 16 years' worth of data taken at 30-second intervals from GPS satellites. They were looking for patterns of glitches in the GPS time data that would indicate the presence of dark matter.

Many of the experiments designed to look for dark matter require technology that is expensive to make, set up, and use—costing millions, or even billions, of dollars. But in this case, the scientists were able to rely on GPS technology that was already in place, so the only cost was for analysis of the data from the project.

The scientists did not find evidence of dark matter, but even a null result is useful information. Dark matter may take a variety of forms consistent with the laws of physics. Dark matter scientists call these forms "models" of dark matter. The analysis was carried out based on particular models, which can now be ruled out, helping scientists decide how to focus future investigations.

COMMUNICATE SCIENTIFIC INFORMATION

Find out about one other experiment designed to search for dark matter. What type of evidence are the scientists looking for? What technology did they use? Communicate your findings in a brief digital slideshow presentation.

MODULE 24
STUDY GUIDE

 GO ONLINE to study with your Science Notebook.

Lesson 1 THE MILKY WAY GALAXY

- The discovery of variable stars aided in determining the shape of the Milky Way.
- RR Lyrae and Cepheid are two types of variable stars used to measure distances.
- Globular clusters of old stars are found in the nuclear bulge and halo of the Milky Way.
- The spiral arms of the Milky Way are made of younger stars and gaseous nebulae.
- Population I stars are found in the spiral arms, while Population II stars are in the central bulge and halo.

- variable star
- RR Lyrae variable
- Cepheid variable
- halo
- Population I star
- Population II star
- spiral density wave

Lesson 2 OTHER GALAXIES IN THE UNIVERSE

- Galaxies can be elliptical, disk-shaped, or irregular.
- Galaxies range in mass from 1 million Suns to more than a trillion Suns.
- Many galaxies seem to be organized in groups called clusters.
- Hubble's law helped astronomers discover that the universe is expanding.
- Quasars are the nuclei of faraway galaxies that are dim and seen as they were long ago, due to their great distances.

- dark matter
- supercluster
- Hubble constant
- active galactic nucleus
- radio galaxy
- quasar

Lesson 3 COSMOLOGY

- The study of the universe's origin, nature, and evolution is cosmology.
- The Big Bang model came from observations of density and acceleration.
- The critical density of the universe, along with the amount of dark energy, will determine if the universe is open or closed.
- Cosmic background radiation supports the Big Bang theory.
- The universe is made mostly of dark matter and dark energy, whose natures are unknown.

- cosmology
- Big Bang theory
- cosmic background radiation

REVISIT THE PHENOMENON

Why did it take so long to discover other galaxies?

CER Claim, Evidence, Reasoning

Explain Your Reasoning Revisit the claim you made when you encountered the phenomenon. Summarize the evidence you gathered from your investigations and research and finalize your Summary Table. Does your evidence support your claim? If not, revise your claim. Explain why your evidence supports your claim.

STEM UNIT PROJECT
Now that you've completed the module, revisit your STEM unit project. You will apply your evidence from this module and complete your project.

GO FURTHER

SEP Data Analysis Lab

How was the Hubble constant derived?

Plotting the distances and speeds for a number of galaxies created the expansion constant for Hubble's law.

Data and Observations Use the data to construct a graph. Plot the distance on the x-axis and the speed on the y-axis. Use a ruler to draw a straight line through the center of the band of points on the graph, so that approximately as many points lie above the line as lie below it. Make sure your line starts at the origin. Measure the slope by choosing a point on the line and dividing the speed at that point by the distance.

CER Analyze and Interpret Data

1. **Claim** What does the slope represent?
2. **Evidence** How accurate do you think your value of H is? Explain.
3. **Reasoning** How would an astronomer improve this measurement of H?

Galaxy Data

Distance (Mpc)	3.0	8.3	10.9	16.2	17.0	20.4	21.9	26.5	33.7	36.8	38.7	43.9	45.1	47.6
Speed (km/s)	210	450	972	1383	1202	1685	1594	2087	2813	2697	3177	3835	3470	3784

STUDENT RESOURCES

Skillbuilder Handbook

Make Comparisons

Why learn this skill?

Suppose you want to buy a portable music player, and you must choose among three different models. You would probably compare the characteristics of the three models, such as price, amount of memory, sound quality, and size to determine which model is best for you.

In the study of Earth science, you often compare the structures and functions of one type of rock or planet with another. You will also compare scientific discoveries or events from one time period with those from a different time period. This helps you gain an understanding of how the past has affected the present.

Learn the Skill

When making comparisons, you examine two or more groups, situations, events, or theories. You must first decide what items will be compared and determine which characteristics you will use to compare them. Then identify any similarities and differences.

For example, comparisons can be made between the two minerals shown on this page. The physical properties of halite can be compared to the physical properties of quartz.

Practice the Skill

Create a table with the title *Mineral Comparison*. Make two columns. Label the first column *Halite*, and the second column *Quartz*. List all of your observations of these two minerals in the appropriate column of your table. Similarities you might point out are that both minerals are solids that occur as crystals, and both are inorganic compounds. Differences might include that halite has a cubic crystal structure, whereas quartz has a hexagonal crystal structure.

When you have finished the table, answer these questions.

1. What items are being compared? How are they being compared?
2. What properties do the minerals have in common?
3. What properties are unique to each mineral?

Apply the Skill

Make Comparisons Read two editorial articles in a science journal or magazine that express different viewpoints on the same issue. Identify the similarities and differences between the two points of view.

Halite

Quartz

(l)Vladislav Gajic/Shutterstock, (r)Siede Preis/Getty Images

Analyze Information

Why learn this skill?

Analyzing, or looking at separate parts of something to understand the entire piece, is a way to think critically about written work. The ability to analyze information is important when determining which ideas are more useful than others.

Learn the Skill

To analyze information, use the following steps:

- Identify the topic being discussed.
- Examine how the information is organized—identify the main points.
- Summarize the information in your own words, and then make a statement based on your understanding of the topic and what you already know.

Practice the Skill

Read the following excerpt from *National Geographic.* Use the steps listed above to analyze the information and answer the questions that follow.

His name alone makes Fabien Cousteau, grandson of the late Jacques, a big fish in the world of underwater exploration. Now he's taking that big-fish status to extremes. The Paris-born, New York-based explorer had become a virtual shark, thanks to his new shark-shaped submarine. He uses the sub to dive incognito among the oceans' top predators, great white sharks.

Created at a cost of more than $100,000, the 4.3-meter-long contraption is designed to look and move as much like the real thing as possible. It carries a single passenger, who fits inside lying down, propped up on elbows to navigate and observe. "This is akin to being the first human being in the space capsule in outer space," Cousteau said. "It's pretty similar. You have no idea what's going to happen; it's a prototype."

Cousteau used the submarine to make a documentary intended to demystify the notion that great white sharks are ruthless, mindless killers. Great whites have been around for more than 400 million years. Anything that has survived that long isn't "stupid," he said.

Cousteau calls the sub Troy, *in reference to the mythical Trojan horse statue, in which Greek soldiers were spirited into the fortress kingdom of* Troy. *Propelled by a wagging tail and covered in a flexible, skinlike material, the sub—created by Cousteau and a team of scientists and engineers—swims silently. The steel-ribbed, womblike*

Fabien Cousteau waves from inside Aquarius Reef Base, a laboratory 63 feet below the surface in the waters off Key Largo, in the Florida Keys national Marine Sanctuary.

interior is filled with water, requiring Cousteau to wear a wet suit and use scuba gear to breathe.

Importantly, Troy *allows Cousteau to be a shark, not shark bait. At the heart of the project is a desire to observe what great white sharks do when people aren't around to watch. Prior to this, most shark observations have come from humans sitting in cages and enticing the predators with bait—conditions that spawn unnatural behaviors, Cousteau said. "Now all of the sudden we can see what they do as white sharks rather than as trained circus animals," he said.*

While Cousteau is reluctant to guess what the sharks thought when Troy *invaded their space, the explorer said they seemed to act naturally. Some even puffed their gills and gaped toward* Troy—*actions thought to be communication signals. And though a few sharks made aggressive gestures, none of the predators attacked the shark-shaped sub.*

1. What topic is being discussed?
2. What are the main points of the article?
3. Summarize the information in this article, and then provide your analysis based on this information and your own knowledge.

Apply the Skill

Analyze Information Find a short, informative article on a new scientific discovery or new application of science technology, such as hybrid-car technology. Analyze the information and make a statement of your own.

Wilfredo Lee/AP Images

Synthesize Information

Why learn this skill?

The skill of synthesizing involves combining and analyzing information gathered from separate sources or at different times to make logical connections. Being able to synthesize information can be a useful skill for you as a student when you need to gather data from several sources for a report or a presentation.

Learn the Skill

Follow these steps to synthesize information:

- Select important and relevant information.
- Analyze the information and build connections.
- Reinforce or modify the connections as you acquire new information.

Suppose you need to write a research paper on global levels of atmospheric carbon dioxide (CO_2) levels. You need to synthesize what you learn to inform others. You can begin by detailing the ideas and information from sources you already have about global levels of atmospheric carbon dioxide. A table such as **Table 1** could help you categorize the facts from these sources.

Table 1 Global Levels of Atmospheric CO_2

Year	Global Atmospheric CO_2 Concentration (ppm)	Year	Global Atmospheric CO_2 Concentration (ppm)
1745	279	1935	307
1791	280	1949	311
1816	284	1958	312
1843	287	1965	318
1854	288	1974	330
1874	290	1984	344
1894	297	1995	361
1909	299	1998	367
1921	302	2005	385

Then you might select an additional article about greenhouse gases, such as the one below.

According to the National Academy of Scientists, Earth's surface temperature has risen about one degree Fahrenheit in the past 100 years. This increase in temperature can be correlated to an increase in the concentration of carbon dioxide and other greenhouse gases in the atmosphere. How might this increase in temperature affect Earth's climate?

Carbon dioxide is one of the greenhouse gases that helps keep temperatures on Earth warm enough to support life. However, a buildup of carbon dioxide and other greenhouse gases such as methane and nitrous oxide can lead to global warming, an increase in Earth's average surface temperature. Since the industrial revolution in the 1800s, atmospheric concentrations of carbon dioxide have increased by almost 30 percent, methane concentrations have more than doubled, and nitrous oxide concentrations have increased approximately 15 percent. Scientists attribute these increases to the burning of fossil fuels for automobiles, industry, and electricity, as well as deforestation, increased agriculture, landfills, and mining.

Practice the Skill

Use the table and the passage on this page to answer these questions.

1. What information is presented in the table?
2. What is the main idea of the passage? What information does the passage add to your knowledge about the topic?
3. By synthesizing the two sources and using your own knowledge, what conclusions can you draw about global warming?

Apply the Skill

Synthesize Information Find two sources of information on the same topic and write a short report. In your report, answer these questions: What kinds of sources did you use? What are the main ideas of each source? How does each source add to your understanding of the topic? Do the sources support or contradict each other?

Take Notes and Outline

Why learn this skill?

One of the best ways to remember something is to write it down. Taking notes—writing down information in a brief and orderly format—not only helps you remember, but also makes studying easier.

Learn the Skill

There are several styles of note-taking, but the goal of every style is to explain information and put it in a logical order. As you read, identify and summarize the main ideas and details that support them and write them in your notes. Paraphrase—that is, state in your own words—the information rather than copying it directly from the text. Use note cards or develop a personal "shorthand"—using symbols to represent words—to represent the information in a compact manner.

You might also find it helpful to create an outline when taking notes. When outlining material, first read the material to identify the main ideas. In textbooks, look at the section headings for clues to main topics. Then identify the subheadings. Place supporting details under the appropriate headings. The basic pattern for outlines is shown below:

```
Main Topic
  I. First Idea or Item
      A. First Detail
          1. Subdetail
          2. Subdetail
      B. Second Detail
  II. Second Idea or Item
      A. First Detail
      B. Second Detail
          1. Subdetail
          2. Subdetail
  III. Third Idea or Item
```

Practice the Skill

Read the following excerpt from *National Geographic*. Use the steps you just read about to take notes and create an outline. Then answer the questions that follow.

Dinosaur fans still have a lot to look forward to. According to a new estimate of dinosaur diversity, the 21st century will bring an avalanche of new discoveries. "We only know about 29 percent of all dinosaurs out there to be found," said study co-author Peter Dodson, a paleobiologist and anatomy professor at the University of Pennsylvania in Philadelphia.

Dodson and statistics professor Steve Wang of Swarthmore College, in Swarthmore, Pennsylvania, made a statistical analysis of an exhaustive database of all known dinosaur genera (the taxonomic group one notch above species). They then used this data to estimate the total number of genera preserved in the fossil record.

The pair predicts that scientists will eventually discover 1,844 dinosaur genera in total—at least 1,300 more than the 527 recognized today from remains other than isolated teeth. What's more, the duo believes that 75 percent of these dinos will be discovered within the next 60 to 100 years and 90 percent within 100 to 140 years, based on an analysis of historical discovery patterns.

The tally applies only to specimens preserved as fossils. Many other types of dinosaurs likely roamed the Earth during the dinosaurs' 160-million-year reign, but remains from these species will never be known to science, the researchers say.

1. What is the main topic?
2. What are the first, second, and third ideas?
3. Name two details for each of the ideas.
4. Name two subdetails for each of the details.

Apply the Skill

Take Notes and Outline Scan a science journal for a short article about a new laboratory technique. Take notes by using shorthand or by creating an outline. Summarize the article using only your notes.

Understand Cause and Effect

Why learn this skill?

In order to understand an event, you should look for how that event or chain of events came about. When scientists are unsure of the cause for an event, they often design experiments. Although there might be an explanation, an experiment should be performed to be certain the cause created the event you observed. This process examines the causes and effects of events.

Learn the Skill

Calderas can form when the summit or side of a volcano collapses into the magma chamber that once fueled the volcano. An empty magma chamber can *cause* the volcano to collapse. The caldera that forms is the *effect*, or result. The figure below shows how one event—the **cause**—led to another—the **effect**.

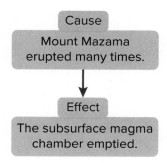

You can often identify cause-and-effect relationships in sentences from clue words such as the following.

because	produced
due to	as a result
so that	that is why
therefore	for this reason
thus	consequently
led to	in order to

Read the sample sentences below.

"The volcano collapsed into the partially empty magma chamber. As a result, a depression was formed where the volcano once stood."

In the example above, the cause is the collapse of the volcano. The cause-and-effect clue words "as a result" tell you that the depression is the effect of the collapsing volcano.

In a chain of events, an effect often becomes the cause of other events. The next chart shows the complete chain of events that occur when a caldera forms.

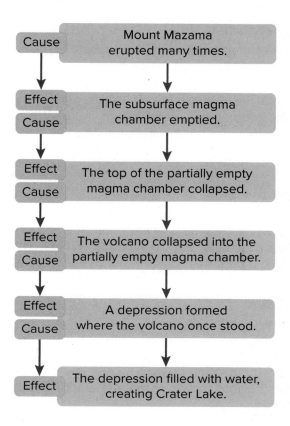

Practice the Skill

Make a chart like the one above showing which events listed below are causes and which are effects.

1. As water vapor rises, it cools and changes back to a liquid.
2. Droplets inside clouds join to form bigger drops.
3. Water evaporates from oceans, lakes, and rivers.
4. Water vapor rises into the atmosphere.
5. Water droplets become heavy and fall as rain or snow.

Apply the Skill

Understand Cause and Effect Read an account of a recent scientific event or discovery in a science journal. Determine at least one cause and one effect of that event. Show the chain of events in a chart.

Analyze Media Sources

Why learn this skill?

To stay informed, people use a variety of media sources, including print media, broadcast media, and electronic media. The Internet has become an especially valuable research tool. It is convenient to use, and the information it contains is plentiful. Whichever media source you use to gather information, it is important to analyze the source to determine its accuracy and reliability.

Learn the Skill

There are a number of issues to consider when analyzing a media source. The most important one is to check the accuracy of the source and content. The author and publishers or sponsors should be credible and clearly indicated. To analyze print media or broadcast media, ask yourself the following questions.

- Is the information current?
- Are the sources revealed?
- Is more than one source used?
- Is the information biased?
- Does the information represent both sides of an issue?
- Is the information reported firsthand or secondhand?

For electronic media, ask yourself these questions in addition to the ones above.

- Is the author credible and clearly identified?
- Are the facts on the Web site documented?
- Are the links within the Web site appropriate and current?
- Does the Web site contain links to other useful resources?

Practice the Skill

To practice analyzing print media, choose two articles on global warming, one from a newspaper and the other from a newsmagazine. Then answer these questions.

1. What points are the authors of the articles trying to make? Were they successful? Can the facts be verified?
2. Did either article reflect a bias toward one viewpoint or another? List any unsupported statements.

3. Was the information reported firsthand or secondhand? Do the articles seem to represent both sides fairly?
4. How many sources can you identify in the articles? List them.

To analyze electronic media, read through the list of links provided by your teacher. Choose one link from the list, read the information on that Web site, and then answer these questions.

1. Who is the author or sponsor of the Web site?
2. What links does the Web site contain? How are they appropriate to the topic?
3. What sources were used for the information on the Web site?

Apply the Skill

Analyze Media Sources Think of a national issue on which public opinion is divided. Read newspaper features, editorials, and Web sites, and monitor television reports about the issue. Which news sources more fairly represents the issue? Which news sources have the most reliable information? Can you identify any biases? Can you verify the credibility of the news source?

Don Mason/Blend Images LLC

Use Graphic Organizers

Why learn this skill?

While you read this textbook, you will be looking for important ideas or concepts. One way to arrange these ideas is to create a graphic organizer. In addition to Foldables®, you will find various other graphic organizers throughout your book. Some organizers show a sequence, or flow, of events. Other organizers emphasize the relationship among concepts. Developing your own organizers while you read will help you better understand and remember what you read.

Learn the Skill

An **events chain concept map** is used to describe a sequence of events, such as a stage of a process or procedure. When making an events-chain map, first identify the event that starts the sequence and add events in chronological order until you reach an outcome.

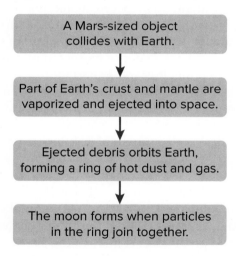

In a **cycle concept map,** the series of events do not produce a final outcome. The event that appears to be the final event relates back to the initiating event. Therefore, the cycle repeats itself.

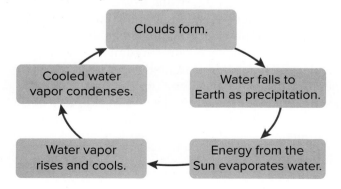

A **network tree concept map** shows the relationship among concepts, which are written in order from general to specific. The words written on the lines between the circles, called linking words, describe the relationships among the concepts. The concepts and the linking words can form sentences.

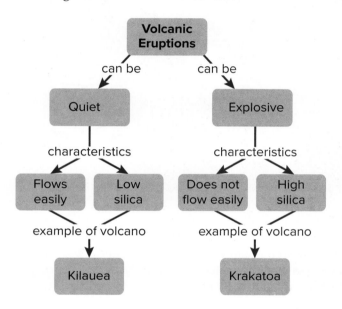

Practice the Skill

1. Create an events chain concept map of the events in sedimentary rock formation.
2. Create a cycle concept map of the nitrogen cycle. Make sure that the cycle shows the event that appears to be the final event relating back to the starting event.
3. Create a network tree concept map with these words: *Cenozoic, trilobites, eras, Paleozoic, mammals, dinosaurs, first land plants, Gondwana, Mesozoic, early Pangaea, late Pangaea.* Add linking words to describe the relationships between the concepts.

Apply the Skill

Use Graphic Organizers Create an events chain concept map of the scientific method. Create a cycle concept map of the water cycle. Create a network tree concept map of pollution that includes air and water, sources of each pollution type, and examples of each type of pollution.

Debate Skills

New research always is leading to new scientific theories. There are often opposing points of view on how this research is conducted, how it is interpreted, and how it is communicated. *The Earth Science and Society* features in your book offer a chance to debate a current controversial topic. Here is an overview on how to conduct a debate.

Choose a Position and Research

First, choose an Earth science issue that has at least two opposing viewpoints. The issue can come from current events, your textbook, or your teacher. These topics could include global warming or fossil fuel use. Topics are stated as affirmative declarations such as "Global warming is not detrimental to the environment."

One speaker will argue the positive position—the viewpoint that supports the statement—and another speaker will argue the negative position—the viewpoint that disputes the statement. Either individually or with a group, choose your position for the debate. The viewpoint that you choose does not have to reflect your personal belief. The purpose of debate is to create a strong argument supported by scientific evidence.

After choosing your position, conduct research to support your viewpoint. Use the Internet, find articles in your library, or use your textbook to gather evidence to support your argument.

A strong argument contains scientific evidence, expert opinions, and your own analysis of the issue. Research the opposing position also. Becoming aware of what points the other side might argue will help you to strengthen the evidence for your position.

Hold the Debate

You will have a specific amount of time, determined by your teacher, in which to present your argument. Organize your speech to fit within the time limit: explain the viewpoint that you will be arguing, present an analysis of your evidence, and conclude by summing up your most important points. Try to vary the elements of your argument. Your speech should not be a list of facts, a reading of a newspaper article, or a statement of your personal opinion, but an organized analysis of your evidence presented in your own manner of speaking. It is also important to remember that you must never make personal attacks against your opponent. Argue the issue. You will be evaluated on your overall presentation, organization and development of ideas, and strength of support for your argument.

Additional Roles There are other roles that you can play in a debate. You can act as the timekeeper. The timekeeper times the length of the debaters' speeches and gives quiet signals to the speaker when time is almost up (usually a hand signal).

You can also act as a judge. There are important elements to look for when judging a speech: an introduction that tells the audience what position the speaker will be arguing, strong evidence that supports the speaker's position, and organization. It is helpful to take notes during the debate to summarize the main points of each side's argument. Then, decide which debater presented the strongest argument for his or her position. You can have a class discussion about the strengths and weaknesses of the debate and other viewpoints on this issue that could be argued.

Experimental data is often expressed using numbers and units. The following sections provide an overview of the common system of units and some calculations involving units.

Measure in SI

The International System of Measurements, abbreviated SI, is accepted as the standard for measurement throughout most of the world. The SI system contains seven base units. All other units of measurement can be derived from these base units.

Table 2 SI Base Units

Measurement	Unit	Symbol
Length	meter	m
Mass	kilogram	kg
Time	second	s
Electric current	ampere	A
Temperature	kelvin	K
Amount of substance	mole	mol
Intensity of light	candela	cd

Some units are derived by combining base units. For example, units for volume are derived from units of length. A liter (L) is a cubic decimeter (dm^3, or $dm \times dm \times dm$). Units of density (g/L) are derived from units of mass (g) and units of volume (L).

When units are multiplied by factors of ten, new units are created. For example, if a base unit is multiplied by 1000, the new unit has the prefix *kilo-*. One thousand meters is equal to one kilometer. Prefixes for some units are shown in **Table 3**.

To convert a given unit to a unit with a different factor of ten, multiply the unit by a conversion factor. A conversion factor is a ratio equal to one. The equivalents in **Table 3** can be used to make such a ratio. For example, 1 km = 1000 m. Two conversion factors can be made from this equivalent.

$$\frac{1000 \text{ m}}{1 \text{ km}} = 1 \quad \text{and} \quad \frac{1 \text{ km}}{1000 \text{ m}} = 1$$

To convert one unit to another factor of ten, choose the conversion factor that has the unit you are converting from in the denominator.

$$1 \text{ km} \times \frac{1000 \text{ m}}{1 \text{ km}} = 1000 \text{ m}$$

A unit can be multiplied by several conversion factors to obtain the desired unit.

Table 3 Common SI Prefixes

Prefix	Symbol	Equivalents
mega-	m	1×10^6 base units
kilo-	k	1×10^3 base units
hecto-	h	1×10^2 base units
deka-	da	1×10^1 base units
deci-	d	1×10^{-1} base units
centi-	c	1×10^{-2} base units
milli-	m	1×10^{-3} base units
micro-	μ	1×10^{-6} base units
micro-	n	1×10^{-9} base units
pico-	p	1×10^{-12} base units

Practice Problem 1 How would you convert 1000 micrometers to kilometers?

Convert Temperature

The following formulas can be used to convert between Fahrenheit and Celsius temperatures. Notice that each equation can be obtained by algebraically rearranging the other. Therefore, you only need to remember one of the equations.

Conversion of Fahrenheit to Celsius

$$°C = \frac{(°F) - 32}{1.8}$$

Conversion of Celsius to Fahrenheit

$$°F = 1.8(°C) + 32$$

Make and Use Tables

Tables help visually organize data so that it can be interpreted more easily. Tables are composed of several components—a title describing the contents of the table, columns and rows that separate and organize information, and headings that describe the information in each column or row.

Table 4 Glacier Movement Rates

Depth (m)	Distance (m)	Average Speed (m/day)
0	13.1	0.198
20	13.1	0.198
60	12.8	0.194
100	12.2	0.185
140	11.2	0.170
180	9.6	0.145

Looking at this table, you should not only be able to pick out specific information, but you should also notice trends.

Practice Problem 2 If scientists drilled another 40 m into the glacier, what would the speed of the glacier's movement be at that depth?

Make and Use Graphs

Scientists often organize data in graphs. The types of graphs typically used in science are the line graph, the bar graph, and the circle graph.

Line Graphs A line graph is used to show the relationship between two variables. The independent variable is plotted on the horizontal axis, called the x-axis. The dependent variable is plotted on the vertical axis, called the y-axis. The dependent variable (y) changes as a result of a change in the independent variable (x).

Suppose your class wanted to collect data about humidity. You could make a graph of the amount of water vapor that air can hold at various temperatures. **Table 5** shows the data.

Table 5 Amount of Water Vapor in Air at Various Temperatures

Air Temperature (°C)	Air (g/m³)
10	10
20	18
30	31
40	50
50	80

To make a graph of the amount of water vapor in air, start by determining the dependent and independent variables. The average amount of water vapor found per cubic meter of air is the dependent variable and is plotted on the y-axis. The independent variable, air temperature, is plotted on the x-axis.

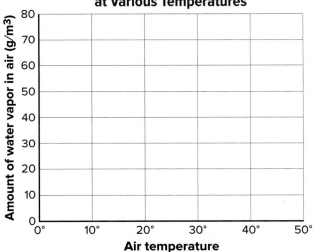

Amount of Water Vapor in Air at Various Temperatures

Plain or graph paper can be used to construct graphs. Draw a grid on your paper or a box around the squares that you intend to use on your graph paper. Give your graph a title and label each axis with a title and units. In this example, label the x-axis *Air temperature*. Because the lowest temperature was 10 and the highest was 50, you know that you will have to start numbers on the y-axis at least at 0 and number to at least 50. You decide to start numbering at 0 and number by equally spaced intervals of ten.

Label the *y*-axis of your graph *Amount of water vapor in air (g/m³)*. Begin plotting points by locating 0°C on the *x*-axis and 5 g/m³ on the *y*-axis. Where an imaginary vertical line from the *x*-axis and an imaginary horizontal line from the *y*-axis meet, place the first data point. Place other data points using the same process. After all the points are plotted, draw a "best fit" straight line through all the points.

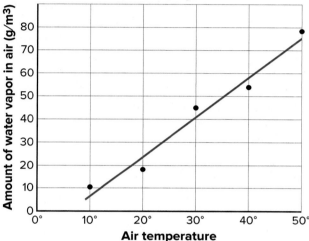

Practice Problem 3 According to the graph, does the amount of water vapor in air increase or decrease with air temperature?

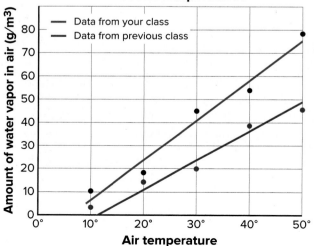

What if you wanted to compare the data about humidity collected by your class with similar data collected a year ago by a different class? The data from the other class can be plotted on the same graph to make the comparison. Include a key with different lines indicating different sets of data.

Practice Problem 4 How did the data from your class compare to the data from the previous class?

Bar Graphs A bar graph displays a comparison of different categories of data by representing each category with a bar. The length of the bar is related to the category's frequency. To make a bar graph, set up the *x*-axis and *y*-axis as you did for the line graph. Plot the data by drawing thick bars from the *x*-axis up to the *y*-axis point.

Net Energy Efficiency

Look at the graph above. The independent variable is the energy efficiency. The dependent variable is the heating method.

Practice Problem 5 Which type of heating method has the second greatest efficiency? Is this more than twice as efficient as the lowest efficiency? Explain.

Bar graphs can also be used to display multiple sets of data in different categories at the same time. A bar graph that displays two sets of data is called a double-bar graph. Double-bar graphs have a legend to denote which bars represent each set of data. The graph below is an example of a double-bar graph.

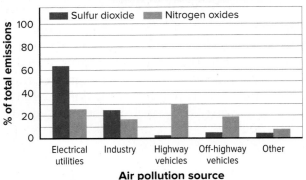

Sources of Acid Precipitation in the United States

Circle Graphs A circle graph consists of a circle divided into sections that represent parts of a whole. When all the sections are placed together, they equal 100 percent of the whole.

Suppose you want to make a circle graph to show the percentage of solid wastes generated by various industries in the United States each year. The total amount of solid waste generated each year is estimated at ten billion metric tons. The whole circle graph will therefore represent this amount of solid waste. You find that 7.5 billion metric tons of waste is generated by mining and oil and gas production. The total amount of solid waste generated each year by mining and oil and gas production makes up one section of the circle graph, as follows.

$$\text{Segment of circle for total waste} = \frac{\text{waste from mining and oil and gas production}}{\text{total waste}}$$

$$= \frac{7.5}{10}$$

$$= 0.75 \times 360°$$

$$= 270°$$

To draw your circle graph, you will need a compass and a protractor. First, use the compass to draw a circle.

Then, draw a straight line from the center to the edge of the circle. Place your protractor on this line, and mark the point on the circle where 270° angle will intersect the circle. Draw a straight line from the center of the circle to the intersection point. This is the section for the waste generated from mining and oil and gas production.

Now, try to perform the same operation for the other data to find the number of degrees of the circle that each represents, and draw them in as well: agriculture, 1.3 billion metric tons; industry, 0.95 billion metric tons; municipal, 0.15 billion metric tons; and sewage sludge, 0.1 billion metric tons.

Complete your graph by labeling the sections of the graph and giving the graph a title. Your completed graph should look similar to the one below.

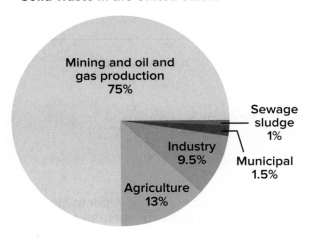

Solid Waste in the United States

Practice Problem 6 There are 25 varieties of flowering plants growing around the high school. Construct a circle graph showing the percentage of each flower's color. Two varieties have yellow blooms, five varieties have blue-purple blooms, eight varieties have white blooms, and ten varieties have red blooms.

Safety in the Laboratory

The Earth science laboratory is a safe place to work if you are careful to observe the following important safety rules. You are responsible for your own safety and for the safety of others. The safety rules given here will protect you and others from harm in the laboratory. While carrying out procedures in any of the activities or GeoLabs, take note of the safety symbols and warning statements.

Safety Rules

1. Always read and complete the lab safety form and obtain your teacher's permission before beginning an investigation.
2. Study the procedure outline in the text. If you have questions, ask your teacher. Make sure that you understand all safety symbols shown on the page.
3. Use the safety equipment provided for you. Safety goggles and an apron should be worn during all investigations that involve the use of chemicals.
4. When heating test tubes, always slant them away from yourself and others.
5. Never eat or drink in the lab, and never use lab glassware as food or drink containers. Never inhale chemicals. Do not taste any substances or draw any material into a tube or pipet with your mouth.
6. If you spill any chemical, wash it off immediately with water. Report the spill immediately to your teacher.
7. Know the location and proper use of the fire extinguisher, eye wash, safety shower, fire blanket, fire alarm, and first aid kit. First aid procedures in the science laboratory are listed in **Table 1.**
8. Keep materials away from flames. Tie back hair and loose clothing when you are working with flames.
9. If a fire should break out in the lab, or if your clothing should catch fire, smother it with the fire blanket or a coat, get under a safety shower, or use the fire department's recommendation for putting out a fire on your clothing: stop, drop, and roll. NEVER RUN.
10. Report any accident or injury, no matter how small, to your teacher.

Clean-Up Procedures

1. Turn off the water and gas. Disconnect electrical devices.
2. Return all materials to their proper places.
3. Dispose of chemicals and other materials as directed by your teacher. Place broken glass and solid substances in the proper containers. Never discard materials in the sink.
4. Clean your work area.
5. Wash your hands thoroughly after working in the laboratory.

Table 1 First Aid in the Science Laboratory

Injury	Safe Response
Burns	Apply cold water. Call your teacher immediately.
Cuts and bruises	Stop any bleeding by applying direct pressure. Cover cuts with a clean dressing. Apply cold compresses to bruises. Call your teacher immediately.
Fainting	Leave the person lying down. Loosen any tight clothing and keep crowds away. Call your teacher immediately.
Foreign matter in eye	Flush with plenty of water. Use an eyewash bottle or fountain.
Poisoning	Note the suspected poisoning agent and call your teacher immediately.
Any spills on skin	Flush with large amounts of water or use safety shower. Call your teacher immediately.

Safety Symbols

Safety symbols in the following table are used in the lab activities to indicate possible hazards. Learn the meaning of each symbol. **It is recommended that you wear safety goggles and apron at all times in the lab. This might be required in your school district.**

Safety Symbols		Hazard	Examples	Precaution	Remedy
Disposal		Special disposal procedures need to be followed.	certain chemicals, living organisms	Do not dispose of these materials in the sink or trash can.	Dispose of wastes as directed by your teacher.
Biological		Organisms or other biological materials that might be harmful to humans	bacteria, fungi, blood, unpreserved tissues, plant materials	Avoid skin contact with these materials. Wear mask or gloves.	Notify your teacher if you suspect contact with material. Wash hands thoroughly.
Extreme Temperature		Objects that can burn skin by being too cold or too hot	boiling liquids, hot plates, dry ice, liquid nitrogen	Use proper protection when handling.	Go to your teacher for first aid.
Sharp Object		Use of tools or glassware that can easily puncture or slice skin	razor blades, pins, scalpels, pointed tools, dissecting probes, broken glass	Practice common-sense behavior and follow guidelines for use of the tool.	Go to your teacher for first aid.
Fume		Possible danger to respiratory tract from fumes	ammonia, acetone, nail polish remover, heated sulfur, moth balls	Be sure there is good ventilation. Never smell fumes directly. Wear a mask.	Leave foul area and notify your teacher immediately.
Electrical		Possible danger from electrical shock or burn	improper grounding, liquid spills, short circuits, exposed wires	Double-check setup with teacher. Check condition of wires and apparatus. Use GFI-protected outlets.	Do not attempt to fix electrical problems. Notify your teacher immediately.
Irritant		Substances that can irritate the skin or mucous membranes of the respiratory tract	pollen, moth balls, steel wool, fiberglass, potassium permanganate	Wear dust mask and gloves. Practice extra care when handling these materials.	Go to your teacher for first aid.
Chemical		Chemicals that can react with and destroy tissue and other materials	bleaches such as hydrogen peroxide; acids such as sulfuric acid, hydrochloric acid; bases such as ammonia, sodium hydroxide	Wear goggles, gloves, and an apron.	Immediately flush the affected area with water and notify your teacher.
Toxic		Substance may be poisonous if touched, inhaled, or swallowed.	mercury, many metal compounds, iodine, poinsettia plant parts	Follow your teacher's instructions.	Always wash hands thoroughly after use. Go to your teacher for first aid.
Flammable		Flammable chemicals may be ignited by open flame, spark, or exposed heat.	alcohol, kerosene, potassium permanganate	Avoid open flames and heat when using flammable chemicals.	Notify your teacher immediately. Use fire safety equipment if applicable.
Open Flame		Open flame in use, may cause fire.	hair, clothing, paper, synthetic materials	Tie back hair and loose clothing. Follow teacher's instruction on lighting and extinguishing flames.	Notify your teacher immediately. Use fire safety equipment if applicable.

 Eye Safety Proper eye protection should be worn at all times by anyone performing or observing science activities.

 Clothing Protection This symbol appears when substances could stain or burn clothing.

 Animal Safety This symbol appears when safety of animals and students must be ensured.

 Radioactivity This symbol appears when radioactive materials are used.

 **Handwashing** After the lab, wash hands with soap and water before removing goggles.

PERIODIC TABLE OF THE ELEMENTS

Key:

Atomic number	1
Symbol	**H**
Element	Hydrogen
Atomic mass	1.008

Legend: Metal · Metalloid · Nonmetal · ⊙ Synthetic

Main table (Group / Period):

Period \ Group	1	2	3	4	5	6	7	8	9	10	11	12	13	14	15	16	17	18
1	1 **H** Hydrogen 1.008																	2 **He** Helium 4.003
2	3 **Li** Lithium 6.941	4 **Be** Beryllium 9.012											5 **B** Boron 10.811	6 **C** Carbon 12.011	7 **N** Nitrogen 14.007	8 **O** Oxygen 15.999	9 **F** Fluorine 18.998	10 **Ne** Neon 20.180
3	11 **Na** Sodium 22.990	12 **Mg** Magnesium 24.305											13 **Al** Aluminum 26.982	14 **Si** Silicon 28.086	15 **P** Phosphorus 30.974	16 **S** Sulfur 32.066	17 **Cl** Chlorine 35.453	18 **Ar** Argon 39.948
4	19 **K** Potassium 39.098	20 **Ca** Calcium 40.078	21 **Sc** Scandium 44.956	22 **Ti** Titanium 47.867	23 **V** Vanadium 50.942	24 **Cr** Chromium 51.996	25 **Mn** Manganese 54.938	26 **Fe** Iron 55.847	27 **Co** Cobalt 58.933	28 **Ni** Nickel 58.693	29 **Cu** Copper 63.546	30 **Zn** Zinc 65.39	31 **Ga** Gallium 69.723	32 **Ge** Germanium 72.61	33 **As** Arsenic 74.922	34 **Se** Selenium 78.971	35 **Br** Bromine 79.904	36 **Kr** Krypton 83.80
5	37 **Rb** Rubidium 85.468	38 **Sr** Strontium 87.62	39 **Y** Yttrium 88.906	40 **Zr** Zirconium 91.224	41 **Nb** Niobium 92.906	42 **Mo** Molybdenum 95.95	43 **Tc** Technetium (98) ⊙	44 **Ru** Ruthenium 101.07	45 **Rh** Rhodium 102.906	46 **Pd** Palladium 106.42	47 **Ag** Silver 107.868	48 **Cd** Cadmium 112.411	49 **In** Indium 114.82	50 **Sn** Tin 118.710	51 **Sb** Antimony 121.757	52 **Te** Tellurium 127.60	53 **I** Iodine 126.904	54 **Xe** Xenon 131.290
6	55 **Cs** Cesium 132.905	56 **Ba** Barium 137.327	57 **La** Lanthanum 138.905	72 **Hf** Hafnium 178.49	73 **Ta** Tantalum 180.948	74 **W** Tungsten 183.84	75 **Re** Rhenium 186.207	76 **Os** Osmium 190.23	77 **Ir** Iridium 192.217	78 **Pt** Platinum 195.08	79 **Au** Gold 196.967	80 **Hg** Mercury 200.59	81 **Tl** Thallium 204.383	82 **Pb** Lead 207.2	83 **Bi** Bismuth 208.980	84 **Po** Polonium 208.982	85 **At** Astatine 209.987	86 **Rn** Radon 222.018
7	87 **Fr** Francium (223)	88 **Ra** Radium (226)	89 **Ac** Actinium (227)	104 **Rf** Rutherfordium * (267) ⊙	105 **Db** Dubnium * (270) ⊙	106 **Sg** Seaborgium * (269) ⊙	107 **Bh** Bohrium * (270) ⊙	108 **Hs** Hassium * (277) ⊙	109 **Mt** Meitnerium * (278) ⊙	110 **Ds** Darmstadtium * (281) ⊙	111 **Rg** Roentgenium * (281) ⊙	112 **Cn** Copernicium * (285) ⊙	113 **Nh** Nihonium * (286) ⊙	114 **Fl** Flerovium * (289) ⊙	115 **Mc** Moscovium * (289) ⊙	116 **Lv** Livermorium * (293) ⊙	117 **Ts** Tennessine * (294) ⊙	118 **Og** Oganesson * (294) ⊙

Lanthanide series

58 **Ce** Cerium 140.115	59 **Pr** Praseodymium 140.908	60 **Nd** Neodymium 144.242	61 **Pm** Promethium (145) ⊙	62 **Sm** Samarium 150.36	63 **Eu** Europium 151.965	64 **Gd** Gadolinium 157.25	65 **Tb** Terbium 158.925	66 **Dy** Dysprosium 162.50	67 **Ho** Holmium 164.930	68 **Er** Erbium 167.259	69 **Tm** Thulium 168.934	70 **Yb** Ytterbium 173.04	71 **Lu** Lutetium 174.967

Actinide series

90 **Th** Thorium 232.038	91 **Pa** Protactinium 231.036	92 **U** Uranium 238.029	93 **Np** Neptunium (237) ⊙	94 **Pu** Plutonium (244)	95 **Am** Americium (243) ⊙	96 **Cm** Curium (247)	97 **Bk** Berkelium (247)	98 **Cf** Californium (251)	99 **Es** Einsteinium * (252) ⊙	100 **Fm** Fermium * (257) ⊙	101 **Md** Mendelevium * (258) ⊙	102 **No** Nobelium * (259) ⊙	103 **Lr** Lawrencium * (262) ⊙

The number in parentheses is the mass number of the longest-lived isotope for that element.

* Properties are largely predicted.

Reference Handbook

Physiographic Map of Earth

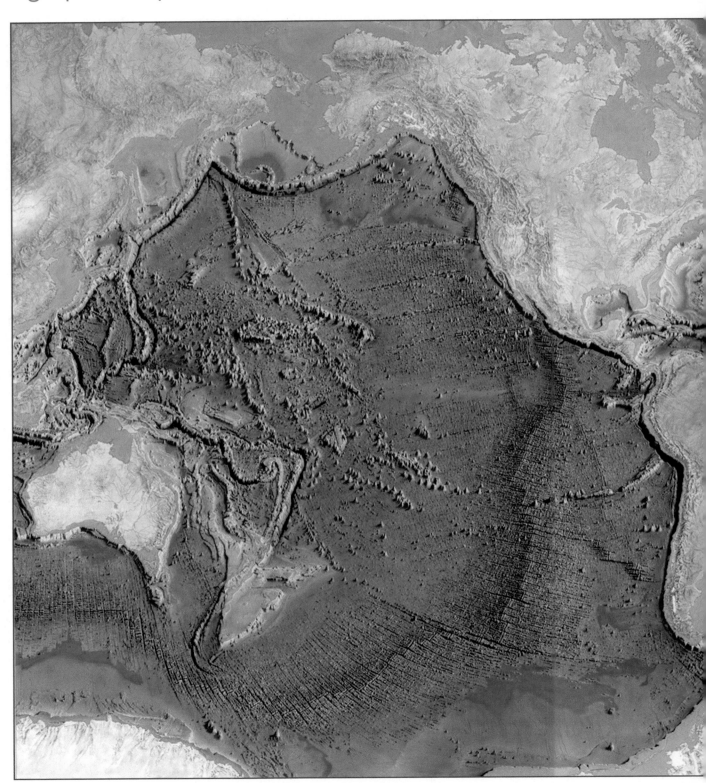

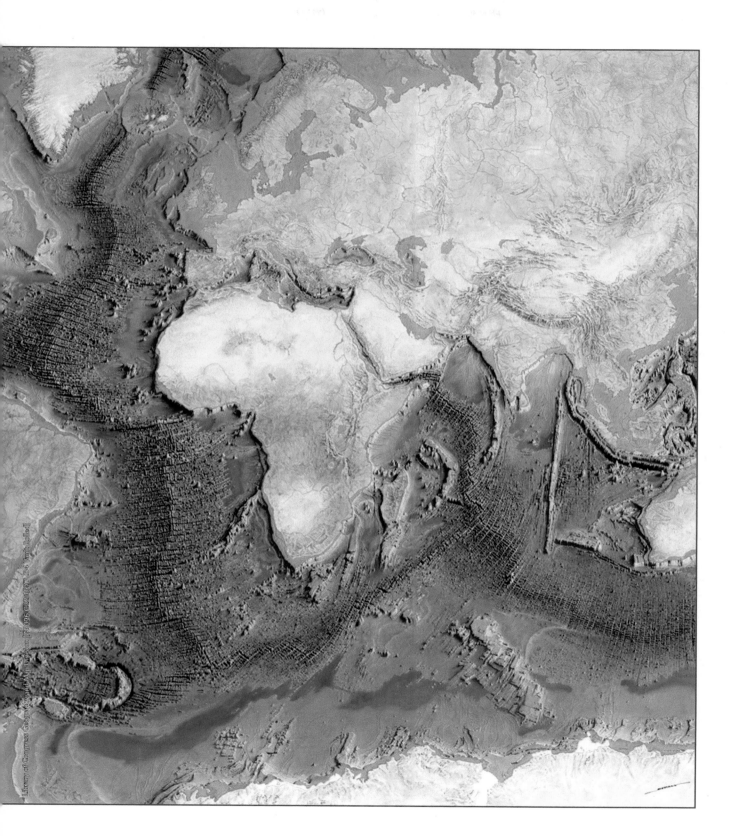

Topographic Map Symbols

Roads and Railroads

Primary highway, hard surface	
Secondary highway, hard surface	
Light-duty road, hard or improved surface	
Unimproved road	
Railroad: single track and multiple track	
Railroads in juxtaposition	

Buildings and Structures

Buildings	
School, church, and cemetery	cem
Barn and warehouse	
Wells, not water (with labels)	o oil o gas
Tanks: oil, water, etc. (labeled if water)	water
Open-pit mine, quarry, or prospect	
Tunnel	
Benchmark	BM Δ 293
Bridge	
Campsite	

Habitats

Marsh (swamp)	
Wooded marsh	
Woods or brushwood	
Vineyard	
Submerged marsh	
Mangrove	
Coral reef, rocks	
Orchard	

Urban area	
Perennial streams	
Elevated aqueduct	
Water well and spring	
Small rapids	
Large rapids	
Intermittent lake	
Intermittent stream	
Glacier	
Large falls	
Dry lake bed	

Surface Elevations

Spot elevation	×7369
Water elevation	670
Index contour	100
Intermediate contour	
Depression contour	

Boundaries

National	
State	
County, parish, municipal	
Civil township, precinct, town, barrio	
Incorporated city, village, town, hamlet	
Reservation, national or state	
Small park, cemetery, airport, etc.	
Land grant	
Township or range line, United States land survey	
Township or range line, approximate location	

Weather Map Symbols

Sample Plotted Report at Each Station

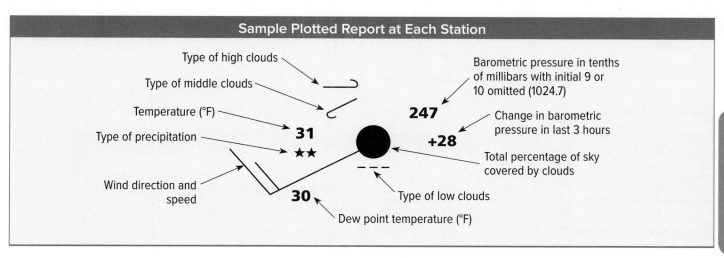

Type of high clouds

Type of middle clouds

Temperature (°F)

Type of precipitation

Wind direction and speed

31

★★

30

247

+28

Barometric pressure in tenths of millibars with initial 9 or 10 omitted (1024.7)

Change in barometric pressure in last 3 hours

Total percentage of sky covered by clouds

Type of low clouds

Dew point temperature (°F)

Symbols Used in Plotting Report

Precipitation	Wind Direction and Speed	Sky Coverage	Fronts and Pressure Systems
☰ Fog	○ 0 calm	○ No cover	(H) or High — Center of high- or
★ Snow	1–2 knots	1/10 or less	(L) or Low — low-pressure system
● Rain	3–7 knots	2/10 to 3/10	▲▲▲▲ Cold front
⦦ Thunderstorm	8–12 knots	4/10	●●● Warm front
❥ Drizzle	13–17 knots	1/2	▲▲●● Occluded front
▽ Showers	18–22 knots	6/10	▲▽▲▽ Stationary front
	23–27 knots	7/10	
	48–52 knots	Overcast with openings	
	1 knot = 1.852 km/h	Completely overcast	

Clouds

Some Types of High Clouds	Some Types of Middle Clouds	Some Types of Low Clouds
Scattered cirrus	Thin altostratus layer	Cumulus of fair weather
Dense cirrus in patches	Thick altostratus layer	Stratocumulus
Veil of cirrus covering entire sky	Thin altocumulus in patches	Fractocumulus of bad weather
Cirrus not covering entire sky	Thin altocumulus in bands	Stratus of fair weather

Table 2 Relative Humidity %

Dry-Bulb Temperature	Dry-Bulb Temperature Minus Wet-Bulb Temperature (°C)									
	1	2	3	4	5	6	7	8	9	10
0°C	81	64	46	29	13					
1°C	83	66	49	33	18					
2°C	84	68	52	37	22	7				
3°C	84	69	55	40	25	12				
4°C	85	71	57	43	29	16				
5°C	85	72	58	45	32	20				
6°C	86	73	60	48	35	24	11			
7°C	86	74	61	49	38	26	15			
8°C	87	75	63	51	40	29	19	8		
9°C	87	76	65	53	42	32	21	12		
10°C	88	77	66	55	44	34	24	15	6	
11°C	89	78	67	56	46	36	27	18	9	
12°C	89	78	68	58	48	39	29	21	12	
13°C	89	79	69	59	50	41	32	22	15	7
14°C	90	79	70	60	51	42	34	26	18	10
15°C	90	80	71	61	53	44	36	27	20	13
16°C	90	81	71	63	54	46	38	30	23	15
17°C	90	81	72	64	55	47	40	32	25	18
18°C	91	82	73	65	57	49	41	34	27	20
19°C	91	82	74	65	58	50	43	36	29	22
20°C	91	83	74	66	59	51	44	37	31	24
21°C	91	83	75	67	60	53	46	39	32	26
22°C	92	83	76	68	61	54	47	40	34	28
23°C	92	84	76	69	62	55	48	42	36	30
24°C	92	84	77	69	62	56	49	43	37	31
25°C	92	84	77	70	63	57	50	44	39	33
26°C	92	85	78	71	64	58	51	46	40	34
27°C	92	85	78	71	65	58	52	47	41	36
28°C	93	85	78	72	65	59	53	48	42	37
29°C	93	86	79	72	66	60	54	49	43	38
30°C	93	86	79	73	67	61	55	50	44	39
31°C	93	86	80	73	67	62	56	50	45	40
32°C	93	86	80	74	68	62	57	51	46	41

Table 3 Minerals with Metallic Luster

Mineral (Formula)	Color	Streak	Hardness	Specific Gravity	Crystal System	Breakage Pattern	Uses and Other Properties
Bornite (Cu_5FeS_4)	bronze, tarnishes to dark blue purple	gray-black	3	4.9–5.4	tetragonal	uneven fracture	source of copper; called "peacock ore" because of the purple shine when it tarnishes
Chalcopyrite ($CuFeS_2$)	brassy to yellow	greenish black	3.5–4	4.2	tetragonal	uneven fracture	main ore of copper
Chromite ((Fe, Mg)(Cr, $Al)_2O_4$)	black or brown	brown to black	5.5	4.6	cubic	irregular fracture	ore of chromium, stainless steel, metallurgical bricks
Copper (Cu)	copper red	copper red	3	8.5–9	cubic	hackly	coins, pipes, gutters, wire, cooking utensils, jewelry; malleable and ductile
Galena (PbS)	gray	gray to black	2.5	7.5	cubic	cubic cleavage perfect	source of lead, used in pipes, shields for X rays, fishing equipment sinkers
Gold (Au)	pale to golden yellow	yellow	2.5–3	19.3	cubic	hackly	jewelry, money, gold leaf, fillings for teeth, medicines; does not tarnish
Graphite (C)	black to gray	black to gray	1–2	2.3	hexagonal	basal cleavage (scales)	pencil lead, lubricants for locks, rods to control some small nuclear reactions, battery poles
Hematite (specular) (Fe_2O_3)	black or reddish brown	red or red-dish brown	6	5.3	hexagonal	irregular fracture	source of iron; roasted in a blast furnace, converted to "pig" iron, made into steel
Magnetite ((Fe, Mg) Fe_2O_4)	black	black	6	5.2	cubic	conchoidal fracture	source of iron, naturally magnetic, called lodestone
Pyrite (FeS_2)	light, brassy yellow	greenish black	6.5	5.0	cubic	uneven fracture	source of iron, "fool's gold," alters to limonite
Pyrrhotite ($Fe_{1-x}S$)* *contains one less atom of Fe than S	bronze	gray-black	4	4.6	hexagonal	uneven fracture	an ore of iron and sulfur; may be magnetic
Silver (Ag)	silvery white, tarnishes to black	light gray to silver	2.5	10–12	cubic	hackly	coins, fillings for teeth, jewelry, silverplate, wires; malleable and ductile

Table 4 Minerals with Nonmetallic Luster

Mineral (Formula)	Color	Streak	Hardness	Specific Gravity	Crystal System	Breakage Pattern	Uses and Other Properties
Augite ((Ca, Na) (Mg, Fe^2, Al) (Al, Si)$_2$O$_6$)	black	colorless	6	3.3	monoclinic	2-directional cleavage	square or 8-sided cross section
Corundum (Al$_2$O$_3$)	colorless, blue, brown, green, white, pink, red	colorless	9	4.0	hexagonal	fracture	gemstones: ruby is red, sapphire is blue; abrasive
Fluorite (CaF$_2$)	colorless, white, blue, green, red, yellow, purple	colorless	4	3–3.2	cubic	cleavage	manufacture of optical equipment; glows under UV light
Garnet ((Mg, Fe^2, Ca, Mn^2)$_3$, (Al, Fe^3, Mn^3, V, Cr)$_2$, (SiO$_4$)$_3$)	deep yellow-red, green, black	colorless	7.5	3.5	cubic	conchoidal fracture	used in jewelry; also used as an abrasive
Hornblende ((Ca, Na)$_{2-3}$ (Mg, Fe^2, Fe^3, Al)$_5$, (Al, Si)$_8$O$_{22}$ (OH)$_2$)	green to black	gray to white	5–6	3.4	monoclinic	cleavage in two directions	will transmit light on thin edges; 6-sided cross section
Limonite (hydrous iron oxides)	yellow, brown, black	yellow, brown	5.5	2.7–4.3	N/A	conchoidal fracture	source of iron; weathers easily, coloring matter of soils
Olivine ((Mg, Fe)$_2$ SiO$_4$)	olive green	colorless	6.5	3.5	orthorhombic	conchoidal fracture	gemstones, refractory sand
Plagioclase feldspar ((Na, Ca) Al(Si, Al) Si$_2$O$_8$)	gray, green, white	colorless	6	2.5	triclinic	two cleavage planes meet at 86° angle	used in ceramics; striations present on some faces
Potassium feldspar (KAlSi$_3$O$_8$)	colorless, white to gray, green, yellow, pink	colorless	6	2.5	monoclinic	two cleavage planes meet at 90° angle	insoluble in acids; used in the manufacture of porcelain
Quartz (SiO$_2$)	colorless, various colors	colorless	7	2.6	hexagonal	conchoidal fracture	glass manufacture, electronic equipment, radios, computers, watches, gemstones
Topaz ((Al$_2$SiO$_4$ (F, OH)$_2$)	colorless, white, pink, yellow, pale blue	colorless	8	3.5	orthorhombic	basal cleavage	valuable gemstone

Table 5 Common Rocks

Rock Type	Rock Name	Characteristics
Igneous (intrusive)	granite	large mineral grains of quartz, feldspar, hornblende, and mica; usually light in color
	diorite	large mineral grains of feldspar, hornblende, and mica; less quartz than granite; intermediate in color
	gabbro	large mineral grains of feldspar, hornblende, augite, olivine, and mica; no quartz; dark in color
Igneous (extrusive)	rhyolite	small or no visible grains of quartz, feldspar, hornblende, and mica; light in color
	andesite	small or no visible grains of quartz, feldspar, hornblende, and mica; less quartz than rhyolite; intermediate in color
	basalt	small or no visible grains of feldspar, hornblende, augite, olivine, and mica; no quartz; dark in color; vesicles may be present
	obsidian	glassy texture; no visible grains; volcanic glass; fracture is conchoidal; color is usually black, but may be red-brown or black with white flecks
	pumice	frothy texture; floats; usually light in color
Sedimentary (clastic)	conglomerate	coarse-grained; gravel- or pebble-sized grains
	sandstone	sand-sized grains 1/16 to 2 mm in size; varies in color
	siltstone	grains smaller than sand but larger than clay
	shale	smallest grains; usually dark in color
Sedimentary (chemical or biochemical)	limestone	major mineral is calcite; usually forms in oceans, lakes, rivers, and caves; often contains fossils; effervesces in dilute HCl
	coal	occurs in swampy, low-lying areas; compacted layers of organic material, mainly plant remains
Sedimentary (chemical)	rock salt	commonly forms by the evaporation of seawater
Metamorphic	gneiss	well-developed banding because of alternating layers of different minerals, usually of different colors; common parent rock is granite
	schist	well-developed parallel arrangement of flat, sheetlike minerals, mainly micas; common parent rocks are shale and phyllite
	phyllite	shiny or silky appearance; may look wrinkled; common parent rocks are shale and slate
	slate	harder, denser, and shinier than shale; common parent rock is shale
Metamorphic (nonfoliated)	marble	interlocking calcite or dolomite crystals; common parent rock is limestone
	soapstone	composed mainly of the mineral talc; soft with a greasy feel
	quartzite	hard and well-cemented with interlocking quartz crystals; common parent rock is sandstone

Solar System Charts The Planets

	Mercury	Venus	Earth	Mars	Jupiter	Saturn	Uranus	Neptune
Mass (kg)	3.3020×10^{23}	4.8685×10^{24}	5.9736×10^{24}	6.4185×10^{23}	1.8986×10^{27}	5.6846×10^{26}	8.6832×10^{25}	1.0243×10^{26}
Equatorial radius (km)	2439.7	6051.8	6378.1	3396.2	71,492	60,268	25,559	24,764
Mean density (kg/m³)	5427	5243	5515	3933	1326	687	1270	1638
Albedo	0.068	0.900	0.306	0.250	0.343	0.342	0.300	0.290
Semimajor axis (km)	5.791×10^{7}	1.0821×10^{8}	1.4960×10^{8}	2.2792×10^{8}	7.7857×10^{8}	1.43353×10^{9}	2.87246×10^{9}	4.49506×10^{9}
Orbital period (Earth days)	87.969	224.701	365.256	686.980	4332.589	10,759.220	30,685.4	60,189
Orbital inclination (degrees)	7.000	3.390	0.000	1.850	1.304	2.485	0.772	1.769
Orbital eccentricity	0.2056	0.0067	0.0167	0.0935	0.0489	0.0565	0.0457	0.0113
Rotational period (hours)	1407.6	5832.5ᴿ	23.9345	24.6229	9.9250	10.656	17.24ᴿ	16.11
Axial tilt (degrees)	0.01	177.36	23.45	25.19	3.13	26.73	97.77	28.32
Average surface temperature (K)	440	737	288	210	165	134	76	72
Number of known moons*	0	0	1	2	79	62	27	14

*Number as of 2018.
ᴿ indicates retrograde rotation.

The Moon	
Mass (kg)	7.349×10^{22}
Equatorial radius (km)	1738.1
Mean density (kg/m³)	3350
Albedo	0.12
Semimajor axis (km)	3.844×10^{5}
Orbital period (Earth days)	27.3217
Lunar period (Earth days)	29.53
Orbital inclination (degrees)	5.145
Orbital eccentricity	0.0549
Rotational period (hours)	655.728

The Sun	
Mass (kg)	1.989×10^{30}
Equatorial radius (km)	6.96×10^{5}
Mean density (kg/m³)	1408
Absolute magnitude	4.83
Luminosity (W)	384.6
Spectral type	G2
Rotational period (hours)	609.12
Average temperature (K)	5778

GLOSSARY/GLOSARIO

Multilingual Glossary

The multilingual science glossary includes Arabic, Bengali, Chinese, English, Haitian Creole, Hmong, Korean, Portuguese, Russian, Tagalog, Urdu, and Vietnamese.

Pronunciation Key

Use the following key to help you sound out words in the glossary.

a . . . . b**a**ck (BAK)	**ew** . . . . f**oo**d (FEWD)
ay . . . . d**ay** (DAY)	**yoo** . . . . p**u**re (PYOOR)
ah . . . . f**a**ther (FAH thur)	**yew** . . . . f**ew** (FYEW)
ow . . . . fl**ow**er (FLOW ur)	**uh** . . . . comm**a** (CAHM uh)
ar . . . . c**ar** (CAR)	**u** (+ con) . . . . r**u**b (RUB)
e . . . . l**e**ss (LES)	**sh** . . . . **sh**elf (SHELF)
ee . . . . l**ea**f (LEEF)	**ch** . . . . na**t**ure (NAY chur)
ih . . . . tr**i**p (TRIHP)	**g** . . . . **g**ift (GIHFT)
i (i + con + e) . . . **i**dea, l**i**fe (i DEE uh, life)	**j** . . . . **g**em (JEM)
oh . . . . g**o** (GOH)	**ing** . . . . s**ing** (SING)
aw . . . . s**o**ft (SAWFT)	**zh** . . . . vi**si**on (VIHZH un)
or . . . . **or**bit (OR but)	**k** . . . . ca**k**e (KAYK)
oy . . . . c**oi**n (COYN)	**s** . . . . **s**eed, **c**ent (SEED, SENT)
oo . . . . f**oo**t (FOOT)	**z** . . . . **z**one, rai**s**e (ZOHN, RAYZ)

Cómo usar el glosario en español:
1. Busca el término en inglés que desees encontrar.
2. El término en español, junto con la definición, se encuentran en la columna de la derecha.

ENGLISH ESPAÑOL

abrasion: process of erosion in which windblown or waterborne particles, such as sand, scrape against rock surfaces or other materials and wear them away.

abrasión: proceso erosivo en que las partículas por el viento o el agua, como la arena, chocan y raspan superficies rocosas u otros materiales y los desgastan.

absolute-age dating: method that enables scientists to determine the actual age of certain rocks and other objects.

datación absoluta: permite a los científicos determinar la antigüedad real de ciertas rocas y objetos.

absolute magnitude: brightness an object would have if it were placed at a distance of 10 pc; classification system for stellar brightness that can be calculated when the actual distance to a star is known.

magnitud absoluta: brillo que tendría un objeto si estuviera a una distancia de 10 pc; sistema de clasificación del brillo estelar que se puede calcular cuando se conoce la distancia verdadera hasta la estrella.

abyssal plain: smooth, flat part of the seafloor covered with muddy sediments and sedimentary rocks that extends seaward from the continental margin.

llanura abisal: parte plana y lisa del fondo del mar cubierta con sedimentos fangosos y rocas sedimentarias y que se extiende desde el margen continental hacia el mar.

acid: solution containing a substance that produces hydrogen ions: (H^+) in water.

ácido: solución que contiene una sustancia que produce iones hidrógeno (H^+) en agua.

acid precipitation: any precipitation with a pH of less than 5.0 that forms when sulfur dioxide and nitrogen oxides combine with moisture in the atmosphere to produce sulfuric acid and nitric acid.

precipitación ácida: toda precipitación con un pH menor que 5.0 que se forma cuando se combinan el dióxido de azufre y óxidos de nitrógeno con la humedad en la atmósfera para producir ácido sulfúrico o ácido nítrico.

active galactic nucleus (AGN): a galaxy's core in which highly energetic objects or activities are located.

núcleo galáctico activo (NGA): centro de la galaxia donde se ubican cuerpos o suceden eventos con gran cantidad de energía.

aggregate: mixture of sand, gravel, and crushed stone that accumulates naturally; found in floodplains, alluvial fans, or glacial deposits.

agregado: mezcla natural de arena, grava y piedra triturada que se acumula naturalmente; se encuentra en llanuras aluviales, abanicos aluviales o depósitos glaciales.

air mass: large volume of air that has the characteristics of the area over which it forms.

masa de aire: gran volumen de aire que tiene las características del área sobre la que se forma.

GLOSSARY/GLOSARIO

air-mass thunderstorm: type of thunderstorm in which air rises because of unequal heating of Earth's surface within a single air mass and is most common during the afternoon and evening.

albedo: percentage of sunlight that is reflected by the surface of a planet or a satellite, such as the Moon.

altered hard part: fossil whose soft organic material has been removed and whose hard parts have been changed by recrystallization or mineral replacement.

amino acid: a building block of proteins.

amniotic (am nee AH tihk) egg: egg with a shell, providing a complete environment for a developing embryo.

amplitude: the height of a wave.

analog forecast: weather forecast that compares current weather patterns to patterns that occurred in the past.

anemometer (a nuh MAH muh tur): weather instrument used to measure wind speed.

apogee: farthest point in an object's orbit to Earth.

apparent magnitude: classification system based on how bright a star appears to be; does not take distance into account so cannot indicate how bright a star actually is.

aquiclude: layer of impermeable material, such as silt, clay, or shale, that is a barrier to groundwater.

aquifer: permeable underground layer through which groundwater flows relatively easily.

artesian well: fountain of water that spurts above the land surface when a well taps a confined aquifer containing water under pressure.

asteroid (AS tuh royd): metallic or silica-rich object, 1 m to 950 km in diameter, that bombarded early Earth, generating heat energy; rocky remnant of the early solar system found mostly between the orbits of Mars and Jupiter in the asteroid belt.

astronomical unit (AU): the average distance between the Sun and Earth, 1.496×10^8 km.

astronomy: study of objects beyond Earth's atmosphere.

atmosphere: blanket of gases surrounding Earth that contains about 78 percent nitrogen, 21 percent oxygen, and 1 percent other gases such as argon, carbon dioxide, and water vapor.

atomic number: number of protons contained in an atom's nucleus.

avalanche: landslide that occurs in a mountainous area when snow falls on an icy crust, becomes heavy, slips off, and slides swiftly down a mountainside.

tormenta eléctrica de masa de aire: tipo de tormenta en que el aire asciende debido al calentamiento desigual de la superficie terrestre bajo una misma masa de aire; esw más común durante la tarde y la noche.

albedo: porcentaje de luz solar que refleja la superficie de un planeta o un satélite, como por ejemplo, la Luna.

partes duras alteradas: fósiles cuya materia orgánica blando ha desaparecido y cuyas partes duras han sido transformadas por recristalización o sustitución de minerales.

aminoácido: unidad básica de las proteínas.

huevo amniótico: huevo con cascarón; provee un ambiente completo para el embrión en desarrollo.

amplitud: la altura de una onda.

pronóstico análogo: pronóstico del tiempo que compara los patrones actuales del clima con patrones ocurridos en el pasado.

anemómetro: instrumento meteorológico que se utiliza para medir la velocidad de viento.

apogeo: punto de la órbita de un objeto en que ésta se encuentra más alejada de la Tierra.

magnitud aparente: sistema de clasificación basado el brillo aparente de una estrella; no toma en cuenta la distancia y por lo tanto no indica el brillo real de la estrella.

acuiclusos: capas impermeables que sirven de barrera a las aguas subterráneas, como por ejemplo limo, arcilla o esquisto.

acuífero: capa subterránea permeable por la cual el agua subterránea fluye de manera relativamente fácil.

pozo artesiano: fuente de agua que brota hacia la superficie terrestre, cuando un pozo conecta con un acuífero confinado que contiene agua bajo presión.

asteroide: cuerpo metálico o rico en sílice que mide de 1 m a 950 km de diámetro y que bombardeó la Tierra primitiva generando energía calórica; restos rocosos del sistema solar primitivo que se hallan principalmente entre las órbitas de Marte y Júpiter, en el cinturón de asteroides.

unidad astronómica (UA): la distancia promedio entre el Sol y la Tierra, equivale a 1.496×10^8 km.

astronomía: el estudio de los cuerpos que se encuentran más allá de la atmósfera de la Tierra.

atmósfera: manto de gases que rodea la Tierra; está compuesta aproximadamente por 78 por ciento de nitrógeno, 21 por ciento de oxígeno y 1 por ciento de otros gases como el argón, el dióxido de carbono y el vapor del agua.

número atómico: número de protones que contiene el núcleo de un átomo.

avalancha: deslizamiento que ocurre en un área montañosa cuando la nieve cae sobre una capa helada, aumenta de peso, se desprende y se resbala rápidamente montaña abajo.

banded-iron formations: alternating bands of iron oxide and chert, an iron-poor sedimentary rock.

barometer: instrument used to measure air pressure.

barrier island: long ridges of sand or other sediment that are deposited or shaped by the longshore current and separated from the mainland.

basaltic rock: rock that is dark colored, has lower silica contents, and is rich in iron and magnesium; contains mostly plagioclase and pyroxene.

base: substance that produces hydroxide ions (OH⁻) in water.

base level: the elevation at which a stream enters another stream or body of water.

batholith: coarse-grained, irregularly-shaped pluton that covers at least 100 km²; generally forms 10–30 km below Earth's surface; and is common in the interior of major mountain chains.

beach: area in which loose sediment is deposited and moved about by waves along the shore.

bedding: horizontal layering in sedimentary rock; layers that can range from a few millimeters to several meters thick.

bed load: describes sediments that are too heavy or large to be kept in suspension or solution and are pushed or rolled along the bottom of a streambed.

bedrock: unweathered, solid parent rock that can consist of limestone, marble, granite, or other quarried rock.

belt: low, warm, dark-colored cloud that sinks and flows rapidly in the Jovian atmosphere.

Big Bang theory: theory that proposes that the universe began as a single point and has been expanding ever since.

binary star: one of two stars that are bound together by gravity and orbit a common center of mass.

biomass fuels: fuels derived from living things; renewable resources.

bioremediation: use of organisms to clean up toxic waste.

biosphere: all of Earth's organisms and the environments in which they live.

bipedal: walking upright on two legs.

black hole: small, extremely dense remnant of a star whose gravity is so immense that not even light can escape its gravitational field.

Bowen's reaction series: sequential, predictable, dual-branched pattern in which minerals crystallize from cooling magma.

breaker: collapsing wave that forms when a wave reaches shallow water and becomes so steep that the crest topples forward.

formaciones de hierro en bandas: bandas alternadas de óxido ferroso y pedernal, roca sedimentaria deficiente en hierro.

barómetro: instrumento que se usa para medir la presión atmosférica.

barrera litoral: grandes lomas de arena u otro sedimento que son depositadas, o que adquieren su forma, por la acción de las corrientes litorales y están separadas del continente.

roca basáltica: roca oscura con bajo contenido en sílice pero rica en hierro y magnesio; contiene principalmente plagioclasa y piroxenos.

base: sustancia que produce iones hidróxido (OH⁻) en agua.

nivel base: elevación a la cual una corriente entra a otra corriente o masa de agua.

batolito: plutones de grano grueso y de forma irregular que cubre por lo menos 100 km²; generalmente se forma de 10 a 30 km bajo la superficie terrestre y es común en el interior de las principales cadenas montañosas.

playa: área en que sedimentos sueltos son depositados y transportados por las olas a lo largo de la costa.

estratificación: capas horizontales de roca sedimentaria; capas que pueden medir de un milímetro a varios metros de grosor.

carga de fondo: término que describe los sedimentos que no se mantienen en suspensión, o en solución, porque son demasiado pesados o grandes y son empujados o arrastrados sobre el fondo del cauce de una corriente.

roca firme: roca madre sólida no meteorizada que puede consistir en piedra caliza, mármol, granito o alguna otra piedra de cantera.

cinturón: nube baja, tibia y oscura que desciende y fluye rápidamente en la atmósfera joviana.

teoría de la Gran Explosión: propone que el universo empezó en un solo punto y se ha estado expan-diendo desde entonces.

estrella binaria: una de dos estrellas unidas por la gravedad que giran alrededor de un centro común de masa.

biocombustible: combustibles derivados de los seres vivos; recursos renovables.

biorremediación: uso de organismos para limpiar desechos tóxicos.

biosfera: incluye a todos los organismos de la Tierra y los ambientes en que éstos viven.

bipedalismo: que camina erguido sobre dos piernas.

agujero negro: restos de una estrella muy densos y pequeños cuya gravedad es tan grande que ni la luz puede escapar de su campo de gravedad.

serie de reacción de Bowen: patrón de dos ramas, predecible y secuencial que siguen los minerales al cristalizarse a partir de magma que se enfría.

rompiente: ola que se colapsa; se forma cuando una ola alcanza aguas poco profundas y se vuelve tan empinada que la cresta de la ola se cae hacia adelante.

Glossary/Glosario

C

caldera: large crater, up to 100 km in diameter, that can form when the summit or side of a volcano collapses into the magma chamber during or after an eruption.

Cambrian explosion: sudden appearance of a diverse collection of organisms in the Cambrian fossil record.

Canadian shield: name given to the Precambrian shield in North America because much of it is exposed in Canada.

carrying capacity: number of organisms that a specific environment can support.

cartography: science of mapmaking.

cast: fossil formed when an earlier fossil of a plant or animal leaves a cavity that becomes filled with minerals or sediment.

cave: a natural underground opening connected to Earth's surface, usually formed when groundwater dissolves limestone.

cementation: process of sedimentary rock formation that occurs when dissolved minerals precipitate out of groundwater and either a new mineral grows between the sediment grains or the same mineral grows between and over the grains.

Cepheid variable: star with pulsation period ranging from 1 to 100 days and varying luminosity.

chemical bond: force that holds the atoms of elements together in a compound.

chemical reaction: change of one or more substances into other substances.

chemical weathering: process by which rocks and minerals undergo changes in their composition due to chemical reactions with agents such as acids, water, oxygen, and carbon dioxide.

chromosphere: layer of the Sun's atmosphere above the photosphere and below the corona that is about 2500 km thick and has a temperature around 30,000 K at its top.

cinder cone: steep-sided, generally small volcano that is built by the accumulation of tephra around the vent.

cirque: deep depression scooped out by a valley glacier.

cirrus (SIHR us): high clouds made up of ice crystals that form at heights of 6000 m.

clastic: composed of rock and mineral fragments produced by weathering and erosion and classified according to particle size and shape.

clastic sedimentary rock: most common type of sedimentary rock, formed from the abundant deposits of loose sediments that accumulate on Earth's surface; classified according to the size of their particles.

cleavage: the manner in which a mineral breaks along planes where atomic bonding is weak.

climate: the long-term averages and variations in weather for a particular area.

caldera: cráter grande, de hasta 100 km de diámetro, que se forma cuando la cumbre o la ladera de un volcán se desploman en la cámara de magma durante o después de una erupción.

explosión del Cámbrico: aparición repentina de un conjunto diverso de organismos en el registro fósil del Cámbrico.

escudo Canadiense: nombre que recibe el escudo Precámbrico en Norteamérica porque la mayor parte está expuesto en Canadá.

capacidad de carga: número de organismos que un ambiente específico puede sustentar.

cartografía: ciencia de la elaboración de mapas.

molde: fósil que se forma cuando un fósil precedente de una planta o un animal forma una cavidad que se rellena con minerales o sedimentos.

caverna: cavidad subterránea abierta a la superficie terrestre, generalmente se forma cuando el agua subterránea disuelve la piedra caliza.

cementación: proceso de formación de roca sedimentaria que ocurre cuando los minerales disueltos del agua subterránea se precipitan y se forma un nuevo mineral entre los granos de sedimento o se acumula el mismo mineral entre y sobre los granos.

variable cefeida: estrella con periodo de pulsación que dura de 1 a 100 días y con mayor o menor luminosidad.

enlace químico: fuerza que mantiene unidos los átomos de los elementos en un compuesto.

reacción química: sucede cuando una o más sustancias se convierten en otras sustancias.

meteorización química: proceso mediante el cual las rocas y los minerales experimentan cambios en su composición, debido a reacciones químicas con agentes como ácidos, agua, oxígeno o dióxido de carbono.

cromosfera: capa de la atmósfera del Sol situada encima de la fotosfera y debajo de la corona; mide aproximadamente 2500 km de ancho y tiene una temperatura cercana a 30,000 K en su parte superior.

cono de carbonilla: volcán empinada cara generalmente pequeño que es construido por a la acumulación de tefrita alrededor de la chimenea.

circo: depresión profunda formada por un glaciar de valle.

cirro: nubes altas formadas por cristales de hielo que se forman a alturas de 6000 m.

clástico: describe los fragmentos de roca y de mineral producidos por la meteorización y la erosión; se clasifican según su tamaño y forma de partícula.

roca sedimentaria clástica: el tipo más común de roca sedimentaria; se forma a partir de los abundantes depósitos de sedimentos sueltos que se acumulan sobre la superficie de la Tierra; se clasifican según el tamaño de sus partículas.

crucero: la forma en la cuál un mineral se rompe a lo largo de los planos donde los enlaces atómicos son débiles.

clima: promedio durante un largo periodo de las variaciones en las condiciones del tiempo de un área determinada.

climatology: study of Earth's climate in order to understand and predict climatic change, based on past and present variations in temperature, precipitation, wind, and other weather variables.

coalescence (ko uh LEH sunts): process that occurs when cloud droplets collide and form larger droplets, which eventually become too heavy to remain aloft and can fall to Earth as precipitation.

cogeneration: production of two usable forms of energy at the same time from the same process, which can conserve resources and generate income.

cold wave: extended period of below-average temperatures caused by large, high-pressure systems of continental polar or arctic origin.

comet: small, eccentrically orbiting body made of rock and ice which have one or more tails that point away from the Sun.

composite volcano: generally cone-shaped with concave slopes; built by violent eruptions of volcanic fragments and lava that accumulate in alternating layers.

compound: substance composed of atoms of two or more different elements that are chemically combined.

compressive force: squeezing force that can cause the intense deformation—folding, faulting metamorphism, and igneous intrusions—associated with mountain building.

condensation: process by which a cooling gas changes into a liquid and releases thermal energy.

condensation nucleus: small particle in the atmosphere around which cloud droplets can form.

conduction: the transfer of thermal energy between objects in contact by the collisions between the particles in the objects.

conduit: a tubelike structure that allows lava to reach the surface.

conic projection: map that is highly accurate for small areas, made by projecting points and lines from a globe onto a cone.

constellation: group of stars that forms a pattern in the sky that resembles an animal, mythological character, or everyday object.

contact metamorphism: local effect that occurs when molten rock meets solid rock.

continental drift: Wegener's hypothesis that Earth's continents were joined as a single landmass, called Pangaea, that broke apart about 200 mya and slowly moved to their present positions.

continental glacier: glacier that forms over a broad, continent-sized area of land and usually spreads out from its center.

climatología: estudio del clima de la Tierra para entender y pronosticar los cambios climáticos; se basa en variaciones pasadas y presentes de temperatura, precipitación, viento y otras variables del tiempo.

coalescencia: proceso que ocurre cuando las gotas de nube chocan entre sí, formando gotas cada vez más grandes; estas gotas puede llegar a ser demasiado pesadas para seguir suspendidas en el aire y entonces caen a la Tierra como precipitación.

cogeneración: producción simultánea de dos formas útiles de energía a partir del mismo proceso; puede ayudar a conservar recursos y obtener ganancias.

onda fría: período prolongado de temperaturas más bajas que el promedio, causado por grandes sistemas de alta presión de origen polar continental o ártico.

cometa: cuerpo pequeño de órbita excéntrica compuesto por roca y hielo y que contiene una o más colas que apuntan hacia el lado opuesto al Sol.

volcán compuesto: volcán que en general tiene forma cónica y laderas cóncavas; se forma por erupciones violentas de fragmentos y lava volcánicos que se acumulan creando capas alternadas.

compuesto: sustancia compuesta por átomos de dos o más elementos diferentes unidos químicamente.

fuerzas de compresión: fuerzas de aplastamiento que pueden causar intensas deformaciones como plegamientos, fallas, metamorfismo e intrusiones ígneas; asociadas con la formación de montañas.

condensación: proceso por el cual un gas enfriador se transforma en un líquido y libera energía térmica.

núcleos de condensación: partículas pequeñas de la atmósfera alrededor de las cuales se pueden formar las gotas de nubes.

conducción: transferencia de energía entre cuerpos en contacto debida a la colisión entre las partículas de los cuerpos.

conducto: estructura tubular que permite que la lava llegue a la superficie.

proyección cónica: mapa de gran exactitud para áreas pequeñas que se elabora mediante la proyección de puntos y líneas de un globo a un cono.

constelación: grupo de estrellas que forman en el firmamento un patrón que semeja un animal, un personaje mitológico o un objeto cotidiano.

metamorfismo de contacto: efecto local que ocurre cuando la roca fundida se encuentra con roca sólida.

deriva continental: hipótesis de Wegener que propone que los continentes de la Tierra estaban unidos en una sola masa terrestre, llamada Pangaea, la cual se separó hace aproximadamente 200 millones de años y que los fragmentos resultantes se movieron lentamente a sus ubicaciones actuales.

glaciar continental: glaciar que se forma sobre una amplia área del tamaño de un continente y que generalmente se extiende a partir de su centro.

Glossary/Glosario

continental margin: area where edges of continents meet the ocean; represents the shallowest part of the ocean that consists of the continental shelf, the continental slope, and the continental rise.

continental rise: gently sloping accumulation of sediments deposited by a turbidity current at the foot of a continental margin.

continental shelf: shallowest part of a continental margin, with an average depth of 130 m and an average width of 60 km, that extends into the ocean from the shore and provides a nutrient-rich home to large numbers of fish.

continental slope: sloping oceanic region found beyond the continental shelf that generally marks the edge of the continental crust and may be cut by sub-marine canyons.

contour interval: difference in elevation between two side-by-side contour lines on a topographic map.

contour line: line on a topographic map that connects points of equal elevation.

control: standard for comparison in an experiment.

convection: the transfer of thermal energy by the movement of heated material from one place to another.

convergent boundary: place where two tectonic plates are moving toward each other; is associated with trenches, islands arcs, and folded mountains.

Coriolis effect: effect of a rotating body that influences the motion of any object or fluid; on Earth, air moving north or south from the equator appears to move right or left, respectively; the combination of the Coriolis effect and Earth's heat imbalance creates the trade winds, polar easterlies, and prevailing westerlies.

corona: top layer of the Sun's atmosphere that extends from the top of the chromosphere and typically ranges in temperature from 3 million to 5 million K.

correlation: matching of rock outcrops of one geographic region to another.

cosmic background radiation: weak radiation that is left over from the early, hot stages of the Big Bang expansion of the universe.

cosmology: study of the universe, including its current nature, origin, and evolution, based on observation and the use of theoretical models.

covalent bond: attraction of two atoms for a shared pair of electrons that holds the atoms together.

crater: bowl-shaped depression that forms around the central vent at the summit of a volcano.

craton (KRAY tahn): continental core formed from Archean or Proterozoic microcontinents; deepest (as far as 200 km into the mantle) and most stable part of a continent.

creep: slow, steady downhill movement of loose weathered Earth materials, especially soils, causing objects on a slope to tilt.

crest: highest point of a wave.

margen continental: área donde los límites de los continentes se unen con el océano; representa la parte menos profunda del océano y consiste en la plataforma continental, el talud continental y el pie del talud continental.

pie del talud continental: acumulación de sedimentos, con pendiente leve, depositados por una corriente de turbidez al pie de un margen continental.

plataforma continental: parte más superficial del margen continental, tiene una profundidad promedio de 130 m y una anchura promedio de 60 km, se extiende hacia el océano desde la costa y proporciona un lugar rico en nutrientes a un gran número de peces.

talud continental: región oceánica inclinada que se encuentra más allá de la plataforma continental; generalmente marca el límite de la corteza continental y puede estar seccionada por cañones submarinos.

intervalo entre curvas de nivel: diferencia en la elevación entre dos curvas de nivel contiguas en un mapa topográfico.

curva de nivel: curva en un mapa topográfico que conecta puntos de igual elevación.

control: estándar de comparación en un experimento.

convección: transferencia de energía térmica debido al movimiento de material caliente de un lado a otro.

límite convergente: lugar donde dos placas tectónicas se mueven aproximándose cada vez más entre sí; está asociado con fosas abisales, arcos insulares y montañas plegadas.

efecto de Coriolis: efecto producido por un cuerpo en rotación que influye en el movimiento de todo cuerpo objeto o fluido; en la Tierra, las corrientes aire que se mueven desde el norte o desde el sur parecen desplazarse hacia la derecha o hacia la izquierda, respectivamente; la combinación del efecto de Coriolis y el desequilibrio térmico de la Tierra originan los vientos alisios, los vientos polares del este y los vientos dominantes del oeste.

corona: capa superior de la atmósfera del Sol que se extiende desde la parte superior de la cromosfera y típicamente tiene un rango de temperatura de 3 a 5 millones K.

correlación: correspondencia entre los afloramientos rocosos de una región geográfica y otra.

radiación cósmica de fondo: radiación residual débil proveniente de las calientes etapas iniciales de la expansión del universo causada por la Gran Explosión.

cosmología: estudio del universo; abarca su naturaleza actual, su origen y evolución y se basa en la observación y el uso de modelos teóricos.

enlace covalente: atracción de dos átomos hacia un par compartido de electrones que mantienen a los átomos unidos.

cráter: depresión en forma de tazón que generalmente se forma alrededor de la abertura central en la cumbre de un volcán.

cratón: zona central de un continente formada a partir de microcontinentes del arcaico o del Proterozoico; son la parte más profunda (penetran hasta 200 km hacia el manto) y estable de un continente.

deslizamiento: movimiento cuesta abajo constante y lento de materia meteorizada suelta de la Tierra, especialmente los suelos, lo que ocasiona que se inclinen los objetos en una ladera.

cresta: punto más alto de una onda.

cross-bedding: depositional feature of sedimentary rock that forms as inclined layers of sediment are carried forward across a horizontal surface.

cross-cutting relationships: the principle that an intrusion or fault is younger than the rock it cuts across.

cryosphere: the frozen portion of water on Earth's surface.

crystal: solid in which atoms are arranged in repeating patterns.

crystalline structure: regular geometric pattern of particles in most solids, giving a solid a definite shape and volume.

cumulus (KYEW myuh lus): puffy, lumpy-looking clouds that usually occur below 2000 m.

cyanobacteria: microscopic, photosynthetic prokaryotes that formed stromatolites and changed early Earth's atmosphere by generating oxygen.

estratificación cruzada: característica de la depo-sitación de roca sedimentaria que se forma a medida que capas inclinadas de sedimento son arrastradas hacia delante, a lo largo de una superficie horizontal.

relaciones de corte transversal: principio que establece que una intrusión o falla es menos antigua que la roca que atraviesa.

crisofera: la parte de agua congelada sobre la superficie de la Tierra.

cristal: sólido cuyos átomos están ordenados en patrones repetitivos.

estructura cristalina: patrón geométrico y regular que tienen las partículas en la mayoría de los sólidos; dan al sólido una forma y volumen definidos.

cúmulo: nubes esponjosas con aspecto de madejas de algodón que generalmente se hallan a alturas menores de 2000 m.

cianobacterias: organismos procariotas fotosintéticos microscópicos que formaron estromatolitos y modificaron la atmósfera primitiva de la Tierra al producir oxígeno.

dark matter: invisible material thought to be made up of a form of subatomic particle that interacts only weakly with other matter.

deep-sea trench: elongated, sometimes arc-shaped depression in the seafloor that can extend for thousands of kilometers; is the deepest part of the ocean basin, and is found primarily in the Pacific Ocean.

deflation: lowering of land surface caused by wind erosion of loose surface particles, often leaving coarse sediments behind.

deforestation: removal of trees from a forested area without adequate replanting, often using clear-cutting, which can result in loss of topsoil and water pollution.

delta: triangular deposit, usually made up of silt and clay particles, that forms where a stream enters a large body of water.

dendrochronology: science of using tree rings to determine absolute age; helps to date relatively recent geologic events and environmental changes.

density current: movement of ocean water that occurs in depths too great to be affected by surface winds and is generated by differences in water temperature and salinity.

density-dependent factor: environmental factor, such as disease, predators, or lack of food, that increasingly affects a population as the population's size increases.

density-independent factor: environmental factor that does not depend on population size, such as storms, flood, fires, or pollution.

dependent variable: factor in an experiment that can change if the independent variable is changed.

materia oscura: sustancia invisible formada por algún tipo de partícula subatómica que interactúa débilmente con otros tipos de material.

fosa abisal: depresión alargada y en algunas ocasiones con forma de arco, que se puede extender miles de kilómetros; es la parte más profunda de la cuenca oceánica y se halla principalmente en el océano Pacífico.

deflación: depresión de la superficie terrestre causada por la erosión eólica de partículas superficiales sueltas; a menudo sólo contiene sedimentos gruesos.

deforestación: eliminación de árboles de un área forestal, sin realizar una adecuada reforestación; a menudo es resultado de una corta a hecho, lo que puede ocasionar la pérdida del mantillo y la contaminación de las aguas.

delta: depósito triangular compuesto generalmente por partículas de limo y arcilla, que se forma en el sitio donde una corriente de agua entra a una gran masa de agua.

dendrocronología: ciencia que usa los anillos de crecimiento anual de los árboles para determinar la edad absoluta; permite datar eventos geológicos y cambios ambientales relativamente recientes.

corriente de densidad: movimiento de las aguas oceánicas que ocurre a grandes profundidades, no se ve afectado por los vientos superficiales y es generado por las diferencias en temperatura y salinidad del agua.

factor dependiente de la densidad: factor ambiental como las enfermedades, los depredadores o la falta de alimento, que afecta con creciente intensidad a una población a medida que aumenta el tamaño de su población.

factor independiente de la densidad: factor ambiental, como las tempestades, las inundaciones, los incendios o la contaminación, que no son afectados por el tamaño de la población.

variable dependiente: factor de un experimento que puede cambiar al variar la variable independiente.

Glossary/Glosario

GLOSSARY/GLOSARIO

deposition: occurs when eroded materials are dropped in another location.

desalination: process that removes salt from seawater in order to provide freshwater.

desertification: process by which productive land becomes desert; in arid areas can occur through the loss of topsoil.

dew point: temperature to which air is cooled at a constant pressure to reach saturation, at which point condensation can occur.

differentiation (dih fuh ren shee AY shun): process in which a planet becomes internally zoned, with the heavy materials sinking toward the center and the lighter materials accumulating near its surface.

digital forecast: weather forecast that uses numerical data to predict how atmospheric variables change over time.

dike: pluton that cuts across preexisting rocks and often forms when magma invades cracks in surrounding rock bodies.

discharge: measure of a volume of stream water that flows over a specific location in a particular amount of time.

divergent boundary: place where two of Earth's tectonic plates are moving apart; is associated with volcanism, earthquakes, and high heat flow, and is found primarily on the seafloor.

divide: elevated land that divides one watershed from another.

Doppler effect: change in the wave frequency that occurs due to the relative motion of the wave as it moves toward or away from an observer.

downburst: violent downdrafts that are concentrated in a local area.

drawdown: difference between the water level in a pumped well and the original water-table level.

drought: extended period of well-below-average rainfall, usually caused by shifts in global wind patterns, allowing high-pressure systems to remain for weeks or months over continental areas.

drumlin: elongated landform that results when a glacier moves over an older moraine.

dune: pile of windblown sand that develops over time, whose shape depends on sand availability, wind velocity and direction, and amount of vegetation present.

dwarf planet: an object that, due to its own gravity, is spherical in shape, orbits the Sun, is not a satellite, and has not cleared the area of its orbit of smaller debris.

depositación: ocurre cuando los materiales erosionados son depositados en otro sitio.

desalinización: proceso de eliminación de la sal del agua marina para obtener agua dulce.

desertificación: proceso mediante el cual las tierras productivas se convierten en desierto; en áreas áridas puede ocurrir debido a la pérdida del mantillo del suelo.

punto de rocío: temperatura a la cual el aire que se enfría a una presión constante alcanza la saturación, punto en el cual ocurre la condensación.

diferenciación: proceso en que un planeta se divide internamente en zonas, los materiales pesados se hunden hacia el centro, mientras que los materiales más ligeros se acumulan cerca de su superficie.

pronóstico digital: pronóstico del tiempo que se basa en datos numéricos para predecir el cambio de las variables atmosféricas con el tiempo.

dique: plutón que atraviesa las rocas preexistentes; suele formarse cuando el magma invade las grietas de los cuerpos rocosos circundantes.

descarga: medida del volumen de agua corriente que fluye sobre una ubicación dada en cierto lapso de tiempo.

límite divergente: lugar donde dos placas tectónicas terrestres se alejan entre sí; se asocia con actividad volcánica, terremotos, un alto flujo de calor y se hallan principalmente en el fondo marino.

divisoria: terreno elevado que separa una cuenca hidrográfica de otra.

efecto Doppler: cambio en la frecuencia de onda que ocurre debido al movimiento relativo de la onda a medida que se acerca o se aleja de un observador.

reventón: violentos chorros de viento descendientes que se concentran en un área local.

tasa de agotamiento: diferencia entre el nivel de agua en un pozo artesanal en uso y el nivel original del manto freático.

sequía: período prolongado con precipitación muy por debajo del promedio, generalmente es causado por cambios en los patrones globales de vientos, lo que permite que los sistemas de alta presión permanezcan sobre áreas continentales durante semanas o meses.

drumlin: formación alargada de tierra que se forma cuando un glaciar se mueve sobre una morrena más antigua.

duna: pila de arena formada a lo largo del tiempo por el arrastre de partículas por el viento, cuya forma depende de la disponibilidad de arena, la velocidad y dirección del viento y la cantidad de vegetación presente.

planeta menor: cuerpo que debido a su propia gravedad tiene forma esférica, tiene una órbita alrededor del Sol, no es un satélite y no ha eliminado restos más pequeños del área de su órbita.

E

eccentricity: ratio of the distance between the foci to the length of the major axis; defines the shape of a planet's elliptical orbit.

ecliptic plane: plane of Earth's orbit around the Sun.

Ediacaran biota (ee dee A kuh ruhn • by OH tuh): fossils of various multicellular organisms from about 635 mya.

ejecta: material that falls back to the lunar surface after being blasted out by the impact of a space object.

elastic deformation: causes materials to bend and stretch; proportional to stress, so if the stress is reduced or returns to zero the strain or deformation is reduced or disappears.

El Niño: a band of anomalously warm ocean temperatures that occasionally develops off the western coast of South America and can cause short-term climatic changes felt worldwide.

electromagnetic spectrum: all types of electromagnetic radiation arranged according to wavelength and frequency.

electron: tiny atomic particle with little mass and a negative electric charge; an atom's electrons are equal in number to its protons and are located in a cloudlike region surrounding the nucleus.

element: natural or artificial substance that cannot be broken down into simpler substances by physical or chemical means.

ellipse: an oval that is centered on two points called foci; the shape of planets' orbits.

energy efficiency: a type of conservation in which the amount of work produced is compared to the amount of energy used.

Enhanced Fujita Tornado Damage scale: classifies tornadoes according to their destruction and estimated wind speed on a scale ranging from EF0 to EF5.

environmental science: study of the interactions of humans with environment.

eon: longest time unit in the geologic time scale.

epicenter (EH pih sen tur): point on Earth's surface directly above the focus of an earthquake.

epoch: time unit in the geological time scale, smaller than a period, measured in hundreds of thousands to millions of years.

equator: imaginary line that lies at 0° latitude and circles Earth midway between the North and South poles, dividing Earth into the northern hemisphere and the southern hemisphere.

equinox: time of year during which Earth's axis is at a 90° angle to the Sun; both hemispheres receive exactly 12 hours of sunlight and the Sun is directly overhead at the equator.

era: second-longest time unit in the geologic time scale, measured in tens to hundreds of millions of years, and defined by differences in life-forms that are preserved in rocks.

excentricidad: razón de la distancia entre los focos y la longitud del eje mayor; define la forma de la órbita elíptica de un planeta.

plano de la eclíptica: plano de la órbita de la Tierra alrededor del Sol.

biota Ediacarana: fósiles de diversos organismos multicelulares de hace cerca de 635 millones de años.

eyecta: material que cae de regreso a la superficie lunar luego de ser expulsado por el impacto de un cuerpo espacial.

deformación elástica: ocasiona que los materiales se doblen y se estiren; es proporcional al grado de tensión, por lo que si la tensión se reduce o desaparece, la deformación también se reduce o desaparece.

El Niño: una banda de agua oceánica que tiene temperaturas anómalamente cálidas que en ocasiones se desarrolla frente a la costa occidental de Sudamérica; puede causar cambios climáticos a corto plazo que afectan a todo el mundo.

espectro electromagnético: clasificación de todos los tipos de radiación electromagnética de acuerdo con su frecuencia y longitud de onda.

electrón: partícula atómica diminuta con masa pequeña y carga eléctrica negativa; los electrones están ubicados en una región con forma de nube que rodea al núcleo del átomo y su número es igual al número de protones del átomo.

elemento: sustancia natural o artificial que no puede separarse en sustancias más simples por medios físicos o químicos.

elipse: óvalo centrado en dos puntos llamados focos; la forma de las órbitas de los planetas.

eficiencia energética: tipo de conservación en el cual la cantidad de trabajo producido se compara con la cantidad de energía utilizada.

escala mejorada de Fujita para daños de tornados: clasifica los tornados según el daño que causan y la velocidad de sus vientos aproximado en una escala que va de EF0 a EF5.

ciencias ambientales: estudio de las interacciones del hombre con su entorno.

eon: unidad más larga de tiempo en la escala de tiempo geológico.

epicentro: punto en la superficie terrestre ubicado directamente encima del foco de un sismo.

época: unidad de tiempo en la escala de tiempo geológico, es más pequeña que un período y se mide en millones a centenares de millares de años.

ecuador: línea imaginaria que yace en la latitud 0° y que circunda la Tierra entre los polos Norte y Sur, dividiendo a la Tierra en dos hemisferios iguales: norte y sur.

equinoccio: epoca del año durante la cual el eje de la Tierra forma un ángulo de 90° con el Sol, ambos hemisferios reciben exactamente 12 horas de luz solar y el Sol se halla exactamente sobre el ecuador.

era: segunda unidad más grande de tiempo en la escala del tiempo geológico; se mide en decenas a centenas de millones de años y se define según las diferencias en las formas de vida preservadas en las rocas.

GLOSSARY/GLOSARIO

erosion: removal and transport of weathered materials from one location to another by agents such as water, wind, glaciers, and gravity.

esker: long, winding ridge of layered sediments deposited by streams that flow beneath a melting glacier.

estuary: coastal area of lowest salinity often occurs where the lower end of a freshwater river or stream enters the ocean.

eukaryote (yew KE ree oht): organism composed of one or more cells each of which usually contains a nucleus; larger and more complex than a prokaryote.

eutrophication: process by which lakes become rich in nutrients from the surrounding watershed, resulting in a change in the kinds of organisms in the lake.

evaporation: vaporization—change of state from a liquid to a gas, involving thermal energy.

evaporite: the layers of chemical sedimentary rocks that form when concentrations of dissolved minerals in a body of water reach saturation due to the evaporation of water; crystal grains precipitate out of solution and settle to the bottom.

evolution (eh vuh LEW shun): the change in species over time.

exfoliation: mechanical weathering process in which outer rock layers are stripped away, often resulting in dome-shaped formations.

exosphere: outermost layer of Earth's atmosphere that is located above the thermosphere with no clear boundary at the top; transitional region between Earth's atmosphere and outer space.

exponential growth: pattern of growth in which a population of organisms grows faster as it increases in size, resulting in a population explosion.

extrusive rock: fine-grained igneous rock that is formed when molten rock cools quickly and solidifies on Earth's surface.

eye: calm center of a tropical cyclone that develops when the winds around its center reach at least 120 km/h.

eyewall: band where the strongest winds in a hurricane are usually concentrated, surrounding the eye.

erosión: eliminación y transporte de materiales meteorizados de un lugar a otro por agentes como el agua, el viento, los glaciares y la gravedad.

ésker: formación larga y sinuosa de sedimentos estratificados, depositados por corrientes que fluyen debajo de un glaciar que se derrite.

estuario: área costera de agua salobre que se forma en el sitio donde la desembocadura de un río o corriente de agua dulce entra al océano; provee una fuente excelente de alimento y refugio para organismos marinos comercialmente importantes.

eucariota: organismo compuesto por unas o más células nucleadas; generalmente es más grande y más complejo que un procariota.

eutroficación: proceso de aumento de la cantidad de nutrientes que contiene un lago, alimentado por los nutrientes provenientes de las cuenca circundante, lo que causa un cambio en los tipos de organismos que habitan el lago.

evaporación: vaporización: cambio de estado de un líquido a gas que implica energía térmica.

evaporita: capas de roca química sedimentaria que se forman cuando la concentración de minerales disueltos en una masa de agua alcanza el punto de saturación debido a la evaporación del agua; los cristales se precipitan de la solución y se asientan en el fondo.

evolución: cambios de las especies a lo largo del tiempo.

exfoliación: proceso de meteorización mecánica que causa la eliminación de los estratos rocosos exte-riores, a menudo produce formaciones en forma de domo.

exosfera: capa más externa de la atmósfera terrestre, está localizada por encima de la termosfera y no tiene un límite definido en su parte más alejada; región de transición entre la atmósfera de la Tierra y el espacio exterior.

crecimiento exponencial: patrón de crecimiento en que una población de organismos crece cada vez más rápido a medida que aumenta de tamaño, causando una explosión demográfica.

roca extrusiva: roca ígnea de grano fino que se forma cuando la roca fundida se enfría rápidamente y se solidifica en la superficie terrestre.

ojo: centro de calma de un ciclón tropical que se desarrolla cuando los vientos a su alrededor alcanzan por lo menos 120 km/h.

pared del ojo de huracán: banda que rodea el ojo de un huracán donde generalmente se concentran los vientos más fuertes.

F

fault: fracture or system of fractures in Earth's crust that occurs when stress is applied too quickly or stress is too great; can form as a result of horizontal compression (reverse fault), horizontal shear (strike-slip fault), or horizontal tension (normal fault).

fault-block mountain: mountain that forms when large pieces of crust are tilted, uplifted, or dropped downward between large normal faults.

falla: fractura o sistema de fracturas en la corteza terrestre que ocurren en sitios donde se aplica tensión rápidamente o donde la tensión es demasiado grande; se puede formar como resultado de una compresión horizontal (falla invertida, un cizallamiento horizontal (falla de transformación) o una tensión horizontal (falla normal).

montañas de bloque de falla: montañas que se forman cuando trozos grandes de corteza se inclinan, se elevan o se hunden entre fallas normales grandes.

feedback: the reaction of one system to the changes in another system.

fission: process in which heavy atomic nuclei split into smaller, lighter atomic nuclei.

fissure: long crack in Earth's crust.

flood: potentially devastating natural occurrence in which water spills over the sides of a stream's banks onto adjacent land areas.

flood basalt: huge amounts of lava that erupt from fissures.

floodplain: broad, flat, fertile area extending out from a stream's bank that is covered with water during floods.

focus: point of the initial fault rupture where an earthquake originates that usually lies at least several kilometers beneath Earth's surface.

foliated: describes metamorphic rock, such as schist or gneiss, whose minerals are squeezed under high pressure and arranged in wavy layers and bands.

fossil fuel: nonrenewable energy resource formed over geologic time from the compression and partial decomposition of organisms that lived millions of years ago.

fractional crystallization: process in which different minerals crystallize from magma at different temperatures, removing elements from magma.

fracture: when a mineral breaks into pieces with arclike, rough, or jagged edges.

front: boundary between two air masses of differing densities; can be cold, warm, stationary, or occluded and can stretch over large areas of Earth's surface.

frontal thunderstorm: type of thunderstorm usually produced by an advancing cold front, which can result in a line of thunderstorms hundreds of kilometers long, or, more rarely, an advancing warm front, which can result in a relatively mild thunderstorm.

frost wedging: mechanical weathering process that occurs when water repeatedly freezes and thaws in the cracks of rocks, often resulting in rocks splitting.

fuel: material, such as wood, peat, or coal, burned to produce energy.

fusion: the combining of lightweight nuclei into heavier nuclei; occurs in the core of the Sun where temperatures and pressure are extremely high.

retroacción: la reacción de un sistema a los cambios de otro sistema.

fisión: proceso mediante el cual los núcleos atómicos pesados se dividen en núcleos más livianos y pequeños.

fisura: grandes grietas en la Tierra.

inundación: acontecimiento natural potencialmente devastador en que el agua se desborda de las riberas de una corriente y cubre los terrenos adyacentes.

basalto de meseta: grandes cantidades de lava que salen por las fisuras.

llanura aluvial: área fértil, plana y ancha que se extiende desde las riberas de una corriente y queda cubierta por agua durante las inundaciones.

foco: punto inicial de ruptura de la falla donde se origina un terremoto; generalmente se halla varios kilómetros debajo de la superficie terrestre.

foliada: describe roca metamórfica, como el esquisto o el gneis, cuyos minerales son comprimidos bajo presiones altas, formando ordenadas capas y bandas onduladas.

combustible fósil: recurso energético no renovable que se forma a lo largo del tiempo geológico, a partir de la compresión y descomposición parcial de organismos que vivieron hace millones de años.

cristalización fraccionaria: proceso en el cual diferentes minerales se cristalizan a diferentes temperaturas a partir del magma, eliminando elementos del magma.

fractura: sucede cuando un mineral se rompe en pedazos con bordes ásperos, arqueados o serrados.

frente: límite entre dos masas de aire con diferentes densidades; puede ser frío, cálido, estacionario u ocluido y puede extenderse sobre grandes áreas de la superficie de la Tierra.

tormenta frontal: tipo de tormenta que es producida generalmente por el avance de un frente frío, pudiendo producir una línea de tormentas de cientos de kilómetros de largo, o en menor frecuencia por el avance de un frente cálido, produciendo tormentas relativamente ligeras.

erosión periglaciar: proceso mecánico de meteorización que ocurre cuando el agua se congela y se descongela, en repetidas ocasiones, en las grietas de las rocas, ocasionando el rompimiento de las mismas.

combustible: materiales como la leña, la turba o el carbón, que se queman para producir energía.

fusión: combinación de núcleos livianos para formar núcleos más pesados: sucede en el núcleo del Sol donde las temperaturas y la presión son extremadamente altas.

G

gas giant planet: large, gaseous planet that is very cold at its surface; has ring systems, many moons, and lacks solid surfaces—Jupiter, Saturn, Uranus, and Neptune.

gem: rare, precious, highly prized mineral that can be cut, polished, and used for jewelry.

gigantes gaseosos: planetas grandes y gaseosos con superficies muy frías; tienen sistemas de anillos, muchas lunas y carecen de superficie sólida: Júpiter, Saturno, Urano y Neptuno.

gema: mineral sumamente valioso, precioso y escaso que se puede cortar, pulir y utilizar en joyería.

GLOSSARY/GLOSARIO

Geographic Information System (GIS): a mapping system that uses worldwide databases from remote sensing to create layers of information that can be superimposed upon each other to form a comprehensive map.

geologic map: map that shows the distribution, arrangement, and types of rocks below the soil, and other geologic features.

geologic time scale: record of Earth's history from its origin 4.6 bya to the present.

geology: study of materials that make up Earth and the processes that form and change these materials, and the history of the planet and its life-forms since its origin.

geosphere: the part of Earth from its surface to its center.

geothermal energy: energy produced by Earth's naturally occurring heat, steam, and hot water.

geyser: explosive hot spring that erupts regularly.

glacier: large, moving mass of ice that forms near Earth's poles and in mountainous regions at high elevations.

glass: solid that consists of densely packed atoms with a random arrangement and lacks crystals or has crystals that are not visible.

Global Positioning System (GPS): satellite-based navigation system that permits a user to pinpoint his or her exact location on Earth.

global climate change: The changes in long-term averages and variations in weather around the world, like temperature, precipitation, frequency and strength of severe weather, acidification of oceans, and glaciation; this is due to increased temperature resulting from increased levels of CO_2 in the atmosphere from human activity.

gnomonic (noh MAHN ihk) projection: map useful in plotting long-distance trips by boat or plane, made by projecting points and lines from a globe onto a piece of paper that touches the globe at a single point.

graded bedding: type of bedding in which particle sizes become progressively heavier and coarser toward the bottom layers.

granitic rock: light-colored, intrusive igneous rock that has high silica content.

greenhouse effect: natural process where certain atmospheric gases absorb heat radiating off the planet's surface and prevents this heat from being lost to space, which helps keep Earth warm enough to sustain life

gully erosion: erosion that occurs when a rill channel widens and deepens.

guyot: large, extinct, basaltic volcano with a flat, submerged top.

Sistema de Información Geográfica (SIG): sistema para la elaboración de mapas que usa bases de datos mundiales obtenidos por sensores remotos, para crear capas de información que se pueden superponer para elaborar mapas que combinen dicha información.

mapa geológico: mapa que muestra la distribución, el orden y los tipos de roca del subsuelo, así como otras características geológicas.

escala del tiempo geológico: registro de la historia de la Tierra desde su origen, hace 4.6 billones de años, hasta el presente.

geología: estudio de los materiales que conforman la Tierra y de los procesos de formación y cambio de estos materiales, así como la historia del planeta y sus formas de vida desde su origen.

geosfera: región que abarca desde la superficie hasta el centro de la Tierra.

energía geotérmica: energía producida naturalmente en la Tierra por el calor, el vapor y el agua caliente.

géiser: manantial termal explosivo que hace erupción regularmente.

glaciar: enormes masas móviles de hielo que se forman cerca de los polos de la Tierra o en grandes elevaciones en regiones montañosas.

vidrio: sólido formado por átomos densamente comprimidos en un ordenamiento aleatorio; carece de cristales o sus cristales no son visibles.

Sistema de posicionamiento global (SPG): sistema de navegación por satélite que permite al usuario localizar su ubicación exacta sobre la Tierra.

cambio climático global: los cambios en los promedios y las variaciones de larga duración en el tiempo por todo el mundo, como la temperatura, la precipitación, la frecuencia y la fuerza del tiempo severo, la acidificación de los océanos, y la glaciación; es producto del aumento en las temperaturas que resulta del aumento en el CO_2 atmosférico a consecuencia de la actividad humana.

proyección gnomónica: mapa útil para trazar viajes de distancias largas por barco o por avión; se elabora proyectando los puntos y las líneas de un globo sobre una hoja de papel que toca el globo en un solo punto.

estratificación graduada: característica de la depositación de rocas sedimentarias en la cual las partículas son progresivamente más pesadas y gruesas hacia las capas inferiores de la estratificación.

roca granítica: roca intrusiva ígnea de color claro que tiene un alto contenido de sílice.

efecto invernadero: el proceso natural en que ciertos gases atmosféricos absorben el calor que radia de la superficie terrestre y previenen que este calor se pierda para el espacio, lo cual calienta la Tierra suficiente para sostener la vida.

erosión en barrancos: erosión que ocurre cuando el cauce de un arroyuelo se ensancha y profundiza.

guyot: grande volcán basáltico extinto con la cima que es plana y está sumergida.

H

---

half-life: period of time it takes for a radioactive isotope, such as carbon-14, to decay to one-half of its original amount.

halo: spherical region where globular clusters are located; surrounds the Milky Way's nuclear bulge and disk.

vida media: período de tiempo que demora un isótopo radiactivo, como el carbono 14, en desintegrarse a la mitad de su cantidad radiactiva original.

halo: región esférica donde se ubican los cúmulos globulares; rodea el disco y el núcleo central de la Vía Láctea.

hardness: measure of how easily a mineral can be scratched, which is determined by the arrangement of a mineral's atoms.

heat island: urban area where climate is warmer than in the surrounding countryside due to factors such as numerous concrete buildings and large expanses of asphalt.

heat wave: extended period of above-average temperatures caused by large, high-pressure systems that warm by compression and block cooler air masses.

Hertzsprung-Russell diagram (H-R diagram): graph that relates stellar characteristics—class, mass, temperature, magnitude, diameter, and luminosity.

highland: light-colored, mountainous, heavily cratered area of the Moon, composed mostly of lunar breccias.

Homo sapiens: species to which modern humans belong.

hot spot: unusually hot area in Earth's mantle where high-temperature plumes of mantle material rise toward the surface.

hot spring: thermal spring with temperatures higher than that of the human body.

Hubble constant: value (*H*) used to calculate the rate at which the universe is expanding; measured in kilometers per second per megaparsec.

humidity: amount of water vapor in the atmosphere at a given location on Earth's surface.

hydrocarbon: molecules with hydrogen and carbon bonds only; the result of the combination of carbon dioxide and water during photosynthesis.

hydroelectric power: power generated by converting the energy of free-falling water to electricity.

hydrogen bond: forms when the positive ends of some water molecules are attracted to the negative ends of other water molecules; causes water's surface to contract and allows water to adhere to and coat a solid.

hydrosphere: all the water in Earth's oceans, lakes, seas, rivers, and glaciers plus all the water in the atmosphere.

hydrothermal metamorphism: occurs when very hot water reacts with rock, altering its mineralogy and chemistry.

hygrometer (hi GRAH muh tur): weather instrument used to measure humidity.

hypothesis: a testable explanation of a situation.

dureza: medida de la facilidad con la que un mineral es rayado; está determinada por el ordenamiento de los átomos del mineral.

isla de calor: área urbana donde el clima es más caliente que en el área rural circundante, debido a factores como los numerosos edificios de concreto y las grandes extensiones de asfalto.

ola de calor: período extenso de temperaturas más altas que el promedio; es causado por grandes sistemas de alta presión que se calientan por compresión y bloquean las masas de aire más frías.

diagrama de Hertzsprung-Russell (diagrama H-R): gráfica que relaciona características estelares: incluyendo la clase, la masa, la temperatura, la magnitud, el diámetro y la luminosidad.

tierras altas: áreas de la Luna de color claro, con muchos cráteres y montañas, compuestas en su mayor parte de brechas lunares.

Homo sapiens: especie a la cual pertenecen los seres humanos modernos.

punto caliente: área muy caliente del manto de la Tierra donde plumas de material del manto a gran temperatura ascienden a la superficie.

fuente caliente: manantial termal con temperaturas más altas que las del cuerpo humano.

constante de Hubble: valor (*H*) que sirve para calcular la velocidad de expansión del universo; se mide en kilómetros por segundo por megaparsec.

humedad: cantidad de vapor de agua en el aire en un sitio determinado de la Tierra.

hidrocarburo: molécula que sólo contiene enlaces entre átomos de hidrógeno y de carbono; es producto de la unión del dióxido de carbono y el agua durante la fotosíntesis.

energía hidroeléctrica: se genera al convertir la energía de una caída de agua en electricidad.

enlace de hidrógeno: se forma cuando el extremo positivo de algunas moléculas de agua son atraídas por el extremo negativo de otras moléculas de agua; ocasiona que la superficie del agua se contraiga y permite al agua adherirse y recubrir un sólido.

hidrosfera: toda el agua en los océanos, los lagos, los mares, los ríos y los glaciares de la Tierra, además de toda el agua en la atmósfera.

metamorfismo hidrotérmico: ocurre cuando agua muy caliente reacciona con la roca, alterando su mineralogía y su química.

higrómetro: instrumento meteorológico que se usa para medir la humedad.

hipótesis: explicación de una situación que se puede poner a prueba.

I

ice age: period of extensive glacial coverage, producing long-term climatic changes, where average global temperatures decreased by 5°C.

igneous rock: intrusive or extrusive rock formed from the cooling and crystallization of magma or lava.

impact crater: crater formed when space material crashes into the surface of a celestial body.

glaciación: período de formación de una amplia cobertura glacial que produce cambios climáticos de largo plazo en que las temperaturas globales promedio desminuyen 5°C.

roca ígnea: roca intrusiva o extrusiva formada a partir del enfriamiento y cristalización del magma o lava.

cráter de impacto: cráter que se forma cuando material proveniente del espacio impacta la superficie de un objeto celeste.

GLOSSARY/GLOSARIO

independent variable: factor that is manipulated by the experimenter in an experiment.

index fossils: remains of plants or animals that were abundant, widely distributed, and existed briefly that can be used by geologists to correlate or date rock layers.

infiltration: process by which precipitation that has fallen on land surfaces enters the ground and becomes groundwater.

interferometry: process that links separate telescopes so they act as one telescope, producing more detailed images as the distance between them increases.

International Date Line: the 180° meridian, which serves as the transition line for calendar days.

intrusive rock: coarse-grained igneous rock that is formed when molten rock cools slowly and solidifies inside Earth's crust.

ion: an atom that has gained or lost an electron.

ionic bond: attractive force between two ions with opposite charge.

iridium (ih RID ee um): metal that is rare in rocks at Earth's surface but is relatively common in asteroids.

isobar: line on a weather map connecting areas of equal pressure

isochron (I suh krahn): imaginary line on a map that shows points of the same age; formed at the same time.

isostasy (I SAHS tuh see): condition of equilibrium that describes the displacement of Earth's mantle by Earth's continental and oceanic crust.

isostatic rebound: slow process of Earth's crust rising as the result of the removal of overlaying material.

isotherm: line on a weather map connecting areas of equal temperature.

isotope: an atom of an element that has a different mass number than the element but the same chemical properties.

variable independiente: factor que es manipulado por el investigador en un experimento.

fósiles guía: restos de plantas o animales que fueron abundantes, tuvieron una amplia distribución y existieron poco tiempo, que sirven a los geólogos para correlacionar o para datar estratos rocosos.

infiltración: proceso mediante el cual la precipi-tación que cae sobre la superficie terrestre entra al suelo y se convierte en agua subterránea.

interferometría: proceso que combina telescopios separados para que funcionen como un solo telescopio, produciendo imágenes más detalladas al aumentar la distancia entre ellos.

línea internacional de cambio de fecha: el meridiano 180°; sirve como la línea de transición para los días del calendario.

roca intrusiva: roca ígnea de grano grueso que se forma cuando la roca fundida se enfría lentamente y se solidifica en el interior de la corteza terrestre.

ion: átomo que ha ganado o perdido un electrón.

enlace iónico: fuerza de atracción entre dos iones con cargas opuestas.

iridio: metal escaso en las rocas de la superficie terrestre, pero relativamente común en los meteoritos y los asteroides.

isobara: línea de un mapa meteorológico que conecta áreas con igual presión.

isocrona: línea imaginaria en un mapa que conecta puntos con la misma antigüedad; que se formaron al mismo tiempo.

isostasia: condición de equilibrio que describe el desplazamiento del manto terrestre por las cortezas continental y oceánica de la Tierra.

rebote isostático: proceso lento de elevación de la corteza terrestre producto de la eliminación del material sobreyacente.

isoterma: línea en un mapa meteorológico que conecta áreas con la misma temperatura.

isótopo: átomo de un elemento que tiene un distinto número de masa que el elemento, pero las mismas propiedades químicas.

J

jet stream: narrow wind band that occurs above large temperature contrasts and can flow as fast as 185 km/h.

corriente de chorro: banda de vientos estrecha situada por encima de áreas con grandes contrastes de temperatura y que puede alcanzar una rapidez de 185 km/h.

K

kame: a conical mound of layered sediment that accumulates in a depression on a retreating glacier.

karst topography: irregular topography with sinkholes, sinks, and sinking streams caused by groundwater dissolution of limestone.

kame: montículo cónico de sedimento estratificado que es depositado por corrientes que fluyen bajo un glaciar que se derrite.

topografía cárstica: topografía irregular con sumideros, hundimientos y corrientes que desaparecen, causada por la disolución de la piedra caliza por el agua subterránea.

kettle: a lake formed when runoff and precipitation filled a kettle hole, which is a depression that formed when an ice block from a continental glacier became covered with sediment and melted.

key bed: a rock or sediment layer that serves as a time marker in the rock record and results from volcanic ash or meteorite-impact debris that spread out and covered large areas of Earth.

kimberlite: rare, ultramafic rock that can contain diamonds and other minerals formed only under very high pressures.

Köppen classification system: classification system for climates, divided into five types, based on the mean monthly values of temperature and precipitation and types of vegetation.

Kuiper (KI pur) belt: region of the space that lies outside the orbit of Neptune, 30 to 50 AU from the Sun, where small solar system bodies that are mostly rock and ice probably formed.

marmita: lago que se forma cuando la escorrentía y la precipitación llenan el hueco de una marmita, que es la depresión que se forma cuando un bloque de hielo de un glaciar continental queda cubierto con sedimento y se derrite.

estrato guía: capa de sedimento que sirve como marcador de tiempo del registro geológico; está formado por cenizas volcánicas o por los restos del impacto de un meteorito que se esparcen y cubren grandes áreas de la Tierra.

kimberlita: roca ultramáfica poco común que puede contener diamantes y otros minerales que sólo se forman bajo presiones muy altas.

sistema de clasificación de Köppen: sistema de clasificación de los climas; los clasifica en cinco tipos básicos en base a los valores mensuales promedio de temperatura y precipitación y a los tipos de vegetación.

cinturón de Kuiper: pequeños cuerpos del sistema solar formados principalmente por roca y hielo, yacen más allá de la órbita de Neptuno, entre 30 a 50 UA del Sol, y es muy probable que se hayan formado en esta región.

L

laccolith (LA kuh lihth): relatively small, mushroom-shaped pluton that forms when magma intrudes into parallel rock layers close to Earth's surface.

lake: natural or human-made body of water that can form when a depression on land fills with water.

Landsat satellite: information-gathering satellite that uses visible light and infrared radiation to map Earth's surface.

landslide: rapid downslope movement of a mass of loose soil, rock, or debris that has separated from the bedrock; can be triggered by an earthquake.

latent heat: stored energy in water vapor that is not released to warm the atmosphere until condensation takes place.

latitude: distance in degrees north and south of the equator.

Laurentia (law REN shuh): ancient continent formed during the Proterozoic that is the core of modern-day North America.

lava: magma that flows out onto Earth's surface.

Le Système International d'Unités (SI): replacement for the metric system; based on a decimal system using the number 10 as the base unit; includes the meter (m), second (s), and kilogram (kg).

liquid metallic hydrogen: form of hydrogen with both liquid and metallic properties that exists as a layer in the Jovian atmosphere.

lithification: the physical and chemical processes that transform sediments into sedimentary rocks.

loess (LUSS): thick, windblown, fertile deposit of silt that contains high levels of nutrients and minerals.

longitude: distance in degrees east and west of the prime meridian.

longshore bar: submerged sandbar located in the surf zone of most beaches.

lacolito: plutón relativamente pequeño con forma de champiñón que se forma cuando se introduce el magma entre estratos rocosos paralelos, cerca de la superficie terrestre.

lago: masa de agua, natural o hecha por el hombre, que se forma cuando una depresión terrestre se llena de agua.

satélite Landsat: satélite que recoge información, usando luz visible y radiación infrarroja para mapear la superficie terrestre.

derrumbe: rápido desplazamiento cuesta abajo de una masa de tierra, rocas o escombros sueltos que se han separado del lecho rocoso; puede ser causado por un terremoto.

calor latente: energía almacenada en el vapor de agua que no es liberada para calentar la atmósfera, hasta que ocurre la condensación.

latitud: distancia en grados hacia el norte o el sur del ecuador.

Laurencia: antiguo continente que se formó durante el Proterozoico y que en la actualidad corresponde al centro de Norteamérica.

lava: magma que fluye por la superficie terrestre.

Le Système Internacional d'Unités/Sistema Internacional de Unidades (SI): sustituto del sistema métrico; se basa en el sistema decimal por lo que usa el número 10 como unidad base: incluye el metro (m), el segundo (s) y el kilogramo (kg).

hidrógeno metálico líquido: forma de hidrógeno con propiedades de líquido y de metal que forma una capa en la atmósfera joviana.

litificación: procesos físicos y químicos que transforman los sedimentos en roca sedimentaria.

loes: amplio depósito fértil de limo que es arrastrado por el viento y contiene niveles altos de nutrientes y minerales.

longitud: distancia en grados hacia el este o el oeste del primer meridiano.

barra litoral: barra de arena sumergida ubicada en la zona de oleaje de la mayoría de las playas.

GLOSSARY/GLOSARIO

longshore current: current that flows parallel to the shore, moves large amounts of sediments, and is formed when incoming breakers spill over a longshore bar.

luminosity: energy output from the surface of a star per second; measured in watts.

lunar eclipse: when Earth passes between the Sun and the Moon, and Earth's shadow falls on the Moon; occurs only during a full moon.

luster: the way that a mineral reflects light from its surface; two types—metallic and nonmetallic.

corriente litoral: corriente que fluye paralela a la costa, transporta grandes cantidades de sedimentos y se forma cuando las olas rompen a lo largo de una larga barra litoral.

luminosidad: energía que irradia la superficie de una estrella por segundo; se mide en vatios.

eclipse lunar: sucede cuando la Tierra pasa entre el Sol y la Luna y la sombra de la Tierra cae sobre la Luna; ocurre sólo durante la luna llena.

lustre: manera en que la superficie de un mineral refleja la luz; existen dos tipos: metálico o no metálico.

M

magnetic reversal: when Earth's magnetic field changes polarity between normal and reversed.

magnetometer (mag nuh TAH muh tur): device used to map the ocean floor that detects small changes in magnetic fields.

magnitude: measure of the energy released during an earthquake, which can be described using the Richter scale.

main sequence: in an H-R diagram, the broad, diagonal band that includes about 90 percent of all stars and runs from hot, luminous stars in the upper-left corner to cool, dim stars in the lower-right corner.

map legend: key that explains what the symbols on a map represent.

map scale: ratio between the distances shown on a map and the actual distances on Earth's surface.

maria (MAH ree uh): dark-colored, smooth plains on the Moon's surface.

mass extinction: occurs when an unusually large number of organisms disappear from the rock record at about the same time.

mass movement: downslope movement of earth materials due to gravity that can occur suddenly or very slowly, depending on the weight of the material, its resistance to sliding, and whether a trigger, such as an earthquake, is involved.

mass number: combined number of protons and neutrons in the nucleus of an atom.

matter: anything that has volume and mass.

meander: curve or bend in a stream formed when a stream's slope decreases, water builds up in the stream channel, and moving water erodes away the sides of the streambed.

mechanical weathering: process that breaks down rocks and minerals into smaller pieces but does not involve any change in their composition.

Mercator projection: map with parallel lines of latitude and longitude that shows true direction and the correct shapes of landmasses but distorts areas near the poles.

inversión magnética: sucede cuando el campo magnético de la Tierra cambia polaridad entre normal e invertida.

magnetómetro: aparato que sirve para mapear el fondo marino; detecta cambios pequeños en los campos magnéticos.

magnitud: medida de la energía liberada durante un sismo; se puede describir usando la escala de Richter.

secuencia principal: la ancha banda diagonal de un diagrama H-R que contiene cerca del 90 por ciento de todas las estrellas; contiene desde estrellas calientes y luminosas en la esquina superior izquierda, hasta estrellas frías de brillo débil en la esquina inferior derecha.

leyenda del mapa: clave que explica los símbolos en un mapa.

escala del mapa: razón entre las distancias que se muestran en un mapa y las distancias reales en la superficie terrestre.

mar: planicie lunar lisa y de color oscuro.

extinción masiva: ocurre cuando un número insólitamente grande de organismos desaparece del registro geológico aproximadamente al mismo tiempo.

movimiento de masa: movimiento cuesta abajo de materiales terrestres debido a la gravedad; puede ocurrir de manera repentina o muy lentamente: dependiendo del peso del material, la resistencia del material a deslizarse y de si ha ocurrido algún evento que lo desencadene, como un sismo.

número de masa: número combinado de protones y neutrones en el núcleo de un átomo.

materia: todo aquello que tiene volumen y masa.

meandro: curva o desviación en una corriente; se forma cuando disminuye la pendiente de la corriente, por lo que el agua se acumula en el cauce y el movimiento del agua erosiona los costados del cauce.

meteorización mecánica: proceso de rompimiento de rocas y minerales en trozos más pequeños que no afecta la composición del material.

proyección de Mercator: mapa con líneas de latitud y longitud paralelas que muestra la dirección real y las formas correctas de las masas terrestres, aunque las áreas cercanas a los polos aparecen distorsionadas.

mesosphere: layer of Earth's atmosphere above the stratopause.

metallic bond: positive ions of metal held together by the negative electrons between them; allows metals to conduct electricity.

meteor: streak of light produced when a meteoroid falls toward Earth and burns up in Earth's atmosphere.

meteorite (MEE tee uh rite): a small fragment of an orbiting body that has fallen to Earth, generating heat; does not completely burn up in Earth's atmosphere and strikes Earth's surface, sometimes causing an impact crater.

meteoroid: piece of interplanetary material that falls toward Earth and enters its atmosphere.

meteorology: the study of the atmosphere, which is the air surrounding Earth.

meteor shower: occurs when Earth intersects a cometary orbit and comet particles burn up as they enter Earth's upper atmosphere.

microclimate: localized climate that differs from the surrounding regional climate.

microcontinent: a small fragment of granite-rich crust formed during the Archean.

mid-ocean ridge: chain of underwater mountains that run throughout the ocean basins, have a total length over 65,000 km, and contain active and extinct volcanoes.

mineral: naturally occurring, inorganic solid with a specific chemical composition and a definite crystalline structure.

mineral replacement: the process where pore spaces of an organism's buried parts are filled in with minerals from groundwater.

modified Mercalli scale: measures earthquake intensity on a scale from I to XII; the higher the number, the greater the damage the earthquake has caused.

mold: fossil that can form when a shelled organism decays in sedimentary rock and is removed by erosion or weathering, leaving a hollowed-out impression.

molecule: combination of two or more atoms joined by covalent bonds.

moment magnitude scale: scale used to measure earthquake magnitude—taking into account the size of the fault rupture, the rocks' stiffness, and amount of movement along the fault—using values that can be estimated from the size of several types of seismic waves.

moraine: ridge or layer of mixed debris deposited by a melting glacier.

mountain thunderstorm: occurs when an air mass rises from orographic lifting, which involves air moving up the side of a mountain.

mudflow: rapidly flowing, often destructive mixture of mud and water that may be triggered by an earthquake, intense rainstorm, or volcanic eruption.

mesosfera: capa de la atmósfera terrestre ubicada encima de la estratopausa.

enlace metálico: iones metálicos positivos que se mantienen unidos debido la carga negativa de los electrones que se encuentran entre ellos; permite a los metales conducir electricidad.

estrella fugaz: rayo luminoso que se produce cuando un meteoroide cae a la Tierra y se quema en la atmósfera terrestre.

meteorito: fragmento pequeño de un cuerpo en órbita que cae a la Tierra generando calor; como no se quema completamente en la atmósfera, choca con la superficie terrestre y produce un cráter de impacto.

meteoroide: trozo de material interplanetario que cae a la Tierra y entra a la atmósfera terrestre.

meteorología: estudio de la atmósfera, la capa de aire que rodea la Tierra.

lluvia de estrellas: ocurre cuando la Tierra interseca la órbita de un cometa y las partículas del cometa se queman al entrar a las capas superiores de la atmósfera terrestre.

microclima: clima localizado que difiere del clima regional circundante.

microcontinentes: trozos pequeños de corteza rica en granito que se formaron durante el Arcaico.

dorsales mediooceánicas: cadenas montañosas submarinas que se extienden a través de las cuencas oceánicas, tienen una longitud total de más de 65,000 km y contienen innumerables volcanes activos y extintos.

mineral: sólido inorgánico natural con una composición química específica y una estructura cristalina definida.

sustitución de minerales: proceso en que los poros de las partes enterradas de un organismo se llenan con los minerales provenientes de aguas subterráneas.

escala de Mercalli modificada: mide la intensidad de un sismo en una escala de I a XII; a medida que aumenta el número, mayor es el daño causado.

molde: fósil que se forma cuando un organismo con concha se descompone en roca sedimentaria y es removido por erosión o meteorización, quedando una impresión hueca.

molécula: combinación de dos o más átomos unidos por enlaces covalentes.

escala de magnitud momentánea: escala que sirve para medir la intensidad de un sismo (tomando en cuenta el tamaño de la ruptura de la falla, la rigidez de la roca y la cantidad del movimiento a lo largo de la falla) usando valores estimados a partir de la magnitud de varios tipos de ondas sísmicas.

morrena: loma o estrato de detritos mezclados que deposita un glaciar al derretirse.

tormenta orográfica: sucede cuando una masa de aire sube por ascenso orográfico, lo que implica el ascenso por la ladera de una montaña.

flujo o corriente de lodo: mezcla de lodo y agua que fluye rápidamente y que a menudo es destructiva; puede ser causada por un terremoto, una lluvia intensa o una erupción volcánica.

Glossary/Glosario

GLOSSARY/GLOSARIO

N

natural resource: resources provided by Earth, including air, water, land, all living organisms, nutrients, rocks, and minerals.

neap tide: tide that occurs during first- or third-quarter Moon, when the Sun, the Moon, and Earth form a right angle; this causes solar tides to diminish lunar tides, causing high tides to be lower than normal and low tides to be higher than normal.

nebula: large cloud of interstellar gas and dust that collapses on itself, due to its own gravity, and forms a hot, condensed object that will become a new star.

neutron: tiny atomic particle that is electrically neutral and has about the same mass as a proton.

neutron star: collapsed, dense core of a star that forms quickly while its outer layers are falling inward, has a radius of about 10 km, a mass 1.4 to 3 times that of the Sun, and contains mostly neutrons.

nitrogen-fixing bacteria: bacteria found in water or soil; can grow on the roots of some plants, capture nitrogen gas, and change into a form that plants use to build proteins.

nonfoliated: describes metamorphic rocks like quartzite and marble, composed mainly of minerals that form with blocky crystal shapes.

nonpoint source: water-pollution source that generates pollution from widely spread areas, such as runoff from roads.

nonrenewable resource: resource that exists in Earth's crust in a fixed amount and can be replaced only by geologic, physical, or chemical processes that take hundreds of millions of years.

normal: standard value for a location, including rainfall, wind speed, and temperatures, based on meteorological records compiled for at least 30 years.

nuclear fission: the process in which a heavy nucleus divides to form smaller nuclei and one or two neutrons and produces a large amount of energy.

nucleus (NEW klee us): positively charged center of an atom, made up of protons and neutrons and surrounded by electrons in energy levels.

recursos naturales: recursos que provee la Tierra: incluyendo el aire, el agua, la tierra, todos los organismos vivos, los nutrientes, las rocas y los minerales.

marea muerta: durante el primero o el tercer cuartos lunares, el Sol, la Luna y la Tierra se encuentran en ángulo recto, causando que las mareas solares reduzcan la intensidad de las mareas lunares, lo que provoca que la marea alta sea menor que lo normal y la marea baja sea mayor que lo normal.

nebulosa: extensa nube de gas y polvo interestelares que se colapsa en sí misma debido a su propia gravedad, formando un cuerpo condensado caliente que se convertirá en una estrella nueva.

neutrón: partícula atómica diminuta, eléctricamente neutra; tiene una masa similar a la de un protón.

estrella de neutrones: núcleo denso y colapsado de una estrella que se forma rápidamente, al mismo tiempo que sus capas exteriores se contraen; tiene un radio aproximado de 10 km, una masa de 1.4 a 3 veces la del Sol y contiene principalmente neutrones.

bacteria fijadora de nitrógeno: bacteria que habita el suelo o el agua; puede crecer en las raíces de algunas plantas, capturar el gas nitrógeno y convertirlo a una forma que las plantas pueden usar para fabricar proteínas.

no foliada: describe roca metamórfica, como la cuarcita y el mármol, compuesta principalmente de minerales que forman bloques cristalinos.

fuente no puntual: fuente de contaminación del agua que genera contaminación a partir de áreas muy extensas, como la escorrentía de los caminos.

recurso no renovable: recurso que existe en la corteza terrestre en una cantidad fija y que sólo puede ser regenerado por procesos geológicos, físicos o químicos que demoran centenas de millones de años.

normales: valores estándar para un sitio: incluyen la lluvia, la velocidad del viento y las temperaturas; se basan en los registros meteorológicos recopilados durante por lo menos 30 años.

fisión nuclear: proceso de división de un núcleo pesado en núcleos más pequeños y uno o dos neutrones, produciendo una gran cantidad de energía.

núcleo: centro del átomo, tiene carga positiva, está compuesto por protones y neutrones y está rodeado por electrones localizados en niveles de energía.

O

oceanography: study of Earth's oceans including the creatures that inhabit its waters, its physical and chemical properties, and the effects of human activities.

ore: mineral or rock that contains a valuable substance that can be mined at a profit.

original horizontality: the principle that sedimentary rocks are deposited in horizontal or nearly horizontal layers.

oceanografía: estudio de los océanos de la Tierra: incluyendo sus propiedades físicas y químicas, los seres que los habitan y los efectos de las actividades humanas sobre ellos.

mena: mineral o roca que contiene una sustancia valiosa que se puede extraer con fines de lucro.

horizontalidad original: principio que establece que las rocas sedimentarias se depositan formando estratos horizontales o casi horizontales.

original preservation: describes a fossil with soft and hard parts that have undergone very little change since the organism's death.

orogeny (oh RAH juh nee): cycle of processes that form all mountain ranges, resulting in broad, linear regions of deformation that you know as mountain ranges but in geology are known as orogenic belts.

orographic lifting: cloud formation that occurs when warm, moist air is forced to rise up the side of a mountain.

outwash plain: area at the leading edge of a glacier, where outwash is deposited by meltwater streams.

oxidation: chemical reaction of oxygen with other substances.

ozone hole: a seasonal decrease on ozone over Earth's polar regions.

preservación de material original: describe un fósil cuyas partes blandas y duras han sufrido muy pocos cambios desde la muerte del organismo.

orogenia: ciclo de procesos que forman todas las cadenas montañosas, dando como resultado grandes regiones lineares de deformación llamadas cadenas montañosas, pero que en geología se conocen como cinturones orogénicos.

ascenso orográfico: formación de nubes que se produce cuando el aire húmedo caliente es forzado a ascender por la ladera de una montaña.

llanura aluvial: área en el borde frontal de un glaciar donde las corrientes del agua que se derrite depositan los derrubios.

oxidación: reacción química del oxígeno con alguna otra sustancias.

agujero de ozono: disminución estacional del ozono sobre las regiones polares de la Tierra.

P

paleogeography (pay lee oh jee AH gruh fee): the ancient geographic setting of an area.

paleomagnetism: study of Earth's magnetic record using data gathered from iron-bearing minerals in rocks that have recorded the orientation of Earth's magnetic field at the time of their formation.

Pangaea (pan JEE uh): ancient landmass made up of all the continents that began to break apart about 200 mya.

parallax: apparent positional shift of an object caused by the motion of the observer.

parsec (pc): the distance equal to 3.26 ly or 3.086×10^{13} km.

partial melting: process in which different minerals melt into magma at different temperatures, changing its composition.

passive margin: edge of a continent along which there is no tectonic activity.

peat: light, spongy, organic fossil fuel derived from moss and other bog plants.

pegmatite: igneous rock with extremely large-grained minerals that can contain rare ores such as lithium and beryllium.

perigee: closest point in the Moon's elliptical orbit to Earth.

period: third-longest time unit in the geologic time scale, measured in tens of millions of years.

permeability: ability of a material to let water pass through; is high in material with large, well-connected pores and low in material with few pores or small pores.

pesticide: chemical applied to plants to kill insects and weeds.

photochemical smog: a type of air pollution; a yellow-brown haze formed mainly from automobile exhaust in the presence of sunlight.

photosphere: innermost layer of the Sun's atmosphere that is also its visible surface, has an average temperature of 5800 K, and is about 400 km thick.

paleogeografía: características geográficas antiguas de un área.

paleomagnetismo: estudio del registro magnético de la Tierra; utiliza la información recogida a partir de minerales ferrosos en las rocas porque este tipo de minerales registran la orientación del campo magnético de la Tierra en el momento en que se forman.

Pangaea: antigua masa terrestre compuesta por todos los continentes, los cuales se empezaron a separar hace cerca de 200 millones de años.

paralaje: cambio aparente de la posición de un cuerpo causado por el movimiento del observador.

parsec: distancia de 3.26 ly o 3.086×10^{13} km.

fundición parcial: proceso en el cual diferentes minerales se funden en el magma a diferentes tempe-raturas, cambiando su composición.

margen pasivo: límite de un continente a lo largo del cual no ocurre actividad tectónica.

turba: combustible fósil liviano, esponjoso y orgánico derivado del musgo y otras plantas de ciénegas.

pegmatita: roca ignea con grano extremadamente grueso que pueden contener minerales raros como el litio y el berilio.

perigeo: punto más cercano a la Tierra en la órbita elíptica de la Luna.

período: tercera unidad de tiempo más grande en la escala del tiempo geológico; se mide en decenas de millones de años.

permeabilidad: capacidad de un material de permitir el paso del agua; es grande en materiales con poros grandes y bien conectados y baja en materiales con pocos poros o con poros pequeños.

pesticida: sustancia química que se aplica a las plantas para eliminar insectos y malas hierbas.

smog fotoquímico: tipo de contaminación del aire; niebla color amarillo marrón que se forma debido principalmente a las emisiones de los autos en presencia de la luz solar.

fotosfera: capa más interior de la atmósfera solar; corresponde a su superficie visible, tiene una temperatura promedio de 5800 K y mide aproximadamente 400 km de ancho.

GLOSSARY/GLOSARIO

photovoltaic cell: thin, transparent wafer that converts sunlight into electrical energy and is made up of two layers of two types of silicon.

phytoplankton: microscopic organisms that are the basis of marine food chains; abundant during the Cretaceous and the remains of their shell-like hard parts are found in chalk deposits worldwide.

planetesimal: space object built of solid particles that can form planets through collisions and mergers.

plasma: hot, highly ionized, electrically conducting gas.

plastic deformation: permanent deformation caused by strain when stress exceeds a certain value.

plateau: a relatively flat-topped area.

pluton (PLOO tahn): intrusive igneous rock body, including batholiths, stocks, sills, and dikes, formed through mountain-building processes and oceanic-oceanic collisions; can be exposed at Earth's surface due to uplift and erosion.

point source: water-pollution source that generates pollution from a single point of origin, such as an industrial site.

polar easterlies: global wind systems that lie between latitudes 60° N and 60° S and the poles and is characterized by cold air.

polar zones: areas of Earth where solar radiation strikes at a low angle, resulting in temperatures that are nearly always cold; extend from 66.5° north and south of the equator to the poles.

pollutant: substance that enters Earth's geochemical cycles and can harm the health of living things or adversely affect their activities.

Population I star: star in the disk or arms that has a small amount of heavy elements.

Population II star: star in the halo or bulge that contains traces of heavy elements.

porosity: percentage of open spaces between grains in a material.

porphyritic (por fuh RIH tihk) texture: rock texture characterized by large, well-formed crystals surrounded by finer-grained crystals of the same or different mineral.

Precambrian (pree KAM bree un): informal unit of geologic time consisting of the first three eons during which Earth formed and became habitable.

Precambrian shield: the top of a craton exposed at Earth's surface

precipitation: all solid and liquid forms of water—including rain, snow, sleet, and hail—that fall from clouds.

prevailing westerlies: global wind system that lies between 30° and 60° north and south latitudes, where surface air moves toward the poles in an easterly direction.

primary wave: seismic wave that squeezes and pushes rocks in the same direction that the wave travels, known as a P-wave.

celdas fotovoltaicas: láminas delgadas y transpa-rentes que convierten la luz solar en energía eléctrica; están compuestas de dos capas con dos tipos de silicio.

fitoplancton: organismos microscópicos que son la base de las cadenas alimenticias marinas; fueron muy abundantes durante el Cretáceo y los restos de sus caparazones se encuentran en depósitos de carbonato de calcio por todo el mundo.

planetesimal: cuerpo espacial formado por partículas sólidas y los cuales pueden formar planetas mediante choques y fusiones.

plasma: gas caliente, altamente ionizado y conductor de electricidad.

deformación dúctil: cuando la presión excede cierto valor; la tensión producida causa una deformación permanente.

altiplanicie: área relativamente plana en la parte más alta.

plutones: cuerpos rocosos ígneos intrusivos: incluye batolitos, macizos magmáticos, intrusiones y diques formados durante los procesos orogénicos y durante la colisión de placas oceánicas; pueden quedar expuestos a la superficie terrestre debido a levantamientos y erosión.

fuente puntual: fuente de contaminación de agua que genera contaminación a partir de un solo punto de origen, por ejemplo, una zona industrial.

vientos polares del este: sistemas globales del viento que se encuentran entre los polos y las latitudes 60°N y 60°S; se caracterizan por tener aire frío.

zonas polares: áreas de la Tierra donde la radiación solar llega con un ángulo bajo, ocasionando que las temperaturas casi siempre sean frías; se extienden desde los 66.5° hasta los polos, en ambos hemisferios.

contaminante: sustancia que entra a los ciclos geoquímicos de la Tierra y puede causar daños a la salud de los seres vivos o afectar adversamente sus actividades.

estrella de la población I: estrella ubicada en el disco o los brazos y que contiene pequeñas cantidades de elementos pesados.

estrella de la población II: estrella ubicada en el halo o en el núcleo y que contiene trazas de elementos pesados.

porosidad: porcentaje de espacios abiertos entre los granos de una roca.

textura porfírica: textura rocosa caracterizada por cristales grandes bien formados, rodeados por cristales de grano más fino del mismo mineral o de uno diferente.

Precámbrico: unidad del tiempo geológico que consiste en los primeros tres eones; periodo durante el cual la Tierra se formó y adquirió condiciones aptas para la vida.

escudo Precámbrico: parte alta de un cratón que está expuesta en la superficie de la Tierra.

precipitación: toda forma líquida o sólida de agua: lluvia, nieve, aguanieve o granizo, que cae de las nubes.

vientos dominantes del oeste: sistema de vientos globales ubicado entre los 30° y los 60° de latitud, en ambos hemisferios, donde el aire superficial se desplaza hacia los polos en dirección este.

onda primaria: onda sísmica que comprime y empuja las rocas en la misma dirección en que viaja la onda; se conocen como ondas P.

prime meridian: imaginary line representing 0° longitude, running from the North Pole, through Greenwich, England, to the South Pole.

principle of inclusion: the principle that fragments, called inclusions, in a rock layer must be older than the rock layer that contains them.

prokaryote (proh KE ree oht): unicellular organism that lacks a nucleus.

prominence: arc of gas ejected from the chromosphere, or gas that condenses in the Sun's inner corona and rains back to the surface, that can reach temperatures over 50,000 K and is associated with sunspots.

proton: tiny atomic particle that has mass and a positive electric charge.

protostar: hot, condensed object at the center of a nebula that will become a new star when nuclear fusion reactions begin.

pulsar: a spinning neutron star that exhibits a pulsing pattern.

pyroclastic flow: swift-moving, potentially deadly clouds of gas, ash, and other volcanic material produced by a violent eruption.

primer meridiano: línea imaginaria que representa la longitud 0°; va desde el Polo Norte hasta el Polo Sur, pasando por Greenwich, Inglaterra.

principio de las inclusiones: principio que establece que los fragmentos, llamados inclusiones, contenidos por un estrato rocoso deben ser más antiguos que la roca que los contiene.

procariota: organismo unicelular que carece de núcleo.

protuberancia solar: arco de gas expulsado de la cromosfera o gas que se condensa en la corona interna del Sol y que se precipita de nuevo sobre su superficie; puede alcanzar temperaturas mayores a los 50,000 K y está asociada a la presencia de manchas solares.

protón: partícula atómica diminuta que tiene masa y una carga eléctrica positiva.

protoestrella: cuerpo condensado, caliente, ubicado en el centro de una nebulosa, que se convertirá en una estrella nueva cuando inicien las reacciones de fusión nuclear.

pulsar: estrella de neutrones giratoria que exhibe un patrón de pulsaciones.

flujo piroclástico: nubes de gas, cenizas y otros materiales volcánicos, potencialmente mortales, que se desplazan rápidamente y que son producidas por una erupción violenta.

Q

quasar: starlike, very bright, extremely distant object with emission lines in its spectra.

cuásares: cuerpos semejantes a estrellas, muy brillantes y extremadamente lejanos, con líneas de emisión en sus espectros.

R

radiation: the transfer of thermal energy by electromagnetic waves; the transfer of thermal energy from the Sun to Earth by radiation.

radioactive decay: emission of radioactive particles and its resulting change into other isotopes over time.

radiocarbon dating: determines the age of relatively young organic objects; objects that are alive or were once alive.

radio galaxy: very bright, often giant, elliptical galaxy that emits as much or more energy in the form of radio wavelengths as it does wavelengths of visible light.

radiometric dating: process used to determine the absolute age of a rock or fossil by determining the ratio of parent isotopes to daughter isotopes within a given sample.

radiosonde (RAY dee oh sahnd): balloon-borne weather instrument whose sensors measure air pressure, humidity, temperature, wind speed, and wind direction of the upper atmosphere.

ray: long trail of ejecta that radiates outward from an impact crater.

recharge: process by which water from precipitation and runoff is added to the zone of saturation.

radiación: transferencia de energía mediante por ondas electromagnéticas; la transferencia de energía térmica del Sol a la Tierra por radiación.

desintegración radiactiva: emisión de partículas atómicas que a lo largo del tiempo produce nuevos isótopos.

datación radiocarbónica: permite determinar la edad de cuerpos orgánicos relativamente recientes, cuerpos que están vivos o que alguna vez estuvieron vivos.

radiogalaxia: galaxia elíptica muy brillante, a menudo gigantesca, cuya emisión de energía en forma de ondas de radio es similar a la que emite como ondas de luz visible.

datación radiométrica: proceso que permite establecer la edad absoluta de una roca o un fósil, al determinar la razón entre los isótopos originales y los isótopos derivados de una muestra dada.

radiosonda: instrumento meteorológico que se monta en un globo y cuyos sensores miden la presión atmosférica, la humedad, la temperatura, así como la velocidad y dirección del viento en la atmósfera superior.

rayo: largo rastro de eyecta que irradia de un cráter de impacto.

recarga: proceso mediante el cual el agua de la precipitación y de la escorrentía entra a la zona de saturación.

Glossary/Glosario

GLOSSARY/GLOSARIO

reclamation: process in which a mining company restores land used during mining operations to its original contours and replants vegetation.

red bed: a sedimentary rock deposit that contains oxidized iron; provides evidence that free oxygen existed in the atmosphere during the Proterozoic.

reflecting telescope: telescope that uses mirrors to focus visible light.

refracting telescope: telescope that uses lenses to focus visible light.

regional metamorphism: process that affects large areas of Earth's crust, producing belts classified as low, medium, or high grade, depending on pressure on the rocks, temperature, and depth below the surface.

regolith: layer of loose, ground-up rock on the lunar surface.

regression: occurs when sea level falls, causing the shoreline to move seaward, and results in shallow-water deposits overlying deep-water deposits.

rejuvenation: process during which a stream resumes downcutting toward its base level, increasing its rate of flow.

relative-age dating: establishing the order of past geologic events.

relative humidity: ratio of water vapor contained in a specific volume of air compared with how much water vapor that amount of air actually can hold; expressed as a percentage.

remote sensing: process of gathering data about Earth from instruments far above the planet's surface.

renewable resource: natural resource, such as fresh air and most groundwater, that can be replaced by nature in a short period of time.

residual soil: soil that develops from parent material which is similar to local bedrock.

retrograde motion: a planet's apparent backward movement in the sky.

return stroke: a branch channel of positively charged ions that rushes upward from the ground to meet the stepped leader.

Richter scale: numerical rating system used to measure the amount of energy released during an earthquake; an increase of 1 in the scale represents an increase in amplitude of a factor of 10.

ridge push: tectonic process associated with convection currents in Earth's mantle that occurs when the weight of an elevated ridge pushes an oceanic plate toward a subduction zone.

rift valley: long, narrow depression that forms when continental crust begins to separate at a divergent boundary.

rill erosion: erosion in which water running down the side of a slope carves a small stream channel.

rille: valleylike structure that meanders across some regions of the Moon's maria.

recuperación: proceso en que una compañía minera restaura los terrenos usados en las actividades mineras a sus contornos originales y reforesta con nueva vegetación.

lecho rojo: depósito de roca sedimentaria que contiene hierro oxidado; es evidencia de que había oxígeno libre en la atmósfera durante el Proterozoico.

telescopio reflector: telescopio que usa espejos para enfocar la luz visible.

telescopio refractor: telescopio que usa lentes para enfocar la luz visible.

metamorfismo regional: proceso que afecta grandes áreas de la corteza terrestre; produce cinturones de bajo, medio o alto grado, dependiendo de la presión sobre las rocas, la temperatura y la profundidad bajo la superficie.

regolito: estrato de roca suelta y molida en la superficie lunar.

regresión: ocurre cuando baja el nivel del mar, provocando que la costa avance hacia el mar, ocasiona que depósitos de agua superficiales cubran depósitos de agua profundos.

rejuvenecimiento: proceso en que una corriente reanuda la erosión hacia su nivel base, aumentando su tasa de flujo.

datación relativa: ordenamiento por antigüedad de eventos geológicos pasados.

humedad relativa: razón del vapor de agua que contiene un volumen específico de aire, en comparación con la cantidad de vapor de agua que ese volumen de aire podría contener, expresado como porcentaje.

percepción remota: proceso de recopilación de datos sobre la Tierra con instrumentos alejados de la superficie del planeta.

recurso renovable: recurso natural, como el aire y la mayoría de las aguas subterráneas, que la naturaleza puede reemplazar en un período corto de tiempo.

suelo residual: suelo que se desarrolla a partir del material original y es similar a la roca madre local.

movimiento retrógrado: movimiento aparentemente en retroceso de un planeta en el cielo.

descarga de retorno: un canal con iones de carga positiva que asciende desde el suelo para encontrarse con la descarga líder o guía escalonada.

escala de Richter: escala numérica que se emplea para medir la cantidad de energía liberada durante un sismo; un aumento de 1 unidad en esta escala representa un aumento en amplitud de un factor de 10.

empuje de la dorsal: proceso tectónico asociado con las corrientes de convección en el manto de la Tierra, que ocurre cuando el peso de una cordillera elevada empuja una placa oceánica hacia una zona de subducción.

valle del rift: depresión larga y estrecha que se forma cuando la corteza continental se empieza a separar en un límite divergente.

erosión por surcos: erosión en la cual el agua que corre cuesta abajo forma un canal pequeño.

surco: formación tipo valle que serpentea a través de algunas regiones de los mares lunares.

rock cycle: continuous, dynamic set of processes by which rocks are changed into other types of rock.

root: thickened areas of continental material that counterbalance the parts of a mountain that rise above Earth's surface; detected by gravitational and seismic studies.

RR Lyrae variable: star with pulsation period ranging from 1.5 hours to 1.2 days; generally have the same luminosity, regardless of pulsation period length.

runoff: water that flows downslope on Earth's surface and may enter a stream, river, or lake; its rate is influenced by the angle of the slope, vegetation, rate of precipitation, and soil composition.

ciclo de las rocas: conjunto de procesos continuos y dinámicos a través de los cuales las rocas se transforman en otros tipos de roca.

raíz: gruesas áreas de material continental que compensan por las partes de la montaña que suben la superficie de la tierra; que son detectadas en estudios sísmicos o gravitacionales.

variable RR Lyrae: estrella con periodo de pulsación que dura de 1.5 horas a 1.2 días; en general tiene la misma luminosidad, independientemente de la duración de la pulsación.

escorrentía: agua que corre cuesta abajo sobre la superficie terrestre y que puede incorporarse a una corriente, río o lago; su tasa de flujo está influida por el ángulo de la pendiente, la vegetación, la tasa de precipi-tación y la composición del suelo.

S

Saffir-Simpson Hurricane Wind scale: classifies hurricanes according to wind speed on a scale ranging from Category 1 to Category 5.

salinity: measure of the amount of salts dissolved in seawater, which is 35 ppt, or 3.5% on average.

saturation: the point at which water molecules leaving the water's surface equals the rate of water molecules returning to the surface.

scarp: a line of cliffs produced by erosion or faulting.

scientific law: a principle that describes the behavior of a natural phenomenon.

scientific methods: a series of problem-solving procedures that help scientists conduct experiments.

scientific model: an idea, a system, or a mathematical expression that represents the idea being explained.

scientific notation: a method used by scientists to express a number as a value between 1 and 10 multiplied by a power of 10.

scientific theory: an explanation based on many observations during repeated experiments; valid only if consistent with observations, can be used to make testable predictions, and is the simplest explanation; can be changed or modified with the discovery of new data.

sea-breeze thunderstorm: local air-mass thunderstorm that commonly occurs along a coastal area because land and water store and release thermal energy differently.

seafloor spreading: the hypothesis that new ocean crust is formed at mid-ocean ridges and destroyed at deep-sea trenches; occurs in a continuous cycle of magma intrusion and spreading.

sea level: level of the oceans' surfaces, which has risen at a rate of about 3 mm per year from 1994 to 2012.

seamount: basaltic, submerged volcano on the seafloor that is more than 1 km high.

escala de Vientos Huracanados Saffir-Simpson: clasifica los huracanes según la velocidad de sus vientos en una escala que va desde la Categoría 1 hasta la Categoría 5.

salinidad: medida de la cantidad de sales disueltas en el agua de mar; en promedio es de 35 ppt ó 3.5%.

saturación: sucede en el punto en el cual la tasa de salida de moléculas de agua en la superficie es igual a la tasa de retorno de las moléculas a la superficie.

escarpes: una línea de acantilados por erosión o fallas.

ley científica: principio que describe el comportamiento de un fenómeno natural.

métodos científicos: serie de procedimientos para resolver problemas que ayudan a los científicos a realizar experimentos.

modelo científico: idea, sistema o expresión matemática que representa la idea que se quiere explicar.

notación científica: método que usan los científicos para expresar un número como un valor entre 1 y 10 multiplicado por una potencia de 10.

teoría científica: explicación basada en muchas observaciones realizadas durante experimentos repetidos; sólo es válida si es consistente con las observaciones, permite hacer predicciones comprobables y es la explicación más sencilla; puede ser modificada debido al descubrimiento de nuevos hechos.

tormenta eléctrica de brisa marina: tormenta local de masa de aire que ocurre comúnmente a lo largo de un área costera; ocurren porque la tierra y el agua almacenan y liberan energía térmica de manera distinta.

expansión del suelo marino: hipótesis que propone que la nueva corteza oceánica se forma en las dorsales medioceánicas y se destruye en las fosas submarinas profundas; ocurre según un ciclo continuo de intrusión y expansión del magma.

nivel del mar: nivel de la superficie del océano; actualmente sube a una velocidad de 3 mm por año de 1994 a 2012.

montaña submarina: volcán basáltico sumergido en el fondo marino que mide más de 1 km de altura.

GLOSSARY/GLOSARIO

season: short-term periods with specific weather conditions caused by regular variations in temperature, hours of daylight, and weather patterns that are due to the tilt of Earth's axis as it revolves around the Sun, causing different areas of Earth to receive different amounts of solar radiation.

secondary wave: seismic wave that causes rock particles to move at right angles to the direction of the wave, known as an S-wave.

sediment: small pieces of rock that are moved and deposited by water, wind, glaciers, and gravity.

seismic gap: place along an active fault that has not experienced an earthquake for a long time.

seismic wave: the vibrations of the ground during an earthquake.

seismogram (SIZE muh gram): record produced by a seismometer that can provide individual tracking of each type of seismic wave.

seismometer (size MAH muh tur): instrument used to measure horizontal or vertical motion during an earthquake.

shield volcano: broad volcano with gently sloping sides built by non-explosive eruptions of basaltic lava that accumulates in layers.

side-scan sonar: technique that directs sound waves at an angle to the seafloor or deep-lake floor, allowing underwater topographic features to be mapped.

silicate: mineral that contains silicon **(Si), oxygen (O)**, and usually one or more other elements.

sill: pluton that forms when magma intrudes parallel rock layers.

sinkhole: depression in Earth's surface formed when a cave collapses or bedrock is dissolved by acidic rain or moist soil.

slab pull: tectonic process associated with convection currents in Earth's mantle that occurs as the weight of the subducting plate pulls the trailing lithosphere into a subduction zone.

slump: mass movement that occurs when earth materials in a landslide rotate and slide along a curved surface, leaving a crescent-shaped scar on a slope.

soil: loose covering of weathered rock and decayed organic matter overlying Earth's bedrock that is characterized by texture, fertility, and color and whose composition is determined by its parent rock and environmental conditions.

soil horizon: distinct layer within a soil profile.

soil liquefaction (lih kwuh FAK shun): process associated with seismic vibrations that occur in areas of sand that is nearly saturated; resulting in the ground behaving like a liquid.

soil profile: a vertical sequence of soil layers.

solar eclipse: when the Moon passes between Earth and the Sun and the Moon casts a shadow on Earth, blocking Earth's view of the Sun; can be partial or total.

solar flare: violent eruption of radiation and particles from the Sun's surface that is associated with sunspots.

estación: períodos de corto plazo con específicas de tiempo causados por variaciones regulares en temperatura, horas de luz solar y patrones meteorológicos, provocadas por la inclinación del eje de la Tierra cuando gira alrededor del Sol, lo que ocasiona que las distintas áreas de la Tierra reciban diferentes cantidades de radiación solar.

onda secundaria: onda sísmica que ocasiona que las partículas de las rocas se muevan en ángulo recto con respecto a la dirección de la onda.

sedimentos: partículas pequeñas de roca que el agua, el viento, los glaciares y la gravedad mueven y depositan.

vacío sísmico: lugar a lo largo de una falla activa que no ha sufrido un terremoto durante mucho tiempo.

onda sísmica: vibraciones del terreno durante un sismo.

sismograma: registro producido por un sismógrafo que proporciona un registro individual de cada tipo de onda sísmica.

sismógrafo: instrumento que sirve para medir los movimientos horizontales y verticales durante un sismo.

volcán de escudo: volcán ancho, de laderas con inclinación suave, formado por erupciones no explosivas de lava basáltica que se acumula en estratos.

sonar de escaneo lateral: técnica que dirige las ondas sonoras en ángulo hacia el fondo del mar o de un lago profundo, lo que permite trazar el relieve topográfico submarino.

silicato: mineral que contiene silicio **(Si), oxígeno (O)** y generalmente uno o más elementos adicionales.

intrusión: plutón que se forma cuando el magma penetra estratos rocosos paralelos.

sumidero: depresión en la superficie terrestre que se forma cuando una caverna se colapsa o cuando el lecho rocoso es disuelto por lluvia ácida o suelo húmedo.

tracción de placa: proceso tectónico asociado con las corrientes de convección del manto de la Tierra, que ocurre cuando el peso de la placa subductora jala la litosfera hacia una zona de subducción.

deslizamiento rotacional: movimiento en masa que ocurre cuando los materiales terrestres de un derrumbe giran y se deslizan a lo largo de una superficie curva, dejando una cicatriz con forma de medialuna en la pendiente.

suelo: cubierta suelta de roca meteorizada y materia orgánica en descomposición que cubre el lecho rocoso terrestre; se caracteriza por su textura, fertilidad y color y su composición está determinada por la roca madre y las condiciones ambientales.

horizonte del suelo: capa distintiva dentro de un perfil del suelo.

licuefacción del suelo: proceso asociado con las vibraciones sísmicas que ocurren en las áreas arenosas casi saturadas; el resultado es que el suelo actúa como un líquido.

perfil del suelo: sucesión vertical de capas del suelo.

eclipse solar: sucede cuando la Luna pasa entre la Tierra y el Sol y la Luna proyecta su sombra sobre la Tierra, bloqueando la luz del Sol; puede ser parcial o total.

erupción solar: violenta erupción de radiación y partículas desde la superficie del Sol que está asociada con las manchas solares.

solar wind: wind of charged particles **(ions)** that flows throughout the solar system and begins as gas flowing outward from the Sun's corona at high speeds.

solstice: period when the Sun is overhead at its farthest distance either north or south of the equator.

solution: homogeneous mixture whose components cannot be distinguished and can be classified as liquid, gaseous, solid, or a combination; the method of transport for materials that are dissolved in a stream's water.

sonar: use of sound waves to detect and measure objects underwater.

source region: area over which an air mass forms.

specific gravity: ratio of the mass of a substance to the mass of an equal volume of H_2O at 4°C.

spiral density wave: spiral regions of alternating density which rotates as a rigid pattern.

spring: natural discharge of groundwater at Earth's surface where an aquifer and an aquiclude come in contact.

spring tide: during full or new moon, the Sun, the Moon, and Earth are all aligned; this causes solar tides to enhance lunar tides, causing high tides to be higher than normal and low tides to be lower than normal.

stalactite: cone-shaped or cylindrical dripstone deposit of calcium carbonate that hangs like an icicle from a cave's ceiling.

stalagmite: mound-shaped dripstone deposit of calcium carbonate that forms on a cave's floor beneath a stalactite.

station model: record of weather data for a specific place at a specific time, using meteorological symbols.

stepped leader: A branched channel of partially-charged air that forms between positive and negative regions; generally moves from the center of the cloud toward the ground.

stock: irregularly shaped pluton that is similar to a batholith but smaller, generally forms 5–30 km beneath Earth's surface, and cuts across older rocks.

storm surge: occurs when powerful, hurricane-force winds drive a mound of ocean water toward shore, where it washes over the land, often causing enormous damage.

strain: deformation of materials in response to stress.

stratosphere: layer of Earth's atmosphere that is located above the tropopause and is made up primarily of concentrated ozone.

stratus (STRAY tus): a layered sheetlike cloud that covers much or all of the sky in a given area.

streak: color a mineral leaves when it is rubbed across an unglazed porcelain plate or when it is broken up and powdered.

stream bank: ground bordering each side of a stream that keeps the moving water confined.

viento solar: viento de partículas cargadas **(iones)** que fluye a través del sistema solar y comienza como un gas que es despedido a gran velocidad por la corona del Sol.

solsticio: sucede cuando el Sol se halla en el horizonte a su mayor distancia al norte o al sur del ecuador.

solución: mezcla homogénea cuyos componentes no se pueden distinguir; puede clasificarse como líquida, gaseosa, sólida o una combinación de éstas; el método de transporte de materiales que están disueltos en las aguas de una corriente.

sonar: uso de ondas sonoras para detectar y medir objetos submarinos.

región fuente: área sobre la cual se forma una masa de aire.

gravedad específica: razón de la masa de una sustancia con relación a la masa de un volumen igual de H_2O a 4°C.

ondas de densidad espirales: regiones en espiral con densidad variable que giran siguiendo un patrón rígido.

manantial: descarga natural de agua subterránea en la superficie terrestre, en el punto donde un acuífero y un acuicluso entran el contacto.

marea viva: durante la luna nueva o la luna llena, el Sol, la Luna y la Tierra se encuentran alineados; esto ocasiona que la marea solar aumente el efecto de la marea lunar y provoca que la marea alta sea más alta que lo normal y que la marea baja sea más baja que lo normal.

estalactita: depósito rocoso de carbonato de calcio, de forma cónica o cilíndrica, que se forma por goteo y que cuelga como un carámbano del techo de una caverna.

estalagmita: depósito de carbonato de calcio, con forma de montículo, que se forma por goteo en el piso de una caverna, debajo de una estalactita.

código meteorológico: registro de los datos del tiempo para un lugar específico en un tiempo dado, usando símbolos meteorológicos.

guía escalonada: el canal ramificado con aire parcialmente cargado que forma entre regiones positivas y negativas; generalmente se mueve del centro de la nube hacia la tierra.

macizo magmático: plutón de forma irregular, similar a un batolito pero más pequeño; generalmente se forma de 5 a 30 km bajo la superficie terrestre y atraviesa rocas más antiguas.

marejada ciclónica: ocurre cuando poderosos vientos huracanados arrojan una gran masa de agua del océano hacia la costa, desparramándose por el terreno y causando a menudo un daño enorme.

tensión: deformación de los materiales en res-puesta a un estrés.

estratosfera: capa de la atmósfera terrestre ubicada por encima de la tropopausa; está compuesta principalmente de ozono concentrado.

estrato: nube con forma de capas delgadas que cubre la mayoría o todo el cielo en cierta área.

veta: color que deja un mineral cuando es frotado contra un plato de porcelana sin barnizar o cuando se rompe y se pulveriza.

margen de una corriente de agua: terreno que limita ambos lados de una corriente, manteniendo confinada la corriente de agua en movimiento.

Glossary/Glosario

GLOSSARY/GLOSARIO

stream channel: narrow pathway carved into sediment or rock by the movement of surface water.

stress: forces per unit area that act on a material—compression, tension, and shear.

stromatolite (stroh MA tuh lite): large mat or mound composed of billions of photosynthesizing cyanobacteria; dominated shallow oceans during the Proterozoic.

subduction: process by which one tectonic plate slips beneath another tectonic plate.

sublimation: process by which a solid slowly changes to a gas without first entering a liquid state.

sunspot: dark spot on the surface of the photosphere; occur in pairs.

supercell: extremely powerful, self-sustaining thunderstorm characterized by intense, rotating updrafts.

supercluster: gigantic threadlike or sheetlike cluster of galaxies that is hundreds of millions of light-years in size.

supernova: massive explosion that occurs when the outer layers of a star are blown off.

superposition: the principle that, in an undisturbed rock sequence, the oldest rocks are on the bottom and each consecutive layer is younger than the layer beneath it.

surface current: wind-driven movement of ocean water that primarily affects the upper few hundred meters of the ocean.

suspension: the method of transport for all particles small enough to be held up by the turbulence of a stream's moving water.

sustainable energy: involves global management of Earth's natural resources to ensure that current and future energy needs will be met without harming the environment.

sustainable yield: replacement of renewable resources at the same rate at which they are consumed.

synchronous rotation: the state at which an orbiting body's orbital and rotational periods are equal.

cauce fluvial: estrecha vía labrada en el sedimento, o en la roca, por el movimiento del agua en la superficie.

estrés: fuerza por unidad de área que actúa sobre un material: puede ser por compresión, tensión o cizallamiento.

estromatolitos: montículos grandes compuestos de billones de cianobacterias fotosintéticas; dominaron los océanos superficiales durante el Proterozoico.

subducción: proceso en que una placa tectónica se desliza por debajo de otra.

sublimación: proceso en que un sólido se convierte lentamente en gas, sin convertirse primero al estado líquido.

mancha solar: mancha oscura en la superficie de la fotosfera; ocurren en pares.

supercelda: tormenta autosostenible extremadamente poderosa, caracterizada por tener intensas cor-rientes ascendentes giratorias.

supercúmulo: cúmulo gigantesco de galaxias con forma de filamento o lámina que mide centenares de millones de años luz.

supernova: enorme explosión que ocurre cuando estallan las capas exteriores de una estrella.

superposición: principio que establece que en una sucesión rocosa no perturbada, los estratos rocosos más antiguos se encuentran en el fondo y que cada capa sucesiva es más reciente que la capa subyacente.

corriente superficial: movimiento de las aguas del océano producido por el viento, que afecta principalmente los primeros cientos de metros superiores de las aguas del océano.

suspensión: método de transporte de todas las partículas que son suficientemente pequeñas como para ser mantenidas en el agua por la turbulencia de la corriente del agua en movimiento.

energía sostenible: implica la administración global de los recursos naturales de la Tierra para asegurar que se satisfagan las necesidades energéticas actuales y futuras, sin causar daños al ambiente.

rendimiento sostenible: regeneración de los recursos renovables a la misma velocidad con que se consumen.

rotación sincronizada: estado en que los periodos de la órbita y de rotación de un cuerpo orbitando son iguales.

T

---

tailings: material left after mineral ore has been extracted from parent rock; can release harmful chemicals into groundwater or surface water.

tectonic plates: huge pieces of Earth's crust that cover its surface and fit together at their edges.

temperate zone: area of Earth that extends between 23.5° and 66.5° north and south of the equator and has moderate temperatures.

temperature inversion: increase in temperature with height in an atmospheric layer, which inverts the temperature-altitude relationship and can worsen air-pollution problems.

escombreras: material que queda después de que se ha extraído la mena de la roca madre; puede liberar sustancias químicas tóxicas hacia las aguas subterráneas y superficiales.

placas tectónicas: enormes fragmentos de corteza que cubren la superficie terrestre; sus límites se corresponden entre sí.

zonas templadas: áreas de la Tierra que se extienden entre los 23.5° y los 66.5°, al norte y al sur del ecuador; experimentan temperaturas moderadas.

inversión de temperatura: aumento de temperatura que ocurre al aumentar la altitud en alguna capa de la atmósfera; invierte la relación entre la altitud y la temperatura y puede empeorar los problemas de contaminación del aire.

temperature profile: plots changing ocean water temperatures against depth, which varies, depending on location and season.

tephra: rock fragments, classified by size, that are thrown into the air during a volcanic eruption and fall to the ground.

terrestrial planets: rocky-surfaced, relatively small, dense inner planets closest to the Sun—Mercury, Venus, Earth, and Mars.

tetrahedron: a geometric solid having four sides that are equilateral triangles.

texture: the size, shape, and distribution of the crystals or grains that make up a rock.

thermocline: transitional ocean layer that lies between the relatively warm, sunlit surface layer and the colder, dark, dense bottom layer and is characterized by temperatures that decrease rapidly with depth.

thermometer: instrument used to measure temperature using either the Faherenheit or Celsius scale.

thermosphere: layer of Earth's atmosphere that is located above the mesopause; oxygen atoms absorb solar radiation causing the temperature to increase in this layer.

tide: periodic rise and fall of sea level caused by the gravitational attraction among Earth, the Moon, and the Sun.

topographic map: map that uses contour lines, symbols, and color to show changes in the elevation of Earth's surface and features such as mountains, bridges, and rivers.

topography: the change in elevation of the crust.

tornado: violent, whirling column of air in contact with the ground that forms when wind direction and speed suddenly change with height; is often associated with a supercell, and can be extremely damaging.

trace fossil: indirect fossil evidence of an organism; traces of worm trails, footprints, and tunneling burrows.

trade winds: two global wind systems that flow between 30° north and south latitudes, where air sinks, warms, and returns to the equator in a westerly direction.

transform boundary: place where two tectonic plates slide horizontally past each another; is characterized by long faults and shallow earthquakes.

transgression: occurs when sea level rises and causes the shoreline to move inland, resulting in deeper-water deposits overlying shallower-water deposits.

transported soil: soil that has been moved away from its parent material by water, wind, gravity, or a glacier.

tropical cyclone: large, low-pressure, rotating tropical storm that gets its energy from the evaporation of warm ocean water and the release of heat.

tropics: area of Earth that receives the most solar radiation, is generally warm year-round, and extends between 23.5° south and 23.5° north of the equator.

perfil de temperatura: diagramas que muestran cómo cambia la temperatura del océano con la profundidad; varía según la ubicación y la temporada.

tefrita: fragmentos rocosos que se clasifican por tamaño; son lanzados al aire durante una erupción volcánica y luego caen al suelo.

planetas terrestres: planetas internos, densos, relativamente pequeños, con superficie rocosa y cercanos al Sol: Mercurio, Venus, la Tierra, y Marte.

tetraedro: sólido geométrico que tiene cuatro lados con forma de triángulo equilátero.

textura: tamaño, forma y distribución de los granos o cristales que forman una roca.

termoclina: capa de transición del océano que se halla entre la capa superficial iluminada por el Sol, que tiene una temperatura relativamente tibia, y la capa inferior, que es densa, oscura y fría; se caracteriza por tener temperaturas que disminuyen rápidamente con la profundidad.

termómetro: instrumento que sirve para medir la temperatura en grados Fahrenheit o Celsius.

termosfera: capa de la atmósfera terrestre ubicada por encima de la mesopausa; los átomos de oxígeno absorben radiación solar, haciendo que la temperatura aumente en esta capa.

marea: ascenso y descenso periódicos del nivel del mar causados por la atracción gravitacional entre la Tierra, la Luna y el Sol.

mapa topográfico: mapa que usa curvas de nivel, símbolos y colores para mostrar los cambios en la elevación de la superficie terrestre, e incluye rasgos como las montañas, los puentes y los ríos.

topografía: el cambio en la elevación de la corteza.

tornado: violenta columna giratoria de aire en contacto con el suelo; se forma cuando la dirección y la velocidad del viento cambian repentinamente con la altura; a menudo está asociada con una supercelda y puede ser extremadamente dañina.

fósiles traza: pruebas fósiles indirectas de un organismo: incluye rastros de gusanos, huellas de pasos y madrigueras.

vientos alisios: dos sistemas globales de vientos que se desplazan entre los 30° de latitud norte y sur, donde el aire desciende, se calienta y regresa al ecuador con dirección oeste.

límite transformante: lugar donde dos placas tectónicas se deslizan horizontalmente, una al lado de la otra y en sentidos opuestos; se caracteriza por presentar grandes fallas y terremotos superficiales.

transgresión: ocurre cuando el nivel del mar aumenta y hace que el litoral retroceda hacia el interior; ocasiona depósitos de agua más profunda que cubren depósitos de agua menos profunda.

suelo transportado: suelo que ha sido transportado lejos de su roca madre por el agua, el viento, gravedad, o un glaciar.

ciclón tropical: gran tormenta giratoria de baja presión que obtiene su energía de la evaporación de las tibias aguas del mar y la liberación de calor.

trópicos: área de la Tierra que recibe la mayor cantidad de radiación solar, generalmente es caliente todo el año y se extiende entre 23.5° sur y 23.5° norte del ecuador.

Glossary/Glosario

GLOSSARY/GLOSARIO

troposphere: layer of the atmosphere closest to Earth's surface, where most of the mass of the atmosphere is found and in which most weather takes place and air pollution collects.

troposfera: capa de la atmósfera más cercana a la superficie terrestre; en ella se halla la mayoría de la masa atmosférica, ocurren la mayoría de los fenómenos meteorológicos y se concentran la mayoría de los contaminantes.

trough: lowest point of a wave.

seno: punto más bajo de una onda.

tsunami (soo NAH mee): large, powerful ocean wave generated by the vertical motions of the seafloor during an earthquake; in shallow water, can form huge, fast-moving breakers exceeding 30 m in height that can damage coastal areas.

tsunami: enorme y poderosa ola marina generada por los movimientos verticales del fondo del mar durante un sismo; en aguas superficiales, puede formar inmensas olas muy rápidas de mas de 30 m de altura que pueden causar daños en las áreas costeras.

turbidity current: rapidly flowing ocean current that can cut deep-sea canyons in continental slopes and deposit the sediments in the form of a continental rise.

corriente de turbidez: corriente oceánica de flujo rápido que puede formar cañones en los taludes continentales y depositar los sedimentos para formar el pie del talud continental.

U

unconformity: gap in the rock record caused by erosion or weathering.

disconformidad: discontinuidad en el registro geológico causada por la erosión o la meteorización.

uniformitarianism: the theory that geologic processes occurring today have been occurring since Earth formed.

uniformitarianismo: este principio establece que los procesos geológicos que ocurren actualmente han estado ocurriendo desde que la Tierra se formó.

uplifted mountain: mountain that forms when large regions of Earth are forced slowly upward without much deformation.

levantamiento montañoso: montañas que se forman cuando grandes regiones de la Tierra son levantadas lentamente sin que ocurra mucha deformación.

upwelling: upward movement of ocean water that occurs when winds push surface water aside and it is replaced with cold, deeper waters that originate below the thermocline.

corriente resurgente: movimiento ascendente de las aguas del océano que ocurre cuando los vientos remueven las aguas superficiales, causando que sean reemplazadas por aguas más frías y profundas prove-nientes de profundidades mayores que la termoclina.

upward streamer: a branching channel of positively charged ions that rushes upward from the ground to meet the stepped leader

rayo de retorno: el canal ramificado de iones con una carga positiva que avanza desde la tierra para encontrarse con la guía escalonada.

V

valley glacier: glacier that forms in a valley in a mountainous area and widens V-shaped stream valleys into U-shaped glacial valleys as it moves downslope.

glaciar de valle: glaciar que se forma en un valle de un área montañosa; al deslizarse cuesta abajo, ensancha los valles de corrientes con forma en V y los convierte en valles glaciales con forma de U.

variable star: star in the giant branch of the Hertzsprung-Russell diagram that pulsates in brightness due to its outer layers expanding and contracting.

estrella variable: estrella en la rama de las gigantes del diagrama Hertzsprung-Russell, cuya luminosidad presenta pulsaciones debidas a la expansión y contracción de sus capas exteriores.

varve: alternating light-colored and dark-colored sedimentary layer of sand, clay, and silt deposited in a lake that can be used to date cyclic events and changes in the environment.

varve: estratos sedimentarios de colores claros y oscuros alternados, compuestos de arena, arcilla y limo, depositados en un lago, que sirven para datar acontecimientos cíclicos y cambios en el ambiente.

vent: opening in Earth's crust through which lava erupts and flows out onto the surface.

chimenea: abertura en la corteza terrestre por la cual fluye lava hacia la superficie.

ventifact: rock shaped by windblown sediments.

ventifacto: roca moldeada por sedimentos arrastrados por el viento.

vesicular texture: characterized by containing vesicles, or holes, formed by gas bubbles.

textura vesicular: caracterizado por que contiene vesículas, o agujeros, formados por burbujas de gas.

viscosity: a substance's internal resistance to flow.

viscosidad: resistencia interna a fluir de una sustancia.

volcanism: describes all the processes associated with the discharge of magma, hot water, and steam.

vulcanismo: describe todos los procesos asociados con la descarga de magma, agua caliente y vapor.

watershed: land area drained by a stream system.

water table: upper boundary of the zone of saturation that rises during wet seasons and drops during dry periods.

wave: rhythmic movement that carries energy through matter or space and, in oceans, is generated mainly by wind moving over the surface of the water.

wave refraction: process in which waves advancing toward shore slow when they encounter shallower water, causing the initially straight wave crests to bend toward the headlands.

weather: short-term variations in atmosphere phenomena that interact and affect the environment and life on Earth.

weathering: chemical or mechanical process that breaks down and changes rocks on or near Earth's surface and whose rate is influenced by factors such as precipitation and temperature.

well: deep hole drilled or dug into the ground to reach a reservoir of groundwater.

wetland: any land area, such as a bog or marsh, that is covered in water a large part of the year and supports specific plant species.

windchill index: measures the windchill factor, by estimating the heat loss from human skin caused by a combination of wind and cold air.

cuenca: área de terreno drenada por un sistema de corrientes de agua.

capa freática: límite superior de la zona de saturación; aumenta durante la temporada de lluvias y disminuye durante los períodos de sequía.

onda (ola): movimiento rítmico que transporta energía a través de la materia o el espacio; en los océanos, es generado principalmente por el movimiento del viento sobre la superficie del agua.

refracción de onda: proceso en que las olas avanzan hacia la costa y reducen su velocidad, cuando llegan a aguas menos profundas, ocasionando que las crestas de las olas, inicialmente rectas, se inclinen hacia los promontorios.

tiempo: variaciones a corto plazo en los fenómenos que suceden en la atmósfera, que interactúan y afectan el entorno de la vida en la Tierra.

meteorización: proceso químico o mecánico que rompe y modifica las rocas que se hallan sobre o cerca de la superficie terrestre; su velocidad se ve influida por factores como la precipitación y la temperatura.

pozo: hoyo profundo perforado o excavado en el suelo para alcanzar un depósito de agua subterránea.

humedal: toda área, como un pantano o una ciénaga, que se encuentra cubierta de agua gran parte del año y que alberga especies específicas de plantas.

índice de sensación térmica: índice que toma en cuenta el efecto del viento en la sensación térmica, al estimar la pérdida de calor de la piel humana causada por la combinación de viento y aire frío.

zircon: very stable and common mineral that scientists often use to age-date old rocks.

zone: high, cool, light-colored cloud that rises and flows rapidly in the Jovian atmosphere.

zone of aeration region above the water table where materials are moist, but pores contain mostly air.

zone of saturation: region below Earth's surface where all the pores of a material are completely filled with groundwater.

circón: mineral sumamente estable que los científicos usan para datar rocas antiguas.

zona: nubes altas, relativamente frías y de color claro, que se elevan y desplazan con rapidez en la atmósfera joviana.

zona de aeración: región sobre el manto freático en que los materiales están húmedos, pero los poros contienen principalmente aire.

zona de saturación: región profunda bajo la superficie terrestre donde todos los poros del material están completamente llenos con agua subterránea.

Index

Index

INDEX

Index

Index

Index

Index

An Earth Scientist's Guide to the Periodic Table

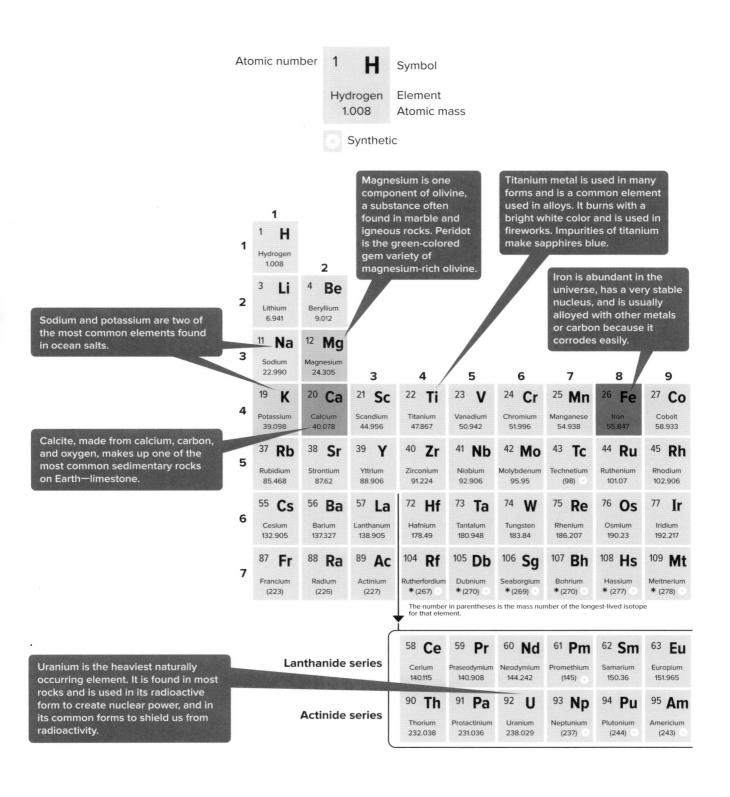

Geologic Time Scale

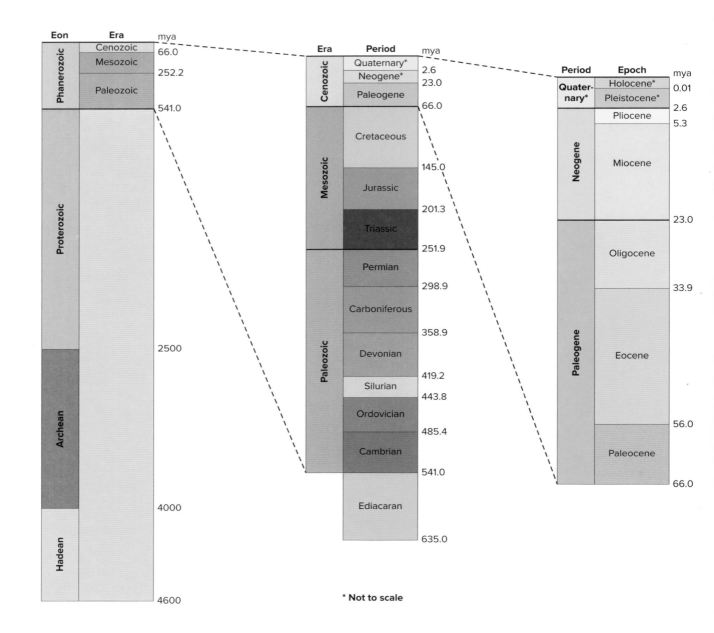

* **Not to scale**